Mobile Radio
Communications

Contributors

Editor: **R. Steele,** *BSc, PhD, DSc, FEng, FIEE, SMIEEE,* University of Southampton and Multiple Access Communications Ltd., UK.

D. G. Appleby, *BSc(Eng), CEng, MIEE,* University of Southampton, UK.

D. Greenwood, *BSc,* University of Southampton, currently with RAM Communications Consultants Ltd., UK.

L. Hanzo, *Dipl Ing, MSc, PhD, MIEEE,* University of Southampton and Multiple Access Communications Ltd., UK.

Y. F. Ko, *BSc, PhD, MIEEE,* University of Southampton, currently with Mercury Personal Communications, UK.

R. A. Salami, *MSc, PhD,* University of Southampton, currently with Communications Research Centre, Department of Electronic Engineering, University of Sherbrooke, Canada.

J. Stefanov, *Dipl Ing, PhD, C.Eng, MIEE,* University of Southampton and Multiple Access Communications Ltd., UK.

I. J. Wassell, *BSc, B.Eng, PhD,* Multiple Access Communications Ltd., currently with Hutchison Personal Communications Ltd., UK.

K. H. J. Wong, *BSc, PhD, C.Eng, MIEE, Eur Ing, CPEng, MIEAust, MIEEE,* University of Southampton, currently with Konwa Communication Ltd, Hong Kong.

K. H. H. Wong, *BSc, PhD, MIEEE,* University of Southampton, currently with Hutchison Telephone Company Ltd, Hong Kong.

Mobile Radio Communications

Raymond Steele (Ed),
Professor of Communications in the Electronics and Computer Science Department, University of Southampton,
and
Managing Director of Multiple Access Communications Ltd., Southampton

PENTECH PRESS
Publishers - London

IEEE PRESS
New York

First published 1992
by Pentech Press Limited
Graham Lodge, Graham Road
London NW4 3DG

North, South and Central America rights assigned
exclusively to IEEE Press, 445 Hoes Lane,
Piscataway, NJ 08855

ISBN 0-7803-1102-7
IEEE Order No. PC04572

Reprinted 1994, 1995

British Library Cataloguing-in-Publication Data.
A catalogue record for this book is available from the
British Library.

ISBN 0-7273-1406-8

First printed and reprinted (1994) in Great Britain by BPC Wheatons Ltd, Exeter

Reprinted (1995) in the United States of America

Preface

Analogue cellular mobile radio systems have taken root in many countries, untethering the telephone and enabling people to conduct conversations away from the home or office and while on the move. The systems are spectrally efficient with the frequency bands assigned by the regulatory bodies being repeatedly reused over countries and even continents.

The fixed networks are becoming digital enabling the introduction of the integrated digital service network (ISDN). No longer are communications to be restricted to voice. Instead a range of services, such as fax, video conferencing, slow-scan tv, and computer data transfer, are becoming increasingly available. Mobile radio, with its paucity of spectrum, is beginning to move towards ISDN. The second generation cellular networks that are in the process of being deployed are all-digital networks having complex radio links to connect the mobile users to their base stations. The mobile voice and data communications are supported by elaborate network protocols that provide registration, location of mobile users, handovers between base stations as the mobiles roam, call initiation and call clear-down, and so forth. In addition there are management, maintenance, and numerous functions unseen by the user that combine to facilitate high quality mobile communications. Although this book does consider some network issues, it principally addresses the radio aspects of mobile communications.

Chapter 1 is a bottom-up approach to cellular radio. Starting with the propagation environment of a single mobile communicating with a base station, it progresses via multiple access methods, first generation and second generation mobile systems, to cordless telecommunications and ends with teletraffic for mobile radio. The Chapter is designed to give the reader a range of concepts that will prepare her or him for the more focused chapters which follow.

Chapter 2 considers mobile radio propagation in a quantitative manner, establishing the background material that is the backbone of mobile radio communications.

A pre-requisite to digital telephony is the selection of an appropriate speech codec to encode the analogue speech signal into a digital format. Chapter 3 commences by describing simple waveform speech codecs, and then the discourse deepens into an exposition of analysis-by-synthesis codecs. Having encoded speech we need to apply forward error correction coding together with interleaving of the coded data in order to combat error bursts that occur due to the fading in the mobile radio channel. Chapter 4 addresses these issues.

The interleaved data are transmitted via a suitable modulator over a mobile radio channel to a distant receiver which recovers the data. There are many different methods of modulation but we opt for those appropriate for mobile communications. In Chapter 5 we consider quaternary frequency shift keying (QFSK) which was a contending modem for the pan-European cellular network. Chapter 6 deals with the more complex modulation known as generalised phase modulation. In this chapter we consider Viterbi equalisation of wideband dispersive mobile radio channels.

Frequency hopping is an important technique in mobile radio whereby a user's channel hops from one frequency carrier to another to avoid being in a deep fade for long periods of time. Chapter 7 is devoted to slow frequency hopping cellular systems, and an estimation of their spectral efficiency is presented. The final chapter describes the pan-European mobile radio system known as GSM. This chapter guides the reader through the complexities of this mobile radio network.

The Book has been written by the staff at Southampton University and at Multiple Access Communications Ltd. The names of the writers of each chapter are presented at the beginning of their chapter. All of the contributors are indebted to our many colleagues who have enhanced our understanding of the subject. In particular, we thank Dr J. C. S. Cheung Dr S. T. Chia, Dr S. El-Dolil, Dr P. M. Fortune, Dr E. Green, Dr W. H. Lam C. C. Lee, Dr M. A. Nofal, R. D. Stewart, Dr W. T. Webb and J. Williams We also acknowledge our valuable associations with Roke Manor Research BT Laboratories at Martlesham Heath, the Department of Trade and Industry, and the Radiocommunications Agency.

Those authors who did not type their own manuscripts thank Jenn Clark and Debbie Sheridan for the work they did on their behalf. We are grateful to Phil Evans for the production of many of the drawings, and the Editor is particularly appreciative of the efforts of Lajos Hanzo in getting the finished manuscript in a suitable form for printing.

Raymond Steele

Contents

Chapter 1

Introduction to Digital Cellular Radio

R.Steele[1]

1.1 The Background to Digital Cellular Mobile Radio

Following the pioneering work of Hertz, the experiments of Marconi at the end of the 19th century demonstrated the feasibility that radio communications could take place between transceivers that were mobile and far apart. Henceforth telegraphic and voice communications were not inherently limited to users' equipment tethered by wires. Instead, a freedom to roam and yet still communicate was possible. However, only relatively few individuals were to enjoy this type of communication during the next several decades.

Morse-coded on-off keying was mainly used for mobile radio communications until the 1920s. It was not until 1928 that the first land mobile radio system for broadcasting messages to police vehicles was deployed. In 1933 a two-way mobile radio voice system was introduced by the Bayonne New Jersey Police Department. The early mobile radio transceivers were by today's standards very simple, noisy, bulky and heavy. They used power-hungry valves, operated in the lower frequency part of the VHF band, and had a range of some ten miles. Reference [1] has interesting photographs of mobile radio equipment in 1936.

The military embraced mobile communications in order to effectively deploy their forces in battle. Vital mobile services, such as those of the police,

---

[1]University of Southampton and Multiple Access Communications Ltd.

ambulance, fire, marine, aviation and so forth, also introduced mobile communications to facilitate their operations. Early mobile radio communications were of poor quality. This was due to the radio propagation characteristics that resulted in the received signal being composed of many versions of the transmitted signal, each with different time delays, amplitudes and phases. The vector sum of these signals gave a received envelope with temporal and spatial variations. As the mobile station travelled, the received signal level often experienced large and rapid variations that caused considerable speech degradation. Of course these propagation characteristics exist today, but they had to be combatted then by a technology that was in a primitive evolutionary state. Whereas today's semiconductor technology can use millions of transistors to compensate for the propagation effects, early transceivers usually had less than ten valves.

The bandwidth that could be utilised by existing technology has always been a scarce commodity in radio communications. The long and medium wave bands were used for broadcasters, while the low frequency (LF) and high frequency (HF) bands were soon occupied with world wide communication services. The technology was inappropriate for good quality mobile communications in the VHF and UHF bands. The concept of frequency reuse was appreciated, but not its application to high user density mobile communications. Thus, for many years the quality of mobile communications was significantly worse than for wire communications as the technology was inadequate, and the operators were unable to utilise bandwidth in the higher frequency bands.

While fixed commercial analogue telephone networks were evolving into digital networks (thanks to the invention of low power, miniscule size, microelectronic devices) the mobile radio scene was also slowly altering. Private land mobile (PLM) radio systems came into use, catering for special groups, rather than individual members of the general public. Although Bell Laboratories conceived cellular radio in 1947, their parent company did not start to deploy a cellular network until 1979. The long gestation period was due to the waiting for necessary developments in technology. It was not until the arrival of custom designed integrated circuits, microprocessors, frequency synthesisers, high capacity fast switches, and so forth, that a cellular radio network could be realised. The 1980s witnessed the introduction of a number of analogue cellular radio systems, often known as public land mobile radio (PLMR) networks. Operating in the UHF band they represented a step-change in the complexity of civil communication systems. They enabled mobile users to have telephone conversations while on the move with any other users who were connected to the public switched telephone networks (PSTNs) or the integrated service digital networks (ISDNs). In the 1990s there will be another leap forward in mobile communications with the deployment of digital cellular networks and digital cordless telecommunication systems. These second generation mobile radio systems will provide a range of services in addition to telephony.

This book is concerned with the principles and techniques of digital mobile radio transmissions, and only marginally with mobile radio network issues. This opening chapter gives an overview of digital mobile radio in a qualitative way, leaving subsequent chapters to treat the subject in greater depth. Rather than a top-down approach, we tackle the subject from the bottom upwards. This means we commence our discourse by considering the propagation phenomena that are responsible for much of the complexity that occurs in mobile radio communication systems. Armed with some understanding of this topic we describe in Section 1.3 how multiple users may be accommodated. Once we have introduced the principles of multiple access, we discuss in Section 1.4 the first generation mobile radio systems which employ analogue modulation. The principles of digital cellular mobile radio transmissions are introduced in Section 1.5, paving the way for an overview of the second generation cellular mobile radio systems in Section 1.6. Section 1.7 considers cordless communications, while Section 1.8 provides useful teletraffic equations.

1.2 Mobile Radio Propagation

Mobile radio communications in cellular radio take place between a fixed base station (BS) and a number of roaming mobile stations (MSs) [2, 3, 4]. The geographical area in which these communications occur is called a cell, and we may consider that the cell boundary marks the maximum distance that a MS can roam from the BS before the quality of communications becomes unacceptably poor. The cells in mobile radio communications vary substantially in size and shape. Traditionally their size is large, up to 30 km radius, when there is rarely a line-of-slight (LOS) between the BS and its MSs. More recently small cells of some 1 km radius have been used where LOS is more probable. Cells, known as microcells, have been proposed [5, 6, 7] whose size may be only 100 m along the side of a city block. In microcells LOS is often a feature. The presence of LOS has a profound effect on radio propagation, and this means that the characteristics of radio propagation are highly dependent on the cell size and shape.

In Section 1.4 we will describe how the cells are organised in clusters that use the entire spectrum assigned by the regulatory bodies, how these clusters are replicated and tessellated using the same radio frequency band to give coverage over an entire country, and how the base stations in the cells communicate with their mobiles and with other users via the PSTN [8]. However, for the present we will describe issues that relate to radio propagation of signals between a base station (BS) and a mobile station (MS).

Mobile radio propagation [2, 3, 4, 9] is considered indepth in Chapter 2. However, it is essential to introduce here some rudimentary notions on propagation in order to proceed to the wider system issues. As the distance between a BS and a MS increases the received mean signal level tends to decrease. The way this occurs is examined in Section 1.2.5. Over relatively short distances the received mean signal is essentially constant, but the received signal level can vary rapidly by amounts typically up to 40 dB. These rapid variations are known as fast fading and we deal with this phenomenon in this Section and in Sections 1.2.2 to 1.2.4.

Let us consider a BS transmitting an unmodulated carrier which pervades the coverage area in which a MS is travelling. The MS does not receive one version of the transmitted carrier, but a number which have been reflected and diffracted by buildings and other urban paraphernalia. Indeed in most environments, each version of the transmitted signal received by the MS is subjected to a specific time delay, amplitude, phase and Doppler shift depending on its path from the BS to the MS. As a consequence the constant amplitude carrier signal transmitted may be substantially different from the signal the MS receives. When the signals from the various paths sum constructively at the MS antenna, the received signal level is enhanced. A serious condition occurs when the multipath signals, i.e., the transmitted signal arriving via many paths, vectorially sum to a small value. When this occurs the received signal is said to be in a fade, and the phenomenon is called multipath fading. As the MS travels it passes through an electromagnetic field that results in the received signal level experiencing fades approximately every half wavelength along its route. When a very deep fade occurs the received signal is essentially zero, and the receiver output is dependent on the channel noise, i.e., the channel signal-to-noise ratio (SNR) can be negative.

The above discussion relates to the transmission of an unmodulated carrier, an event that does not occur in practice. We are concerned here with digital mobile communications, where the propagation phenomena are highly dependent on the ratio of the symbol duration to the delay spread of the time variant radio channel. The delay spread may be considered as the length of the received pulse when an impulse is transmitted. We can see that if we transmit data at a slow rate the data can easily be resolved at the receiver. This is because the extension of a data pulse due to the multipaths is completed before the next impulse is transmitted. However, if we increase the transmitted data rate a point will be reached where each data symbol significantly spreads into adjacent symbols, a phenomenon known as intersymbol interference (ISI). Without the use of channel equalisers to remove the ISI the bit error rate (BER) may become unacceptably high.

Suppose that we continue to transmit at the high data rate which caused ISI, but move the MS closer to the BS while decreasing the radiated power to allow for the smaller BS to MS separation distance. If this distance is sufficiently small the delay spread will have decreased as delays of the multipaths components are, in general, smaller. The ISI will cease to be significant, removing the need for channel equalisation. The communications are still subjected to fading, but these fades may be very deep. The fading is said to be flat as it occurs uniformly across the frequency band of the channel. This is not so when ISI occurs as frequency selective fading is exhibited, i.e., some frequencies fade relative to others over the channel bandwidth.

The consequence is that cellular radio networks using large cells, where the excess delay spread may exceed 10 μs, need equalisers when the bit rate is relatively low, say 64kb/s, while cordless communications in buildings where the excess delay spread is often significantly below a microsecond may exhibit flat fading when the bit rate exceeds a megabit/sec. Small cells are not just smaller, they have different propagation features [7]. Very small cells, sometimes referred to as picocells may support many megabits/sec without equalisation because the delay spread is only tens of nanoseconds.

1.2.1 Gaussian Channel

The simplest type of channel is the Gaussian channel. It is often referred to as the additive white Gaussian noise (AWGN) channel. Basically it is the noise generated in the receiver when the transmission path is ideal. The noise is assumed to have a constant power spectral density over the channel bandwidth, and a Gaussian amplitude probability density function (PDF). This type of channel might be considered to be unrealisable in digital mobile radio, but this is not so. In microcells it is possible to have a LOS with essentially no multipath, giving a Gaussian channel. Even when there is multipath fading, but the mobile is stationary and there are no other moving objects, such as vehicles, in its vicinity, the mobile channel may be thought of as Gaussian with the effects of fading represented by a local path loss.

The Gaussian channel is also important for providing an upper bound on system performance. For a given modulation scheme we may calculate, or measure in a laboratory, the BER performance in the presence of a Gaussian channel. When multipath fading occurs the BER will increase for a given channel SNR. By using techniques to combat multipath fading, such as diversity, equalisation, channel coding, data interleaving, and so forth, techniques to be described throughout the book, we can observe how close the BER approaches that for the Gaussian channel.

1.2.2 Rayleigh Fading Channel

If each multipath component in the received signal is independent then the PDF of its envelope is Rayleigh. A typical received signal's fading envelope and phase as a function of time is shown in Figure 1.1. The Rayleigh PDF of the envelope is presented in Figure 1.2. The probability of experiencing a

RAYLEIGH FADING ENVELOPE

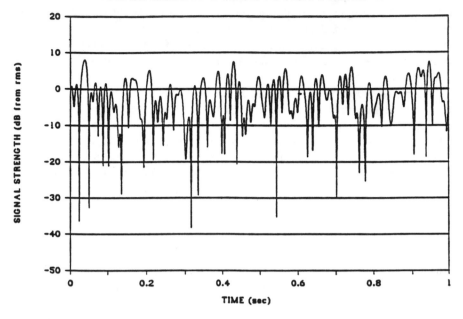

RAYLEIGH FADING PHASE

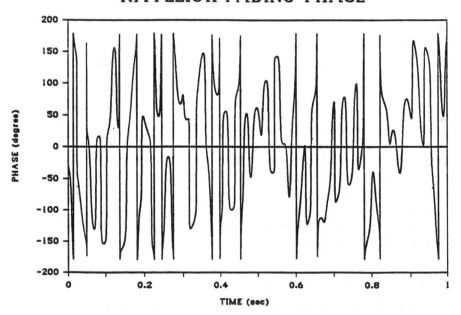

Figure 1.1: Typical Profile of the received signal's Rayleigh fading envelope and phase. Vehicular MS speed of 30 mph, carrier frequency of 900 MHz.

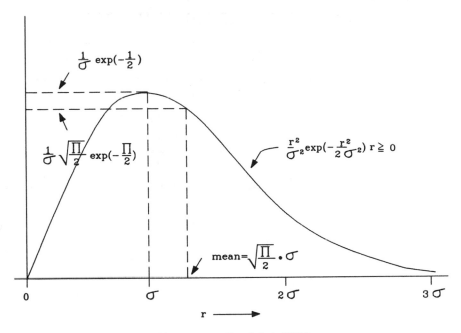

Figure 1.2: Rayleigh PDF.

deep fade below, say, 3σ, is the area in the tail of the PDF between 3σ and infinity, where σ is the rms value of the received signal envelope $r(t)$. Notice that the Rayleigh PDF applies to positive values, namely the magnitude of the received signal envelope, and that σ and the mean of the distribution are similar.

The impulse response of the flat Rayleigh fading mobile radio channel consists of a single delta function whose weight has a Rayleigh PDF. This occurs because all the multipath components manifest themselves in a bunch with negligible delay spread between them, and when modelled as a single delta function they combine to have a Rayleigh PDF. Thus as the MS travels the received signal fades in a manner similar to that shown in Figure 1.1, while the weight of the delta function in the impulse response also changes according to a Rayleigh PDF. When the MS experiences a deep fade the weight of the delta function is small, and vice versa when the received signal is enhanced.

Notice that the Gaussian channel may be represented by an impulse response having a constant weight delta function, i.e., an ideal channel, to which is added an AWGN source.

Representation of mobile radio channels is required for both mathematical analysis and computer simulation of mobile radio systems. A Rayleigh fading profile channel can be modelled using the arrangement shown in Fig-

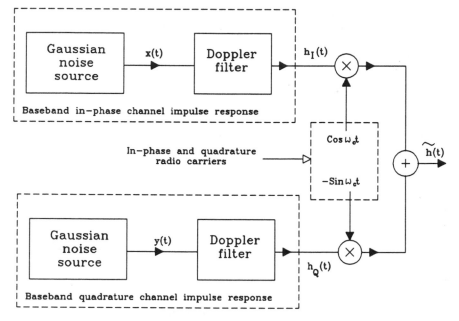

Figure 1.3: Model to generate a Rayleigh fading profile.

ure 1.3. Observe that there are two quadrature channels in the model. The outputs from the Gaussian noise sources are applied to filters that represent the effects of Doppler frequency shifts. So before continuing with the description of the model we must say a few words regarding the Doppler changes to the transmitted signal as perceived by the MS. As throughout this Chapter we confine ourselves to basic concepts. Let us again consider the transmission of an unmodulated carrier from a BS. A MS travelling in a direction making an angle α_i with respect to the signal received on the i-th path has its carrier frequency f_c modified to $f_c + f_m \cos \alpha_i$, where $f_m = v/\lambda$ $= v f_c / c$,and v is the speed of the MS, $\lambda = c/f_c$ is the wavelength of the carrier, and c is the velocity of light. Notice that a Doppler frequency can be positive or negative depending on α_i, and that the maximum and minimum Doppler frequencies are $\pm f_m$. These extreme frequencies correspond to the $\alpha_i = 0°$ and $180°$, when the ray is aligned with the street that the MS is travelling along, and corresponds to the ray coming towards or from behind the MS, respectively. It is analogous to the change in the frequency of a whistle from a train perceived by a person standing on a railway line when the train is bearing down or receding from the person, respectively.

Assuming that α_i is uniformly distributed, the Doppler frequency has a random cosine distribution. The Doppler power spectral density $S(f)$ can

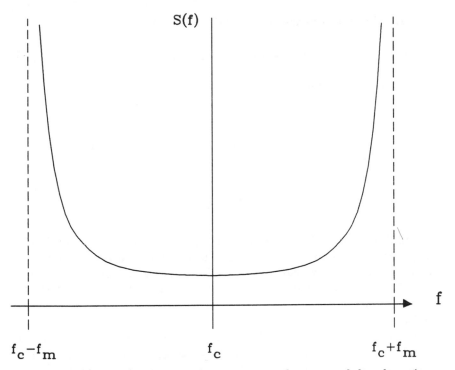

Figure 1.4: Sketch of a Doppler spectrum of an unmodulated carrier.

be computed by equating the incident received power in an angle $d\alpha$ with the Doppler power $S(f)df$, where df is found by differentiating the Doppler frequency term $f_m \cos \alpha$ with respect to α. The incident received power at the MS depends on the power gain of the antenna and the polarisation used. Thus the transmission of an unmodulated carrier is received as a multipath signal whose spectrum is not a single carrier frequency f_c, but contains frequencies up to $f_c \pm f_m$. A typical spectrum is sketched in Figure 1.4. In general we can express the received RF spectrum $S(f)$ for a particular MS speed, antenna and polarisation as,

$$S(f) = \frac{A}{\sqrt{1 - (f/f_m)^2}} \tag{1.1}$$

where A is a constant. Observe that f_m depends on the product of the speed of the MS and the propagation frequency.

Let us now return to Figure 1.3. If the Doppler filters are absent such that $h_I(t) = x(t)$ and $h_Q(t) = y(t)$, then the output is

$$\tilde{h}(t) = x(t) \cos \omega_c t - y(t) \sin \omega_c t \tag{1.2}$$

where $x(t)$ and $y(t)$ are independent Gaussian random variables. This equation is simply that of band-limited white noise where

$$\tilde{h}(t) = R(t) \cos\left(\omega_c t + \psi(t)\right) \tag{1.3}$$

and we infer in Chapter 2 that

$$R(t) = \left(x^2(t) + y^2(t)\right)^{\frac{1}{2}} \tag{1.4}$$

is Rayleigh distributed, and

$$\psi(t) = \tan^{-1}\left(\frac{y(t)}{x(t)}\right) \tag{1.5}$$

has a uniform PDF. We observe that we have obtained a signal $\tilde{h}(t)$ having a Rayleigh envelope, and with a white power spectral density (PSD) as both $x(t)$ and $y(t)$ have flat PSDs. By introducing the Doppler filters in Figure 1.3 we do not change the Rayleigh envelope statistics of $\tilde{h}(t)$, but we do introduce the necessary correlation between frequency components in $\tilde{h}(t)$.

The Rayleigh fading signal profile $\tilde{h}(t)$ is composed of two quadrature channels, a feature we will utilise in connection with the transmission of continuous phase modulated (CPM) signals addressed in Chapter 6. We also notice that if the quadrature carriers are removed we are left with a quadrature baseband representation of the Rayleigh fading channel, a representation that is essential when doing computer simulations because to simulate the many radio frequency (RF) cycles per data symbol is impracticable. The signal $\tilde{h}(t)$ may be represented as

$$\tilde{h}(t) = Re\left[h(t)e^{j\omega_c t}\right] \tag{1.6}$$

where $h(t)$ is the complex baseband representation of $\tilde{h}(t)$,

$$h(t) = h_I(t) + jh_Q(t) \tag{1.7}$$

with $h_I(t)$ and $h_Q(t)$ marked on Figure 1.3, ω_c is the angular carrier frequency and $Re[(\cdot)]$ is the real part of $(\cdot)$. If the output from the transmitter is $\tilde{s}(t)$, the received RF signal is

$$\tilde{r}(t) = \tilde{h}(t) * \tilde{s}(t) + \tilde{n}(t) \tag{1.8}$$

where $\tilde{n}(t)$ is the additive receiver noise at RF, and $*$ means convolution. Observe that $\sim$ above a symbol indicates that it is an RF and not a baseband signal.

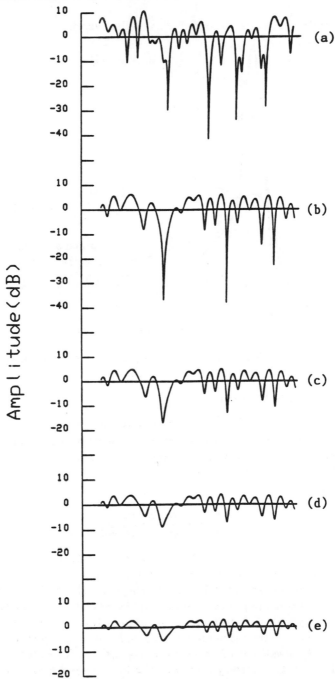

Figure 1.5: Rician fading profiles for MS travelling at 30 mph. Sub-figures (a), (b), (c), (d) and (e) refer to a K value of 0,4,8,16 and 32, respectively.

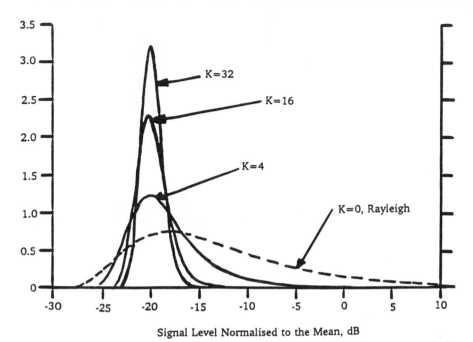

Signal Level Normalised to the Mean, dB

Figure 1.6: Rician PDFs normalised to their local means.

1.2.3   Rician Channel

In microcellular mobile radio a dominant path, which may be a line-of-sight (LOS) path, often occurs at the receiver, in addition to the many scattered paths. This dominant path may significantly decrease the depth of fading. The PDF of the received envelope is said to be Rician. We introduce a Rician parameter,

$$K = \frac{power\ in\ the\ dominant\ path}{power\ in\ the\ scattered\ paths} \qquad (1.9)$$

and emphasise that sometimes this parameter is defined as the ratio of the power in the scattered path to the power in the dominant path. Notice that when K is zero the channel is Rayleigh, whereas if K is infinite the channel is Gaussian. There are mobile radio channels that do not conform to either Gaussian, Rayleigh or Rician fading statistics. However, usually one does apply, and the Rician channel may be considered as the general case.

Figure 1.5 shows a set of envelope fading profiles for different values of K recorded over an arbitrary interval. The fades have a high probability of being very deep when K=0 (Rayleigh fading) to being very shallow when K = 32 (approaching Gaussian). When the received signal is in a deep fade below the average level of channel noise an error burst occurs. However,

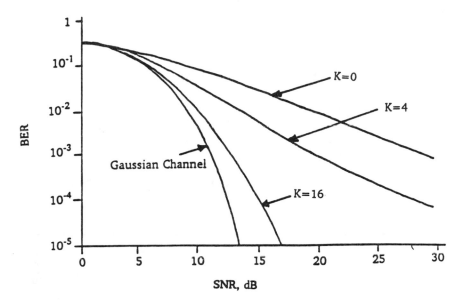

Figure 1.7: BER versus channel SNR for different Ks. Non-coherent FSK.

the same average noise level will not cause as many error bursts when K is higher.

The Rician PDFs for different values of K normalised with respect to their local mean values are displayed in Figure 1.6. It is evident that the Rayleigh PDF has the highest probability of being in a deep fade below the mean, while the Gaussian PDF has the lowest. This is also apparent in Figure 1.5. As an example of how Rician fading provides a superior performance to Rayleigh fading we show in Figure 1.7 the curves of BER as a function of signal-to-noise ratio (SNR) for different values of K [10]. The modulation, is non-coherent frequency shift keying (FSK). For a BER of 10^{-3} an SNR of 30 dB is required for a Rayleigh fading channel, because even for this low channel noise power there are occasionally deep fades that go below the noise floor of the receiver inducing an error burst. The Rayleigh fading channel is a feature of large cells. Macrocells of some 2 km diameter can exhibit both Rayleigh fading and Rician fading, but with usually low values of K. When microcells are used K can vary widely, but is often above five and values of 30 are not uncommon, see Section 1.2.6. When K is sufficiently high that it approaches a Gaussian channel an SNR of only 11 dB is required to achieve a BER of 10^{-3}.

Another by-product of the Rician channel is the improvement in cochannel interference performance. For the same FSK modulation, the BER

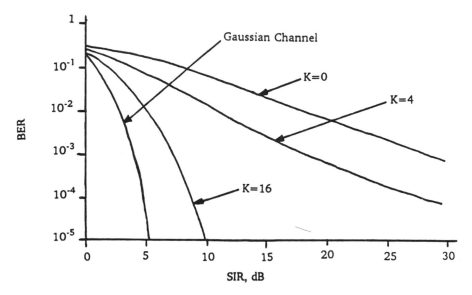

Figure 1.8: BER versus SIR for different KS. Non-Coherent FSK.

performance in the presence of a single cochannel interferer is displayed in Figure 1.8 for different values of K. Again we observe that the higher the value of K the lower the signal-to-interference ratio (SIR) required to achieve a particular BER. This improvement in SIR performance due to the presence of a Rician mobile radio channel having a high K factor means that handover of communications to another BS due to an unacceptably high BER is required to be rapid as the MS needs only lower its SIR level by a few dBs for the BER to rapidly increase. This increasing BER with decreasing SIR is much slower in Rayleigh fading channels.

The Rician channel representation is shown in Figure 1.9. The diagram is essentially that shown in Figure 1.3 for the Rayleigh channel, but with the dominant radio path represented by

$$\tilde{r}_D(t) = (I_D^2 + Q_D^2)^{1/2} \cos\{(\omega_C + \omega_D)t + \phi_D\} \qquad (1.10)$$

where I_D and Q_D are amplitudes of the quadrature components, ϕ_D is $\arctan(Q_D/I_D)$ and $\omega_C + \omega_D$ is the angular frequency of the dominant path. If this path intersects the MS at an angle of α_D when the MS is travelling at velocity v, the input RF spectrum shown in Figure 1.4 has an additional delta function at the frequency $f_c + (v/\lambda) \cos \alpha_D$, i.e., $\omega_D = (2\pi v/\lambda) \cos \alpha_D$. By calculating the ratio of the mean square value of $\tilde{r}_D(t)$ to the mean square value of $\tilde{h}(t)$ for the Rayleigh fading model shown in Figure 1.3, we obtain the value of K.

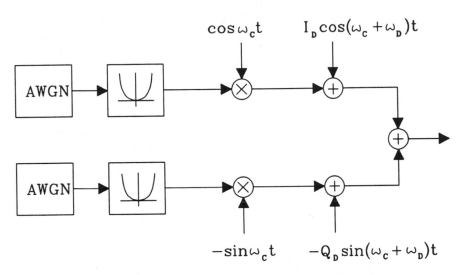

Figure 1.9: Model to generate a Rician Channel.

1.2.4 Wideband Channels

We have argued that the effect of multipath propagation is to spread the received symbols. In wideband channels the symbol rate is sufficiently high that each symbol is spread over adjacent symbols causing intersymbol interference (ISI). In order for the receiver to remove the ISI and regenerate the symbols correctly it must determine the impulse response of the mobile radio channel. This response must be frequently remeasured as the mobile channel may change rapidly both in time and space. Channel sounding is the means by which the impulse response of the time variant mobile radio channel is estimated. We will show in Chapter 6 that an estimate of the channel impulse response at the receiver is vital if equalisation of the channel is to be accomplished in order to operate the link at an acceptable BER.

Viewing the mobile radio channel as a time varying linear network we can measure its impulse response by a number of techniques. Suffice to say here is that if a pseudo random binary sequence (PRBS) modulates an RF carrier and is transmitted over an ideal channel to a receiver having a quadrature demodulator, then the use of correlators on each quadrature output will yield a narrow pulse whose width is approximately two bits wide. When the same PRBS is transmitted over a mobile radio channel the combined correlator outputs yields the impulse response of the channel, or more accurately, the response to the narrow pulse obtained for the ideal channel. Notice that as a quadrature demodulator is used the impulse response is determined in both magnitude and phase.

The magnitude of a typical impulse response is shown in Figure 1.10(a). If we partition the time delay axis as shown in Figure 1.10(b) into equal delay segments, usually called delay bins, then there will be, in general, a number of received signals in each bin corresponding to the different paths whose times of arrival are within the bin duration. These signals when vectorially combined can be represented by a delta function occurring in the centre of the bin having a weight that is Rayleigh distributed, see Figure 1.10(c). As the smaller impulses are of less significance we may introduce a threshold T and discard all components whose weight is below T. This leads to the simplified discrete impulse response shown in Figure 1.10(d). The discrete impulse in Figure 1.10(c) can be represented at RF by

$$\tilde{h}(t) = \left\{ \sum_{i=1}^{L} \beta_i e^{j\phi_i} \delta(t - \tau_i) \right\} e^{j\omega_c t} \tag{1.11}$$

where β_i and ϕ_i are the weight and the phase of the component in the i-th bin, respectively, which occurs at time $t = \tau_i$, and L is the last component in the response. The weight of the delta function $\delta(t - \tau_i)$ is β_i, and the complex baseband representation of $\tilde{h}(t)$ is

$$
\begin{aligned}
h(t) &= \sum_{i=1}^{L} \beta_i e^{j\phi_i} \delta(t - \tau_i) \\
&= \sum_{i=1}^{L} \beta_i \cos\phi_i \, \delta(t - \tau_i) + j\sum_{i=1}^{L} \beta_i \sin\phi_i \delta(t - \tau_i) \\
&= h_I(t) + jh_Q(t).
\end{aligned}
\tag{1.12}
$$

To model the discretised baseband impulse response $h(t)$ of the wideband channel shown in Figure 1.10(c) we need to formulate L narrow-band baseband Rayleigh fading channels. Figure 1.3 represents a narrow-band bandpass Rayleigh fading model. By removing the quadrature carriers and the adder we obtain the baseband inphase (I) and quadrature (Q) channels, each consisting of an AWGN source in cascade with a filter representing the effects of Doppler shifts. For the wideband channel we need L of these baseband narrow band channels for both the I and Q components.

Suppose a data signal $\alpha(t)$ is applied to a modulator to give a high frequency signal $\tilde{s}(\alpha, t)$, where again the $\sim$ above the symbol indicates that it is a bandpass radio signal. In Section 6.1.1 we describe a continuous phase modulation signal

$$\tilde{s}(\alpha, t) = A\cos\left(2\pi f_o t + \phi(t, \alpha)\right) \tag{1.13}$$

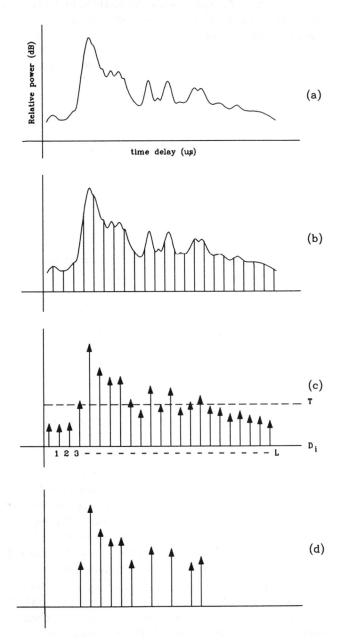

Figure 1.10: Wideband channel impulse response: (a) actual, (b) partitioned into equal delay segments, (c) discretised and (d) simplified discretised response.

where A is a constant, f_o is the carrier frequency, and $\phi(t, \alpha)$ is the phase angle of the phasor that carries $\alpha(t)$. We consider this signal as an example, to show that by writing it as

$$
\begin{aligned}
\tilde{s}(\alpha, t) &= Re\left[A \exp\left(\phi(t, \alpha)\right) . \exp(2\pi f_o t)\right] \\
&= Re\left[s(t, \alpha) \exp(2\pi f_o t)\right]
\end{aligned}
\tag{1.14}
$$

we can identify the baseband signal component of $\tilde{s}(\alpha, t)$ as

$$
s(t, \alpha) = s_I(t, \alpha) + j s_Q(t, \alpha)
\tag{1.15}
$$

where

$$
s_I(t, \alpha) = A \cos \phi(t, \alpha)
\tag{1.16}
$$

and

$$
s_Q(t, \alpha) = A \sin \phi(t, \alpha).
\tag{1.17}
$$

When this signal $s(t, \alpha)$ is convolved with the channel impulse response $h(t, \alpha)$ we obtain the baseband received signal

$$
r(t, \alpha) = s(t, \alpha) * h(t, \alpha)
\tag{1.18}
$$

where $*$ means convolution. Substituting in the complex values of $s(t, \alpha)$ and $h(t, \alpha)$ yields

$$
\begin{aligned}
r(t, \alpha) &= (s_I(t, \alpha) + j s_Q(t, \alpha)) * (h_I(t) + j h_Q(t)) \\
&= \{s_I(t, \alpha) * h_I(t) - s_Q(t, \alpha) * h_Q(t)\} \\
&+ j\{s_I(t, \alpha) * h_Q(t) + s_Q(t, \alpha) * h_I(t)\} \\
&= r_I(t, \alpha) + j r_Q(t, \alpha).
\end{aligned}
\tag{1.19}
$$

This equation implies a block diagram of the baseband mobile radio channel of the form shown in Figure 1.11. To formulate each of the $r_I(t, \alpha)$ and $r_Q(t, \alpha)$ signals we use an L-stage shift register, whose delay D is equal to the duration between the delay bins. Consider the inphase modulated signal $s_I(t)$ applied to this register as shown in Figure 1.12. As $s_I(t)$ passes along the register, the outputs from each delay stage in the register are applied to two sets of multipliers. The signals $h_I(t)$ and $h_Q(t)$ are produced having components $h_{I,i}$ and $h_{Q,i}$; $i = 1, 2, ...L$, respectively, where each $h_{(),i}$ is a Doppler filtered AWGN signal, i.e., a signal having a Rayleigh envelope. Hence $h_{I,i}$, and $h_{Q,i}$ constitute the wideband channel impulse response where each component has a Rayleigh fading envelope. An identical arrangement to that in Figure 1.12 is used for $s_Q(t)$, and by formulating the convolutional terms in Equation 1.19 we obtain $r_I(t)$ and $r_Q(t)$ in Figure 1.11. Notice that all the Doppler filters in the Figure 1.12 are the same, although they would all need to be modified for different MS speeds.

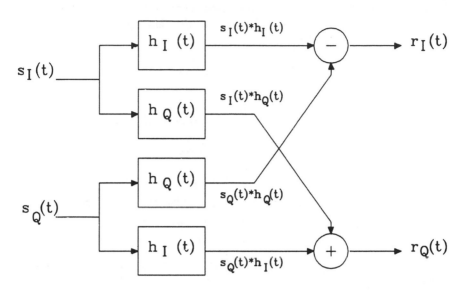

Figure 1.11: Quadrature representation of a baseband mobile radio channel, where $h_I(t)$ and $h_Q(t)$ are the inphase and quadrative channel impulse responses, respectively. (Additive noise is not shown)

The AWGN sources are statistically independent. This type of modelling is useful for computer simulation (with appropriate modifications).

In general the independent fading components in the wideband channel impulse response are not all Rayleigh distributed. However, Rayleigh is usually a reasonable assumption. When dealing with a specific type of environment, cell size, and so forth, it is better to use experimental results in modelling the channel. By averaging a set of experimental channel impulse responses an average channel impulse response is obtained and divided into delay bins. Again, each bin contains a coefficient ($h_{I,i}$ or $h_{Q,i}$) whose mean is adjusted to the value of the average impulse response in that bin.

1.2.4.1   GSM Wideband Channels

In formulating the specification for the pan-European digital mobile radio standard, the GSM Committee identified a number of channel impulse response models [11]. These models have impulse responses with either 6 or 12 components. In general the classical Doppler spectrum defined by Equation 1.1 is used, but they also employ the Rician Doppler spectrum as previously described. For rural areas (RA) the impulse response has 6 components, the first component h_1 has a Rician-type Doppler spectrum, while $h_2 \cdots h_6$ have the classical Rayleigh one. The other channel impulse

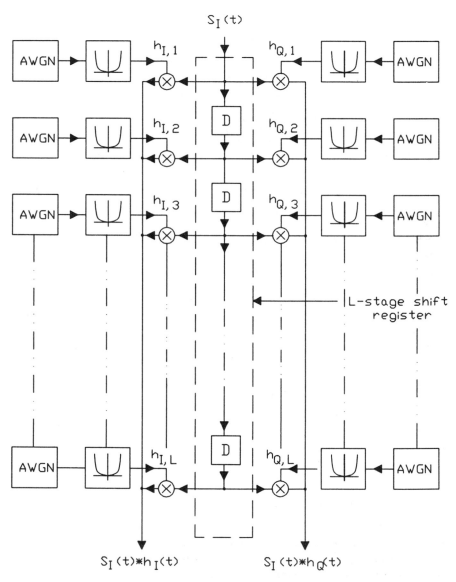

Figure 1.12: Generation of the baseband inphase channel output signals.

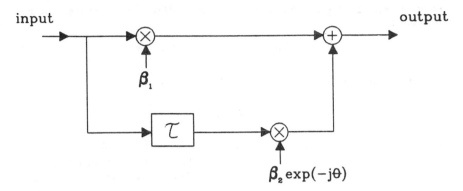

Figure 1.13: Two-ray channel model.

responses specified by GSM have all their components subjected to the classical Doppler spectrum. We emphasise that measured Doppler spectra may be significantly different from the shape shown in Figure 1.4.

Chapter 8 is concerned exclusively with the GSM system, and Figure 8.25 displays the impulse responses specified for different environments. Notice that rural response has uniformly spaced components. The transmission bit rate of GSM is 271 kb/s, giving a bit duration of 3.7 μs which is long compared to the maximum delay of 0.5 μs. Consequently there is little dispersion with this channel. However, if the bit rate is, say, 2 Mb/s, then there would be significant ISI. The channel response for the hilly terrain (HT) is a group of components with a delay $< 0.5\mu s$, followed by other components at much greater delays that are the consequence of long paths resulting from reflection and diffraction from the hills. The typical urban channel (TU) impulse response spans 5 μs and, as for the HT response, equalisation is essential at transmitted rates of 271 kb/s. There is also a channel impulse response specified for testing equalisers having six components equally spaced by 3.2 μs and spanning 16 μs. This type of channel response ensures considerable ISI and tests the equaliser's ability to remove it.

1.2.4.2 The Two-ray Rayleigh Fading Channel

The block diagram of the two-ray channel model is shown in Figure 1.13. The input signal is separated into two components or rays, one weighted by β_1 and the other delayed by τ and weighted by $\beta_2 exp(-j\theta)$. Both β_1 and β_2 are random variables from Rayleigh distributions, while θ is the relative phase between the two rays and is uniformly distributed between 0 and 2π. The outputs from the two rays are summed at the receiver input. In using this model, τ is held constant and system performance parameters, such as bit error rate (BER), determined. For a particular τ, the channel impulse response is

$$h(t) = \beta_1 \delta(t) + \beta_2 exp(-j\theta)\delta(t - \tau) \qquad (1.20)$$

corresponding to a channel frequency response of

$$H(f) = \beta_1 + \beta_2 exp(-j\theta)exp(-j2\pi f\tau). \qquad (1.21)$$

We observe that $H(f)$ exhibits frequency selective fading due to the variations in β_1 and β_2. Further, the model is based on the assumption that enough paths occur in the two bins to justify that only two rays need be considered, and that both rays are subjected to independent Rayleigh fading.

It is usual to determine the system parameters as a function of τ. For example, if the receiver had an equaliser that operated by using a window of W seconds of the estimated channel impulse response where the energy was greatest, then when τ was zero the BER performance would be poor as the channel would be a single path flat fading Rayleigh channel. As no dispersion occurs there is nothing the equaliser can do to improve performance. When τ is larger and both paths occur within W, the equaliser performs well, making use of the information in both paths. However, when τ is so large that only one ray resides within W, dispersion occurs, but the equaliser cannot exploit the diversity that inherently resides in the second path. Hence the graph of BER versus τ has a minimum when the equaliser can make best use of the two rays residing within W.

1.2.4.3 Real Channel Impulse Responses

The above wideband channel impulse responses are models. They relate to average conditions. What is often required are the sequences of actual channel impulse responses in circumstances that have a profound effect on system performance. One of these is when a MS enters a handover region. Handover is when the communications to the MS are switched from one BS to a more suitable one. Approaching a handover condition can be gradual when a MS is proceeding along a street or in a rural area. However, in a microcell the received signal level can rapidly decrease by some 20 dBs due to diffraction losses as the MS turns a corner, coupled with a radically different impulse response. In these conditions the channel may rapidly deteriorate and fast hand-over is required. For those involved in studying handover procedures, it is not only generalised models of channels that are required in the simulation, but experimental soundings of worst case scenarios that occur near and at the cell boundaries. This means that soundings at a rapid rate are required at crucial locations.

1.2.5 Path Loss

So far we have discussed the fast fading phenomena that occurs when a MS travels over distances where the mean signal level is approximately constant. In addition to these rapid fluctuations in the received signal level, less signal power is received as the BS to MS distance increases. Even in free space the power captured by an antenna decreases with increasing distance as there is less power on the surface of a larger than a smaller sphere whose radii are the transmitter to receiver separation. In cities there are losses in power due to reflection, diffraction around structures and refraction within them. We refer to this diminution of received power with distance as path loss(PL). There are many treatises dealing with the fundamentals of radio propagation and how to estimate PL [2, 3, 9]. They tend to start with PL for propagation in free space, and then in the presence of different types of surfaces, followed by diffraction and reflection around obstacles, and so forth. Techniques are available to replace the undulations in the terrain relative to a flat earth by the use of knife edges. For example [9, 12], an equivalent knife edge can be positioned where the optical paths from each terminal to its horizon intersect. Allowing for many obstacles by knife edge diffraction techniques can become complex. Terrain-based techniques [13] determine the terrain profile between the BS and MS. A check is made to see if there is a LOS path, and whether there is Fresnel-zone clearance over the path implying "free-space" propagation characteristics. If this is true the free-space and plane-earth loss calculations are made and the higher one is adopted. Should this situation not prevail multiple knife edge calculations are performed (up to three edges) to find the PL.

The above approaches are used for BS antennas located at high elevations providing coverage over tens of kilometers. In urban environments the BS antennas for large cells are located on tops of tall buildings. As the MSs move along the streets, rarely within the line-of-sight (LOS) of their BS antennas, the calculation of the PL for any particular MS is, in general, analytically impossible. This means that network designers must site their BSs in a sub-optimal way; disregarding the other legal and financial difficulties involved in acquiring cell site locations. They must site the BSs to provide radio coverage over most of the geographical area they serve, which results in some areas receiving considerable overlapping radio coverage by a number of BSs. This increases the infrastructure costs.

However, the network designers do have a number of techniques for wisely locating their BSs. These are usually company confidential, built up with a mixture of published information, company experience and computer modelling. We will not stray into this specialised field. However, the important contributions of Okumura et al [14] and Hata [15] should be mentioned. Faced with intractable analytical calculations, an empirical approach of estimating PL was used by Okumura et al who carried out a detailed propagation measurement programme in the Tokyo area. Their results provide an estimate of the PL in dBs as the free-space path loss in dBs for a BS to MS distance d, to which is added the urban loss in dBs for a quasi-smooth terrain when the BS antenna is at $h_t = 200$ m while the MS antenna height is $h_r = 3$ m. From this are subtracted correction factors in dBs to allow for the actual BS and MS antenna heights in a particular situation. In addition, the PL calculations can be modified for different terrains, types of urbanisation, and so forth. This excellent work was presented in graphical form, and it was Hata [14] who turned it into a set of readily usable equations. These equations are presented in Chapter 2, and are defined in terms of the propagation frequency f in MHz ($150 < f < 1500$), h_t in m($30 < h_t < 300$) and d in km ($1 < d < 20$). We now know that the frequency range can be extended to at least 2 GHz and that providing the BS antenna clears the local building line, h_t can be reduced below 30 m, by adding a constant factor of some 6 dBs to PL. The Hata equations are primarily concerned with estimating PL in large cells. They yield straight-line PL(dB) versus log d curves.

When the received signal at the MS is averaged out to remove the effects of fast fading the local mean signal level at a particular distance d is obtained. The statistics of the local means are log-normally distributed, and seem to be dependent on the local environment. The standard deviations are typically 6 to 9 dB, and may be as small as 3 dB. The signal received at the MS is attenuated as the distance d increases due to increasing PL. At a particular d we may compute the PL using an empirical formula, such as Hata's, and then allow for the log-normally distributed mean, and the Rayleigh distributed fast fading. Identifying the receiver's noise floor level (NFL), we allow for there to be a low probability of the signal level rapidly fading below it, i.e., the NFL intersects the Rayleigh PDF such that there is only a small area in the tail of the PDF below NFL. Next we construct both the Rayleigh and log-normal PDFs. The mean of the log-normal PDF coincides with the path loss curve, while the mean of the Rayleigh PDF overlays the log-normal PDF at its two or three sigma point. Hata's straight-line curve enables the transmitted power to be estimated for the maximum cell distance d_{max}. Increasing the transmitter power moves the curve upwards pro rata and increases d_{max} accordingly.

1.2.6 Propagation in Microcells for Highways and City Streets

We will discuss microcellular mobile radio in later sections. Suffice to say here that microcells are small cells which may be a small segment of a highway, a street along the side of a city block, part of a park, an office floor and so forth. They are small areas where the teletraffic is high. Indeed, microcells are the most effective means of providing high user density mobile radio communications.

As we are still setting the propagation framework for system concepts, we consider in a qualitative way propagation in highway and street microcells [16]. Vehicular mobile stations (MSs) are applicable to both these types of microcells, but pedestrian MS are expected to be more numerous in city street microcells than vehicular ones. To establish a microcell we must contain the radiation so that the frequency band allotted by the regulatory authority can be re-used far more frequently than in conventional large cells. Accordingly the BS antenna is not mounted at a high elevation to get wide area coverage. Instead it is some 5 to 12 m above the ground, and the radiated power is generally in the milliwatt range [7]. We will commence by considering the path loss characteristics of vehicular highway microcells.

Path Loss

Early conceptual and theoretical investigations into highway microcells were made by Steele and Prabhu [6]. Having no experimental data, and being concerned with interference from other highway microcells, they assumed that free space propagation applied. Chia et al [17] later made measurements at 900 MHz along a number of highways in Southern England. An 18-element Yagi BS antenna having a gain of 15 dB and a front-to-back ratio of 25 dB was used. The power delivered to the antenna was 16 mW. During each experiment the antenna was mounted on a bridge that crossed above a motorway, with the antenna pointing along the motorway. As the MS drove along the motorway the received signal level was recorded.

Figure 1.14 shows a set of received signal level versus distance curves. When driving near the BS the MS receiver was saturated as shown by the flat top curves. Because the radiated power from the back of the antenna was reduced by 25 dB compared to that from the front, the highway did not lose line-of-sight (LOS) with the antenna in the backward direction before the received signal level decreased to the noise floor. It was found that an inverse fourth power propagation law was always a good fit to the data. The radiation in the forward direction went farther than in the backward direction. Again the inverse fourth power law was appropriate in the forward direction, although the variation in the received signal level was as much as ±10 dB. Notice that the curve for the M25 motorway at junction 21A initially decreased rapidly. This was due to the motorway turning sharply into a cutting, losing LOS and causing the signal to experience a diffraction loss. However, having sustained this loss the propagation conformed to the inverse fourth power law.

The countryside in Southern England is essentially flat and the inverse fourth power propagation law observed in highway microcells can be explained by a two-path model consisting of a LOS path and a path reflected

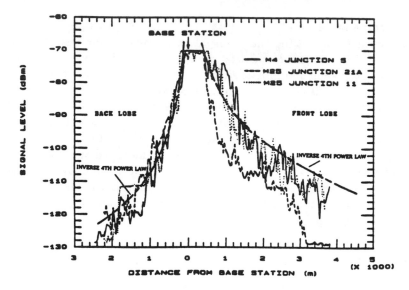

Figure 1.14: Received signal profiles at three different motorway locations (Chia et al. [17]).

from the road surface. This simple model for highway microcells has been investigated by both Rustako et al [18] and Green [10]. The theoretical curves of normalised signal level versus distance of the MS from the BS are shown in Figure 1.15. The transmissions are at 900 MHz and the BS and MS antenna heights are 5 and 1.5 m, respectively. The received signal level for the two-path model is relatively constant for distances close to the BS. The signal level then decreases in a fluctuating manner until the MS is some 200 m from the BS when the curve becomes smoother and decreases according to an inverse fourth power law.

Vehicular MSs in cities travel significantly slower than on countryside highways. Indeed in some cities their average speed is comparable with 19-th century horse transportation speeds! The high density of vehicles on city streets means that vehicular microcells in cities should, in general, be shorter than highway microcells as the mobiles spend longer on a given length of road. A useful design criterion is to arrange for mobiles to spend similar times in microcells in order to harmonise the handover rates of communications between adjacent microcells. A vehicular MS driving through a city effectively proceeds along a canyon or trench. The BS antenna is at lamp-post elevation and the buildings restrain the radiation to be within the canyon. There is also penetration of electromagnetic energy into buildings which provides useful radio coverage there. Even ignoring the presence of other vehicles, there are a number of radiation paths between a BS and a MS, but the predominant paths are the LOS path, the ground reflected

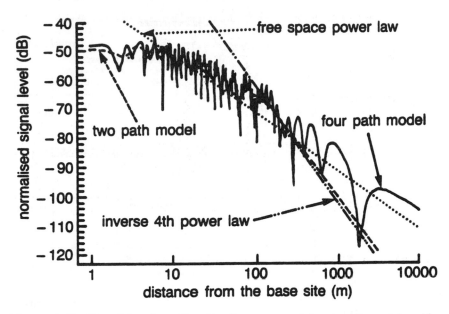

Figure 1.15: Signal level profiles for the two- and four-path models. Also shown are the free space and inverse 4th power laws (Green [10]).

path, and the two single reflected wall paths. When theoretical calculations are made based on these four most significant paths, and a perfectly smooth canyon and road having homogeneous properties are assumed, we obtain the curves shown in Figure 1.15.

The received signal level fluctuates significantly more than for the two-path model as we are now vectorially summing two additional paths at the MS antenna. Up to distances of 200 to 300 m the averages of the two-path and the four path curves are similar, but after these distances the curve for the four-path model becomes closer to the free space characteristic than to the curve for the inverse fourth power law.

Another way of considering the curves in Figure 1.15 is that they can be modelled as a single pole circuit, rather like a resistance-capacitance integrator, having a break frequency. However, the 3 dB down break distance is well within 100 m, and a break distance of more interest was proposed by Green [10]. He divided his measured data into two parts and applied regression analysis to each part to obtain two path loss straight line curves. The distance b where the data divided was varied, curves obtained, and the goodness of fit to the data found for both curves. The value of b that gave the best fit was identified and the two path loss laws established. From measurements in Central London the double regression analysis at 900 MHz yielded variations from 1.7 to 2.14 for the exponent of the curve for the data closest to the BS. For the data after the break distance b, the exponent var-

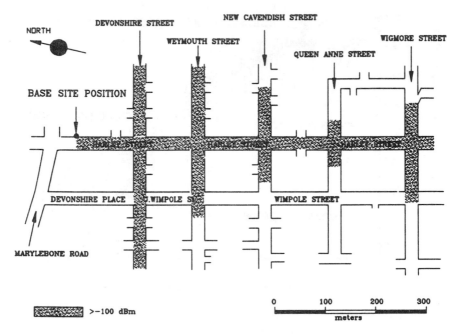

Figure 1.16: Street plan of the Harley Street area, Central London.

ied from 4.46 to 9.19, while *b* varied from 207 to 316 m. Chia [19] making measurements at 1.7 GHz reported the first and second exponents as 1.0 and 4.8, respectively, and *b*=115 m.

The above experimental results do not agree in detail to those predicted by the four-path model. This is not surprising as the buildings are not homogeneous and large vehicles can cause multiple reflections and diffractions. However, there are some general conclusions that can be made. Close to the BS the path loss is less than in free space, then it decreases with an exponent of two or less, and eventually it can plummet at a rapid rate. By arranging for the microcell to be sufficiently small, eg., 100 to 300 m, the signal level is reasonably well maintained throughout the microcell. Outside the boundary of the microcell the signal level falls rapidly, offering less interference to adjacent microcells.

Another important design parameter is the loss of signal level as a MS turns a street corner. To find this parameter measurements were made in the Harley Street area of Central London at 900 MHz [17]. This area was characterised by a near rectilinear grid pattern of roads, as shown in Figure 1.16. The power into the antenna was again 16 mW. A 10 dB gain Yagi antenna with a front-to-back ratio of 15 dB, a 5 dB gain omnidirectional colinear array, and a 9 dB gain corner reflector antenna with a 25 dB front-to-back ratio were used.

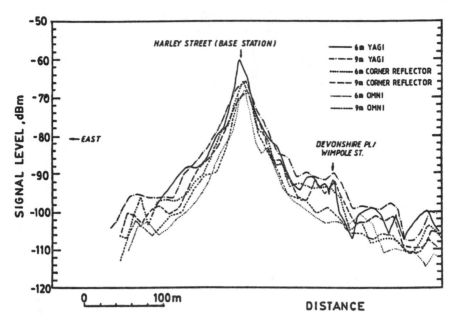

Figure 1.17: Received signal level profiles along Weymouth Street for different BS antennas (Chia et al. [17]).

With the BS antenna located at 6 or 9 m elevation in the position shown in Figure 1.16, the signal level was measured by the MS as it drove around the area. Figure 1.17 shows the received signal level along Weymouth Street for different BS antenna heights. Whether the antenna was at 6 or 9 m made negligible difference, although the effect of antenna gains was evident. The slopes of the curves in the vicinity of the corners varied from 0.55 to 1.3 dB/m. This rapid loss of signal level as a MS turns a corner means that unless care is taken in siting microcellular BSs, calls may be forced to prematurely terminate if the received signal level goes below the receiver noise floor before the MS transmissions can be transferred to another BS that can provide a higher signal level.

The shaded area in Figure 1.16 shows the region where acceptable communications can probably be maintained. The two-dimensional microcell has a christmas tree appearance with the BS near the foot of the tree. (A mirror image of the tree about its base is expected for an omni-directional antenna allowing for a concomitant increase in radiated power). The height of the tree is about 620 m, and its width at the first branch level is 350 m. A propagation model by Chia based on diffraction theory offers path loss predictions which are in reasonable agreement with these measurements [17].

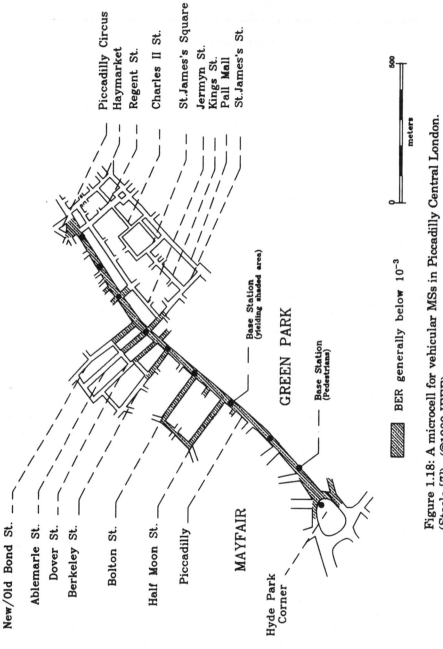

Figure 1.18: A microcell for vehicular MSs in Piccadilly Central London. (Steele [7]). (©1989 IEEE).

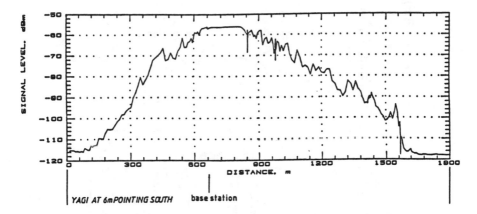

Figure 1.19: Received signal level profile along Piccadilly (Steele [7]).
(©1989 IEEE)

Figure 1.18 shows a microcell for vehicular MSs in Central London, while
Figure 1.19 shows the received signal level along Piccadilly for this microcell [7]. Similar observations to those made in the Harley Street area apply
here. We also show in Figure 1.18 a number of microcell BSs for pedestrian MSs. Typically we would arrange for the handover times between
pedestrian microcell BSs to be similar to the handover times for the larger
vehicular microcell BSs. Consequently there may be many pedestrian microcells within a vehicular microcell. A key point to emphasise is that by
deploying BS antennas on lamp posts or clamped to the sides of buildings
at elevations of 5 to 12 m allows the buildings and the streets they frame to
control the radio propagation. The building heights are unimportant, but
not the ability of the buildings to absorb or duct electromagnetic radiation.

Fading in Street Microcells

Although anticipated by Steele and Prabhu [6], Green [10] demonstrated
that microcells are not merely smaller, but have better propagation properties than large cells. The following experiment was performed along the
A33 road in Southampton. A microcell antenna was located on a pedestrian cross-walk at 7.5 m above the road. Omni-directional transmissions
at 1 mW and 900 MHz were radiated to the vehicular MS travelling along
the A33 road. Another BS antenna was located at a height of 35 m on the
tallest building on the Southampton University campus. This macrocellular BS was a mile from the microcellular BS and radiated 8 W at a similar
frequency. The vehicular MS was also able to receive the macrocellular BS
transmissions.

Figure 1.20 shows the received signal level and Rician K-factor distance

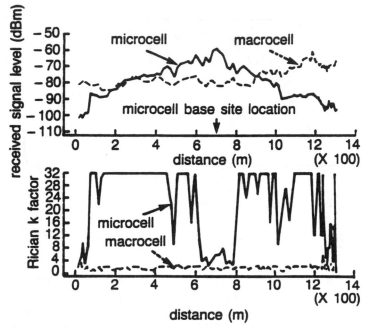

Figure 1.20: Received signal and K-factor profiles for the microcellular BS and the oversailing macrocell BS (Green [10]).

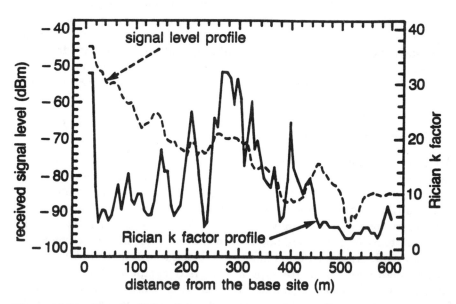

Figure 1.21: Received signal level and K-factor profiles for Harley Street, Central London (Green [10]).

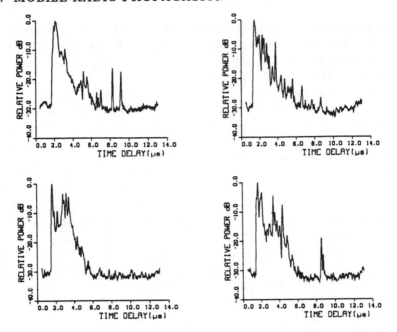

Figure 1.22: Some worst-case average impulse response envelopes for conventional size cells (Bultitude [20]). (©1989 IEEE)

profiles at the MS for transmissions from the microcellular and macrocellular BSs. The received signal level for the macrocellular BS was relatively constant throughout the microcell, and K was 1 or 2, i.e., close to Rayleigh fading. By contrast, the LOS between the microcellular BS and the MS often produced high K-factors, virtually Gaussian channel conditions, due in part to the tree-line nature of the road. Near the transmitter the K values were low, and this was probably due to the antenna radiation pattern, and path loss nulls associated with the two-and four-path models. As expected, the received signal level from the microcellular BS decreased with distance along the road that formed the microcell, but because of the higher K factors the microcell will have a better performance for digital transmissions than can be achieved by the macrocellular BS.

The path loss is the important factor in determining the microcell boundary. The K-factors only become influential as the edge of the microcell is approached. If the K-factors are low in this region, the channel is poor with occasional deep fades causing bursts of digital errors. Should the K-factor be high, few or no errors will occur, even though the received signal level is becoming low. In this situation the signal level could descend below the noise floor level causing the quality of transmission to deteriorate rapidly and catastrophically. The same effect can happen when the MS experiences cochannel interference from another BS and the cochannel is operating with a low K. Thus being able to predict the value of K is desirable. It is also difficult. Further, how to arrange for K to be relatively high at the microcell boundary is unknown at the present time.

Figure 1.21 shows the variations of signal level and K-factors for Harley Street made by Green at 900 MHz [10]. Chia [19] has measured K-distance profiles at 1.7 GHz showing high K values near the BS, but for distances from 20 to 300 m away from the BS the K values varied in a relatively random way between 2 and 20. He also found that the K factors differed significantly with BS location. Our own experiments confirm the widely differing K-profiles along similar streets, and until more is known about their generation the designer should work on worse case scenarios. We also note that the absolute value of K quoted by research workers depends on the windowing used, see Section 2.7.

Figures 1.22 and 1.23 show worst-case average impulse responses for small conventional size cells i.e., macrocells and for street microcells, obtained by Bultitude [20] at 910 MHz in urban conditions. For the macrocells the transmitter antenna was mounted 78 m above the ground and the cell radius was approximately 1 km. The microcell had a length of three blocks, with the BS antenna at heights of 8.2 or 3.7 m. The MS antenna height was 3.5 m. The impulse responses for the macrocells are shown in Figure 1.22 and relate to responses at the MS in the vicinity of different city blocks. The microcell channel impulse responses are shown in Figure 1.23. Each row corresponds to a specific location, while the first and second columns relate to BS antenna heights of 8.2 and 3.7 m, respectively. Although the maximum rms delay spreads for the microcells exceeded those of the macrocells the power of the long-delay components was not greater than -23 dB. However, the average rms delay spread was 3.7 times lower for microcells, and in general the microcell impulse responses indicated an ability to support higher bit rate transmissions without the need for equalisation compared to those for the macrocells. When typical average impulse responses were examined the microcell was clearly superior. From Bultitude's results we can extract suitable channel impulse responses for both street and for office microcells [21]. These are displayed in Figure 1.24.

1.2.7 Indoor Radio Propagation

Mobile communications within an office-type environment are often referred to as cordless telecommunications (CT). They can be achieved by introducing an optical or cable network within the building to which small fixed stations (FSs) are connected. These FSs communicate with portable stations (PSs) via radio. Notice that FS and PS correspond to base station (BS) and mobile station (MS) in cellular radio terminology. When cables are used it is possible to make them into leaky feeders, radiating electromag-

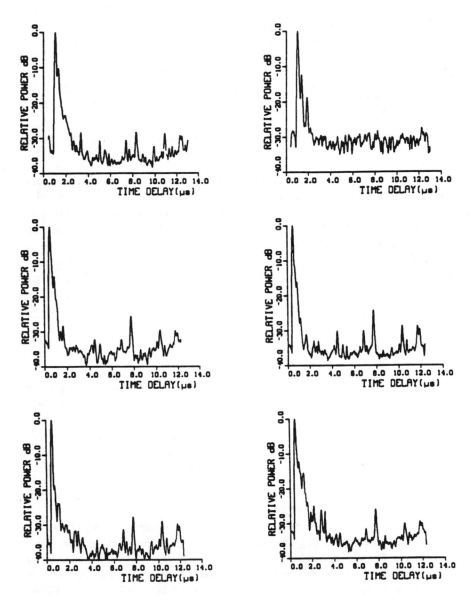

Figure 1.23: Some worst-case average impulse response envelopes for micro-cells (Bultitude [20]). (©1989 IEEE)

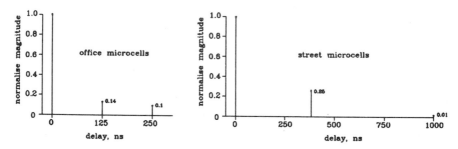

Figure 1.24: Discretised channel impulse responses.

netic energy directly to the PSs. Introducing an optical or cable local area network (LAN) into a building may be too expensive, and an alternative approach is to use a radio LAN, which, although easier to install, must be carefully designed to accommodate local movement of people and changes in the radio path due to resiting of office equipment.

Buildings exhibit great variation in size, shape and type of materials employed, making propagation prediction difficult. Compounding the problem is that radiation can exit from a building and return via buildings close by. Indoor propagation in a building as a function of its construction and its proximity to other buildings is very complex.

Path Loss

Keenan and Motley [22] made measurements in a modern multistory office block at 864 and 1728 MHz. The building was 500 m by 15 m, steel-framed with reinforced concrete and had plasterboard internal walls. The receiver was located at one end of the sixth floor and the transmitter moved on this and the next lower two floors. Adopting the traditional starting point, they opted for the straight line representation of path loss, namely,

$$PL = L(\nu) + 10n \log_{10} d, (dB) \qquad (1.22)$$

where d was the straight-line distance between the transmitter and receiver, and therefore passed through floors and walls. The propagation loss exponent was n. An additional loss was observed which was accommodated by a so-called clutter loss $L(\nu)$, which was easy to introduce but difficult to quantify its physical meaning. From Equation 1.22, $L(\nu)$ was the PL at d = 1 m. The distribution of $L(\nu)$ was lognormal with variance ν. For both propagation frequencies n was close to 4, but the regression lines provided a poor fit to the data. The PL equation was then modified to

$$PL = L(\nu) + 20 log_{10} d + n_f a_f + n_w a_w \qquad (1.23)$$

i.e., n=2, where the attenuation in dBs of the floors and walls was a_f and a_w, and the number of floors and walls along the line d were n_f and n_w, respectively. The regression fit for this expression was much better. The values of $L(\nu)$ at 864 and 1728 MHz were 32 and 38 dB, with standard deviations of 3 and 4 dB, respectively. This law is elegant in its simplicity.

Owen and Pudney [23] found that the Keenan and Motley model provided a good fit to their data at short range, except they used the horizontal range rather than the three-dimensional straight line distance d. This was because for signals on floors adjacent to the one housing the transmitter there was only a small path loss in the vertical plane, as the lower signal levels on these floors was essentially due to the attenuation by the floors. Energy also arrived on other floors by means of stairwells and lift shafts. At 1650 MHz the floor loss factor was 14 dB, while the wall losses were 3 to 4 dB for the double plasterboard and 7 to 9 dB for breeze block or brick. The parameter $L(\nu)$ was 29 dB. When the propagation frequency was 900 MHz the floor loss factor was 12 dB and $L(\nu)$ was 23 dB. The higher $L(\nu)$ at 1650 MHz was due to the reduced antenna aperature at this frequency compared to that at 900 MHz. For a 100 dB path loss the BS to MS distance exceeded 70 m on the same floor, was 30 m for the floor above, and 20 m for the floor above that, when the top propagation frequency was 1650 MHz. The corresponding distances at 900 MHz were 70 m, 55 m and 30 m. The propagation decay exponent was approximately 3.5 for both frequencies. These results apply to a particular building, and they can be expected to vary significantly with the type and construction of the building, the furniture and equipment it houses, and the number and deployment of the people who populate it.

Bultitude et al [24] found that for line-of-sight (LOS) along a corridor, the exponent n was 1 for 1.7 GHz and 1 to 3 for 860 MHz transmissions. The average loss by a single partition into an office was 5 dB at both 860 MHz and 1.7 GHz. A large variation in loss between floors occurred which depended on the building structure. For example, at 1.7 GHz the losses to the adjacent floor varied from 26 to 41 dB, while over a distance of two floors the variation reduced from 45 to 52 dBs. For an open office, free space propagation plus a lumped attenuation that was both frequency and distance dependent occurred. At 1.7 GHz, this lumped attenuation was -1 dB for d < 3 m, and -9 dB for d > 20 m. The attenuation tended to be a couple of dBs less at 860 MHz. For FS and PS on the same floor, the furniture in an open office contributed 5 to 7 dB loss when LOS was achieved, and 12 to 17 dB when there was no LOS. The building penetration loss was found to be some 24 to 37 and 17 to 29 dBs depending on the angle of incidence for 1.7 GHz and 860 MHz, respectively.

Cox, a pioneer in mobile radio propagation, and his associates studied the problem of low level transmissions into and around suburban houses when the BS was located at a height of 27 ft and when the BS to MS distance varied up to 2500 ft [25]. This was equivalent to having a street microcell that also included the rooms in houses. However, the measurements were done within and in the immediate vicinity of the houses. The exponent n was found to vary from 3 to 6.2. The slow fading, e.g., between rooms, and between houses, was log normal.

Fading Properties

Based on measurements at 1.5 GHz in a narrow two-storey building, Saleh and Valenzuela[26] noted that the indoor channel impulse response changed slowly as it was related to people's movement, and in the absence of LOS it was independent of the polarisation of the transmitting and receiving antennas. The maximum delay spreads were up to 200 ns in rooms, and values in excess of 300 ns were found in hallways. The rms delay spread within rooms had a median value of 25 ns and a maximum value of 50 ns.

Of particular interest is their model of the channel impulse response. They observed that rays arrive in clusters, and the rays that trigger a cluster are due to the building superstructure. When a particular ray arrives in the vicinity of the transceiver multiple reflections occur in the local environment generating a sequence of received rays. The later the arrival of a ray the more reflections it has probably experienced and the smaller its magnitude is likely to be. While these rays are still arriving, albeit of negligible magnitude, another ray arrives to initiate the next cluster, and so on.

Figure 1.25 shows spatially averaged power profiles within four different rooms. The dashed lines are the responses generated by their proposed model, namely, that the baseband channel impulse response is

$$h(t) = \sum_{l=0}^{\infty} \left[\sum_{k=0}^{\infty} \beta_{k,l} e^{j\theta_{k,l}} \delta(t - T_l - \tau_{k,l}) \right]. \tag{1.24}$$

In the square bracket we see the representation of an infinite number of rays in the ℓ-th cluster. The arrival time of the ray that initiated the cluster is T_ℓ, the arrival time of the kth ray measured from T_ℓ is $\tau_{k,\ell}$. Both T_ℓ and $\tau_{k,\ell}$ are independent of each other, and their inter-arrival times have exponential PDFs. Each ray is represented by a Dirac function $\delta(t - T_\ell - \tau_{k,\ell})$ whose weight is $\beta_{k,\ell}$ with phase $\theta_{k,\ell}$. The weights $\beta_{k,\ell}$ are independent Rayleigh variables whose variances decay exponentially for both the rays that initiate a cluster and for rays within a cluster. Their phase angles are independent uniformly distributed random variables over $(0, 2\pi]$. Typical parameter values are: cluster arrival rate = 1/300 ns; ray arrival rate = 1/5 nsec, cluster delay time constant = 60 nsec, ray delay time constant = 20 nsec.

Point-to-point radio links, as distinct from BS to MS links, may be used in office environments to avoid the use of fixed fibre or cable links. The

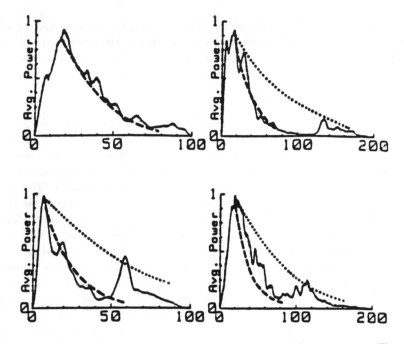

Figure 1.25: Spatially averaged power profiles within different rooms. The time axis in nanoseconds. (Saleh and Velanzuela [26]). (©1987 IEEE)

radio channel for these fixed radio links may be considered to be a nearly time invariant channel cascaded with a time variant channel, the latter due to the movement of people within the immediate environment of the terminals, while the former exists in the absence of movement. It appears that a notch may exist in the relatively static channel frequency response for long periods, regardless of the multi-path spread [24]. This means that even if the time variant channel does not require equalisation, the static channel does. For transmissions between fixed terminals, the average fading below and above the mean at 1.7 GHz was found to be 27 to 35 dB and 6 to 12 dB, respectively, for non-LOS transmission. The corresponding figures (only one set of measurements) for a LOS experiment was 25 and 4 dB. These large variations indicate that point-to-point communication channels within buildings will require complex channel conditioning circuits.

Devasirvatham [27] making measurements in a large building at 850 MHz observed a median rms time delay spread of 125 ns and a maximum delay spread of 250 ns. Bultitude [28] compared the power delay profiles in two buildings at 910 MHz and 1.75 GHz. The average profiles differed signif-icantly within the two buildings, although the standard deviation of the rms delay spreads was greater for 1.7 GHz. The coverage was less uniform at 1.75 GHz than at 910 MHz. Interestingly the median rms delay spread was about 30 ns, significantly less than the larger building measured in reference [27]. Even lower values of median rms delay spreads, 11 ns, were reported by Davies et al [29] at 1.7 GHz.

1.2.7.1 60 GHz Propagation

Electromagnetic radiation is partially absorbed by oxygen molecules in the atmosphere. Resonant absorption lines occur in the band 50 to 70 GHz [30, 31]. They are resolvable at high altitudes where the molecular density is low, but broaden at the earth's surface inflicting significant attenuation on electromagnetic radiation. The spectrum from 51.4 to 66 GHz is the absorption band A_1. Another absorption band, A_2, stretches from 105 to 134 GHz. There are peaks in the attenuation of electromagnetic radiation due to water vapour absorption at 22 and 200 GHz. Figure 1.26 shows the oxygen and water vapour absorption curves. We notice that when the absorption in A_1 reaches a peak the attenuation due to water vapour is near a minimum [32]. The oxygen absorption is lower in A_2 than in A_1, while the water vapour attennuation is higher. These observations suggest that band A_1 is more suitable for communications than band A_2.

Alexander and Pulgliese [33] examined the prospects of using 60 GHz communications in buildings. Like them, Steele was attracted by the large bandwidths in band A_1, and the characteristic that the oxygen absorption would limit the cell size and hence improve frequency re-use [5]. The path loss at ground level due to oxygen absorption is some 14 dB/km, and there is an additional loss due to rain (that can in bad weather exceed that due to oxygen absorption). These losses are in addition to the usual path loss values discussed previously. The diffraction of 60 GHz radiation around a corner is only a metre or so, which means that a microcell is strictly defined by line-of-sight (LOS). This is both an advantage and a disadvantage. If microcells are to be sharply defined by building structures to limit co-channel interference; or to act as a communication node providing high bit rate LOS communications, then the use of band A_1 is desirable. If the signal is required to propagate beyond LOS, to provide coverage for an irregular shape microcell, then transmissions in band A_1 are unsuitable.

A study [34] of 60 GHz propagation for microcells, revealed that even along corridors in buildings free-space propagation applied as the wavelength is only 0.5 cm. Satisfactory communications were found to occur without the need to align the transmitter and receiver in offices and in lecture theatres of differing constructions. The transmissivity and reflectivity of aluminium or brass; wood; plasterboard; and glass; were measured in percentages as < 0.06, > 99; 6.3, 2; 63, 3; and 25, 16. The attenuation in signal strength was typically 8 dB for pedestrians, 10 to 14 dB for cars, 4 dB for bicycles and motorcycles, and 16 dB for buses, the measurements being made as they crossed the LOS path between the transmitter and receiver.

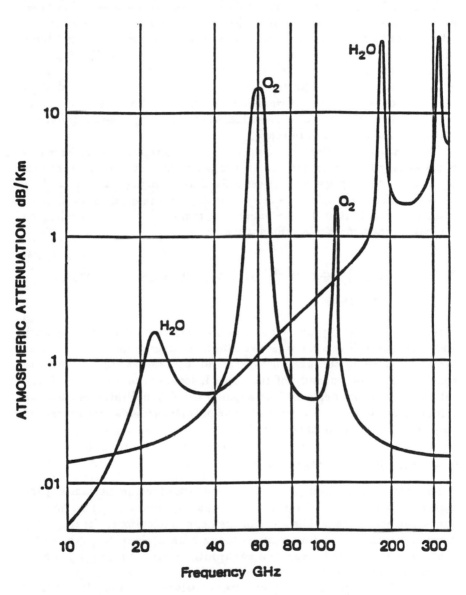

Figure 1.26: Oxygen and water vapour absorption curves.

An attractive use for transmissions in band A_1 is as a microcellular point-to-point link, where the transmitters and receivers are above the height of pedestrians but below the urban sky-line. Because of the absorption properties of band A_1, the same frequencies can be repeatedly re-used across a city to give numerous point-to-point links [34]. These links can distribute 2B+D ISDN channels to each dwelling, or link microcellular BSs together. We note that these applications can also use lower frequencies, say 15-50 GHz, as the urban infrastructure will absorb the radiation and allow frequency re-use. 60 GHz transmissions are of particular value when buildings are widely spaced, e.g., in suburbia, as then the oxygen absorption will effectively truncate the path range.

This concludes our discourse on mobile radio propagation, and the reader is advised to consult Chapter 2 for an in-depth treatment. Our deliberations so far relate to propagation between a fixed and moving transceiver. Now we consider how multiple users can access the radio channel. After that we will describe in Section 1.4 the notion of a radio cell, clusters of cells, and the basic arrangement of a cellular network.

1.3 Principles of Multiple Access Communications

Mobile radio networks provide mobile radio communications for many users who must be able to access or receive calls from other mobile users, or from users connected to the fixed networks. Ideally, they should be able to communicate independently of their speed, location, or the time of day, subject to teletraffic demand. The way in which the multiple users access a communication system is typically by one of three methods based on either frequency, time or code allocation [35].

Frequency Division Multiple Access

In frequency division multiple access (FDMA) the spectrum provided by the regulatory bodies is sub-divided into contiguous frequency bands, and the bands assigned to the mobile users for their communications. For frequency division duplex (FDD) transmissions using FDMA there is a group of n contiguous sub-bands occupying a bandwidth W Hz for forward or down-link radio transmissions from a BS to its MSs, and a similar group of n sub-bands for the reverse or up-link transmissions from the MSs to their BS. A band of frequencies separates the two groups. The arrangement is shown in Figure 1.27. Interference in adjacent sub-bands due to non-perfect channel filtering is designed to be below an acceptable threshold.

Each MS is allocated a sub-band, or channel, in both of the FDD bands of W Hz for the duration of its call. All the first generation analogue cellular

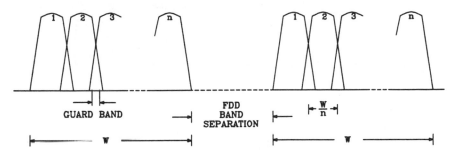

Figure 1.27: FDMA/FDD channel arrangement.

mobile radio systems in service use FDMA/FDD with the speech conveyed by analogue frequency modulation (FM), while the control is performed digitally with the data transmitted via frequency shift keying (FSK). A serious disadvantage of FDMA is that a separate transceiver is required at the BS for each MS in its coverage area. Although each BS does not use all n channels, it may use many tens of them. Further, high power antenna combining networks are required to handle the simultaneous transmissions of many channels. A significant advantage of using FDMA in the first generation systems is that as each user's transmissions are over a narrow channel of bandwidth W/n the fading is flat, and this is easier to handle than transmissions over dispersive channels.

FDMA can also be used with time division duplex (TDD). Here only one band is provided for mobile transmissions, so a time frame structure is used allowing transmissions to be done during one half of the frame while the other half of the frame is available to receive signals. For the same number of mobile users the bandwidth required is the same as for FDD, namely, $2W$ Hz. FDMA/TDD is used in the cordless telecommunications (CT), see Section 1.7.

Time Division Multiple Access

Time division multiple access (TDMA) is a method to enable n users to access the assigned bandwidth W on a time basis. Each user accesses the full bandwidth W (not W/n as in FDMA) but for only a fraction of the time and on a periodic basis. Instead of requiring n radio carriers to convey the communications of n users as in FDMA, only one carrier is required in TDMA. Each user gains access to the carrier for $1/n$ of the time and generally in an ordered sequence.

In TDMA a framing structure is used as shown in Figure 1.28. Typically, a user is given a slot in a frame of duration T having n slots. If a user generates continuous data at a rate of R bits/sec, it must be transmitted

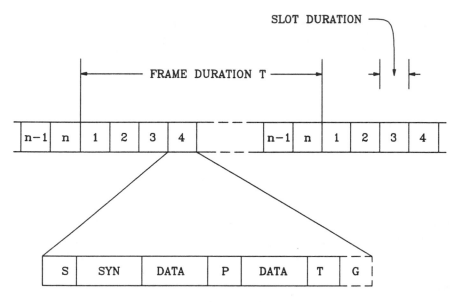

Figure 1.28: TDMA framing structure

in a burst at a higher rate of at least nR during each frame when it is the users' turn to transmit. In practice the actual transmitted rate R_t of a user is considerably in excess of nR because the data is transmitted in each slot in a packetised format, as shown in Figure 1.28, with extra bits to aid the receiver. Starting bits (S) and tailing bits (T) are often added to the data, such as all zero sequences, to assist in data recovery. Sometimes the header includes bits (SYN) to assist the receiver in bit timing recovery and frame synchronisation. A sequence (P) may be inserted to estimate the impulse response of the radio channel. A guard space (G), which may be viewed as a number of guard bits, is located between packets to ease synchronisation at the receiver and to accommodate the different propagation delays between MSs close to the BS and those far away.

In general, the higher transmission rate of TDMA compared to FDMA means that the channel often becomes dispersive, adding considerable complexity in signal processing at the receiver. However, there are advantages in using TDMA. In TDMA/FDD there are two frequency bands, one for the transmission and the other for reception. When the BS transmits the MSs are switched to reception, and when the MSs transmit the BS receives their signals. This significantly simplifies transceiver design compared to FDMA where transmission and reception are performed simultaneously. Further, the BS needs only one transceiver to accommodate its n TDMA users, as it only processes them one at a time at instants according to their slot position.

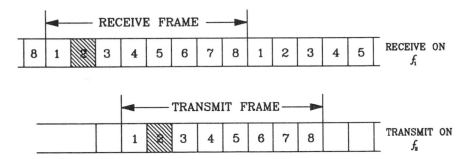

Figure 1.29: Transmitting and receiving slot assignments in TDMA.

Another feature of TDMA is the transceiver's ability to monitor other channels during slots when it is neither transmitting nor receiving signals. Figure 1.29 is drawn for a MS operating on channel 2, i.e., slot 2, with a frame of 8 slots, where the transmitting frame is three slot positions displaced from the receiving frame. The receiving and transmitting carriers are f_1 and f_2, respectively. Monitoring of another BS's signal can be done, say, in receiver time slot 7, but with the MS retuned to the frequency of the monitored BS. The MS transceiver will re-synthesise its oscillator as it changes from receiver to transmitter to monitor modes in this periodic sequence.

In mobile radio communications a number of TDMA carriers, each carrying n users, may be assigned unique carriers to produce a TDMA/FDMA multiple access arrangement. This approach where each TDMA/FDMA operates with FDD is employed by the pan-European GSM and the American IS-54 systems, see Section 1.6. The digital European cordless telecommunications (DECT) network described in Section 1.7.2 uses TDMA/TDD/-FDMA.

Code Division Multiple Access

Spread spectrum multiple access (SSMA) is another method that allows multiple users to access the mobile radio communications network. Each user is allowed to use all the bandwidth, like TDMA, and for the complete duration of the call, like FDMA. The presence of all the mobile users having their signals occupying the entire bandwidth at the same time appears at first sight to be a nightmare, with each user interfering with every other user. So it is perhaps surprising to realise that spread spectrum communications was originally conceived to provide covert communications with an innate robustness to jamming.

There are two basic types of SSMA. One is direct sequence SSMA (DS/SSMA), which is more frequently referred to as code division multiple access

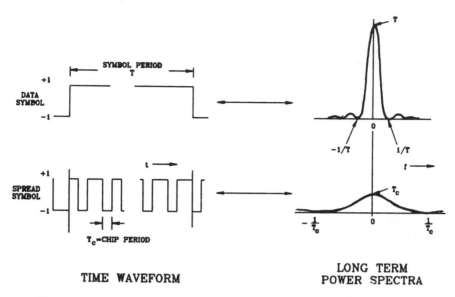

Figure 1.30: The long term spectra of the data and spread signals.

(CDMA) [36, 37, 38], and the other is called frequency hopped SSMA (FH/SSMA). The latter scheme was proposed [39] as an alternative to FDMA in the first generation cellular radio systems. CDMA is a second generation digital cellular scheme whose parameters are given in Section 1.6 [40, 41]. As CDMA is currently proving to be a successful multiple access method for cellular radio we will describe its salient points, and leave the reader to pursue FH/SSMA. CDMA is conceptually more complex than FDMA and TDMA, but not necessarily more difficult to implement because of the advances in microelectronics.

Let us consider a MS generating data at a rate of R bits/sec. We "spread" each bit by representing it by a sequence of N_c pulses, known as chips, within the bit period T. Each chip has a duration T_c, and $T = 1/R = N_c \cdot T_c$. The bandwidth of the spread signal is much greater than the bandwidth of the data signal. Figure 1.30 shows the long term power spectra of the data and spread signals. Suppose there are M users, such that when a logical 1 bit is generated each MS uses its unique PN code of N_c chips, and when a logical 0 is formed the inverse of this PN code is transmitted. Suppose these M mobile users transmit to a BS using the same frequency band via binary phase shift keying (BPSK). This modulation means that the polarity of the chips controls the phase of the transmitted signal. For the kth user the transmitted CDMA signal is

$$s_k(t) = \sqrt{2P_s}.a_k(t).b_k(t)\cos(2\pi f_c t + \theta_k) \qquad (1.25)$$

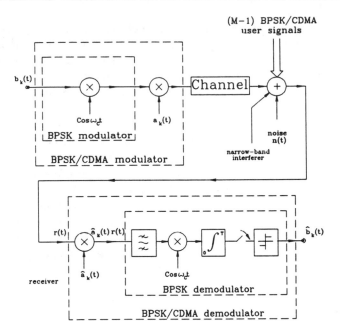

Figure 1.31: Basic BPSK/CDMA link

where P_s is the transmitted power, $a_k(t)$ is the spreading code signal for the kth user, $b_k(t)$ is the kth users data signal, f_c is the common carrier frequency used by all mobiles and θ_k is the phase of the kth user.

Figure 1.31 shows the arrangement of a BPSK modulator and demodulator. We may therefore view the CDMA signal $s_k(t)$ as the multiplication of the BPSK signal $b_k(t)\cos\omega_c t$ by the spreading signal $a_k(t)$. The recovery of the BPSK signal is achieved by multiplying the received signal r(t) by $\hat{a}_k(t)$, where $\hat{a}_k(t)$ is $a_k(t)$ suitably delayed such that the $a_k(t)$ that arrives at the receiver is synchronised with $\hat{a}_k(t)$. Observe that the BPSK signal was spread at the modulator due to the multiplication by $a_k(t)$, and consequently all components of r(t), other than the wanted component, will experience spreading due to the multiplication by $\hat{a}_k(t)$. For those unwanted components the BPSK/CDMA demodulator acts like a CDMA modulator! Only the wanted component is de-spread into a narrowband BPSK signal, as the product of $\hat{a}_k(t)\,a_k(t)$ is unity.

The power spectral density (PSD) of the received signal r(t) is shown in Figure 1.32(a). The components of this PSD are the PSD of the M BPSK/CDMA users, i.e., the received kth BPSK/CDMA signal plus the (M-1) other BPSK/CDMA users, and the PSD of the receiver noise. Included is the PSD of an arbitrary narrow-band interferer, such as a point-to-point radio signal whose carrier is f_I. After multiplication of r(t) by

$\hat{a}(k)$ the wanted component in r(t), namely the BPSK signal is obtained. The PSD of the signal components applied to the BPSK demodulator in Figure 1.31 are displayed in Figure 1.32(b). The PSD of the wanted signal is a signal occupying a bandwidth of 2/T about the carrier. The narrowband interference has been spread by $\hat{a}(t)$ into a wideband signal, and the M-1 other CDMA signals, and the receiver noise, remain wideband.

The signal $\hat{a}_k(t)r(t)$, whose PSD is shown in Figure 1.32(b), is applied to the band pass filter shown in Figure 1.31. This filter has a bandwidth from $f_c - (1/2T)$ to $f_c + (1/2T)$, which accepts the wanted component, and rejects most of the interference from the (M-1) other CDMA users and from the narrow band interference, as well as from the receiver noise. In keeping with the spirit of this introductory chapter we will keep the mathematics to a minimum. Rather than proving the reduction in noise power, we can see that as the PSD of $r(t)\hat{a}_k(t)$ is composed of $sinc^2()$ functions, then if the filter has a bandwidth sufficient to pass the wanted signal, the noise and interference components are reduced by (T/T_c). This quotient is called the processing gain

$$G = \frac{T}{T_c} \qquad (1.26)$$

i.e., the ratio of the chip rate to the data rate. The greater the amount of spreading by having more chips per data bit, the larger G becomes, and the lower the noise power at the multiplier in the BPSK demodulator.

The signal at the output of the filter is multiplied by a coherent carrier, and the resulting signal integrated over a bit period T. The output of the integrator is sampled at the end of each integration period, and the polarity of the sample specifies the logical state of the recovered bit.

We notice that the receiver was faced with M BPSK/CDMA users and recovered the wanted signal as it knew $\hat{a}_k(t)$. A base station receiver must have $\hat{a}_k(t)$; $k = 1, 2, \ldots, M$, in order to recover the signals from all M mobile stations. The separation of the received signal into the M user signals is done on a code basis, and hence the term code division multiple access.

The description given has been simplified. For a mobile channel impulse response given by Equation 1.12, the received signal at a CDMA BS is

$$r(t) = \sum_{k=1}^{M}\left[\sum_{i=1}^{L}\beta_i b_k(t - \tau_i - \Gamma_k)a_k(t - \tau_i - \Gamma_k)\right.$$

$$\left. \cdot \cos\{\omega_c t - \omega_c(\tau_i + \Gamma_k) + \Psi_k + \phi_i\}\right] + n_k(t) \qquad (1.27)$$

where for the kth MS, Γ_k is the delay when transmissions start, ψ_k is the phase of the carrier, and $n_k(t)$ is the receiver noise. If the BS is recovering

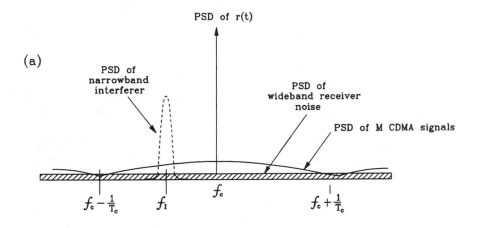

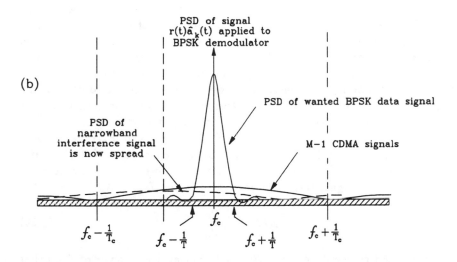

Figure 1.32: Power spectral density of, (a) received signal r(t), and (b) after de-spreading

the kth MS signal, it forms

$$Z_k^j = \int_0^T r(t)a_k(t - \tau_j - \Gamma_k)$$
$$\cdot \cos\{\omega_c t - \omega_c(\tau_j + \Gamma_k) + \psi_k + \phi_j\}\, dt \qquad (1.28)$$

where the receiver is locked onto the j-th path assuming β_j is the largest component. It is common for CDMA receivers to have the ability to repeat this process for other paths, say u and v, to give outputs Z_k^j, Z_k^u, Z_k^v. By combining these signals coherently we make our decision by sampling the combination and generating a logical one if the sample is positive, otherwise a logical zero is formulated. The procedure is known as correlation diversity or path diversity, or the RAKE process.

The ability to synchronise the locally generated version of the users code prior to cross-correlation is essential if good performance is to be achieved [40]. Asynchronous CDMA has a poor performance unless different forms of diversity reception are added [37].

The ratio of the bandwidth $W(\simeq 1/T_c)$ of the spread signal to the bit rate $R=(1/T)$ of the data is the processing gain G. The more chips in the code the more unique it is and the higher G becomes. To maximise the number of users it is important that all MSs transmit at power levels such that the received power at the BS from each of them is, to a good approximation, the same. If one MS transmits at too high a power level the quality of the link for all MSs deteriorates. To ensure link quality is maintained in the presence of a rogue mobile it is necessary to reduce the number of mobile users. Consequently every effort is made to control the radiated power from each MS, no matter where it is in the cell, such that the received signal power at the BS is of the required value.

This signal power is the product of R (bits/sec) and E_b (energy/bit), while the interference from the other MSs is the product of W(Hz) and $(J_o + N_o)$ (watts/Hz), where J_0 and N_o are the interference and receiver noise PSDs. The signal-to-interference ratio is, for $J_0 >> N_o$,

$$SIR = \frac{RE_b}{WJ_o}. \qquad (1.29)$$

If the wanted received power is P_R, and the received power of the other (M-1) users is $P_R(M$-1), we have an alternative expression for SIR of

$$SIR = \frac{P_R}{P_R(M - 1)}. \qquad (1.30)$$

Hence,

$$M \simeq \frac{W/R}{E_b/J_o} = \frac{G}{(E_b/J_o)} \qquad (1.31)$$

and M increases with the processing gain. The ability to control the power P_R from each MS so that it is the same at the BS in the presence of Rayleigh fading is a major problem in CDMA, and essential if Equation (1.31) is to be valid.

CDMA transmissions in neighbouring cells using the same carrier frequency will cause interference which we can allow for by introducing a factor F. This factor reduces the number of users as the interference due to users in other cells is added to the interference caused by the other mobiles in the users cell. When one speaker is listening in a conversation his bit rate can be significantly reduced to allow only the back-ground noise to be transmitted. As on average people speak in conversations for only 40% of the time the interference they generate can be significantly decreased when they are not speaking. We designate the reduction in interference due to speaker inactivity by introducing a factor d. Notice that this inactivity results in the overall interference being reduced and this benefits all users. This is radically different to the situation in TDMA and FDMA where only cochannel users benefit. It is also common place in cellular radio to introduce sectorisation of cells where instead of using omnidirectional antennas at a base site, S antennas are used each radiating into a sector of $(360/S)^o$. These interference mitigating factors, lead to an increase in the number of CDMA users to

$$ M \simeq \frac{W/R}{E_b/J} \cdot \frac{1}{d} . F.S. \qquad (1.32) $$

Imperfect power control can be allowed for by multiplying the right handside of the Equation 1.29 by another factor which is less than unity, and better than 0.5. Typically we may anticipate in a well run system that $d = 0.4$, $F = 0.6$, $S = 3$ to 6.

1.4 First Generation Mobile Radio Systems

These systems are essentially concerned with the transmission of speech signals, although they are able to transmit data at relatively low bit rates. They are usually referred to as 'analogue' systems as the speech signals are not digitally encoded prior to transmission on a radio frequency (RF) carrier. However, all the command and control of the network is digital. The user accesses the systems by means of frequency division multiple access (FDMA). When a call is connected, the mobile user is assigned a frequency band exclusively for his or her use until the call is completed.

There are numerous analogue mobile radio networks [3, 4]. These include the Nordic Mobile Telephone (NMT) system, the American Advanced Mobile Phone Service (AMPS), the British Total Access Communications System (TACS), the German Netz C and D networks, and the Japanese

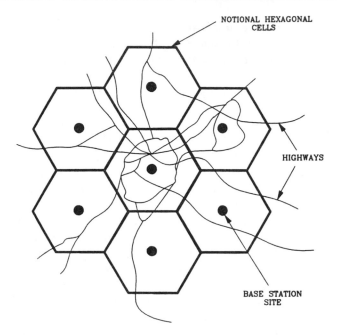

Figure 1.33: A cluster of hexagonal cells overlaying geographical area.

Nippon Advanced Mobile Telephone System (NAMTS). There are only minor differences between these analogue mobile radio systems.

Communications between a roving mobile station (MS) and another MS, or between a MS and a fixed station, such as a telephone connected to the public switched telephone network (PSTN) or an integrated services digital network (ISDN), generally involves the use of cellular techniques. The basic concept in FDMA (and TDMA) cellular mobile radio is to divide the spectrum W assigned by the regulatory body into equal parts B_c, and to allocate B_c to each base station (BS) in a cluster of N BSs until all the bandwidth $W(=NB_c)$ is used. Conceptually, each BS may be viewed as being located at the centre of hexagonal cell (which is not physically realised) and the hexagons are tessellated to provide a continuous mosaic over a geographical area as illustrated in Figure 1.33. The radiated power from a BS is sufficient to provide adequate radio coverage for all the MSs travelling in its domain, or cell.

As the assigned frequency band W is totally used by the cluster of N BSs, it must be repeatedly reused if contiguous radio coverage is to be nationally provided. This means that N-cell clusters must be tessellated as shown in Figure 1.34 where seven cell clusters are shown. The consequence of tessellating clusters is that a MS travelling in, say, cell 4 of a particular cluster will experience interference from BSs located in cell 4 of other clusters that are

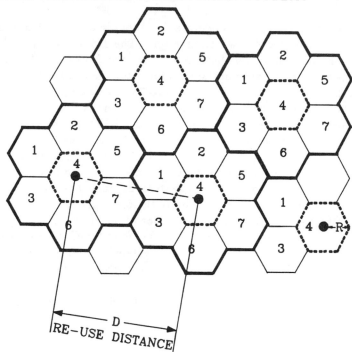

Figure 1.34: Tessellated clusters of cells; seven cells per cluster.

transmitting to MSs in their cells using the same frequency. For the seven-cell cluster there are six significant interfering BSs, although only four are shown in 1.34. This interference is known as cochannel interference. The BSs that use the same radio frequencies must be sufficiently separated for the cochannel interference to be acceptably small. As a consequence mobile radio communications are often referred to as being interference limited in that the BSs are moved close together to increase the density of users, but not so close that the cochannel interference results in unacceptable speech quality to the mobile user. The relationship between the reuse distance D, the cell radius R and the number of cells N is $D = (3N)^{1/2}R$.

The same bandwidth B_c, but not the same frequency band, is assigned to each cell. If each user is allocated a bandwidth B_u, then the maximum number of channels per cell is B_c/B_u. As the cell covers an area A the density of channels is $B_c/(AB_u)$. By making A small this density increases. Consequently in city centres where the density of MSs is high the cell radius may be only 1 km, while in remote rural areas the radius may be some 35 km and yet still be sufficient to accommodate user demand.

The actual shapes of the cells are significantly different from the notional hexagonal ones. They are determined by the terrain, buildings, directivity of the antennas, radiated power level, and so forth. The BSs in the first generation systems occupy a moderate size room in an office block as each radio channel requires its own transceiver. The BS antennas tend to be mounted on the tops of tall buildings or on the peaks of hills, and the radiated power may be substantial ($\simeq$ 100watts) when the cells are large.

As the MSs are generally beyond the line-of-sight (LOS) of the BS the received signal envelope in the first generation systems experiences fast fading having a Rayleigh probability density function (PDF). As the bandwidth of the mobile radio channel B_u is is either 25 or 30 kHz, the fading is said to be flat. i.e., all the frequencies across the band fade by the same amount. As a consequence the channel is devoid of dispersion and data transmissions do not experience intersymbol interference (ISI). The PDF of the slow fading signal level is log-normal. The path loss (PL) experienced by the MS decreases with an exponent of the order of 3.5 rather than the inverse square law and two-path plane earth exponents of 2 and 4, respectively.

The radio links in the first generation systems are simple compared to those found in the second-generation system. Essentially the link is the transmission of frequency modulated (FM) speech, or frequency shift keying (FSK) data. The band used for TACS is 935-950 MHz for the forward band or downlink, and 890-950 MHz for the reverse band or up-link. The complexity of the first generation system lies in their control procedure.

1.4.1 Network Aspects

We have discussed the behaviour of mobile radio propagation in Section 1.2, emphasising the fading nature of the channel, that it can be dispersive and noisy. When describing cellular clusters we introduced cochannel interference from other cells and there is adjacent channel interference from users of adjacent channels. Much of the emphasis of this book is devoted to the methods of establishing reliable communications of acceptable quality over these hostile channels. However, this is only part of the problem of enabling users to make and receive calls while on the move. We need to have a network that will establish and terminate calls with mobiles, track the mobiles as they travel, enable their calls to be switched between BSs to maintain call quality, and so forth. To achieve all these requirements we have the cellular network.

The control of the first-generation cellular network is digital. Although this control is simpler than the control procedures adopted for the second-generation digital networks, e.g., see those described in Chapter 8, it does have all the basic features found in these more complex networks. We will therefore describe the control aspects of the British TACS network.

In TACS the BSs are connected by permanent links to mobile switching centres (MSCs) which are computer controlled telephone exchanges specifically designed for handling cellular services. The MSCs in turn are connected to the PSTN and to other MSCs. The arrangement is shown in

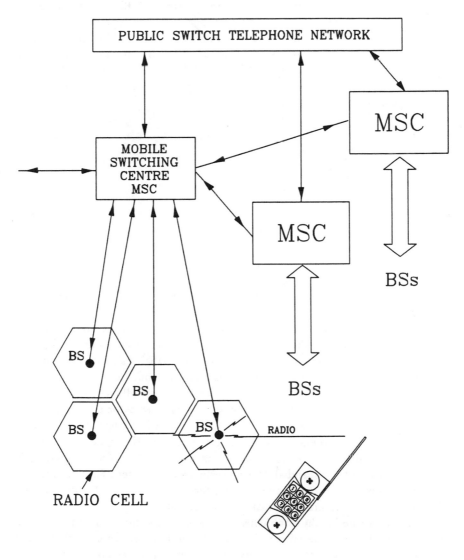

Figure 1.35: Network arrangement for TACS.

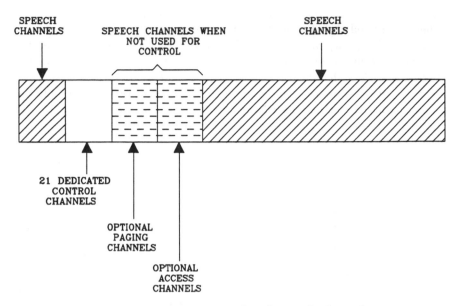

Figure 1.36: TACS control and speech channels.

Figure 1.35. It enables MSs to communicate with other MSs and to non-mobile users. It also allows calls to be connected to MSs who are not in their home area.

The cellular network must keep track of all the MSs that are subscribers to the network. It does this by forming traffic areas which consist of groups of cells. The MSCs log the current locations of all MSs through a process known as registration. When the MS is switched on, but is not making a call, it constantly listens to one of the common signalling channels transmitted by the BSs to determine the traffic area in which it is currently residing. Should the quality of the current signalling channel deteriorate as the MS travels it scans the signalling channels until it finds one with an acceptable quality. If the traffic area has changed the MS automatically, and unknown to the user, calls the new BS, gives its identification number, and is thereby registered in the new traffic area. The old BS de-registers the MS.

When a call has been established and a conversation is in progress the MS may travel near the edge of its cell. Should the communications quality become unsatisfactory the network is required to ensure that the MS changes its BS to one that can provide a better quality channel. The process of switching from one BS to another BS is called handover or hand-off. The handover process is required to be automatic, with no interruption of the call.

To facilitate the handover procedure the BS constantly monitors the received signal levels from all mobiles with speech transmissions in progress. When a received signal level drops below a threshold it informs the MSC that a hand-off may be required. The MSC commands all the surrounding base stations to measure the signal strength of the mobile, and then chooses the BS with the strongest received signal to handle the call. Once the new base station has been informed of the handover request, and the radio channel allocation has been made, the original base station is commanded to send a control message to the mobile to have it re-tune to the newly assigned channel. This happens automatically within a few seconds, and the mobile user is aware of only a very brief break in transmission (about 400 ms) when the actual handover takes place.

The TACS system has both control and speech radio channels. There are 21 channels reserved for control which cannot be used as speech channels. Of the remaining channels, some may be used for control where traffic demands extra control channels; otherwise, they can all be used as speech channels. Figure 1.36 shows the TACS control and speech channels.

1.4.1.1 Control Channels

The TACS control channels are used for the co-ordination of mobiles and for all call set-up procedures. Functionally, there are three types of control channel: dedicated control channels, paging channels and access channels. All mobiles are permanently programmed with the channel numbers of the dedicated control channels and they scan these channels at switch-on. The dedicated control channels carry basic information about the network and inform the mobiles about the channel numbers of the paging channels.

The paging channels are used both to transmit messages to specific mobiles, for example, to alert them of incoming calls, and for general network information, such as traffic area identity, channel numbers of the access channels, access methods to be used by mobiles, etc. The access channels are used by mobiles for accessing the network to initiate out-going calls, to register their location and to respond to paging calls.

All three types of control channel carry status information in the sequences of data blocks called overhead messages. The type of overhead message depends on the function of the control channel. Where the functions of dedicated control channel, paging channel and access channel are combined, the overhead messages contain all the relevant information as one message train.

TACS has a total of four signalling channels as shown in Figure 1.37. The Forward Control Channel (FOCC) and the Reverse Control Channel (RECC) are used for setting-up calls, as well as maintaining contact between BS and MS when no calls are in progress. The Forward Voice Channel (FVC) and the Reverse Voice Channel (RVC) must be compatible with a standard voice channel. In order to maintain system flexibility, FOCC and

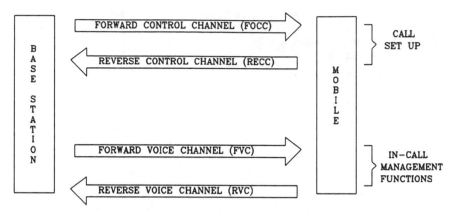

Figure 1.37: The TACS signalling channels.

RECC are also compatible with a standard voice channel. Data messages on the FVC and RVC are kept short because they are inserted periodically into the users voice channel. The FOCC is transmitted at all times so that a mobile entering the system can acquire system specific information, e.g., which channels are allocated for voice traffic. The RECC is only active intermittently, e.g., when the MS moves from one BS to another (no conversation in progress) or the MS initiates a call.

Signalling between the mobile and the base station for call set-up, hand-off and other similar control functions is carried digitally by using frequency-shift keying (FSK) of the radio-frequency carrier with a deviation of 6.4 kHz. The basic data rate used is 8 kbit/s, but to facilitate clock recovery at the receiver the data is Manchester encoded before transmission to give a rate of 16 kbit/s.

It is necessary for all signalling information to be correctly received. Accordingly robust error protection is used having the combined techniques of repeated transmission with majority decision and forward error correction coding. The message to be sent is coded into a block and a parity word is generated from it by using Bose-Chaudhuri-Hocquenghem (BCH) coding. The parity word is appended to the message to form a signalling block. A complete signalling frame is then formed which starts with a bit- and word-synchronisation sequences, followed by the combined message and parity word repeated several times.

When the message is received, a bit-by-bit majority decision is carried out on the repeated blocks, and many of the errors that may have occurred are corrected. The parity word is then used to correct up to one remaining error, and to detect if more than one error is present. If there are two or more errors remaining after majority decision, they cannot be corrected and the message is rejected.

Supervision

Whilst a call is in progress, a supervisory audio tone (SAT) is transmitted by the base station and looped back by the mobile. Both base station and mobile require the presence of the SAT on the received signal to enable the audio path. Three different frequencies are used for the SAT, all around 6 kHz. During call set-up, the base station informs the mobile which SAT to expect on the speech channel. If the SAT is incorrect, the mobile does not enable the audio path, but starts a timer which, on expiry, returns the mobile to stand-by. Similarly, the base station expects to see its transmitted SAT returned and takes this as confirmation that the mobile is operating on the correct channel.

The three SAT frequencies are allocated to the BSs to give a three-cluster repeat pattern. Consequently a cell in the adjoining cluster using the same radio channel uses a different SAT. The effect of this arrangement is to reduce the probability of a co-channel interfering signal being stronger than the wanted signal at a mobile.

Call Origination

When a user makes an outgoing call, the number required is keyed into the mobile, or the required number is extracted from the short-code memory in the mobile. A SEND key is then pressed. This causes the mobile to perform a system access. To do this the mobile first scans the network's access channels. The channel numbers of the access channels are transmitted as part of the overhead information on the paging channels. The mobile chooses the two channels with the highest signal level and attempts to receive the overhead messages being transmitted. Should the mobile fail to receive these messages on the strongest channel, it tries again on the other channel. Contained in the overhead messages are parameters which inform the mobile of the access procedure. Once the mobile has received these parameters, it checks to see if the access channel is in use. If it is available the mobile transmits its message.

When the mobile has sent its message to the network, it turns off its transmitter and remains on the access channel awaiting a message from the base station. For MS call originations, the received message from the BS is normally a speech-channel allocation and contains the channel number and the SAT code. On receipt of the message, the mobile tunes to the required channel and starts to transpond the SAT. If the correct SAT is received at the BS, the audio paths are enabled and the user can hear the call being set-up.

If the access was as a result of a registration, the message received from the BS on the access channel is normally a registration confirmation. On receipt of this message, the mobile returns to the IDLE condition.

Call Receipt

When an incoming call for a MS is received, the MSC checks the current location of the mobile which has been obtained through the registration procedures. A paging call is then transmitted on the paging channel of all BSs in the mobile's current traffic area. When the MS receives a paging call, it accesses the network in the same way as for call originations, but the message sent to the BS informs the network that the access is as a result of receiving a page. The MS receives a speech-channel allocation from the BS, tunes to the new channel, and checks the SAT received. The BS then transmits an alert message to the MS, causing it to alert the user of the incoming call and to transmit a continuous 8 kHz signalling tone. When the user answers the call, the signalling tone is turned off, the audio paths are enabled and the call proceeds.

Power Levels and Power Control

The transmit power levels used at a BS are chosen to give the required coverage area. The maximum effective radiated power (ERP) per channel is limited to 100W. Mobile stations have a nominal power class, namely 1, 2, 3 and 4 with ERPs of 10W, 4W, 1.6W and 0.6W, respectively. However, Classes 2 and 4 are used by the majority of MSs. In addition, the BS instructs the MSs within its coverage area to adjust their transmitted power to the minimum level required for acceptable performance. The power control facility reduces interference within the system and improves spectral occupancy. A maximum of eight different power levels may be emitted by a MS. The minimum ERP is -22 dBW and the step changes of ERP are 4 dB. The ERP to be used is transmitted to the MS using a three-bit attenuation code. On receipt of the power control message from the BS, the mobile sends an acknowledgement to the BS and selects the appropriate power level.

Call Termination

When a mobile user completes a call and replaces the handset, the mobile transmits a long burst (1.8s) of 8 kHz signalling tone to the BS, and then re-enters the control-channel scanning procedure. If the originator of a call on the PSTN clears down, a release message is sent to the mobile, which responds by sending a burst of 8 kHz signalling tone, after which it re-enters the control-channel scanning procedure.

We have shown how the TACS network operates to facilitate the ability of mobiles to make and receive calls while on the move. This is not the full story, but it does indicate that perhaps the major complexity of cellular radio lies in its network organisation. However, the quality of the communications is dependent on the radio links. We now examine the basic digital mobile radio link.

1.5 Digital Cellular Mobile Radio Systems

The second generation cellular mobile radio systems are all digital, and once more the primary service is mobile telephony. FDMA is abandoned in favour of time division multiple access (TDMA), although one American system employs code division multiple access (CDMA). The consequence is that the transmitted bit rate per carrier is significantly increased compared to transmitting digital speech via FDMA, and this usually results in time variant, dispersive mobile radio channels.

1.5.1 Communication Sub-systems

Of paramount importance is the speech codec which encodes the speech into a digital format for transmission, and decodes the regenerated bits at the receiver output to provide the recovered speech signal. We have discussed the time varying nature of the mobile radio channel, emphasising the distortions it can impose on the transmitted signal. Consequently we condition the signals prior to transmission in order to assist the receiver in its difficult role of regenerating the data with an acceptably low bit error rate (BER). Conditioning sub-systems include forward error correcting (FEC) codecs that operate on the encoded speech signal, increasing the bit rate while allowing the receiver to perform bit error correction. We also need to introduce interleavers to scramble the FEC data. Scrambling the data at the transmitter enables the descrambling process at the receiver to convert burst errors into random errors, and this improves the performance of the FEC decoding. The RF modem is another important sub-system that determines spectral occupancy and battery power drain. We also need to consider equalisers to remove the effects of channel dispersion, and the use of diversity in order to receive a number of versions of the transmitted signal and to combine them in such a way that the BER is decreased. We will now briefly describe some of these important sub-systems although they are considered in depth in subsequent chapters.

Speech Codec

There are a range of speech codecs that enable us to decrease the bit rate for an increase in complexity, while maintaining speech quality [42, 43, 44]. Logarithmic pulse code modulation (log-PCM) has long been used by the wire networks and can be used in mobile radio when there is sufficient channel bandwidth [45, 46, 47, 48, 49, 50]. Adaptive delta modulation

(ADM) [51] has long been favoured by the military for mobile communications, and is well suited to cordless telecommunication (CT) applications, i.e., short range mobile communications as used in offices. The speech quality of ADM at 32 kb/s is not as good as that of adaptive differential pulse code modulation (ADPCM) at the same bit rate when the transmissions are over ideal channels. However, over mobile radio channels where burst errors occur it generally provides better quality speech. In addition the battery consumption of ADM is significantly lower than that of ADPCM. Nevertheless, ADPCM is preferred in cordless communications, see Section 1.7, as it is and international standard, and ADM does not perform well when multiple tandem links are used in the network.

Sub-band coding (SBC) [52] at 16 kb/s is more appropriate for cellular radio where bandwidth is usually at premium compared to CT applications. The lower bit rate of 16 kb/s is achieved with considerable extra complexity and delay compared to ADM and ADPCM. Analysis-by-synthesis (ABS) techniques are currently in vogue for cellular radio [53, 54]. They operate effectively (after some channel coding) from 8 to 13 kb/s, and yield near toll quality speech. A popular ABS scheme is the code excited linear predictive codec (CELP) [53, 54]. The regular pulse excited linear predictive codec (RPE-LPC) had been adopted for the pan-European GSM network [55]. Low bit rate vocoders at 2.4 kb/s and below [56] that are used by the military are not currently employed in digital cellular mobile radio. Speech codecs are discussed in-depth in Chapter 3.

Channel Codec

The digitally encoded speech is channel coded [57] to enable the receiver to correct many of the symbol errors that occur during transmission [52]. As the symbol errors are associated with deep fades in the signal level they tend to occur in bursts. In order to correct symbol errors at the receiver we can either use a long block channel code that spans a number of error bursts, or use a short block code and interleave the data prior to transmission. The long codes work well because for most of the time there are few deep fades within the code words, and although error bursts occur during these fades the resulting erroneous symbols are only a small fraction of the total symbols in the code words. The consequence is that the channel decoder is able to correct the relatively few symbol errors. The penalty of using long codes is the delay, and more importantly, the complexity of the channel codec.

Short block codes do not have the high error correcting power of the long block codes. However, by interleaving the data prior to transmission, the errors in the bursts are randomised at the receiver during the de-interleaving process. Consequently during decoding there are relatively few errors in a block and satisfactory error correction is achieved. Short codes are usually preferred to long codes because of the lower complexity, but the overall delay is similar due to the delay incurred by the interleaving process. Important block codes are the Bose-Chaudhuri-Hocquenghem (BCH) and Reed Solomon (RS) codes that use one bit per symbol or many bits per symbol, respectively. The RS codes are more complex than the BCH codes with a more powerful and reliable error detection and error correction capability. The abilities of these codes are known with mathematical certainty.

Another class of channel coding used in mobile ratio is convolutional coding (CC). Interleaving must be used to combat error bursts, and the complexity of the codec increases exponentially with the error correcting power. CCs do not have the reliable error detecting capabilities of block codes, and the designer cannot guarantee the number of errors that will be corrected. The CCs are also subjected to error propagation effects.

However, CCs are often preferred to block codes in mobile radio applications. They can employ soft decoding whereby the signal applied for CC decoding is not a sequence of bits from the demodulator, but a multilevel signal. This approach yields an enhanced performance. Another feature of CCs is the use of 'puncturing', a process which provides a high rate code (i.e., the ratio of the information bits to the total bits) by periodically deleting some of the coded bits from the coder output. Puncturing reduces the complexity of the CC decoder compared to an identical rate non-punctured CC, while weakening its correcting power. Speech can be encoded, followed by convolutional coding. By puncturing the CC bits, control data can be inserted without effecting the speech encoding. The transmitted bit rate is unaltered, although the correcting power is marginally decreased.

In Chapter 4 we present the detailed operation of BCH and Reed Solomon block codes, convolutional codes, and the operation of different interleavers. The importance of channel coding and interleaving will be demonstrated.

Modulation

The modulation process converts the channel coded speech into a format suitable for transmission over the mobile radio channel [35]. It is desirable that the digital modulation technique employed has a high bandwidth efficiency, i.e., the bit rate per channel bandwidth is high for a given power expenditure and at a specific bit error rate (BER). The implementation costs should also be low. Constant-envelope modulated signals find favour in both digital cellular and cordless telecommunications (CT) radio links. They have acceptably good bandwidth efficiencies and relatively low battery power drain due to their use of class-C amplifiers.

Of particular importance in mobile radio is Gaussian minimum shift keying (GMSK), which is considered in detail in Chapter 6. The modulator is basically a Gaussian filter followed by a voltage controlled oscillator (VCO). The modulated signal has a spectrum that is narrow due to the deliberate introduction of ISI into the transmitted data stream. The ISI is generated by passing the input data through a digital filter having a Gaussian shaped impulse response which causes each bit to be spread over a number of bit intervals. The greater the bit spreading the narrower the spectrum of the GMSK signal and the better the adjacent channel interference performance. However, the clock recovery and symbol detection at the receiver becomes more difficult.

In fast frequency shift keying (FFSK) or minimum shift keying (MSK), logical 0 and logical 1 data bits are conveyed by assigning distinct carrier frequencies to them with a suitable frequency band between them to avoid detection ambiguities. To achieve a spectral occupancy that is less than that of MSK each data bit is filtered, to yield smooth transitions in the data signal before it modulates a VCO. We notice that filtering of the data is also done in GMSK, but an essential difference between MSK and GMSK is that in MSK the shaping of each data bit is done over only one bit period. When the shaping is done over one bit period it is referred to as a full response system, as distinct from a partial response system when it is implemented over more than one bit period. The consequence of MSK being a full response system is that the bandwidth of the modulated signal is larger than that for GMSK. However, the detection of MSK signals is simpler as its eye-pattern is more open.

The tighter spectrum of GMSK means that carrier spacings can be closer than in MSK for the same adjacent channel interference, producing a higher spectral efficiency. With the oscilloscope synchronised to symbol timing, the display of the demodulated signal has and eye-like appearance due to the data being filtered. The GMSK parameter BT is related to the number of bit periods over which each data bit is spread. Common values employed are BT = 0.3 and 0.5, e.g., BT = 0.5 means the spreading is over two bit periods. As this spreading does not occur in MSK its eye-pattern is wide open, and the bit regenerator samples the signal at the instants corresponding to the widest openings of the eye. For moderate amounts of signal noise, it is easy to identify the polarity of the signal and thereby regenerate the bits correctly. By contrast the GMSK eye-patterns are less open, making it more difficult in the presence of noise to identify the correct polarity of the received bit. By spreading each bit over a number of bit periods a multilevel signal and hence a multilevel eye-pattern is formed which depends on the variations in the patterns of logical ones and zeros in the data. Some patterns are easier to regenerate than others.

In contrast to these constant envelope modulation methods supporting one bit per symbol, we have multi-level modulators having n bits per symbol. For example, the modulated signal can have different discrete amplitudes, as in multi-level amplitude shift keying, known as m-ary ASK, where $m = 2^n$ signifies the number of levels. Sometimes the amplitude of the modulated phasor is constant, but its phase has one of m distinct values as in multi-level phase shift keying, called m-ary PSK. By combining the two types of modulation we obtain quadrature amplitude modulation (QAM) where the data is conveyed by both the magnitude and phase of the transmitted phasor [49, 50]. Many other types of multi-level modulation are possible. The twin phase modulation, known as star-QAM [58], has the desirable property that when used with differentially encoded data it effectively avoids the need for automatic gain control (AGC) and carrier recovery procedures at the receiver.

Figure 1.38 shows the constellations for these multi-level modulation signals. The dots or constellation points, correspond to the tips of phasors whose other extremities are at the centre of the constellation. Each constellation point is associated with a data word. For example, a 16-level QAM constellation supports 4-bit words on each constellation point or phasor. So during any symbol period the transmitted phasor, suitably filtered, will correspond to one point and convey 4-bits. By having 256-levels, 8 bits are conveyed by each transmitted symbol. Another multi-level modulation method is m-ary FSK where n bits of data are transmitted during each symbol period as one of m unique carrier frequencies.

The desirable feature of multi-level modulation is that each symbol carries n-bits, $n \geq 2$, and as the symbol rate determines bandwidth occupancy, the signal bandwidth is lower compared to binary modulation methods where $n = 1$. However, the channel SNR is required to be higher for multi-level modulators as the regenerator does not make simple binary decisions. QAM modems also require linear ampliers which although having lower efficiencies can be used in microcellular mobile radio networks as the radiated power levels are much lower than those used with large cells.

1.5.2 FDMA Digital Link

Having briefly discussed some of the communication sub-systems, we will now describe how they are cascaded to form a basic FDMA digital mobile radio link. The transmitter contains a speech and channel codec, an interleaver to provide time diversity, and a modulator prior to amplification for transmission over the mobile radio channel. At the receiver demodulation ensues followed by clock recovery and bit regeneration. The resulting data stream is de-interleaved causing the burst errors to be partially randomised, and channel decoding and speech decoding executed to recover the speech signal. The block diagram of the basic FDMA digital link is shown in Figure 1.39. Also shown in the Figure is a switched diversity arrangement. The two radio channels are associated with each of the two receiver antennas. These antennas are spaced sufficiently far apart that their received signals are uncorrelated. The signal applied to the demodulator switches from one antenna to the other whenever the received signal level falls below a system threshold or above and acceptable BER. We emphasise that there

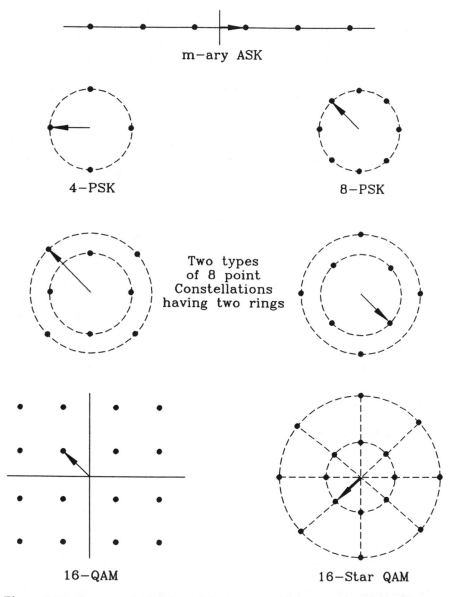

Figure 1.38: Some multi-level modulation constellations. Arbitrary phasors are drawn on the constellations.

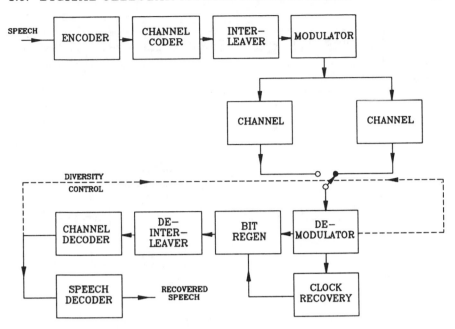

Figure 1.39: Basic FDMA digital link showing second-order switched diversity.

are a variety of space diversity techniques [2]. The one shown in Figure 1.37 is the simplest and is known as second-order switched diversity. By using more antennas the order of the diversity increases pro rata. As the signal applied to the demodulator is frequently switched in order to provide a good signal level the statistics of the signal being demodulated changes from, say, Rayleigh when only a single antenna is continuously used, to near-Gaussian when the order of diversity is high. A Gaussian channel is the best we can achieve as all the fading is eliminated. The general case of switched diversity is known as selection diversity (SD), where each antenna has its own receiver and the one with highest baseband SNR is selected to be the demodulated signal.

The weakness of selection diversity is that only one antenna is used at any instant, while all the others are disregarded. Maximal ratio combining diversity (MRCD) seeks to exploit the signals from each antenna by weighting each signal in proportion to their SNRs and then summing them. Accordingly in MRCD the individual signals in each diversity branch are cophased and combined, exploiting all the received signals, even those with poor SNRs. However, MRCD is more difficult to implement than SD.

Space diversity, i.e., where the receiving antennas are spaced apart, is usually employed in flat fading environments as the deep fades can be ef-

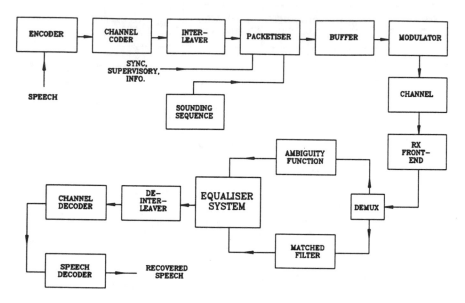

Figure 1.40: Basic TDMA radio link.

fectively combatted at the expense of system complexity. However, space diversity is less useful with dispersive channels, although it can be effectively employed but not in the simple ways discussed above. In dispersive systems the diversity procedures must be integrated into the equalisation techniques, or into the adaptive correlation diversity methods used in spread spectrum receivers. The system shown in Figure 1.39 assumes that the symbol rate is sufficiently low that dispersion has not occurred. We now consider TDMA links where dispersion is usually prevalent.

1.5.3 TDMA Digital Link

In time division multiple access (TDMA) systems a number of users signals are transmitted on a single RF carrier. Their transmissions are synchronised to occur in a particular time slot in each TDMA frame and hence the data is sent in packets where the packet duration is marginally shorter than the slot duration. If there are n slots in a TDMA frame, then it follows that if a user's coded data are generated at a rate R it will be transmitted at a rate in excess of nR, whereas in FDMA it is transmitted at R. The actual TDMA transmitted rate is above nR as described in connection with Figure 1.28.

Figure 1.40 shows the basic arrangement of a TDMA transmitter-receiver link. At the transmitter a packetiser accepts the coded interleaved digital speech at a rate R, forms a packet over a period of a frame and releases it as a data burst to the modulator for transmission. The high speed data on the RF carrier suffers considerable distortion during its transmission over the mobile radio channel and this must be compensated for at the receiver. After the receiver has demodulated the RF signal, demultiplexing of the baseband signal ensues to yield the channel data and the propagation channel sounding data. Referring to Figure 1.40, the sounding channel data is applied to a matched filter whose impulse response is matched to the sounding sequence inserted into the packet prior to transmission. Should the sounding sequence at the transmitter be passed through the matched filter (i.e., an ideal channel) a sharp pulse of some two bits width is produced. If this sharp pulse were passed through the equivalent baseband channel we would obtain the impulse response of the channel, slightly degraded by the sharp pulse not being an ideal delta function. We see that the cascading of the sounding sequence with the matched filter and radio channel yields an estimate of the baseband channel impulse response $h'(t)$. Because the mobile radio channel is essentially linear, we obtain the same $h'(t)$ by transmitting the sounding sequence over the mobile the channel, followed by the matched filtering, i.e., the arrangement shown in Figure 1.40.

Having obtained a measure of the complex baseband channel impulse response $h'(t)$ we are able to perform channel equalisation on the traffic data. Equalisation is a process which attempts to remove the ISI introduced by the channel [21,59-61]. In linear equalisers, or decision directed equalisers, the knowledge of $h'(t)$ enables the equalisers coefficients to be found. In Viterbi equalisation, which is really a maximum likelihood sequence detection process, a baseband modulator is used at the receiver to generate a wide range of possible signals that could have been received over the real channel. The data signal and the locally generated estimates of the data signal are combined to form mean square error signals, called metrics, which are used in a Viterbi processor (VP) to identify the most probable sequence of data transmitted. The operation of the VP is explained in detail in Chapter 6. The ambiguity function in Figure 1.40 is a filter whose impulse response is the convolution of the sounding sequence with the matched filter impulse response. It is introduced to compensate for the channel being sounded by an impulse of this type and not a delta function, i.e., both the data and the locally generated estimates of the signal are subjected to the same distortion.

Having regenerated the data in the packet, de-interleaving is performed (this may be over a number of packets) followed by channel decoding and speech decoding.

1.6 Second Generation Cellular Mobile Systems

In the USA the system to be introduced, known at the time of writing as IS-54, has the TDMA carriers spaced by 30 kHz to align with those in their analogue advanced mobile phone service (AMPS). Each carrier supports three users at the TDMA rate of 48.6 kb/s. The dual-mode transmissions are in the bands 824-849 MHz and 869-894 MHz. Vector sum excited linear prediction speech encoding (see Section 3.5) is used operating at 7.95 kb/s. After channel coding the rate becomes 13 kb/s, and allowing for control information the effective rate per user is 16.2kb/s. The 48.6 kb/s TDMA rate is transmitted using $\pi/4$ shifted DQPSK modulation at 2 bits/symbol [62].

The Japanese are taking a similar approach. As their first generation analogue system has carrier spacings of 25 kHz, they use this bandwidth to introduce a TDMA rate of 42 kb/s, again using $\pi/4$ shifted DQPSK. The speech channel coding rate is 11.2 kb/s.

The pan-European digital cellular mobile system is called GSM. These initials originally stood for, 'Groupe Speciale Mobile', after the name of the committee responsible for its specification, but it now represents, 'Global System for Mobile Communications'. The GSM network began to be deployed in Europe in July 1991. It is a much more ambitious network than IS-54. Operating in a TDMA mode, a regular pulse excited linear predictive coder (RPE-LPC) encodes the speech at 13 kb/s. This is followed by channel coding and bit interleaving to yield a voice rate of 22.8 kb/s. The data are assembled into packets with a propagation sounding sequence located in the centre of the packets, and transmitted via GMSK at a TDMA rate of 270.8 kb/s. The carrier spacing is 200 kHz. The frequency bands are 935-960 and 890-915 MHz. Because of the high transmission rate and the large cells that may be used (up to 35 km radius), the mobile radio channel is often dispersive and this requires receivers to employ channel equalisation as outlined in the previous section. Chapter 8 deals exclusively with the GSM system and we will refrain from discussing it further here.

In 1989 the British Government announced that service providers would be licensed to operate so-called personal communication networks (PCNs). This has now been done and these PCNs are likely to be operational by 1993. They will use a modified form of GSM, called DCS 1800, meaning a digital cellular system at 1800 MHz. Duplex bands of 75 MHz with a 20 MHz guard band will be used. Frequencies assigned are 1805-1880 MHz for the down-link (BS to MSs) and 1710-1785 MHz for the up-link (MSs to BS). Each service provider has a contiguous block of spectrum. The DCS 1800 specification is essentially that of GSM, with minor modifications as DCS 1800 is expected to operate with smaller cells than those used in GSM.

This means that the radiated power levels are lower for DCS 1800, and we may anticipate that the complexity of the equaliser will be significantly decreased. Both these features should decrease battery power drain in hand-held portables.

1.6.1 Qualcomm CDMA

Qualcomm Incorporated in the USA opted for CDMA as the multiple access method for mobile radio. Although CDMA has been well understood for a long time, its use in cellular radio had been avoided due mainly to the problems associated with power control. If the standard deviation of the received power from each mobile at the BS is not controlled to an accuracy of approximately ± 1 dB relative to the target received power the number of users supported by the system can be significantly curtailed. Other problems sited were whether there were sufficient codes available for a large number of mobile users, and difficulties of synchronisation. These major and many other minor problems have been successfully addressed by this CDMA system.

Qualcomm CDMA operates at the top of the AMPS band. The CDMA bandwidth required for each up-and down-link is 1.23 MHz, equivalent to 41 AMPS channels ($41 \times 30\ KHz = 1.23\ MHz$). This CDMA network also operates in the 1.7 to 1.8 GHz band.

Down-link

There is one pilot channel, one synchronisation channel, and 62 other channels. All of the 62 channels can be used for traffic, but up to 7 can be used for paging. The 64 Walsh codes of length 64 are used for each of these channels. The first 64 Walsh codes are shown in Figure 1.41. Walsh code W_0, an all-one code is used for the pilot, the alternating polarity W_{32} is used for the synchronisation channel, while the paging and traffic data use the other 62 Walsh codes. It is important to notice that the Walsh codes are used to identify the channel. Their modus operanti is very different on the reverse or up-link.

Figure 1.42 shows a block diagram of the BS transmitter. The pilot channel consists of a pair of pseudo random binary sequences (PRBS) at 1.2288 Mchip/sec. The synchronisation channel data at 1200 b/s is convolutionally encoded to 2400 b/s, repeated to 4800 b/s and interleaved over the period of the pilot PRBS. Each of these interleaved symbols spans 4 Walsh symbols; so that when W_{32} generated at 1.2288 Mchip/sec is exclusive-ORed with the sync data, a signal of 1.2288 Mchips/sec is produced.

The speech is encoded by a variable rate vocoder that generates forward traffic channel data at rates of 1.2, 2.4, 4.8 or 9.6 kb/s, depending on speaker activity. As the frame duration is fixed at 20 ms, the number of bits per

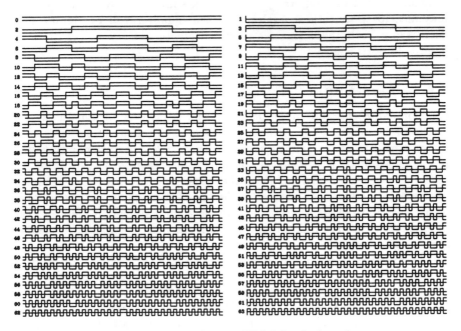

Figure 1.41: The first 64 Walsh basis functions.

frame varies according to the traffic rate. Half rate convolutional encoding with a constraint length of 9 doubles the traffic rate to give rates from 2.4 to 19.2 k symbols/s. To ensure the rate is always 19.2 k symbols/s, data repetition is appropriately used at the lower speech rates. Interleaving is performed over 20 ms, and the higher the data repetition used, the lower is the transmission power of the symbols.

A long code of 2^{12}-1($=4.4 \times 10^{12}$) is generated containing the user's electronic serial number embedded in the mobile stations long code mask. This code, suitably processed, scrambles the output stream from the block interleaver. The scrambled data is multiplexed with the power control information which essentially steals bits from the scrambled data. The multiplex signal remains at 19.2 kb/s and is changed to 1.2288 Mchip/s by the Walsh code W_i assigned to the users traffic channel. This signal is spread at 1.2288 Mchip/s by the pilot quadrature PRBS signals, and the resulting quadrature signals are then weighted.

The last set of channels are the paging channels. They provide the MSs with system information and instructions, in addition to acknowledgement messages following access requests made on the MSs' access channels. Essentially the paging channel data is processed in a similar way to the traffic channel data; but with the following exceptions. There is no variation in

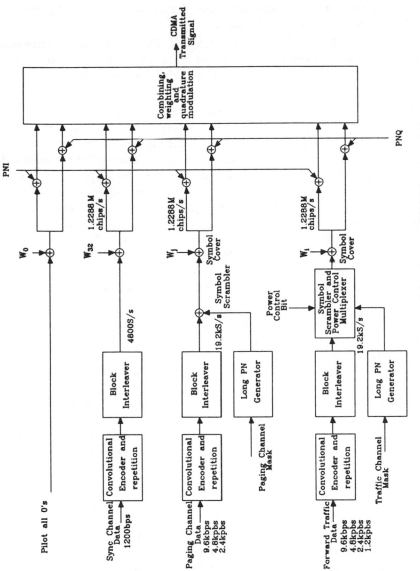

Figure 1.42: Block diagram of Qualcomm CDMA transmitter.

the power level on a per frame basis, and the 42 bit mask used to generate the long code contains different data.

All the 64 CDMA channels are combined to give single I and Q channels. These signals are applied to quadrature modulators and the resulting signals summed to form a CDMA/QPSK signal. The resulting CDMA signal is linearly amplified.

The pilot CDMA signal transmitted by a BS provides a coherent carrier reference for all MSs to use in their demodulation process. The transmitted pilot signal level for all BSs is some 4 to 6 dB higher than a traffic channel and is of constant value. The pilot signals are quadrature PRBS signals with a period of 32768 chips. As the chip rate is 1.2288 Mchip/s($= 128 \times 9600$ where 9600, is the maximum bit rate of the speech codec) the pilot PRBS corresponds to a period of 26.66 ms, equivalent to 75 pilot channel code repetitions every two seconds. The pilot signals from all BSs use the same PRBS, but each BS is characterised by a unique time offset of its PRBS. These offsets are in increments of 64 chips providing 511 unique offsets relative to the zero offset code. These large numbers of offsets ensure that unique BS identification can be performed, even in dense microcellular environments.

A MS processes the pilot channel and finds the strongest signal components. The processed pilot signal provides an accurate estimation of the time delay, phase and magnitude of three of the multipath components. These components are tracked in the presence of fast fading, and coherent reception with combining is used. The chip rate on the pilot channel, and on all channels is locked to precise system time, e.g., by using the Global Positioning System (GPS). Once the MS identifies the strongest pilot offset by processing the multipath components from the pilot channel correlator, it examines the signal on its synchronisation channel which is locked to the PRBS signal on the pilot channel.

All synchronisation channels use the same code W_{32} to spread their data. The information rate on the synchronisation channel is 1200 b/s (although its chip rate is 1.2288 Mchips/sec). Because the synchronisation channel is time aligned with its BS's pilot channel, the MS finds on the synchronisation channel the information pertinent to this particular BS. The synchronisation channel message contains time-of-day and long code synchronisation to ensure that the long code generators at the BS and MS are aligned and identical.

The MS now attempts to access the paging channel, and listens for system information. The MS enters the idle state when it has completed acquisition and synchronisation. It listens to the assigned paging channel and is able to receive and initiate calls. When told by the paging channel that voice traffic is available on a particular channel, the MS recovers the speech data by applying the inverse of the scrambling procedures shown in Figure 1.42.

Up-link

The up-link from MS to BS uses the same 32768 chip short code employed on the down-link, and the MS also uses its unique code embedded in the long 2^{42}-1 PRBS. Speech is again convolutionally coded, this time using a rate 1/3 code of constraint length 9, and data is repeated depending on speech activity. Interleaving is then performed over the vocoder block length of 20 ms. However, the repeated symbols are not transmitted giving rise to a variable transmission duty cycle. Whereas the down-link uses one-bit symbols, the up-link groups the data into 6-bit symbols. Each symbol generates an appropriate 64-chip Walsh code which is then combined with both the long PRBS to bring the rate up to 1.2288 Mchip/s, and the short code to launch it onto the quadrature modulation channels. Notice that the Walsh codes are used in a totally different way in the two links. In the down-link the Walsh codes label the channels, while on the up-link they convey data to their only destination, namely the BS.

In the traffic operating mode, the Walsh coded signals at a MS are modulated by the long 2^{42}-1 PRBS with a specific time offset that is unique to a particular MS, enabling the BS to distinguish signals arriving from different MSs. This long code is the same long PRBS that was used on the down-link. Further modulation by the quadrature 32768 chips PRBSs ensues, but with a fixed zero offset, followed by quadrature modulation. The reverse link does not use a pilot CDMA as to give each MS a pilot channel would be impracticable. The receiver at the BS has a tracking receiver and four receivers that each locks on to a significant path in the channel impulse response. The outputs of 64 correlators, one associated with each Walsh functions, are examined for each receiver. The outputs of the four correlation receivers are combined and the correlator number having the maximum output selected to identify the recovered 6-bit symbol. Progressive 6-bit symbols are serialised, de-enterleaved and convolutionally decoded.

The reverse CDMA transmission can accommodate up to 62 traffic channels and up to 32 access channels per paging channel. The access channel enables the MS to communicate non-traffic information, such as originating calls and responding to paging. The access rate is fixed at 4.8 kb/s. The output duty cycle is 100%; and the access channel is identified by a long PN mask having an access number, a paging channel number (on the forward or down-link) associated with the access channel, and other system data.

Each BS transmits on the same frequency (initially there will only be one carrier frequency) a pilot signal of constant power. As mentioned above, this pilot signal is sent as a CDMA signal whose code identifies the BS to the MS. However, the received power level of the received pilot also enables the MS to estimate the BS to MS path loss (PL). Knowing the PL the MS adjusts its transmitted power such that the BS will receive the signal at the requisite power level. However, the MS transmits and receives in duplex frequency bands which although having similar average PL values, have different instantaneously received signal levels due to the independent fading in each transmit and receive band. To allow for this independent fading, the BS measures the MS received power and informs the MS to make the appropriate fine adjustment to its transmitter power. One command every 1.25 ms adjusts the transmitted power from the MS in steps of ± 0.5dB. The dynamic range of the transmitted power is 85 dB.

The MS measures the SIR by comparing the desired signal power with the total interference and noise powers. If the SIR is below a threshold the MS requests the BS to increase its transmitter power, and vice versa when the SIR is above the threshold. The changes in MS transmitter power are small for this situation, $\simeq 0.5$dB over a range of ± 6dB, and are made at the vocoder frame rate.

An interesting feature of CDMA is that it can operate with single cell clusters. Neighbouring BSs transmit to their MSs using the same carrier frequency, although different codes are used for the pilot, set-up and traffic channels. As a MS moves to the edge of its single cell, i.e., cluster, the adjacent BS assigns a modem to the call, while the current BS continues to handle the call. The call is then handled by both BSs on a make-before-break basis. In effect, handover diversity occurs with both BSs handling the call until the MS moves sufficiently close to one of the BSs which then exclusively handles the call. This handover procedure is called a 'soft handover', as distinct to the more conventional break-before-make 'hard handover' method. Soft handovers are also made between sectors in a cell.

The Qualcomm CDMA system operates with a low E_b/N_o ratio, exploits voice activity, and uses sectorisation of cells. Each sector has 64 CDMA channels as previously described. It is a synchronised system, and there are three receivers to provide path diversity at the MS and four receivers are used at the cell site. The single cell cluster enables cells to be easily replicated along streets and into buildings. Complex frequency reassignment procedures required in TDMA and FDMA systems when small cells are introduced to alleviate teletraffic hot-spots are avoided. Nevertheless, CDMA/FDMA systems can be deployed to increase capacity if additional spectrum is available. Although conceived for the current range of conventional cell sizes, it is able to operate in microcells.

1.7 Cordless Telecommunications

Cordless telecommunications (CT) networks are designed for mobile radio coverage over relatively small distances, such as in office environments. Because of the small cells, or microcells, in which CTs operate the networks are basically much simpler than cellular ones.

The propagation environments in CT networks have less average delay spreads, but greater variability than cells in cellular networks due to the widely differing types of building construction. A discussion of the propagation in CT environments is provided in Section 1.2.7. Suffice to say that microcells and picocells are used in CT enabling very high bit rates per square kilometer or large values of Erlangs/MHz/km^2 to be achieved. This means that the complex speech codecs used in cellular radio can often be discarded in favour of simpler codecs that are inherently more robust to channel errors, have negligible encoding delays, and consume relatively little battery power. The higher bit rates that can be accommodated in CT also allow more bits to be used for synchronisation and control. The microcells yield other advantages. The excess path delay of the received radio signals is much lower than in cellular radio, and often much higher TDMA bit rates can be transmitted without the need for equalisation. Channel coding can often be avoided, a wider range of services accommodated, and so forth. We again emphasise that the most crucial factor determining system capacity in mobile radio system design is the cell size, and from a radio point of view, small is best!

1.7.1 CT2 System

The first operational digital CT is the British CT2 system [63]. CT2 has three principle applications; cordless PABX, cordless telephony, and telepoint in which the user can only call into the network. However, it must be stressed that in its non-telepoint mode it enables the cordless user either to be called, or to dial into the network.

CT2 operates in the band 864.1 to 868.1 MHz. The channel spacing is 100 kHz enabling 40 time division duplex (TDD or ping-pong) channels to be accommodated. In TDD one RF carrier supports transmissions on both the up-link and the down-link. To do this a frame structure shown in Figure 1.43 is used. ADPCM speech at 32 kb/s is sent and received as a B channel or bearer channel. The D channel or signalling channel is used to control the link. This arrangement, known as Multiplex One, is used to transfer signalling and data across an established link. The data is transmitted at rates 72 kb/s, with the first half of the frame for fixed station (FS) to portable station (PS) while the second half is for PS to FS transmissions. Notice that in CT terminology BS and MS becomes FS and PS, or cordless fixed part (CFP) and cordless portable part (CPP), respectively. Two or four bits on each link are assigned to the signalling channel at rates of 1 or 2 kb/s. All CT2 equipment must be able to operate using the 66-bit burst format, whereas the 68-bit burst is optional. When the shorter burst format is used the guard time is extended by two bits and hence the transmission rate remains at 72 kb/s, irrespective of the signalling rate. The frame length is 144 bits, corresponding to 2 ms.

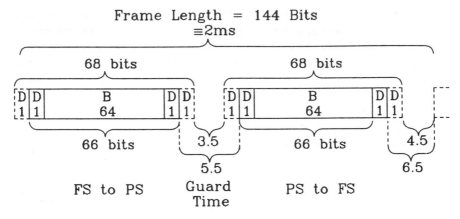

Figure 1.43: CT2 Multiplex One frame structure.

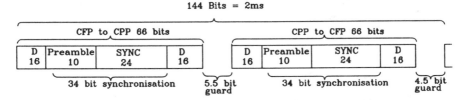

Figure 1.44: CT2 Multiplex Two frame structure.

Multiplex One is used once a link is established. However, in setting-up the link between the FS and the PS, Multiplex Two is used whose format is shown in Figure 1.44. The 34-bit synchronisation channel consists of a 10-bit alternate zero and one sequence, followed by a 24-bit sequence to facilitate burst synchronisation. The D channel has 32 bits resulting in a 16 kb/s control link. When the FS has a call for the PS it selects Multiplex Two, using its down-link part of the frame to call the PS with the information in the D channels, while the PS responds using the up-link part of the frame. The FS controls the timing of the PS transmissions.

When the PS initiates a call it uses Multiplex Three. The alternate up-link and down-link transmissions of Multiplexes One and Two are discarded. Instead the PS transmits five consecutive frames each having 144 bits. In the first frame, the first sub-multiplex group consists of a 6-bit preamble which is followed by a 10-bit D channel, by an 8-bit preamble, by a 10-bit D channel. The D channel is composed of 20 bits and contains the identity of the PS. Next come two sub-multiplex groups, each having an 8-bit preamble, followed by a 10-bit D channel, by another 8-bit preamble and 10-bit D channel. The last sub-multiplex group is the same as the two centre ones, except that it has a final 2-bit preamble. The same format of four sub-multiplex groups is repeated for the next three 144-bit frames. The fifth frame has a 12-bit preamble, followed by a 24-bit channel marker synchronisation word that informs the FS that a PS is calling. This arrangement is repeated three times to give a 144-bit fame. During the next two 144-bit frames the PS listens for a response from the FS. Multiplex Three continues to be used until a link is established, whence Multiplex One is used and speech transmissions commence.

1.7.2 Digital European Cordless Telecommunications (DECT) Systems

Again TDD is used, but with 12 channels per carrier and a carrier spacing of 1728 MHz in the 1880-1900 MHz band [63]. The TDMA rate is 1152 kb/s. DECT has a number of attractive features. It employs ADPCM at 32 kb/s; and GMSK modulation with the higher normalised bandwidth of 0.5 compared to the 0.3 used in GSM, thereby simplifying clock recovery procedures. No channel coding is used, neither is channel equalisation as the microcells are in general small and in office environments. Whereas GSM and CT2 have control channel rates of 967 and 2000 b/s, and control channel delays of 480 and 32 ms, respectively, the corresponding values for DECT are 6400 b/s and 5 ms. Thus DECT has a powerful control capability. A system similar to DECT is the Ericsson CT3 system. The DECT network arrangement is shown in Figure 1.45. The telepoint FSs are connected directly to the public switched telephone network (PSTN), as are the FSs installed in residential and office buildings. In larger offices, or complexes, e.g., airports, a number of FSs are connected to a radio link exchange (RLE) which distributes calls from neighbouring PABXs to the PSs they serve.

The DECT architecture consists of a concentrator which trunks N PSTN lines to L DECT lines. The link conversion unit (LCU) reformats the PSTN/DECT data to the new form. For example, incoming PSTN 64 kb/s A-law PCM speech is transcoded to 32 kb/s ADPCM for transmissions in the DECT network. The L lines carrying ADPCM speech are then conveyed to the RLE for distribution to the FSs and hence to the PSs. The transmission rate of the standard 2B+D ISDN channel is 144 kb/s, which is converted in the DECT system to 2 x 32 + 6.4 = 70.4 kb/s. It can also support the full ISDN rate of 144 kb/s.

The structure of the DECT TDMA frame is shown is Figure 1.46. The frame duration is 10 ms, and the TDMA slot length is 0.417 ms. In each slot a burst of 416 bits is transmitted at 1152 kb/s. At the commencement of a burst there is a 16-bit preamble followed by a 16-bit synchronisation code. Next comes 64 control bits conveyed on the C-channel, followed by 320 information bits on the I-channel. Notice that the information rate is 320 bits in 10 ms, i.e., 32 kb/s, the rate of the ADPCM speech encoder. The

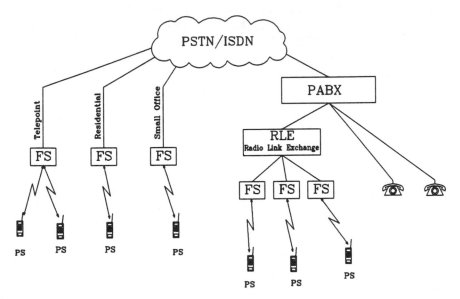

Figure 1.45: The DECT network.

control rate is relatively high, 64 bits/10 ms = 6.4 kb/s. The control data is partitioned into control (C), paging (P) and broadcasting (Q), depending on the activity. A guard period of 64 bits equivalent to 52.1 μs is used. The preamble and sync code are present in every burst (the first 32 bits) to guarantee synchronisation of the time slot. The 16-bit synchronisation code on the down-link is inverted on the up-link transmissions.

A free channel is defined as one with a signal strength below a system threshold, or the channel with the lowest signal strength. This must be the situation on both links. When the FS has a call for a PS, it uses one or more free channels for signalling. On detecting its handshake code the PS responds. For calls initiated by the PS, one or more free channels are used to convey the PS handshake code to the FS for a period up to five seconds. Upon detection of the code the FS responds to the PS. The DECT system allows for handovers. The time between handovers must be at least three seconds, and one FS can use up to 75% of all available channels.

Other features include multiple rate transmissions using more than one time slot per frame, and uni-directional transmissions, i.e., when the time slots in both halves of the frame can be used to receive or transmit data.

The data link layer is divided into two layers. Layer 2a comprises error protection, link quality assessment and handover control, while Layer 2b is concerned with the data link layer function. Layers 2a and 2b are called the Medium Access Control layer and the Logical Link Control Layer, respectively. The 64-bit control field consists of an 8-bit header, followed by

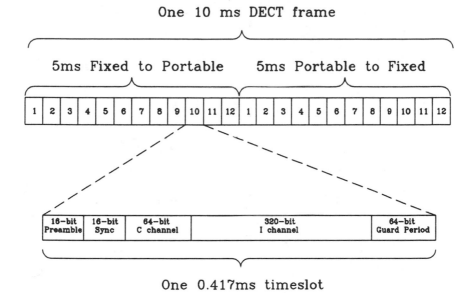

Figure 1.46: DECT TDMA frame structure.

40-bit content and finally a 16-bit cyclic redundancy check (CRC) code to combat transmission errors. The header defines the type of message, and whether the system is residential, business or telepoint. One header bit is used for paging and identifying the portable. The 40 bits of content are shared between layers 2a and 2b depending on the message type, and the service at the network being extended by DECT. The CRC code is a BCH (63,48,2) code.

1.7.3 Parameters of CTs and Cellular Systems

Both the digital cellular systems and the digital CT networks enable mobile users to access the national and international PSTNs and ISDNs. The first-generation cellular systems were conceived as mobile telephones, and much of this thinking prevails in the second-generation. However, hand-portables are increasingly used, and non-telephony services are incorporated to a limited extent. Nevertheless, second-generation cellular networks are designed for mobiles operating in large cells. The equipment is accordingly relatively bulky, complex and expensive, with a reliance on signal processing to enable each user to operate at low bit rates. The service costs are much higher than those of CTs, but they do offer near ubiquitous communications.

CTs are designed for short-range communications in buildings and their immediate environments and are essentially consumer products, being low cost, light-weight, and inexpensive to buy and operate. Table 1.1 presents

System	TACS	GSM	DCS-1800	Qualcomm CDMA	IS-54	CT2	DECT	Units
Forward Band	935-950	935-960	1805-1880	869-894	869-894	864-868	1880-1900	MHz
Reverse Band	890-905	890-915	1710-1785	824-849	824-849	864-868	1880-1900	MHz
Multiple Access	FDMA	TDMA	TDMA	CDMA	TDMA	FDMA	TDMA	
Duplex	FDD	FDD	FDD	FDD	FDD	TDD	TDD	
Carrier Spacing	25	200	200	1250	30	100	1728	kHz
Channels/carrier	1	8	8	55-62	3	1	12	
Bandwidth/channel	50	50	50	21	20	100	144	kHz
Modulation	FM	GMSK	GMSK	QPSK	$\frac{\pi}{4}$-DQPSK	FSK	GMSK	
Modulation Rate	N/A	271	271	1228	48.6	72	1152	kb/s
Voice Rate	N/A	22.8	22.8	9.6	13	32	32	kb/s
Speech codec	N/A	RPE-LTP	RPE-LTP	CELP	VSELP	ADPCM	ADPCM	
Uncoded Voice Rate	N/A	13	13	1.2-9.6	8	32	32	kb/s
Control Channel Name	-	SACCH	SACCH	SACCH	SACCH	D	C	
Control Channel Rate	-	967	967	800	600	2000	6400	b/s
Control Message Size	-	184	184	1	65	64	64	bits
Control Delay	-	480	480	1.25	240	32	10	ms
Peak Power (Mobile)	0.6-10	2-20	0.25-2	0.6-3	0.6-3	10mW	250mW	W
Mean Power (Mobile)	0.6-10	0.25-2.5	0.03-0.25	0.6-3	0.6-3	5mW	10mW	W
Power Control	Yes	Yes	Yes	Yes	Yes	Yes	No	
Voice Activity Detection	Yes	Yes	?	Yes	?	No	No	
Handover	No	Yes	No	Yes	No	Yes	Yes	
Dynamic Channel Allocation	No	No	No	N/A	No	Yes	Yes	
Min. Cluster Size	7	3	3	1	7	N/A	N/A	
Capacity/cell/MHz	2.8	6.7	6.7	16.5†	7	N/A	N/A	Duplex chs/cell/Mhz

†Assumes $\frac{2}{3}$ frequency re-use

N/A means not applicable.

Table 1.1: Comparison of key system parameters

some parameters of the systems we have considered.

1.8 Teletraffic Considerations

Networks have far fewer mobile radio channels than users. This is because only a small percentage of subscribers make call attempts at any time. The network operator usually designs his network such that at the busy hour only a small percentage, say 2%, will have their call attempts blocked because all the channels are in use. The probability P_n of a request for a new call being denied must be significantly greater than the probability P_h of an existing call being forced to terminate due to a handover failure between two BSs.

There are many books on general traffic theory [64-66], but the literature dealing with the application of traffic theory to cellular radio is relatively sparse [67-69] [2]. As this chapter attempts an overview of mobile radio we will only mention some of the salient points of the subject.

In conventional cellular radio where the cells are large, and handovers between BSs are relatively rare, we can assume that the number of MSs in a cell is so much greater than the number of BS channels N, that to a good approximation the number of users made be considered as infinite. In this situation the probability of a call attempt being blocked is given by

$$B = \frac{(\lambda/\mu)^N/N!}{\sum_{k=0}^{N}(\lambda/\mu)^k/k!} \qquad (1.33)$$

where λ is the mean call arrival rate, and μ is the mean rate at which calls are terminated, i.e., cleared from the system. Equation (1.33) is known as the Erlang B formula or Erlang's first formula. It is derived on the basis that no queueing occurs so that if all N channels are busy the call is blocked and the user may try again. The offered traffic to the BS is

$$A = \lambda/\mu = \lambda T_c; \text{Erlangs} \qquad (1.34)$$

where T_c is the mean value of the call durations, or the mean channel holding time. The blocked traffic is AB, while the traffic carried by the BS is $A(1 - B)$. The call requests arriving at the BS, and being cleared by the network, are independent events. It is interesting to notice that the arrival of call requests at the BS, and the number of motor vehicles and pedestrians (both potential MSs) passing a point on the road-side are Poisson distributed, while the separation between calls, motor vehicles and pedestrians are exponentially distributed. This observation is important when simulating mobile radio systems with a view to obtaining teletraffic results.

---

[2]There will be a book published in this series dealing exclusively with this subject

In microcells the number of MSs, M, may not be significantly different from N. If the total offered traffic to the microcellular BS is A, where A is small and the blocking probability is zero, then the average offered traffic per user is $a = (A/M) = (\alpha/\mu)$, where α is the reciprocal of the average inter-arrival time. When j of the N BS channels are in use the mean call arrival rate is $\lambda_j = (M - j)\alpha$, and $\mu_j = j\mu$. The probability of N channels being busy is

$$P_N = \frac{\binom{M}{N} a^N}{\sum_{k=0}^{N} \binom{M}{k} a^k} \tag{1.35}$$

which is the fraction of time during which all channels are in use.

Another important probability is the probability that a call is attempted when all channels are in use. This probability is given by

$$P_B = \frac{\binom{M-1}{N} a^N}{\sum_{k=0}^{N} \binom{M-1}{k} a^k} \tag{1.36}$$

and a can also be expressed as

$$a = \frac{A}{M - A(1 - B)}. \tag{1.37}$$

These equations are known as Engset formulae [64].

Determining the number of channels required at a BS in a conventional large cell system is straight forward. Having decided on the offered traffic per BS, i.e., the total offered traffic $A = \lambda/\mu$ in Erlangs, and the blocking probability B, the value of N is next determined using the Erlang B formula. This formula is tabulated [65] in terms of A, N and B, enabling N to be found quickly. It is a straight forward procedure to then calculate the effects on A, N and B of sectorisation of cells using directional antennas, and the splitting of a large cell into smaller cells.

Applying teletraffic theory to microcells is much more difficult due to MSs frequently making handovers [68,69]. This means that the channel holding time in the microcell becomes a fraction of the call duration. The average channel holding time is

$$\overline{T}_H = \gamma_n \overline{T}_{Hn} + \gamma_h \overline{T}_{Hh} \tag{1.38}$$

where $\overline{T}_{Hn}$ and $\overline{T}_{Hh}$ are the average holding times of the new call and the handover call, respectively, and

$$\gamma_n = (1 - \gamma_h) \tag{1.39}$$

is the ratio of the carried new call rate to the total call rate, while γ_h is the ratio of carried handover rate to the new call rate. The total mean arrival rate at a BS is

$$\lambda_T = \lambda_N + \lambda_H \tag{1.40}$$

where λ_N and λ_H are new call rate and the handover request rate, respectively.

The traffic carried by the microcell BS is

$$A_{cm} = A_T(1 - P_{bn}) \tag{1.41}$$

where P_{bn} is the probability of a new call attempt being blocked. This traffic formula is applicable for the case where no priority assignment of channels is provided for handover requests. In this situation P_{bn} is the same as the probability of a call being forced to terminate when requesting a handover. To decrease the probability of handover failure we use a macrocell to oversail a cluster of microcells. The traffic carried by one macrocell BS is

$$A_{CM} = A_M(1 - P_{fhM}) \tag{1.42}$$

where A_M is the traffic offered by the cluster of microcellular BSs to the macrocellular BS due to the former BSs not having sufficient channels to accommodate the demand for handovers, and P_{fhM} is the probability of handover failure in the macrocell. The total traffic carried by the network is

$$A_{CT} = C_m A_{cm} + C_M A_{CM} \tag{1.43}$$

where C_m and C_M are the total number of microcells and macrocells, respectively.

The channel utilisation is defined as the carried traffic per channel. With each microcellular BS having N channels, and each macrocellular BS having N_o channels, the average channel utilisation is

$$\rho = \frac{A_{cm} + A_{CM}}{N + N_o}. \tag{1.44}$$

If we define the spectral efficiency η in terms of Erlangs per Hz per m^2, and we assume that to a first approximation that the number of channels carried by the macrocells is very small compared to those carried by the microcells, then

$$\eta = \frac{A_{CT}}{S_T W} \simeq \frac{\rho}{S_m M_{mc} B_c} \tag{1.45}$$

where S_T is the total area covered by the network, S_m is the average area of each microcell, W is the total available bandwidth allocated to the network, B_c is the equivalent bandwidth per channel, and M_{mc} is the number of microcells per cluster. Notice that for high η: S_m should be small, i.e., deploy microcells, B_c should be small and depends on the modulation and multiple access method used; while the number of microcells per cluster should be small. One of the virtues of CDMA is that it meets these requirements, particularly $M_{mc} = 1$. It might appear that to increase ρ the number of channels should be decreased. However, if the grade-of-service

(GOS) is to be maintained, the blocking probability must remain low. As a consequence it is more appropriate to increase N which will increase ρ, and therefore η, for the same blocking probability.

*   *

This opening chapter has attempted to introduce the reader to digital cellular radio using a bottom-up approach. The mathematics has been kept to a minimum, with the emphasis on concepts. We now embark on a series of chapters in which the subject treatment is more detailed and focused.

Bibliography

[1] "Car radio telephones", *The Autocar*, pp.A22-A23, 30 October 1936.

[2] **W.C.Jakes**, "Microwave mobile communications", *John Wiley & Sons*, New York, 1974.

[3] **W.Y.C.Lee**, "Mobile cellular communications", *McGraw Hill*, New York, 1989.

[4] **J.D.Parsons** and **J.G.Gardiner**, "Mobile communication systems", *Blackie*, London 1989.

[5] **R.Steele**, "Towards a high capacity digital cellular mobile radio system", *Proc. of the IEE*, PtF, No.5, pp.405-415, August 1985.

[6] **R.Steele** and **V.K.Prabhu**, "Mobile radio cellular structures for high user density and large data rates", *Proc. of the IEE*, Pt F, No.5, pp.396-404, August 1985.

[7] **R.Steele**, "The cellular environment of lightweight hand-held portables", *IEEE Communications Magazine*, pp.20-29, July 1989.

[8] **V.H.MacDonald**, "The cellular concept", *Bell System Tech. J.*, Vol 58, No.1, pp.15-41, January 1979.

[9] **J.D.Parson** "The mobile radio propagation channel", Pentech Press, London, 1992.

[10] **E.Green**, "Radio link design for microcellular systems", *British Telecom Technology J*, Vol 8, No.1, pp.85-96 January 1990.

[11] GSM Recommendation 05.05, Annex 3, pp.13-16, November 1988.

[12] **K.Bullington**, "Radio propagation at frequencies above 30 Mc/s", *Proc. IRE 35*, pp.1122-1136, (1947).

[13] **R.Edwards** and **J.Durkin**, "Computer prediction of service area for VHF mobile radio networks", *Proc IRE 116 (9)* pp.1493-1500, 1969.

[14] **Y.Okumuma, E.Ohmori, T.Kawano and K.Fukuda**, "Field strength and its variability in VHF and UHF land-mobile radio service", *Review of the Elec. Comm. Lab.*, Vol 16, No.9 and 10, pp.825-873, 1968.

[15] **M.Hata**, 'Empirical formula for propagation loss in land mobile radio services", *IEEE Trans. Veh. Technol.*, Vol VT-29, No.3, pp.317-325, August 1980.

[16] **R.Steele**, "The importance of propagation phenomena in personal communication networks". *IEE 7-th International Conference on Antennas and Propagation*, ICAP, York, No.333, Pt I, pp.1-5, April 1991.

[17] **S.T.S.Chia, R.Steele, E.Green and A.Baran**, "Propagation and bit error ratio measurements for a microcellular system", *J IERE*, Vol 57, No.6 (Supplement), pp.S255-S266, November/December 1987.

[18] **A.J.Rustako, N.Amitay, G.J.Owens and R.S.Roman**, "Radio propagation measurements at microwave frequencies for microcellular mobile and personal communications", *IEEE ICC'89*, pp.482-488, Boston, USA, 1989.

[19] **S.T.S.Chia**, "Radiowave propagation and handover criteria for microcells", *British Telecom Tech J*, Vol 8, No.4, pp.50-61, October 1990.

[20] **R.Bultitude and G.Bedal**, "Propagation characteristics on microcellular urban mobile radio channels at 910 MHz", *IEEE J-SAC*, Vol 7, No.1, pp.31-39, January 1989.

[21] **W.T.Webb and R.Steele**, "Equaliser techniques for QAM transmission over dispersive mobile radio channels", *IEE Proc.-I*, Vol 138, No.6, pp.566-576, December 1991.

[22] **J.M.Keenan and A.J.Motley**, "Radio coverage in buildings", *British Telecom Technol J*, Vol 8, No.1, pp.19-24, January 1990.

[23] **F.C.Owen and C.D.Pudney**, "In-building propagation at 900 MHz and 1650 MHz for digital cordless telephones", *6th Int Conf on Antennas and Propagation*, ICAP'89, Pt 2: Propagation, Conf Pub No.301, pp.276-280, 1989.

[24] **R.J.C.Bultitude, P Melancon and J.LeBel**, "Data regarding indoor radio propagation", *Wireless'90*, Calgary, Canada, July 1990.

[25] **D.C.Cox**, "Universal digital portable radio communications", *Proc IEEE*, Vol 75, No.4, pp.436-477, April 1987.

[26] **A.A.M.Saleh and R.A.Valenzula**, "A statistical model for indoor multipath propagation", *IEEE J-SAC*, pp.128-137, February 1987.

[27] **D.M.J.Devasirvatham**, "Time delay spread measurements of wideband radio signals within a building", *Electronics Letters*, Vol 20, No.23, pp.951-952, 8 November 1984.

[28] **R.J.C.Bultitude, S.A.Mahmoud** and **W.A.Sullivan**, "A comparison of indoor radio propagation characteristics at 910 MHz and 1.75 GHz", *IEEE JSAC*, Vol 7, No.1, pp.20-30, January 1989.

[29] **R.Davies, A.Simpson** and **J.P.McGeehan**, "Propagation measurements at 1.7 GHz for microcellular urban communications", *Electronic Letters*, Vol 26, No.14, pp.1053-1054, 5 July 1990.

[30] **E.Damosso, L.Stola** and **G.Brussaard**, "Characterisation of the 50-70 GHZ band for space communications", *ESA J*, 7, pp.25-43, 1983.

[31] **R.H.Ott** and **M.C.Thompson**, "Atmospheric amplitude spectra in an absorption region", *Proceedings of IEEE AP-S symposium*, Amherst, MA, USA, pp.594-597, 1976.

[32] **O.E.De Lange, A.F.Dietrich** and **D.C.Hogg**, "An experiment on propagation of 60 GHz waves through rain", *Bell Syst. Tech.J.*, 54, pp.165-176, 1975.

[33] **S.E.Alexander** and **G.Pulgliese**, "Cordless communication within buildings: results of measurements at 900 MHz and 60 GHz", *Br Telecom Tech.,J*, 1, pp.99-105, 1983.

[34] **S.Chia, D.Greenwood, D.Rickard, C.R.Shephard** and **R.Steele**, "Propagation studies for a point-to-point 60 GHz microcellular system for urban environments", *IEE Communications '86*, Birmingham, pp.28-32, 13-15 May 1986.

[35] **J.G.Proakis**, "Digital communications", *McGraw Hill*, 1989.

[36] **W.C.Y.Lee**, "Overview of cellular CDMA", *IEEE Trans. on Veh. Technol*, Vol 40, No.2, pp.291-302, May 1991.

[37] **W.H.Lam** and **R.Steele**, "Performance of direct-sequence spread spectrum multiple access systems in mobile radio", *IEE Proc-I*, Vol 138 No.1, pp.1-14, February 1991.

[38] **J.K.Holmes**, "Coherent spread spectrum systems", *John Wiley*, New York, 1981.

[39] **G.R.Cooper** and **R.W.Nettleton**, "A spread-spectrum technique for high capacity mobile communications", *IEEE Trans. Veh. Technol*, Vol VT-27, pp.264-275, November 1978.

[40] **K.S.Gilhousen,** **I.M.Jacobs,** **R.Padovani** and **A.J.Viterbi**, L A Weaver and C E Wheatley, "On the capacity of a cellular CDMA system", *IEEE Trans. Veh. Technol*, Vol 40, No.2, pp.303-312, May 1991.

[41] **A.J.Viterbi**, "Wireless digital communication: a view based on three lessons learned", *IEEE Communications Magazine*, pp.33-36, September 1991.

[42] **B.G.Haskell** and **R.Steele**, "Audio and video bit rate reduction", *Proc. IEEE*, Vol 69, No.2, pp.252-262, February 1981.

[43] **N.S.Jayant** and **P.Noll**, "Digital coding of waveforms", *Prentice-Hall*, 1984.

[44] **T.Aoyama, W.R.Daumer** and **G.Modena**, "Voice coding for communications", *Special Issue of IEEE JSAC*, Vol 6, No.2, pp.225-452, February 1988.

[45] **W.C.Wong, R.Steele, B.Glance** and **D.Horn**, "Time diversity with adaptive error detection to combat Rayleigh fading in digital mobile radio", *IEEE Trans*, COM-31, pp.378-387, March 1983.

[46] **C-E.Sundberg, W.C.Wong** and **R.Steele**, "Weighting strategies for companded PCM transmitted over Rayleigh fading and Gaussian channels", *Bell Syst. Tech. J.*, Vol 63, No.4, pp.587-626, April 1984.

[47] **R.Steele, C-E.Sundberg** and **W.C.Wong**, "Transmission errors in companded PCM over Gaussian and Rayleigh fading channels", *Bell Syst. Tech. J.*, Vol 63, No.6, pp.955-990, July/ August 1984.

[48] **W.C.Wong, R.Steele** and **C-E.W.Sundberg**, "Soft decision demodulation to reduce the effect of transmission errors in logarithmic PCM transmitted over Rayleigh fading channels", *Bell Syst. Tech. J.*, Vol 63, No.10, pp.2193-2213, December 1984.

[49] **R.Steele, C-E.W.Sundberg** and **W.C.Wong**, "Transmission of log-PCM via QAM over Gaussian and Rayleigh fading channels", *IEE Proc.* Vol 134, Pt.F, No.6, pp.539-556, October 1987.

[50] **C-E.W.Sundberg, W.C.Wong** and **R.Steele**, "Logarithmic PCM weighted QAM transmission over Gaussian and Rayleigh fading channels", *IEE proc.*, Vol 134, Pt.F, No.6, pp.557-570, October 1987.

[51] **R.Steele**, "Delta modulation systems", *Pentech Press*, London, 1975.

[52] **L.Hanzo, R.Steele** and **P-M.Fortune**, "A subband coding, BCH coding and 16-QAM system for mobile radio speech applications", *IEEE Trans. Veh. Technol.*, Vol 39, No.4, pp.327-339, November 1990.

[53] **R.C.Cox** "Robust CELP coders for noisy background and noisy channels", *IEEE Proc ICASSP'89*, Glasgow, pp.739-742, 23-26 May 1989.

[54] **L.Hanzo, R.Salami** and **R.Steele**, " A 2.1 kBd speech transmission system for Rayleigh fading channels", *IEE Colloquium on Speech Coding*, London, Digest No.1989/112, pp.10/1-10/5, 9 October 1989.

[55] **P.Vary**, "GSM speech codec", *Digital Cellular Radio Conference*, Hagen, Germany, paper 2a, October 1988.

[56] **J.L.Flanagan**, "Speech analysis, synthesis and perception", *Springer-Verlag*, Berlin, 1972.

[57] **R.E.Blahut**, "Theory and practice of error control codes", *Addison-Wesley*, 1983.

[58] **W.T.Webb, L.Hanzo** and **R.Steele**, "Bandwidth efficient QAM schemes for Rayleigh fading channels", *IEE Proc.-I*, Vol 138, No.3, pp.169-175, June 1991.

[59] **A.P.Clark**, "Advanced data-transmission systems", *Pentech Press*, London, 1977.

[60] **A.Duel-Hallen** and **C.Heegard**, "Delayed decision-feedback sequence estimation," *IEEE Trans. on Comms.*, Vol COM-37, pp.428-436, May 1989.

[61] **J.C.S.Cheung**, "Adaptive equalisers for wideband time division multiple access mobile radio", *PhD Thesis*, University of Southampton, England, 1992.

[62] **D.J.Goodman**, "Second generation wireless information networks", *IEEE Trans. Veh. Technol.*, No.2, pp.336-374, May 1991.

[63] **W.H.W.Tuttlebee** (Ed), "Cordless telecommunication in Europe", *Springer-Verlag*, 1990.

[64] **M.Schwartz**, "Telecommunication Networks: protocols, modeling and analysis", *Addison-Wesley*, 1987.

[65] **J.R.Boucher**, "Voice teletraffic systems engineering", *Artech House*, 1988.

[66] **L.Kleinrock**, " Queueing systems volume 1: theory", *John Wiley & Sons, New York*, 1975

[67] **D.Hong** and **S.S.Rappaport**, "Traffic model and performance analysis for cellular mobile radio telephone systems with prioritized and nonprioritized hand-off procedures", *IEEE Trans. Veh. Technol.*, Vol VT-3, pp.77-92, August 1986.

[68] **S.A.Dolil, W.C.Wong** and **R.Steele**, "Teletraffic performance of highway microcells with overlay macrocell", *IEEE JSAC*, Vol 7, No.1., pp.71-78, January 1989.

[69] **R.Steele** and **M.Nofal**, "Teletraffic performance of microcellular performance of microcellular personal communications networks", *IEE Proc.-I*, Vol. 139, No.4, August 1992.

Chapter 2

Characterisation of Mobile Radio Channels

D. Greenwood[1] and L. Hanzo[2]

Modern society is continually demanding more and better communications services [1]. There is a very real market for a global network allowing voice and data communications between any two points on the earth's surface, no matter how remote. A key factor in the realisation of such a network is the ability to provide multimedia services via cellular mobile radio systems. Unfortunately, mobile radio channels are extremely harsh media for information transmission. It is the intention of this chapter to introduce the reader to the behaviour of mobile radio channels and to their characterisation.

In Section 2.2 several mobile radio channel types are defined, and their usage is discussed. The characteristics of each of these channel are examined in terms of their physical structure in Section 2.3. By considering the bandwidth and duration of the information signals they carry, channels can be classified according to their invariance properties. This concept is discussed in Section 2.4 along with the resulting simplifications.

Section 2.5 introduces the Bello system functions—a set of functions which describe general linear time-variant channels with a powerful mathematical elegance. These functions are commonly used to describe mobile radio channels not only for their simplicity and ease of manipulation, but also because they assist in the intuitive understanding of channel behaviour. Simplifications to the general theory are discussed as statistical constraints are placed on the channel, leading to the development of the QWSSUS (Quasi Wide Sense Stationary Uncorrelated Scattering) channel.

---

[1] University of Southampton
[2] University of Southampton and Multiple Access Communications Ltd.

Section 2.6 discusses the application of the QWSSUS model to the characterisation of mobile radio channels, while Section 2.7 portrays a practical approach to their description for the system-designer.

The chapter begins with a review of the complex notation often used in the study of time-variant linear channels. This notation will be used extensively throughout this chapter.

2.1 Complex Baseband Representation of Bandpass Signals and Systems

A useful tool for studying bandpass communication systems is the complex lowpass equivalent notation. It provides a mathematical shorthand which bypasses the tedious trigonometry that generally accompanies the mathematical manipulation of signals modulated onto a sinusoidal carrier. The following description pursues the approach of Stein and Jones [2].

2.1.1 Bandpass Signals

The general form of a bandpass signal, $x(t)$, having a carrier frequency f_c is

$$x(t) = A(t)\cos[2\pi f_c t + \phi(t)], \qquad (2.1)$$

where either, or both, of the amplitude, $A(t)$, and phase, $\phi(t)$, are used to carry the message information, such as digitally encoded speech. Trigonometric expansion of this equation yields

$$x(t) = u_I(t)\cos 2\pi f_c t - u_Q(t)\sin 2\pi f_c t, \qquad (2.2)$$

where

$$u_I(t) = A(t)\cos\phi(t) \qquad (2.3)$$

and

$$u_Q(t) = A(t)\sin\phi(t) \qquad (2.4)$$

are the envelopes of the two quadratic carrier frequency components. Using complex notation, we may write

$$x(t) = \Re\left\{x_+(t)\right\}, \qquad (2.5)$$

where $x_+(t)$ is referred to as the pre-envelope of $x(t)$ [3] or the analytic signal. The pre-envelope of a bandpass signal can be represented as a phasor, with the bandpass signal given by the image of the pre-envelope along the real axis. Figure 2.1 shows a phasor representation of $x_+(t)$.

It is seen from Equations 2.2 and 2.5 above that

$$x_+(t) = u(t)\exp(j2\pi f_c t), \qquad (2.6)$$

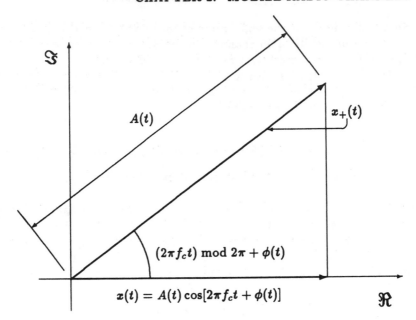

Figure 2.1: Phasor representation of the pre-envelope of a bandpass signal

where

$$u(t) = u_I(t) + ju_Q(t). \tag{2.7}$$

The waveform $u(t)$ is referred to as the complex envelope, or complex lowpass equivalent, of $x(t)$. It is also a phasor, as shown in Figure 2.2. The pre-envelope, $x_+(t)$, is obtained by rotating the phasor $u(t)$ with an angular velocity $2\pi f_c$. In many modulators, the modulation waveform is generated from its complex envelope. Figure 2.3 shows how this is accomplished.

Equations 2.5 and 2.7 reveal that

$$x(t) = \Re \left\{ u(t) \exp(j2\pi f_c t) \right\}. \tag{2.8}$$

This equation shows that knowledge of $u(t)$ and f_c completely describes the signal $x(t)$. As all the message information is represented by $u(t)$, it is common to describe $x(t)$ by its complex lowpass equivalent alone—the presence of a carrier frequency being implied.

To establish how the frequency spectra of $x(t)$ and $u(t)$ [$X(f)$ and $U(f)$ respectively] are related, consider the spectrum of the bandpass signal $x(t)$. That is,

$$X(f) = \int_{-\infty}^{\infty} x(t) \exp(-j2\pi f t) \, dt. \tag{2.9}$$

Substituting for $x(t)$ in the above equation from Equation 2.8 gives

$$X(f) = \int_{-\infty}^{\infty} \Re \left\{ u(t) \exp(j2\pi f_c t) \right\} \exp(-j2\pi f t) \, dt. \tag{2.10}$$

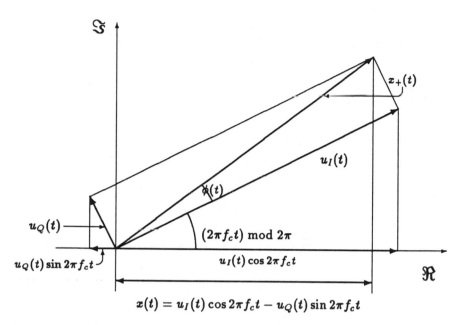

$$x(t) = u_I(t) \cos 2\pi f_c t - u_Q(t) \sin 2\pi f_c t$$

Figure 2.2: Phasor representation of the complex envelope of a bandpass signal

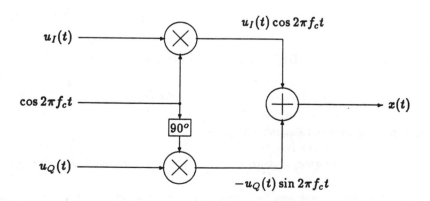

Figure 2.3: Generation of a bandpass signal from its complex envelope

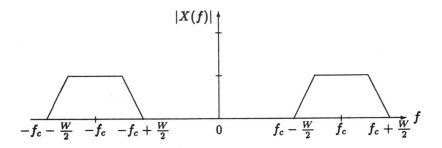

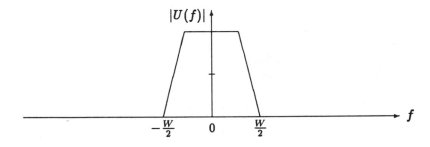

Figure 2.4: Amplitude spectra relationships

It is easily shown that the real part of a complex variable, z, can be written as

$$\Re\{z\} = \frac{1}{2}\{z + z^*\}, \qquad (2.11)$$

where z^* is its complex conjugate. Hence

$$X(f) = \frac{1}{2}\int_{-\infty}^{\infty} [u(t)\exp(j2\pi f_c t) + u^*(t)\exp(-j2\pi f_c t)]\exp(-j2\pi f t)\,dt. \qquad (2.12)$$

Defining the spectrum $U(f)$ as the Fourier transform of $u(t)$,

$$U(f) = \int_{-\infty}^{\infty} u(t)\exp(-j2\pi f t)\,dt, \qquad (2.13)$$

enables us to express Equation 2.12 as

$$X(f) = \frac{1}{2}[U(f - f_c) + U^*(-f - f_c)]. \qquad (2.14)$$

Figure 2.4 shows an example amplitude spectrum for a bandpass signal $X(f)$ and the corresponding lowpass spectrum of $U(f)$.

In the cellular mobile radio environment the bandwidth of information carrying signals is always very much less than the carrier frequency. Thus, as shown in Figure 2.4, the two components of X(f) in Equation 2.14 do not overlap in frequency. If they did overlap then the complex notation could

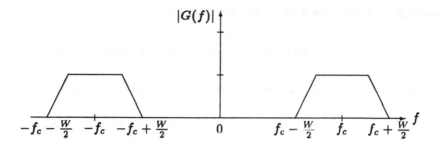

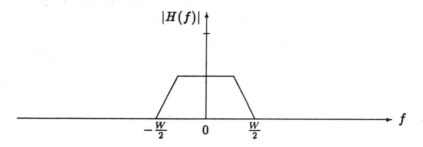

Figure 2.5: Frequency response of a bandpass system and its lowpass complex representation

still be employed, but would require the explicit use of Hilbert transforms [3], [4], [5].

2.1.2 Linear Bandpass Systems

Linear bandpass systems, exemplified here by the mobile radio channel, can also be described using complex notation. The impulse response of such a bandpass system, $g(t)$, can be written as the inverse Fourier transform of its frequency response, $G(f)$,

$$g(t) = \int_{-\infty}^{\infty} G(f) \exp(j2\pi ft) \, df. \qquad (2.15)$$

A representative frequency response for $G(f)$ is shown in Figure 2.5. It is seen to comprise identical components centered at $f = \pm f_c$. In order to derive a complex lowpass equivalent representation for a bandpass system we must first establish $g(t)$ in terms of just one of these components, say that centered at $f = f_c$. This component can then be frequency translated to baseband, resulting in the frequency response for the complex lowpass equivalent system.

This is accomplished in the following manner. The range of integration of the integral in Equation 2.15 is split as,

$$g(t) = \int_0^\infty G(f) \exp(j2\pi ft) df + \int_{-\infty}^0 G(f) \exp(j2\pi ft) \, df, \qquad (2.16)$$

and the variable $f' = -f$ is substituted into the second integral to give

$$g(t) = \int_0^\infty G(f) \exp(j2\pi ft) df + \int_0^\infty G(-f') \exp(-j2\pi f't) \, df. \quad (2.17)$$

It is noted that a physical system must possess a real impulse response, and that for a real $g(t)$ the following equality holds,

$$G(-f) = G^\star(f). \qquad (2.18)$$

Equation 2.17 then becomes,

$$g(t) = \int_0^\infty G(f) \exp(j2\pi ft) df + \int_0^\infty G^\star(f') \exp(-j2\pi f't) \, df, \qquad (2.19)$$

which when compared with Equation 2.11 reveals that

$$g(t) = 2\Re \left\{ \int_0^\infty G(f) \exp(j2\pi ft) \, df \right\}. \qquad (2.20)$$

Having obtained $g(t)$ in terms of the positive frequency component of $G(f)$, the bandpass to complex lowpass equivalent can be effected. The spectrum of a lowpass equivalent system is defined such that it is equal to the positive frequency component of $G(f)$ centred on zero frequency. That is,

$$H(f - f_c) = \begin{cases} G(f) & f > 0 \\ 0 & f < 0. \end{cases} \qquad (2.21)$$

Equation 2.20 is then equivalent to

$$g(t) = 2\Re \left\{ \int_{-\infty}^\infty H(f - f_c) \exp(j2\pi ft) \, df \right\}, \qquad (2.22)$$

or

$$g(t) = 2\Re \left\{ h(t) \exp(j2\pi f_c t) \right\}, \qquad (2.23)$$

where

$$h(t) = \int_{-\infty}^\infty H(f) \exp(j2\pi ft) \, dt \qquad (2.24)$$

is the complex lowpass equivalent impulse response of the system. As with $u(t)$, $h(t)$ can be written in terms of its inphase and quadrature components

$$h(t) = h_I(t) + jh_Q(t). \qquad (2.25)$$

It has thus been shown that in much the same way as bandpass signals, bandpass systems may be fully described by knowledge of their complex lowpass equivalent, and their center frequency.

It is easily shown from Equations 2.18 and 2.21 that the frequency response of the system is described by

$$G(f) = H(f - f_c) + H^*(-f - f_c). \qquad (2.26)$$

This equation differs in form from that of Equation 2.14 by a factor of a half (compare Figures 2.4 and 2.5). The reason for this difference will become apparent after reading the next section.

2.1.3 Response of a Linear Bandpass System

The response of a linear bandpass system, $y(t)$, must be a bandpass signal, even if the input to the system is not bandlimited. Referring to Section 2.1.1 we see that this response can therefore be represented in the manner of Equation 2.8, i.e.,

$$y(t) = \Re\{z(t)\exp(j2\pi f_c t)\}, \qquad (2.27)$$

where $z(t)$ is the complex lowpass equivalent signal, namely, the complex envelope of $y(t)$. Furthermore, it is seen from Equation 2.14 that the signal's spectrum is given by,

$$Y(f) = \frac{1}{2}[Z(f - f_c) + Z^*(-f - f_c)]. \qquad (2.28)$$

A bandpass system, described in the frequency domain by the transfer function $G(f)$, with an input signal spectrum of $X(f)$, has an output

$$Y(f) = G(f)X(f). \qquad (2.29)$$

Using Equations 2.14, and 2.26, Equation 2.29 is expanded to give,

$$Y(f) = \frac{1}{2}[H(f - f_c) + H^*(-f - f_c)][U(f - f_c) + U^*(-f - f_c)]. \qquad (2.30)$$

As indicated on page 96, the bandwidth of signals encountered in the mobile radio environment are small compared with the carrier frequency. Hence the terms in the product $H(f - f_c)U^*(-f - f_c)$ do not overlap in frequency, and the product equates to zero. Similarly, the product $H^*(-f - f_c)U(f - f_c)$ is equal to zero.

The spectrum of the received RF signal is therefore

$$Y(f) = \frac{1}{2}[H(f - f_c)U(f - f_c) + H^*(-f - f_c)U^*(-f - f_c)]. \qquad (2.31)$$

This Equation may be compared with Equation 2.28 to obtain the complex lowpass equivalent relationship,

$$Z(f) = H(f)U(f), \qquad (2.32)$$

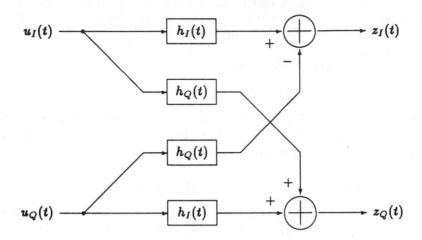

Figure 2.6: Complex lowpass equivalent of a bandpass system

from which the complex envelope of $y(t)$ may be deduced as,

$$z(t) = \int_{-\infty}^{\infty} h(\xi) u(t - \xi)\, d\xi. \qquad (2.33)$$

This may also be written, as

$$z(t) = h(t) \star u(t), \qquad (2.34)$$

where $\star$ represents convolution.

Substituting for $h(t)$ and $u(t)$ into Equation 2.34 from Equations 2.25 and 2.7, respectively, gives the complex envelope of the output signal as the sum of four convolutions

$$z(t) = h_I(t) \star u_I(t) - h_Q(t) \star u_Q(t) + j[h_I(t) \star u_Q(t) + h_Q(t) \star u_I(t)], \quad (2.35)$$

which in terms of the inphase and quadrature components of $z(t)$ yield

$$z_I(t) = h_I(t) \star u_I(t) - h_Q(t) \star u_Q(t) \qquad (2.36)$$

$$z_Q(t) = h_I(t) \star u_Q(t) + h_Q(t) \star u_I(t). \qquad (2.37)$$

This reveals that the structure of an equivalent complex lowpass system has the form shown in Figure 2.6.

Let us now return to the question of why Equation 2.26 differs from Equation 2.14 by a factor of two. The answer is that, it is simply to ensure that Equation 2.32 has the same form as Equation 2.29. Had Equation 2.21 been defined such that the frequency domain relationship for bandpass systems was analogous to that of a bandpass signal, i.e.,

$$G(f) = \frac{1}{2}[H(f - f_c) + H^\star(-f - f_c)],$$

then Equation 2.32 would have read

$$Z(f) = \frac{1}{2}H(f)U(f).$$

If this were the case, complex lowpass equivalent signals and systems theory would not mirror conventional theory. That is, the response of a complex lowpass system would not be the convolution of its impulse response and the complex lowpass input signal.

Conceptually, the bandpass to complex lowpass transformation of signals can be viewed as mapping both positive and negative frequency components of the signals' spectra to baseband and summing them. Systems, however, do not possess frequency spectra, they respond to the spectra of signals (i.e., they are described by frequency responses). Hence, the reason that a linear bandpass system has a positive and a negative component to its frequency response is so that it will shape both components of the input signal.

After a bandpass to complex lowpass transformation, the two components of a signal's spectrum are then centred on zero frequency. Only one component of the system's frequency response needs to be mapped to baseband, because this will then shape both of the signal's components at once.

The results derived in this section illustrate that we can reduce problems involving high frequency bandpass-type radio signals and systems to baseband schemes using their complex lowpass equivalents. This is essential if computer simulations of mobile radio channels and equipment are to be carried out, as it is impractical to simulate a high frequency radio carrier.

Nevertheless, in order to be able to apply the complex lowpass equivalent representation to the simulation of practical bandpass communication systems, there is one further equivalence that must be derived. That is, the baseband representation of the ubiquitous additive white Gaussian noise (AWGN).

2.1.4   Noise in Bandpass Systems

In communications systems noise from all sources are referenced to the receiver input and are represented by a single noise source added directly to the received signal. This is illustrated in Figure 2.7. The dominant noise source is the Gaussian distributed thermal noise generated within the receiver. As the spectrum of thermal noise extends to frequencies of the order of 10^{13} Hz, the additive noise source is assumed to be white (possessing all frequencies).

AWGN has an infinite spectrum. It therefore follows from the earlier discussion (see page 96) that the use of Hilbert transforms is required in order to derive a complex lowpass equivalent representation. Not only is this difficult, it is also unnecessary.

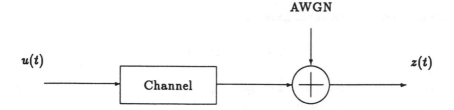

Figure 2.7: The arrangement assumed when analysing the noise properties of a bandpass system

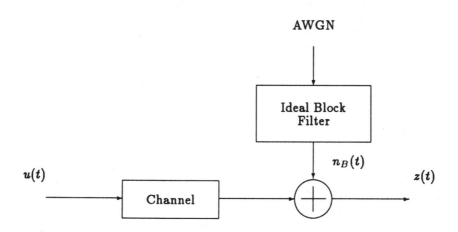

Figure 2.8: The noise model used when analysing a bandpass system

A simplified approach is adopted based on the assumption that bandpass systems are unaware of signals that lie outside their frequency band. The AWGN can therefore be regarded as having been passed through an ideal block filter prior to its addition to the received signal. Figure 2.8 illustrates this modified model. The block filter has the ideal frequency response,

$$H(f) = \begin{cases} 1 & f_c - \frac{B}{2} \leq f \leq f_c + \frac{B}{2} \\ 0 & \text{otherwise,} \end{cases} \qquad (2.38)$$

where B is greater than the bandwidth of the system, but not large when compared to the center frequency.

To noise falling in band, the filter is transparent. Noise power outside the system bandwidth receives infinite attenuation. The resulting bandpass Gaussian noise process, $n_B(t)$, is then added to the received signal.

The bandpass noise process, $n_B(t)$, can be represented in the form of

Equation 2.8. That is,

$$n_B(t) = \Re\left\{n(t)\exp(j2\pi f_c t)\right\}. \tag{2.39}$$

The complex lowpass equivalent noise signal, $n(t)$, remains Gaussian distributed, because the signal statistics are unaffected by the frequency translation.

It is possible to represent $n(t)$ as the sum of two quadrature Gaussian noise processes,

$$n(t) = n_I(t) + jn_Q(t). \tag{2.40}$$

The quadrature components, $n_I(t)$ and $n_Q(t)$, are independent Gaussian variables both with the same mean and variance as $n(t)$ [6].

2.2 Mobile Radio Channel Types

The meaning of a communications channel is not universally agreed, and it is often used in an imprecise way. We may view a particular channel as the link between two points along a path of communications. When defining a specific channel we shall indicate under what conditions the channel exhibits either or both of the properties of linearity and reciprocity. Linearity is often described as follows.

If signals x_1 and x_2 applied to a channel give rise to the output signals y_1 and y_2, respectively, then the channel is said to be linear if an input signal $x = x_1 + x_2$ produces an output signal $y = y_1 + y_2$.

Often a channel behaves in a linear fashion only over certain regions of input voltage, temperature, supply voltage, etc.. When this is the case, we refer to the regions of linear operation of the channel. The linearity of a channel is important when amplitude sensitive modulation schemes, such as quadrature amplitude modulation (QAM) are employed.

A channel is called a reciprocal channel if its behaviour is identical regardless of the direction of information flow. It follows that a reciprocal channel need only be investigated in one direction. Figure 2.9 shows the channels we consider to be of use to the systems engineer.

2.2.1 The Propagation Channel

The propagation channel is the physical medium that supports electromagnetic wave propagation between a transmit and a receive antenna. In other words, it consists of everything that influences propagation between two antennas.

It is assumed that in the mobile radio environment propagating waves will only encounter media which are both bilateral and linear (an example of when this is not the case is in an ionized plasma). This assumption implies that mobile radio propagation channels are both linear and reciprocal . The channel is also time-variant due to the movement of the mobile.

TRANSMITTER   RECEIVER

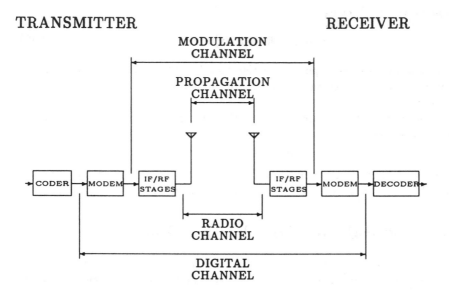

Figure 2.9: Channel types arising in radio communications

2.2.2   The Radio Channel

The transmitter antenna, propagation channel and receiver antenna viewed collectively constitute the radio channel. As the propagation channel is reciprocal, so reciprocity of the radio channel depends on the antennas used.

It can be shown [7] that antennas exhibit the same transmit and receive radiation patterns in free space if they are bilateral, linear and passive. Under these circumstances the antennas are reciprocal, and therefore so is the radio channel.

Nonlinearities can occur in antenna systems due to rust, ice, and mounting structures, but they are usually small and we shall assume they can be neglected.

2.2.3   The Modulation Channel

The modulation channel extends from the output of the modulator to the input of the demodulator, and is composed of the transmitter front-end, receiver front-end, and the radio channel. It is of particular interest to designers of modulation schemes, and trellis coding systems.

Assuming a linear radio channel, the linearity of the modulation channel is determined by the transfer characteristics of the front-ends of the transmitter and receiver. Modulation systems that employ multilevel amplitude modulation, such as quadrature amplitude modulation (QAM), require the modulation channel to be approximately linear.

To achieve linearity, amplifiers are biased to operate in their linear regions, low distortion mixers are used, and linear phase filters employed. Because linear phase filters (Bessel or Gaussian) have a slow attenuation role off, more stages are required to obtain the same selectivity as that of steeper, non-linear-phase filter families. Linear amplifiers are more expensive than non-linear versions having the same output power.

Although on a one off basis the cost of having a linear front-end may not present a problem, in a commercial cellular system where every base station and mobile has to be equipped, it becomes a major consideration.

Power efficiency is an additional problem. Amplifiers operated in their linear region (Class A) are inefficient, compared with non-linear (for example Class C) amplifiers. In a mobile environment power efficiency is of paramount concern, because the size and weight of a hand-held portable is governed largely by the batteries it uses.

Understandably, system designers avoid using linear front-ends unless it is justified by the need for high bit rate transmission in microcellular environments where the radiated power levels are relatively low.

The modulation channel is non-reciprocal, since amplifiers and other front-end components are non-reciprocal. This is not generally a problem, because a transceiver uses separate front-end equipment for the transmit and receive operations. The two radio sections are then connected to the antenna via a duplexer. Hence, in a cellular radio system, the modulation channel from base station to mobile is different than that from the mobile to base station.

2.2.4   The Digital Channel

A further channel has been proposed by Aulin [8] for the case of digital transmissions. Called the digital channel, it consists of all the system components (including the radio channel) linking the unmodulated digital sequence at the transmitter to the regenerated sequence at the receiver. The digital channel is of value to source coding and channel coding engineers.

The digital channel is nonlinear because the output can only take on certain fixed values. Reciprocity does not hold for the same reasons as described in Section 2.2.3.

2.2.5   A Channel Naming Convention

The propagation channel in the mobile radio environment is called the mobile radio propagation channel. Formally the radio channel is called the mobile radio radio channel. Similarly, the modulation and digital channels are referred to as the mobile radio modulation channel and the mobile radio digital channel, respectively.

Nevertheless, we will often refer to just the mobile radio channel. Unless otherwise stated, whenever the mobile radio channel is refered to, the mobile radio radio channel is implied.

2.3   Physical Description of the Channels

A prerequisite to combating the impairments experienced by a radio signal when it is transmitted over a mobile radio channel is to understand how these impairments originate. In the following section we examine the modifications to the transmitted signal by each of the above channels.

2.3.1   The Propagation Channel

The transmitted signal follows many different paths before arriving at the receiving antenna, and it is the aggregate of these paths that constitutes the mobile radio propagation channel. Figure 2.10 shows (in two dimensions only) two simplified propagation scenarios.

Each path may support a unique combination of propagation phenomena, nevertheless, the effect of an individual path when viewed by the receiver is to attenuate, delay and phase shift the transmitted signal. The receiver antenna has a voltage induced in it that is the superposition of many scaled and phased echoes of the transmitted waveform.

Motion of the mobile and nearby scatterers, such as trucks and buses, may cause Doppler frequency shifts in each received signal component.

Maxwell's equations tell us that wherever there exists a time varying electric field there must also be a time varying magnetic field and vice versa. In the following theory the signal $x_p(t)$ is a generic symbol which can be used to represent either an electric or a magnetic field component of the transmitted signal. Similarly $y_p(t)$ is a generic symbol representing a field component of the received signal.

2.3.1.1   The Received Signal

Consider a signal $x_p(t)$, of bandwidth B_x and centre frequency f_c radiated from a perfect isotropic radiator into a mobile radio propagation channel (the subscript p indicates an association of the variable with the propagation channel).

If B_x is small, relative to the centre frequency, the characteristics of each propagation path may be regarded as being independent of frequency (even though the propagation channel itself may be frequency dispersive). However if B_x is large compared to f_c, then the propagation phenomena (e.g. reflection and diffraction) can no longer be regarded as frequency independent over the band, and may cause appreciable signal distortion over individual paths.

It is assumed that signals of practical interest have bandwidths sufficiently narrow for the channel to be nondispersive.

The component, $y_{pi}(t)$, of the received signal due to the ith path will then be a replica of the transmitted waveform, delayed by $\tau_i(t)$ seconds, attenuated by a factor $a_i(t)$, and phase retarded (due to reflections and

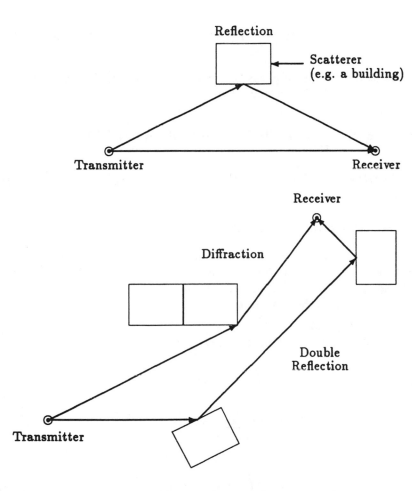

Figure 2.10: Two possible propagation channel scenarios

diffractions) by $\theta_i(t)$ radians. That is,

$$y_{pi}(t) = \Re\{a_i(t)x_p[t - \tau_i(t)]\exp j[\omega_c(t - \tau_i(t)) - \theta_i(t)]\}. \qquad (2.41)$$

Summing the received components over all the propagation paths supported by the channel yields the total received signal,

$$y_p(t) = \Re\left\{\sum_{i=0}^{I-1} a_i(t)x_p[t - \tau_i(t)]\exp j[\omega_c(t - \tau_i(t)) - \theta_i(t)]\right\} \qquad (2.42)$$

where I is the number of paths comprising the channel. Using the complex notation of Section 2.1 we can write,

$$z_p(t) = \sum_{i=0}^{I-1} a_i(t)u_p[t - \tau_i(t)]\exp -j[\omega_c\tau_i(t) + \theta_i(t)], \qquad (2.43)$$

where $u_p(t)$ and $z_p(t)$ are the complex envelopes of $x_p(t)$ and $y_p(t)$, respectively.

2.3.1.2   The Impulse Response of the Channel

In the mobile radio environment it is normally impossible to establish the exact value of I. The summation over the number of paths is therefore replaced by the integral over all the possible delays. This allows the complex envelope of $y_p(t)$ to be written as,

$$z_p(t) = \int_0^\infty a_{p\tau}(t)u_p(t - \tau)\exp -j[\omega_c\tau + \theta_{p\tau}(t)]\,d\tau. \qquad (2.44)$$

The variables $a_{p\tau}(t)$ and $\theta_{p\tau}(t)$ are given by,

$$a_{p\tau}(t) = |\sum_{\tau_i(t)=\tau} a_i(t)\exp -j\theta_i(t)| \qquad (2.45)$$

and

$$\theta_{p\tau}(t) = \arg\left(\sum_{\tau_i(t)=\tau} a_i(t)\exp -j\theta_i(t)\right), \qquad (2.46)$$

respectively.

Both Equation 2.43 and Equation 2.44 provide accurate descriptions of the received signal. However, Equation 2.44 is in a mathematically much more convenient form.

The function, $h_p(t, \tau)$ is defined as,

$$h_p(t, \tau) \triangleq a_{p\tau}(t)\exp -j[\omega_c\tau + \theta_{p\tau}(t)] \qquad (2.47)$$

so that Equation 2.44 can be written as,

$$z_p(t) = \int_0^\infty h_p(t, \tau) u_p(t - \tau) \, d\tau. \tag{2.48}$$

Comparing Equation 2.48 with Equation 2.33 in Section 2.1.3 $h_p(t, \tau)$ is identified as the time varying impulse response of the propagation channel. Specifically, $h_p(t, \tau)$ is the response of the lowpass equivalent channel at time t to a unit impulse τ seconds in the past. It is known as the input delay-spread function, and is one of eight system functions described by Bello [9] which can be used to fully characterise linear time-variant channels. These functions are discussed in Section 2.5.

2.3.1.3   The Effect of Time Variations on the Channel

To examine the effects of the time dependence of the channel, we return to Equation 2.43 and express all the channel parameters at instant t_0 as linear functions of time, that is,

$$a_i(t_0 + \delta t) = \dot{a}_i(t_0)\delta t + a_i, \tag{2.49}$$

$$\tau_i(t_0 + \delta t) = \dot{\tau}_i(t_0)\delta t + \tau_i, \tag{2.50}$$

and

$$\theta_i(t_0 + \delta t) = \dot{\theta}_i(t_0)\delta t + \theta_i, \tag{2.51}$$

where a_i , τ_i and θ_i are values taken at time $t = t_0$, δt is measured from time t_0, and a dot above a symbol signifies differentiation with respect to time. Without loss of generality we can choose t_0 to equal the time origin $t = 0$. Substituting Equations 2.49 through 2.51 into Equation 2.43 gives,

$$
\begin{aligned}
z_p(t_0 + \delta t) &= \sum_{i=0}^{I-1} [\dot{a}_i(t_0)\delta t + a_i] u_p[t - \dot{\tau}_i(t_0)\delta t - \tau_i] \cdot \\
&\quad \exp -j\{\omega_c[\dot{\tau}_i(t_0)\delta t + \tau_i] + \dot{\theta}_i(t_0)\delta t + \theta_i\},
\end{aligned} \tag{2.52}
$$

an equation illustrating the complexity of the time dependence of the mobile radio propagation channel.

For very small δt, we can write

$$\dot{a}_i(t_0)\delta t \approx 0, \quad \dot{\tau}_i(t_0)\delta t \approx 0, \quad \dot{\theta}_i(t_0)\delta t \approx 0. \tag{2.53}$$

In the cellular mobile radio environment, however, $\omega_c \gg 10^6 \text{Hz}$. Therefore, in spite of the approximations given above, the product $\omega_c \dot{\tau}_i(t_0)\delta t$ in Equation 2.52 is not negligible.

Equation 2.52 can be simplified for small time periods as,

$$z_p(t_0 + \delta t) = \sum_{i=0}^{I-1} a_i u_p(t - \tau_i) \exp j(2\pi\nu_i(t_0)\delta t - \phi_i), \tag{2.54}$$

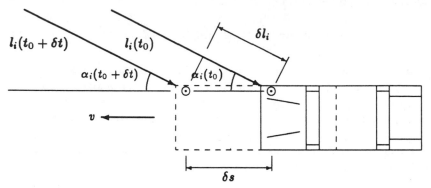

Figure 2.11: Signal geometry for small movements of the mobile

where
$$2\pi\nu_i(t_0) = -\omega_c\dot{\tau}_i(t_0) \tag{2.55}$$
and
$$\phi_i = \omega_c\tau_i + \theta_i. \tag{2.56}$$

The variable $\nu_i(t_0)$ represents the Doppler frequency shift due to changes in the electrical length, $l_i(t_0)$, of the ith path at time t_0. Figure 2.11 illustrates (in plan) the geometry associated with a small movement of the mobile. Referring to this figure it is seen that if the difference between $\alpha_i(t_0 + \delta t)$ and $\alpha_i(t_0)$ is small, then the two arrival paths are approximately parallel and the change in path length in time δt is

$$\delta l_i = l_i(t_0 + \delta t) - l_i(t_0) = -\delta s \cos \alpha_i(t_0). \tag{2.57}$$

The rate of change of the delay associated with the path is then

$$\dot{\tau}_i(t_0) = \lim_{\delta t \to 0} -\frac{\delta s}{\delta t}\frac{1}{c}\cos\alpha_i(t_0), \tag{2.58}$$

where c is the velocity of electromagnetic waves in free space.

In the limit Equation 2.58 becomes

$$\dot{\tau}_i(t_0) = -\frac{v}{c}\cos\alpha_i(t_0), \tag{2.59}$$

where v is the velocity of the mobile. Combining this result with Equation 2.55 gives the Doppler frequency shift

$$\nu_i(t_0) = \frac{\omega_c v}{2\pi c}\cos\alpha_i(t_0) = \frac{v}{\lambda}\cos\alpha_i(t_0), \tag{2.60}$$

where $\lambda = c/f_c$ is the propagation wavelength. Movement of the scatterers effecting the ith path (we assume that the base station is stationary) will also cause Doppler shifting.

Once more (c.f. Section 2.3.1.2) it is noted that the number of paths comprising the propagation channel is generally indeterminate. Delay and Doppler shifts are therefore represented as continuous domains. It is then deduced from Equations 2.54 to 2.56 that,

$$z_p(t) = \int_0^\infty \int_{-\infty}^\infty S_p(\tau, \nu) u_p(t - \tau) \exp j2\pi\nu t \, d\nu \, d\tau, \qquad (2.61)$$

where

$$S_p(\tau, \nu) = \sum_{\substack{\tau_i(t)=\tau \\ \nu_i(t)=\nu}} a_i(t) \exp -j\theta_i(t). \qquad (2.62)$$

$S_p(\tau, \nu)$ is another of the Bello system functions, called the delay-Doppler-spread function. Equations 2.45 through 2.48 with Equation 2.61 and Equation 2.62 reveal the relationship

$$h_p(t, \tau) = \int_{-\infty}^\infty S_p(\tau, \nu) \exp j2\pi\nu t \, d\nu. \qquad (2.63)$$

That is, the delay-Doppler-spread and input delay-spread functions form a Fourier transform pair over the time and Doppler shift variables, while τ is a fixed parameter. This relationship exemplifies the elegance with which the Bello functions are related. The delay-Doppler-spread function is interesting in that it explicitly illustrates both the time and frequency dispersion of a channel.

2.3.1.4 Channel Effects on Systems of Finite Delay Resolution

All radio communications systems have a finite delay resolution related to the reciprocal of their transmission bandwidths. Two propagation paths separated by less than the system's delay resolution will appear to the receiver as one path. This is illustrated in Figure 2.12. The actual channel impulse response shown comprises three impulses at delays τ_a, τ_b and τ_c, nevertheless the system of low delay resolution only sees two signals, with the apparent delays τ_a' and τ_b'.

To investigate the effect of finite delay resolution on the received signal we represent all the paths arriving with delays in the range

$$\tau_n - \frac{\Delta\tau}{2} \leq \tau < \tau_n + \frac{\Delta\tau}{2} \qquad (2.64)$$

as a single path of delay τ_n. The delay range can then be partitioned such that,

$$\tau_{n+1} = \tau_n + \Delta\tau, \qquad (2.65)$$

for $n \in 0, 1, \ldots$ and arbitrarily,

$$\tau_0 = \frac{\Delta\tau}{2}. \qquad (2.66)$$

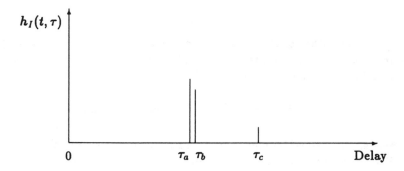

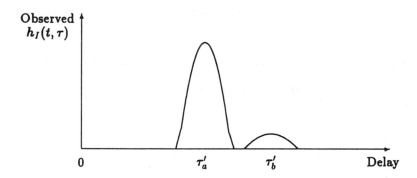

Figure 2.12: Inphase component of the channel Impulse Response Observed by a System of Low Delay Resolution

The delay range defined in Equation 2.64 is called the nth delay bin. The width, $\Delta\tau$, of the delay bin is chosen to be less or equal to the delay resolution of the system using the channel.

From Equation 2.48 the received signal due to the nth delay bin is then

$$z_{pn}(t) \approx \int_{\tau_n - \frac{\Delta\tau}{2}}^{\tau_n + \frac{\Delta\tau}{2} - \epsilon} u_p(t - \tau) h_p(t, \tau)\, d\tau, \qquad (2.67)$$

where ϵ is vanishingly small but not equal to zero. Adopting a bin width at least as narrow as the delay resolution of the system, ensures that the receiver cannot register the difference between a signal arriving with delay $\tau_n + \Delta\tau/2 - \epsilon$ and one arriving with delay $\tau_n - \Delta\tau/2$. The approximation

$$u_p(t - \tau_n + \xi) \approx u_p(t - \tau_n), \qquad (2.68)$$

where

$$|\xi| \leq \frac{\Delta\tau}{2}, \qquad (2.69)$$

can therefore be applied to Equation 2.67 to give

$$z_{pn}(t) \approx u_p(t - \tau_n) \int_{\tau_n - \frac{\Delta\tau}{2}}^{\tau_n + \frac{\Delta\tau}{2} - \epsilon} h_p(t, \tau)\, d\tau. \qquad (2.70)$$

Defining the channel response for the nth delay bin as

$$h_{p\Delta\tau}(t, \tau_n) \triangleq \int_{\tau_n - \frac{\Delta\tau}{2}}^{\tau_n + \frac{\Delta\tau}{2} - \epsilon} h_p(t, \tau)\, d\tau \qquad (2.71)$$

provides the approximation to the complex envelope of the received signal as

$$z_p(t) \approx \int_0^{\infty} h_{p\Delta\tau}(t, \tau) u_p(t - \tau)\, d\tau, \qquad (2.72)$$

where the function

$$h_{p\Delta\tau}(t, \tau) = \sum_{n=0}^{\infty} h_{p\Delta\tau}(t, \tau_n) \delta(\tau - \tau_n) \qquad (2.73)$$

may be regarded as the band-limited input delay-spread function of the propagation channel, with the factor $\Delta\tau$ determining the response bandwidth.

Figure 2.13 shows how the channel impulse response considered in Figure 2.12 maps into its band-limited version.

If $\Delta\tau$ is allowed to tend towards zero, the bandwidth over which the channel is observed increases until in the limit, $h_{p\Delta\tau}(t, \tau)$ is identical to the theoretical impulse response of the channel, $h_p(t, \tau)$.

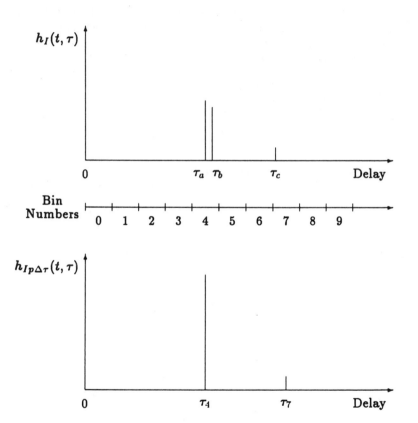

Figure 2.13: The inphase component of the band-limited impulse response of a channel

As a vector sum of several phasors, $h_{p\Delta\tau}(t, \tau_n)$ is itself a phasor, and can be represented as,

$$h_{p\Delta\tau}(t, \tau_n) = a_{p\Delta\tau}(t, \tau) \exp{-j[\omega_c \tau_n + \theta_{p\Delta\tau}(t, \tau)]}, \qquad (2.74)$$

where $a_{p\Delta\tau}(t, \tau)$ and $\theta_{p\Delta\tau}(t, \tau)$ are random variables. The distribution of $\theta_{p\Delta\tau}(t, \tau)$ will be over $[0, 2\pi]$ if $\Delta\tau \geq 1/f_c$ because the phase error in approximating a particular signal from the nth delay bin as having delay τ_n may be as great as $\pm\pi$. Furthermore, since the path delay $\tau \gg \Delta\tau$, $\theta_{p\Delta\tau}(t, \tau)$ is uniformly distributed.

Fitting an accurate distribution to the random variable $a_{p\Delta\tau}(t, \tau)$ is not straightforward, and problems of this nature have taxed many minds over the last century. The most successful distribution, in terms of popularity, was proposed by Lord Rayleigh in 1880. His distribution, known as the Rayleigh distribution, refers to a specific situation, and may not always accurately describe observed distributions of $a_{p\Delta\tau}(t, \tau)$. Nevertheless, it has maintained its position in mobile radio theory by virtue of representing a worst case scenario.

2.3.1.5 Channel Effects on Systems of Finite Doppler Resolution

In a manner similar to the delay domain, the Doppler domain may be partitioned into Doppler bins, such that,

$$\nu_{m+1} = \nu_m + \Delta\nu, \qquad (2.75)$$

for $m \in \dots, -1, 0, 1, \dots$ and

$$\nu_0 = 0. \qquad (2.76)$$

The width of the Doppler delay bins, $\Delta\nu$, is chosen to be less than the frequency resolution of the system. Therefore all signals subject to a Doppler shift falling in the mth Doppler bin can be considered to be shifted by ν_m Hz.

The function $S_{p\Delta\tau\Delta\nu}(\tau_n, \nu_m)$ can then be defined as,

$$S_{p\Delta\tau\Delta\nu}(\tau_n, \nu_m) \triangleq \int_{\tau_n - \frac{\Delta\tau}{2}}^{\tau_n + \frac{\Delta\tau}{2} - \epsilon} \int_{\nu_m - \frac{\Delta\nu}{2}}^{\nu_m + \frac{\Delta\nu}{2} - \epsilon} S(\tau, \nu) \, d\nu \, d\tau, \qquad (2.77)$$

where both ϵ and ϵ are vanishingly small but not equal to zero. We therefore obtain the approximation,

$$z_p(t) \approx \int_0^\infty \int_{-\infty}^\infty S_{p\Delta\tau\Delta\nu}(\tau, \nu) u_p(t - \tau) \exp{j2\pi\nu t} \, d\nu \, d\tau, \qquad (2.78)$$

where

$$S_{p\Delta\tau\Delta\nu}(\tau, \nu) = \sum_{n=0}^\infty \sum_{m=-\infty}^\infty S_{p\Delta\tau\Delta\nu}(\tau_n, \nu_m)\delta(\tau - \tau_n)\delta(\nu - \nu_m). \qquad (2.79)$$

2.3.2   The Radio Channel

Antennas provide the means for interfacing communication equipment with
the propagation channel. Currents in the transmitter antenna generate
electromagnetic radiation. This radiation travels via the propagation channel to the receiver where electromagnetic coupling generates currents in the
receiver antenna.

It is impossible to build an antenna which radiates equally in all directions (an isotropic radiator). Real antennas radiate more strongly in certain
directions than in others. The radiation pattern of an antenna is the gain of
that antenna as a function of the direction of radiation. When the antenna
is reciprocal, the radiation pattern for transmission is identical to that of
reception. In spherical coordinates the direction of radiation is defined by
the zenith angle, θ, and the azimuthal angle, ϕ, as shown in Figure 2.14.

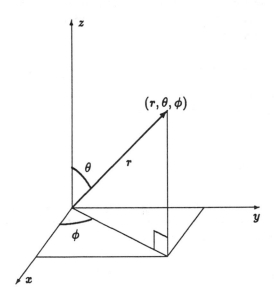

Figure 2.14: The spherical polar coordinates system

The directive gain function of an antenna is often defined relative to that
of an isotropic radiator as

$$G(\theta, \phi) \triangleq \frac{K(\theta, \phi)}{W/4\pi}, \tag{2.80}$$

where W is the total radiated power, and 4π is the total number of stera-
dians in a sphere. The radiation intensity $K(\theta, \phi)$ is defined as the power

radiated in a given direction per unit solid angle such that,

$$W = \int_0^\pi \int_0^{2\pi} K(\theta, \phi) \sin \theta \, d\theta \, d\phi. \qquad (2.81)$$

$G(\theta, \phi)$ may also be defined relative to other standard antennas, such as the quarter-wave monopole or the short dipole [7].

Real antennas may also produce a phase shift which is once more a function of the direction of transmission. Thus a continuous equiphase surface (a wavefront) propagating in all directions from the antenna need not be equal to the surface of a sphere.

Denoting the complex envelope of a signal transmitted across a mobile radio radio channel by $u_r(t)$, we can write the complex envelope of the input to the ith path of the propagation channel as,

$$u_{pi}(t) = \sqrt{G_t(\theta_{ti}, \phi_{ti})} \, u_r(t) \exp j\psi_t(\theta_{ti}, \phi_{ti}), \qquad (2.82)$$

where θ_{ti} and ϕ_{ti} define the direction of transmission of the radiation following the ith path relative to the zenith and the azimuthal angles respectively, $G_t(\theta, \phi)$ is the directive gain of the transmitter antenna, and $\psi_t(\theta, \phi)$ is its directive phase shift.

The complex envelope of the signal at the output of the radio channel due to the ith propagation path is

$$z_{ri}(t) = \sqrt{G_r(\theta_{ri}, \phi_{ri})} \, z_{pi}(t) \exp j\psi_r(\theta_{ri}, \phi_{ri}), \qquad (2.83)$$

where θ_{ri} and ϕ_{ri} identify the angle of arrival of the ith path, $G_r(\theta, \phi)$ is the directive gain of the receiver antenna and $\psi_r(\theta, \phi)$ is its directive phase shift.

Replacing the real signals in Equation 2.41 by their complex lowpass equivalents gives the complex envelope of the received signal for the ith path of the *propagation* channel as:

$$z_{pi}(t) = a_i(t)u_{pi}[t - \tau_i(t)] \exp -j[\omega_c \tau_i(t) + \theta_i(t)]. \qquad (2.84)$$

In the above equation $u_{pi}(t)$ is used, because the inputs to each propagation path are not necessarily identical. Substituting for $z_{pi}(t)$ into Equation 2.83 from Equation 2.84 yields,

$$\begin{aligned} z_{ri}(t) &= \sqrt{G_r(\theta_{ri}, \phi_{ri})} \\ &\quad a_i(t)u_{pi}[t - \tau_i(t)] \exp -j[\omega_c \tau_i(t) + \theta_i(t)] \exp j\psi_r(\theta_{ri}, \phi_{ri}). \end{aligned} \qquad (2.85)$$

The combination of Equations 2.82 and 2.85 yields,

$$\begin{aligned} z_{ri}(t) &= \sqrt{G_r(\theta_{ri}, \phi_{ri})G_t(\theta_{ti}, \phi_{ti})} \, a_i(t)u_r[t - \tau_i(t)] \\ &\quad \exp -j[\omega_c \tau_i(t) + \theta_i(t)] \exp j[\psi_r(\theta_{ri}, \phi_{ri}) + \psi_t(\theta_{ti}, \phi_{ti})], \end{aligned} \qquad (2.86)$$

which can then be summed over all the paths to give the output of the
radio channel as

$$z_r(t) = \sum_{i=0}^{I-1} \sqrt{G_r(\theta_{ri}, \phi_{ri})G_t(\theta_{ti}, \phi_{ti})}\, a_i(t)u_r[t - \tau_i(t)]$$
$$\exp -j[\omega_c\tau_i(t) + \theta_i(t)] \exp j[\psi_r(\theta_{ri}, \phi_{ri}) + \psi_t(\theta_{ti}, \phi_{ti})].$$

$$(2.87)$$

Integrating over the delay domain instead of summing over all the propagation paths allows $z_r(t)$ to be written in the form

$$z_r(t) = \int_0^\infty h_r(t, \tau)u_r(t - \tau)\, d\tau, \qquad (2.88)$$

where

$$h_r(t, \tau) = \sum_{\tau_i(t)=\tau} w_i a_i(t) \exp -j\theta_i(t) \qquad (2.89)$$

is the input delay-spread function of the radio channel, and the antenna
weighting function for the i_{th} path is

$$w_i = \sqrt{G_r(\theta_{ri}, \phi_{ri})G_t(\theta_{ti}, \phi_{ti})}\, \exp j[\psi_r(\theta_{ri}, \phi_{ri}) + \psi_t(\theta_{ti}, \phi_{ti})]. \qquad (2.90)$$

The antenna weighting function can be measured for a given pair of transmit and receive angles. However, it is difficult, if not impossible to measure
all the angles of transmission and reception associated with each propagation path [10].

A measurement system can seldom measure the true response of a propagation channel. It must measure a radio channel, which can at best only
be an approximation to the propagation channel. Further, unless $a_i(t)$ and
$\theta_i(t)$ appearing in Equation 2.89 can be established for each path comprising the propagation channel, $h_r(t, \tau)$ cannot be accurately deduced for any
other radio channel (that is, for any other combination of transmitter and
receiver antennas).

2.3.3   The Modulation Channel

The modulation channel combines the front-end of the radio equipment
and the radio channel as shown in Figure 2.9. It represents the complete
signal path between the output of the modulator and the input to the demodulator. If both front-ends are linear, then the complex lowpass impulse
response of the modulation channel is

$$h_m(t, \tau) = h_T(\tau) \star h_r(t, \tau) \star h_R(\tau), \qquad (2.91)$$

where $h_T(\tau)$, $h_r(t, \tau)$ and $h_R(\tau)$ represent the complex lowpass equivalent
impulse responses of the transmitter front-end, the radio channel and the

receiver front-end, respectively, and $\star$ represents convolution. As the transmitter and receiver front-ends are assumed to be time-invariant, they are not considered functions of t. If either or both of the front-ends are nonlinear then the modulation channel cannot be described fully by an impulse response.

2.3.4 The Digital Channel

The digital channel was introduced by Aulin [8]. It consists of all the system components (including the radio channel) linking the unmodulated digital sequence at the transmitter to the regenerated sequence at the receiver. It is characterized by bit error patterns.

This channel is of great use to engineers working at baseband. For example a knowledge of the digital channel will enable a speech codec to be designed for a particular set of bit error statistics, without the designer needing to know the complexities of the propagation channel, modem and transceiver behaviour.

The digital channel is nonlinear, so the relationship between it and the modulation channel is not simply described. It is also non-reciprocal, because modems are non-reciprocal.

2.4 Classification of Channels

The impulse response of a mobile radio channel generally exhibits both delay and Doppler spreading. Delay spreading results in two effects, time dispersion and frequency-selective fading. Doppler spreading leads to frequency dispersion and time-selective fading.

Although all four effects are displayed by mobile radio channels, whether they are apparent to systems operating over a channel is dependent on the nature of the transmitted signal. The channel *as perceived by the system* can therefore be classified according to which effects are dominant.

In order to develop a system of classification we shall first examine each of the effects mentioned above.

2.4.1 Time Dispersion and Frequency-Selective Fading

Time dispersion and frequency-selective fading are both manifestations of multipath propagation with delay spread. The presence of one effect perforce implies the presence of the other.

Time dispersion stretches a signal in time so that the duration of the received signal is greater than that of the transmitted signal. Frequency-selective fading filters the transmitted signal, attenuating certain frequencies more than others. Two frequency components closely spaced receive approximately the same attenuation, however, if they are far apart they often receive vastly different attenuations.

Time dispersion is a result of the signals taking different times to cross the channel by different propagation paths. Frequency selective fading occurs because the electrical length of each propagation path can be expressed as a function of frequency.

If the bandwidth, B_x, of the transmitted signal is sufficiently narrow, then all the transmitted frequency components will receive about the same amount of attenuation, and the signal will be passed undistorted, without frequency selective fading.

As the transmission bandwidth is increased, the frequency components at the extremes of the transmitted spectrum will start to be attenuated by different amounts. Thus the channel is having a filtering effect, and is distorting the transmitted waveform, that is frequency selective fading is experienced. The distortion increases as B_x is increased.

For very large transmission bandwidths the receiver may be able to observe distinct echoes of the transmitted waveform. At this point the system is able to recognize time dispersion, since the delay spread of the channel is greater than the delay resolution of the receiver. In digital systems this results in intersymbol interference.

The minimum transmission bandwidth at which time dispersion is observable is inversely proportional to the maximum excess delay of the channel, τ_m, where the excess delay is the actual delay minus the delay of the first arrival path. The constant of proportionality is system dependent, but shall be taken here to be $\frac{1}{4}$.

There are thus two observable effects of delay spread; distortion and dispersion.

A measure of the transmission bandwidth at which distortion becomes appreciable is often based on the channel's coherence bandwidth. The coherence bandwidth $B_c(t)$ indicates the frequency separation at which the attenuation of the amplitudes of two frequency components becomes decorrelated such that the envelope correlation coefficient, $\rho(\Delta f, \Delta t)$, reaches a predesignated value. This value has in the past been taken as 0.9 [11], 0.5 [12, 13] and $1/e$ (0.37) [14], [15], [16].

The amount of signal distortion required before a specific system's performance is effected is heavily dependent on the modulation and demodulation techniques employed. A particular system may start to have problems when the transmission bandwidth corresponds to a value of 0.9 for the envelope correlation coefficient, whereas a more robust system may perform perfectly satisfactorily up to a transmission bandwidth corresponding to an envelope correlation coefficient of 0.37.

In line with the two major works on mobile radio communications Lee [12] and Jakes [13] the coherence bandwidth is taken to correspond to an envelope correlation coefficient of 0.5. That is

$$\rho(B_c(t), 0) = 0.5, \tag{2.92}$$

where [13]

$$\rho(\Delta f, \Delta t) \triangleq \frac{\langle a_1 a_2 \rangle - \langle a_1 \rangle \langle a_2 \rangle}{\sqrt{[\langle a_1^2 \rangle - \langle a_1 \rangle^2][\langle a_2^2 \rangle - \langle a_2 \rangle^2]}}. \tag{2.93}$$

In the above definition $\langle\rangle$ denotes the ensemble average. Variables a_1 and a_2 represents the amplitudes of signals at frequencies f_1 and f_2, respectively, and at times t_1 and t_2, respectively, where $|f_2 - f_1| = \Delta f$ and $|t_2 - t_1| = \Delta t$.

To derive a value for $B_c(t)$, we shall employ the approximation that in the mobile radio environment the amplitude of each received signal is unity and that the probability of receiving a signal with delay τ is given by

$$p(\tau) = \frac{1}{2\pi\sigma(t)} \exp \frac{-\tau}{\sigma(t)}, \tag{2.94}$$

where $\sigma(t)$ is the delay spread of the channel. This may at first appear a widely inaccurate approximation for the mobile radio environment, since we realise that signals of different delays rarely arrive with equal amplitudes. It should, however, be remembered that this is a mathematical model. The same results would be produced from a model that uses signals arriving at fixed delays with exponentially distributed amplitudes—intuitively a more acceptable scenario.

The delay spread, $\sigma(t)$ is equal to the square root of the second central moment of the channel's power-delay profile, $P_h(\tau)$. That is

$$\sigma(t) = \frac{\int_0^\infty (\tau - d(t))^2 P_h(\tau) \, d\tau}{\int_0^\infty P_h(\tau) \, d\tau}, \tag{2.95}$$

where $d(t)$ is the mean propagation delay, given by

$$d(t) = \frac{\int_0^\infty \tau P_h(\tau) \, d\tau}{\int_0^\infty P_h(\tau) \, d\tau}. \tag{2.96}$$

The power-delay profile of the channel, to be described in Section 2.6.3, is given as

$$\begin{aligned} P_h(\tau) &= |h(t,\tau)|^2 \\ &= h_I(t,\tau)^2 + h_Q(t,\tau)^2. \end{aligned} \tag{2.97}$$

Approximating $p(\tau)$ as shown in Equation 2.94 does not of course describe all mobile radio channels, since the specific environment of each system varies. However, results indicate that it is not an unreasonable assumption [17, 18]. It can then be shown [12, 13] that the envelope correlation coefficient for two signals separated by Δf Hz and Δt seconds is equal to

$$\rho(\Delta f, \Delta t) = \frac{J_0^2(2\pi f_m \Delta t)}{1 + (2\pi \Delta f)^2 \sigma^2}, \tag{2.98}$$

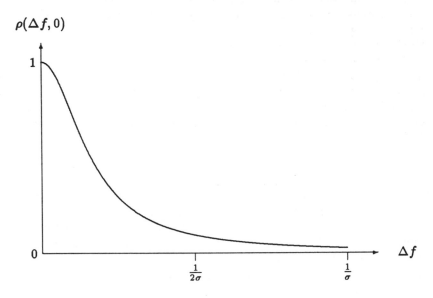

Figure 2.15: Envelope correlation with frequency separation

where $J_0()$ is the zero order Bessel function of the first kind and $f_m = v/c$ is the maximum Doppler shift for a vehicular velocity of v, with c representing the velocity of light. To observe the decorrelation of two signals as their frequency separation is increased, Δt is set equal to zero in Equation 2.98. This gives the frequency correlation function as

$$\rho(\Delta f, 0) = \frac{1}{1 + (2\pi\Delta f)^2\sigma^2}. \tag{2.99}$$

The graphical representation of Equation 2.99 is shown in Figure 2.15, where the envelope correlation decreases with the frequency separation of the signals. The correlation bandwidth is obtained from Equations 2.92 and 2.99 as

$$B_c(t) = \frac{1}{2\pi\sigma(t)}. \tag{2.100}$$

A typical delay spread value of $2\mu s$ for conventional size cells in an urban environment [19] results in a coherence bandwidth of about 80kHz. In practice, the correlation function $\rho(\Delta f, 0)$ does not decrease monotonically with increasing frequency separation [20] and the presence of a strong echo with excess delay τ_{ex}, will lead to an oscillatory component in the correlation coefficient of frequency $1/\tau_{ex}$ [21]. In this case, as Δf is increased the first occurrence of the envelope correlation coefficient dropping below 0.5 is taken as the channel coherence bandwidth.

2.4.2   Frequency Dispersion and Time-Selective Fading

When a channel is time variant it is referred to as possessing time-selective fading. Time-selective fading can cause signal distortion, because the channel may change its characteristics whilst the transmitted signal is in flight. The channel seen by the leading edge of the signal is not the same as that seen by the trailing edge.

In Section 2.3.1.3 it was shown that when the response of a channel is time-variant, Doppler spreading (frequency dispersion) occurs. Frequency dispersion results in the signal bandwidth being stretched so that the received signal's bandwidth is different (greater or less) from that of the transmitted signal.

If a signal has a short duration then it is passed through the channel before any significant change in the channel characteristics can take place. As the signal's duration is increased, the channel is able to change whilst the signal is still in flight, thereby causing distortion. The distortion increases as the signal duration is increased. At the same time, Doppler spreading of the signal increases relative to the transmission bandwidth until it is possible to observe significant widening of the received spectrum. That is, when the maximum Doppler frequency is larger than the Doppler resolution of the receiver,

$$f_m > \Delta \nu. \tag{2.101}$$

The minimum signal duration at which frequency dispersion becomes noticable is inversely proportional to the magnitude of the maximum Doppler shift experienced by the signal, f_m. The constant of proportionality is again somewhat arbitrary, but will be taken as being equal to $\frac{1}{4}$.

In a manner similar to that of Section 2.4.1, we can estimate at what transmitted signal duration distortion becomes noticeable by referring to the channel's coherence time, $T_c(t)$. Analogous to the channel's coherence bandwidth (Equation 2.92) the coherence time is defined as,

$$\rho(0, T_c(t)) = 0.5. \tag{2.102}$$

Setting $\Delta f = 0$ in Equation 2.98 gives

$$\rho(0, \Delta t) = J_0^2 (2\pi f_m \Delta t), \tag{2.103}$$

which is plotted in Figure 2.16.   From the previous two equations

$$T_c(t) = \frac{J_0^2 (2\pi f_m \Delta t)}{2} \approx \frac{9}{16\pi f_m}. \tag{2.104}$$

For example, for a vehicular speed of 30 ms^{-1} (67.5 mph), a channel centered on 1.7 GHz exhibits a coherence time of approximately 1 ms. This corresponds to a transmitted bit rate of 1 kb/s. Signals with bit rates in excess of 1 kb/s can therefore assume the channel to be non-distorting in time.

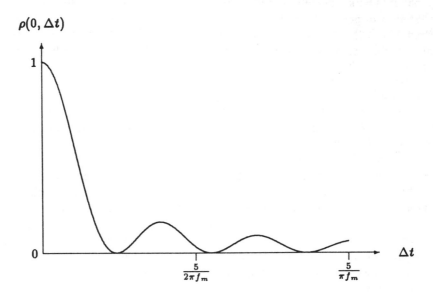

Figure 2.16: Envelope correlation with time separation

2.4.3   Channel Classifications

The coherence bandwidth and coherence time are properties of a channel which may be used to assess how it will appear to transmitted signals. If the bandwidth of a transmission is less than the coherence bandwidth of the channel, the frequency-selective fading, and therefore the time dispersion of the channel appear to be transparent to the signal. The channel is viewed by the system as having a flat response across the transmission band, and is therefore referred to as being frequency-flat.

Similarly, if the duration of the received waveform is less than the coherence time, the channel will appear to the signal to be time-invariant. Notice that we have specified the *received* waveform duration, since this is the time for which the signal is in flight. It is generally taken as the transmitted signal duration (the symbol period for digital transmissions) plus the channel delay spread, $\sigma(t)$ (defined in Equation 2.95). As the channel response appears to be constant for the duration of the signal's flight, the channel is referred to as time-flat.

When a channel is flat in both frequency and time, it is called a flat-flat channel. When a channel is flat neither in frequency nor in time, it is often referred to as a doubly dispersive channel. This nomenclature, however, is somewhat misleading because such a channel need only cause signal distortion not dispersion. Hence we shall refer to a channel that is neither time-flat nor frequency-flat as a non-flat channel.

Figure 2.17 shows the classification of channels following the above approach. The shaded region of the figure indicates the physical restriction that it is impossible for the time bandwidth product of a signal to be less than 1/2 [22]. A more rigorous system of classification, emphasising the differences between distorting and dispersive channels, is shown in Figure 2.18.

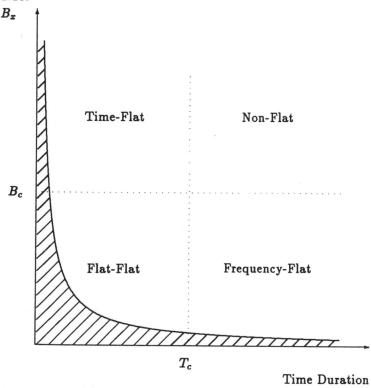

Figure 2.17: Channel Classifications

The flat-flat, or doubly-flat channel does not fade with either time or frequency. Using the approximate values derived above, it is seen that a signal of bandwidth less than $B_c(t)$ Hz and duration less than $T_c(t)$ seconds will observe a flat-flat channel at time t. For example, in this category resides the uplink from the mobile stations (MS) to the base station (BS) of the MATS-D system [23]. MATS-D is a hybrid mobile radio system that was put forward as a contender for the pan-European cellular radio system. It employs a narrowband frequency division multiple access (FDMA) scheme for the uplink. The transmissions employ generalised tamed frequency modulation (GTFM) to give a bandwidth of approximately 25 kHz for a bit rate of 19.5 kb/s.

The frequency-flat fading channel is observed by narrowband channel

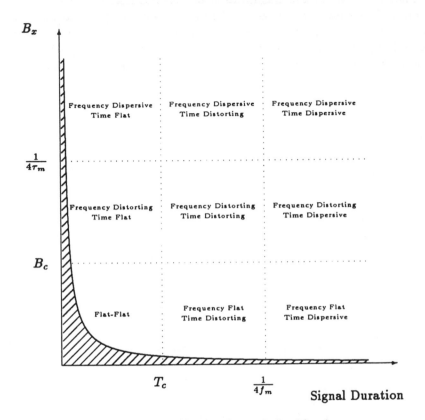

Figure 2.18: Detailed channel classifications

sounders. These sounders transmit a monochromatic (single tone) signal continuously, and so approximate to a signal of infinitesimal bandwidth and infinite duration. The envelope of the received process therefore varies in sympathy with the channel.

For a delay spread of $2\mu s$, the time-flat fading channel applies to all mobile radio systems using digital transmissions with bit rates in excess of 80 kb/s (see page 120). Many of the systems put forward for the pan-European cellular system (e.g., S900D [24], DMS90 [25], MATS-D (downlink) [23], and CD900 [26]) including the GSM system that has been adopted [27], fall into this category.

The situation where the channel is flat with neither time nor frequency does not occur with mobile radio channels. This is because of the short delay spreads and low mobile speeds. Such a channel may be encountered in satellite to aircraft communications where greater mobile (aircraft) speeds are combined with large excess delays due to ground reflections.

Variable	Notation	Units
Time	t	Seconds
Frequency	f	Hertz
Delay	τ	Seconds
Doppler Shift	ν	Hertz

Table 2.1: Variables used to describe linear time-variant channels

Although $2\mu s$ is a typical delay spread value for example in New York, it may not be appropriate for all mobile radio environments. Suburban environments tend to show less delay spread [20, 28], whilst some measurements in cities have yielded delay spreads of $5\mu s$ and greater [29, 30]. In extreme environments, e.g., hilly terrain, delay spreads of up to $17\mu s$ have been recorded [31]. In order to measure the characteristics of a time-flat channel, wideband sounding techniques must be employed.

2.5 A Systems Approach to Linear Time-Variant Channels

Bello [9] proposed a set of eight system functions to describe linear time-variant channels. Each function embodies a complete description of the channel, and full knowledge of one function allows calculation of any of the others. Each one uses two of the four variables shown in Table 2.1.

2.5.1 The Variables Used For System Characterisation

The familiar time and frequency variables, t and f, are by definition [32] dual network variables, whilst the delay and Doppler shift variables, τ and ν, are dual operators describing time and frequency translation. The concept of duality has been discussed at length by Bello [32], however, for the purposes of this discussion, it is sufficient to understand that two operators (functions, elements, or systems) are dual when the behaviour of one with reference to a time-related domain (the time or the delay domain) is identical to the behaviour of the other referenced to the corresponding frequency-related domain (i.e., the frequency or the Doppler shift domain, respectively).

A common mistake when first encountering the delay variable is to assume that since τ is measured in seconds it is linearly dependent on the time variable t. This is not the case (i.e., $\tau \not\propto t$). The delay variable is orthogonal to the time variable, and as such can be drawn on a set of Cartesian coordinates. It is easiest to appreciate that the variables t and τ are independent, if the electrical lengths of propagation paths are considered.

A channel's electrical length is one of its physical properties, and is related to τ by

$$l_e = \frac{\tau}{c}. \tag{2.105}$$

Even if it were possible to freeze time, a path would still possess an electrical length of l_e, and therefore an associated delay. Although the electrical length of a particular path may vary with time, perhaps due to the motion of the mobile, in general there may exist a path at any instant in time possessing any positive electrical length. That is, the two variables are independent, and from Equation 2.105 we deduce that τ is also independent of t.

It is perhaps more difficult to understand that ν is orthogonal to f, because the Doppler shift associated with a particular path is a function of the frequency of transmission and physically it is caused by a change in the delay or electrical length of a path, as evidenced by Equation 2.55. The rate of change of l_e with respect to time is expressed in ms^{-1}. Equations 2.55 and 2.105 show that the Doppler shift is actually the rate of change of the physical length of the path, dl/dt, scaled by the signal's frequency. The frequency scaling occurs because a signal perceives length in terms of wavelengths, not absolute measures. As dl/dt is independent of frequency, it is theoretically possible for a path to exist possessing any Doppler shift value at any particular frequency.

2.5.2 The Bello System Functions

Two of the Bello system functions were encountered in Section 2.3.1 when the propagation channel was analysed in the time domain. The first one is the **input delay-spread function**, $h(t, \tau)$, defined by

$$z(t) = \int_{-\infty}^{\infty} h(t, \tau) u(t - \tau) \, d\tau \tag{2.106}$$

and interpreted as the response of the channel at time t to a unit impulse input τ seconds in the past. This function describes the channel in terms of the t–τ domain.

It is helpful to visualise Equation 2.106, as the output from a densely tapped delay line, where $h(t, \tau) d\tau$ is the tap weighting for delay τ, as seen in Figure 2.19. Notice that $h(t, \tau)$ is called the *input* delay-spread function because the delay is associated with the *input* port of the channel. A further Bello function, $g(t, \tau)$, will be introduced which has the delay associated with the output port of the channel. This function is then called the output delay-spread function.

Computer simulations generally employ the tapped delay line approach to model mobile radio channels, using the complex baseband equivalent system of Figure 2.6.

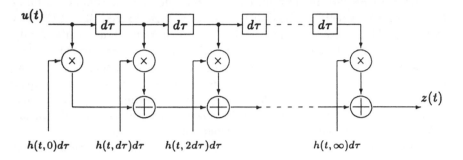

Figure 2.19: Tapped delay line representation of the input delay-spread function

The second Bello function, which has already been presented is the **delay-Doppler-spread function**, $S(\tau, \nu)$. This function was defined by Equation 2.61 as:

$$z(t) \triangleq \int_{-\infty}^{\infty} \int_{-\infty}^{\infty} S(\tau, \nu) u(t - \tau) \exp j2\pi\nu t \, d\nu \, d\tau \qquad (2.107)$$

and is interpreted as the gain experienced by signals suffering first delay in the range $[\tau, \tau + d\tau]$ then Doppler shift in the range $[\nu, \nu + d\nu]$.

$S(\tau, \nu)$ uses the τ–ν domain to describe the channel. It is seldom used in the simulation of channels because the double integral in the above equation requires more computation than the single integral of Equation 2.106. Nevertheless, the delay-Doppler-spread function has found favour as a means of displaying the dispersive characteristics of a channel, because it explicitly shows both time and frequency dispersion.

Equation 2.63 shows that $h(t, \tau)$ and $S(\tau, \nu)$ form a complex Fourier pair over the variables ν and t, with the common variable τ. It may come as somewhat of a surprise that the time domain Fourier transformation transforms into the Doppler shift domain and not into the frequency domain. This is because it is a change in the channel's behaviour as a function of time that causes a Doppler shift, whilst it is the frequency response of the channel at a specific time as a function of the delay variable that determines the channel's spectrum. As ν is the dual of τ, and t is the dual of f, we can deduce from the above relationship that the delay domain Fourier transforms with the frequency domain.

From these two Fourier relationships it is possible to deduce why Bello functions require two variables, and also why there are only eight Bello functions.

Fourier transform pairs contain the same information, presented in different forms. The time and Doppler shift domains contain information describing frequency dispersion, whilst the frequency and delay domains contain time dispersion information. To *fully* describe a channel, a function must use at least two variables, one from each Fourier pair. Should a third variable be used, its information would be redundant, while if only one variable were to be used, the description would be incomplete.

The number of ways of permuting two variables from four is 12. However, Bello has only defined eight functions. This is because four of these permutations are due to variables from the same Fourier pair (i.e., t, ν; ν, t; τ, f; and f, τ;). Functions defined for these permutations would contain information about only one type of dispersion.

The six Bello functions that have yet to be presented will be discussed below.

The time-variant transfer function, $T(f, t)$ is defined by the equation

$$z(t) = \int_{-\infty}^{\infty} T(f, t) U(f) \exp j2\pi f t \, df. \qquad (2.108)$$

It is interpreted as the complex envelope of the received signal for a cissoidal input at the carrier frequency. As the name implies, $T(f, t)$ is the time-variant equivalent of the conventional (time-invariant) system transfer function. Equations 2.106 and 2.108 may be manipulated as follows to reveal that $T(f, t)$ and $h(t, \tau)$ form a Fourier pair with the common variable t.

Replacing $u(t - \tau)$ in Equation 2.106 with its Fourier transform gives

$$z(t) = \int_{-\infty}^{\infty} h(t, \tau) \int_{-\infty}^{\infty} U(f) \exp j2\pi f(t - \tau) \, df \, d\tau, \qquad (2.109)$$

which on rearranging yields

$$z(t) = \int_{-\infty}^{\infty} \left\{ \int_{-\infty}^{\infty} h(t, \tau) \exp -j2\pi f\tau \, d\tau \right\} U(f) \exp j2\pi f t \, df. \qquad (2.110)$$

Equating the above equation to Equation 2.108 shows the required Fourier relationship, that is:

$$h(t, \tau) = \int_{-\infty}^{\infty} T(f, t) \exp j2\pi f\tau \, df. \qquad (2.111)$$

The output Doppler-spread function, $H(f, \nu)$ describes the channel in the f–ν domain, and is defined by

$$Z(f) = \int_{-\infty}^{\infty} U(f - \nu) H(f - \nu, \nu) \, d\nu. \qquad (2.112)$$

The interpretation of the function $H(f, \nu)$ is as the spectral response of the channel at a frequency ν Hz above a cissoidal input at f Hz.

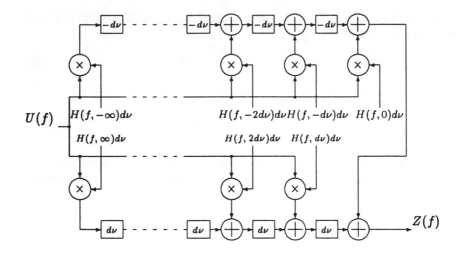

Figure 2.20: Tapped frequency conversion chain representation of the output doppler-spread function

A differential circuit representation of Equation 2.112 is given in Figure 2.20. The circuit is a densely tapped frequency conversion chain.

The Doppler shift $d\nu$ is associated with the output of the channel, hence the name *output* Doppler-spread function. $H(f,\nu)$ offers an alternative approach to computer simulation instead of using $h(t,\tau)$. The two models (Figures 2.19 and 2.20) are equally easy to translate into program code, allowing a worker to deal with channel inputs in either the time or frequency domains.

It is possible to show that $H(f,\nu)$ and $T(f,t)$ are a Fourier pair by examining the received spectrum when a cissoidal input at $f = f'$ is present. From the previous equation we then have:

$$Z(f) = \int_{-\infty}^{\infty} \delta(f - f' - \nu)H(f - \nu, \nu)\,d\nu. \qquad (2.113)$$

Taking the Fourier transform of both sides of this equation with respect to f, we get:

$$z(t) = \int_{-\infty}^{\infty}\int_{-\infty}^{\infty} \delta(f - f' - \nu)H(f - \nu, \nu)\exp j2\pi ft\,d\nu\,df, \qquad (2.114)$$

which reduces to

$$z(t) = \int_{-\infty}^{\infty} H(f', \nu)\exp j2\pi(f' + \nu)t\,d\nu. \qquad (2.115)$$

Now from Equation 2.108, the received signal for a cissoidal input at $f = f'$ may also be written as,

$$z(t) = T(f', t) \exp j2\pi f't. \tag{2.116}$$

Equating the above two equations gives the required relationship,

$$T(f, t) = \int_{-\infty}^{\infty} H(f, \nu) \exp j2\pi\nu t \, d\nu. \tag{2.117}$$

$H(f, \nu)$ also forms a Fourier transform pair with $S(\tau, \nu)$, this time with the common variable ν. This may be derived as follows. Fourier transformation of both sides of Equation 2.107 with respect to t gives

$$Z(f) = \int_{-\infty}^{\infty} \int_{-\infty}^{\infty} \int_{-\infty}^{\infty} S(\tau, \nu) u(t - \tau) \exp j[2\pi(\nu - f)t] \, d\nu \, d\tau \, dt, \tag{2.118}$$

which can be rearranged as

$$Z(f) = \int_{-\infty}^{\infty} \int_{-\infty}^{\infty} S(\tau, \nu) \left\{ \int_{-\infty}^{\infty} u(t - \tau) \exp j[2\pi(\nu - f)t] \, dt \right\} \, d\nu \, d\tau. \tag{2.119}$$

The term in braces is evaluated using the shifting property of Fourier transform, so that

$$Z(f) = \int_{-\infty}^{\infty} \int_{-\infty}^{\infty} S(\tau, \nu) \exp j[2\pi(f - \nu)\tau] U(f - \nu) \, d\nu \, d\tau. \tag{2.120}$$

Comparing this equation with Equation 2.106 reveals the Fourier relationship

$$S(\tau, \nu) = \int_{-\infty}^{\infty} H(f, \nu) \exp j2\pi\tau f \, d\tau. \tag{2.121}$$

Figure 2.21 shows how the four Fourier relationships derived so far (Equations 2.63, 2.111, 2.117 and 2.121) allow the functions to be arranged in a symmetric pattern.

Just as the delay process is associated with the input of the channel for $h(t, \tau)$, we can derive a further Bello function by associating the Doppler shift with the input of the channel. The resulting function is called the **input Doppler-spread function**, $G(f, \nu)$. The model describing this situation is shown in Figure 2.22, from which we see that

$$Z(f) = \int_{-\infty}^{\infty} G(f, \nu) U(f - \nu) \, d\nu. \tag{2.122}$$

Comparison of this equation and Equation 2.106 shows that they are identical in form and both represent a convolution. Equation 2.122 is in fact the dual relationship to that of Equation 2.106, and $G(f, \nu)$ is the dual function of $h(t, \tau)$. Duality was introduced at the start of this section. To

LEGEND :   *F*   Fourier Transform
           F^{-1} Inverse Fourier Transform

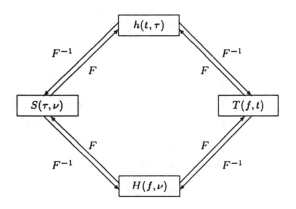

Figure 2.21: Fourier relationships amongst the first set of Bello functions

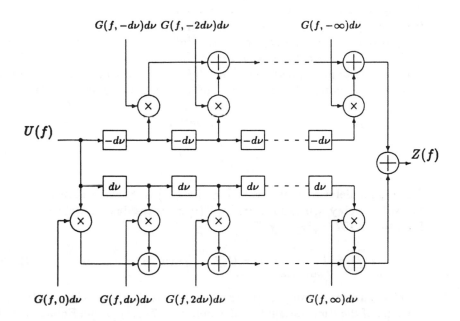

Figure 2.22: Tapped frequency conversion chain representation of the input doppler-spread function

Original notation	Dual notation
t	f
f	t
τ	ν
ν	τ
$\exp(\cdot)$	$\exp -(\cdot)$
$\exp -(\cdot)$	$\exp(\cdot)$
$x(t)$	$X(f)$
$X(f)$	$x(t)$

Table 2.2: Notational changes to establish dual relations

reiterate, dual systems, or operators, for example, behave in an identical manner, however one exists in a dual domain to the other.

Bello [32] has presented the techniques used to manipulate time-frequency duality relationships. In brief, to obtain a dual relationship replace each function by its dual and make the substitutions shown in Table 2.2.

By applying duality relations to the defining equations of the functions in Figure 2.21, the definitions of the remaining system functions are found. The input Doppler-spread function, $G(f, \nu)$, has already been introduced, as seen in Equation 2.122 above. The **output delay-spread function**, $g(t, \tau)$, is the dual of the output Doppler-spread function, $H(f, \nu)$ and its defining equation is obtained from Equation 2.112 applying duality as:

$$z(t) = \int_{-\infty}^{\infty} u(t - \tau)g(t - \tau, \tau)\, d\tau. \tag{2.123}$$

The tapped delay line representation of $g(t, \tau)$ is shown in Figure 2.23. The **Doppler-delay-spread function**, $V(\nu, \tau)$, is the dual of $S(\tau, \nu)$, and from Equation 2.107 is given by

$$Z(f) = \int_{-\infty}^{\infty} \int_{-\infty}^{\infty} V(\nu, \tau)U(f - \nu)\exp -j2\pi\tau f\, d\tau\, d\nu. \tag{2.124}$$

Finally, the **frequency-dependent modulation function**, $M(t, f)$, is the dual of the time-variant transfer function, $T(f, t)$. From Equation 2.108

$$Z(f) = \int_{-\infty}^{\infty} M(t, f)u(t)\exp -j2\pi ft\, dt. \tag{2.125}$$

Figure 2.24 shows that this second set of Bello functions can also be

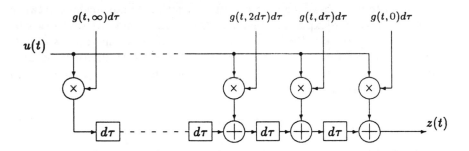

Figure 2.23: Tapped delay line representation of the output delay-spread function

LEGEND : F Fourier Transform
           F^{-1} Inverse Fourier Transform

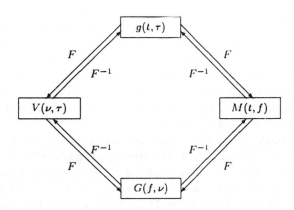

Figure 2.24: Fourier relationships amongst the second set of Bello functions

arranged symmetrically with respect to complex Fourier transforms.

A summary of the definitions and interpretations of all the eight Bello functions are presented in Table 2.3.

In order to move between the two sets of functions, relationships between sister functions are exploited. Sister functions are functions that describe the channel in terms of the same domain. For example, the input and output delay-spread functions, $h(t, \tau)$ and $g(t, \tau)$, are sister functions because they both describe the channel in the t-τ domain. From Equations 2.106 and 2.123 it is seen that

$$h(t, \tau) = g(t - \tau, \tau). \qquad (2.126)$$

The relationships between the other three pairs of sister functions are easily derived from their defining equations. They are summarised in Table 2.4. As sister functions describe the channel in the same domain, it is usual to employ just one set of functions. The most commonly used set [33], [34], [35] is the one illustrated in Figure 2.21. The functions in this set generally occur as shown in Table 2.5.

So far, we have introduced a family of system functions which fully describe a time-variant channel, and have provided techniques for manipulating these functions. Attention will now be given to randomly time-variant channels. Such channels are particularly useful to study, because, as will be seen in Section 2.6, mobile radio channels can be regarded as being randomly time-variant.

2.5.3　Description of Randomly Time-Variant Channels

A general linear time-variant channel can be viewed as the superposition of a deterministic channel and a purely random, zero ensemble average channel, shown in Figure 2.25.

The deterministic channel may be fully characterised by applying directly the system functions described above. However, the functions become stochastic processes, when they are used to describe the randomly varying component of the channel.

Unfortunately, a full statistical description of the system functions requires the determination of multidimensional pdf's for the functions, and this is not a trivial task. A less stringent, but practical approach [9, 36] to characterising purely random channels involves the determination of the correlation functions for any one of the Bello system functions.

2.5.3.1　Autocorrelation of a Bandpass Stochastic Process

The autocorrelation $R_y(t_1, t_2)$ of a stochastic process, $y(t)$, is the ensemble average, or expected value, of the product $y(t_1)y(t_2)$ and may be written

Name	Notation	Definition	Interpretation	Dual
Input delay-spread function	$h(t,\tau)$	$z(t) = \int h(t,\tau)u(t-\tau)d\tau$	The channel response at time t to a unit impulse input τ seconds in the past	$G(f,\nu)$
Time-variant Transfer function	$T(f,t)$	$z(t) = \int T(f,t)U(f)\exp j2\pi ft df$	The complex envelope of the received signal due to a cisoidal input at the carrier frequency	$M(t,f)$
Output Doppler-spread function	$H(f,\nu)$	$Z(f) = \int H(f-\nu,\nu)u(f-\nu)d\nu$	The spectral response of the channel at a frequency ν Hz above the cisoidal input at frequency f Hz	$g(t,\tau)$
Delay-Doppler-spread function	$S(\tau,\nu)$	$z(t) = \int\int S(\tau,\nu)u(t-\tau) \exp j2\pi\nu t d\nu d\tau$	The gain afforded signals suffering first delay in $[\tau,\tau+d\tau]$ then Doppler shift in $[\nu,\nu+d\nu]$	$V(\nu,\tau)$
Ouput delay-spread function	$g(t,\tau)$	$z(t) = \int g(t-\tau,\tau)u(t-\tau)d\tau$	The channel response τ seconds in the future to a unit impulse input at time t	$H(f,\nu)$
Frequency-dependent modulation function	$M(t,f)$	$Z(f) = \int M(t,f)u(t) \exp -j2\pi t f dt$	The complex amplitude spectrum of the received signal for a unit impulse input at time $t = 0$	$T(f,t)$
Input Doppler-spread function	$G(f,\nu)$	$Z(f) = \int G(f,\nu)u(f-\nu)d\nu$	The spectral response of the channel at a frequency f Hz due to a cisoidal input ν Hz below f	$h(t,\tau)$
Doppler-delay-spread function	$V(\nu,\tau)$	$Z(f) = \int\int V(\nu,\tau)U(f-\nu) \exp -j2\pi\tau\nu d\tau d\nu$	The gain afforded signals suffering first Doppler shift in $[\nu,\nu+d\nu]$ then delay in $[\tau,\tau+d\tau]$	$S(\tau,\nu)$

Table 2.3: The Bello functions

Function	Sister function	Relationship
$h(t,\tau)$	$g(t,\tau)$	$h(t,\tau) = g(t - \tau, \tau)$
$T(f,t)$	$M(t,f)$	$\displaystyle\int\int M(t',f') \exp{-j2\pi(f - f')(t - t')}\,df'\,dt'$
$H(f,\nu)$	$G(f,\nu)$	$H(f,\nu) = G(f + \nu, \nu)$
$S(\tau,\nu)$	$V(\nu,\tau)$	$S(\tau,\nu) = V(\nu,\tau) \exp{-j2\pi\nu\tau}$

Table 2.4: Sister functions

Function	Normal Occurrence
$h(t,\tau)$	Measured directly by time domain wideband sounders. Used in computer simulations.
$T(f,t)$	Measured directly by narrowband (single tone) sounders.
$H(f,\nu)$	Measured by frequency domain wideband sounders. Used in computer simulations.
$S(\tau,\nu)$	Used to display the time and frequency dispersion of the channel simultaneously.

Table 2.5: Occurrences of the first set of Bello functions

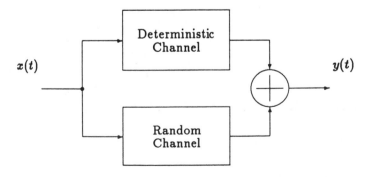

Figure 2.25: The decomposition of a channel into its deterministic and random components

as [37]

$$R_y(t_1, t_2) = \int_{-\infty}^{\infty} \int_{-\infty}^{\infty} y_1 y_2 f(y_1, y_2; t_1, t_2) \, dy_1 \, dy_2, \qquad (2.127)$$

where $f(y_1, y_2; t_1, t_2)$ is the second order density of the process $y(t)$, given by

$$f(y_1, y_2; t_1, t_2) \triangleq \frac{\partial^2 F(y_1, y_2; t_1, t_2)}{\partial y_1 \partial y_2}. \qquad (2.128)$$

The cumulative densiy function $F(y_1, y_2; t_1, t_2)$ is defined as the joint probability,

$$F(y_1, y_2; t_1, t_2) \triangleq \Pr\left\{ y(t_1) < y_1, y(t_2) < y_2 \right\}. \qquad (2.129)$$

The ensemble average of a quantity, $(\cdot)$, is denoted by angle brackets, like $\langle (\cdot) \rangle$. The autocorrelation of $y(t)$ can therefore be written as,

$$R_y(t_1, t_2) = \langle y(t_1) y(t_2) \rangle. \qquad (2.130)$$

This autocorrelation may be expressed in terms of the signal's complex envelope, $z(t)$, namely

$$R_y(t_1, t_2) = \langle \Re\left\{ z(t_1) \exp j2\pi f_c t_1 \right\} \Re\left\{ z(t_2) \exp j2\pi f_c t_2 \right\} \rangle, \qquad (2.131)$$

or, applying Equation 2.11 we have:

$$R_y(t_1, t_2) = \frac{1}{2} \Re\left\{ \langle z(t_1) z(t_2) \rangle \exp j2\pi f_c(t_2 + t_1) \right\}$$
$$+ \frac{1}{2} \Re\left\{ \langle z^{\star}(t_1) z(t_2) \rangle \exp j2\pi f_c(t_2 - t_1) \right\}, \qquad (2.132)$$

where $\star$ identifies a complex conjugate, and the exponentials have been taken outside the averaging process, as they are deterministic. The autocorrelation of a real bandpass stochastic process is seen to be the sum of two autocorrelation functions that depend upon the complex envelope of the process $y(t)$ at instants t_1 and t_2. Bello has reported [9] that most narrowband processes are constituted such that

$$\langle z(t_1) z(t_2) \rangle = 0. \qquad (2.133)$$

If the channel exhibits wide-sense stationarity (WSS) in the time variable, which will be discussed later in Section 2.5.3.3, the above equation must be true. This is, because the WSS criterion implies that the process's time-domain characteristics, such as the autocorrelation, cannot be dependent on absolute time, only on the time difference, $t_2 - t_1$. It is assumed that Equation 2.133 is applicable for all mobile radio channels, enabling Equation 2.132 to be simplified to

$$R_y(t_1, t_2) = \frac{1}{2} \Re\left\{ R_z(t_1, t_2) \exp j2\pi f_c(t_2 - t_1) \right\}, \qquad (2.134)$$

where

$$R_z(t_1, t_2) = \langle z^{\star}(t_1) z(t_2) \rangle. \qquad (2.135)$$

Autocorrelation Function	Input-Output Correlation Function Relationships
$R_h(t_1, t_2; \tau_1, \tau_2) = \langle h^*(t_1, \tau_1) h(t_2, \tau_2) \rangle$	$R_z(t_1, t_2) = \iint \mathcal{F}_u(t_1 - \tau_1, t_2 - \tau_2) R_h(t_1, t_2; \tau_1, \tau_2) d\tau_1 d\tau_2$
$R_T(f_1, f_2; t_1, t_2) = \langle T^*(f_1, t_1) T(f_2, t_2) \rangle$	$R_z(t_1, t_2) = \iint \mathcal{F}_U(f_1, f_2) R_T(f_1, f_2; t_1, t_2) \exp j2\pi(f_2 t_2 - f_1 t_1) df_1 df_2$
$R_H(f_1, f_2; \nu_1, \nu_2) = \langle H^*(f_1, \nu_1) H(f_2, \nu_2) \rangle$	$R_z(f_1, f_2) = \iint \mathcal{F}_U(f_1 - \nu_1, f_2 - \nu_2) R_H(f_1 - \nu_1, f_2 - \nu_2; \nu_1, \nu_2) d\nu_1 d\nu_2$
$R_S(\tau_1, \tau_2; \nu_1, \nu_2) = \langle S^*(\tau_1, \nu_1) S(\tau_2, \nu_2) \rangle$	$R_z(t_1, t_2) = \iiiint \mathcal{F}_u(t_1 - \tau_1, t_2 - \tau_2) R_S(\tau_1, \tau_2; \nu_1, \nu_2) \exp j2\pi(\nu_2 t_2 - \nu_1 t_1) d\nu_1 d\tau_1 d\nu_2 d\tau_2$
$R_g(t_1, t_2; \tau_1, \tau_2) = \langle g^*(t_1, \tau_1) g(t_2, \tau_2) \rangle$	$R_z(t_1, t_2) = \iint \mathcal{F}_u(t_1 - \tau_1, t_2 - \tau_2) R_g(t_1 - \tau_1, t_2 - \tau_2; \tau_1, \tau_2) d\tau_1 d\tau_2$
$R_M(t_1, t_2; f_1, f_2) = \langle M^*(t_1, f_1) M(t_2, f_2) \rangle$	$R_z(f_1, f_2) = \iint \mathcal{F}_u(t_1, t_2) R_M(t_1, t_2; f_1, f_2) \exp -j2\pi(t_2 f_2 - t_1 f_1) dt_1 dt_2$
$R_G(f_1, f_2; \nu_1, \nu_2) = \langle G^*(f_1, \nu_1) G(f_2, \nu_2) \rangle$	$R_z(f_1, f_2) = \iint \mathcal{F}_U(f_1 - \nu_1, f_2 - \nu_2) R_G(f_1, f_2; \nu_1, \nu_2) d\nu_1 d\nu_2$
$R_V(\nu_1, \nu_2; \tau_1, \tau_2) = \langle V^*(\nu_1, \tau_1) V(\nu_2, \tau_2) \rangle$	$R_z(f_1, f_2) = \iiiint \mathcal{F}_U(f_1 - \nu_1, f_2 - \nu_2) R_V(\nu_1, \nu_2; \tau_1, \tau_2) \exp -j2\pi(\tau_2 f_2 - \tau_1 f_1) d\tau_1 d\nu_1 d\tau_2 d\nu_2$

Table 2.6: Correlation functions of the first set of Bello's system functions

2.5.3.2 General Randomly Time-Variant Channels

The Bello functions are defined in terms of the complex lowpass equivalent notation. Hence, in line with Equation 2.135 above, the correlation function for the input delay-spread function is given by

$$R_h(t_1, t_2; \tau_1, \tau_2) = \langle h^*(t_1, \tau_1) h(t_2, \tau_2) \rangle . \qquad (2.136)$$

Expressions relating the autocorrelation functions of the channel output to the correlation functions of the Bello system functions are easily derived from the functions' defining equations given in Table 2.3. For example, from Equation 2.106

$$
\begin{aligned}
z^*(t_1) z(t_2) &= \int_{-\infty}^{\infty} h^*(t_1, \tau_1) u^*(t_1 - \tau_1) \, d\tau_1 \int_{-\infty}^{\infty} h(t_2, \tau_2) u(t_2 - \tau_2) \, d\tau_2 \\
&= \int_{-\infty}^{\infty} \int_{-\infty}^{\infty} h^*(t_1, \tau_1) h(t_2, \tau_2) u^*(t_1 - \tau_1) u(t_2 - \tau_2) \, d\tau_1 d\tau_2 .
\end{aligned}
$$
$$(2.137)$$

Taking the ensemble average of both sides of the above equation gives

$$\langle z^*(t_1) z(t_2) \rangle = \int_{-\infty}^{\infty} \int_{-\infty}^{\infty} \langle h^*(t_1, \tau_1) h(t_2, \tau_2) u^*(t_1 - \tau_1) u(t_2 - \tau_2) \rangle \, d\tau_1 d\tau_2 .$$
$$(2.138)$$

If the channel input is deterministic, it can be removed from the ensemble average to yield

$$R_z(t_1, t_2) = \int_{-\infty}^{\infty} \int_{-\infty}^{\infty} R_h(t_1, t_2; \tau_1, \tau_2) u^*(t_1 - \tau_1) u(t_2 - \tau_2) \, d\tau_1 d\tau_2 . \quad (2.139)$$

However if $u(t)$ is a random process which is assumed to be independent of the channel characteristics, Equation 2.137 becomes

$$R_z(t_1, t_2) = \int_{-\infty}^{\infty} \int_{-\infty}^{\infty} R_h(t_1, t_2; \tau_1, \tau_2) R_u(t_1 - \tau_1, t_2 - \tau_2) \, d\tau_1 d\tau_2 , \quad (2.140)$$

where

$$R_u(t_1, t_2) = \langle u^*(t_1) u(t_2) \rangle . \qquad (2.141)$$

Defining

$$\mathcal{F}_u(t_1, t_2) = \begin{cases} u^*(t_1) u(t_2) & u(t) \text{ deterministic} \\ \langle u^*(t_1) u(t_2) \rangle & u(t) \text{ random,} \end{cases} \qquad (2.142)$$

Equations 2.139 and 2.140 can be combined as

$$R_z(t_1, t_2) = \int_{-\infty}^{\infty} \int_{-\infty}^{\infty} R_h(t_1, t_2; \tau_1, \tau_2) \mathcal{F}_u(t_1 - \tau_1, t_2 - \tau_2) \, d\tau_1 d\tau_2 . \quad (2.143)$$

Table 2.6 details the relations between the channel output autocorrelation functions and the correlation functions of the remaining Bello functions. Notice that the dual function of $\mathcal{F}_U$ is employed in the table, that is,

$$\mathcal{F}_U(f_1, f_2) = \begin{cases} U^*(f_1)U(f_2) & U(f) \text{ deterministic} \\ \langle U^*(f_1)U(f_2)\rangle & U(f) \text{ random.} \end{cases} \tag{2.144}$$

As one would expect, since the system functions can be arranged symmetrically with respect to their Fourier relationships, the correlation functions can also be arranged symmetrically, this time with respect to their double Fourier relationships. As an example consider the relationship between the functions $T(f, t)$ and $H(f, \nu)$. From Equation 2.117

$$T^*(f_1, t_1)T(f_2, t_2) = \int_{-\infty}^{\infty}\int_{-\infty}^{\infty} H^*(f_1, \nu_1)H(f_2, \nu_2)$$

$$\exp j2\pi(\nu_2 t_2 - \nu_1 t_1)\, d\nu_1\, d\nu_2. \tag{2.145}$$

Taking ensemble averages of both sides

$$R_T(f_1, f_2; t_1, t_2) = \int_{-\infty}^{\infty}\int_{-\infty}^{\infty} R_H(f_1, f_2; \nu_1, \nu_2)\exp j2\pi(\nu_2 t_2 - \nu_1 t_1)\, d\nu_1\, d\nu_2. \tag{2.146}$$

The above equation shows that $R_T(f_1, f_2; t_1, t_2)$ is the two-dimensional Fourier transform of the correlation function $R_H(f_1, f_2; \nu_1, \nu_2)$ with the convention that when transforming from a pair of time variables to a pair of frequency variables a positive exponential connects the first variable in each pair and a negative the second.

Figure 2.26 illustrates the symmetric double Fourier transform relationships of the correlation functions of the first set of Bello functions. If required, a table showing the relationships between the correlation functions of sister Bello functions can easily be derived from Tables 2.4 and 2.6.

The relationships derived above for the correlation functions of Bello's system functions may be applied to any time-variant linear channel. However, if the statistical behaviour of a channel obeys certain constraints, then it is possible to simplify them. Specifically, if a channel is wide-sense-stationary in the time domain and/or the frequency domain then its correlation functions can be simplified. The following sections investigate how.

2.5.3.3   Wide-Sense Stationary Channels

Firstly, we shall examine a channel that exhibits wide-sense (second-order) stationarity. A process is called wide-sense stationary with respect to time if its first two moments (mean and autocorrelation) are independent of absolute time. That is the correlation function for a wide-sense stationary

LEGEND : DF Double Fourier Transform
          DF^{-1} Inverse Double Fourier Transform

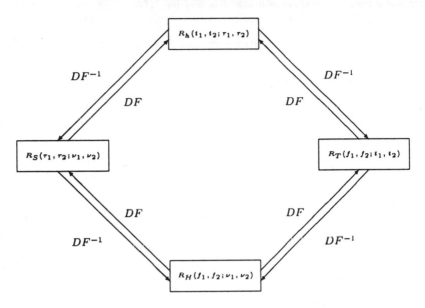

Figure 2.26: Fourier relations of the correlations for the first set of Bello functions

(WSS) channel depends on time difference, and not absolute time. For example, the correlation of the input delay-spread function for a WSS channel becomes

$$R_h(t_1, t_2; \tau_1, \tau_2)\Big|_{\text{WSS}} \equiv R_h(\Delta t; \tau_1, \tau_2), \tag{2.147}$$

where

$$\Delta t = t_2 - t_1. \tag{2.148}$$

Substituting for $R_h(t_1, t_2; \tau_1, \tau_2)$ from Equation 2.147 into Equation 2.143 gives

$$R_z(t_1, t_2) = \int_{-\infty}^{\infty} \int_{-\infty}^{\infty} R_h(\Delta t; \tau_1, \tau_2)\mathcal{F}_u(t_1 - \tau_1, t_2 - \tau_2)\, d\tau_1 d\tau_2. \tag{2.149}$$

The autocorrelation function of the channel output is still a function of absolute time even though the channel correlation function is expressed in terms of differential time. This is because the channel output is of course dependent on its input, which need not have any statistical constraints placed upon it.

Earlier discussion indicated that the Doppler shift domain contains the same information as the time domain. This implies that wide-sense stationarity in the time variable, t, must also manifest itself in the Doppler domain. To illustrate the way in which this occurs we shall look at the correlation function of the delay-Doppler-spread function $R_S(\tau_1, \tau_2; \nu_1, \nu_2)$. Applying the information contained in Figure 2.26, we can write

$$R_S(\tau_1, \tau_2; \nu_1, \nu_2) = \int_{-\infty}^{\infty} \int_{-\infty}^{\infty} R_h(t_1, t_2; \tau_1, \tau_2) \exp j2\pi(\nu_1 t_1 - \nu_2 t_2)\, dt_1\, dt_2.$$

(2.150)

Substituting Equations 2.147 and 2.148 for the WSS channel gives

$$R_S(\tau_1, \tau_2; \nu_1, \nu_2)\Big|_{\text{WSS}} = \int_{-\infty}^{\infty} \exp j2\pi t_1(\nu_1 - \nu_2)\, dt_1$$

$$\cdot \int_{-\infty}^{\infty} R_h(\Delta t; \tau_1, \tau_2) \exp -j2\pi\nu_2 \Delta t\, d(\Delta t).$$

(2.151)

The second integral may be recognised as the Fourier transform of an autocorrelation function, which from the Wiener-Khinchine theorem results in a power spectral density function. The other integral is zero except for the case $\nu_1 = \nu_2$, when it is infinite. This is recognised as the definition of a unit impulse at $\nu_2 = \nu_1$. We can thus write,

$$R_S(\tau_1, \tau_2; \nu_1, \nu_2)\Big|_{\text{WSS}} \equiv P_S(\tau_1, \tau_2; \nu_2)\delta(\nu_2 - \nu_1),$$

(2.152)

where $P_S(\tau_1, \tau_2; \nu_2)$ is the cross power spectral density of $h(t, \tau_1)$ and $h(t, \tau_2)$. The impulse function in Equation 2.152 implies that, for a WSS channel, signals arriving with different Doppler shift values are uncorrelated.

As ν_1 does not feature in the function $P_S(\tau_1, \tau_2; \nu_2)$ we can drop the suffix on ν_2 to get,

$$R_S(\tau_1, \tau_2; \nu_1, \nu_2)\Big|_{\text{WSS}} \equiv P_S(\tau_1, \tau_2; \nu)\delta(\nu_2 - \nu_1).$$

(2.153)

Column 2 in Table 2.7 lists the correlation functions of the Bello system functions for the WSS channel. The dual functions for a general time-variant linear channel are no longer duals for a WSS channel. This is because we have applied statistical constraints to variables t and ν, but not to their dual variables, f and τ respectively. Figure 2.27 shows, how the correlation functions for the first set of Bello functions are related on a WSS channel.

In order to fully characterise a WSS channel, the correlation functions must be established for all frequencies.

2.5.3.4 Uncorrelated Scattering Channels

The dual of the WSS channel is the uncorrelated scattering (US) channel. For this channel, the statistics describing signals arriving with different delays are uncorrelated.

General Channel	WSS Channel	US Channel	WSSUS Channel
$R_h(t_1, t_2; \tau_1, \tau_2)$	$R_h(\Delta t; \tau_1, \tau_2)$	$P_h(t_1, t_2; \tau_2)\delta(\tau_2 - \tau_1)$	$P_h(\Delta t; \tau_2)\delta(\tau_2 - \tau_1)$
$R_T(f_1, f_2; t_1, t_2)$	$R_T(f_1, f_2; \Delta t)$	$R_T(\Delta f; t_1, t_2)$	$R_T(\Delta f; \Delta t)$
$R_H(f_1, f_2; \nu_1, \nu_2)$	$R_H(f_1, f_2; \nu_2)\delta(\nu_2 - \nu_1)$	$R_H(\Delta f; \nu_1, \nu_2)$	$R_H(\Delta f; \nu_2)\delta(\nu_2 - \nu_1)$
$R_S(\tau_1, \tau_2; \nu_1, \nu_2)$	$R_S(\tau_1, \tau_2; \nu_2)\delta(\nu_2 - \nu_1)$	$R_S(\tau_2; \nu_1, \nu_2)\delta(\tau_2 - \tau_1)$	$R_S(\tau_2; \nu_2)\delta(\nu_2 - \nu_1)\delta(\tau_2 - \tau_1)$
$R_g(t_1, t_2; \tau_1, \tau_2)$	$R_g(\Delta t; \tau_1, \tau_2)$	$P_g(t_1, t_2; \tau_2)\delta(\tau_2 - \tau_1)$	$P_g(\Delta t; \tau_2)\delta(\tau_2 - \tau_1)$
$R_M(t_1, t_2; f_1, f_2)$	$R_M(\Delta t; f_1, f_2)$	$R_M(t_1, t_2; \Delta f)$	$R_M(\Delta t; \Delta f)$
$R_G(f_1, f_2; \nu_1, \nu_2)$	$R_G(f_1, f_2; \nu_2)\delta(\nu_2 - \nu_1)$	$R_G(\Delta f; \nu_1, \nu_2)$	$R_G(\Delta f; \nu_2)\delta(\nu_2 - \nu_1)$
$R_V(\nu_1, \nu_2; \tau_1, \tau_2)$	$R_V(\nu_2; \tau_1, \tau_2)\delta(\nu_2 - \nu_1)$	$R_V(\nu_1, \nu_2; \tau_2)\delta(\tau_2 - \tau_1)$	$R_V(\nu_2; \tau_2)\delta(\tau_2 - \tau_1)\delta(\nu_2 - \nu_1)$

Table 2.7: Correlation functions for different channel types

LEGEND : $(D)F$ (Double) Fourier Transform
            $(D)F^{-1}$ Inverse (Double) Fourier Transform

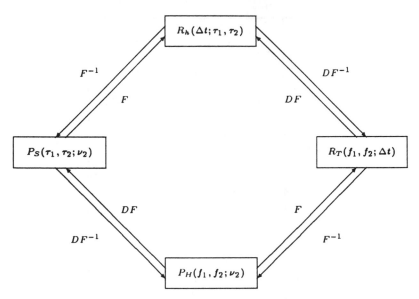

Figure 2.27: Fourier relationships amongst the correlation functions of the first set of Bello functions for WSS channels

The WSS channel was seen to be wide-sense stationary in the time domain and to possess uncorrelated scattering in the Doppler shift domain. By applying duality we can state immediately that a channel possessing uncorrelated scattering in the delay domain will be wide-sense stationary in the frequency domain. Hence, for the output Doppler-spread function, we can write,

$$R_H(f_1, f_2; \nu_1, \nu_2)\Big|_{US} \equiv R_H(\Delta f; \nu_1, \nu_2), \qquad (2.154)$$

where

$$\Delta f = f_2 - f_1. \qquad (2.155)$$

The correlation function of the delay-Doppler-spread function is derived from $R_H(f_1, f_2; \nu_1, \nu_2)$ by double Fourier transform. (See Figure 2.26.) Applying Equation 2.154 we have:

$$R_S(\tau_1, \tau_2; \nu_1, \nu_2)\Big|_{US} = \int_{-\infty}^{\infty} \exp j2\pi f_1(\tau_1 - \tau_2)\, df_1$$
$$\cdot \int_{-\infty}^{\infty} R_{II}(\Delta f; \nu_1, \nu_2) \exp -j2\pi\tau_2 \Delta f\, d(\Delta f). \qquad (2.156)$$

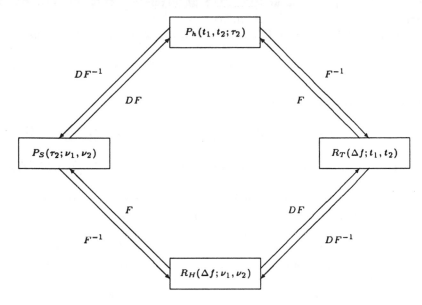

Figure 2.28: Fourier relationships amongst correlation functions of the first set of Bello functions for US channels

Then expressing $R_S(\tau_1, \tau_2; \nu_1, \nu_2)$ in terms of the cross-power spectral density of $H(f, \nu_1)$ and $H(f, \nu_2)$ reveals that

$$R_S(\tau_1, \tau_2; \nu_1, \nu_2)\Big|_{\text{US}} \equiv P_S(\tau; \nu_1, \nu_2)\delta(\tau_2 - \tau_1), \qquad (2.157)$$

which illustrates the uncorrelated scattering in delay domain. The remaining correlation functions are listed in Column 3 of Table 2.7, and the relationships between the functions for the first set of Bello functions are displayed in Figure 2.28. As with a WSS channel, dual functions for a general time-variant linear channel are not dual for a US channel.

A full description of the US channel requires the correlation functions to be established for all time.

2.5.3.5 Wide-Sense Stationary Uncorrelated Scattering Channels

The most useful channel as far as the mobile radio engineer is concerned is a hybridisation of the above channels. Referred to as the wide-sense stationary uncorrelated scattering (WSSUS) channel, its first and second order statistics are invariant under translation in time and frequency. This means that the correlation functions for a WSSUS channel need only be worked out once, since they apply for all time and all frequency.

The correlation functions are easily deduced by applying first the wide-sense stationary criteria and then the uncorrelated scattering criteria, or vice versa.

Consider $R_h(t_1, t_2; \tau_1, \tau_2)$. Under wide sense stationary conditions,

$$R_h(t_1, t_2; \tau_1, \tau_2)\Big|_{\text{WSS}} \equiv R_h(\Delta t; \tau_1, \tau_2)\delta(\nu_2 - \nu_1), \qquad (2.158)$$

then adding the uncorrelated scattering restriction,

$$R_h(t_1, t_2; \tau_1, \tau_2)\Big|_{\text{WSSUS}} \equiv P_h(\Delta t; \tau)\delta(\tau_2 - \tau_1), \qquad (2.159)$$

where $P_h(\Delta t; \tau)$ is the cross-power spectral density of $T(f, t_1)$ and $T(f, t_1 + \Delta t)$.

Table 2.7 lists the correlation functions for a WSSUS channel in Column 4, and Figure 2.29 illustrates their inter-relations for the first set of Bello functions. Notice from this table that dual functions under the general time-variant linear channel are still duals under a WSSUS channel. This is because although we have applied certain statistical constraints to t and ν, we have also applied the dual constraints to f and τ.

WSSUS channels are of particular significance because they are the simplest channels to analyse, that exhibit both time and frequency fading. Workers are therefore disposed to approximate real channels by WSSUS channels.

2.5.3.6 Quasi-Wide-Sense Stationary Uncorrelated Scattering Channels

In order to utilise the benefits of a WSSUS channel in the characterisation of real channels, the Quasi-WSSUS (QWSSUS) channel was introduced [9]. A QWSSUS channel behaves as a WSSUS channel for a restricted interval of time T and a band of frequencies B. Outside this region, the channel correlation functions can no longer be assumed invariant with time, frequency or both.

Bello suggested in [9] that a useful method of describing real channels is to work out the correlation functions over time and frequency intervals small enough for the channel to be described by a hypothetical WSSUS channel. That is, successively apply a QWSSUS model. Then determine the statistics of these correlation functions over longer time periods to fully characterise the channel.

2.6   Channel Description by Bello Functions

This section discusses how the Bello functions introduced previously are applied to the characterisation of mobile radio channels. It explains that mobile radio channels are purely random, and describes how a practical

LEGEND : F Fourier Transform
           F^{-1} Inverse Fourier Transform

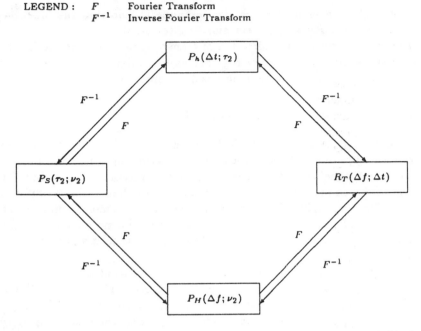

Figure 2.29: Fourier relationships between correlation functions of the first set of Bello functions for WSSUS channels

approach to their characterisation has evolved based on the QWSSUS concept.

2.6.1   Space-variance

Consider a mobile station (MS) roaming through an area illuminated by a fixed base station (BS) transmitting a constant single tone. As it moves the MS will see random variations in the amplitude and phase of the signal it receives. Assuming that all scatterers comprising the channel are stationary, then whenever the MS stops, the amplitude and phase of the received signal both remain constant. That is, the channel appears to be time-invariant. When the MS starts to move again the channel once more appears time-variant.

The channel characteristics are therefore dependent on the position of the MS. In the case of single tone transmission, the MS is seen to move through a standing electromagnetic field of random amplitude and phase.

When the MS transmits to the BS, a different physical scenario exists. The receiver is stationary and the field at the BS is changing due to movement of the transmitter. Nevertheless, consideration of the reciprocity of the channel reveals that the effect can be modelled in the same way as before. That is, the BS can be regarded as being a mobile terminal, "moving through" the same standing electromagnetic field that the MS would experience if the BS were the transmitting terminal.

Hence, regardless of the direction of transmission, the characteristics of the mobile radio channel can be regarded as being dependent on the spatial position of the MS, rather than on absolute time.

2.6.2   Statistical Characteristics

The earlier discussion of the physical structure of mobile radio channels (see Section 2.3) outlined the statistical nature of the amplitude and phase of the received signal component corresponding to a particular delay bin. The phase of the component was reasoned to be a random variable uniformly distributed over $[0, 2\pi]$. This being the case, the ensemble average of any one of the Bello functions when used to describe the mobile radio channel must be zero. Put another way, the mobile radio channel is a purely random channel.

In Section 2.5.3.5 it was explained that it is advantageous to be able to describe a random channel in terms of a WSSUS channel, and that to achieve this with practical channels the QWSSUS concept was proposed.

For the case of the mobile radio channel an approach based upon the QWSSUS method has evolved, where instead of restricting the frequency and time intervals over which a hypothetical WSSUS channel will adequately describe the real channel, the bounds are defined in terms of a frequency interval and a physical area.

The original QWSSUS approach described in Section 2.5.3.6 partitions a channel in frequency and time such that the second-order channel statistics are invariant to frequency translations within a bandwidth B, and time translations within an interval of duration T. That is, the channel must be WSS in both frequency[2] and time.

That mobile radio channels are WSS in the frequency variable was stated earlier in Section 2.3.1.1. This stationarity is a result of the physical properties of the propagation media.

As described above, for the mobile radio channel, wide-sense stationarity in the time variable may be commuted to wide-sense stationarity with respect to the position of the MS. In line with the original QWSSUS approach we want to partition the area through which the MS will roam into small areas within each of which the channel characteristics are WSS with position.

Small changes in the position of the MS can cause dramatic changes in the amplitude and phase of the received signal corresponding to a particular delay bin. This effect is referred to as fast fading. Although the characteristics of each path (i.e., $\{a_i(t), \tau_i(t), \theta_i(t)\}$) will have changed almost imperceptibly, the ω_c multiplicative factor described on page 109 may

---

[2] Wide-sense stationarity in the frequency variable is generally referred to as uncorrelated scattering (US).

produce a relatively large phase shift in the received signal component due to a particular propagation path. The interference of the received signal components from all the propagation paths comprising the channel can therefore change from say, predominantly constructive to predominantly destructive over a very short distance.

For the fast fading statistics to be WSS there should be no change in the mean and variance of the fading process. This implies that the characteristics of each path, such as amplitude, delay and phase retardation, remain unchanged. Hence for small areas, the channel can be considered WSS with respect to position.

Furthermore, combining the two stationarity criteria shows that mobile radio channels, partioned as small areas, are QWSSUS and can therefore be described in terms of the functions shown in Figure 2.29, and listed in Column 4 of Table 2.7.

Motion of the MS over large areas results in a second fading effect, called slow fading. This is the result of significant changes in any of the three propagation path characteristics. It could be due to the obscuring of one building by another, or by a change in the position of a scatterer relative to the MS.

A mobile roaming over a large area will experience both types of fading, the fast fading being superimposed on the slow fading.

It should be understood that this method of applying QWSSUS channels to mobile radio channels derives from the assumption that changes in the channel characteristics are due essentially to movement of the mobile, and that variations due to moving scatterers are a second order effect. In support of this premise, Cox has reported that in New York City cars and trucks generally produce only minor multipath effects [19].

2.6.3   Small-Area Characterisation

Small-areas are generally taken to have a radius approximately equal to a few tens of wavelengths [19, 36]. In general such areas are arbitrarily chosen, however care must be exercised as features of the local topology may cause significant changes over relatively short distances. This may be the case close to road junctions in urban environments.

In Section 2.5.3 it was noted that a practical approach to channel characterisation is adopted that involves taking the mean and correlation of one of the Bello functions. However, as the mobile radio channel is purely random it possesses zero ensemble average. Thus, the approach reduces to the establishment of the correlation function for any one of the eight Bello functions.

Consider the input delay-spread function, $h(t, \tau)$. Commuting time to position, this function can be rewritten as $h(p, \tau)$, where p denotes the position of the MS. The correlation function of $h(p, \tau)$ for a small-area is $P_h(\Delta s, \tau)$ where Δs is the distance between the points at which samples of

$h(p, \tau)$ are taken. If Δs is set to zero, $P_h(0, \tau)$ (or just $P_h(\tau)$) is obtained. $P_h(\tau)$ is called the power-delay profile of the channel, and is the power spectral density of the channel as a function of delay.

Each measurement of $P_{hp}(\tau)$ is a sample value of the product $h^\star(p, \tau)h(p, \tau)$ for a specific position of the MS. To establish the channel correlation function applicable to the small-area, the ensemble average, $< h^\star(p, \tau)h(p, \tau) >$, taken across the whole area, must be evaluated.

Hence the correlation function is the average power-delay profile, given by

$$P_h(\tau) = \frac{1}{K} \sum_{k=1}^{K} P_{hp}(\tau), \qquad (2.160)$$

where K is the number of samples of the power-delay profile taken over the small-area.

The statistics of each delay bin comprising the mobile radio channel are often assumed to be Gaussian in nature. This is because the local scatterers around the terminals give rise to many propagation paths of virtually identical delay. The central limit theorem [12] can then be applied to reach the assumption of Gaussian statistics.

$P_h(\tau)$ provides a complete description of the channel over the small-area if the channel statistics are Gaussian, because in this case WSS implies strict-sense stationarity (SSS) [12].

Analysis of the statistical distribution of $P_{hp}(\tau)$ across small-areas has in general supported the Gaussian assumption [12, 13], although this is not always the case [38]. The amplitude distribution, $\sqrt{P_{hp}(\tau)}$, for a given delay has often been found to fit either a Rayleigh or a Ricean distribution, thereby implying Gaussian statistics.

In narrowband propagation studies, it is often $R_{Tp}(0, 0)$ which is measured. From Figure 2.29, it is seen that,

$$R_T(\Delta f; \Delta t) = \int_{-\infty}^{\infty} P_h(\Delta t; \tau) \exp -j2\pi f \tau d\tau. \qquad (2.161)$$

From which it is seen that evaluation of $R_T(0, 0)$ for the narrowband channel is identical to that of $P_h(\tau)$.

2.6.4 Large-Area Characterisation

By analysing the results obtained within small-areas over large-areas, a description of the slow fading process results. Large-areas are taken as covering geographical districts of similar constitution, such as suburban, or rural. Large-area characterisation takes one of two forms.

In the first form, characterisation is by means of statistical analysis of the variation of channel descriptors derived from the small-area results. For example, for wideband channels, large-areas may be described by the distribution of the delay-spread and mean delay of the average power-delay profile of the channel, see Equations 2.95 and 2.96, respectively. Another descriptor often used is the coherence bandwidth. Narrowband channel descriptors include, level crossing rates and fade durations.

The second approach is to analyse the variation in the Bello functions over large-areas. In the case of the average power-delay profile this involves determining the probability of occupancy of a delay bin, and the distribution of the amplitude and phase retardation associated with each bin [39, 40, 41]. For narrowband channels, this reduces to measuring the variation in the mean signal strength.

Further analysis of large-areas over grossly dissimilar propagation areas results in gross channel descriptors useful for the prediction of radio coverage during cellular system planning. Gross descriptors are used to modify the basic free space path loss equation, in order to be able to predict the gross channel characteristics for a given propagation environment. "Rule of thumb" descriptors have evolved this way. For instance, a loss factor representing the degree of urbanisation. Fast fading and slow fading effects are still superimposed upon the predicted channel response.

2.7 Practical Description of Mobile Channels

Having portrayed the mobile radio environment in theoretical terms in time, frequency, delay and Doppler-shift domains by means of Bello functions, we now embark upon the practical characterisation of mobile channels for the practising engineer. Our main goal in this section is to describe the wave-propagation environment as simply as possible, while deriving a set of relevant parameters for power budget and system designers. In harmony with this ambition we characterise the channel by the help of Figure 2.30 in terms of:

1. Propagation pathloss law,

2. Slow fading statistics,

3. Fast fading statistics,

which in general will vary as a function of the propagation frequency, surrounding natural and man-made objects, vehicular speed, etc. Clearly, a deterministic treatment is not possible due to the unpredictable variation of channel features, hence we resort to statistical methods. The general approach is to develop theoretical models and check their validity by statistical methods against various real propagation environments.

In this chapter we cannot elaborate on the propagation channels of each existing and perspective mobile radio system, spanning maritime mobile satellite systems, public land mobile radio (PLMR) services, private mobile radio (PMR) schemes, high capacity personal communications networks

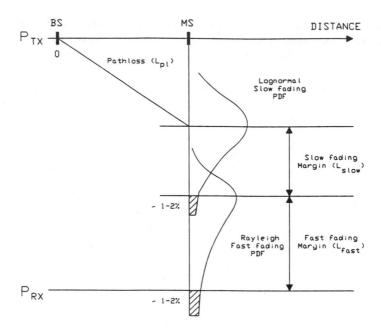

Figure 2.30: Power-budget design

(PCN) penetrating buildings, halls, etc. The channel and terrain features, multiple access, modulation and signal detection methods, bandwidths, vehicular speeds, etc. are so vastly different that a generic model would be extremely complicated, yet inaccurate. For an all-encompassing up-to-date reference Parson's excellent book published in this series is recommended [56].

This book is concerned mainly with PLMR whilst offering outlook to the emerging PCN. We therefore restrict our discourse to channels in the 900-1800 MHz band, at the same time recognising that the methodology used can be applied to all mobile channels. These frequencies fall in the so-called Ultra High Frequency (UHF) band where convenient antenna sizes are associated with power efficient transmitters and compact receivers. Also the wave propagation is coveniently curtailed by the horizon thereby limiting cochannel interference, when the frequencies are reused in neighbouring cell clusters. At these frequencies, even if there is no line-of-sight path between transmitter and receiver, by means of wave scattering, reflection and diffraction generally sufficient signal power is received to ensure communications.

The prediction of the expected mean or median received signal power plays a crucial role in determining the coverage area of a specific base station and for known interference tolerance also determines the closest acceptable reuse of the propagation frequency deployed. For high antenna elevations and large rural cells a more slowly decaying power exponent is expected than for low elevations and densely built-up urban areas. As suggested by Figure 2.30, the received signal is also subjected to slow or shadow fading which is mainly governed by the characteristic terrain features in the vicinity of the mobile receiver. When designing the system's power budget and coverage area pattern, the slow fading phenomenon is taken into account by including a shadow fading margin as demonstrated by Figure 2.30.

Statistically speaking this requires increasing the transmitted power P_{tx} by the shadow fading margin L_{slow} which is usually chosen to be the $1-2\%$ quantile of the slow fading probability density function (PDF) to minimise the probability of unsatisfactorily low received signal power P_{rx}. Additionally, the short term fast signal fading due to multipath propagation is taken into account by deploying the so-called fast fading margin L_{fast}, which is typically chosen to be also a few percent quantile of the fast fading distribution. In the worst-case scenario both of these fading margins are simultaneously exceeded by the superimposed slow and fast envelope fading. This situation is often referred to as 'fading margin overload', resulting in a very low-level received signal almost entirely covered in noise. The probability of these cases can be taken to be the sum of the individual margin overload probabilities, when the error probability is close to 0.5, since the received signal is essentially noise. Clearly, the system's error correction codec must be designed to be able to combat this worst-case average bit error probability. This reveals an important trade-off in terms of designed fading margin overload probability, transmitted signal power and error correction coding 'power', which will be made more explicit in the concluding part of this section.

2.7.1 Propagation Pathloss Law

In our probabilistic approach it is difficult to give a worst-case pathloss exponent for any mobile channel. However, it is possible to specify the most optimistic scenario. That is propagation in free space. The free-space pathloss, L_{pl} is given by [56]:

$$L_{pl} = -10\log_{10} G_T - 10\log_{10} G_R + 20\log_{10} f^{Hz} + 20\log_{10} d^m - 147.6dB,$$
$$(2.162)$$

where G_T and G_R are the transmitter and receiver antenna gains, f^{Hz} is the propagation frequency in Hz and d^m is the distance from the BS antenna in km. Observe that the free-space pathloss is increased by 6 dB every time, the propagation frequency is doubled or the distance from the mobile is doubled. This corresponds to a 20 dB/decade decay and at d=1 km, f=1 GHz and $G_T = G_R = 1$ a pathloss of $L_F = 92.4$ dB is encountered. Clearly, not only technological difficulties, but also propagation losses discourage the deployment of higher frequencies. Nevertheless, spectrum is usually only available in these higher frequency bands.

In practice, for UHF mobile radio propagation channels of interest to us, the free-space conditions do not apply. There are however a number of useful pathloss prediction models that can be adopted to derive other prediction bounds. One such case is the 'plane earth' model. This is a two-path model constituted by a direct line of sight path and a ground-reflected one, as discussed in Section 1.2.6 and depicted in Figure 1.15, which ignores the curvature of the earth's surface. Assuming transmitter base station (BS) and receiver mobile station (MS) antenna heights of $h_{BS}^m, h_{MS}^m \ll d$, respectively, the plane earth pathloss formula [56] can be derived:

$$L_{pl} = -10 \log_{10} G_T - 10 \log_{10} G_R - 20 \log_{10} h_{BS}^m - 20 \log_{10} h_{MS}^m + 40 \log_{10} d^m,$$
$$(2.163)$$

where the dependence on propagation frequency is removed. Observe that a 6 dB pathloss reduction is resulted, when doubling the transmitter or receiver antenna elevations, and there is an inverse fourth power law decay with increasing the BS-MS distance d. In the close vicinity of the transmitter antenna, where h_{BS} or $h_{MS} \ll d$ does not hold, Equation 2.163 is no longer valid. Instead, distance-dependent periodic received signal level maxima and minima are experienced, as suggested by Figure 1.15 [42].

The urban microcellular channels are more realistically described by a four-path model including two more reflected waves from building walls along the streets [42], [57], [58]. In this scenario it is assumed that the transmitter antenna is below the characteristic urban skyline. The four-path model of Figure 1.15 used by Green [42] assumed smooth reflecting surfaces yielding specular reflections with no scattering, finite permittivity and conductivity, vertically polarised waves and half-wave dipole antennas. The resultant pathloss profile vs. distance becomes rather erratic with received signal level variations in excess of 20 dB, which renders pathloss modelling by a simple power exponent rather inaccurate, however attractive it would appear due to its simplicity.

There exists a wide variety of further refined models with different strengths, weaknesses and applicability, which take into account other channel imperfections neglected so far. The most widely used of these in the mobile radio environment is the Hata pathloss model [63], which will be discussed in the following section. Parsons [56] gives a detailed comparative study of how various multiple diffractions can be taken into account in illuminating shadowed or obstructed areas, highlighting a number of published pathloss prediction models. Further pathloss model comparisons are readily found in [59], [60], [61]. Here we use the comprehensively tabulated summary of [60] to provide a quick overview in Table 2.8.

It is quite plausible that more sophisticated models guarantee generally better predictions but are more difficult to evaluate. Irrespective of the prediction model deployed, estimated values always have to be verified by measurements other than those utilised to derive the empirical model invoked. If necessary, correction factors have to be derived and introduced in further predictions. This is the approach we will adopt in our further

discussions. For examplary purposes we propose to verify the applicability
of the Hata model [63] to the 1.8 GHz microcellular environment. In doing
so we fit minimum mean squared error regression lines to our measurement
data, compare this model to Hata's predictions and derive appropriate
correction factors to generate further pathloss estimates.

2.7.1.1   The Hata Pathloss Models

Hata developed three pathloss models described below. These were developed from an extensive data base derived by Okumura [62] from measurements in and around Tokyo. *The typical urban Hata model* is defined
as:

$$L_{Hu} = 69.55 + 26.16\log_{10}f - 13.82\log_{10}h_{BS} - a(h_{MS}) +$$
$$(44.9 - 6.55\log_{10}h_{BS})\log_{10}d \ [dB], \tag{2.164}$$

where f is the propagation frequency in MHz, h_{BS} and h_{MS} are the BS
and MS antenna elevations in terms of metres, respectively, $a(h_{MS})$ is a
terrain dependent correction factor, while d is the BS-MS distance in km.
The correction factor $a(h_{MS})$ for small and medium sized cities was found
to be

$$a(h_{MS}) = (1.1\log_{10}f - 0.7)h_{MS} - (1.56\log_{10}f - 0.8), \tag{2.165}$$

while for large cities is frequency-parametrized:

$$a(h_{MS}) = \begin{cases} 8.29[\log_{10}(1.54h_{MS})]^2 - 1.1 & \text{if } f \leq 200MHz \\ 3.2[\log_{10}(11.75h_{MS})]^2 - 4.97 & \text{if } f \geq 400MHz \end{cases}. \tag{2.166}$$

The typical suburban Hata model applies a correction factor to the urban
model yielding:

$$L_{Hsuburban} = L_{Hu} - 2[\log_{10}(f/28)]^2 - 5.4 \ [dB]. \tag{2.167}$$

The rural Hata model modifies the urban formula differently, as seen below:

$$L_{Hrural} = L_{Hu} - 4.78(\log_{10}f)^2 + 18.33\log_{10}f - 40.94 \ [dB]. \tag{2.168}$$

Before we try to interpret these formulae in terms of power-loss exponents,
the fundamental limitations of its parameters have to be listed:

$$f: \quad 150 - 1500MHz$$
$$h_{BS}: \quad 30 - 200m$$
$$h_{MS}: \quad 1 - 10m$$
$$d: \quad 1 - 20km.$$

For a 900 MHz PLMR system these conditions can be usually satisfied
but for a 1.8 GHz typical PCN urban microcell all these limits have to be
slightly stretched.

Name	Ref	Date	Frequency	Environment	Remarks
Allesbrook & Parsons	[70]	1977	85-441MHz	urban	Correction factor for VHF
BBC	[66]	1974	UHF(VHF)	urban/rural	Correction factor for VHF
Blomquist & Ladell	[78]	1974	30-900MHz	urban	Deterministic model for pathloss
Bullington	[75]	1947	>30MHz	rural	Equivalent knife-edge for diffraction
Deygout	[77]	1966	VHF	hilly	Main knife-edge diffraction model
Egli	[68]	1957	40-900MHz	rural	Correction factor of irregular terrain
Edwards & Durkin	[65]	1969	30-300MHz	urban/suburban	Plane earth model
Epstein & Peterson	[76]	1953	850MHz	hilly	Multiple knife-edge diffraction
Hata	[63]	1980	100MHz-3GHz	urban/suburban	Computerised Okumura's graphical model
Ibrahim & Parsons	[71]	1981	150-450MHz	urban	Empirical & Semi-empirical models
Japanese Atlas	[79]	1957	30MHz-10GHz	hilly	Multiple knife-edge diffraction
Kessler & Wiggins	[72]	1977	VHF/UHF	urban/rural	Semi-empirical model
Lee	[74]	1982	UHF	urban/suburban	Based on inverse-square law, simple
Leubbers	[80]	1984	VHF	rural	Based on Uniform theory of diffraction
Longley & Rice	[64]	1968	VHF/UHF	rural	Two-ray model + correction factors
Lustgarten & Madison	[73]	1977	VHF/UHF	rural	For quick computation of pathloss
Murphy	[69]	1970	VHF/UHF	rural	Statistical model for irregular terrain
Okumura	[62]	1968	100MHz-3GHz	urban/suburban	Graphical results, mainly for Japan
Palmer	[67]	1979	UHF	urban/rural	Terrain data-base computer model

Table 2.8: Comparison of various pathloss models [60]

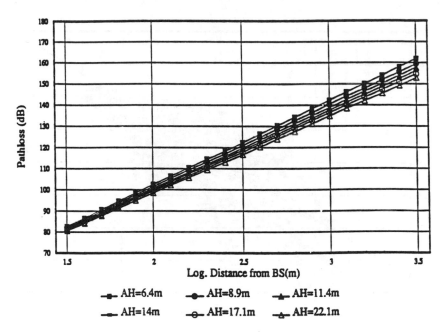

Figure 2.31: Hata pathloss in urban environment at various antenna heights

In what follows we now evaluate the Hata prediction for a specific set of values used in our experiments to check its applicability and accuracy in urban microcells. The measured and predicted values are then compared for a large set of measurements to derive relevant correction factors to Hata's model allowing its deployment in microcellular environments. The measurements were carried out in typical urban environments in Southern England at a propagation frequency of 1.8 GHz, BS antenna heights (AH) of 6.4 m, 8.9 m, 11.4 m, 14 m, 17.1 m and 22.1 m, and MS antenna height of 2 m. Using the urban Hata model and the above mentioned parameters the predicted pathlosses are plotted as a function of logarithmic distance in Figure 2.31, where, for example, abscissa values 2 and 3 correspond to $10^2 = 100$ and $10^3 = 1000$ m, respectively. As expected, the higher the antenna elevation, the less steep the pathloss prediction. The power-loss exponents for these parameters vary between 3.962 (39.62 dB/decade) and 3.61 (36.1 dB/decade) for AH=6.4 m and AH=22.1 m, respectively. These exponent values are reasonably close to the inverse fourth power law of the two-path 'plane earth' model. However, they provide a better approximation of the expected measured pathloss in various propagation environments.

The measurement results were collected by sampling and logging the received signal strength at the MS at distances of 3.22 mm and averaging these samples over 2000-sample long windows to remove the effects of fast fading. This delivered a received signal value every 6.44 m. We then fitted

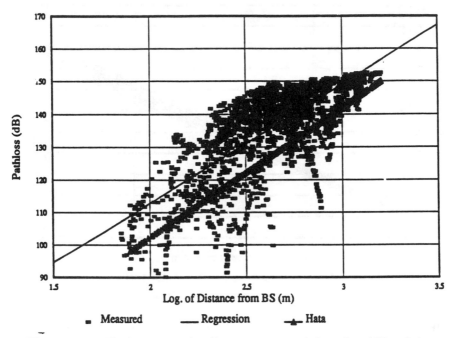

Figure 2.32: Fitting regression line to measured data for $AH = 6.4m$

a minimum mean squared error regression line to the averaged measured data points and compared it to the appropriate Hata model, as seen in Figures 2.32- 2.35 for the antenna heights of AH=6.4 m, 8.9 m, 11.4 m and AH=14 m.

Observe that as the antenna height is increased from $AH = 6.4m$ to $AH = 14m$, the regression lines fitted to the measured data become increasingly more optimistic than the corresponding Hata estimates, which is attributable to the fact that the antenna is gradually elevated beyond the urban skyline. Naturally, local building and terrain features do influence these findings, but the larger the measured data-base, the more consistent the predictions become. The pathloss regression lines for the antenna elevations AH=6.4 m, 8.9 m, 11.4 m and AH=14 m are summarised in Figure 2.36 along with the two extreme Hata models corresponding to AH=6.4 m and AH=14 m. The regression and Hata pathloss law gradients for our experiments are summarised in Table 2.9, which show reasonable agreement only for the lower antenna elevations, below the urban skyline. For the higher elevations clearly different propagation phenomena dominate. We have to point out that the regression line fitting is inevitably biased towards measurements between 500 m and 1000 m, since there are more streets to be measured further away from the BS than in its immediate vicinity, as inferred from the clustering of measurement points in Figures 2.32-2.35. Nevertheless, from Figure 2.36 we see that in the most important

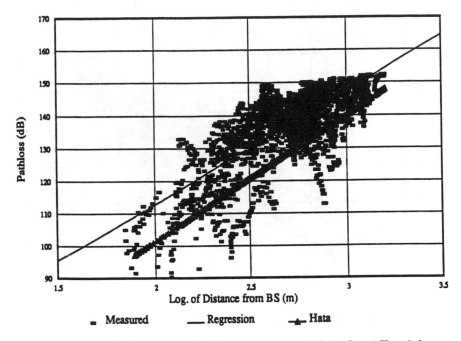

Figure 2.33: Fitting regression line to measured data for $AH = 8.9m$

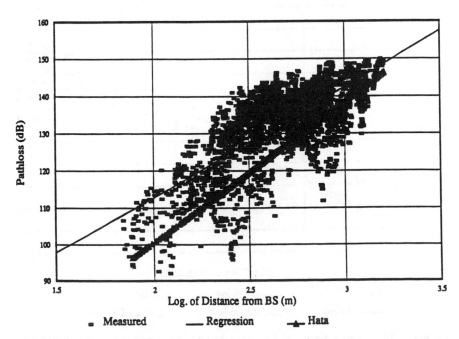

Figure 2.34: Fitting regression line to measured data for $AH = 11.4m$

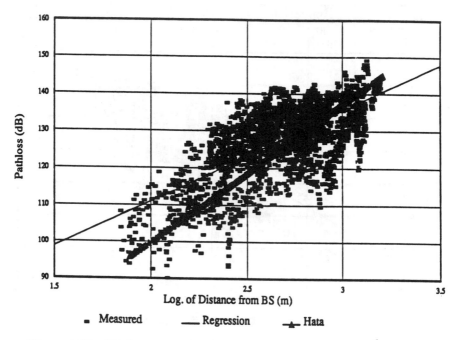

Figure 2.35: Fitting regression line to measured data for $AH = 14m$

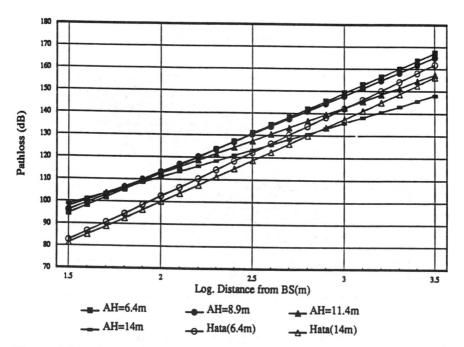

Figure 2.36: Comparison of pathloss regression lines with corresponding Hata models

AH(m)	Regression gradient (dB/decade)	Hata gradient (dB/decade)	Gradient difference (dB/decade)
14	24.6	37.4	-12.8
11.4	29.8	38	-8.2
8.9	34.5	38.7	-4.2
6.4	36.3	39.6	-3.6

Table 2.9: Measured and predicted pathloss gradients

d=100-1000 m region the Hata model gives approximately 10 dB more pessimistic estimates for the parameters considered than the fitted regression lines, if this extremely simplistic model is acceptable.

2.7.2 Slow Fading Statistics

Having derived the propagation pathloss law from our measurements we briefly focus our attention on the characterisation of the slow fading phenomena, which constitutes the second component of the overall power budget design of mobile radio links, as portrayed in Figure 2.30. In slow fading analysis the effects of fast fading and pathloss have to be ignored. The fast fading fluctuations have already been removed for pathloss modelling by averaging over 6.4 m distances. The slow fading fluctuations are simply separated by subtracting the best-fit pathloss regression estimate from each individual 6.4 m-spaced averaged received signal value. The slow fading histograms derived this way from Figures 2.32-2.35 for the previously used four antenna heights of AH=6.4 m, 8.9 m, 11.4 m and AH=14 m are depicted in Figures 2.37- 2.40. As expected, these figures suggest a lognormal distribution in terms of dBs due to normally distributed random shadowing effects. Indeed, when subjected to rigorous Kolmogorov-Smirnov and χ^2 (Chi-square) distribution fitting techniques (see later in Section 2.7.3.4) using the lognormal hypothesis, the hypothesis is confirmed at a high confidence level. The associated standard deviations are 6.5 dB, 6.8 dB, 7.3 dB and 7.8 dB for AH=6.4 m, 8.9 m, 11.4 m and 14 m, respectively. When amalgamating all four slow fading histograms, Figure 2.41 is derived, which has an even smoother lognormal distribution due to the higher number of measured points.

2.7.3 Fast Fading Evaluation

2.7.3.1 Analysis of Fast Fading Statistics

Irrespective of the distribution of the numerous individual constituent propagation paths of both quadrature components (a_i, a_q) of the received signal, their distribution is normal due to the central limit theorem. Then the com-

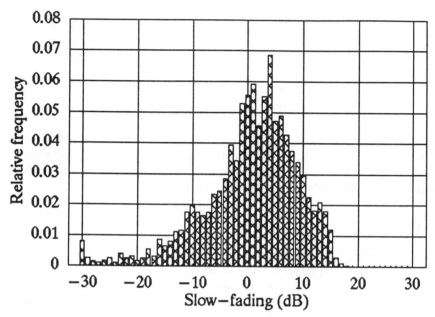

Figure 2.37: Typical microcellular slow fading histogram for $AH = 6.4m$

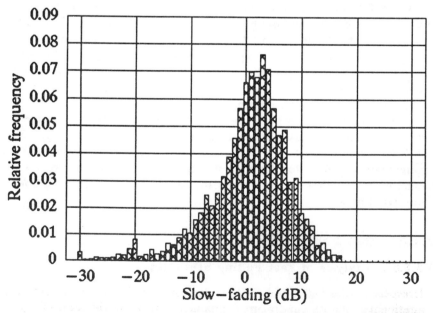

Figure 2.38: Typical microcellular slow fading histogram for $AH = 8.9m$

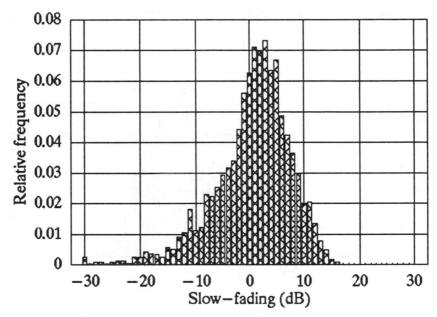

Figure 2.39: Typical microcellular slow fading histogram for $AH = 11.4m$

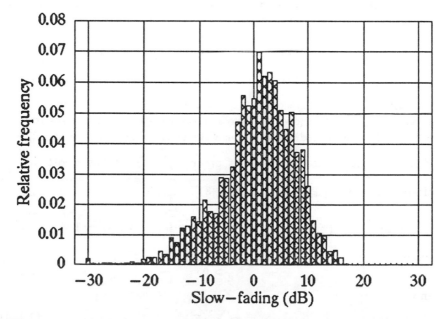

Figure 2.40: Typical microcellular slow fading histogram for $AH = 14m$

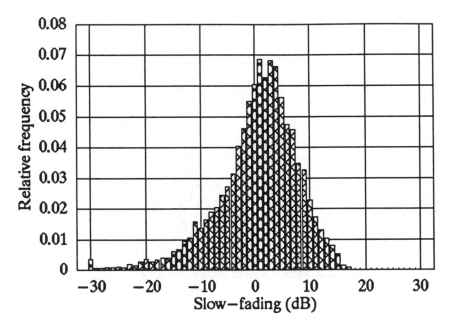

Figure 2.41: Typical microcellular slow fading histogram for various antenna elevations

plex baseband equivalent signal's amplitude and phase characteristics are given by:

$$a(k) = \sqrt{a_i^2(k) + a_q^2(k)} \tag{2.169}$$

$$\phi(k) = \arctan[a_i(k)/a_q(k)]. \tag{2.170}$$

Our aim is now to determine the distribution of the amplitude $a(k)$, if $a_i(k)$ and $a_q(k)$ are known to have a normal distribution. In general, for n normally distributed random constituent processes with means $\overline{a_i}$ and identical variances σ, the resultant process $y = \sum_{i=1}^{n} a_i^2$ has a so-called χ^2 distribution with a PDF given below [81]:

$$p(y) = \frac{1}{2\sigma^2} \left(\frac{y}{s^2}\right)^{(n-2)/4} \cdot e^{-(s^2+y)/2\sigma^2} \cdot I_{(n/2)-1}\left(\sqrt{y}\frac{s}{\sigma^2}\right) \tag{2.171}$$

where

$$y \geq 0 \tag{2.172}$$

and

$$s^2 = \sum_{i=1}^{n} (\overline{a_i})^2 \tag{2.173}$$

is the so-called non-centrality parameter computed from the first moments of the component processes $a_1 \cdots a_n$. If the constituent processes have zero means, the χ^2 distribution is central, otherwise non-central. Each of these processes have a variance of σ^2 and $I_k(x)$ is the modified k-th order Bessel-function of the first kind, given by

$$I_k(x) = \sum_{j=0}^{\infty} \frac{(x/2)^{k+2j}}{j! \Gamma(k+j+1)} \, , \quad x \geq 0. \qquad (2.174)$$

The Γ function is defined as

$$\begin{aligned}
\Gamma(p) &= \int_0^{\infty} t^{p-1} e^{-t} dt \quad \text{if } p > 0 \\
\Gamma(p) &= (p-1)! \quad \text{if } p > 0 \text{ integer} \qquad (2.175) \\
\Gamma(\tfrac{1}{2}) &= \sqrt{\pi} \, , \quad \Gamma(\tfrac{3}{2}) = \frac{\sqrt{\pi}}{2}.
\end{aligned}$$

In our case we have two quadrature components, i.e. $n = 2$, $s^2 = (\overline{a_i})^2 + (\overline{a_q})^2$, the envelope is computed as $a = \sqrt{y} = \sqrt{a_i^2 + a_q^2}, a^2 = y, p(a)da = p(y)dy$, and hence $p(a) = p(y)dy/da = 2ap(y)$ yielding the Rician PDF

$$p_{\text{Rice}}(a) = \frac{a}{\sigma^2} e^{-(a^2+s^2)/2\sigma^2} I_o\left(\frac{as}{\sigma^2}\right) \quad a \geq 0. \qquad (2.176)$$

Formally introducing the Rician K-factor as

$$K = s^2/2\sigma^2 \qquad (2.177)$$

renders the Rician distribution's PDF to depend on one parameter only:

$$p_{\text{Rice}}(a) = \frac{a}{\sigma^2} \cdot e^{-\frac{a^2}{2\sigma^2}} \cdot e^{-K} \cdot I_o\left(\frac{a}{\sigma} \cdot \sqrt{2K}\right), \qquad (2.178)$$

where K physically represents the ratio of the power received in the direct line-of-sight path, to the total power received via indirect scattered paths. Therefore, if there is no dominant propagation path, $K = 0$, $e^{-K} = 1$ and $I_0(0) = 1$ yielding the worst-case Rayleigh PDF:

$$p_{\text{Rayleigh}}(a) = \frac{a}{\sigma^2} e^{-\frac{a^2}{2\sigma^2}}. \qquad (2.179)$$

Conversely, in the clear direct line-of-sight situation with no scattered power, $K = \infty$, yielding a 'Dirac-delta shaped' PDF, representing a step-function-like CDF. The signal at the receive antenna then has a constant amplitude with a probability of one. Such a channel is referred to as a Gaussian channel. This is because although there is no fading present, the receiver will still see the additive white Gaussian noise (AWGN) referenced to its input, as seen in section 2.1.4. Clearly, if the K-factor is known, the fast fading envelope's distribution is described perfectly.

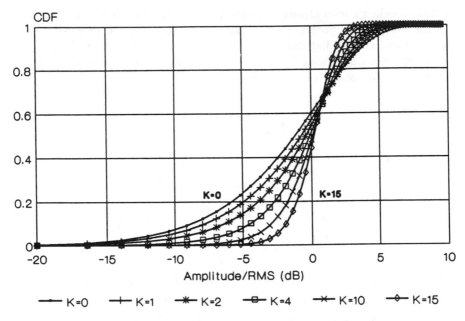

Figure 2.42: Rician CDFs on linear scale

The Rician CDF takes the shape of [81]

$$
\begin{aligned}
C_{\mathrm{Rice}}(a) &= 1 - e^{-\left(K + \frac{a^2}{2\sigma^2}\right)} \sum_{m=0}^{\infty} \left(\frac{s}{a}\right)^m \cdot I_m\left(\frac{a\,s}{\sigma^2}\right) \\
&= 1 - e^{-\left(K + \frac{a^2}{2\sigma^2}\right)} \sum_{m=0}^{\infty} \left(\frac{\sigma\sqrt{2K}}{a}\right)^m \cdot I_m\left(\frac{a\sqrt{2K}}{\sigma}\right).
\end{aligned}
$$

(2.180)

Clearly, this formula is more difficult to evaluate than the PDF of Equation 2.178 due to the summation of an infinite number of terms, requiring double or quadruple precision and it is avoided in numerical evaluations, if possible. However, in practical terms it is sufficient to increase m to a value, where the last term's contribution becomes less than 0.1 %.

A range of Rician CDFs evaluated from Equation 2.180 are plotted on a linear scale in Figure 2.42 for $K = 0, 1, 2, 4, 8$ and 15. Figure 2.43 shows the same Rician CDFs plotted on a more convenient logarithmic scale, which reveals the enormous difference in terms of deep fades for the K values considered. When choosing the fading margin overload probability, Figure 2.43 expands the high-attenuation tails of the CDFs, where for example for a Rician CDF with $K = 1$ the $15dB$ fading margin overload probability is seen to be approximately 10^{-2}. Lastly, a set of Rician PDFs computed from Equation 2.178 are seen in Figure 1.6, while the corresponding fading

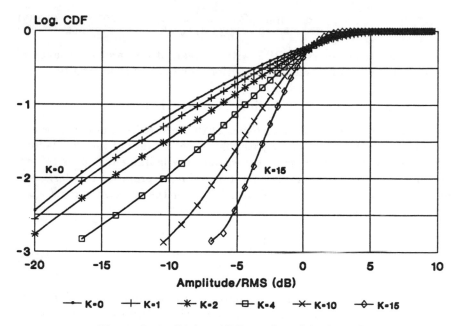

Figure 2.43: Rician CDFs on logarithmic scale

envelopes are depicted in Figure 1.5.

2.7.3.2   The Relation of Rician and Gaussian PDFs

In this subsection we show that the Rician PDF tends to the Gaussian one, as the K factor tends to infinity. To do so we simplify Equation 2.178 by introducing the transformation [55]

$$\alpha = \frac{a}{\sigma\sqrt{2K}}, \tag{2.181}$$

which yields

$$p(\alpha) = p(a)\, da/d\alpha = p(a)\, \sigma\sqrt{2K}. \tag{2.182}$$

Substituting $p_{\text{Rice}}(a)$ from Equation 2.178 in Equation 2.182 we have:

$$p(\alpha) = 2K\alpha \cdot e^{-K(\alpha^2+1)} \cdot I_0(2K\alpha). \tag{2.183}$$

For large x values

$$I_0(x) \approx \frac{e^x}{\sqrt{2\pi x}}\left[1 + \frac{1^2}{1!\, 8x} + \frac{1^2 \cdot 3^2}{2!\, (8x)^2} + \ldots\right] \approx \frac{e^x}{\sqrt{2\pi x}}. \tag{2.184}$$

Hence from Equation 2.183 we have:

$$\lim_{K\to\infty} p(\alpha) \approx \lim_{K\to\infty} 2K\alpha \cdot e^{-K(\alpha^2+1)} \cdot \frac{e^{2K\alpha}}{\sqrt{2\pi}\sqrt{2K\alpha}}$$

$$\approx \lim_{K \to \infty} \sqrt{\alpha} \frac{1}{\sqrt{2\pi} \cdot 1/\sqrt{2K}} \cdot e^{-\frac{(\alpha-1)^2}{2 \cdot 1/2K}} , \qquad (2.185)$$

which tends to a Gaussian PDF with a mean of one and a variance of $1/2K \approx 0$, yielding a Dirac delta function when K tends to infinity.

2.7.3.3   Extracting Fast Fading Characteristics

To determine the fast fading statistics of the received signal envelope one has to remove the effects of the path loss as well as that of the slow fading. The standard recognised method to extract the fading envelope is to normalise the received signal to its local RMS value, as proposed by Clarke [43] and used by other authors since then [42]. For the received sample $r(x_i)$ the local RMS is given by

$$\text{RMS} = \left(\frac{1}{W} \sum_{i-W/2}^{i+W/2} [r(x_i)]^2 \right)^{\frac{1}{2}} \qquad (2.186)$$

where W represents the window-length for the computation. This local RMS estimate is computed for each individual received sample in a sliding window and the normalised samples $r(x_i)/\text{RMS}$ are subjected to distribution fitting algorithms. The adequacy of this normalisation depends on the appropriate selection of W. Lee [44] suggested a window of 40 wave-lengths (λ) for conventional cell sizes, but in agreement with other authors [42], we found that in microcells the local RMS received signal level undergoes quite large fluctuations in such wide windows. This could effect the local statistics. Our experiments with a range of W values suggested that for any signal envelope sampling rate the window size must 'cover' a computation interval of about $4\lambda - 10\lambda$. This gives a sound RMS estimate and does not distort the fast fading statistics, hence we opted for $W = 200$ samples. After this smoothing-normalisation the fast fading envelope is stored in the computer ready for distribution fitting.

Knowledge of the expected Rician K-factor is important in system design, as it allows estimation of the fast fading margin required in the link budget calculation. A long term average of the K-factor gives an estimation of the average performance that can be expected. However, more efficient system design requires knowledge of the variation in K-factor as the mobile moves. In extreme, an individual K-factor could be calculated for each fade. This would require a very short computational window but as the computational window becomes shorter, the K-factor variations become more erratic. That is, since the distribution gradually changes, the confidence in the goodness-of-fit reduces. Furthermore, such very short windows would result in such a vast amount of information that computation and analysis become impractical.

In this paragraph we set out to determine the optimum window size of the fast fading distribution fitting. To achieve this we synthesized a file

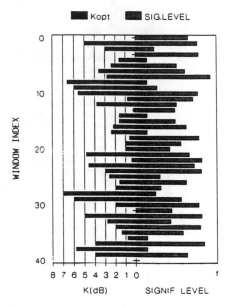

Figure 2.44: K-factor and significance-level profiles for $D = 2000$

of 80,000 fading envelope values with an overall K=3 dB or $K \approx 2$ using
Figure 1.9 in Chapter 1. This represents moderately severe fading, as can
be seen from the CDFs in Figures 2.42 and 2.43. Observe that localised
K-factors will have increased variation, as the window size is reduced, re-
vealing the true fast fading profile. Accordingly, the K-factors were then
evaluated by the distribution fitting algorithms to be highlighted in the
next subsection using computation windows of length D=1000, 2000, 4000,
8000 and 80,000 samples. As expected, using D=80,000, the overall K-
factor was measured to be K=3 dB or $K \approx 2$, with a very high degree of
confidence, i.e. the significance level in the goodness-of-fit test was almost
unity. As the computation window size, D, was decreased, the variation in
the K-factors increased. The derived K-factor profiles and their associated
significance levels are shown for window sizes of D=2000, 4000 and 8000 in
Figures 2.44–2.46. where the whole file of 80,000 points is represented in
each case and a significance level in excess of 0.1 implies a high degree of
confidence in the fit. Analysis of these figures and a variety of similar profiles using different window sizes show that the variation in the K-factors
does not increase significantly for window sizes below $D = 1000 - 2000$,
suggesting that the window is sufficiently short to track the change in the
K-factor. However, reducing the window size further significantly lowers
the confidence measures associated with the K-factor, implying a poor fit.
Observe that although the file of 80,000 samples gives a K-factor of 2, the
average of the K values over the windows in Figures 2.44–2.46 is generally
a different value. Analysis of other data files revealed similar results con-

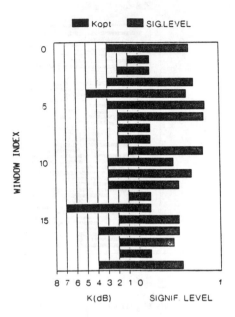

Figure 2.45: K-factor and significance-level profiles for $D = 4000$

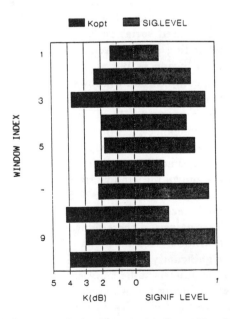

Figure 2.46: K-factor and significance-level profiles for $D = 8000$

firming our choice of $D = 1000 - 2000$ for distribution fitting when tracking K-profiles.

The erratic K-factor variations seen in Figure 2.44 are fairly typical, making their efficient use in power budget design rather difficult. Clearly, the K-variations must be statistically characterized in terms of their distributions, which were found to have 'near-normal' PDFs for a variety of scenarios, but the expected value and variance of the K-factors depended on the local paraphernalia. The interested reader is referred to Figures 1.20 and 1.21 for some measured K-profiles.

With the analysis parameters settled, one can proceed to analyse the measured data in a spirit similar to that of references [45]–[47].

2.7.3.4 Goodness-of-fit Techniques

K-profiles of measured propagation data are conveniently evaluated by computing the PDFs and CDFs of the measured samples for windows of, for example, $D = 1000 \ldots 2000$ smoothed values and then comparing them with a set of hypothesis PDFs and/or CDFs. Each comparison yields a probability, confidence measure or significance level proportional to the likelihood that the measured distribution is really a representative of the assumed hypothesis distribution.

The specific comparison giving the highest significance level is checked against the minimum acceptance level allowing rejection or acceptance of the initial hypothesis.

There is an abundance of goodness-of-fit methods for testing the statistical relevance of a match between a measured and a hypothesis distribution, all of which have different strengths and weaknesses [49], [50]. Here we briefly consider the Chi-square (χ^2) and the Kolmogorov-Smirnov (KS) methods.

2.7.3.4.1 Chi-square Goodness-of-fit Test The method devised in 1900 by Pearson is based on a normalised quadratic sum of the deviations of the observed occurrences (n_i) from their hypothesised expected values ($n \cdot p_i$), where n_i and p_i represent the observed number of occurrences in bin i and their expected probabilities, respectively. There are k bins and a total of n samples in the experiment. Then the degree of freedom is $(k-1)$, since the only linear restriction present is that

$$\sum_{j=1}^{k} n_j = n. \tag{2.187}$$

Other restrictions can be introduced for distributions where some parameters, such as mean or variance, have to be estimated. Each maximum likelihood parameter estimate reduces the degree of freedom by one. The measure of deviation is then computed as:

$$\chi^2 = \sum_{i=1}^{k} \frac{(n_i - n \cdot p_i)^2}{n \cdot p_i}, \tag{2.188}$$

which can be shown to have Chi-square distribution if the differences between n_i and $n \cdot p_i$ are non-deterministic. Therefore the confidence measure associated with a specific χ^2 value in Equation 2.188 gives the probability that such a χ^2 value could have come from the χ^2 distribution of Equation 2.171, which is also given in the Tables of [49] and [50].

2.7.3.4.2 Kolmogorov-Smirnov (KS) Goodness-of-fit Test In contrast to the Chi-square fitting, where the binned measured PDF was used, the KS-test uses the CDFs. According to Kolmogorov and Smirnov the limiting distribution of the maximum CDF deviation D_n between the measured distribution $C_n(x)$ and hypothesis $C(x)$

$$\sqrt{n} \cdot D_n = \sqrt{n} \cdot \mathrm{Max}|C_n(x) - C(x)| \tag{2.189}$$

is characterised by [50]

$$
\begin{aligned}
H(x) &= \lim_{n \to \infty} [CDF_{D_n}(x)] = \lim_{n \to \infty} P[\sqrt{n}D_n \le x] \\
&= \left[1 - 2\sum_{j=1}^{\infty}(-1)^{(j-1)}e^{-2j^2x^2}\right] \cdot I_{0,\infty}(x)
\end{aligned}
\tag{2.190}
$$

where the indicator function $I_{(0,\infty)}(x)$ is

$$I_{(0,\infty)}(x) = \begin{cases} 1 & \text{if } 0 < x < \infty \\ 0 & \text{otherwise} \end{cases} \tag{2.191}$$

Observe that the CDF $H(x)$ in Equation 2.190 does not depend on the distribution $C_n(x)$, which explains the versatility of the method. On the other hand, the KS method does not allow testing of composite hypothesises, where some parameters have to be estimated, as this would reduce the degree of freedom.

In practical hypothesis testing the measured data of each window of D samples has to be sorted in ascending order, which actually gives a fine but non-uniformly spaced representation of its CDF. Then for each individual measured and sorted sample the corresponding CDF value has to be found from the hypothesis CDF and the maximum deviation over the window must be remembered for every hypothesis CDF. The hypothesis CDF with the smallest maximum deviation D_n has the highest confidence, the value of which is computed from Equation 2.190 or from tables given in [51].

The KS test is computationally more demanding than the χ^2 test, because the computation of the Rician CDFs from Equation 2.180 implies the

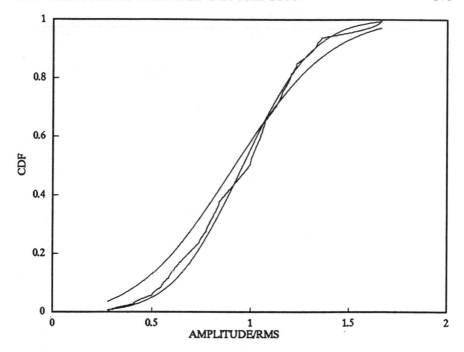

Figure 2.47: KS CDF-fitting using a window size of $D = 1000$

evaluation of a high number of summation terms, necessitating double or quadruple precision computations. Furthermore, as opposed to the χ^2 test, where the hypothesis PDFs can be prestored, here the hypothesis CDF of Equation 2.180 must be evaluated 'on-line' for every single, non-uniformly spaced abscissa value, constituted by the measured fading samples. Hence for large-scale measurement programmes the χ^2 test is preferred.

2.7.3.4.3 Goodness-of-fit of the Hypothesis Distribution In distribution fitting the higher the number of samples used to compute the measured CDF and PDF, the smaller is the tolerable discrepancy for a specific significance level between the best-fit theoretical PDF or CDF and its experimental counterpart. This can be inferred from Equations 2.188 and 2.190 for both the χ^2 and the KS test, since the error terms in these equations increase, as the number of samples n in a fitting window is increased. The reason for this is plausible, since for larger number of samples the statistical relevance of the experimental data is enhanced, allowing diminishing random differences only between the measured and hypothesis statistics. This is clearly illustrated by Figures 2.47–2.49, where the measured CDFs of a $K = 3\,dB$ Rician channel generated by theory are depicted using window sizes of D=1000, 2000 and 4000 samples, together with their best matching theoretical counterparts. Specifically, Figure 2.47 displays the goodness-of-fit to be expected, if a window of D=1000 samples is used.

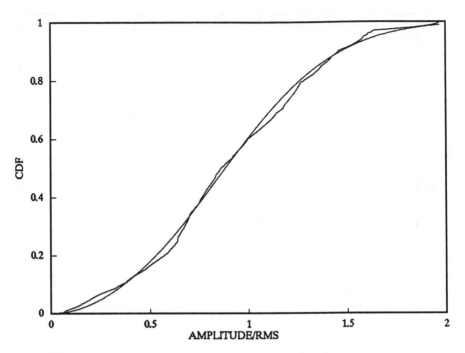

Figure 2.48: KS CDF-fitting using a window size of $D = 2000$

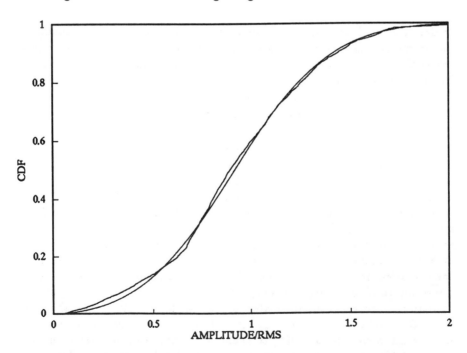

Figure 2.49: KS CDF-fitting using a window size of $D = 4000$

As expected, the KS distribution fitting procedure favours the CDF, representing $K = 3\,dB$ from the preselected set of 80 hypothesis CDFs, to yield the best fitting theoretical CDF. For comparison we also depicted the $K = 7\,dB$ CDF, which exhibits a consistently different shape, while the differences with respect to the $K = 3\,dB$ CDF appear to be random. When the window size is increased to D=2000 and 4000 samples, as seen in Figures 2.48 and 2.49, the differences become less significant, as the higher number of samples encountered gives a more adequate representation of the random process.

Figure 2.50 displays the significance level as a function of the Chi-square distribution parameter χ^2 computed from Equation 2.171. Here we utilised a window-size of 1000 samples and 80 'bins' for categorising the data between 0 and $4 \cdot RMS$. Using the χ^2 error term computed from Equation 2.171 the significance level, referred to also as confidence measure (CM), is read from Figure 2.50a as the intercept of the CM-curve with the appropriate vertical grid line at the specific χ^2-value. For CMs lower than 10^{-2} the logarithmically scaled curve of Figure 2.50b is more preferable. Observe that for a χ^2-value of 60 the confidence level is near-unity, while a value of 120 only achieves a CM of approximately 10^{-3}. In practice any hypothesis PDF having $\chi^2 > 120$, i.e. $CM < 10^{-3}$ has to be rejected.

Similarly, the linear and logarithmic Kolmogorov-Smirnov (KS) significance levels are depicted in Figure 2.51, where the abscissa values are scaled by the square-root of the sample number (SAMPNO). Again, the sharply decaying curves ensure a well-defined acceptance or rejection of the hypothesis distribution and the logarithmically scaled curve in Figure 2.51b conveniently expands the lower end of the CM-scale.

2.7.4 Summary

Below we summarise the practical characterisation of microcellular mobile radio channels by way of an example using our three-step approach portrayed in Figure 2.30 and the results from our previous deliberations.

1. We estimate the pathloss using the Hata-model and deploy a correction factor corresponding to the antenna elevation, deduced from measurements: $L_{pl} = L_{Hata} + L_{corr}$.

2. Using the characteristic slow fading variance of, say $\sigma = 7dB$, assuming lognormal slow fading PDF and allowing for a 1.4% slow 'fading margin overload' probability we introduce a 'slow fading margin' of $L_{slow} = 2 \cdot \sigma = 14dB$.

3. Assuming a typical Rician fading with $K = 10$ and a fast 'fading margin overload' probability of 1% a 'fast fading margin' of $L_{fast} = 7dB$ is inferred from Figure 2.42.

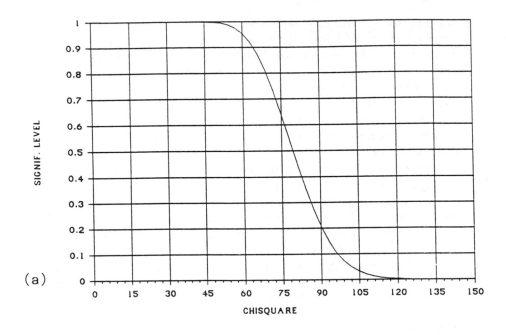

(a)

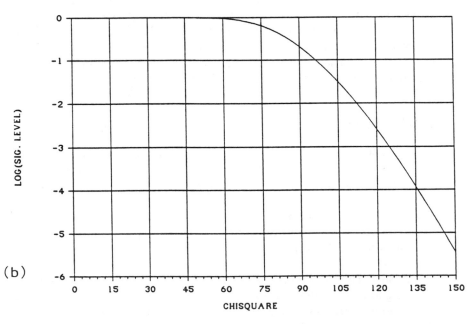

(b)

Figure 2.50: Chi-square significance level variation (a) on linear scale, or (b) on logarithmic scale

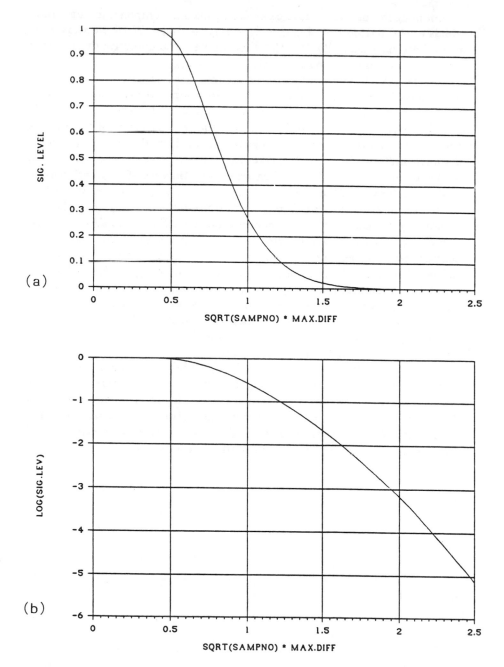

Figure 2.51: Kolmogorov-Smirnov significance level variation (a) on linear scale (b) on logarithmic scale

Summing the pathloss and the two fading margin components from above yields a total pathloss of: $L_{total} = L_{pl} + L_{slow} + L_{fast}$. With the knowledge of the receiver sensitivity P_{rec} this then allows us to compute the required minimum transmitted power as: $P_{tx} = P_{rec} + L_{total}$.

For the sake of illustration let us compare the transmitted power requirements for an urban microcellular environment with cell radii of 300 m and 100 m, using the above mentioned $L_{slow} = 14\,dB$ and $L_{fast} = 7\,dB$ fading margins. From Figure 2.36 we find that our extensive measurement programme in this environment suggests a pathloss of $L_{pl} = 130\,dB$ at a distance of 300 m from the BS, when the antenna elevation is AH=6.4 m, while at 100 m $L_{pl} = 110\,dB$. Then we have pathlosses of $L_{300} = 130\,dB + 14\,dB + 7\,dB = 151\,dB$ and $L_{100} = 131\,dB$. Assuming a receiver sensitivity of -104 dBm, as in the Pan-European digital mobile radio system, the corresponding transmitted power requirements are: $P_{300} = 47\,dBm \approx 50\,W$ and $P_{100} = 27\,dBm \approx 0.5\,W$. Clearly, for this low antenna elevation microcellular scenario a cell radius of larger than 100 m is becoming unrealistic in terms of transmitted power. Increasing the antenna height substantially reduces the transmitted power requirement or extends the cell radius. When the BS antenna is elevated beyond the urban sky-line, substantially higher coverage area can be achieved at the same transmitted power. However, the fading gradually becomes Rayleigh, requiring a higher fading margin and thereby reducing the gains achieved.

Observe that the power budget design highlighted outlines the requirements imposed on the error correction codec as well. Namely, in the event when the slow and fast fading exceed the total fading margin of $L_{fading} = L_{slow} + L_{fast} = 21\,dB$, the channel codec is faced with an error probability of 50 %, hence it must be designed to correct an error rate of about half of the fading margin overload probability.

Bibliography

[1] **D. C. Cox.** "Universal digital portable radio communications". *Proc. IEEE,* *vol.75, no.4,* pp.436-477, April 1987.

[2] **S. Stein** and **J.J. Jones.** *Modern Communication Principles,* McGraw-Hill, 1967.

[3] **J. Dugundji.** "Envelopes and pre-envelopes of real waveforms". *IRE Trans., vol.IT-4, no.1,* pp.53-57, March 1958.

[4] **M. Schwartz, W.R. Bennett,** and **S. Stein.** *Communications systems and techniques,* McGraw-Hill, 1966.

[5] **D.J. Sakrison.** *Communications theory: Transmission of waveforms and digital information,* John Wiley & Sons, 1968.

[6] **W.B. Davenport, Jr.,** and **W.L. Root.** *An introduction to the theory of random signals and noise,* McGraw-Hill, New York, 1958

[7] **S.Y. Liao.** *Engineering applications of electromagnetic theory,* West Publishing Company, 1988.

[8] **T. Aulin.** "Characteristics of a digital mobile radio channel". *IEEE Trans., vol.VT-30, no.2,* pp.45-53, May 1981.

[9] **P.A. Bello.** "Characterization of randomly time-variant linear channels". *IEEE Trans., vol.CS-11, no.4,* pp.360-393, December 1963.

[10] **W.C.Y. Lee.** "Finding the approximate angular probability density function of wave arrival by using a directional antenna". *IEEE Trans., vol.AP-21, no.3,* pp.328-334, May 1973.

[11] **D.C. Cox** and **R.P. Leck.** "Correlation bandwidth and delay spread multipath propagation statistics for 910-MHz urban mobile radio channels". *IEEE Trans., vol.COM-23, no.11,* pp.1271-1280, November 1975.

[12] **W.Y.C. Lee.** *Mobile communications engineering,* McGraw Hill, 1982.

[13] **W.C. Jakes, ed.** *Microwave mobile communications,* Wiley, New York, 1974.

[14] **P.A. Bello** and **B.D. Nelin.** "The effect of frequency selective fading on the binary error probabilities of incoherent and differentially coherent matched filter receivers". *IEEE Trans., vol.CS-11, no.2,* pp.170-186, June 1963.

[15] **H.F. Schmid.** "A prediction model for multipath propagation of pulse signals at VHF and UHF over irregular terrain". *IEEE Trans., vol.AP-18, no.2,* pp.253-258, March 1970.

[16] P. Melançon and J. Le Bel. "A characterization of the frequency selective fading of the mobile radio channel". *IEEE Trans.*, vol. VT-35, no.4, pp.153-161, November 1986.

[17] W.R. Young, Jr., and L.Y. Lacy. "Echoes in transmission at 450 megacycles from land-to-car radio units". *Proc. IRE*, vol.38, pp.255-258, March 1950.

[18] R.H. Clarke. "A statistical theory of mobile-radio reception". *B.S.T.J.*, vol.47, pp. 957-1000, July-August 1968.

[19] D.C. Cox. "910 MHz urban mobile radio propagation: Multipath characteristics in New York city". *IEEE Trans.*, vol. VT-22, no.4, pp.104-110, November 1973.

[20] A.S. Bajwa and J.D. Parsons. "Small-area characterisation of UHF urban and suburban mobile radio propagation". *IEE Proc.*, vol.129, pt.F, no.2, pp.102-109, April 1982.

[21] J. Zander. "A stochastical model of the urban UHF radio channel". *IEEE Trans.*, vol. VT-30, no.4, pp.145-155, November 1981.

[22] D. Gabor. "Theory of communications". *Journal of the IEE, Part III*, vol.93, pp. 429-457, November 1946.

[23] D.G. Appleby, P.M. Fortune, Y.F. Ko, W.H. Lam, R. Steele, I.J. Wassell, and K.H.H. Wong. "The propose multiple access methods for the pan-European mobile radio systems". *IEE Colloquium on "Multiple Access Techniques in Radio Systems"*, Savoy Place, London, pp.1.1-1.26, October 1986.

[24] GSM Doc. No. 85/85. *"Description of the Experimental S-900D for Digital Radiotelephone"*.

[25] J. Udenfeldt. "DMS90–An experimental TDMA digital mobile telephone system". *Ericsson Radio System AB, Stockholm, Sweden*.

[26] K.D. Eckert and G. Höfgen. "The fully digital cellular radio telephone system CD900". *Nordic Seminar on Digital Land Mobile Radio Communication, Espoo, Finland*, pp.249-259, February 1985.

[27] *GSM Specifications*, ETSI, 1988.

[28] D.C. Cox. "Delay doppler characteristics of multipath propagation at 910 MHz in a suburban mobile radio environment". *IEEE Trans.*, vol. AP-20, no.5, pp. 625-635, September 1972.

[29] T. Takeuchi, F. Ikegami, S. Yoshida, and N. Kikuma. "Comparison of multipath delay spread characteristics with BER performance of high speed digital mobile transmission". 38th *IEEE Vehicular Technology Conference, Philadelphia, Pennsylvania*, pp. 119-126, 15-17 June 1988.

[30] M. Nilson. "Measurements of some characteristics of multipath propagation at 900 MHz in the city of Stockholm". *COST 207 TD (85)14*, pp. 117-132, 11 February 1985.

[31] P. Lo Muzio, G. Guidotti, N. Benvenuto, and S. Pupolin. "Experimental characterization of VHF propagation in a mountain environment". *Alta Frequenza*, vol.56, no.6, pp.273-282, August 1987.

[32] **P. A. Bello.** "Time-frequency duality". *IEEE Trans., vol.IT-10, no.1,* pp.18-33, January 1964.

[33] **J.D. Parsons and A.S. Bajwa.** "Wideband characterisation of fading mobile radio channels". *IEE Proc., vol.129, pt.F, no.2,* pp.95-101, April 1982.

[34] **P.W. Huish and E. Gurdenli.** "Propagation measurements and planning requirements for digital cellular systems". *Second Nordic Seminar on Digital Land Mobile Radio Communications, Stockholm,* pp.199-204, 14-16 October 1986.

[35] **P. Olivier and J. Tiffon.** "Transfer function measurement as a characterisation of the urban mobile radio channel". *IEE Fifth Int. Conf. on Antennas and Propagation, ICAP 87, IEE Conf. Pub. no.274, pt.2,* pp.95-98, 30 March-2 April 1987.

[36] **J.D. Parsons and A.S. Bajwa.** "Wideband characterisation of fading mobile radio channels". *IEE Proc., vol.129, pt.F, no.2,* pp.95-101, April 1982.

[37] **A. Papoulis.** *Probability random variables and stochastic processes,* McGraw-Hill, 1981.

[38] **IEEE Vehicular Technology Society Committee on Radio Propagation.** "Coverage prediction for mobile radio systems operating in the 800/900 MHz frequency range". *IEEE Trans., vol.VT-37, no.1,* pp.3-72, February 1988.

[39] **G.L. Turin, F.D. Clapp, T.L. Johnston, S.B. Fine, and D. Lavry.** "A statistical model of urban multipath propagation". *IEEE Trans., vol.VT-21, no.1,* pp.1-9, February 1972.

[40] **H. Suzuki.** "A statistical model for urban radio propagation". *IEEE Trans., vol.COM-25, no.7,* pp.673-680, July 1977.

[41] **H. Hashemi.** "Simulation of the urban radio propagation channel". *IEEE Trans., vol.VT-28, no.3,* pp.213-225, August 1979.

[42] **E. Green.** "Radio link design for microcellular systems". *BT Technology J., vol.8, no.1,* January 1990.

[43] **R.H. Clarke.** "A statistical theory of mobile-radio reception". *BSTJ, vol.47,* pp. 957-1000, 1968.

[44] **W.C.Y. Lee.** "Estimate of local average power of a mobile radio signal". *IEEE Trans. on Veh. Techn., vol.VT-34, no.1,* pp. 22-27, 1985.

[45] **R.J.C. Bultitude.** "Measurement, characterisation and modelling of indoor 800/900 MHz radio channels for digital communications". *IEEE Comms Magazine, vol.25, no.6,* pp. 5-12, June 1987.

[46] **R.J.C. Bultitude, S.A. Mahmoud and W.A. Sullivan.** "A comparison of indoor propagation characteristics at 910 MHz and 1.75 GHz". *IEEE JSAC, vol.7, no.1,* pp. 20-30, January 1989.

[47] **R.J.C. Bultitude and G.K. Bedal.** "Propagation characteristics on microcellular urban mobile radio channels". *IEEE JSAC, vol.7, no.1,* pp. 31-39, January 1989.

[48] **J.G. Proakis.** *Digital Communications*, McGraw-Hill, 1983.

[49] **R.B. D'Agostino and M.A. Stephens (Ed).** *Goodness-of-fit Techniques*, Marcel Dekker Inc., 1986.

[50] **A.M. Mood, F.A. Graybill and D.C. Boes.** *Introduction to the Theory of Statistics*, McGraw-Hill, 1985.

[51] **F.J. Massey Jr.** "The Kolmogorov-Smirnov test for goodness-of-fit". J. Amer Statist. Ass., pp. 46-70, 1951.

[52] **S. Siegel.** *Non-parametric statistics for the behavioural sciences*, McGraw Hill, 1956.

[53] **C. Leach.** *Introduction to statistics*, John Wiley & Sons.

[54] **J. Cheung.** "PhD Thesis". *Univ of Southampton*, 1991

[55] **L. Pap.** *Private Communication*, 1992

[56] **J. D. Parsons.** "The Mobile Radio Propagation Channel". *Pentech Press*, 1991

[57] **A. Rustako, N. Amitay, G.J. Owens, R.S. Roman.** "Propagation Measurments at Microwave Frequencies for Microcellular Mobile and Personal Communications". *Proc. of 39th IEEE VTC*, pp. 316-320, 1989

[58] **A.J. Rustako, N. Amitay, G.J. Owens, R.S. Roman.** "Radio Propagation Measurements at Microwave Frequencies for Microcellular Mobile and Personal Communications". *IEEE Int. Conf. on Comm's (ICC'89)*, pp. 482-488, Boston, U.S.A.

[59] **J.F. Aurand, R.E. Post.** "A Comparison of Prediction Methods for 800 MHz Mobile Radio Propagation". *IEEE Tr. VT-34, No 4*, pp. 149-153, Nov. 1985

[60] **S.T.S. Chia.** "Propagation Studies for Microcellular Mobile Radio". *PhD. Thesis*, University of Southampton, U.K., 1987

[61] **G.Y. Delisle, J. Lefevre, M. Lecours, J. Chouinard.** "Propagation Loss Prediction: A Comparative Study with Application to Mobile Radio Channel". *IEEE Trans. VT-34, No 2*, pp. 86-96, May 1985

[62] **Y. Okumura, E. Ohmori, T. Kawano, K. Fukuda.** "Field Strength and its Variability in VHF and UHF Land Mobile Service". *Review of the Electrical Communication Laboratory, Vol 16*, pp. 825-873, Sept.-Oct. 1968

[63] **M. Hata.** "Empirical Formula for Propagation Loss in Land Mobile Radio". *IEEE Trans. VT-29*, pp. 317-325, August 1980

[64] **A.G. Longley, P.L. Rice.** "Predictions of Tropospheric Radio Transmission Loss over Irregular Terrain - A Computer Method". *ESSA Tech. Rep. ERL79-ITS67*, 1968

[65] **R. Edwards, J. Durkin.** "Computer Prediction of Service Areas for VHF Mobile Radio Networks". *Proc. IEE 116, 9*, pp. 1493-1500, 1969

[66] **J.H. Causebrook.** "Computer Prediction of UHF Broadcast Service Areas". *BBC Research Report RD 1974-4*

[67] **F.H. Palmer.** "VHF/UHF Path-Loss Calculations Using Terrain Profiles Deduced from a Digital Topographic Data Base". *AGARD Conf. Proc., No 269*, pp. 26-1-26-11, 1979

[68] **J.J. Egli.** "Radio Propagation above 40 Mc over Irregular Terrain". *Proc. IRE*, pp. 1383-1391, 1957

[69] **J.D. Murphy.** "Statistical Propagation Model for Irregular Terrain Paths between Transportable and Mobile Antennas". *AGARD Conf. Proc., No 70*, pp. 49-1-49-20, 1970

[70] **K. Allesbrook, J.D. Parsons.** "Mobile Radio Propagation in British Cities at Frequencies in the VHF and UHF Bands". *IEEE Trans VT-26, No 4*, pp. 313-323, 1977

[71] **M.F. Ibrahim, J.D. Parsons.** "Signal Strength Prediction in Built Up Area". *Part I, Proc IEE, Vol 130, pt F*, pp. 377-384, August 1983

[72] **W.J. Kessler, M.J. Wiggins.** "A Simplified Method for Calculating UHF Base-to-Mobile Statistical Coverage Contours over Irregular Terrain". *27th IEEE Veh. Techn. Conf.*, pp. 227-236, 1977

[73] **M.N. Lustgarten, J.A. Madison.** "An Empirical Propagation Model". *IEEE Trans. EMC-19, No 3*, pp. 301-309, 1977

[74] **W.C.Y. Lee.** "Mobile Communications Engineering". *New York: McGrawHill, Chapter 3 and 4*, 1982

[75] **K. Bullington.** "Radio Propagation at Frequencies above 30 Mc". *Proc. IRE, 35*, pp. 1122-1136, 1947

[76] **J. Epstein, D.W. Peterson.** "An Experimental Study of Wave Propagation at 850 Mc". *Proc IRE, 41, 5*, pp. 595-611, 1953

[77] **J. Deygout.** "Multiple Kinfe-Edge Diffraction of Microwaves". *IEEE Trans. AP-14*, pp. 480-489, 1966

[78] **A. Blomquist, L. Ladell.** "Prediction and Calculation of Transmission Loss in Different Types of Terrain". *NATO-AGARD Conf. Publ. CP-144*, 1974

[79] "Atlas of radio Waves Propagation Curves for Frequencies between 30 and 10000 Mc/s". *Radio Research Laboratory, Ministry of Postal Services, Tokyo, Japan*, pp. 172-179, 1957

[80] **R.J. Leubbers.** "Finite Conductivity Uniform GTD versus Knife Edge Diffraction in Prediction of Propagation Path Loss". *IEEE Trans. AP-32, No 1*, pp. 7-76, January 1984

[81] **J.G. Poakis.** "Digital Communications". *McGraw-Hill*, 1983

Chapter 3

Speech Coding

R.A. Salami[1], L. Hanzo[2], R. Steele[3], K.H.J. Wong[4], and I. Wassell[5]

3.1 Introduction

A recurrent theme in digital cellular mobile radio is spectral efficiency, which is generally taken to mean the user density for the allotted spectrum. We have argued in Chapter 1 that the most influential factor in determining the spectral efficiency is the cell size. Microcells reuse the allotted spectrum over a smaller geographical area and thereby produce a more spectrally efficient system. For a given cell size and bandwidth allocation there are a set of sub-systems, such as the speech encoders, channel coders, interleavers, modulators, and so forth, that are influential in determining the number of mobile users that can be accommodated. By reducing the bit rate of the speech encoders, or the amount of channel coding required, the number of users and therefore the spectral efficiency can be increased.

The bit rate of the speech encoder for the pan-European digital cellular mobile radio system was initially set at 16 kb/s, a value which contained a minimal of channel coding and was considered to be the lowest bit rate to provide near-toll quality speech, and yet to have low enough complexity for implementation on a single monolithic chip. The main contenders for the

[1] University of Southampton
[2] University of Southampton and Multiple Access Communications
[3] University of Southampton and Multiple Access Communications
[4] University of Southampton
[5] Multiple Access Communications Ltd.

pan-European system speech encoder was the regular pulse excited (RPE) linear predictive codec [1], the multipulse excited (MPE) linear predictive codec [2] and a number of sub-band coders (SBC) [3]. The speech encoder selected, as well as the other system details of the GSM pan-European system are presented in Chapter 8. The next phase for the GSM system will be to determine the most suitable so-called half-rate codec, i.e., a codec giving toll quality speech at 8 kb/s. The leading contender is the code-book excited linear predictive coder (CELP) [4], to be described in a later section.

Other speech encoders may also be employed in mobile radio systems. When microcells or picocells [5] are used more bandwidth becomes available and simpler waveform encoders such as delta modulation (DM) [6], pulse code modulation (PCM) [7], and differential pulse code modulation (DPCM) [8, 9] may be appropriate. These codecs are inexpensive to implement, have minimal encoding delay and have relatively low power consumption. By employing adaptive DM, (ADM), at 32 kb/s the codec displays the virtues of toll quality, low power consumption and robustness to transmission errors.

Speech encoders are vital in determining the transmitted bit rate and the quality of the recovered speech. For a full treatise on speech encoding we refer the reader to the excellent books of Jayant and Noll [10], Flanagan [11] as well as that of Rabiner and Schafer [12]. Our intention is to glide over DM, PCM and DPCM as it has been extensively treated in the above references. We will mainly concentrate on speech codecs operating at 4-16 kb/s that are in vogue for mobile radio applications at the present time. Of the simple codecs that are suitable for small cells, only the embedded delta modulator (EDM) will be described [13].

As this book is devoted to mobile radio and not to speech encoding, we will not discuss the basic properties and categories of speech. Instead we will commence by briefly reviewing major waveform encoding techniques in Section 3.2 as they may have a role to play in microcellular mobile radio systems. Sections 3.3 and 3.4 deal with EDM and SBC, respectively. The remainder of the chapter is concerned with hybrid encoding techniques that are finding favour in current mobile radio applications.

3.2 Brief Review of Major Waveform Encoding Techniques

There are basically three classes of speech encoders, namely waveform, source, and hybrid encoders. Waveform encoders accept continuous analogue speech signals and encode them into digital signals prior to transmission. Decoders at the receivers reverse the encoding process to recover the speech signals. In the absence of transmission errors, the recovered speech waveforms have a close resemblance to the original speech waveforms. The basis of waveform encoding is that if the recipient can obtain a replica of

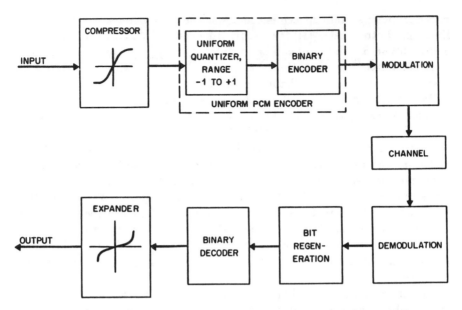

Figure 3.1: Block diagram of a basic PCM transmission link

the original speech waveform the speech quality will be excellent. In practice the encoding process introduces quantisation noise (which is really a waveform distortion), but this is usually sufficiently small for the speech quality to be unaffected. Waveform codecs are of low complexity and are therefore inexpensive. Both the power consumption and the encoding delay are low. They can usually be employed to encode other signals, such as signalling tones, voice band data, and with certain provisos, music signals.

The simplest waveform codec is pulse code modulation (PCM), where the speech is sampled at the Nyquist rate, and each sample is represented by a binary number thereby introducing quantisation noise. Logarithmic PCM (log-PCM) has been standardised by the CCITT to operate at 64 kb/s. The samples are near-logarithmically compressed and binary encoded prior to transmission. At the receiver binary decoding is performed followed by logarithmic expansion to recover the speech signal. The basic PCM arrangement is shown in Figure 3.1. An in-depth study of transmitting log-PCM signals over mobile radio channels has been made [14, 15, 16, 17, 18, 19] where it is demonstrated that log-PCM can be employed, in spite of its vulnerability to transmission errors, provided sufficient radio spectrum is available.

The quasi-periodic vibrations of the vocal cords, and the constrained movements of the vocal articulators, such as the jaw and tongue, cause the samples in voiced speech signals to be highly correlated. By contrast, unvoiced speech tends to be relatively uncorrelated. In conversational speech, a typical talker will speak for 40% of the time, and for the active speech

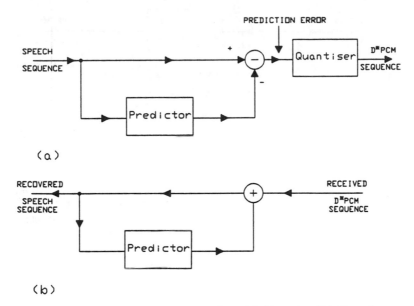

Figure 3.2: Differential-star PCM codec: (a) Encoder (b) Decoder

periods voiced speech occurs some four times more frequently than unvoiced speech. The dominance of voiced speech means that the correlation in speech signals can be effectively exploited by predicting the speech
sequence, forming an error between the actual speech and the predicted
speech, and quantizing the error sequence prior to transmission. This
method is known as differential-star pulse code modulation D*PCM, [10].
The decoder has the same predictor as used in the encoder and operates
on the previously decoded samples to form a predicted sample which when
added to the received error sample gives the output speech sample. The
recovered speech is obtained by low pass filtering. The codec arrangement
is shown in Figure 3.2. Notice that the role of the predictor at the encoder
is to form an error sequence whose variance is significantly less than the
variance of the input speech sequence. By decreasing the variance of a
signal applied to a quantiser, the step-size is reduced for a given number of
quantization levels, and the quantization noise decreases with decreasing
step-size.

A superior form of differential encoding is differential pulse code modulation (DPCM). Figure 3.3 shows the block diagram of a DPCM encoder.
Like D*PCM, the error signal is quantised, and binary encoded (not shown)
for transmission. However, the predicted signal is not formed from the input speech. Instead the predictions are based on the decoded sequence
(identical to the sequence at the output of the decoder in Figure 3.3 in the
absence of transmission errors) that consists of the input speech sequence
contaminated by quantization noise. As the predictors in the encoder and

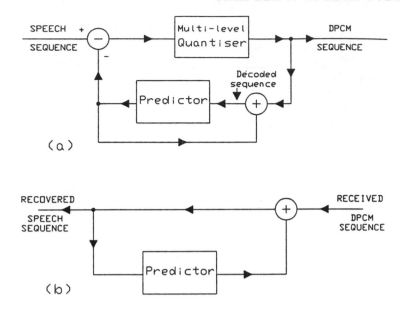

Figure 3.3: Differential Pulse Code Modulation Codec: (a) Encoder (b) Decoder

decoder both operate on the recovered sequence the performance is superior to D*PCM whose predictors make predictions from different sequences. The quantitative performance of DPCM and D*PCM are given by Jayant and Noll [10].

The quantisers used in DPCM can be uniform (fixed step-size), non-uniform (un-equal step-sizes), or adaptive where the step-size changes depending on the previous samples. The quantisers may operate on one sample when they are scalar quantisers, or they may treat a sequence of samples as a vector to perform vector quantisation. The predictors may have fixed coefficients, adaptive coefficients, or the coefficients may be organised as vectors [10], [20]. Two predictors may be employed, one to accommodate the variations in the vocal tract, and the other to predict the excitation rate associated with the vocal cord vibrations. The role of the prediction is to decrease the variance of the error signal that is quantised and transmitted. As the prediction is made on noisy samples the order of the predictor is limited to 3 or 4, and the predictor gain is typically 15 dB [21]. When adaptive quantisers or predictors are used the encoding process is known as adaptive DPCM (ADPCM). Toll quality ADPCM codecs can operate at 24 kb/s [22], although the CCITT ADPCM codec is specified at 32 kb/s. Cordless telecommunication systems, such as CT2 and DECT operate with this CCITT ADPCM codec (see Chapter 1).

The bit rate can be decreased to 16 kb/s for essentially the same speech quality by using transform coding (TC) [23]. In TC the speech samples are

placed in contiguous blocks and transformed into the frequency domain.
Many transforms can be used, but the most popular is the discrete cosine
transform (DCT) [10]. Each transform coefficient is encoded with a number
of bits dependent on its perceptual importance. Some coefficients are discarded as inconsequential. The number of bits assigned to each coefficient
and the decision to discard a coefficient can be made to depend on the block
statistics. The arrangement is known as adaptive DCT (ADCT). Sub-band
coding (SBC) is a special case of ADCT where the speech signal is filtered
into bands, and an adaptive coder, such as adaptive PCM, (APCM) or
ADPCM, is used to encode each sub-band signal. The encoded sub-band
signals are multiplexed prior to transmission. At the receiver demultiplexing of the sub-band encoded signal is performed, followed by decoding. The
sub-band signals are combined to give the recovered speech. SBC has been
well researched [3, 10, 24] and can be implemented on a digital signal processing (DSP) chip. Of the six contending encoders for the pan-European
system, four were varieties of SBCs. We therefore give special attention to
SBC in Section 3.4.

3.3 Embedded Delta Modulation

Radio transmissions for digital cordless telecommunication (CT) systems
are made over short distances, say, 100 to 300 m, with a small base station
at an elevation of some 6 to 12 m. The hand-held portable is required to be
light-weight, inexpensive, and operate with low battery power drain. The
bit rate for the CT2 system is 32 kb/s, and delta modulation (DM) codecs
operate satisfactorily at this rate with a power consumption that is four to
six times lower than that for ADPCM. We will refrain from a discussion of
DM as it has been well documented by Steele [6]. Instead we will examine
embedded delta modulation (EDM) as it offers trade-offs between bit rate
and speech quality, while having a relatively low power consumption. Its
embedded characteristics allow the network to efficiently discard bits in
network overload situations. We will commence by considering the concept
of an embedded code.

Suppose an encoder generates R_{max} bits per sample, and the network
decides to delete R_{del} bits in each sample, such that the decoder decodes
samples having

$$R_{dec} = R_{max} - R_{del} \qquad (3.1)$$

bits. On the proviso that the quality of the decoded output is essentially
the same as for a R_{dec} bits per sample codec, then the encoder is referred
to as an embedded one. We observe that PCM is an embedded code if the
R_{del} bits are the least significant bits. Differential encoders do not produce
embedded codes. For example, if every sixth bit, say, is discarded from
the bit stream of a delta modulator (DM), the added noise in the decoded
output is significantly higher than if the DM had operated with a clock

Figure 3.4: Generalised embedded codec

rate reduced by 5/6. This is particularly so if an adaptive DM (ADM) is used. The generalised block diagram of an embedded codec is shown in Figure 3.4.

In order to form embedded differential codecs we need to multiplex more than one codec. For example, in multi-stage delta modulation (MSDM), N DM encoders are cascaded with each encoding the error signal of the previous encoder. The outputs from the N DM decoded outputs are suitably delayed and combined to give the recovered speech. The overall error is that of the last DM in the N cascaded chain of DM encoders in the MSDM encoder. The MSDM is an embedded codec because if the transmitted bit rate is reduced by discarding the bits from the last DM encoder, the MSDM has the performance of a $(N - 1)$ stage MSDM.

Another example is that of embedded DPCM [25] where the DPCM encoder operates on the speech signal, and the quantization error is PCM encoded. The combination behaves as an embedded codec if the deleted bits are taken from the PCM encoder.

Embedded coding offers the ability to provide a variable bit rate, and for the network to trade more users for lower speech quality, and vice versa. It can also be used over mobile radio channels by arranging for the secondary PCM codec to be replaced by a forward error correction (FEC) codec when the BER of the channel increases [26].

We emphasize that the power consumption of DM codecs is comparatively low, and for microcells where bit rate is not at so high a premium as in large cells, a lightweight hand-held portable can be achieved using a DM codec. An embedded DM codec, a combination of DM and PCM, can be used in such a portable, having higher quality speech than the basic DM, and an extension of radio coverage when its PCM component is replaced by a FEC codec. Although practical embedded DM (EDM) would use ADM rather than DM codecs, we will consider for exemplary reasons the less complex case of DM.

3.3.1   Theory of EDM

The block diagram of the EDM codec is shown in Figure 3.5. The speech signal is bandlimited to BHz, sampled at $2BHz$ to yield the input sequence $x(n)$. The DM is clocked at $2 BR_{min}$, where R_{min} are the bits generated per Nyquist interval. The DM error sequence $q(n)$ is generated at the rate of $2 BR_{min}$, and is low pass filtered to BHz. As the filtered signal $q_f(n)$

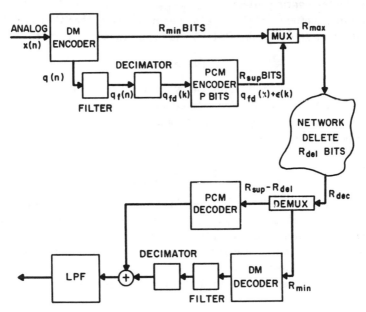

Figure 3.5: Embedded delta modulation codec (Wassell, et al. [13]).
(©1988 IEEE)

is over sampled by R_{min} it is decimated by this amount to yield

$$q_{fd}(k) = q_f(kR_{min}) \tag{3.2}$$

where k is the sample index at the Nyquist rate. The $q_{fd}(k)$ sequence is encoded by a supplemental PCM encoder which yields R_{sup} bits/Nyquist interval, where R_{sup} is P, the number of bits used in the quantisation process. The DM and PCM data are multiplexed to form a transmission rate of

$$R_{max} = R_{min} + P \quad \text{bits/Nyquist interval.} \tag{3.3}$$

Assuming that the network deletes R_{del} bits/Nyquist interval, then on demultiplexing at the receiver the PCM decoder is supplied with $R_{sup} - R_{del}$ bits per Nyquist interval as it is the PCM bits that are discarded by the network. The shortened PCM words are next decoded into a sequence of samples at the Nyquist rate. The DM data from the multiplexer is at $2BR_{min}$, and this is decoded into a sequence at the same rate. Filtering to BHz ensues to remove out-of-band noise, and the over-sampled sequence is decimated by R_{min} to give samples at the Nyquist rate. The PCM samples and the DM samples are now both occurring at the Nyquist rate, with the PCM samples representing the filtered DM error sequence. By adding the DM samples with their error samples we obtain the recovered speech

sequence which is converted into a continuous speech signal by the final
low pass interpolating filter.

The error sequence $q(n)$ has a power $< q^2(n) >$ that depends on both
the DM step-size δ and the PSD of the input speech sequence,

$$< q^2(n) >= a\delta^2/3 \qquad (3.4)$$

where $a = 1$ or 1.3, depending on whether the DM is operating in the
granular region or if its SNR is a maximum. The $q(n)$ sequence is filtered
and decimated to yield $q_{fd}(n)$, a sequence having a Gaussian PDF. The
mean square value of $q_{fd}(n)$ is

$$< q_{fd}(n) >= \frac{< q^2(n) >}{bR_{min}} \qquad (3.5)$$

where b is determined by the shape of the power spectral density (PSD) of
$q(n)$. If $q(n)$ is flat, a condition that occurs when the slope overload noise
power is approximately equal to the granular noise power, b is unity. When
the DM is operating with a sufficiently large step-size δ the DM noise is
granular and b is close to two.

The $q_{fd}(n)$ sequence is applied to the P-bit quantiser having overload
points at $\pm q_o$, where

$$q_o = \Delta 2^{P-1} \qquad (3.6)$$

and $\triangle$ is the step-size of the PCM quantiser. The PCM output sequence
is $q_{fd}(n)$ plus an error sequence $\epsilon(k)$. If the latter has a uniform amplitude
distribution over $\pm\Delta/2$,

$$< \epsilon^2(k) >= \Delta^2/12 \qquad (3.7)$$

and this is the noise power that occurs at the output of the EDM decoder,
it being appreciated that the only source of noise in the EDM codec orig-
inates in the PCM codec (ignoring transmission errors). From the above
equations,

$$< \epsilon^2(k) >= \frac{q_o^2 2^{-2P}}{3}. \qquad (3.8)$$

The overload level q_o is expressed in terms of a loading factor λ such that

$$q_o = \lambda(< q_{fd}^2(k) >)^{\frac{1}{2}} \qquad (3.9)$$

giving

$$< \epsilon^2(k) >= \left(\frac{\lambda^2 2^{-2P}}{3}\right)\left(\frac{a\delta^2}{3bR_{min}}\right) \qquad (3.10)$$

If we follow the path of $x(n)$ through the DM chain in Figure 3.5, we
observe that it is subjected to both filtering and decimation. We therefore
define SNR in terms of the filter decimated $x(n)$, viz:-

$$SNR \triangleq \frac{< x_{fd}^2(k) >}{< \epsilon^2(k) >} \qquad (3.11)$$

and from Equations 3.10,

$$SNR = \frac{< x_{fd}^2(k) > 9bR_{min}2^{2P}}{a\delta^2\lambda^2}; R_{sup} > 0. \tag{3.12}$$

If $P = 0$, i.e., the PCM codec is not used, although decimation is still performed at the receiver, giving

$$SNR_{min} = \frac{< x_{fd}^2(k) >}{< q_{fd}^2(k) >} = \frac{< x_{fd}^2(k) > 3bR_{min}}{a\delta^2} \tag{3.13}$$

We may therefore express the SNR for the EDM in terms of the SNR_{min} due to the DM acting alone, namely,

$$SNR = SNR_{min} \left[\frac{3(2^{2P})}{\lambda^2} \right] \tag{3.14}$$

where $3(2^{2P})/\lambda^2$ is the gain in SNR resulting from the supplemental PCM codec.

3.3.2 Performance of EDM

An RC-filtered test sequence at a rate of 8,000 samples/sec was generated according to

$$x(n) = \beta x(n - 1) + (1 - \beta^2)^{1/2}u(n) \tag{3.15}$$

where $u(n)$ was a zero-mean, white Gaussian noise sequence that was bandlimited by a finite impulse response low-pass filter, and the parameter β was

$$\beta = \exp(-2\pi f_c/f_s) \tag{3.16}$$

with

$$f_c = \frac{1}{2\pi RC} \tag{3.17}$$

and a sampling frequency of f_s. We selected $f_c = 800$ Hz, $f_s = 16$ kHz, giving $R_{min} = 2$, a signal power $< x^2(n) >$ of unity, and an adjacent sample correlation of $x(n)$ as 0.9. We opted for the generation of $x(n)$ as its PSD was similar to the long term PSD of speech. The predictor coefficient of the DM integrator was 0.99, while the DM step sizes to minimise $< q^2(n) >$, and to give negligible slope overload was 0.72, and 1.2, respectively. The PSD of the DM error sequence $q(n)$ for these two values of δ are displayed in Figure 3.6.

The variation of signal-to-noise ratio (SNR), given by Equation 3.11, as a function of bit rate is displayed in Figure 3.7. The DM codec operated at 16 kb/s and the variation in the bit rate was a consequence of assigning more bits to the PCM supplemental codec. The sequence $q_{fd}(k)$ applied to the PCM encoder was unchanged by the EDM bit rate, and maintained a Gaussian-like amplitude distribution. As the bit rate increases so did the

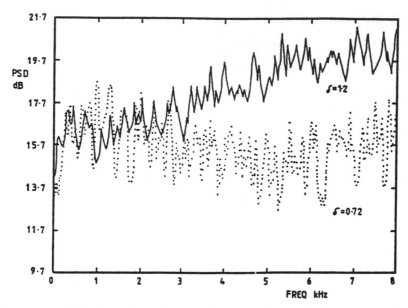

Figure 3.6: PSD functions of the DM error signal q(n) for two different operating conditions

number of PCM quantisation levels, and to ensure the PCM quantisation noise was minimised the overload points were extended, in spite of the PCM step-size $\triangle$ decreasing. This meant that for optimum EDM performance the loading factor λ increased with the EDM bit rate. The curve related to the triangles in Figure 3.7 shows the performance when λ was optimised for each value of P. By operating with a constant λ that was optimised for 40 kb/s transmissions, i.e., R_{sup} was 3 bits/Nyquist sample, we obtained the curve associated with the small squares. The circles in Figure 3.7 give the performance when a logarithmic quantiser having $u = 5, \lambda = 2.5$, was used.

Figure 3.8 compares the performance of the EDM codec relative to 4σ linear PCM, DM and embedded DPCM. As expected, embedded DPCM had the superior performance, but its SNR gain compared to EDM was not large (particularly at low bit rates). Where EDM scored over embedded DPCM was in its low complexity and power consumption.

3.4   Sub-band Coding

In sub-band coding (SBC) [3] the speech signal is filtered into a number of sub-bands, and each sub-band signal is adaptively encoded. The number of bits used in the encoding process differs for each sub-band signal, with bits assigned to quantisers according to perceptual criteria. By encoding

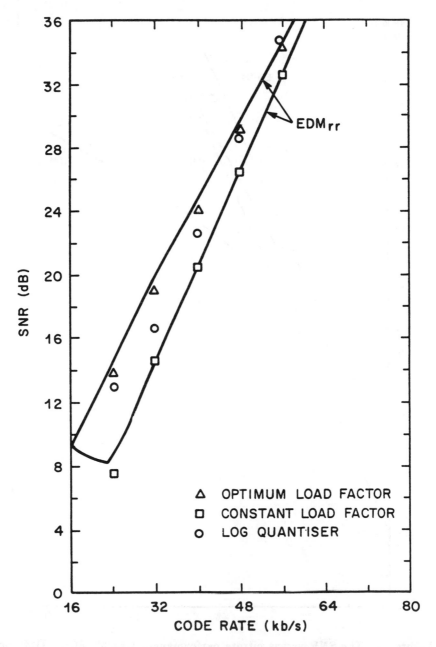

Figure 3.7: SNR variation with bit rate for EDM

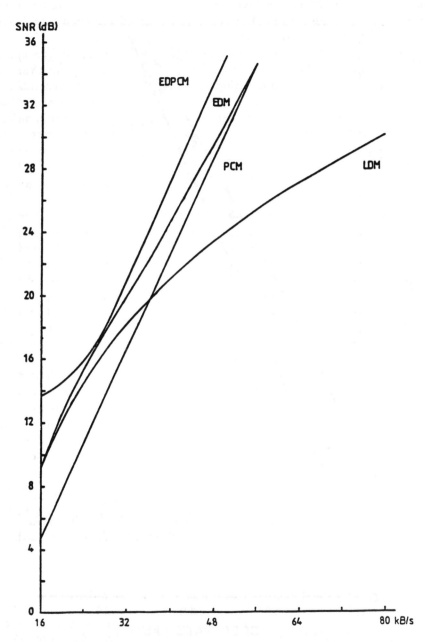

Figure 3.8: The SNR versus bitrate performance of EDM, PCM, DM and DPCM

each sub-band signal individually, the quantization noise is confined within its sub-band. The output bit streams from each encoder are multiplexed and transmitted. At the receiver demultiplexing is performed followed by decoding of each sub-band data signal. The sampled sub-band signals are then combined to yield the recovered speech signal.

At the outset we will assure ourselves that if speech is passed through a set of filters whose bandwidths do not overlap, yet are contiguous over the speech bandwidth, that the resulting sub-band signals may be recombined to recover the speech without perceptual distortion. By demonstrating that sub-band filtering can be performed, and the resulting signals combined to yield a very close replica to the original speech, then perceptual distortion in the recovered SBC speech derives only from the encoding process and channel impairments.

Suppose the speech signal is bandlimited to B Hz and sampled at the Nyquist rate f_s. The samples emanating from the sub-band filters are also at this rate, although the bandwidth of the filters is significantly below the bandwidth of the original speech. If over-sampling of the sub-band signals is to be eliminated, down-sampling must occur according to the ratio of the original bandwidth B to the sub-band bandwidth. As we have mentioned, we need to reconstruct the original speech signal from the sub-band signals with imperceptible distortion. The production of sub-bands using conventional digital filter banks cannot meet this requirement because of the finite width of the bandpass filter transition bands. If the bandpass filters overlap in the frequency domain, sub-sampling causes aliasing which, in a speech system, destroys the harmonic structure of voiced sounds and results in unpleasant perceptual effects. If the bandpass filters do not overlap the speech signal cannot be perfectly reconstructed because the gaps between the channels introduce audible echos. The requirements for alias-free and gap-free reconstruction are particularly stringent in the sub-band coding of speech because even small amounts of aliasing noise or channel gap distortion may cause significant degradation in speech quality. In addition, the order of the filter (N) has to be determined precisely to meet the specific frequency response criterion. An effective method of accommodating these stringent filter requirements is to use quadrature mirror filter (QMF) banks, which enable the sub-band processes and its inverse to provide virtually an overall flat frequency response with near zero aliasing, regardless of the value of N.

3.4.1 Theory of Quadrature Mirror Filtering

The Quadrature Mirror Filter (QMF) was proposed by Esteban and Galand [24]. Johnston [27] designed the filter coefficients for a family of QMF filter banks that were appropriate for a wide range of applications. The QMF analysis/synthesis technique is best described by considering the two-channel system shown in Figure 3.9. The input speech signal is band-

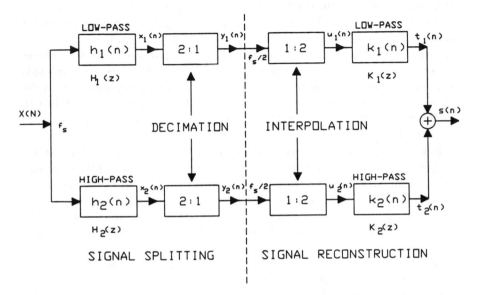

(a) Block Diagram of Two Sub-band qmf Filter Banks

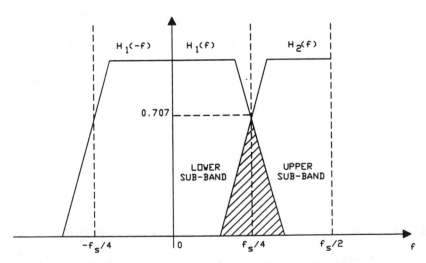

(b) Frequency Response of $H_1(f)$ and $H_2(f)$

Figure 3.9: QMF analysis/synthesis scheme

limited to $f_s/2$, and sampled at $f_s = 1/T = \omega_s/2\pi$, to yield $x(n)$. Filtering by the low-pass filter $H_1(z)$ and the high-pass filter $H_2(z)$ ensues to give the low frequency half-band signal $x_1(n)$ and the high frequency half-band signal $x_2(n)$, respectively. Because the spectrum of $x_1(n)$ and $x_2(n)$ occupy only half the bandwidth of the original signal $x(n)$, the sampling rate in each band can be halved by ignoring every second sample to give the signals $y_1(n)$ and $y_2(n)$.

In the reconstruction process shown in Figure 3.9, the signals $y_1(n)$ and $y_2(n)$ are interpolated by inserting one zero-valued sample between each sample, to give $u_1(n)$ and $u_2(n)$. These sequences are filtered by $K_1(z)$ and $K_2(z)$ to give the signals $t_1(n)$ and $t_2(n)$. Signals $t_1(n)$ and $t_2(n)$ are added together to give the recovered speech $s(n)$.

Let the filter $H_1(z)$ be the sampled, half-band, low-pass filtered z--transform function with an impulse response $h_1(n)$; and $H_2(z)$ be the z-transform function of the corresponding half-band, high-pass, mirror response filter with impulse response $h_2(n)$; such that they satisfy the following magnitude relationship,

$$\left| H_1\left(e^{j\omega T}\right) \right| = \left| H_2\left(e^{j\left(\frac{\omega_s}{2}-\omega\right)T}\right) \right| \tag{3.18}$$

where ω is the angular frequency. The frequency response of $H_1(\cdot)$ and $H_2(\cdot)$ are shown in Figure 3.9b having a 3 dB point at $f_s/4$. Furthermore, if $H_1(z)$ has a symmetrical frequency response function, i.e., the negative frequency response is the mirror image of the positive frequency response as shown in Figure 3.9b, then

$$\left| H_2\left(e^{j\omega T}\right) \right| = \left| H_1\left(e^{-j\left(\frac{\omega_s}{2}-\omega\right)T}\right) \right|. \tag{3.19}$$

Now,

$$
\begin{aligned}
e^{-j\left(\frac{\omega_s}{2}-\omega\right)T} &= e^{-j(\pi-\omega T)} \\
&= \cos(\pi-\omega T) - j\sin(\pi-\omega T) \\
&= -\cos(\omega T) - j\sin(\omega T) \\
&= -e^{j\omega T} \tag{3.20}
\end{aligned}
$$

and therefore Equation 3.19 becomes

$$\left| H_2\left(e^{j\omega T}\right) \right| = \left| H_1\left(-e^{j\omega T}\right) \right| \tag{3.21}$$

The filters $K_1(z)$ and $K_2(z)$ are the half-band, low-pass and high-pass z-domain transfer functions with impulse responses of $k_1(n)$ and $k_2(n)$, corresponding to $H_1(z)$ and $H_2(z)$. Hence they also obey the relationship specified by Equation 3.21.

The z-transform of $x_1(n)$ is

$$X_1(z) = H_1(z)X(z) \tag{3.22}$$

which may be expressed as

$$X_1(z) = a_0 + a_1 z^{-1} + a_2 z^{-2} + a_3 z^{-3} + a_4 z^{-4} + \ldots \qquad (3.23)$$

where $a_i, i = 1, 2, \ldots$, are the z-transform coefficients. When $x_1(n)$ is decimated to give $y_1(n)$, its z-transform is

$$
\begin{aligned}
Y_1(z) &= a_0 + a_2 z^{-1} + a_4 z^{-2} + \ldots \\
&= \frac{1}{2}\left[a_0 + a_1 z^{-\frac{1}{2}} + a_2(z^{-\frac{1}{2}})^2 + a_3(z^{-\frac{1}{2}})^3 + a_4(z^{-\frac{1}{2}})^4 + \ldots\right] \\
&+ \frac{1}{2}\left[a_0 + a_1(-z^{-\frac{1}{2}}) + a_2(-z^{-\frac{1}{2}})^2 + a_3(-z^{-\frac{1}{2}})^3 + a_4(-z^{-\frac{1}{2}})^4 \ldots\right] \\
&= \frac{1}{2}\left[X_1(z^{\frac{1}{2}}) + X_1(-z^{\frac{1}{2}})\right].
\end{aligned}
$$

$$(3.24)$$

The Reverse Process

When $y_1(n)$ is interpolated to give $u_1(n)$. In the z-domain we have:

$$
\begin{aligned}
U_1(z) &= a_0 + 0 \cdot z^{-1} + a_2 z^{-2} + 0 \cdot z^{-3} + a_4 z^{-4} + \ldots \\
&= Y_1(z^2).
\end{aligned}
\qquad (3.25)
$$

If $T_1(z)$, $K_1(z)$ represent the z-transforms of $t_1(n)$ and $k_1(n)$ respectively, then after the final filtering,

$$T_1(z) = K_1(z)U_1(z) \qquad (3.26)$$

Combining Equations 3.22 to 3.26, yields

$$
\begin{aligned}
T_1(z) &= K_1(z)U_1(z) \\
&= K_1(z)Y_1(z^2) \\
&= K_1(z)\frac{1}{2}[X_1(z) + X_1(-z)] \\
&= \frac{1}{2}K_1(z)[H_1(z)X(z) + H_1(-z)X(-z)]. \qquad (3.27)
\end{aligned}
$$

Similarly the high-pass filtered signal $t_2(n)$ has a z-transform

$$T_2(z) = \frac{1}{2}K_2(z)[H_2(z)X(z) + H_2(-z)X(-z)]. \qquad (3.28)$$

We are now able to formulate the z-transform of the recovered signal $s(n)$ as

$$
\begin{aligned}
S(z) &= T_1(z) + T_2(z) \\
&= \frac{1}{2}K_1(z)[H_1(z)X(z) + H_1(-z)X(-z)] \\
&\quad + \frac{1}{2}K_2(z)[H_2(z)X(z) + H_2(-z)X(-z)]
\end{aligned}
$$

and on rearranging

$$S(z) = \frac{1}{2}[H_1(z)K_1(z) + H_2(z)K_2(z)]X(z)$$
$$+ \frac{1}{2}[H_1(-z)K_1(z) + H_2(-z)K_2(z)]X(-z)$$

(3.29)

The second term of this summation represents the aliasing effect due to decimation and can be eliminated if we set

$$K_1(z) = H_1(z)$$

(3.30)

$$K_2(z) = -H_1(-z)$$

(3.31)

and exploit from Equation 3.21 that

$$H_2(z) = H_1(-z)$$

(3.32)

then Equation 3.29 becomes

$$S(z) = \frac{1}{2}[H_1(z)H_1(z) - H_1(-z)H_1(-z)]X(z)$$
$$+ \frac{1}{2}[H_1(-z)H_1(z) - H_1(z)H_1(-z)]X(-z)$$

whence

$$S(z) = \frac{1}{2}[H_1^2(z) - H_1^2(-z)]X(z).$$

(3.33)

Evaluating $S(z)$ on the unit circle of the z-plane gives

$$S\left(e^{j\omega T}\right) = \frac{1}{2}[H_1^2\left(e^{j\omega T}\right) - H_1^2\left(-e^{j\omega T}\right)]X\left(e^{j\omega T}\right)$$

and on noting that

$$-e^{-j\omega T} = e^{j\left(\frac{\omega_t}{2}+\omega\right)T}$$

(3.34)

we have

$$S\left(e^{j\omega T}\right) = \frac{1}{2}\left[H_1^2\left(e^{j\omega T}\right) - H_1^2\left(e^{j\left(\frac{\omega_t}{2}+\omega\right)T}\right)\right]X\left(e^{j\omega T}\right).$$

(3.35)

In order to ensure that the signal $s(n)$ is not subjected to aliasing we need to identify the restrictions on $H_1(z)$. Suppose we select the filter to be a symmetrical FIR filter then $H_1(z)$ may be expressed as

$$H_1(z) = \sum_{n=0}^{N-1} h_1(n)z^{-n}$$

(3.36)

where N is the order of the filter. Because $H_2(z)$ is the mirror response of $H_1(z)$, its impulse response is formed by inverting every second sample of $h_1(n)$. From Equation 3.32 we have

$$
\begin{aligned}
H_2(z) &= H_1(-z) \\
&= \sum_{n=0}^{N-1} h_1(n)(-z)^{-n} \\
&= \sum_{n=0}^{N-1} h_1(n)(-1)^{-n} z^{-n} \\
&= \sum_{n=0}^{N-1} h_1(n)(-1)^{n} z^{-n}. \quad (3.37)
\end{aligned}
$$

Returning to $H_1(z)$, we may express its frequency response by a magnitude response $H_1(\omega)$ and a linear phase function, viz:-

$$
H_1\left(e^{j\omega T}\right) = H_1(\omega)e^{-j(N-1)\pi(\omega/\omega_s)}. \quad (3.38)
$$

Substituting this into Equation 3.35 gives

$$
\begin{aligned}
S\left(e^{j\omega T}\right) &= \frac{1}{2}\left[H_1^2(\omega)e^{-j2(N-1)\pi(\omega/\omega_s)} \right. \\
&\left. \qquad -H_1^2\left(\omega+\frac{\omega_s}{2}\right)e^{-j2(N-1)\pi\left(\frac{\omega}{\omega_s}+\frac{1}{2}\right)} \right] X\left(e^{j\omega T}\right) \\
S\left(e^{j\omega T}\right) &= \frac{1}{2}\left[H_1^2(\omega) - H_1^2\left(\omega+\frac{\omega_s}{2}\right)e^{-j(N-1)\pi} \right] \\
&\qquad e^{-j(N-1)2\pi(\omega/\omega_s)} X\left(e^{j\omega T}\right). \quad (3.39)
\end{aligned}
$$

We now have to consider whether N is even or odd.

1. **N is even**

   By noting that

   $$
   e^{-j(N-1)\pi} = -1 \quad (3.40)
   $$

   we may write Equation 3.39 as

   $$
   S\left(e^{j\omega T}\right) = \frac{1}{2}\left[H_1^2(\omega) + H_1^2(\omega+\frac{\omega_s}{2}) \right] e^{-j(N-1)\omega T} X\left(e^{j\omega T}\right). \quad (3.41)
   $$

   For a perfect all-pass system

   $$
   H_1^2(\omega) + H_1^2(\omega+\frac{\omega_s}{2}) = 1 \quad (3.42)
   $$

   and hence we get,

   $$
   S\left(e^{j\omega T}\right) = \frac{1}{2}e^{-j(N-1)\omega T} X\left(e^{j\omega T}\right) \quad (3.43)
   $$

or in the time domain,

$$s(n) = \frac{1}{2}x(n - N + 1) \qquad (3.44)$$

The output signal is therefore the input speech, subject to a delay of $(N - 1)$ samples and an attenuation of $1/2$. All aliasing components are absent.

2. **N is odd**

For odd values of N the exponential term now becomes,

$$e^{-j(N-1)\pi} = 1 \qquad (3.45)$$

with the result that

$$S\left(e^{j\omega T}\right) = \frac{1}{2}\left[H_1^2(\omega) - H_1^2(\omega + \frac{\omega_s}{2})\right]e^{-j(N-1)\omega T}X\left(e^{j\omega T}\right). \qquad (3.46)$$

However, the filtered $H_1(\omega)$ is symmetrical,

$$H_1(\omega) = H_1(-\omega)$$

and at $\omega = -\omega_s/4$, the term $H_1^2(\omega) - H_1^2(\omega + \frac{\omega_s}{2})$ is zero, and so is $S\left(e^{j\omega T}\right)$. As a consequence, when N is odd, at this frequency there is no output signal. It is therefore essential to employ filters with even values of N.

Summarizing

For perfect signal reconstruction, the following conditions must be satisfied.

1. $H_1(z)$ is a symmetrical FIR filter of even order

2. $H_2(z) = H_1(-z)$

3. $K_1(z) = H_1(z)$

4. $K_2(z) = -H_2(z)$

5. $H_1^2(\omega) + H_1^2(\omega + \frac{\omega_s}{2}) = 1$

We have described a QMF known as a real QMF filter as it deals with real time signals. Although we will not pursue the matter further, a complex quadrature mirror filter (CQMF) having the ability to half the computational complexity has been introduced by Nussbaumer [28] and Galand [29].

In general, the QMF filter banks can be arranged in the form of a tree structure to achieve finer band-splitting. The circuit can be arranged in such a way that either uniform or non-uniform spacing of the split bands can be done, as illustrated in Figures 3.10a and 3.10b. In Figure 3.10a the speech input $x(n)$ is divided into four sub-bands of equal bandwidth, whereas in Figure 3.10b the two lower sub-bands are each a quarter of the speech bandwidth, while the highest sub-band is a half.

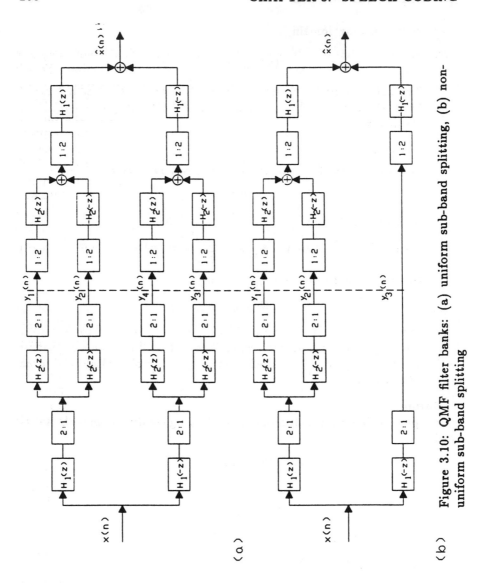

Figure 3.10: QMF filter banks: (a) uniform sub-band splitting, (b) non-uniform sub-band splitting

3.4.2 The SBC Algorithm

Numerous varieties of SBC codecs have been realised, although they are variations of the basic scheme briefly outlined in our introduction. We will consider one variety, namely the one due to Hanes [35], for exemplary purposes. The block diagram of the codec is given in Figure 3.11. The codec is seen to possess a symmetrical tree-structured QMF filter bank that divides the input signal bandwidth into equal bandwidths. The input speech signal $x(n)$ is processed by the first QMF filter stage, QMF_1, to yield a low-frequency component $u_1(n)$ and high-frequency component $u_2(n)$, both with a sampling frequency of $f_s/2$. We are assuming the decimation process is housed within QMF_1. Each signal component is further divided into low- and high-frequency components by the QMF_2 stages. The resulting signals are $v_1(n), v_2(n), v_3(n)$ and $v_4(n)$, and the sampling frequency is now at $f_s/4$. Finally, each of these signals have their bandwidths divided by a third stage of QMF filters to yield signals $w_1(n), w_2(n), \cdots, w_8(n)$, each having a sampling frequency of $f_s/8$.

The signals $w_k(n), k = 1, 2, \cdots, 8$, are individually SBC coded by Jayant's one word memory quantiser [30]. The codewords from each sub-band are multiplexed together and transmitted over the 16 kbit/s mobile radio channel. As most of the energy in telephonic speech is in the frequency range from 200 Hz to 3.2 kHz, we can inhibit the transmissions from the 7th and 8th sub-bands (covering the frequency range from 3.0 kHz to 4.0 kHz) without causing significant perceptual degradation. This also enables us to contain the multiplexed data from the sub-band encoders to 15 kb/s. We will see shortly that side-information is also transmitted to give a total bit rate of 16 kb/s.

At the receiver, the demultiplexed signals are decoded and applied to the third stage of the inverse QMF filter bank. The 7th sub-band signal (frequency range from 3.0 kHz to 3.5 kHz) is regenerated by replicating the 6th sub-band signal in this 7th sub-band. However, this procedure is only invoked when unvoiced speech is present, and is known as high band regeneration. The signals from the QMF_3^{-1} are added in pairs as shown in Figure 3.11, and applied to the QMF_2^{-1} stages. The outputs from the two QMF_2^{-1} stages are summed, filtered by QMF_1^{-1} to yield the recovered speech signal $\hat{x}(n)$. Notice that the interpolation at each filter stage in the decoder is assumed to reside within the QMF^{-1} stages, and that the gain of two at each stage is to compensate for the half factor in Equation 3.44.

In the lower part of the encoder block diagram is shown the sub-system that provides the semi-adaptive bit allocation algorithm. It accepts the signals from the QMF_1 and QMF_2 stages and determines from them the bit-assignment for the sub-band encoders and decoders. The bit-assignment selected is sent to the receiver as side-information. Being of critical importance, this information is protected against channel errors prior to its transmission. It is the bit-assignment that controls the bit rate of the codec for each sub-band, and we will discuss how the bit assignment is made in

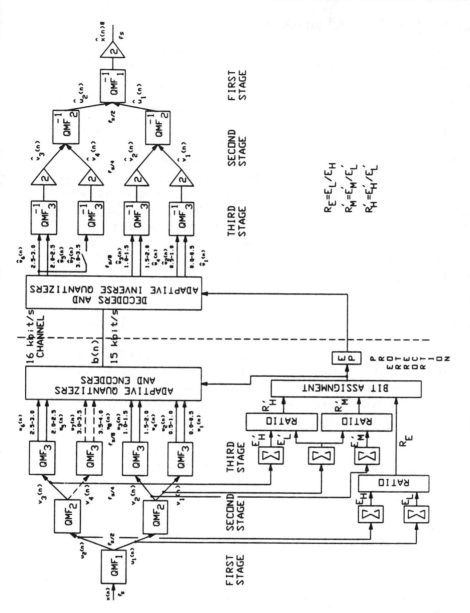

Figure 3.11: Sub-band codec block diagram

Section 3.4.2.3.

3.4.2.1 QMF Filter Bank

The theory of the QMF was given in Section 3.4.1. We now consider its deployment in the SBC codec. Although we have seen that the QMFs can be selected to avoid aliasing effects, there are practical limitations in achieving this feature. In practice compromises must be made. For example, a longer QMF FIR filter is necessary to give a sharper transition in order to reduce the interband aliasing effects. However, a longer QMF filter causes a longer coding delay. To achieve better performance more subbands are needed and this requires longer QMF filters to provide sufficiently sharp transitions. Otherwise the performance improvement due to more subbands will be offset by increased aliasing effects.

In the configuration of the QMF filter band shown in Figure 3.11, with its primary sampling frequency of f_s (=8 kHz), the resultant subband bandwidths are 500 Hz. As a consequence we have to choose the transition band of the filters at every stage to be less than 500 Hz, if across-the-band aliasing distortion is to be avoided. Table 3.1 summarizes the requirement for the normalized transition band at every stage, and selects Johnston's filters [27] having a transition band that is closest to the calculated one.

QMF stage	Sampling frequency	Normalized transition band	Chosen transition band	Transition code letter
1	8 kHz	500/8000=0.0625	0.043	D
2	4 kHz	500/4000=0.1250	0.100	B
3	2 kHz	500/2000=0.2500	0.140	A

Table 3.1: Johnston's QMF transition-band specifications

Next we choose a compromise between the number of taps and the stopband rejection attenuation that achieves a minimum delay. Quoting from Johnston's filters [27], we present the relevant parameters in Table 3.2.

Stage	Transition code letter	Pass-band ripple (dB)	Stop-band rejection (dB)	Chosen taps
1	D	0.025	38	$N_1 = 32$
2	B	0.020	44	$N_2 = 16$
3	A	0.025	48	$N_3 = 12$

Table 3.2: Johnston's QMF stop-band and pass-band specifications

Next, the low-pass filter coefficients are determined as displayed in Table 3.3. From the theory of QMF filters given in Section 3.4.1, the delay in the recovered signal is $(N - 1)$ samples with respect to the pre-decimation sampling rate. Thus, the first, second and third stage delays are $(N_1 - 1)$, $(N_2 - 1)$, and $(N_3 - 1)$ samples with respect to their sampling frequencies,

Filter, h(n)	First stage $N_1 = 32$	Second stage $N_2 = 16$	Third stage $N_3 = 12$
h(1)	2.2451390E-3	2.8981630E-3	-3.8096990E-3
h(2)	-3.9711520E-3	-9.9722520E-3	1.8856590E-2
h(3)	-1.9696720E-3	-1.9209360E-3	-2.7103260E-3
h(4)	8.1819410E-3	3.5968530E-2	-8.4695940E-2
h(5)	8.4268330E-4	-1.6118690E-2	8.8469920E-2
h(6)	-1.4228990E-2	-9.5302340E-2	4.8438940E-1
h(7)	2.0694700E-3	1.0679870E-1	h(6)
h(8)	2.2704150E-2	4.7734690E-1	h(5)
h(9)	-7.9617310E-3	h(8)	h(4)
h(10)	-3.4964400E-2	h(7)	h(3)
h(11)	1.9472180E-2	h(6)	h(2)
h(12)	5.4812130E-2	h(5)	h(1)
h(13)	-4.4524230E-2	h(4)	
h(14)	-9.9338590E-2	h(3)	
h(15)	1.3297250E-1	h(2)	
h(16)	4.6367410E-1	h(1)	
h(17)	h(16)		
h(18)	h(15)		
h(19)	h(14)		
h(20)	h(13)		
h(21)	h(12)		
h(22)	h(11)		
h(23)	h(10)		
h(24)	h(9)		
h(25)	h(8)		
h(26)	h(7)		
h(27)	h(6)		
h(28)	h(5)		
h(29)	h(4)		
h(30)	h(3)		
h(31)	h(2)		
h(32)	h(1)		

Table 3.3: The Johnston-QMF's low-pass filter coefficients

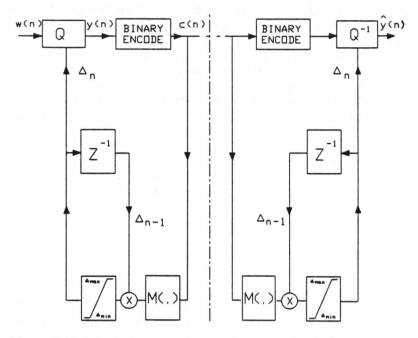

Figure 3.12: Block diagram of Jayant's one-word memory quantiser

f_s, $f_s/2$ and $f_s/4$, respectively. When all these delays are reflected back to the primary sampling frequency f_s, the total delay (D_T) is

$$D_T = (N_1 - 1) + 2(N_2 - 1) + 4(N_3 - 1) \qquad (3.47)$$

In our experiment, $N_1 = 32$, $N_2 = 16$, $N_3 = 12$, giving $D_T = 105$ samples. As the primary sampling frequency f_s is 8 kHz, the total delay is 13 ms.

3.4.2.2   Quantization

To contain the transmitted bit rate we must efficiently encode each sub-band signal. A suitable approach is to use adaptive pulse code modulation (APCM) of the type due to Jayant [30]. Each sub-band signal is encoded by Jayant's adaptive one-word memory quantiser, and the number of bits assigned to each quantiser are either 1, 2, 3, or 4, according to perceptual criteria. Mid-riser quantisers are implemented because of their efficient utilization of the codeword representation of the quantization levels of the sub-band signals. The Jayant quantiser adapts its step-size Δ_n at each sampling instant. The block schematic of the quantiser is displayed in Figure 3.12. The output $y(n)$ of a B-bit quantiser $(B > 1)$ is of the form,

$$y(n) = c(n)\frac{\Delta_n}{2} \qquad (3.48)$$

where $|c(n)| = 1, 3, \cdots, 2^B - 1$; and $\Delta_n > 0$. The step-size Δ_n is given by the previous step-size multiplied by a time-invariant function of the code-word magnitude $|c(n)|$,

$$\Delta_n = \Delta_{n-1} M(|c(n)|) \tag{3.49}$$

Because of hardware limitations, the step-size Δ_n can vary only over a limited dynamic range, from the minimum step-size Δ_{min} to the maximum step-size Δ_{max}, i.e.:

$$\Delta_n = \begin{cases} \Delta_{min} & ; & \Delta_n < \Delta_{min} \\ \Delta_{max} & ; & \Delta_n > \Delta_{max} \\ \Delta_n & ; & \text{otherwise} \end{cases} \tag{3.50}$$

The ratio of the Δ_{max} to Δ_{min}

$$R = \frac{\Delta_{max}}{\Delta_{min}} \tag{3.51}$$

determines the dynamic range of the codec, which is typically 42 dB. The multiplier function $M(\cdot)$ controls the adaption rate of the step-size. For telephonic speech the values of these multipliers for 2, 3 and 4 bit quantisers, are shown in Table 3.4.

B	*Multiplier $M(\cdot)$*
2	0.85, 1.90
3	0.85, 1.00, 1.00, 1.50
4	0.90, 0.90, 0.90, 0.90, 1.20, 1.60, 2.00, 2.40

Table 3.4: Jayant quantiser multipliers for B-bit quantisers

3.4.2.3 Semi-adaptive Bit Allocation

To obtain toll quality speech, i.e., a quality equivalent to 56 kb/s logarithmic PCM, the number of bits B assigned to the adaptive PCM quantisers must vary according to the speech signal statistics. However, a fully adaptive bit allocation strategy introduces considerable delay and complexity that renders it inappropriate for many speech applications. By employing a semi-adaptive bit allocation procedure the only delay is that due to the QMF filter banks. In order to allocate the bits to the bank of quantisers we classify the input speech signal into voiced, unvoiced, and intermediate, and also make provision for the SBC codec to handle signalling/voice-band data (VBD). The latter is designed mainly for transmitting signalling tones such as AC9/AC15 (2280 Hz) and MF5 dual tone inter-register signalling between 540 Hz and 1140 Hz, and 1380 Hz and 1980 Hz.

The classification is performed by first measuring the spectral envelopes of some sub-band signals over a 6 ms window. This period is equivalent to

48 samples for a sampling frequency of 8 kHz, and it approximately corresponds to the delay introduced by the 2nd and 3rd stage QMF filters. The sub-band signals $u_1(n)$ and $u_2(n)$ from QMF_1 in Figure 3.11 are processed to yield

$$E_L = \sum_{n=1}^{24} |u_1(n)| \tag{3.52}$$

$$E_H = \sum_{n=1}^{24} |u_2(n)| \tag{3.53}$$

and their ratio

$$R_E = \frac{E_L}{E_H} \tag{3.54}$$

is formed. Next the short-term-average magnitude estimates from the QMF_2 filters are computed, viz:-

$$E_L' = \sum_{n=1}^{12} |v_1(n)| \tag{3.55}$$

$$E_M' = \sum_{n=1}^{12} |v_2(n)| \tag{3.56}$$

$$E_H' = \sum_{n=1}^{12} |v_3(n)| \tag{3.57}$$

and the ratios

$$R_M' = \frac{E_M'}{E_L'} \tag{3.58}$$

and

$$R_H' = \frac{E_H'}{E_L'} \tag{3.59}$$

determined. The state diagram and the optimized values of R_E, R_M' and R_H' for classifying the four states are shown in Figure 3.13, while Table 3.5 shows the bit allocation for each quantiser as a function of the classification. The classification decision is also transmitted as protected side-information to enable the receiver to employ the correct APCM decoders.

Perceptually improved speech can be obtained by artificially widening the bandwidth of the recovered speech when the unvoiced bit assignment strategy is selected. At the receiver, a high frequency regeneration process couples energy from the 2.5–3.0 kHz into the 3.0–3.5 kHz band. The situation is represented by H in Table 3.5. By this technique unvoiced sounds are reproduced with improved subjective crispness.

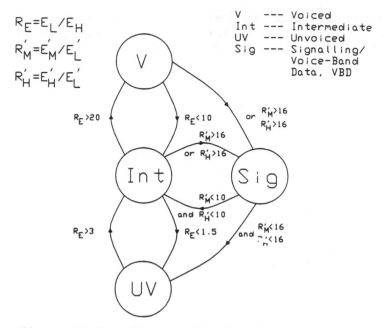

$$R_E = E_L / E_H$$
$$R'_M = E'_M / E'_L$$
$$R'_H = E'_H / E'_L$$

V   --- Voiced
Int --- Intermediate
UV   --- Unvoiced
Sig --- Signalling/
         Voice-Band
         Data, VBD

Figure 3.13: State diagram of the bit-assignment strategy

3.4.3   Simulation Results

The maximum and minimum quantization step-sizes are Δ_{max} and Δ_{min}, and their ratio determines the dynamic range of input signal levels that the adaptive quantiser can accommodate while maintaining a relatively high segmental SNR. The minimum step-size Δ_{min} for each sub-band is selected relative to the energy of its sub-band signal, and each sub-band codec is designed to operate within its dynamic range. The three-dimensional isometric plot in Figure 3.14 shows the segmental-SNR values of each sub-band signal for various input power levels, which is linearly interpolated for intermediate frequencies in case of a female speaker. When these sub-band SEG-SNR values are averaged over all sub-bands we obtain the SBC SEG-SNR as a function of input power as shown for a female speaker in Figure 3.15. The maximum SEG-SNR occurs at the relative input power level of -20 dB, irrespective of whether the speech is male or female. The results are for speech spoken by English people, and each speech signal lasts for 12.8 sec, and is band-limited from 200 Hz to 3.2 kHz.

3.4.4   Increasing the Robustness of SBC to Channel Errors

The basic SBC algorithm with its adaptive quantisation with backward estimation (AQB) quantisers (that do not require the transmission of side-

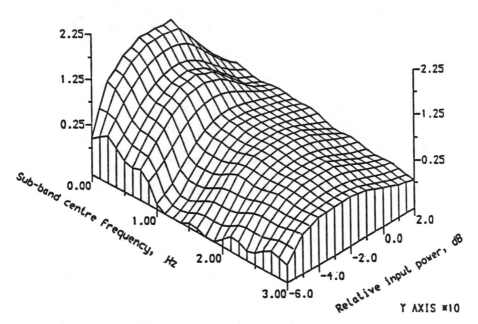

Figure 3.14: The ordinate segmental-SNR in dBs as a function of sub-band centre frequency and relative input power

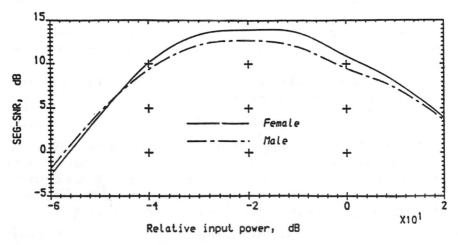

Figure 3.15: Variation of the overall SEG-SNR with input signal power for female and male speakers

BAND, kHz	0.0 ↓ 0.5	0.5 ↓ 1.0	1.0 ↓ 1.5	1.5 ↓ 2.0	2.0 ↓ 2.5	2.5 ↓ 3.0	3.0 ↓ 3.5	3.5 ↓ 4.0
BIT ASSIGNMENT	BIT ALLOCATION PATTERN							
STRATEGY								
VOICED	4	4	3	2	2	0	0	0
INTERMEDIATE	4	3	2	2	2	2	0	0
UNVOICED	2	3	3	3	2	2	H	0
SIGNALLING/VBD	0	0	4	4	4	3	0	0

H indicates high band regeneration.

Table 3.5: SBC Bit-Allocation Table

information) works very well in the absence of transmission errors. However, when the BER exceeds 10^{-3} the step-size at the decoder begins to deviate from the one at the encoder. The step-sizes will reconverge, but only when they both acquire their maximum or minimum step-sizes over many sampling intervals. This is equivalent to forcing the step-size to known values, when the encoding and decoding restarts in agreement.

It was Jayant who pointed out [31] that the adaptive quantisation with forward estimation (AQF) quantiser has a significantly better performance than the AQB type in poor channel conditions, such as those found in mobile radio communications. However it is necessary to regularly transmit the AQF step-size to the receiver. This is not attractive in SBC as there are many quantisers and the deployment of AQFs for each sub-band signal would greatly increase the bit rate. The compromise solution is to put some robustness into the algorithm, while avoiding the transmission of quantiser side-information. We can do this by either making the step-size leaky, or by periodically forcing the step-size to a known value at both the transmitter and receiver.

Let us consider the leakage approach. From Equation 3.49 we see that the step-size Δ_n is dependent on its past history,

$$\Delta_n = M(|c(n)|)M(|c(n-1)|)\cdots M(|c(n-k+1)|)\Delta_{n-k} \qquad (3.60)$$

If, for example, Δ_{n-k} is in error, then all the following step-sizes are in error, and the whole segment of decoded speech is corrupted. To make this adaptation mechanism gradually forget the error, a leakage exponent factor β is introduced, where $\beta < 1$. We compensate for the decrease in the step-size by using a gain factor G to give the quantiser step-size

$$\Delta_n = GM(|c(n)|)\Delta_{n-1}^\beta. \qquad (3.61)$$

If this adaptation mechanism is expanded into the past history of the step-sizes, we get

$$\Delta_n = GM(|c(n)|)G^\beta M^\beta(|c(n-1)|)\cdots$$

$$\cdots G^{\beta^{k-1}} M^{\beta^{k-1}} (|c(n-k+1)|) \Delta^{\beta^k}_{n-k}. \qquad (3.62)$$

As β is less, but close to unity, β^{k-1} is very small, and the history of the multipliers has virtually no effect on the current adaptation of the step-size Δ_n. Thus whenever there are errors in the received codeword the step-size begins to converge to its correct value within a short period of time. The corresponding values of the modified gain factor G, with the leakage factor β, has been optimized by Crochiere [32]. They are listed in Table 3.6.

Leakage factor, β	Modified gain factor, G
63/64	1.08
31/32	1.18
15/16	1.50

Table 3.6: SBC leakage- and associated gain-factors

Now we consider the second method of enhancing the robustness of the quantiser. Here the step-size is forced to a known value at both the encoder and decoder at the same clock instant. This method is more effective at terminating the propagation of incorrect step-sizes compared to making the step-size leaky. Clearly we need to know when to force the step-sizes, and what values the forced step-sizes should be. We recall that the SBC codec has its bit-allocation updated every 6 ms, and we consider this to be the period of one SBC frame. Utilising this frame structure we arrange for the forcing of the step-size to be done at multiples of 6 ms, and if N_f is the number of frames between updates, we have an up-date period of

$$T_u = 6N_f \quad ms. \qquad (3.63)$$

At the beginning of each up-date period, the step-size is forced to a known value Δ_F that resides within the range of Δ_{min} to Δ_{max}. We define the step-size forcing factor F_z in terms of the expression

$$\Delta_F \stackrel{\Delta}{=} F_z (\Delta_{max} - \Delta_{min}) + \Delta_{min}. \qquad (3.64)$$

The optimized values of N_f and F_z are found by simulation using male and female speech signals. In Figure 3.16 we plot the SEG-SNR of each sub-band signal as a function of the forcing factor, F_z, for different T_u and N_f values when female speech was used. The plots for male speech are similar. The sub-band frequency range in each sub-figure is indicated in the bottom left corner. Only the lower six sub-bands are considered as the other two high frequency sub-bands (3.0–4.0 kHz) are not transmitted. The top horizontal line in each sub-figure indicates the performance when no step-size forcing is imposed.

For the AQF quantisers used (see Table 3.4) the coefficients are not optimum. Nevertheless, we see from Figure 3.16 that the general trend is that the less often the forcing is applied, the higher is the SEG-SNR in

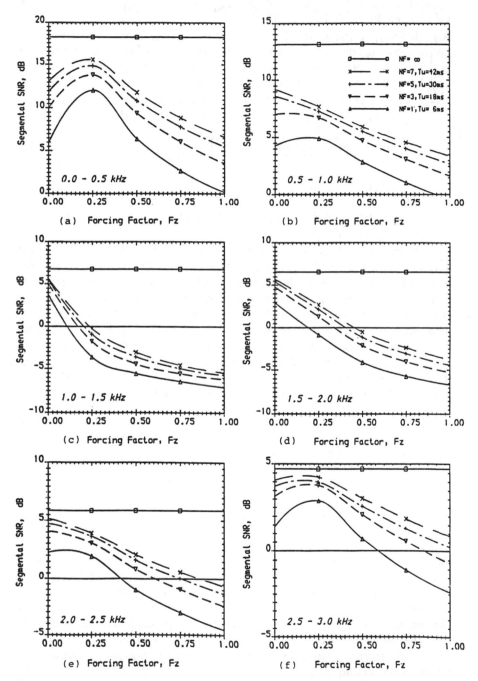

Figure 3.16: Sub-band SEG-SNR against step-size forcing factor F_z for female speech sequence

each sub-band. This is because the quantiser often experiences over-load shortly after step-size forcing, resulting in additional quantization over-loaded noise. The less frequently the step-size forcing occurs, the smaller is the over-loaded noise and the higher the SEG-SNR. The maximum gain in SEG-SNR occurs in sub-bands 0.0–0.5 kHz and 2.5–3.0 kHz when their step-sizes are forced to one-quarter of the range, ie, $F_z = 0.25$, while the other sub-bands are forced to minimum step-size where F_z is zero. The effect of forcing time T_u in terms of the number of frames N_f on the SEG-SNR of each sub-band signal for different values of F_z are displayed in Figure 3.17 for female speech. Similar curves were obtained for male speech. The sub-band frequency range is indicated at the bottom right hand corner of each box. The solid top curve is the upper bound that each sub-band can attain in the absence of forcing. The curves exhibit the same trend as before, namely, the less often the forcing is done, the better is the SEG-SNR value. The rate of improvement in SEG-SNR is much steeper with N_f for F_f in the range 1 to 3 than for N_f in the range 3 to 7. There is not much improvement after N_f equals five. The segmental SNR values due to forcing are all below the unforced case, a sacrifice that will be seen to be justified when transmissions are over radio channels.

Having selected the optimized forcing factor F_z for the different sub-bands we obtain the simulation results shown in Figures 3.18a and 3.18b, the former representing the overall SBC codec performance, while the latter the individual subband performances. The corresponding results for the step-size leakage algorithm are displayed in Figures 3.18c and 3.18d for the leakage factor β over the range from 15/16 to 63/64. In both sub-figures (a) and (d), the top horizontal curves are the SEG-SNR values without the robust algorithm. The step-size forcing algorithm degrades the speech quality by 2 dB for male and 2.2 dB for female speech sequences, respectively, while the corresponding degradation with the step-size leakage algorithm is 0.30 dB and 0.33 dB. However, informal listening tests do not reveal perceptual difference between the two robust algorithms. The step-size leakage method is more complicated to implement because of the calculation of the previous step-size to the power of β. No calculation is required in the step-size forcing approach, which proved also more robust against channel errors in the mobile radio environment.

3.4.5 Codec Performance in Acoustic Environments

Our discussion has assumed that the input speech is uncontaminated by background noise. This is not usually the case in mobile radio environments. Usually the speech is accompanied by background conversations, traffic noise, engine noise and so forth. These background noises combine with the speech and if sufficiently large may affect the coding performance. To simulate a noisy acoustic background, we used a speech correlated noise source called Modulated Noise Reference Unit (MNRU). This was proposed

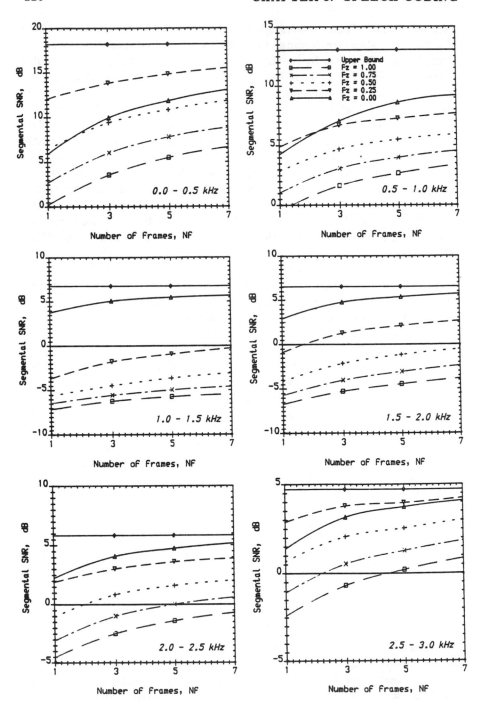

Figure 3.17: Sub-band SEG-SNR against step-size forcing frequency N_f for female speech sequence

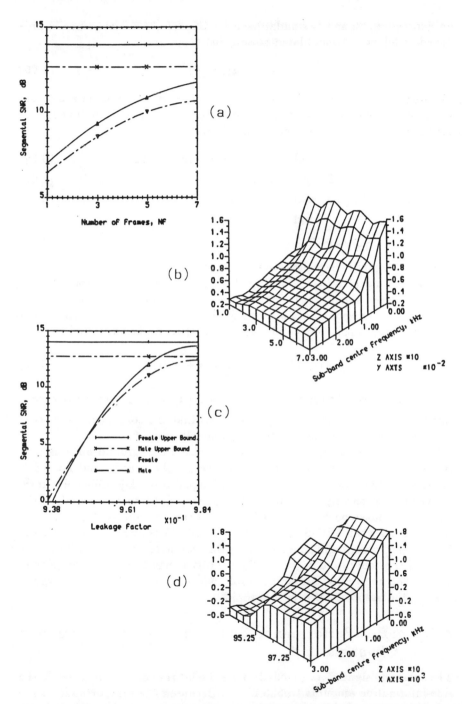

Figure 3.18: SEG-SNR of the SBC codec and that of the individual sub-bands using step-size forcing as well as step-size leakage

by Schroeder [33] and is standardized by CCITT [34]. The source is produced as follows. A modulated noise $N(n)$

$$N(n) = x(n)\epsilon(n) \qquad (3.65)$$

is formed, where $x(n)$ is the original clean speech, and $\epsilon(n)$ is a discrete stochastic process with a Kronecker delta autocorrelation function δ_{n_1,n_2}, which is uncorrelated with the speech, where:

$$
\begin{align}
\epsilon(n) &= \pm 1; \text{with equal probability} & (3.66) \\
E[\epsilon(n)] &= 0; & (3.67) \\
E[\epsilon(n_1)\epsilon(n_2)] &= \delta_{n_1,n_2}; & (3.68) \\
&\text{and} \\
E[\epsilon(n_1)x(n_2)] &= 0 & (3.69)
\end{align}
$$

The $N(n)$ signal is bandpass filtered over the band of 200 Hz to 3.2 kHz, to give the signal $\tilde{N}(n)$. The signal $s(n)$ applied to the encoder is the original speech signal $x(n)$ combined with a scaled version of the noise $\tilde{N}(n)$. The scaling provides a range of speaking environments from quiet to noisy background. The combined signal, $s(n)$ can be expressed as

$$s(n) = \left(1 + \alpha^2\right)^{-\frac{1}{2}} \left[x(n) + \alpha\tilde{N}(n)\right]. \qquad (3.70)$$

The factor α gives the ratio of the power of $\tilde{N}(n)$ to the power in the speech signal, $x(n)$. The term $\left(1 + \alpha^2\right)^{-\frac{1}{2}}$ ensures that the combined signal $s(n)$ has the same power as that of the clean speech $x(n)$.

The SBC codec's performance under acoustic background noise is portrayed in Figure 3.19, where the ordinates are the decoder output SEG-SNR, while the abscissa is the segmental signal-to-background noise ratio (SEG-SBNR), namely the ratio of the power of the speech signal to the background noise power. The 3 dB point of the non-robust step-size algorithm occurs when SEG-SBNR is about 15 dB. When SEG-SBNR exceeds 15 dB, the background noise is negligible compared to the quantization noise of the codec, and so the coding ability is unaffected by background noise. The noise plays a dominant role when SEG-SBNR becomes small, making the output SEG-SNR fall rapidly.

3.4.6   SBC Performance over Rayleigh and Gaussian Channels

The sub-band signals are encoded to yield a bit rate of 15 kbit/s, while the side-information amounts to 333 bit/s. Because of the importance of the side-information, it is protected by repeating it three times and majority logic is used at the receiver. The side-information is therefore transmitted at 1 kbit/s, and consequently the codec bit rate is 16 kbit/s. Suppose

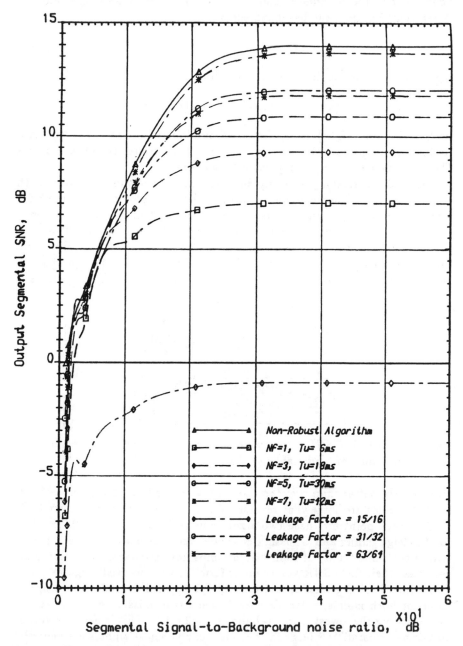

Figure 3.19: Codec performance under noisy acoustic background conditions in terms of output SEG-SNR vs. SEG-SBNR with female speech

the bit stream $b(n)$ modulates a carrier using binary phase shift keying (BPSK), and the modulated signal is conveyed over a Rayleigh channel. After demodulation, the baseband signal is

$$a(n) = R(n)b(n) + I(n) \qquad (3.71)$$

where $R(n)$ is the amplitude of the envelope having a Rayleigh PDF,

$$I(n) = C_x(n)\cos\theta(n) - C_y(n)\sin\theta(n) \qquad (3.72)$$

with $C_x(n)$ and $C_y(n)$ being Gaussian random variables, and $\theta(n)$ is a uniform random phase variable on $(-\pi, \pi)$. If the transmissions are over a Gaussian channel, $R(n)$ in Equation 3.72 is unity.

The SBC algorithms previously described have been simulated over both Rayleigh and Gaussian channels, and the results are plotted in Figure 3.20. The two most robust algorithms are the step-size leakage algorithm with $\beta = 63/64$, and the step-size forcing algorithm with $T_u = 42$ *ms*. Their improvement in segmental SNR compared to the basic non-robust algorithm for the two channels are itemised in Table 3.7 for BERs of 10^{-2} and 10^{-3}. In the high BER region, the step-size forcing algorithm with $T_u = 42$ ms is

	Leakage alg	*Forced alg*
BER	$\beta = 63/64$	$T_u = 42$ ms
	RAYLEIGH CHANNEL	
10^{-3}	7 dB	6 dB
10^{-2}	8.8 dB	11 dB
	GAUSSIAN CHANNEL	
10^{-3}	7.5 dB	6.5 dB
10^{-2}	5.8 dB	8.6 dB

Table 3.7: SBC SEGSNR improvements using stepsize leakage and stepsize forcing via Gaussian and Rayleigh channels

significantly better than the step-size leakage algorithm having $\beta = 63/64$, while the roles are somewhat reversed for low BERs. The cross-over BER performance point for these two algorithms is 3.2×10^{-3} and 2.7×10^{-3} for Rayleigh fading and Gaussian channels, respectively. The other curves within the same family of robust algorithms have similar performance, except that their SEG-SNR values are inferior in the low BER regions.

Informal listening tests reveal that the perceptual degradation is significant for both male and female speech when transmission errors occur to bits encoding the lowest four sub-bands (0.0–2 kHz). This is because voiced speech has most of its energy in these sub-bands whose signals are encoded by higher bit rate quantisers which tend to have an inferior performance in high BER environments. Thus if channel coding is introduced, these encoded sub-band signals require more protection against bit corruption than the encoded bits from bands in the 2 to 3 kHz range.

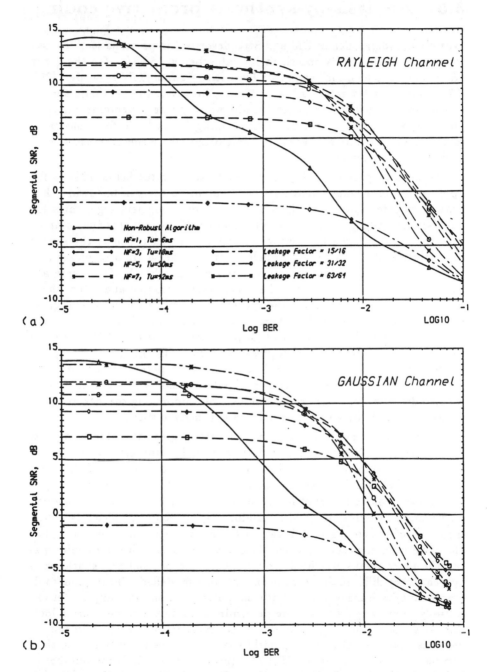

Figure 3.20: SBC performance using various algorithms over Rayleigh and Gaussian channels with female speech sequence

3.5   Analysis-by-synthesis predictive coding

Waveform coders, such as DM and SBC described in previous sections, fail to produce high quality speech at bit rates below 16 kb/s. On the other hand, source coders, such as *Linear Predictive Coding (LPC)* vocoders [36], [37], operate at bit rates as low as 2 kb/s, however, the synthetic speech quality of vocoded speech is not broadly appropriate for commercial telephone applications. Linear predictive coding in its basic form has been mainly used in secure military communications where speech must be carried at very low bit rates.

The need to produce toll-quality speech at bit rates below 10 kb/s for applications on channels that have inherent limitations in bandwidth has drawn the interest of researchers to look at more efficient algorithms for LPC speech coding. The main limitation of LPC vocoding is the assumption that speech signals are either voiced or unvoiced, hence, the source of excitation of the synthesis all-pole filter is either a train of pulses (for voiced speech) or random noise (for unvoiced speech). In fact, there are more than two modes in which the vocal tract is excited and often, these modes are mixed. Even when the speech waveform is voiced, it is a gross simplification to assume that there is only one point of excitation in the entire pitch period. In 1982, Atal [2] proposed a new model for the excitation which is known as *multi-pulse excitation*. In this model, no prior knowledge of a voiced/unvoiced decision or pitch period is needed. The excitation is modelled by a number of pulses (usually 4 per 5 ms) whose amplitudes and positions are determined by minimizing the perceptually weighted error between the original and synthesized speech.

The introduction of this model has generated a great deal of interest, and it was the first of a new generation of analysis-by-synthesis speech coders capable of producing high quality speech at bit rates around 10 kb/s and down to 4.8 kb/s. This new generation of coders use the same all-pole synthesis filter (source model of speech production) as used by LPC vocoders. However, the excitation signal is carefully optimized and efficiently coded using waveform coding techniques. All analysis-by-synthesis coders share the same basic structure in which the excitation is determined by minimizing the perceptually weighted error between the original and synthesized speech. They differ in the way the excitation is modelled. The original multipulse approach assumes that both the pulse positions and amplitudes are initially unknown, then they are determined inside the minimization loop one pulse at a time. The *Regular Pulse Excitation* (RPE) approach [1] assumes that the pulses are regularly spaced and the amplitudes are then computed by solving a set of $M \times M$ equations where M is the number of pulses. In the *Code-Excited Linear Prediction* (CELP) [4], the excitation signal is an entry of a very large stochastically populated codebook. The complexity of these coders increases as the bit rate is reduced. For example, CELP can produce good quality speech at bit rates as low as 4.8 kb/s at

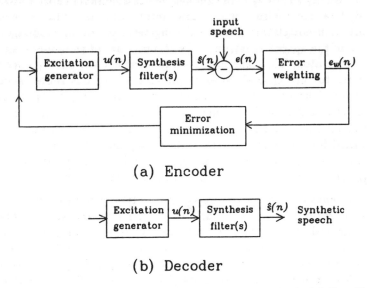

Figure 3.21: General model for analysis-by-synthesis LPC coding

the expense of high computational demands due to the exhaustive search of the large excitation codebook (usually 1024 entries) for determining the optimum innovation sequence.

In the next section, a general model for high quality analysis-by-synthesis speech coding is described. The different parts of the model will be discussed in details. These parts include: The LPC synthesis filter (for modelling the short-term spectral envelope of speech), the pitch predictor (for modelling the spectral fine structure), the error weighting filter and the error minimization procedure. The definition of the excitation sequence plays a dominant role in determining the coder performance and complexity. Different excitation models will be described in details including the multi-pulse, regular-pulse, CELP, self-excited, and binary-pulse excitation. The performance and complexity of these excitation models will be addressed.

3.5.1   General model for analysis-by-synthesis speech coding

The basic structure of the general model for analysis-by-synthesis predictive coding of speech is shown in Figure 3.21. The model consists of three main parts. The first part is the synthesis filter which is an all-pole time-varying filter for modelling the short-time spectral envelope of the speech waveform. It is often called *short-term correlation* filter because its coefficients are computed by predicting a speech sample from few previous samples (usually previous 8-16 samples, hence the name short term). The synthesis filter could also include a long-term correlation filter cascaded to the short-term correlation filter. The long-term predictor models the fine structure of the speech spectrum. The second part of the model is the excitation generator. This generator produces the excitation sequence which is to be fed to the synthesis filter to produce the reconstructed speech at the receiver. The excitation is optimized by minimizing the perceptually weighted error between the original and synthesized speech. As it is shown in Figure 3.21, a local decoder is present inside the encoder, and the analysis method for optimizing the excitation uses the difference between the original and synthesized speech as an error criterion, and it chooses the sequence of excitation which minimizes the weighted error.

The efficiency of this analysis-by-synthesis method comes from the closed loop optimization procedure, which allows the representation of the prediction residual using a very low bit rate, while maintaining high speech quality. This explains the superiority of analysis-by-synthesis predictive coding over other predictive coders which have open-loop structures such as *Residual Excited Linear Prediction* (RELP) coders [39]. The key point in the closed-loop structure is that the prediction residual is quantised by minimizing the perceptually weighted error between the original and reconstructed speech rather than minimizing the error between the residual and its quantised version as in open-loop structures.

The third part of this model is the criterion used in the error minimization. The most common error minimization criterion is the mean squared error (mse). In this model, a subjectively meaningful error minimization criterion is used, where the error $e(n)$ is passed through a perceptual weighting filter which shapes the noise spectrum in a way to make the power concentrated at the formant frequencies of the speech spectrum so that the noise is masked by the speech signal.

The encoding procedure includes two steps: firstly, the synthesis filter parameters are determined from the speech samples (10–30 ms of speech) outside the optimization loop. Secondly, the optimum excitation sequence for this filter is determined by minimizing the weighted error criterion. The excitation optimization interval is usually in the range of 4–7.5 ms which is less than the LPC parameter update frame. The speech frame is therefore divided into sub-blocks, or subframes, where the excitation is determined individually for each subframe. The quantised filter parameters and the quantised excitation are sent to the receiver. The decoding procedure is performed by passing the decoded excitation signal through the synthesis filters to produce the reconstructed speech.

In the following subsections, we will discuss the LPC synthesis and pitch synthesis filters and the computation of their parameters, as well as the error weighting filter and the selection of the error criterion. The definition of every excitation method will be discussed in separate sections.

3.5.1.1 The short-term predictor

The short term predictor models the short-time spectral envelope of the speech. The spectral envelope of a speech segment of length L samples can be approximated by the transmission function of an all-pole digital filter of the form

$$H(z) = \frac{1}{1 - P_s(z)} = \frac{1}{1 - \sum_{k=1}^{p} a_k z^{-k}}, \tag{3.73}$$

where

$$P_s(z) = \sum_{k=1}^{p} a_k z^{-k} \tag{3.74}$$

is the short-term predictor. The coefficients $\{a_k\}$ are computed using the method of Linear Prediction (LP). The set of coefficients $\{a_k\}$ is called the *LPC parameters* or the *predictor coefficients*. The number of coefficients p is called the *predictor order*. The basic idea behind linear predictive analysis is that a speech sample can be approximated as a linear combination of past speech samples (8-16 samples). i.e.

$$\tilde{s}(n) = \sum_{k=1}^{p} a_k s(n - k), \tag{3.75}$$

where $s(n)$ is the speech sample and $\tilde{s}(n)$ is the predicted speech sample at sampling instant n. The prediction error, $e(n)$, is defined as

$$
\begin{aligned}
e(n) &= s(n) - \tilde{s}(n) \\
&= s(n) - \sum_{k=1}^{p} a_k s(n - k)
\end{aligned}
\tag{3.76}
$$

Taking the z-transform of Equation (3.76), we get

$$E(z) = S(z)A(z), \tag{3.77}$$

where

$$A(z) = 1 - \sum_{k=1}^{p} a_k z^{-k}. \tag{3.78}$$

$A(z)$ is the inverse of $H(z)$ in Equation (3.73), hence, $A(z)$ is called the *inverse filter*.

Because of the time-varying nature of the speech, the prediction coefficients should be estimated from short segments of speech signal (10-20 ms). The basic approach is to find a set of predictor coefficients that will minimize the *mean-squared prediction error* over a short segment of speech waveform. The resulting parameters are then assumed to be the parameters of the system function $H(z)$ in the model of speech production in

Equation (3.73). The short-time average prediction error is defined as

$$E = \sum_n e^2(n)$$

$$= \sum_n \left[s(n) - \sum_{k=1}^p a_k s(n-k) \right]^2 . \tag{3.79}$$

To find the values of $\{a_k\}$ that minimize E, we set $\partial E / \partial a_i = 0$ for $i = 1, \ldots, p$. Then

$$\frac{\partial E}{\partial a_i} = -2 \sum_n \left\{ \left[s(n) - \sum_{k=1}^p a_k s(n-k) \right] s(n-i) \right\} = 0, \tag{3.80}$$

which gives

$$\sum_n s(n)s(n-i) = \sum_n \sum_{k=1}^p a_k s(n-k)s(n-i). \tag{3.81}$$

Changing the order of the summation in the right-hand side of Equation (3.81)

$$\sum_n s(n)s(n-i) = \sum_{k=1}^p a_k \sum_n s(n-k)s(n-i), \quad i = 1, \ldots, p. \tag{3.82}$$

If we define

$$\phi(i, k) = \sum_n s(n-i)s(n-k), \tag{3.83}$$

then Equation (3.82) can be written as

$$\sum_{k=1}^p a_k \phi(i, k) = \phi(i, 0), \quad i = 1, \ldots, p. \tag{3.84}$$

This set of p equations in p unknowns can be solved efficiently for the unknown predictor coefficients $\{a_k\}$. First, we must compute $\phi(i, k)$ for $i = 1, \ldots, p$ and $k = 0, \ldots, p$. To compute $\phi(i, k)$ from Equation (3.83), the limits of the summation must be specified. For short-time analysis, the limits must be over a finite interval. Two known methods for linear prediction analysis emerge out of a consideration of the limits of the summation, the *autocorrelation method* and the *covariance method* [40, 41]. There is a third method which computes the reflection coefficients, that are an equivalent representation of the filter parameters, directly from the speech samples, by-passing an estimate of the autocorrelation coefficients. This approach is termed the *covariance lattice method* [42] or the *Burg algorithm* [43]. Although the Burg algorithm has different applications, commonly, the detection of sinusoidal signals in additive noise, Gray et. al. [43] showed that in speech analysis applications, Burg's method does not appear any more useful than the other techniques. In the following two subsections, we will discuss the more commonly used autocorrelation and covariance methods.

3.5.1.1.1   The Autocorrelation Method In this approach, we assume that the error in Equation (3.79) is computed over the infinite duration $-\infty < n < \infty$. Since this can not be done in practice, it is assumed that the waveform segment is identically zero outside the interval $0 \le n \le L_a - 1$, where L_a is the LPC analysis frame length. This is equivalent to multiplying the input speech by a finite length window $w(n)$ that is identically zero outside the interval $0 \le n \le L_a - 1$. Considering Equation (3.79), $e(n)$ is nonzero only in the interval $0 \le n \le L_a + p - 1$. Thus

$$\phi(i, k) = \sum_{n=0}^{L_a + p - 1} s(n - i)s(n - k), \qquad \begin{array}{l} i = 1, \dots, p, \\ k = 0, \dots, p. \end{array} \qquad (3.85)$$

Setting $m = n - i$, Equation (3.85) can be expressed as

$$\phi(i, k) = \sum_{m=0}^{L_a - 1 - (i-k)} s(m)s(m + i - k). \qquad (3.86)$$

Therefore, $\phi(i, k)$ is the short-time autocorrelation of $s(m)$ evaluated for $(i - k)$. That is

$$\phi(i, k) = R(i - k), \qquad (3.87)$$

where

$$R(j) = \sum_{n=0}^{L_a - 1 - j} s(n)s(n + j) = \sum_{n=j}^{L_a - 1} s(n)s(n - j). \qquad (3.88)$$

Therefore, the set of p equations in (3.84) can be expressed as

$$\sum_{k=1}^{p} a_k R(|i - k|) = R(i), \qquad i = 1, \dots, p. \qquad (3.89)$$

Equation (3.89) is expressed in matrix form as

$$\begin{pmatrix} R(0) & R(1) & R(2) & \dots & R(p-1) \\ R(1) & R(0) & R(1) & \dots & R(p-2) \\ R(2) & R(1) & R(0) & \dots & R(p-3) \\ \vdots & \vdots & \vdots & \ddots & \vdots \\ R(p-1) & R(p-2) & R(p-3) & \dots & R(0) \end{pmatrix} \begin{pmatrix} a_1 \\ a_2 \\ a_3 \\ \vdots \\ a_p \end{pmatrix} = \begin{pmatrix} R(1) \\ R(2) \\ R(3) \\ \vdots \\ R(p) \end{pmatrix}$$
$$(3.90)$$

The $p \times p$ matrix of autocorrelation values is a symmetric Toeplitz matrix, i.e., all the elements along a given diagonal are equal. This special property can be exploited to obtain an efficient algorithm for the solution of Equation (3.90). The most efficient solution is a recursive procedure known as Durbin's algorithm, which can be stated as follows [40]:

$$E(0) = R(0)$$

For $i = 1$ to p do

$$k_i = \left[R(i) - \sum_{j=1}^{i-1} a_j^{(i-1)} R(i-j) \right] / E(i-1) \qquad (3.91)$$

$$a_i^{(i)} = k_i$$

For $j = 1$ to $i-1$ do

$$a_j^{(i)} = a_j^{(i-1)} - k_i a_{i-j}^{(i-1)} \qquad (3.92)$$

$$E(i) = (1 - k_i^2) E(i-1). \qquad (3.93)$$

The final solution is given as

$$a_j = a_j^{(p)} \qquad j = 1, \ldots, p. \qquad (3.94)$$

The quantity $E(i)$ in Equation (3.93) is the prediction error of a predictor of order i. The intermediate quantities k_i are known as the *reflection coefficients*. They are the same coefficients which appear in the lossless tube model of the vocal tract [40]. The value of k_i is in the range

$$-1 \leq k_i \leq 1. \qquad (3.95)$$

This condition imposed on the parameter k_i is necessary and sufficient for all the roots of the polynomial $A(z)$ to be inside the unit circle, thereby guaranteeing the stability of the system $H(z)$. It is found that the autocorrelation method always leads to a stable filter $H(z)$.

As mentioned earlier in this section, the speech samples, $s(n)$, are identically set to zero outside the interval $0 \leq n \leq L_a - 1$. However, a sharp truncation of the speech segment is likely to create a large increase in the prediction error at the beginning and the end of the segment being analysed. This problem is avoided by using tapered windows such as a Hamming window whose amplitude falls to zero in a gradual manner. The Hamming window is given by

$$w(n) = 0.54 - 0.46\cos(2\pi n/(L_a - 1)), \qquad 0 \leq n \leq L_a - 1, \quad (3.96)$$

where L_a is the LPC analysis frame length. The length L_a of the Hamming window (i.e. the length of LPC analysis frame) is usually chosen longer than the length of the speech update frame L. The overlapped windowing gives a smoothing effect in LPC analysis, that is, it alleviates abrupt changes in LPC coefficients between analysis blocks.

3.5.1.1.2   The Covariance Method   In contrast to the autocorrelation method, here, we assume that the error E in Equation (3.79) is minimized over a finite interval, $0 \leq n \leq L - 1$. Therefore ϕ k) of Equation (3.83) is given by

$$\phi(i, k) = \sum_{n=0}^{L-1} s(n-i)s(n-k), \qquad \begin{array}{l} i = 1, \ldots, p \\ k = 0, \ldots, p. \end{array} \qquad (3.97)$$

The set of equations in (3.84) can be written in matrix form as

$$
\begin{pmatrix}
\phi(1,1) & \phi(1,2) & \phi(1,3) & \cdots & \phi(1,p) \\
\phi(2,1) & \phi(2,2) & \phi(2,3) & \cdots & \phi(2,p) \\
\phi(3,1) & \phi(3,2) & \phi(3,3) & \cdots & \phi(3,p) \\
\vdots & \vdots & \vdots & \ddots & \vdots \\
\phi(p,1) & \phi(p,2) & \phi(p,3) & \cdots & \phi(p,p)
\end{pmatrix}
\begin{pmatrix}
a_1 \\ a_2 \\ a_3 \\ \vdots \\ a_p
\end{pmatrix}
=
\begin{pmatrix}
\phi(1,0) \\ \phi(2,0) \\ \phi(3,0) \\ \vdots \\ \phi(p,0)
\end{pmatrix}.
$$

(3.98)

Since $\phi(i,k) = \phi(k,i)$, the $p \times p$ matrix is symmetric but it is not Toeplitz. Cholesky decomposition [44] can be used to solve Equation (3.98) where the $p \times p$ matrix $\boldsymbol{\Phi}$ is decomposed into

$$
\boldsymbol{\Phi} = \mathbf{VDV}^T,
$$

(3.99)

where $\mathbf{V}$ is a lower triangular matrix with diagonal elements equal to 1, $\mathbf{D}$ is a diagonal matrix, and T denotes transpose.

Another form of Cholesky decomposition is

$$
\boldsymbol{\Phi} = \mathbf{UU}^T,
$$

(3.100)

where $\mathbf{U}$ has a lower triangular structure. $\mathbf{U}$ in Equation (3.100) and $\mathbf{V}$ in Equation (3.99) are related by

$$
\mathbf{U} = \mathbf{VD}^{1/2}.
$$

(3.101)

The square root in Equation (3.101) requires the matrix $\boldsymbol{\Phi}$ to be positive definite if the decomposition form in Equation (3.100) is to be used.

Unlike the autocorrelation method, the covariance method does not always guarantee the stability of the all-pole filter $H(z)$. To guarantee the stability of $H(z)$, the stabilized covariance method can be used [45]. In the stabilized covariance method, the matrix of covariances $\boldsymbol{\Phi}$ is decomposed according to Equation (3.100). Therefore Equation (3.98) becomes

$$
\mathbf{UU}^T \mathbf{a} = \mathbf{c}
$$

(3.102)

where $\mathbf{a} = [a_1 \ldots a_p]^T$ and $\mathbf{c} = [\phi(1,0) \ldots \phi(p,0)]^T$. The elements of $\mathbf{U}$ are computed from the elements of $\boldsymbol{\Phi}$ by the following recursion:

For $j = 1$ to p do

$$
u_{jj} = \sqrt{\phi_{jj} - \sum_{k=1}^{j-1} u_{jk}}
$$

For $i = j+1$ to p do

$$
u_{ij} = \frac{1}{u_{jj}} \left(\phi_{ij} - \sum_{k=1}^{j-1} u_{ik} u_{jk} \right).
$$

(3.103)

Let

$$g = U^T a,$$ (3.104)

then Equation (3.102) is reduced to

$$Ug = c.$$ (3.105)

Since the matrix U has a lower triangular structure, Equation (3.105) is easily solved for g by

$$g_i = \frac{1}{u_{ii}} \left(c_i - \sum_{k=1}^{i-1} u_{ik} g_k \right), \qquad i = 1, \ldots, p.$$ (3.106)

The reflection coefficients are found from the elements of the vector g by

$$k_i = \frac{g_i}{\sqrt{\phi_{00} - \sum_{k=1}^{i-1} g_k^2}} \qquad i = 1, \ldots, p.$$ (3.107)

The predictor coefficients a_i can be now computed from the reflection coefficients k_i using the same recursive relation as in Durbin's algorithm, namely Equation (3.92). i.e.

$$
\begin{aligned}
&\text{For } i = 1 \text{ to } p \text{ do} \\
&\quad a_i^{(i)} = k_i \\
&\quad \text{For } j = 1 \text{ to } i-1 \text{ do} \\
&\qquad a_j^{(i)} = a_j^{(i-1)} - k_i a_{i-j}^{(i-1)}.
\end{aligned}
$$ (3.108)

Atal has further improved his originally proposed stabilized covariance method by introducing frequency compensation [38] and error weighting [89].

3.5.1.1.3 Considerations in the Choice of LPC Analysis Conditions

The variables in the LPC analysis are: the analysis method, the predictor order and the update frame length. We used the Multi-pulse LPC (to be discussed in a later section) to get some insight into the effect of the above mentioned analysis conditions on the speech quality. Regarding the analysis method, we found that both the autocorrelation method and the stabilized covariance method discussed earlier lead to very similar results. It is difficult to distinguish between both methods under the same analysis conditions. The autocorrelation method is used throughout our investigations. The second variable in the LPC analysis is the number of prediction coefficients p. To reduce the number of bits needed to encode the LPC parameters, it is desirable to use the minimum number of parameters necessary to accurately model the short-term spectral envelope of the speech. It was shown in [41], Chapter 4, that to adequately represent the vocal tract under ideal circumstances, the memory of the model

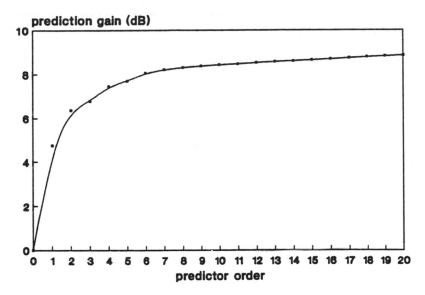

Figure 3.22: Prediction gain versus predictor order in LPC analysis

$A(z)$ must be equal to twice the time required for sound waves to travel from the glottis to the lips, that is, $2\ell/c$, where ℓ is the length of the vocal tract and c is the speed of sound. For example, the representative values $c = 34\,\mathrm{cm/ms}$ and $\ell = 17\,\mathrm{cm}$ result in a necessary memory of $1\,\mathrm{ms}$. When the sampling frequency is f_s samples/sec, the period of $1\,\mathrm{ms}$ corresponds to $f_s/1000$ samples. At $8\,\mathrm{kHz}$ sampling rate, the predictor order p must be at least 8 for this ideal model. It is generally necessary to add several more coefficients (4 or 5) to accommodate other factors excluded from the ideal acoustic tube model (the glottal and lip radiation and the fact that the digitized speech waveforms are not exactly all-pole waveforms). Figure 3.22 shows the average prediction gain against the predictor order p for $20\,\mathrm{s}$ of speech uttered by two males and two females. The prediction gain for a speech frame is given by

$$G = \frac{\sum_n s^2(n)}{\sum_n r^2(n)}, \tag{3.109}$$

where $r(n)$ is the prediction residual. In case of the autocorrelation method the prediction gain can be written as

$$G = \frac{R(0)}{R(0) - \sum_{i=1}^{p} a_i R(i)} = \frac{1}{\prod_{i=1}^{p}(1 - k_i^2)}. \tag{3.110}$$

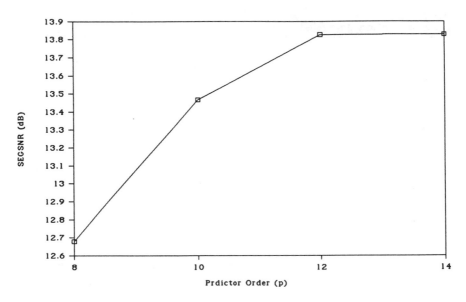

Figure 3.23: SEGSNR versus predictor order for MPE-LPC

The average prediction gain in dB is given by

$$G_{av} = \frac{1}{N_f} \sum_{i=0}^{N_f-1} 10 \log G_i \qquad \text{dB} \qquad (3.111)$$

where N_f is the number of frames. It is clear from Figure 3.22 that the prediction gain starts to saturate for predictor orders larger than 8. Figure 3.23 shows the SEGSNR of the MPE-LPC by varying the predictor order from 8 to 14. The degradation in speech quality becomes noticeable when p is reduced to 8 and the quality saturates when p is above 12. We found that the choice of $p = 10$ is very reasonable to adequately represent the vocal tract.

The third variable in LPC analysis is the updating interval L. Like the predictor order p, the choice of the updating interval is a trade-off between quality and bit rate. It is usually desirable to perform spectral analysis within an interval where the vocal tract movement is negligible. For most vowels, this interval is on the order of 15-20 ms, and it is usually shorter for unvoiced sounds. In fact, the burst associated with the release of an unvoiced stop consonant in the initial position such as $/t/$ may exist for only a few ms. Asynchronous analysis (arbitrary placement of the time interval regardless of the pitch period) will often extend the averaging into the voiced portion following the $/t/$ or the silence preceding the $/t/$ release. Therefore, for accurate analysis of transient sounds, an interval on the order of 10 ms is desirable, while for quasi-periodic sounds like most of the vowels, a 15-20 ms interval is adequate. When 10 predictor coefficients are used, they are usually quantised with 40 bits using scalar quantization of the so-called Log Area Ratios. If the updating interval is 20 ms, 2 kb/s are needed to quantise the LPC parameters. On the other hand, if the coefficients are updated every 10 ms, their bit rate rises to 4 kb/s. This means an increase of 2 kb/s, which is very significant in low bit rate coding of speech. Figure 3.24 shows the decrease in the segmental SNR of MPE with increasing the updating interval. We can conclude that for low bit rate speech coding, a 20 ms interval is sufficient to maintain good speech quality, although this would introduce a little degradation in some sounds which have fast changing spectral characteristics, like transient sounds.

3.5.1.1.4 Quantization of the LPC parameters The spectral envelope represented by the set of LPC parameters a_i can be quantised using either scalar or vector quantization methods. In scalar quantization, the LPC parameters are quantised individually using either uniform or nonuniform quantization. In vector quantization (VQ) [48, 49] the set of LPC parameters is considered as one entity, and they are quantised using a large trained codebook by minimizing a specified spectral distortion measure [50]. Generally speaking, vector quantisers yield smaller quantization error than scalar quantisers at the same bit rate, however, the high complexity associated with VQ algorithms has hindered their use in real time implementations. Conventional vector quantisers use algorithms to design a codebook of LPC parameters based on a long training sequence of LPC vectors [51, 52]. These VQs usually lack robustness when speakers outside the training sequence are tested. Using 10 bit codebooks, conventional vector quantization results into consistently noticeable spectral distortions. Increasing the number of bits used to encode the LPC parameters causes the codebook size to grow exponentially. Accordingly, the storage needs and processing time make such large codebooks impractical in real time applications. Therefore vector quantization of LPC parameters has been limited to applications where coarse quantization of the spectral envelope is sufficient [53]. VQ methods have not proved to be useful in high quality speech coding due to their excessive complexity and poor performance with practically small book sizes. VQ of LPC parameters was mostly used with very low bit rate vocoders which inherently limit the achievable speech quality. Another disadvantage of VQs is their vulnerability to transmission errors specially when vector prediction is used. Therefore the attention in the rest of this section will be focused on scalar quantization of LPC parameters since it is the most useful method in high quality speech coding.

The set of LPC parameters $\{a_k\}$ represents the coefficients of the LPC synthesis filter $H(z) = 1/A(z)$. When quantizing the set of prediction coefficients, one has to insure the stability of the synthesis filter. In other words, the poles of the quantised synthesis filter should lie inside the unit circle, a task which is hard to achieve if the set of parameters $\{a_k\}$ is to be

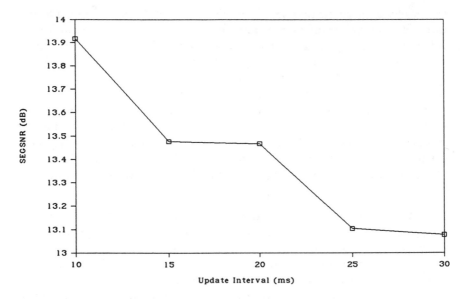

Figure 3.24: SEGSNR versus updating interval of LPC parameters for MPE-LPC

quantised directly. $\{a_k\}$ has to be quantised with 10 bits per parameter to insure the synthesis filter stability. Therefore, it is necessary to transform the LPC coefficients into another set of parameters which are related in a one-to-one manner to the coefficients of a stable synthesis filter. The new set of parameters should possess well behaved statistical properties, and there should be a criterion for guaranteeing the stability of the quantised synthesis filter.

3.5.1.1.4.1   Reflection Coefficients   Such a transformation is the set of reflection coefficients or partial correlation (PARCOR) coefficients. For a stable LPC synthesis filter, the reflection coefficients have the property

$$|k_i| < 1. \tag{3.112}$$

The reflection coefficients are computed as a byproduct of solving the set of $p \times p$ equations in (3.84) using either Durbin's algorithm or the stabilized covariance method. Because for values $|k_i|$ approaching 1, the poles of $H(z)$ approach the unit circle, small changes in k_i can result in large changes in the spectrum. Previous studies have shown that the spectral sensitivity function of the reflection coefficients is approximately U-shaped, with increasing sensitivity as the magnitude of the reflection coefficient approaches unity [54]. Therefore uniform quantization of the reflection coefficients is not efficient. For quantization purposes, the reflection coefficients are transformed to another set of coefficients that exhibit lower spectral sensitivity as k_i approaches 1. Two popular transformations of the reflection coefficients are the inverse sine transformation

$$S_i = \sin^{-1}(k_i) \qquad (3.113)$$

and the log-area ratios

$$LAR_i = \log \frac{1 - k_i}{1 + k_i}. \qquad (3.114)$$

The name *log-area ratio* as the name *reflection coefficient* is obtained from the acoustic tube analogy of the vocal tract [40].

The probability density functions (PDF) of the LARs of an 8th order filter are shown in Figure 3.25. It is clear that the dynamic range of the parameters LAR_i decreases as the index i increases. Therefore more bits are allocated for quantizing the first LARs. The 8 LARs are usually quantised with 6, 6, 5, 5, 4, 4, 3 and 3, respectively. A total of 36 bits are used in this case. For 20 ms LPC parameter updating frames, 1.8 kb/s are needed to quantise the filter coefficients, and the bit rate is reduced to 1.2 kb/s in case of 30 ms updating frames.

To measure the efficiency of a certain LPC parameter quantiser, the log spectral distortion measure is usually used. The log spectral distortion of a speech frame is given by [47]

$$
\begin{aligned}
SD &= \frac{1}{2\pi} \int_{-\pi}^{\pi} \left(10 \log |H(\omega)|^2 - 10 \log \left| \hat{H}(\omega) \right|^2 \right)^2 d\omega \qquad (\text{dB})^2 \\
&= \frac{1}{2\pi} \int_{-\pi}^{\pi} \left(10 \log \frac{\left| \hat{A}(\omega) \right|^2}{|A(\omega)|^2} \right)^2 d\omega \qquad (\text{dB})^2, \qquad (3.115)
\end{aligned}
$$

where $\hat{H}(z)$ and $\hat{A}(z)$ are the quantised synthesis filter and inverse filter, respectively. The log spectral distortion is then averaged over a large number of speech frames. A spectral deviation of 1 dB is considered as the perceptual difference limen for coding the LPC parameters [55]. Utilizing nonuniform quantization reduces the number of bits needed to quantise the LARs to about 36 bits per frame with maintaining a spectral distortion less than 1 dB. Decreasing the number of bits below 30 bits per LPC frame results into noticeable spectral distortion (larger than 1 dB).

3.5.1.1.4.2   Line Spectrum Pairs   Besides the LARs, another important transformation of LPC parameters is the set of *line spectrum pairs* (LSP) [56] or *line spectrum frequencies* [57]. The inverse filter $A(z)$ associated with nth order LPC analysis satisfies the following recursive relation [58]

$$A_n(z) = A_{n-1}(z) - k_n z^n A_{n-1}(z^{-1}), \quad n = 1, \ldots, p, \qquad (3.116)$$

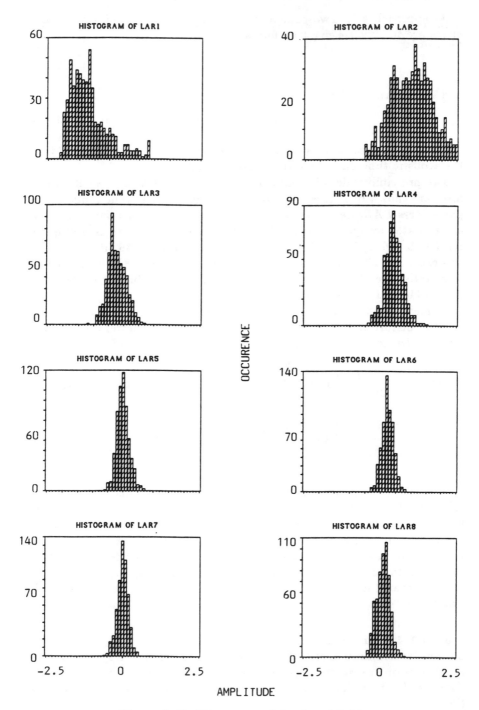

Figure 3.25: Histogram of the first 8 LARs

with $A_0(z) = 1$ and k_n being the $n^t h$ reflection coefficient. Extending the filter order to $n = p + 1$ Equation (3.116) becomes

$$A_{p+1}(z) = A_p(z) - k_{p+1}z^{-(p+1)}A_p(z^{-1}). \qquad (3.117)$$

Consider the two extreme artificial boundary conditions $k_{p+1} = 1$ and $k_{p+1} = -1$. These two conditions correspond, respectively, to a complete closure and complete opening at the glottis in the acoustic tube model. Under these conditions, we obtain the two following polynomials

$$
\begin{aligned}
P(z) &= A(z) - z^{-(p+1)}A(z^{-1}) \\
&= 1 + p_1 z^{-1} + p_2 z^{-2} + \cdots - p_2 z^{-(p-1)} - p_1 z^{-p} - z^{-(p+1)}
\end{aligned}
$$
$$(3.118)$$

for $k_{p+1} = 1$, and

$$
\begin{aligned}
Q(z) &= A(z) + z^{-(p+1)}A(z^{-1}) \\
&= 1 + q_1 z^{-1} + q_2 z^{-2} + \cdots + q_2 z^{-(p-1)} + q_1 z^{-p} + z^{-(p+1)}
\end{aligned}
$$
$$(3.119)$$

for $k_{p+1} = -1$. Notice that the polynomials $P(z)$ and $Q(z)$ are, respectively, antisymmetric and symmetric. It can be shown that the polynomials $P(z)$ and $Q(z)$ possess the following important properties [59]

1. All roots of $P(z)$ and $Q(z)$ are on the unit circle.

2. The roots of $P(z)$ and $Q(z)$ alternate each other on the unit circle.

3. Minimum phase property of $A(z)$ (or the stability of $H(z)$) is easily preserved after quantizing the roots of $P(z)$ and $Q(z)$.

Since the roots of $P(z)$ and $Q(z)$ are on the unit circle, they are given by $e^{j\omega_i}$, and it is easily shown that for the case of even predictor order p, $P(z)$ and $Q(z)$ can be expressed as

$$P(z) = (1 - z^{-1}) \prod_{i=2,4,\dots,p} (1 - 2\cos(2\pi f_i)z^{-1} + z^{-2}), \qquad (3.120)$$

and

$$Q(z) = (1 + z^{-1}) \prod_{i=1,3,\dots,p-1} (1 - 2\cos(2\pi f_i)z^{-1} + z^{-2}). \qquad (3.121)$$

The frequencies f_i in Equations (3.120) and (3.121) are normalized by the sampling frequency f_s. The parameters f_i, $i = 1, \dots, p$, are defined as the *line spectrum frequencies* or the *line spectrum pairs*. It is important to note that $f_0 = 0$ and $f_{p+1} = 0.5$ are always fixed corresponding to the fixed roots $z = 1$ and $z = -1$ of $P(z)$ and $Q(z)$ respectively. Therefore these two fixed roots are excluded from the LSP parameter set. The LSFs can be interpreted as the resonant frequencies of the vocal tract under the two extreme boundary conditions at the glottis (complete closure and complete opening). The second property of the LSFs can be stated as

$$f_0 < f_1 < f_2 < \cdots < f_{p-1} < f_p < f_{p+1}, \qquad (3.122)$$

where $f_0 = 0$ and $f_{p+1} = 0.5$. This relation is known as the *ordering property* of the LSFs. As far as the ordering property is preserved while quantizing the LSFs, the stability of the synthesis filter $H(z)$ is insured.

Several methods for efficient computation of the LSFs can be found in the literature [59, 57, 60, 61, 62]. It has been reported that quantizing the LSP parameter set gives 25% reduction in bit rate compared with the LARs when straight uniform quantization of the LSFs is used [63], and 30% reduction has been obtained when quantizing the LSF differences [59]. However, Atal et al. [64] concluded that quantizing the LSFs does not offer any advantage over quantizing the LARs or the inverse sines. In general, the LSP parameters possess the following properties which allow them to be quantised more efficiently than the LARs

1. The LSFs have well-behaved statistical properties, and the stability of the quantised synthesis filter is easily insured by preserving the ordering property of Equation (3.122). Further, the ordering property can be efficiently used to detect transmission errors in the LSP parameters, and accordingly, substitution algorithms can be utilized to overcome channel errors in LSFs with zero-redundancy.

2. There is evidence of the existence of a direct relation between the LPC spectrum and LSFs. Past studies have shown that a concentration of line spectrum frequencies in a certain frequency band approximately corresponds to a resonance in that band [65]. Further investigation showed that it is not trivial to derive the resonance frequencies from the LSFs [66, 67].

3. The LSP parameters between adjacent frames are highly correlated. This can be exploited to reduce the bit rate by employing predictive quantization. It also leads to efficient interpolation procedures in which the LSFs are transmitted every odd frame, which dramatically reduces their bit rate.

Figure 3.26 shows the histograms of the LSFs of a 10*th* order LPC filter.

Since their introduction by Itakura, the LSP parameters have been intensively studied, and various approaches have been proposed for efficient quantization of the LSFs. These approaches attempted to exploit the previously mentioned properties, namely, the ordering property, the relation between the LSFs and the LPC spectrum, and the correlation between adjacent LSP frames. Recently, Soong and Juang have described a dynamic programming procedure for globally optimizing the allocation of

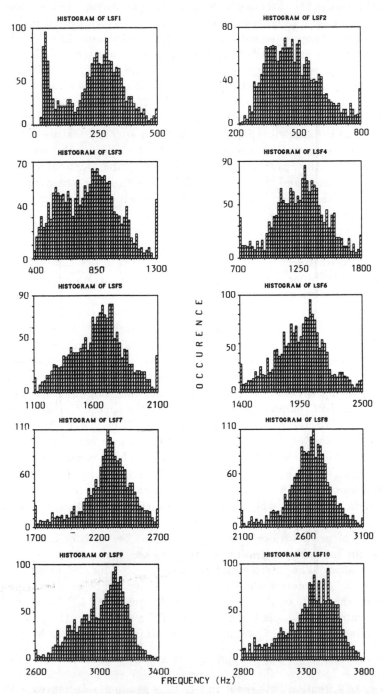

Figure 3.26: Histograms of the LSFs of a 10th order filter

bits as well as the distribution of levels for nonuniform quantization of
LSF differences [68]. Starting from zero bits per frame, they successively
assigned increasing number of bits to parameters which provided the minimum marginal improvement in quantization distortion. LSP-based LPC
vocoders have been formally tested by Kang and Fransen [69]. In 800 b/s
vocoding, they used vector quantization of the LSFs, and in their 4800 b/s
vocoder, they quantised each pair scalarly in terms of its center frequency
$(f_i + f_{i+1})/2$ and offset frequency $(f_{i+1} - f_i)/2$. Crosmer et al. [70] tried to
exploit the relation between the LSFs and the formant frequencies in quantizing the LSP parameters taking into account spectral features known to
be important in perceiving speech signals. In their approach they assumed
that the odd LSFs approximately correspond to the locations of formant
center frequencies, and the closer two LSFs are together, the narrower the
bandwidth of the corresponding pole of the vocal tract. Based on this interpretation they described an approach for quantizing the LSFs. Sugamura
et al. [55] described and compared several schemes to quantise the LSFs
utilizing the ordering property. They obtained the 1 dB difference limen
of spectral distortion with 32 bits/frame. Wong et al [71] obtained similar
results (32 bits/frame with 1 dB spectral distortion) by quantizing the LSF
differences between adjacent frames.

3.5.1.1.4.3   Interpolation of LPC parameters   As we mentioned
earlier, the excitation frame length is usually smaller than the LPC frame.
The LPC frame is, therefore, divided into several subframes, and the excitation parameters are updated every subframe. Figure 3.27 demonstrates
the relationship between the frame, subframe, and Hamming window used
to derive the LPC parameters. In Figure 3.27 the speech frame is of length
160 samples (20 ms), the subframe 40 samples (5 ms), and the Hamming
window 200 samples (25 ms). In this example a new set of LPC parameters is transmitted every 20 ms. Interpolation of LPC parameters between
adjacent frames can be used to obtain a different set of parameters for every subframe. In our example, interpolation enables the updating of filter
parameters every 5 ms while transmitting them every 20 ms, i.e. without
needing the higher bit rate associated with shorter updating frames.

The set of predictor coefficients $\{a_i\}$ cannot be used for interpolation,
because the interpolated parameters in this case do not guarantee a stable
synthesis filter. The interpolation is, therefore, performed using a transformed set of parameters where the filter stability can be easily guaranteed,
e.g. using the LARs or the LSFs. If f_n is the quantised LPC vector in the
present frame and f_{n-1} is the quantised LPC vector from the past frame,
then the interpolated LPC vector sf_k in a subframe k is given by

$$sf_k = \delta_k f_{n-1} + (1 - \delta_k) f_n, \tag{3.123}$$

where δ_k is a fraction between 0 and 1. δ_k is gradually decreased with the
subframe index. For our specific example, a good choice of the values of δ_k

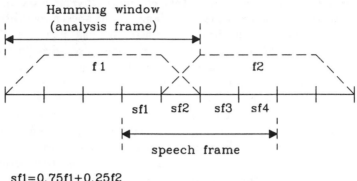

sf1=0.75f1+0.25f2
sf2=0.50f1+0.50f2
sf3=0.25f1+0.75f2
sf4=f2

Figure 3.27: Relationship between the frame, subframe, and Hamming window

is 0.75, 0.5, 0.25, and 0 for $k = 1, \ldots, 4$, respectively. Using these values, the interpolated LPC vectors in the 4 subframes are given by

$$
\begin{aligned}
\mathbf{sf}_1 &= 0.75\mathbf{f}_{n-1} + 0.25\mathbf{f}_n \\
\mathbf{sf}_2 &= 0.5\mathbf{f}_{n-1} + 0.5\mathbf{f}_n \\
\mathbf{sf}_3 &= 0.25\mathbf{f}_{n-1} + 0.75\mathbf{f}_n \\
\mathbf{sf}_4 &= \mathbf{f}_n.
\end{aligned}
$$

Interpolation has been recently used to reduce the bit rate associated with the spectral parameters by transmitting them every odd frame [72] or even every third frame [73]. For the 20 ms speech frame in our example, if the LSFs, are quantised with 36 bits, their corresponding bit rate is 1.8 kb/s. If the LSFs are transmitted every odd frame their bit rate is reduced to 0.9 kb/s. Similarly, if the LSFs are transmitted every third frame their bit rate is further reduced to 0.6 kb/s. In the frames where the LSFs are not transmitted, they are interpolated using the available LSFs similar to Equation (3.123) with a proper weighting fraction δ. This reduction in bit rate offered by interpolation is crucial for encoding at bit rates below 4.8 kb/s. The disadvantage of interpolation is increasing the codec delay. By using interpolation in every odd frame the speech frame size is effectively doubled, and for our example, the end-to-end delay is increased from about 40 ms to 80 ms.

3.5.1.2 The Long-Term Predictor (LTP)

While the short-term predictor models the spectral envelope of the speech segment being analysed, the long-term predictor, or the pitch predictor, is used to model the fine structure of that envelope. Inverse filtering of the speech input removes some of the redundancy in the speech by subtracting from the speech sample its predicted value using the past p samples. This is called short-term prediction since only the previous p samples (usually 10) are used to predict the present sample of speech. The short-term prediction residual, however, still exhibits, to a lesser extent, some periodicity (or redundancy) related to the pitch period of the original speech when it is voiced. This periodicity is on the order of 20–160 samples (50–400 Hz pitch frequencies). Adding the pitch predictor to the inverse filter further removes the redundancy in the residual signal and turns it into a noise like process. It is called *pitch predictor* since it removes the pitch periodicity, or *long-term predictor* since the predictor delay is between 20 and 160 samples. The LTP is not an essential part in medium bit rate LPC coding (e.g. MPE-LPC and RPE-LPC), although including a pitch predictor in these coders improves their performance. However, the long-term predictor is very essential in low bit rate speech coders, as in the CELP, where the excitation signal is modelled by a Gaussian process, therefore, long-term prediction is necessary to insure that the prediction residual is very close to random Gaussian noise process.

The general form of a long-term correlation filter is

$$\frac{1}{P(z)} = \frac{1}{1 - P_l(z)} = \frac{1}{1 - \sum_{k=-m_1}^{m_2} G_k z^{-(\alpha+k)}} \qquad (3.124)$$

where

$$P_l(z) = \sum_{k=-m_1}^{m_2} G_k z^{-(\alpha+k)} \qquad (3.125)$$

is the long-term predictor. For $m_1 = m_2 = 0$, we have a one-tap predictor, and for $m_1 = m_2 = 1$, we have a three-tap predictor. The delay α usually represents the pitch period (or a multiple of it).

The parameters α and $\{G_m\}$ are determined by minimizing the mean-squared residual error after short-term and long-term prediction over a period of N samples. For a one-tap predictor, the long-term prediction residual $e(n)$ is given by

$$e(n) = r(n) - Gr(n - \alpha) \qquad (3.126)$$

where $r(n)$ is the residual signal after short-term prediction. The mean-squared residual E is given by

$$E = \sum_{n=0}^{N-1} e^2(n) = \sum_{n=0}^{N-1} [r(n) - Gr(n - \alpha)]^2 . \qquad (3.127)$$

Setting $\partial E / \partial G = 0$ yields

$$G = \frac{\sum_{n=0}^{N-1} r(n)r(n - \alpha)}{\sum_{n=0}^{N-1} [r(n - \alpha)]^2} \qquad (3.128)$$

and substituting G into Equation (3.127) gives

$$E = \sum_{n=0}^{N-1} r^2(n) - \frac{\left[\sum_{n=0}^{N-1} r(n)r(n-\alpha) \right]^2}{\sum_{n=0}^{N-1} [r(n-\alpha)]^2}. \quad (3.129)$$

Minimizing E means maximizing the second term in the right-hand side of Equation (3.129), which represents the normalized correlation between the residual $r(n)$ and its delayed version. This term is computed for all possible values of α over its specified range, and the value of α which maximizes this term is chosen. The energy $\mathcal{E}$ in the denominator can be easily updated from delay $(\alpha - 1)$ to α instead of recomputing it afresh by

$$\mathcal{E}_\alpha = \mathcal{E}_{\alpha-1} + r^2(-\alpha) - r^2(-\alpha + N) \quad (3.130)$$

which requires 2 instructions (addition plus multiplication). The updating of the term to be maximized requires $N + 4$ instructions for each new delay.

If one-tap LTP is used, then Equation (3.128) is used to compute the gain G. For K-tap LTP, the LTP delay α is first determined by maximizing the second term of Equation (3.129) and then a set of $K \times K$ equations is solved to compute the K predictor gains. For example, if 3-tap LTP is used, the gains are computed by solving the matrix equation

$$\begin{pmatrix} \psi(\alpha - 1, \alpha - 1) & \psi(\alpha - 1, \alpha) & \psi(\alpha - 1, \alpha + 1) \\ \psi(\alpha, \alpha - 1) & \psi(\alpha, \alpha) & \psi(\alpha, \alpha + 1) \\ \psi(\alpha + 1, \alpha - 1) & \psi(\alpha + 1, \alpha) & \psi(\alpha + 1, \alpha + 1) \end{pmatrix}$$

$$\begin{pmatrix} G_{-1} \\ G_0 \\ G_1 \end{pmatrix} = \begin{pmatrix} \psi(0, \alpha - 1) \\ \psi(0, \alpha) \\ \psi(0, \alpha + 1) \end{pmatrix} \quad (3.131)$$

where

$$\psi(i, j) = \sum_{n=0}^{N-1} r(n - i)r(n - j). \quad (3.132)$$

The stability of the pitch synthesis filter $1/P(z)$ is not always guaranteed. For one-tap predictor, the stability condition is $|G| \le 1$. Therefore, the stabilization can be easily carried out by setting $|G| = 1$ whenever $|G| > 1$. For 3-tap predictor, another stabilization procedure can be used [75]. However, the instability of the pitch synthesis filter is not that harmful to the quality of the reconstructed speech. The unstable filter will persist for a few frames (increasing the energy), but eventually, periods of stable filters are encountered, so that the output does not continue to increase with time. Figure 3.28 shows an example of a segment of voiced speech, the residual signal after short-term prediction, and the remaining signal after 1-tap and 3-tap long-term prediction. From Figure 3.28, it is seen that it is enough to use 3-tap filtering to remove the quasi-periodicity in the short-term prediction residual. Usually, to reduce the number of bits

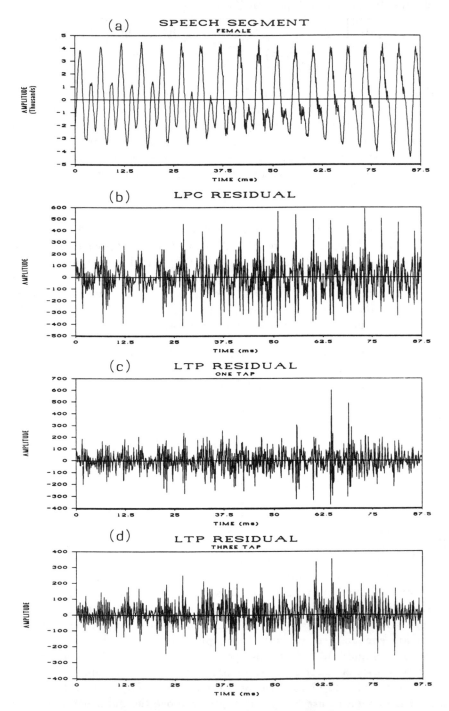

Figure 3.28: (a) 87.5 ms of speech   (b) LPC residual
(c) 1-tap LTP residual   (d) 3-tap LTP residual

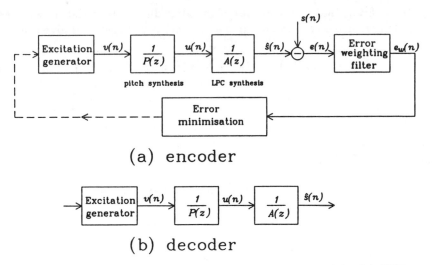

Figure 3.29: General analysis-by-synthesis LPC model with LTP

needed to encode the predictor gains, it is sufficient to use 1-tap LTP. When using LTP, the general codec schematic seen in Figure 3.21 is modified, as portrayed in Figure 3.29.

3.5.1.2.1 Computing the LTP parameters inside the loop: the adaptive codebook approach   In the block diagram of Figure 3.21, the LTP parameters can be determined outside the error minimization loop (directly from the LPC residual signal) as in equations (3.129) and (3.131). However, a significant improvement is achieved when the LTP parameters are optimized inside the analysis-by-synthesis loop [46]. In this case, the computation of the parameters contributes directly to the weighted error minimization procedure. Assuming one-tap long-term prediction, the output of the pitch synthesis filter is given by

$$u(n) = v(n) + Gu(n - \alpha). \qquad (3.133)$$

We first assume that no excitation has been determined, so that Equation (3.133) reduces to

$$u(n) = Gu(n - \alpha). \qquad (3.134)$$

The weighted synthesized speech is given by

$$\hat{s}_w(n) = \sum_{i=0}^{n} u(i)h(n - i) + \hat{s}_0(n), \qquad (3.135)$$

where $h(n)$ is the impulse response of the weighted synthesis filter $1/A(z/\gamma)$ (see Section 3.5.1.3) and $\hat{s}_0(n)$ is the zero-input response of the weighted

synthesis filter, that is, the output of the filter due to its initial states. The
weighted error between the original and synthesized speech is given by

$$e_w(n) = x'(n) - \sum_{i=0}^{n} u(i)h(n-i), \qquad (3.136)$$

where

$$x'(n) = s_w(n) - \hat{s}_0(n) \qquad (3.137)$$

and $s_w(n)$ is the weighted input speech. Substituting Equation (3.134) into
Equation (3.136) gives

$$e_w(n) = x'(n) - Gy_\alpha(n), \qquad (3.138)$$

where

$$y_j(n) = u(n-j) * h(n) = \sum_{i=0}^{n} u(i-j)h(n-i). \qquad (3.139)$$

The mean squared weighted error is given by

$$E_w = \sum_{n=0}^{N-1} [x'(n) - Gy_\alpha(n)]^2. \qquad (3.140)$$

Setting $\partial E_w / \partial G = 0$ leads to

$$G = \frac{\sum_{n=0}^{N-1} x'(n)y_\alpha(n)}{\sum_{n=0}^{N-1} [y_\alpha(n)]^2}. \qquad (3.141)$$

Substituting Equation (3.141) into Equation(3.140) gives

$$E_w = \sum_{n=0}^{N-1} [x'(n)]^2 - \frac{\left[\sum_{n=0}^{N-1} x'(n)y_\alpha(n)\right]^2}{\sum_{n=0}^{N-1} [y_\alpha(n)]^2}. \qquad (3.142)$$

The pitch delay α is selected as the delay which maximizes the second term
in Equation (3.142), and G is then computed from Equation (3.141). Significant speech quality improvement over the open loop solution is achieved
when the long-term predictor parameters are computed inside the optimization loop. The disadvantage of the closed loop solution is the extra
computational load needed to compute the convolution in Equation (3.139)
over the range of the delay α. A fast procedure to compute this convolution
$y_\alpha(n)$ for all the possible delays is to compute it for the first value in the
range and then update it by

$$\begin{aligned}
y_j(0) &= u(-j)h(0) \\
y_j(n) &= u(-j)h(n) + y_{j-1}(n-1), \qquad n = 1, \ldots, N-1.
\end{aligned}$$

$$(3.143)$$

Equation (3.143) requires N operations to determine the convolution $y_j(n)$, while $N(N + 1)/2$ operations are needed when Equation (3.139) is used. The term to be maximized requires $3N + 2$ instructions for each new delay. Another approach (the autocorrelation approach) can be used to update the energy in the denominator in Equation (3.142) with less number of instructions than in the case of the convolution approach [110] (specially for large frame sizes). This approach will be described in a later section while discussing the CELP overlapping codebook approach.

The past synthetic excitation $u(n)$ is stored in an adaptive shift-storage register from $-L_p$ to -1, where L_p is the register or buffer length (usually 147). The contents of this buffer are updated every subframe (excitation frame) by introducing N new samples and dropping the last N samples, that is

$$u(n) \leftarrow u(n + N), \qquad n = -L_p, \ldots, -1. \qquad (3.144)$$

The shift-storage register can be represented by an adaptive codebook, where each codeword is obtained by shifting the previous codeword to the left by one sample. The codewords are given by

$$c_j(n) = u(-j + n), \qquad \begin{matrix} n = 0, \ldots, N - 1, \\ j = N, \ldots, L_p. \end{matrix} \qquad (3.145)$$

For pitch delays less than the excitation frame length N, only the first j values of the codeword $c_j(n)$ are available. In natural speech the pitch delay varies from 20 to 160 samples (from 50 to 400 Hz) and the subframe length N is usually larger than 20 ($N = 60$ is commonly used in 4.8 kb/s coding). For these delays which are less than N the codewords $c_j(n)$ are constructed by repeating the available values until the codeword is completed. That is, for $j < N$

$$c_j(n) = \begin{cases} u(-j + n), & n = 0, \ldots, j - 1, \\ u(-2j + n) & n = j, \ldots, 2j - 1, \end{cases} \qquad (3.146)$$

and so on until the codeword is completed. The delay range 20–147 is commonly used (7 bits). For delays in the range 20 to $N - 1$, the relation in Equation (3.143) has to be modified to accommodate for these delays. For $j < N$ the codeword $c_j(n)$ can be expressed by (assuming $j \geq N/2$)

$$c_j(n) = c_j^{(1)}(n) + c_j^{(2)}(n), \qquad (3.147)$$

where

$$c_j^{(1)}(n) = \begin{cases} u(-j + n), & n = 0, \ldots, j - 1, \\ 0 & n = j, \ldots, N - 1, \end{cases} \qquad (3.148)$$

and

$$c_j^{(2)}(n) = \begin{cases} 0 & n = 0, \ldots, j - 1, \\ u(-2j + n), & n = j, \ldots, N - 1 \end{cases} \qquad (3.149)$$

Accordingly, the filtered codeword is given by

$$y_j(n) = \left(c_j^{(1)}(n) + c_j^{(2)}(n) \right) * h(n)$$
$$= y_j^{(1)}(n) + y_j^{(2)}(n). \tag{3.150}$$

From Equations (3.148) and (3.149)

$$c_j^{(2)}(n) = c_j^{(1)}(n-j), \qquad n = j, \ldots, N-1,$$

which yields

$$y_j^{(2)}(n) = y_j^{(1)}(n-j), \qquad n = j, \ldots, N-1. \tag{3.151}$$

$y_j^{(1)}(n)$ can be updated using the relation in (3.143) from $j = 21$ to 147. For delays $j < N$ $y_j(n)$ is computed from $y_j^{(1)}(n)$ by

$$y_j(n) = y_j^{(1)}(n), \qquad \text{for } n = 0, \ldots, j-1, \tag{3.152}$$
$$= y_j^{(1)}(n) + y_j^{(1)}(n-j), \qquad \text{for } n = j, \ldots, 2j-1, \tag{3.153}$$
$$= y_j^{(1)}(n) + y_j^{(1)}(n-j) + y_j^{(1)}(n-2j), \quad n = 2j, \ldots, N-1. \tag{3.154}$$

Notice that Equation (3.154) is only applied when $j < N/2$. A simpler approach for accommodating delays less than the frame length N is to extend the excitation buffer by the short term prediction residual. That is

$$u(n) = r(n), \qquad n = 0, \ldots, N - \alpha_{\min} - 1, \tag{3.155}$$

where $\alpha_{\min}$ is the minimum value in the range of the pitch delay. In this case, the delays $\alpha < N$ are not treated separately.

The pitch predictor performance can be improved by utilizing noninteger pitch delays. It often happens that the pitch delay does not coincide with the sampling instant. In this case the integer delay nearest to the real pitch delay, or a multiple of it, will be chosen. To find a delay closer to the real one, higher sampling resolution is needed [77, 78, 119] . This is done by upsampling the synthetic excitation signal $u(n)$ (the content of the adaptive buffer) to obtain interpolated codewords in the adaptive codebook. The upsampling factor is determined by the required resolution. Upsampling by a factor m is accomplished by inserting $m - 1$ zeros between each two samples and then low-pass filtering using a filter with cut-off frequency at π/m. A Hamming windowed truncated sinc function is commonly used for FIR low-pass filtering. The standard 4.8 kb/s DoD (Department of Defence, U.S.A.) CELP coder [80] uses 128 integer delays and 128 noninteger delays with the noninteger delays nonuniformly distributed among the integer delays (higher resolution at smaller pitch delays to obtain improvement in speech quality for typical female speakers). The use of nonintegers delays introduces a substantial amount of complexity to the pitch search. In fact, the size of the adaptive codebook is doubled, and interpolation is used to find the codewords corresponding to noninteger delays. To avoid this increase in complexity, the following search procedure can be used [81]

- The match score function is determined for the integer delays only. The match score is the square root of the second term on the right-hand side of Equation (3.142), and it is given by

$$\chi(\alpha) = \frac{\left|\sum_{n=0}^{N-1} x'(n)y_\alpha(n)\right|}{\sqrt{\sum_{n=0}^{N-1}[y_\alpha(n)]^2}}, \qquad \alpha = \alpha_{\min}, \ldots, \alpha_{\max}. \qquad (3.156)$$

- The interpolation is performed on the match score function and its maximum values is searched. This avoids the need to filter the interpolated codewords and compute their match score according to Equation (3.156). In fact, the interpolated codewords are not determined in the first place. Only if the maximum of the interpolated match score function corresponded to a noninteger delay, then the excitation buffer is interpolated to determine the required codeword.

- Only a few interpolated points of the match score need to be computed. The interpolated points are determined around the integer delay which maximizes the score function and its submultiples. The delay (or the fractional delay) which maximizes the match score is selected. If the match score at a submultiple of chosen delay is larger than 0.95 of the maximum match score, then the submultiple delay is favoured. This avoids the selection of the multiples of the actual pitch value and results in a smooth pitch contour [82] which is useful in delta coding of the delay. It is also useful for detecting transmission errors in the delay.

To take advantage of both the simplicity of the open loop and the high performance of the closed loop, the pitch delay can be determined using the open loop solution by maximizing the second term in Equation (3.129) and then the gain can be computed inside the loop using Equation (3.141) [76]. In this case the convolution $y_\alpha(n)$ is computed only once for the value of α determined outside the loop. Nonexhaustive search of the adaptive codebook can be also utilized to reduce the search complexity. This can be done by subsampling, delta coding, and/or hierarchical search.

Table 3.8 shows the SEGSNRs of a CELP coder obtained using the different approaches for determining the LTP parameters. A CELP coder is used as the LTP is an essential part to the coder. A 20 ms speech frame is used with 5 ms excitation frames (subframes). A 9-bit ternary codebook is used (to be explained in a later section). The LTP gain is quantised

LTP method	SEGSNR (dB)	LTP bit rate (kb/s)
IN1	13.3181	2.00
IN2	12.7172	1.80
OUT1	10.3038	2.00
OUT2	9.1432	0.50
OUTIN1	11.0102	2.00
OUTIN2	11.3637	2.00
OUTIN3	10.5513	0.95
OUTIN4	11.3780	1.35

Table 3.8: The effect of long-term prediction on the SEGSNR of CELP

with 3 bits and the delay with 7 (only noninteger delays). IN1 denotes computing the LTP parameters inside the loop (the adaptive codebook approach). IN2 denotes computing the parameters inside the loop with delta coding the delay in even subframes, where the delay in every even subframe is encoded with 5 bits. This reduces the bit rate by 0.2 kb/s at the expense of reducing the SNR by 0.5 dB. The important advantage of delta coding, however, is reducing the search complexity as the adaptive codebook size in even subframes is reduced to 32. OUT1 denotes computing the LTP parameters outside the analysis-by-synthesis loop with 5 ms updating interval (every subframe). This resulted in significant degradation in the CELP performance (3 dB drop in SNR). OUT2 denotes computing the parameters outside the loop with 20 ms updating interval (every speech frame). Only 0.5 kb/s transmission rate is needed to quantise the parameters but the speech quality is reduced dramatically (4 dB drop in SNR). OUTIN1 denotes computing the delay outside the loop and the gain inside the loop and OUTIN2 is the same except that the delay range $\alpha \pm 2$ is searched inside the loop, where α is the delay determined outside the loop. This approach resulted in 1 dB improvement over OUT1 with keeping the low complexity of the open loop search. OUTIN3 and OUTIN4 are similar to OUTIN1 and OUTIN2, respectively, except that the delay is updated every speech frame. In OUTIN4 a delay is determined every speech frame outside the loop and then in every subframe, the delay range $\alpha - 1, \ldots, \alpha + 2$ is searched inside the loop (the delay offset is encoded with 2 bits every subframe). Using OUTIN4 the bit rate is reduced by 0.6 kb/s and the search complexity is significantly reduced at the expense of 2 dB reduction in SNR.

3.5.1.2.2 Quantization of LTP parameters The LTP parameters are the delay α and gain G (or the adaptive codebook index and gain). The delay is quantised with 6–8 bits depending on the range used. Most

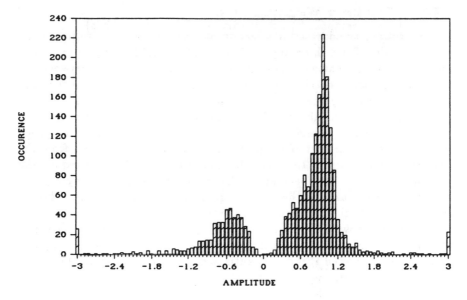

Figure 3.30: Histogram of LTP gain

commonly a 7 bits range is utilized where 128 possible values are used in the range 20–147. To reduce the number of bits the LTP delays can be delta coded in even subframes with 5 bits. Another way to reduce the delay as well as the encoding rate (and also encoder complexity) is to determine a delay outside the loop and then to search for a few codewords around that delay in every subframe. When noninteger delays are used, 8 bits are usually needed (256 entry adaptive codebook). The histogram of the LTP gain is shown in Figure 3.30. It is enough to quantise the gain with 3 or 4 bits (the GSM full-rate coder uses only 2 bits as the gain is restricted to be positive). Due to the nonuniform distribution of the gain, nonuniform quantization has to be used. The quantization levels are determined from a large data base using a Lloyd-Max quantiser. The absolute value of the gain sometimes exceeds 10, so when designing the quantiser the gain value can be truncated. In the DoD 4.8 kb/s CELP coder [80], the range from −1 to 2 is used. In voiced speech segments, the gain value is very close to 1. Negative gains are usually obtained in unvoiced speech frames. Restricting the gain to the range 0–1.2 was found satisfactory. Notice that the gain values larger than 1 or less than -1 correspond to an unstable pitch synthesis filter (poles outside the unit circle). However, the speech quality is not affected by these short unstable periods as the LTP parameters are usually updated every 5 ms. These large gain values are obtained in the transient periods from silence to speech. A detailed treatment of long-term predictors can be found in a series of papers by Kabal and Ramachandran [75, 83, 84].

3.5.1.3   The Error Weighting Filter

In this section, we address the selection of a suitable error criterion in the general model of speech coding of Figure 3.21. Traditionally, speech coding algorithms have attempted to minimize the rms difference between the original and coded speech waveforms. However, it is now well recognized that the subjective perception of the signal distortion is not based on the rms error alone. The theory of auditory masking suggests that noise in the formant regions would be partially or totally masked by the speech signal. Thus, a large part of the perceived noise in a coder comes from the frequency regions where the signal level is low. Therefore, to reduce the perceived noise, its flat spectrum is shaped so that the frequency component in the noise around the formant regions are allowed to have higher energy relative to the components in the inter-formant regions. Now comes the question of how to choose this error shaping (or weighting) filter which appears in Figure 3.21. Incorporating noise shaping in APC [74], Atal and Schroeder showed that the quantization noise appearing in the reconstructed speech signal is given by

$$|N(f)|^2 = |\hat{S}(f) - S(f)|^2 = |\Delta(f)|^2 \left| \frac{1 - F(f)}{1 - P_s(f)} \right|, \qquad (3.157)$$

where $|\Delta(f)|^2$ is the power spectrum of the noise at the output of the quantiser, $P_s(z)$ is the short-term predictor, and $F(z)$ is a feedback filter. In [85], Atal and Schroeder described an efficient method for determining the weighting filter by minimizing the subjective loudness of the quantization noise. In the model of Figure 3.21, the weighting filter $W'(z)$ can be expressed as

$$W'(z) = \frac{1 - P_s(z)}{1 - F(z)} = \frac{A(z)}{B(z)}. \qquad (3.158)$$

Equation (3.158) is derived from Equation (3.157) where

$$\begin{aligned} \Delta(f) &= |\hat{S}(f) - S(f)| \frac{1 - P_s(f)}{1 - F(f)} \\ &= N(f)W(f). \end{aligned} \qquad (3.159)$$

Details about the selection of $B(z)$ are found in [74]. An appropriate choice was found to be $B(z) = A(z/\gamma)$ which gives

$$W'(z) = \frac{A(z)}{A(z/\gamma)} = \frac{1 - \sum_{k=1}^{p} a_k z^{-k}}{1 - \sum_{k=1}^{p} a_k \gamma^k z^{-k}} \qquad (3.160)$$

where γ is a fraction between 0 and 1. The value of γ is determined by the degree to which one wishes to deemphasize the formant regions in the error spectrum. Note that decreasing γ increases the bandwidth of the poles of $W'(z)$. The increase in the bandwidth ω is given by

$$\omega = -\frac{f_s}{\pi} \ln(\gamma), \qquad (3.161)$$

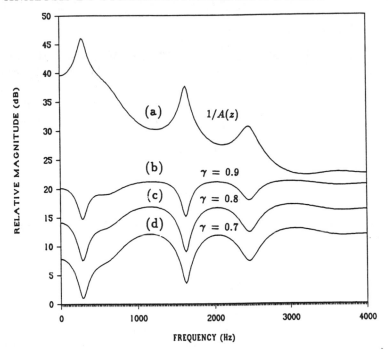

Figure 3.31: Spectrum of $1/A(z)$ and $A(z)/A(z/\gamma)$ for different values of γ

where f_s is the sampling frequency. The choice $\gamma = 0$ gives $W'(z) = A(z)$. In this case, the coder output noise has the same envelope as the original speech. On the other hand, the choice $\gamma = 1$ gives $W'(z) = 1$ which is equivalent to no weighting. A good choice is to use a value of γ between 0.8 and 0.9, which corresponds to an increase in the bandwidth of the poles of $W'(z)$ by 570 down to 270 Hz approximately. Figure 3.31 shows an example of the spectrum of $1/A(z)$ and $A(z)/A(z/\gamma)$ for different values of γ. Makhoul and Berouti [86] discussed other choices of $W'(z)$, where they assumed $W'(z)$ as an all-pole filter or an all-zero filter, but this choice was found to be inferior to the pole-zero filter of Equation (3.160).

Using the error weighting filter given in Equation (3.160), and weighting the original speech and the synthesized one separately before their subtraction, the configuration of Figure 3.21 is reduced to the form shown in Figure 3.32. In this new configuration, the synthesis filter is combined with the error weighting filter to produce the filter

$$W(z) = \frac{1}{A(z/\gamma)} = \frac{1}{1 - \sum_{k=1}^{p} a_k \gamma^k z^{-k}}. \tag{3.162}$$

We will refer to $W(z)$ as the *weighted synthesis filter*. From now on, the configuration of Figure 3.32 will be used as the basic structure for high-quality analysis by synthesis predictive coding.

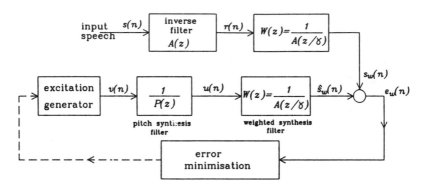

Figure 3.32: Basic structure for analysis-by-synthesis predictive coding using $W'(z) = A(z)/A(z/\gamma)$.

3.5.2 Multi-pulse and Regular-pulse Excitation

In the previous sections we described an efficient basic structure of a model of speech coding which is capable of producing near toll quality speech in the range 4.8–16 kb/s. We showed how the model removes the redundancy in the speech signal by employing short-term and long-term predictors. We discussed also the utilization of an efficient perceptually weighted error minimization criterion. The crucial part of this model which has not been discussed yet is determining an appropriate excitation signal to drive the synthesis filters to produce the synthesized speech. The excitation signal should be defined in a clever way to make the number of bits needed to encode it as small as possible. In this section we discuss the multi-pulse and regular-pulse approaches, and in a later sections we will discuss the code-excited approach. The multi-pulse excited (MPE) and regular-pulse excited (RPE) approaches assume that the excitation signal is modelled by a definite number of pulses in a short time period (5–15 ms). We will denote this time period the excitation frame of length N. The LPC parameters update frame L is usually divided into subframes of length N (L is a multiple of N), and the excitation is determined for every subframe. The difference between MPE and RPE is the way the positions of the pulses are determined. In the next subsection, a general mathematical formulation for determining the pulse positions and amplitudes will be given. In the later subsections, the MPE and RPE algorithms will be discussed and evaluated individually.

3.5.2.1 Formulation of the pulse amplitudes and positions computation

In multipulse excitation, an excitation frame of length N contains M pulses with amplitudes β_i at positions m_i. Therefore, the excitation signal $v(n)$

is defined as

$$v(n) = \sum_{k=0}^{M-1} \beta_k \delta(n - m_k), \qquad n = 0, \ldots, N - 1, \qquad (3.163)$$

where β_k are the pulse amplitudes, m_k are the pulse positions and M is the number of pulses modelling an excitation sequence of length N samples. The pulse amplitudes and positions are determined by minimizing the mean squared weighted error between the original and synthesized speech.

The residual signal after short-term prediction, $r(n)$, is found by filtering the original speech through the inverse filter $A(z)$. The residual $r(n)$ is given by

$$r(n) = s(n) - \sum_{k=1}^{p} a_k s(n - k). \qquad (3.164)$$

The weighted input speech is found by recursive filtering the residual signal $r(n)$ through the weighted synthesis filter $W(z)$ of Equation (3.162). This can be expressed as

$$s_w(n) = r(n) + \sum_{k=1}^{p} a_k \gamma^k s_w(n - k). \qquad (3.165)$$

The weighted input speech, $s_w(n)$, can also be found by convolving the residual signal $r(n)$ with $h(n)$, the impulse response of the weighted synthesis filter $W(z)$, that is

$$s_w(n) = \sum_{i=-\infty}^{n} r(i)h(n - i) = \sum_{i=0}^{n} r(i)h(n - i) + s_0(n), \qquad (3.166)$$

where $s_0(n)$ is the zero-input response of the filter $W(z)$ in the upper branch of Fig. (3.1), i.e. the output of $W(z)$ due to its initial states. The memoryless convolution

$$r(n) * h(n) = \sum_{i=0}^{n} r(i)h(n - i) \qquad (3.167)$$

is referred to as the zero-state response of $W(z)$ to the input $r(n)$. It is preferred to use the recursive relation in Equation (3.165) to compute $s_w(n)$ since it requires only p multiply/add operations per speech sample.

The impulse response, $h(n)$, of the weighted synthesis filter is deduced directly from Equation (3.162) and is given by

$$h(n) = \delta(n) + \sum_{k=1}^{p} a_k \gamma^k h(n - k), \qquad n = 0, \ldots, N - 1, \qquad (3.168)$$

where $h(0) = 1$ and $h(n) = 0$ for $n < 0$.

The weighted synthesized speech $\hat{s}_w(n)$ is computed by convolving the excitation signal $v(n)$ with the impulse response of the combination of the pitch synthesis filter $1/P(z)$ and the weighted synthesis filter $W(z)$. The combined filter $C(z)$ is given by

$$C(z) = \frac{1}{P(z)} \cdot W(z) = \frac{1}{(1 - Gz^{-\alpha})(1 - \sum_{k=1}^{p} a_k \gamma^k z^{-k})}. \qquad (3.169)$$

The impulse response $h_c(n)$ of the combined filter $C(z)$ is given by

$$h_c(n) = f(n) * h(n), \qquad (3.170)$$

where $f(n)$ is the impulse response of the pitch synthesis filter $1/P(z)$ given by

$$\begin{aligned} f(n) &= \delta(n) + Gf(n - \alpha) \\ &= \sum_{i=0}^{n_p} G^i \delta(n - i \cdot \alpha), \qquad n = 0, \ldots, N - 1, \qquad (3.171) \end{aligned}$$

where n_p is the number of pitch periods in the excitation frame of length N and it usually varies from 0 to 3. Substituting Equation (3.171) into Equation (3.170) gives

$$\begin{aligned} h_c(n) &= \sum_{i=0}^{n_p} G^i h(n - i \cdot \alpha) \\ &= h(n) + Gh(n - \alpha) + \cdots + G^{n_p} h(n - n_p \cdot \alpha). \qquad (3.172) \end{aligned}$$

It is plausible from Equation (3.172) that for pitch periods α larger than the the excitation frame length N, $h_c(n)$ is equal to $h(n)$ for the values $n < N$. Therefore if the LTP delay α is restricted to be larger than the excitation frame length N, the pitch synthesis filter will not contribute to the impulse response of the combined filter $C(z)$ and it will only be considered when the zero-input response of the combined filter is computed. Also when the adaptive codebook concept is used $h_c(n)$ becomes equal to $h(n)$ as the pitch synthesis filter is replaced by a codebook. From now on we will assume that the pitch synthesis filter is replaced by an adaptive codebook. In this case the excitation at the input of the LPC synthesis filter is given by

$$u(n) = v(n) + Gc_\alpha(n), \qquad (3.173)$$

where G is the adaptive codebook gain (or the LTP gain) and $c_\alpha(n)$ is the codeword selected from the adaptive codebook (or α is the LTP delay). At the beginning the parameters α and G are determined as was described in section 3.5.1.2. The weighted synthesized speech can now be expressed by

$$\hat{s}_w(n) = v(n) * h(n) + Gc_\alpha(n) * h(n) + \hat{s}_0(n), \qquad (3.174)$$

where the convolution is a memoryless process {as in Equation (3.167)} and $\hat{s}_0(n)$ is the zero-input response of the weighted synthesis filter in the lower branch of Figure 3.32. If the adaptive codebook concept is not used $c_\alpha(n)$ is replaced by $u(n - \alpha)$ and $h(n)$ is the impulse response defined in Equation (3.168). The zero-input response $\hat{s}_0(n)$ is given by

$$\hat{s}_0(n) = \sum_{k=1}^{p} a_k \gamma^k \hat{s}_0(n - k), \qquad n = 0, \ldots, N - 1. \qquad (3.175)$$

where initially,

$$\hat{s}_0(n) = \hat{s}_w(N + n) \qquad \text{for} \qquad n = -p, \ldots, -1. \qquad (3.176)$$

which involves buffering the states in the taps of the filter $W(z)$ at the end of the previous excitation frame. Note that $\hat{s}_0(n)$ has already been determined while computing the adaptive codebook parameters (see Equation (3.135)).

Substituting Equation (3.163) into Equation (3.174) leads to

$$\begin{aligned}
\hat{s}_w(n) &= \sum_{i=0}^{n} \left(\sum_{k=0}^{M-1} \beta_k \delta(n - m_k) \right) h(n - i) + G c_\alpha(n) * h(n) + \hat{s}_0(n), \\
&= \sum_{k=0}^{M-1} \beta_k h(n - m_k) + G y_\alpha(n) + \hat{s}_0(n), \qquad (3.177)
\end{aligned}$$

where

$$y_\alpha(n) = c_\alpha(n) * h(n)$$

is the zero-state response of the weighted synthesis filter to the codeword c_α chosen from the adaptive codebook. Now, the weighted error between the original speech and the synthesized speech is given by

$$\begin{aligned}
e_w(n) &= s_w(n) - \hat{s}_w(n), \\
&= s_w(n) - G y_\alpha(n) - \hat{s}_0(n) - \sum_{k=0}^{M-1} \beta_k h(n - m_k), \\
&= x(n) - \sum_{k=1}^{M} \beta_k h(n - m_k), \qquad (3.178)
\end{aligned}$$

where

$$x(n) = s_w(n) - G y_\alpha(n) - \hat{s}_0(n). \qquad (3.179)$$

The signal $x(n)$ is computed by updating $x'(n)$ of Equation (3.137), that is

$$x(n) = x'(n) - G y_\alpha(n). \qquad (3.180)$$

The mean squared weighted error is given by using Equation 3.178 as

$$
\begin{aligned}
E_w &= \sum_{n=0}^{N-1} e_w^2(n), \\
&= \sum_{n=0}^{N-1} \left[x(n) - \sum_{k=0}^{M-1} \beta_k h(n - m_k) \right]^2.
\end{aligned}
\tag{3.181}
$$

The task now is to find the pulse amplitudes β_k and the pulse positions m_k which minimize the mean squared weighted error in Equation (3.181). By setting $\partial E_w / \partial \beta_i = 0$ for $i = 0, \ldots, M-1$ we get

$$
\frac{\partial E_w}{\partial \beta_i} = -2 \sum_{n=0}^{N-1} \left[x(n) - \sum_{k=0}^{M-1} \beta_k h(n - m_k) \right] h(n - m_i) = 0.
\tag{3.182}
$$

Therefore

$$
\sum_{n=0}^{N-1} x(n) h(n - m_i) = \sum_{n=0}^{N-1} \left[\sum_{k=0}^{M-1} \beta_k h(n - m_k) \right] h(n - m_i).
\tag{3.183}
$$

Reordering the summations in the right-hand side of Equation (3.183) yields

$$
\sum_{k=0}^{M-1} \beta_k \sum_{n=0}^{N-1} h(n - m_k) h(n - m_i) = \sum_{n=0}^{N-1} x(n) h(n - m_i), \qquad i = 0, \ldots, M-1.
\tag{3.184}
$$

Define

$$
\phi(i, j) = \sum_{n=0}^{N-1} h(n - i) h(n - j)
\tag{3.185}
$$

to be the autocorrelation of the impulse response $h(n)$, and

$$
\psi(i) = \sum_{n=0}^{N-1} x(n) h(n - i)
\tag{3.186}
$$

to be the cross-correlation between $x(n)$ and $h(n)$, then the set of M equations in (3.184) can be written as

$$
\sum_{k=0}^{M-1} \beta_k \phi(m_i, m_k) = \psi(m_i), \qquad i = 0, \ldots, M-1.
\tag{3.187}
$$

The set of M equations in (3.187) can be written in matrix form as

$$
\begin{pmatrix}
\phi(m_0, m_0) & \phi(m_0, m_1) & \cdots & \phi(m_0, m_{M-1}) \\
\phi(m_1, m_0) & \phi(m_1, m_1) & \cdots & \phi(m_1, m_{M-1}) \\
\vdots & \vdots & \ddots & \vdots \\
\phi(m_{M-1}, m_0) & \phi(m_{M-1}, m_1) & \cdots & \phi(m_{M-1}, m_{M-1})
\end{pmatrix}.
$$

$$\begin{pmatrix} \beta_0 \\ \beta_1 \\ \vdots \\ \beta_{M-1} \end{pmatrix} = \begin{pmatrix} \psi(m_0) \\ \psi(m_1) \\ \vdots \\ \psi(m_{M-1}) \end{pmatrix} \qquad (3.188)$$

We will now proceed in deriving an expression for the weighted mean-squared error using the optimal pulse positions and amplitudes from Equation (3.187). Using Equation (3.181)

$$
\begin{aligned}
E_w &= \sum_{n=0}^{N-1} x^2(n) - 2 \sum_{n=0}^{N-1} x(n) \sum_{k=0}^{M-1} \beta_k h(n - m_k) \\
&\quad + \sum_{n=0}^{N-1} \left[\sum_{k=0}^{M-1} \beta_k h(n - m_k) \right]^2, \\
&= \sum_{n=0}^{N-1} x^2(n) - 2 \sum_{k=0}^{M-1} \beta_k \psi(m_k) + \sum_{i=0}^{M-1} \sum_{k=0}^{M-1} \beta_i \beta_k \phi(m_i, m_k)
\end{aligned}
$$

$$(3.189)$$

Using the optimum solution for the pulses in Equation (3.187) we have

$$
\begin{aligned}
\sum_{i=0}^{M-1} \beta_i \psi(m_i) &= \sum_{i=0}^{M-1} \beta_i \sum_{k=0}^{M-1} \beta_k \phi(m_i, m_k), \\
&= \sum_{i=0}^{M-1} \sum_{k=0}^{M-1} \beta_i \beta_k \phi(m_i, m_k). \qquad (3.190)
\end{aligned}
$$

Substituting the relation of Equation (3.190) in Equation (3.189), the minimum mean squared weighted error between the original and the synthesized speech is given by

$$E_{\min} = \sum_{n=0}^{N-1} x^2(n) - \sum_{k=0}^{M-1} \beta_k \psi(m_k). \qquad (3.191)$$

To find the optimum pulse positions and amplitudes, Equation (3.188) has to be solved. In Equation (3.188), we have a set of M equations with $2M$ unknowns, expressed in a matrix form. The unknowns are M pulse positions and M pulse amplitudes. Therefore, it is very difficult to find an optimal solution for the pulse positions and amplitudes. In fact, the optimal solution is to solve Equation (3.188) for all the possible combinations of pulse positions, and select the pulse amplitudes which minimize the error in Equation (3.191). The number of pulse position combinations is $^N C_M = N!/((N - M)!M!)$, and for typical values of $N = 40$ and $M = 4$ the number of combinations is 91390. This shows the complexity of an optimal solution. In the following sections, we will discuss two suboptimal approaches for solving Equation (3.188) using the minimum error expression in Equation (3.191). The so-called multipulse excited approach (MPE) determines one pulse at a time and the regular pulse excited method (RPE) assumes predefined regularly spaced positions. We will start with the MPE approach.

3.5.2.2 The Multi-pulse Approach

The multipulse algorithm is a sub-optimal approach for solving Equation (3.188). The algorithm determines one pulse at a time in an M-stage process. At every stage j, a new pulse amplitude β_j and position m_j are computed by using the previously determined pulse positions and amplitudes at stages less than j. At the beginning, we assume that there is only one pulse with amplitude β_0 at position m_0. Now, Equation (3.188) is reduced to

$$\beta_0 = \frac{\psi(m_0)}{\phi(m_0, m_0)}. \tag{3.192}$$

With only one pulse, the minimum error expression in Equation (3.191) is reduced to

$$E_{\min} = \sum_{n=0}^{N-1} x^2(n) - \beta_0 \psi(m_0). \tag{3.193}$$

Substituting Equation (3.192) in (3.193)

$$E_{\min} = \sum_{n=0}^{N-1} x^2(n) - \frac{\psi^2(m_0)}{\phi(m_0, m_0)}. \tag{3.194}$$

To find the first pulse, we search for the value m_0 which minimizes $E_{\min}$ or, equivalently, the value which maximizes the second term in the right-hand side of Equation (3.194). Having determined the first pulse position m_0, the first pulse amplitude β_0 is computed from Equation (3.192). Introducing a second pulse, the mean-squared weighted error is now given by

$$
\begin{aligned}
E_w^{(1)} &= \sum_{n=0}^{N-1} [x(n) - \beta_0 h(n - m_0) - \beta_1 h(n - m_1)]^2 \\
&= \sum_{n=0}^{N-1} \left[x^{(1)}(n) - \beta_1 h(n - m_1) \right]^2 \tag{3.195}
\end{aligned}
$$

where

$$x^{(1)}(n) = x(n) - \beta_0 h(n - m_0). \tag{3.196}$$

Setting $\partial E_w^{(1)}/\partial \beta_1$ to zero leads, similar to Equations (3.192) and (3.194), to the relations

$$\beta_1 = \frac{\psi^{(1)}(m_1)}{\phi(m_1, m_1)}, \tag{3.197}$$

and

$$E_{\min}^{(1)} = \sum_{n=0}^{N-1} \left[x^{(1)}(n) \right]^2 - \frac{\left[\psi^{(1)}(m_1) \right]^2}{\phi(m_1, m_1)} \qquad (3.198)$$

where

$$\begin{aligned}
\psi^{(1)}(n) &= \sum_{i=0}^{N-1} x^{(1)}(i) h(i-n) \\
&= \sum_{i=0}^{N-1} \left[x(i) - \beta_0 h(i - m_0) \right] h(i - n) \\
&= \psi(n) - \beta_0 \phi(m_0, n). \qquad (3.199)
\end{aligned}$$

Therefore, the value of $\psi(n)$ is updated, as in Equation (3.199), by removing the effect of the first pulse, and the second pulse position is determined by the value which maximizes the second term in Equation (3.198). This process is continued until all the pulses are determined. We first initialize $\psi^{(0)}(n) = \psi(n)$. At every stage j, $j = 1, \ldots, M - 1$, the value $\psi^{(j)}(i)$ is found from

$$\psi^{(j)}(i) = \psi^{(j-1)}(i) - \beta_{j-1}\phi(m_{j-1}, i), \qquad i = 0, \ldots, N - 1, \qquad (3.200)$$

and the position of the jth pulse, m_j, is determined by the value i which maximizes the normalized correlation, given by

$$T(i) = \frac{\left[\psi^{(j)}(i) \right]^2}{\phi(i, i)}. \qquad (3.201)$$

The pulse amplitude is then computed by

$$\beta_j = \frac{\psi^{(j)}(m_j)}{\phi(m_j, m_j)}. \qquad (3.202)$$

When the speech segment is voiced, the multipulse algorithm tends to locate more than one pulse at the same position, which virtually reduces the number of pulses in the entire period. This can be avoided by setting $\psi(m_k) = 0$ for $k = 0, \ldots, i - 1$, where i is the index of the pulse being searched, so that no more than one pulse is located at the same position.

Figure 3.33-a shows the signal power with time for the sentence *"to reach the end he needs much courage"* uttered by a female speaker. Figure 3.33-b shows the variation of SEGSNR with time for this speech sequence when the multipulse algorithm is utilized. 4 pulses are placed in an excitation search frame of 40 samples (5 ms) and the LPC parameters update frame is 20 ms. The upper curve is obtained when long-term prediction is utilized and the lower curve without LTP. It is observed that the SNR is high when the speech power is high. The SNR is high in the periods where the speech is quasi-periodic (voiced) compared with the unvoiced or transient periods.

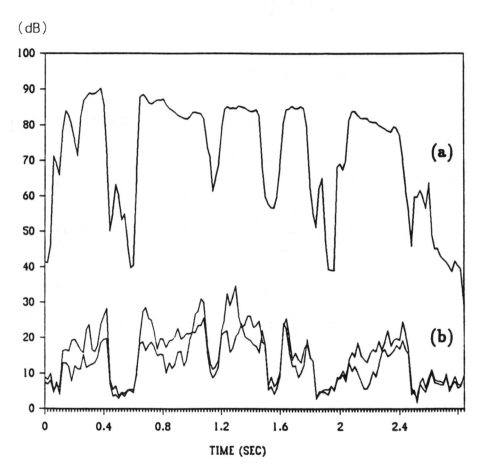

Figure 3.33: (a) Power variation of the sentence *"to reach the end he needs much courage."* (b) Variation of SEGSNR vs. time for MPE using 4 pulses/5 ms with and without LTP (see upper and lower curves, respectively)

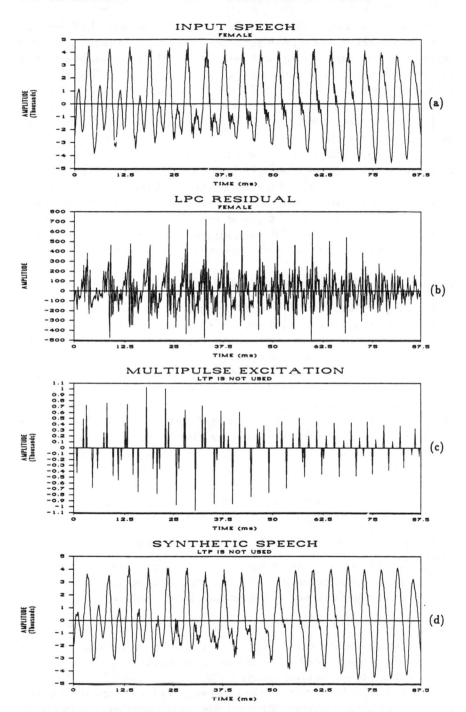

Figure 3.34: (a) 87.5 ms of voiced speech
(b) corresponding LPC residual
(c) multipulse excitation without LTP (4 pulses/5 ms)
(d) reconstructed speech

Figure 3.34 shows how the multipulse algorithm models the LPC residual signal in the absence of a pitch predictor. Figure 3.34-a shows a 87.5 ms segment of female speech, Figure 3.34-b displays the LPC residual for this speech segment, and the multi-pulse excitation is shown in Figure 3.34-c for the case where no LTP is used. The reconstructed speech is depicted in Figure 3.34-d. It can be clearly seen how the quasi-periodicity in the LPC residual is preserved by the multi-pulse algorithm without having any prior knowledge of the pitch period or whether the speech is voiced or unvoiced. Figure 3.35 shows the excitation and reconstructed speech for the same speech segment when LTP is utilized. The excitation pulses are interpolated inside the taps of the pitch synthesis filter, and the excitation of the LPC synthesis filter does not contain zeros as in the multipulse without LTP. The excitation signal is much closer to the LPC residual than in the case where the LTP is not utilized. It is also clear that the fine details in the speech waveform are preserved when LTP is used.

3.5.2.3   Modification of the MPE Algorithm

The multi-pulse algorithm described earlier is suboptimal because the algorithm implicitly assumes that the amplitudes of the past pulses remain constant during the search for the location of the present pulse. Thus the determined pulse amplitudes and positions do not satisfy the set of equations in (3.188). This results in speech quality degradation specially when the pulses are closely spaced [88]. The pulse amplitudes can be reoptimized, after the last pulse position has been determined, by solving the set of equations in (3.188) using the determined pulse positions. However, the pulse positions remain suboptimal as they have been determined using the nonoptimal amplitudes. A better solution is obtained if the pulse amplitudes are reoptimized at every stage of the search. Thus after determining the second pulse position m_1, the amplitudes of the first two pulses β_0 and β_1 are recomputed using Equation (3.188) (2×2 matrix equation), and the correlation ψ is updated using these reoptimized pulses, that is

$$\psi^{(2)}(n) = \psi(n) - \beta_0^{(1)}\phi(m_0, n) - \beta_1^{(1)}\phi(m_1, n).$$

$\psi^{(2)}(n)$ is used to search for the position of the third pulse m_2 and then another set of reoptimized pulses is determined using Equation (3.188) (3×3 matrix equation), and the correlation is updated by

$$\psi^{(3)}(n) = \psi(n) - \beta_0^{(2)}\phi(m_0, n) - \beta_1^{(2)}\phi(m_1, n) - \beta_2^{(2)}\phi(m_2, n).$$

This process is continued until the last pulse position is found, then the pulse amplitudes are finally recomputed using Equation (3.188). For M pulses, this approach requires solving 2×2, 3×3, ..., and $M \times M$ matrix equations in order to reoptimize the pulse amplitudes at each stage of the search. The method becomes rather complex as the number of pulses is increased. Singhal [89] developed a computationally efficient algorithm for reoptimizing the pulse amplitudes without needing to solve Equation (3.188)

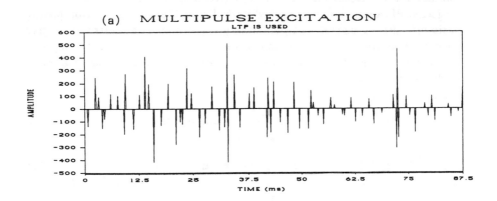

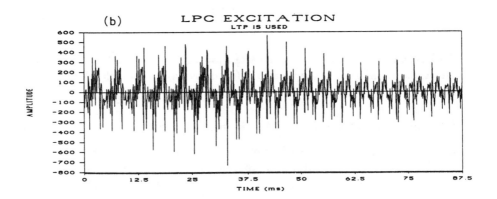

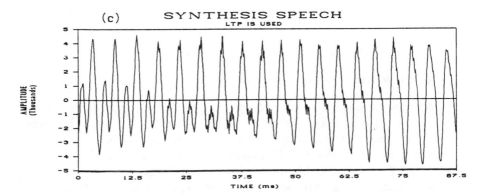

Figure 3.35: (a) Multipulse excitation utilizing LTP.
(b) Excitation of LPC synthesis filter.
(c) reconstructed speech.

at every stage. His algorithm is based on the Cholesky decomposition of the matrix of autocorrelations and it is detailed in [89, 88].

The MPE algorithm is simplified by utilizing the autocorrelation formulation [90, 91]. This is done by extending the summation limits in the error minimization to $-\infty$ and ∞, and windowing with a window having zero values outside the range 0 to $N - 1$. Doing this will reduce the expression in Equation (3.185) to

$$\phi(i, j) = \phi(|i - j|) = \sum_{n=|i-j|}^{N-1} h(n)h(n - |i - j|). \qquad (3.203)$$

The autocorrelation approach will be explained in more detail later while describing the CELP. Using the autocorrelation approach, the term to be maximized in Equation (3.201) is reduced to

$$\mathcal{T}(i) = \frac{\left[\psi^{(j)}(i)\right]^2}{\phi(0)}. \qquad (3.204)$$

This is maximized by maximizing the absolute value of $\psi^{(j)}(i)$, which reduces the number of multiplications (or divisions) to search M pulses by $2MN$.

3.5.2.4   Evaluation of the Multi-pulse Algorithm

In this section, we study the effect of the different MPE encoder parameters on the quality of the synthesized speech. For the results in this section, the LPC parameters are quantised with 36 bits using LSPs. The adaptive codebook gain is quantised with 4 bits and the index with 7 bits in odd subframes and 5 bits in even subframes. The pulse amplitudes are not quantised. The parameters which we are going to take into consideration are

1. The number of pulses per excitation frame.

2. The length of the excitation frame.

We have already shown, in Section 2.3.4, the effect of changing the predictor order p, and the considerations in choosing the LPC analysis method and the updating frame length. Table 3.9 shows the default analysis conditions used in this chapter. This choice of frame lengths is suitable for the bit rate of 9.6 kb/s. At bit rates below 8 kb/s, larger LPC and excitation update frames will be necessary. The selection of predictor order $p = 10$ is satisfactory as we discussed earlier.

3.5.2.4.1   Number of pulses per excitation frame   As we are aiming to achieve low bit rate while maintaining a high synthesized speech quality,

sampling frequency	8000 Hz
LPC analysis frame	200 samples (25 ms)
LPC parameter update frame	160 samples (20 ms)
predictor order	10
analysis method	autocorrelation
LPC quantization	LSPs with 36 bits
excitation frame	40 samples (5 ms)
LTP predictor taps	1
LTP parameter update	40 samples (5 ms)
LTP analysis	adaptive codebook integer delays (20-147) delta coding (7,5,7,5)

Table 3.9: Default analysis conditions used in this chapter

it is desired to use as few pulses as possible for modelling the excitation signal. Using the analysis conditions stated in Table 3.1, and excluding the pitch predictor from the coder, the SEGSNR has been computed with number of pulses varying from 2 to 20 pulses per 5 ms excitation frame, and the results are shown Figure 3.36. Figure 3.37 shows the SEGSNR against number of pulses when long-term prediction is utilized. It is noticed from the lower curves in Figures 3.36 and 3.37 that after few pulses have been placed, the SEGSNR tends to saturate with the increased number of pulses. This is due to the suboptimal solution of the algorithm. The upper curves in the figures show the SEGSNR with the pulses recomputed at every stage, and the middle curve with the pulses recomputed at the last stage. The improvement becomes more significant when the number of pulses is increased. For 9.6 kb/s MPE-LPC coding, at most 4 pulses per 5 ms are used, and deploying amplitude reoptimization does not introduce any significant improvement in speech quality in this case. It is sufficient to reoptimize the pulse amplitudes at the last stage.

Figure 3.38 shows the SEGSNR against number of pulses with and without long-term prediction using last stage reoptimization. A gap of 4–6 dB is noticed between the two curves. When 4 pulses are used, the SNR is increased by 4.5 dB if LTP is deployed. Quantizing the LTP parameters is almost equivalent to quantizing one pulse (amplitude and position) in terms of number of bits used. However, with LTP less number of pulses are needed to model the excitation signal. Using 2 pulses per 5 ms frame with LTP gives similar quality to using 4 pulses without LTP. Therefore deploying LTP results in improved quality at lower bit rates.

Figure 3.39 shows a comparison between the covariance method and autocorrelation method in determining the pulses with and without LTP. The

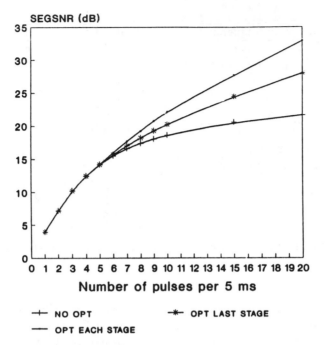

Figure 3.36: SEGSNR versus number of pulses per 5 ms with and without pulse amplitudes reoptimization (without LTP)

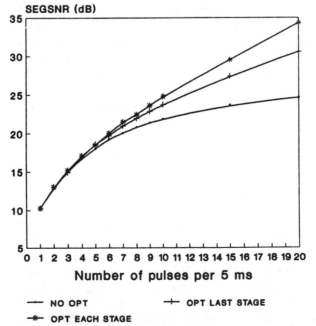

Figure 3.37: SEGSNR versus number of pulses per 5 ms subsegment with and without pulse amplitude reoptimization utilizing LTP

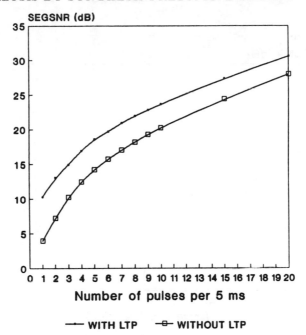

Figure 3.38: SEGSNR versus number of pulses per 5 ms with and without long-term prediction (pulse amplitudes are reoptimized at last stage)

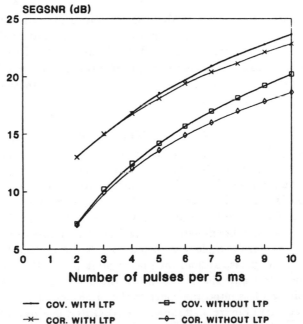

Figure 3.39: SEGSNR versus number of pulses per 5 ms with and without LTP for the covariance and autocorrelation approaches (pulses are reoptimized at last stage)

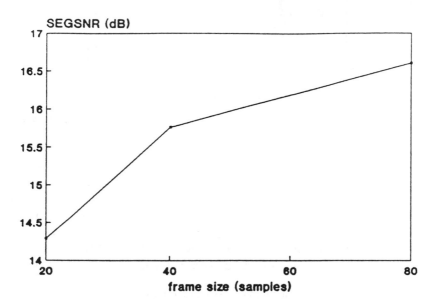

Figure 3.40: SEGSNR vs search frame size

covariance approach gives better SNRs as the number of pulses is increased. For practical values of 3 or 4 pulses per 5 ms the degradation due to using the autocorrelation approach is negligible (about 0.1 dB). Bearing in mind that the autocorrelation approach is computationally more efficient than the covariance approach, it is preferred to be used in practice where the number of pulses is less than 5.

3.5.2.4.2 The length of the excitation frame In the multipulse approach, M pulses are located in a frame of N samples. The choice of the frame size is a trade-off between quality and complexity. It can be shown that the number of operations (multiplication/addition) needed per speech sample is proportional to NM. Therefore, to reduce the complexity, one wishes to reduce the frame size N. On the other hand, the excitation frame size N can not be chosen too small (< 5 ms) to avoid nonoptimal pulse allocation. For example, for voiced speech the multipulse algorithm tends to locate the pulses around the major pitch pulse. If the pitch period is greater than the frame size N (as in low pitch frequency voiced speech), we are modelling a less important part of the pitch period with more pulses than it is necessary. Figure 3.40 shows the SEGSNR against the frame size for 10% pulse rate using multipulse excitation. We can see the increase in SNR for larger frame sizes. To reduce the complexity the choice of 5 ms excitation search frame is reasonable.

Another concern with choosing the search frame size is related to *block*

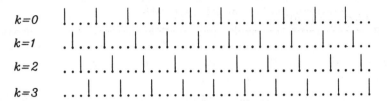

$k=0$ |...|...|...|...|...|...|...|...|...|...|...

$k=1$.|...|...|...|...|...|...|...|...|...|...|..

$k=2$ ..|...|...|...|...|...|...|...|...|...|...|.

$k=3$ ...|...|...|...|...|...|...|...|...|...|...|

Figure 3.41: Candidate excitation patterns in RPE for $N = 40$ and $D = 4$

edge effects [92]. The elements $\phi(i, j)$ of the correlation matrix become small for values of i or j close to N (due to the existence of few terms in the summation in Equation (3.185)). Therefore for pulses located towards the end of the frame the solution in (3.187) becomes ill-conditioned resulting in artificially high pulse amplitudes [88]. To avoid this effect the search frame is made larger than the excitation frame by overlapping with the beginning of the next frame, and the pulses falling in the overlap region are recomputed in the next frame. An overlap region of 2.5 ms was found adequate to prevent ill-conditioned solutions [88].

3.5.2.5   Regular Pulse Excitation Approach (RPE)

Instead of determining one pulse (β_j, m_j) at every stage j assuming that the pulses up to stage $j - 1$ have been determined, Kroon et. al [93, 1] suggested another suboptimal approach for the solution of Equation (3.188) by assuming predefined pulse positions, regularly spaced by distance D. The same approach was also proposed, at a similar time, by Adoul et al. [94], and was called generalized decimation.

In the RPE approach, the excitation sequence for a frame of length N consists of M pulses, regularly spaced by a distance D, where $M = N$ DIV D and DIV denotes integer division (N is not necessarily a multiple of D). Depending on where the first pulse is positioned, D different excitation patterns are obtained. The pulse positions are given by

$$m_i^{(k)} = k + iD, \qquad \begin{aligned} k &= 0,\ldots,D-1, \\ i &= 0,\ldots,M-1, \end{aligned} \qquad (3.205)$$

where k is the position of the first pulse or the *initial phase*. As an example, Figure 3.41 shows the possible excitation patterns for $N = 40$ and $D = 4$. The RPE algorithm consists of solving Equation (3.188) D times (for every possible excitation pattern) to obtain D sets of amplitudes $\{\beta_i^{(k)}\}$ at initial phase k. The mean squared weighted error of Equation (3.191) is then evaluated for every set of computed amplitudes and the set which minimizes the error is chosen. The RPE algorithm requires solving a set of M simultaneous linear equations D times. Typically, $M = 10$ and $D = 4$.

The solution can be performed using Cholesky decomposition. This is the main computational load in the RPE algorithm.

Figure 3.42 shows the regular pulse excitation signal and reconstructed speech for the speech segment shown in Figure 3.34-a without LTP, and Figure 3.43 shows the RPE excitation, the LPC excitation, and the reconstructed speech when LTP is deployed.

It should be noted that the multipulse algorithm needs less number of pulses than the RPE algorithm to achieve the same speech quality. This is because the pulse positions in the MPE algorithm are optimized, unlike the RPE where the positions are predefined. However, although less pulses are used in the MPE case, the pulse positions have to be quantised, while in the RPE case, only the position of the first pulse is quantised with 2 bits usually. Therefore, both the MPE and RPE approaches lead to similar bit rates for the same speech quality. The complexity of the RPE algorithm is higher than that of the MPE. This is because the RPE approach requires the solution of D $(M \times M)$ matrix equations (typically, $M = 10$ and $D = 4$). In later sections, we will look at some methods to reduce the complexity of the RPE algorithm. In the next section, we will examine the effect of different analysis parameters on the quality of the synthesized speech.

3.5.2.6   Evaluation of the RPE Algorithm

The effect of changing the coder parameters has already been studied with the multipulse approach. The same conclusions can be drawn in the RPE case. The only parameters which we will consider in this section are the pulse spacing D and the excitation search frame length N.

3.5.2.6.1   Pulse spacing   Increasing the pulse spacing reduces the number of excitation pulses. Similar to the multipulse approach, the choice of the number of excitation pulses is a trade-off between quality and bit rate. Figure 3.44 shows the SEGSNR for different pulse rates from 1600 to 4000 pulses/sec corresponding to pulse spacing from 5 down to 2. For speech coding at 9.6 kb/s, a good choice of the pulse spacing is $D = 4$ or 5. At this bit rate, the multipulse algorithm usually needs 3–4 pulses per 5 ms. The RPE approach normally gives slightly better quality than the MPE at the same bit rate, at the cost of more complexity. For 5 ms excitation frame and pulse spacing of $D = 4$, we have 10 excitation pulses every 40 samples, and to compute the optimum pulse amplitudes, the RPE algorithm requires solving a 10×10 matrix equation 4 times.

Using LTP gives 2 dB improvement in SNR when $D = 4$. At pulse spacing $D = 2$ (20 pulses per 40 samples) the LTP does not give any improvement as the number of pulses is large enough to model the LPC excitation.

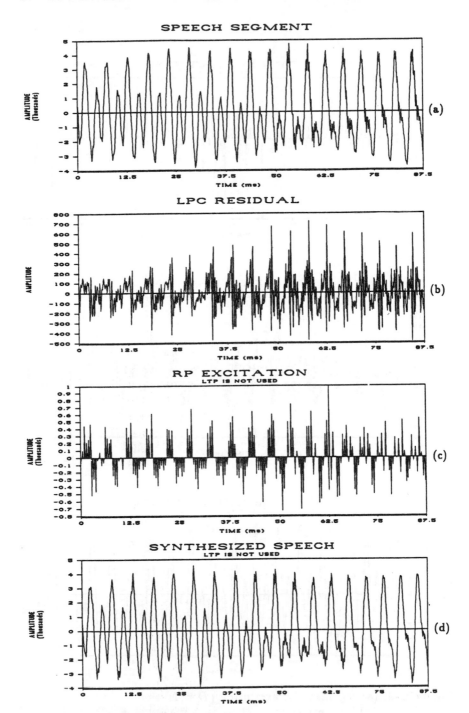

Figure 3.42: (a) 87.5 ms of voiced speech
(b) corresponding short-term prediction residual
(c) regular pulse excitation ($D = 4$)
(d) reconstructed speech

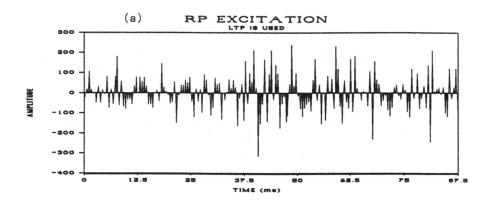

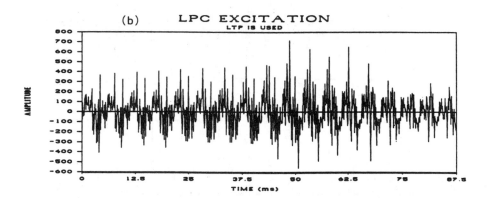

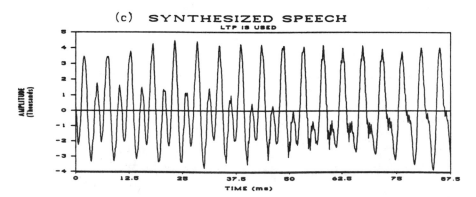

Figure 3.43: (a) regular pulse excitation ($D = 4$)
        (b) LPC excitation
        (c) reconstructed speech

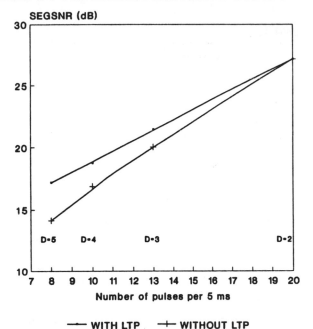

Figure 3.44: Variation of SNR with pulse rate ($D = 2, \ldots, 5$) for RPE without and with LTP inside the optimisation loop

3.5.2.6.2 Excitation search frame length

In the MPE algorithm, we have seen that longer search frames give better results because the pulse positions would be optimized more efficiently. In the RPE algorithm, however, the opposite is true. Since the pulses are regularly spaced, the only factor which controls the positions is the initial phase, or the position of the first pulse, and this changes in every search frame. Therefore, shorter search frames give more flexibility in selecting the pulse positions (higher position updating rate). Figure 3.45 shows the SNR of the RPE with pulse spacing $D = 4$ for the search frame sizes of 20, 40, 52 and 80 (2.5 to 10 ms). The higher SNR is obtained at $N = 20$, and it is not considerably better than the other frame sizes. Another consideration in choosing the excitation frame size is the coder complexity. The longer the frame is, the more complex the coder becomes. For example, at LPC parameters frame of 160 and $D = 4$, if $N = 20$, we have 8 subframes, and we have to solve a 5×5 matrix equation 4 times in each frame. For $N = 80$, we have 2 subframes, and we have to solve a 20×20 matrix equation 4 times in each frame. keeping in mind that solving an $M \times M$ matrix equation is proportional to M^3, then for $N = 20$, the complexity is proportional to 5^3 while for $N = 80$, it is proportional to 20^3. Therefore, shorter excitation frames means less complexity, but on the other hand, we need more bits to encode the positions of the first pulse.

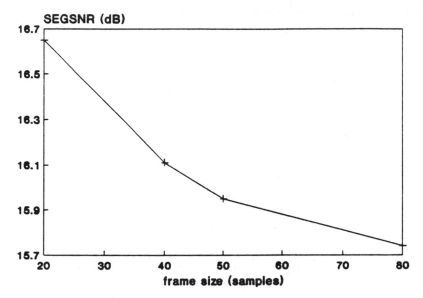

Figure 3.45: SEGSNR variation of the RPE codec with different excitation frame lengths

3.5.2.7   Simplification of the RPE Algorithm

From the evaluation of the RPE algorithm, we have shown that it delivers slightly better quality than the MPE at the same bit rate. The disadvantage of the RPE is the high computational load arising from the necessity to solve 10 simultaneous linear equations every 5 ms (when the pulse spacing $D = 4$). A gross simplification can be achieved by using the autocorrelation method for defining the limits of the summation in computing $\phi(i, j)$. This will have a crucial role in simplifying the matrix in Equation (3.188) due to the regularity of the pulse spacing. We will discuss the autocorrelation approach in the next subsection. Further simplification is obtained by eliminating the matrix inversion in Equation (3.188) and by employing a fixed error weighting filter.

3.5.2.7.1   The autocorrelation approach   Recalling Equation (3.203) and using the autocorrelation approach, $\phi(i, j)$ can be substituted by

$$\phi(i, j) = \phi(|i - j|) = \sum_{n=|i-j|}^{N-1} h(n)h(n - |i - j|). \qquad (3.206)$$

Now, the elements of the matrix in Equation (3.188) are reduced to

$$\phi(m_i^{(k)}, m_j^{(k)}) = \phi(|m_i^{(k)} - m_j^{(k)}|). \qquad (3.207)$$

Using the relation in the definition of the pulse positions in Equation (3.205)

$$
\begin{aligned}
m_i^{(k)} - m_j^{(k)} &= k + iD - (k + jD), \\
&= (i - j)D, \qquad i, j = 0, \ldots, M - 1. \quad (3.208)
\end{aligned}
$$

Exploiting the result in Equation (3.208), Equation (3.188) becomes

$$
\begin{pmatrix}
\phi(0) & \phi(D) & \cdots & \phi([M-1]D) \\
\phi(D) & \phi(0) & \cdots & \phi([M-2]D) \\
\vdots & \vdots & \ddots & \vdots \\
\phi([M-1]D) & \phi([M-2]D) & \cdots & \phi(0)
\end{pmatrix}.
$$
$$
\begin{pmatrix}
\beta_0^{(k)} \\
\beta_1^{(k)} \\
\vdots \\
\beta_{M-1}^{(k)}
\end{pmatrix}
=
\begin{pmatrix}
\psi(m_0^{(k)}) \\
\psi(m_1^{(k)}) \\
\vdots \\
\psi(m_{M-1}^{(k)})
\end{pmatrix}. \quad (3.209)
$$

In Equation (3.209) two complexity reductions can be observed. Firstly, only M values of $\phi(i)$, $i = 0, D, 2D, \ldots, (M-1)D$, are computed. The total number of operations needed is $M(N - D)/2$. The second simplification is that the matrix in Equation (3.209) is independent of the initial phase, thus it is inverted only once every LPC frame rather than D times. Further, the matrix is Toeplitz and it can be solved more efficiently than the symmetric matrix in Equation (3.188) by the use of Levinson's algorithm.

3.5.2.7.2 Eliminating the matrix inversion A closer look at the matrix of correlations in Equation (3.209) suggests that the matrix is strongly diagonal, where the off-diagonal elements become smaller as we go far from the diagonal. i.e. $\phi(0) > |\phi(D)| > \cdots > |\phi([M-1]D)|$. If $g(n)$ is the impulse response of the synthesis filter $1/A(z)$ then it is related to $h(n)$, the impulse response of $1/A(z/\gamma)$, by

$$
h(n) = \gamma^n g(n). \quad (3.210)
$$

The impulse response $g(n)$ is already a decaying function, and the presence of the factor γ^n in Equation (3.210) causes $h(n)$ to decay even faster as γ is less than 1. Figure 3.46 shows an example of the autocorrelation of the impulse response $h(n)$. For a spacing $D = 4$, the diagonals of the autocorrelation matrix are equal to $\phi(0)$, $\phi(4)$, $\phi(8)$, $\ldots$, $\phi(4[M-1])$.

If all the off-diagonal elements of the autocorrelation matrix are set to zero, the matrix is reduced to $\phi(0)\mathbf{I}$, where $\mathbf{I}$ is the identity matrix. In this case Equation (3.209) becomes

$$
\beta_i^{(k)} = \frac{1}{\phi(0)} \psi(m_i^{(k)}), \qquad i = 0, \ldots, M - 1. \quad (3.211)
$$

Autocorrelation

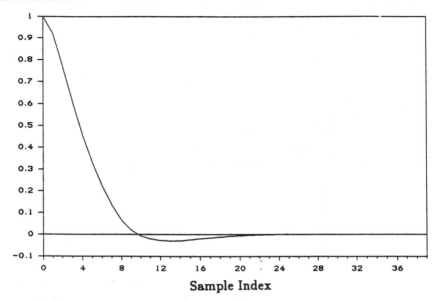

Figure 3.46: Autocorrelation of the typical weighted synthesis filter's impulse response

Recall Equation (3.186) for computing $\psi(n)$

$$\psi(n) = \sum_{i=n}^{N-1} x(i)h(i-n) = x(n) * h(-n),\qquad(3.212)$$

where, from Equation (3.179), and assuming that $\alpha \geq N$

$$x(n) = s_w(n) - Gu(n-\alpha) * h(n) - \hat{s}_0(n),\qquad(3.213)$$

and $\hat{s}_0(n)$ is the zero-input response of the weighted synthesis filter $1/A(\frac{z}{\gamma})$ in the lower branch of Figure 3.32. From Equations (3.166) and (3.213), $x(n)$ can be expressed as

$$x(n) = r(n) * h(n) - Gu(n-\alpha) * h(n) + s_0(n) - \hat{s}_0(n).\qquad(3.214)$$

If we assume that the zero-input responses of the weighted synthesis filters $W(z)$ in both branches of Figure 3.32 are equal, then Equation (3.214) is reduced to

$$
\begin{aligned}
x(n) &= [r(n) - Gu(n-\alpha)] * h(n)\\
&= d(n) * h(n).
\end{aligned}\qquad(3.215)
$$

where

$$d(n) = r(n) - Gu(n - \alpha). \tag{3.216}$$

The signal $d(n)$ can be viewed as the residual after both short-term and long-term prediction. Using the result of Equation (3.215), $\psi(n)$ in (3.212) can be now written as

$$\psi(n) = d(n) * h(n) * h(-n). \tag{3.217}$$

Note that

$$\phi(n) = \sum_{i=n}^{N-1} h(i)h(n-i) = h(n) * h(-n). \tag{3.218}$$

Therefore

$$\psi(n) = d(n) * \phi(n) \tag{3.219}$$

$$= \sum_{i=0}^{N-1} d(i)\phi(|n-i|). \tag{3.220}$$

Note that $\phi(n) = \phi(-n)$, $n = -(N-1), \ldots, N-1$. Let us define

$$z(n) = \frac{\phi(n)}{\phi(0)}, \qquad n = -(N-1), \ldots, N-1, \tag{3.221}$$

to be the normalized autocorrelation of the impulse response $h(n)$. $z(n)$ is a double sided symmetric function where $z(n) = z(-n)$. The pulse amplitudes in Equation (3.211) are now given by

$$\beta_i^{(k)} = \frac{\psi(m_i^{(k)})}{\phi(0)} = d(n) * z(n), \qquad \text{at} \quad n = m_i^{(k)} \tag{3.222}$$

and the mean squared weighted error of Equation (3.191) can be now written as

$$E^{(k)} = \sum_{n=0}^{N-1} x^2(n) - \sum_{i=0}^{M-1} \beta_i^{(k)} \psi(m_i^{(k)})$$

$$= \sum_{n=0}^{N-1} x^2(n) - \phi(0) \sum_{i=0}^{M-1} [\beta_i^{(k)}]^2. \quad k = 0, \ldots, D-1. \tag{3.223}$$

Equations (3.222) and (3.223) are the key equations of an efficient and simple method for RPE coding without the need of solving the set of M linear equations in (3.209). The coder structure of this simplified RPE method is shown in the schematic diagram of Figure 3.47. The method can be described as follows: the short-term prediction residual $r(n)$ is obtained by inverse filtering the original speech through $A(z)$. The residual after

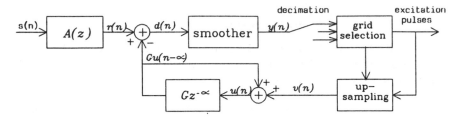

Figure 3.47: Schematic diagram of the simplified RPE structure

long term prediction, $d(n)$, is formed by subtracting from $r(n)$ its estimated value $Gu(n - \alpha)$ (the previously quantised excitation) as in (3.216). The LTP residual $d(n)$ is convolved with the smoothing function $z(n)$. The smoothed LTP residual is given by

$$
\begin{aligned}
y(n) &= d(n) * z(n) \\
&= \sum_{i=0}^{N-1} d(i)z(|n - i|).
\end{aligned}
\tag{3.224}
$$

Now, the smoothed LTP residual $y(n)$ is decomposed into D sets of M amplitudes given by

$$
\{\beta_i^{(k)}\} = \{y(m_i^{(k)})\}, \qquad k = 0, \dots, D - 1.
\tag{3.225}
$$

The energy $T^{(k)}$ is then computed for every set by

$$
T^{(k)} = \sum_{i=0}^{M-1} [\beta_i^{(k)}]^2.
\tag{3.226}
$$

According to Equation (3.223), the set $\{\beta_i^{(k)}\}$ which has maximum energy minimizes the error, thus is chosen to be the excitation signal, where the first pulse β_0 starts at position k and the pulses are separated by a distance D.

As we discussed earlier in this section, the autocorrelation value $\phi(n)$ drops very significantly as n increases. Therefore $z(n)$ can be truncated at $|n| = Q$, where $Q \ll N$, to further reduce the number of terms in the summation of Equation (3.224). This simplified RPE method is detailed in Algorithm 3.1. The closed analysis-by-synthesis loop is broken in this simplified RPE approach. The LTP parameters are determined by minimizing the mean-squared error

$$
E = \sum_{n=0}^{N-1} \left(r(n) - Gu(n - \alpha) \right)^2,
\tag{3.227}
$$

and α is limited to be larger than $N - 1$.

Algorithm 3.1 (simplified RPE) *This algorithm determines M RPE pulses with a simplified method which does not need solving a set of M linear equations. The M pulses β_i are regulary spaced by a distance D in an excitation frame of length N, with the first pulse positioned at k_0.*

1. Compute the short-term prediction residual
$$r(n) = s(n) - \sum_{i=1}^{p} a_i s(n - i).$$

2. Determine the LTP parameters.

3. Compute the LTP residual.
$$d(n) = r(n) - Gu(n - \alpha).$$

4. Compute the smoothed LTP residual $y(n)$
$$y(n) = \sum_{i=0}^{N-1} d(i) z(|n - i|), \quad |n - i| \le Q.$$

5. Decompose $y(n)$ into D sets and compute the energy of every set
    for $k = 0$ to $D - 1$ do
    $$\beta_i^{(k)} = y(k + iD), \ i = 0, \dots, M - 1$$
    $$E^{(k)} = \sum_{i=0}^{M-1} [\beta_i^{(k)}]^2$$

6. Choose the set of pulses with maximum energy
    ref $= 0$
    for $k = 0$ to $D - 1$ do
        if $E^{(k)} >$ ref then
            ref $= E^{(k)}$
            $\beta_i = \beta_i^{(k)}, \ i = 0, \dots, M - 1$
            $k_0 = k$

From the definition of the smoothing function $z(n)$ in Equation (3.221), it is the time-varying normalized autocorrelation of the impulse response of the weighted synthesis filter. Further simplification is obtained when the smoothing function is made fixed. From Figure 3.47 the similarity between the simplified RPE structure and the RELP structure is evident. It is natural therefore to choose the fixed smoothing function as a low-pass filter with cut-off frequency $f_s/(2D)$ where D is the decimation factor. Notice that the smoothing function $z(n)$, where $z(n) = z(-n)$, $n = -Q, \ldots, Q$, is noncausal. A proper approach for designing the low-pass filter is to use a windowed sinc function. For a cut-off frequency $f_s/(2D)$ and a Hamming windowed sinc function the coefficients of the fixed smoothing function are given by [96]

$$z(n) = \frac{D}{n\pi} \sin\left(\frac{n\pi}{D}\right) \left(0.54 + 0.46 \cos\left(\frac{n\pi}{Q}\right)\right), \qquad -Q \leq n \leq Q. \quad (3.228)$$

Usually Q is less than 10. If $z(n)$, $-Q \leq n \leq Q$, is shifted to the right by Q positions, an FIR filter $f(n)$, $0 \leq n \leq 2Q$, is obtained. The resulting FIR filter $f(n)$ is a linear phase filter where $f(n) = f(2Q - n)$. The relation between the double-sided smoother $z(n)$ and the FIR filter $f(n)$ is given by

$$f(n) = z(n - Q), \qquad n = 0, \ldots, 2Q. \quad (3.229)$$

The smoothed residual is given by

$$y(n) = d(n) * z(n) = \sum_{i=-Q}^{Q} z(i)d(n - i), \qquad (n - i) \geq 0, \quad (3.230)$$

or, equivalently,

$$\begin{aligned}
y(n) &= d(n) * f(n + Q), \\
&= \sum_{i=-Q}^{Q} f(i + Q)d(n - i), \qquad n = 0, \ldots, N - 1, \qquad n \geq i, \\
&= \sum_{i=0}^{2Q} f(i)d(n + Q - i), \qquad n = 0, \ldots, N - 1. \quad (n + Q) \geq i,
\end{aligned}$$

$$(3.231)$$

Note that when convolving the segment $d(n)$ of length N with the FIR filter $f(n)$ of length $2Q + 1$, the number of resulting output samples is $N + 2Q$. The N samples resulting from the convolution in (3.231) are the central samples of the convolution $d(n) * f(n)$. Equation (3.231) suggests, therefore, that when the FIR filter is used as a smoother, block filtering is deployed where the first Q samples of the output are discarded.

An FIR low-pass filter with 11 taps is given in [97] for a pulse spacing of $D = 3$. The smoother is a low pass filter with a cut-off frequency at 1333 Hz ($D = 3$) and the taps are given by

$$z(\pm 5) = f(0) = f(10) = -0.016356$$
$$z(\pm 4) = f(1) = f(9) = -0.045649$$
$$z(\pm 3) = f(2) = f(8) = 0$$
$$z(\pm 2) = f(3) = f(7) = 0.250793$$
$$z(\pm 1) = f(4) = f(6) = 0.70079$$
$$z(0) = f(5) = 1$$

With decimation $D = 4$ the filters cut-off frequency is 1000 Hz. The coefficients of a Hamming windowed low-pass filter at this cut-off frequency, and with $Q = 7$, are given by

$$z(0) = 1$$
$$z(\pm 1) = 0.859303$$
$$z(\pm 2) = 0.5263605$$
$$z(\pm 3) = 0.1927755$$
$$z(\pm 4) = 0$$
$$z(\pm 5) = -0.045591$$
$$z(\pm 6) = -0.0319721$$
$$z(\pm 7) = -0.0102893$$

Using the fixed low pass filter as a smoothing function has given better results than using the changing smoothing function which is equal to the normalized autocorrelation of the weighted synthesis filter as in Equation (3.221). This is due to the gross simplification in deriving the structure with changing smoother where the off-diagonal elements of the autocorrelation matrix were set to zero. Using a low pass filter at cut-off frequency $f_s/(2D)$ is more sensible as the simplified structure bears a close similarity with the baseband coder, or residual excited linear predictive coder (RELP) [39]. In RELP coders the LPC residual is low pass filtered, decimated, and the extracted baseband residual is quantised and used to excite the LPC synthesis filter after using interpolation to recover the full band residual (regenerating the residual high frequencies is usually accomplished by spectral folding where zeros are inserted between the baseband samples). The main advantage of the RPE over the RELP is its flexibility in choosing the position of the first excitation pulse, which produces a more appropriate excitation signal. Another difference is the presence of the pitch predictor in the RPE (although pitch prediction has been suggested in RELP coders to reduce the effect of tonal noise [98]). Finally, when the residual signal is smoothed with an FIR low-pass filter of length $2Q + 1$ the residual subframes are smoothed individually (there is no continuous filtering of the residual) using block filtering where only the central N samples of the resulting $N + 2Q$ samples are considered. With these advantages, the simplified RPE delivers better speech quality than the RELP, but it is still inferior to the original RPE because the analysis-by-synthesis optimization loop is broken in the simplified structure. The number of operations (add/multiply) needed to determine the optimum excitation is $2QN$ for the convolution (smoothing) and DM for the energy computation. For the

GSM coder [99], $N = 40$, $Q = 5$ (11-tap), $D = 3$ and $M = 13$. The total
number of operations needed in this case is 11 operations per speech sample. This illustrates the simplicity of the RPE coder with fixed smoothing
filter. In fact, the coder simplicity was a decisive factor in choosing this
coder for the pan-European digital mobile radio system. Figures 3.48-(a)
and (b) show the SEGSNR obtained by the covariance, correlation and simplified RPE approaches described earlier for pulse spacings between $D = 2$
and $D = 5$ with and without LTP, respectively. At $D = 4$ the SNR using
the simplified RPE structure is 2.5 dB less than the covariance (original)
approach, 2 dB less than the autocorrelation approach, and it is similar to
that of the MPE with 4 pulses. For decimation values of 4 or 5, using
the autocorrelation approach is a good choice. Some degradation in speech
quality results when the simplified structure is used with the great advantage of significantly reducing the coder complexity. Another conclusion is
that using a decimation factor of $D = 3$ does not give any improvement over
using $D = 4$. Therefore it is neither necessary nor desirable to deploy this
lower decimation factor, as it would inevitably increase the transmission
bit rate.

3.5.2.8 Quantization of the excitation in MPE and RPE coders

In MPE and RPE coders, the excitation consists of the pulse amplitudes
and pulse positions.

In RPE the pulse positions are defined by the initial phase, or the position of the first pulse, as the pulses are regularly spaced. A decimation
factor of 3 or 4 is usually used, and the initial phase is quantised with 2 bits
in this case. The RPE pulse amplitudes are quantised using adaptive block
quantization. The M pulses in a subframe are scaled by their rms value,
or maximum value. The histograms of the maximum pulse and rms value
of the pulses are shown in Figure 3.49 (a) and (b), respectively. In this
case $D = 4$, $N = 40$, and $M = 10$. It is clear from the histograms that
the scaling value can not be efficiently quantised using a uniform quantiser.
The scaling value is quantised either logarithmically or by using nonuniform quantisers. The histograms of the logarithms of the maximum pulse
and the pulses rms value are shown in Figure 3.50 (a) and (b), respectively. It is clear that uniform quantization of the logarithm is adequate.
When nonuniform quantization is used the quantization and decision levels
are usually designed from a training data set using a Lloyd-Max quantiser [100, 101]. The scaling value is quantised with 5 or 6 bits. Using less
number of bits results in noticeable degradation in speech quality. The histograms of the RPE pulses scaled by their maximum value and by their rms
value are shown in Figure 3.51 (a) and (b), respectively. The normalized
pulses are adequately quantised with 3 bits using nonuniform quantization.
For a subframe of length 40 (5 ms) and 10 RPE pulses, the number of bits
needed to quantise the excitation is: 2 for the initial phase, 6 for the scaling
value, and 30 for the normalized pulses. The bit rate associated with the

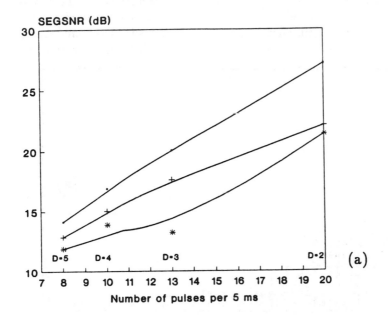

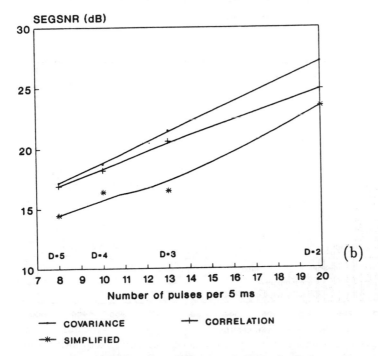

Figure 3.48: SEGSNR for different RPE approaches.
(a) with LTP (b) without LTP

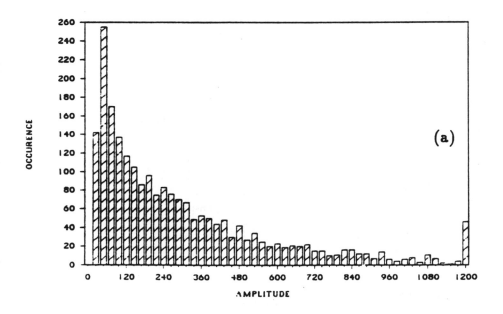

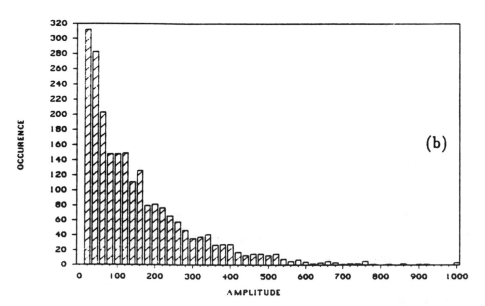

Figure 3.49: (a) Histogram of the maximum RPE pulse
(b) Histogram of the rms value of the RPE pulses

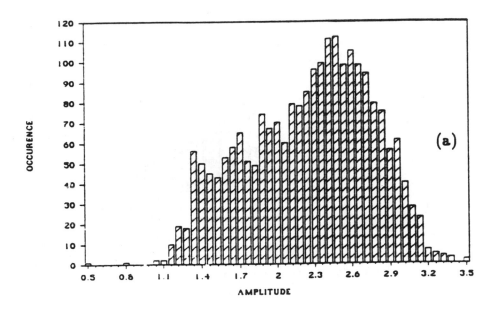

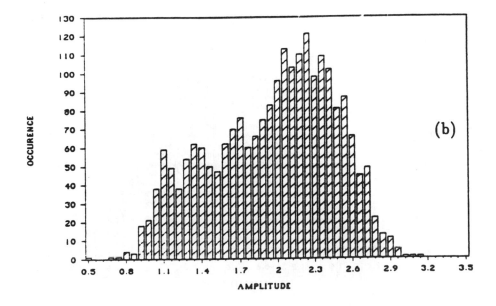

Figure 3.50: (a) Histogram of the logarithm of the maximum RPE pulse
(b) Histogram of the logarithm of the RPE pulses' rms

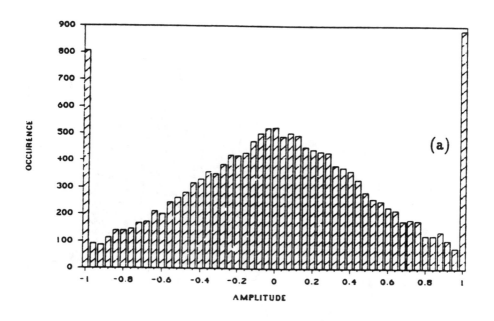

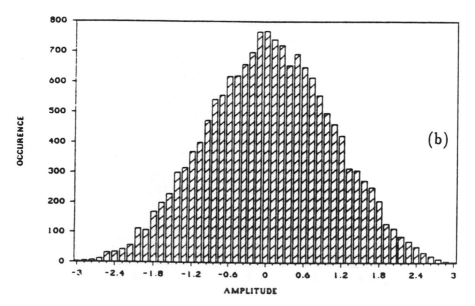

Figure 3.51: Histogram of the normalized RPE pulses
(a) scaled by the maximum pulse (b) by the rms value of the pulses

excitation in this case is 7.6 kb/s. Reducing the bit rate can be achieved by reducing the number of pulses (increasing the decimation factor) and/or reducing the bits needed to quantise the pulses. Increasing the subframe length also reduces the bit rate as the scaling factor is updated less frequently. Using a subframe of $N = 60$ (7.5 ms) and $D = 5$ (12 pulses), the bits needed are: 2 for the initial phase (by limiting it to 4 positions rather than 5), 6 for the maximum pulse, and 36 for the normalized pulses. The excitation bit rate is reduced to 6 kb/s in this case. Taking into account the bit rate needed for LTP and LPC parameters, it is difficult to maintain high quality speech below 9.6 kb/s using the RPE.

In case of MPE a lower number of pulses is needed to obtain the same speech quality as compared to the RPE, since the pulse positions are also optimized. However, the major bit rate contribution in quantizing the excitation is allocated to the pulse positions. If M pulses are allocated in an excitation frame of length N then the total number of position combinations is

$$L_p = \left(\begin{array}{c} N \\ M \end{array} \right) = \frac{N!}{(N-M)!M!}. \tag{3.232}$$

The minimum number of bits needed to quantise M pulse positions is

$$N_{\text{bit}} = \log_2 L_p. \tag{3.233}$$

For the typical values $N = 40$ and $M = 4$ at least 17 bits are needed to quantise the positions of the 4 pulses. In this example if the positions are separately quantised 6 bits are needed for each pulse which results in total of 24 bits. Using differential encoding the pulses can be quantised each with 5 bits for this specific example. The pulses are reordered such that $m_0 < m_1 < m_2 < m_3$ and the quantised quantities are m_0, $m_1 - m_0$, $m_2 - m_1$, and $m_3 - m_2$. In this case the distance between adjacent pulses is restricted to be less than 33, and the positions of the four pulses are encoded with 20 bits.

The most efficient method, which needs the minimum number of bits (N_{bit}) to quantise the pulse positions, is a combinatorial coding scheme [91]. The scheme is given in [102] and is referred to as an enumerative source coding technique. The total number of pulse positions L_p in Equation (3.232) can be represented by an imaginary list from 0 to $L_p - 1$, and the value of the index corresponds to the required pulse positions. The encoding is done by scanning the excitation sequence and incrementing the index every time a pulse is encountered, where the final value of the index is sent. The increment is given by

$$\begin{aligned} \mathcal{I} &= \left(\begin{array}{c} n \\ m \end{array} \right) = \frac{n!}{(n-m)!m!} \quad \text{for } n \geq m \\ &= 0 \quad \text{for } n < m, \end{aligned} \tag{3.234}$$

where n is the remaining number of samples and m is the number of pulses yet to be encoded including the current pulse. The encoding and decoding procedures are described in [102].

Similar to the RPE the pulse amplitudes in MPE are quantised using adaptive block quantization. The pulses are scaled by their rms value, or maximum value, which is quantised with 5–6 bits, and the normalized pulses are quantised with 3–4 bits each. Figure 3.52-(a) shows the histogram of the rms value of the pulses while Figure 3.52-(b) shows the histogram of its logarithm. The histogram of the MPE pulses, normalized by their rms value is shown in Figure 3.53. The normalized pulses are adequately quantised with 3 bits using nonuniform quantization, while the scaling gain is quantised with 5 or 6 bits. Using 4 pulses in an excitation frame of 40 samples, the bits needed are: 6 for the scaling value, 12 bits for the normalized pulses, and 17 bits for the positions using the combinatorial coding scheme. The bit rate needed to quantise the excitation in this case is 7 kb/s. When pitch prediction is used, less number of pulses is needed to represent the excitation signal. Using 3 pulses in a 60 samples subframe reduces the excitation bit rate to 4 kb/s. More efficient approaches can also be used to quantise the amplitudes of the excitation pulses, however, the overall bit rate is not significantly reduced as the quantization of the pulse positions reserves a substantial proportion of the total bit rate. Soheili et al. [103] described several adaptation schemes in quantizing the pulse magnitudes which do not need side information. They found that the most promising adaptation technique is to scale the pulses by the pitch filter memory energy. In this way the scaling gain, which was previously quantised with 5 or 6 bits, is not transmitted and this reduces the bit rate by about 1 kb/s.

3.5.3   Code-Excited Linear Prediction (CELP)

There is currently a high demand for speech coding techniques which are able to produce high quality speech at bit rates below 8 kb/s. Since the full rate GSM speech codec recommendation has been finalized, there has been an intensive research activity devoted to half-rate codecs around 6.5 kb/s encoding rates. The MPE and RPE coders discussed in the previous section can be used to produce good quality speech at bit rates as low as 9.6 kb/s. When the bit rate is reduced below 9.6 kb/s, the MPE and RPE fail in maintaining good speech quality. This is due to the large number of bits needed to encode the excitation pulses, and the quality is deteriorated when these pulses are coarsely quantised, or when their number is reduced, to reduce the bit rate. Therefore, if the analysis-by-synthesis structure is to be used for producing good quality at bit rates below 8 kb/s, more subtle approaches have to be used for defining the excitation signal. The implementation of a long-term predictor in the analysis-by-synthesis loop becomes of prominent importance to remove the redundancy of the speech as much as possible. The residual signal after short-term and long-term prediction becomes noise-like, and it is assumed that the residual can be modelled by a zero-mean Gaussian process with slowly varying power

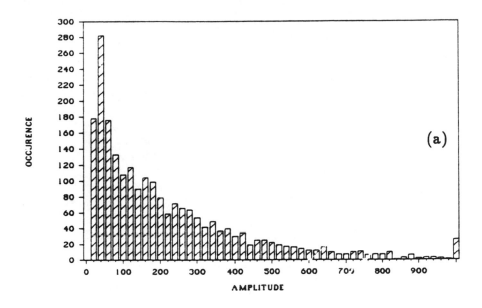

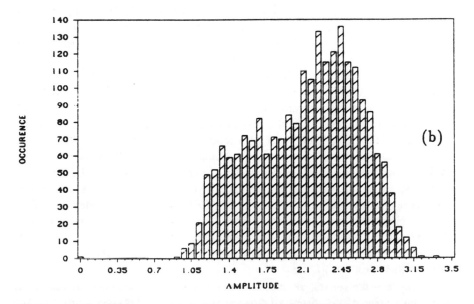

Figure 3.52: (a) Histogram of the rms value of the MPE pulses
(b) Histogram of the logarithm of the MPE pulses' rms value

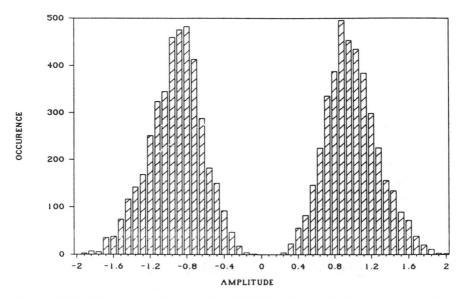

Figure 3.53: Histogram of normalized MPE pulses scaled by the rms value

spectrum. This is the key point in implementing stochastic coders, where
the excitation frame is vector quantised using a large stochastic codebook.
Stochastic coding, or code-excited linear prediction (CELP) coding was first
introduced by Atal and Schroeder in 1984 [104, 4]. A similar approach was
proposed by Copperi and Sereno in 1985 [105]. In the CELP approach, a
5 ms (40 samples) excitation frame is modelled by a Gaussian vector chosen
from a large Gaussian codebook by minimizing the perceptually weighted
error between the original and synthesized speech. Usually, a codebook
of 1024 entries is needed, and the optimum innovation sequence is chosen
by the exhaustive search of the codebook. Using the CELP approach, an
excitation frame (5–7.5 ms) can be encoded with 15 bits only (10 bits for
the book address and 5 bits for the scaling gain). This illustrates the dramatic reduction in bit rate compared, for example, to the GSM RPE-LTP
codec where 47 bits are needed to encode the same excitation. However,
until recently, the high complexity of the CELP algorithm hindered its real
time implementation. The complexity comes from the exhaustive search of
the excitation codebook, where the weighted synthesized speech has to be
computed for all possible codebook entries and compared with the weighted
original speech.

    In the last few years, research activity has been focussed on reducing
the complexity of the originally proposed CELP coder and achieving its
real time implementation using the current DSP technology. Significant
simplification of the CELP encoder has been achieved by using sparse excitation codebooks [106, 107], or center-clipped codebooks [108], in which

most of the excitation samples in the Gaussian random vector are set to zero. Another significant simplification in the codebook structure is to use ternary excitation codebooks where the elements of the excitation vector are set to −1, 0, or 1 [108, 109]. Overlapping sparse or ternary codebooks [108] have also been used to reduce the computational complexity and storage requirements in CELP systems. Binary pulse codebooks [111, 112] are ternary codebooks whereby the excitation pulses are regularly spaced. This specific structure allows for efficient nonexhaustive search procedures to be used. Another efficient way of defining the excitation signal is using the vector sum excitation (VSELP) [113], where the excitation vector is a linear combination of a number of basis vectors weighted with -1 or 1. This special excitation structure yields a significant reduction in the computation needed to identify the optimum excitation vector. An 8 kb/s VSELP coder was recently selected for the future American digital mobile radio system. Algebraic codebooks have also been utilized to reduce the CELP complexity [115] where the codebook is generated using special binary error-correcting codes. Another simplified approach has been proposed [116] in which a CELP system operates on the baseband of the LPC residual signal. The structure of this coder is similar to the GSM RPE coder, and the bit rate reduction is achieved by vector quantizing the smoothed residual with a CELP codebook. The self-excitation concept [117] (or backward excitation recovery [118]) has also been used for producing high quality speech at bit rates below 6.4 kb/s. In this approach, the excitation is obtained by searching through the past excitation signal, and the segment which minimizes the perceptually weighted error between the original and synthesized speech is chosen. The self-excited LPC can be seen as another variant of the CELP in which the codebook is changing, and it has the advantage of less computational demand and less storage requirement with the disadvantage of lacking robustness over noisy transmission environments.

In this chapter we describe the CELP coder and the different approaches used for generating the excitation codebook. In the next section we give a detailed description to the CELP encoding algorithm, we then discuss methods to reduce the complexity of the codebook search procedure and we discuss the use of sparse excitation, ternary excitation, overlapping codebooks, algebraic codebooks, and binary pulse excitation.

3.5.3.1 CELP Principle

After short-term prediction and long-term prediction of the speech signal, the redundancies in the speech signal are almost removed, and the residual signal has very little correlation. A Gaussian process with slowly varying power spectrum can be used to represent the residual signal, and the speech waveform is generated by filtering white Gaussian innovation sequences through the time-varying linear long-term and short-term synthesis filters. The optimum innovation sequence is selected from a codebook of random

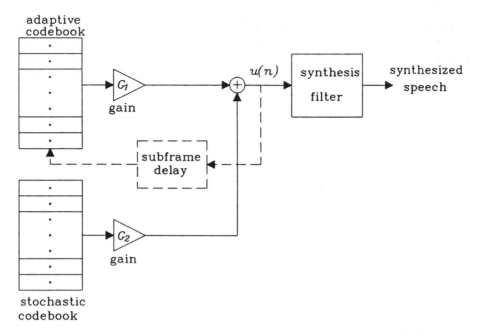

Figure 3.54: Schematic diagram of the CELP synthesis model

white Gaussian sequences by minimizing the subjectively weighted error between the original and the synthesized speech. The schematic diagram of the CELP synthesis model is shown in Figure 3.54. The pitch correlation filter is replaced here by an adaptive overlapping codebook as was discussed in Section 2.4.1. The address selected from the adaptive codebook and the corresponding gain (the pitch delay and gain) along with the address selected from the stochastic codebook and the corresponding scaling gain are sent to the decoder, which uses the same codebooks (in the absence of channel errors) to determine the excitation signal at the input of the LPC synthesis filter to produce the synthesized speech.

The excitation codebook contains L codewords (stochastic vectors) of length N samples (typically $L = 1024$ and $N = 40$ corresponding to a 5 ms excitation frame). The excitation signal of a speech frame of length N is chosen by the exhaustive search of the codebook after scaling the Gaussian vectors by a gain factor β.

The filter $W(z)$ is the weighted synthesis filter given by

$$W(z) = \frac{1}{A(z/\gamma)} = \frac{1}{1 - \sum_{k=1}^{p} a_k \gamma^k z^{-k}}. \qquad (3.235)$$

Having determined the adaptive codebook parameters (pitch delay and gain) as was described in Section 2.4.1, the weighted synthesized speech can be written as

$$\hat{s}_w(n) = \beta c_k(n) * h(n) + G y_\alpha(n) + \hat{s}_0(n), \qquad (3.236)$$

where the convolution is memoryless, $c_k(n)$ is the excitation codeword at index k, β is a scaling factor, $h(n)$ is the impulse response of the weighted synthesis filter $W(z)$, $\hat{s}_0(n)$ is the zero input response of the weighted synthesis filter, G is the adaptive codebook gain and $y_\alpha(n) = c'_\alpha(n) * h(n)$ is the zero-state response of the weighted synthesis filter to the codeword $c'_\alpha(n)$ selected from the adaptive codebook.

The weighted error between the original and synthesized speech is given by

$$
\begin{aligned}
e_w(n) &= s_w(n) - \hat{s}_w(n) \\
&= x(n) - \beta c_k(n) * h(n),
\end{aligned}
\tag{3.237}
$$

where

$$
x(n) = s_w(n) - Gy_\alpha(n) - \hat{s}_0(n).
\tag{3.238}
$$

The signal $x(n)$ is computed by updating $x'(n)$ of Equation (2.76), that is

$$
x(n) = x'(n) - Gy_\alpha(n),
\tag{3.239}
$$

as $x'(n)$ has already been determined while searching the adaptive codebook.

The mean squared weighted error is given by

$$
E = \sum_{n=0}^{N-1} [e_w(n)]^2 = \sum_{n=0}^{N-1} [x(n) - \beta c_k(n) * h(n)]^2.
\tag{3.240}
$$

Setting $\partial E / \partial \beta = 0$, we get

$$
\beta = \frac{\sum_{n=0}^{N-1} x(n)[c_k(n) * h(n)]}{\sum_{n=0}^{N-1} [c_k(n) * h(n)]^2},
\tag{3.241}
$$

and substituting β in Equation (3.240) gives

$$
E = \sum_{n=0}^{N-1} x^2(n) - \frac{\left[\sum_{n=0}^{N-1} x(n)[c_k(n) * h(n)]\right]^2}{\sum_{n=0}^{N-1} [c_k(n) * h(n)]^2}
\tag{3.242}
$$

Equations (3.241) and (3.242) can be written in matrix form as

$$
\beta = \frac{\mathbf{x}^T \mathbf{H} \mathbf{c}_k}{\mathbf{c}_k^T \mathbf{H}^T \mathbf{H} \mathbf{c}_k}
\tag{3.243}
$$

and

$$
\begin{aligned}
E &= \|\mathbf{x} - \beta \mathbf{H} \mathbf{c}_k\|^2 \\
&= \mathbf{x}^T \mathbf{x} - \frac{\left(\mathbf{x}^T \mathbf{H} \mathbf{c}_k\right)^2}{\mathbf{c}_k^T \mathbf{H}^T \mathbf{H} \mathbf{c}_k},
\end{aligned}
\tag{3.244}
$$

where $\mathbf{x}$ and $\mathbf{c}_k$ are N-dimensional vectors given by

$$\mathbf{x}^T = (\begin{array}{cccc} x_0 & x_1 & \ldots & x_{N-1} \end{array}) \tag{3.245}$$

$$\mathbf{c}^T = (\begin{array}{cccc} c_0 & c_1 & \ldots & c_{N-1} \end{array}) \tag{3.246}$$

and $\mathbf{H}$ is a lower triangular convolution matrix of the impulse response $h(n)$ given by

$$\mathbf{H} = \begin{pmatrix} h_0 & 0 & 0 & \ldots & 0 \\ h_1 & h_0 & 0 & \ldots & 0 \\ h_2 & h_1 & h_0 & \ldots & 0 \\ \vdots & \vdots & \vdots & \ddots & \vdots \\ h_{N-1} & h_{N-2} & h_{N-3} & \ldots & h_0 \end{pmatrix}. \tag{3.247}$$

Let

$$\mathbf{\Phi} = \mathbf{H}^T \mathbf{H} \tag{3.248}$$

then $\mathbf{\Phi}$ is a symmetric matrix containing the correlations of the impulse response $h(n)$ given by

$$\phi(i, j) = \sum_{n=\max(i,j)}^{N-1} h(n - i)h(n - j), \qquad i, j = 0, \ldots, N - 1. \tag{3.249}$$

and let

$$\mathbf{\Psi}^T = \mathbf{x}^T \mathbf{H} \tag{3.250}$$

be a vector with elements

$$\psi(i) = x(i) * h(-i) = \sum_{n=i}^{N-1} x(n)h(n - i), \qquad i = 0, \ldots, N - 1. \tag{3.251}$$

The mean squared weighted error can now be minimized by maximizing the second term of Equation (3.244), which is given by

$$T_k = \frac{(C_k)^2}{\mathcal{E}_k} = \frac{(\mathbf{x}^T \mathbf{H} \mathbf{c}_k)^2}{\mathbf{c}_k^T \mathbf{H}^T \mathbf{H} \mathbf{c}_k} = \frac{(\mathbf{\Psi}^T \mathbf{c}_k)^2}{\mathbf{c}_k^T \mathbf{\Phi} \mathbf{c}_k} \tag{3.252}$$

where C_k is the cross-correlation between $\mathbf{x}$ and the filtered codeword $\mathbf{H}\mathbf{c}_k$ and it is given by

$$C_k = \sum_{n=0}^{N-1} x(n) [c_k(n) * h(n)] = \sum_{n=0}^{N-1} \psi(n)c_k(n) \tag{3.253}$$

and $\mathcal{E}_k$ is the energy of the filtered codeword c_k and it is given by

$$\begin{aligned} \mathcal{E}_k &= \sum_{n=0}^{N-1} [c_k(n) * h(n)]^2 \\ &= \sum_{i=0}^{N-1} c_k^2(i)\phi(i, i) + 2 \sum_{i=0}^{N-2} \sum_{j=i+1}^{N-1} c_k(i)c_k(j)\phi(i, j). \end{aligned} \tag{3.254}$$

$\psi(i)$ and $\phi(i,j)$ are computed outside the optimization loop, and the term T_k in Equation (3.252) is evaluated for $k = 0$ to $L - 1$, where L is the codebook size. The codeword with index k which maximizes this term is chosen, and the scalar gain β is then computed from Equation (3.243). In this approach the codeword $c_k(n)$ and the gain β are not jointly optimized since the gain has to be quantised, and the term in Equation (3.252) has been derived using the value of the unquantised gain. The gain and the excitation vector can be jointly optimized as follows: for the codeword with index k the cross-correlation C_k and the energy $\mathcal{E}_k$ are computed from Equations (3.253) and (3.254), respectively. The gain is computed as in Equation (3.243) by

$$\beta_k = \frac{C_k}{\mathcal{E}_k}. \tag{3.255}$$

The gain is then quantised to obtain the value $\hat{\beta}_k$, and this quantised value is substituted in Equation (3.240) to obtain the minimum error

$$
\begin{aligned}
E &= \sum_{n=0}^{N-1} [x(n) - \hat{\beta}_k c_k(n) * h(n)]^2 \\
&= \mathbf{x}^T \mathbf{x} - 2\hat{\beta}_k \mathbf{x}^T \mathbf{H} \mathbf{c}_k + \hat{\beta}_k^2 \mathbf{c}_k^T \mathbf{H}^T \mathbf{H} \mathbf{c}_k \\
&= \mathbf{x}^T \mathbf{x} - 2\hat{\beta}_k C_k + \hat{\beta}_k^2 \mathcal{E}_k
\end{aligned}
\tag{3.256}
$$

Thus the term to be maximized is now given by

$$T_k = \hat{\beta}_k(2C_k - \hat{\beta}_k \mathcal{E}_k). \tag{3.257}$$

This term is computed for every codeword and the one which maximizes the term is chosen along with the quantised gain. This joint optimization approach does not introduce any considerable complexity as the correlation C and the energy $\mathcal{E}$ are computed once for every codeword similar to the case when Equation (3.252) is used. The extra computational load is that the the gain has to be quantised for every possible codeword.

The number of instructions needed to evaluate the expression in Equation (3.257) is approximately N^2 (when Equations (3.253) and (3.254) are used to compute C and $\mathcal{E}$). For a codebook with 1024 entries and an excitation frame of length 40 samples, around 40000 multiplications per speech sample are needed to search the codebook. When the convolution is computed by recursive filtering, the codewords $c_k(n)$ are filtered through the zero-state filter $1/A(z/\gamma)$, where the convolution needs Np instructions, the energy computation in $\mathcal{E}_k$ requires N, and the cross-correlation evaluation in C_k also N instructions, yielding a total of $N(p+2)$ operations. For a 1024 size codebook and a predictor of order 10, around 12000 multiplications per speech sample are required to search the codebook.

It can be seen from the previous discussion that the exhaustive search of the excitation codebook is a computationally demanding procedure, which is difficult to implement in real time. We will now look at some methods which simplify the codebook search procedure without affecting the quality of the output speech.

3.5.3.2   Simplification of the CELP Search Procedure Using the Autocorrelation Approach

Different approaches have been introduced to simplify the codebook search procedure. The frequency domain can be used [119] so that the convolution $c_k(n) * h(n)$ which appears in Equation (3.242) is reduced to the multiplication $C(i)H(i)$, where $C(i)$ is a Gaussian vector and $H(i)$ is the DFT of the impulse response $h(n)$. The number of operations is reduced this way, but we need to compute the DFTs of $h(n)$ and $x(n)$. Another method similar to the frequency domain approach is also proposed in [119]. In this method, the singular value decomposition (SVD) is used to reduce the matrix $\mathbf{H}$ which appears in Equation (3.244) to a diagonal form by expressing it as $\mathbf{H} = \mathbf{U}\mathbf{D}\mathbf{V}^T$, where $\mathbf{D}$ is diagonal, while $\mathbf{U}$ and $\mathbf{V}$ are orthogonal matrices. The properties of orthogonal matrices can be used to reduce the mean squared weighted error in Equation (3.244) to a form where only $4N$ multiplications are needed to evaluate the term to be maximized. This reduces the multiplications needed to search a 1024 entries codebook to 4000 multiplications/sample. However, this approach requires the extra burden of computing the SVD of the matrix $\mathbf{H}$ for every new set of filter parameters, which is proportional to N^3 operations, and for the typical value $N = 40$, more than 1600 operations per speech sample are introduced, which can not be neglected.

A common approach to simplify the search procedure is to use the autocorrelation method [119]. In this approach, the matrix of covariances $\mathbf{\Phi} = \mathbf{H}^T\mathbf{H}$ is reduced to a Toeplitz form by modifying the summation limits in Equation (3.249) so that

$$\phi(i,j) = \phi(|i-j|) = \sum_{n=|i-j|}^{N-1} h(n)h(n-|i-j|). \qquad (3.258)$$

The autocorrelation approach results from modifying the $N \times N$ convolution matrix of Equation (3.247) into a $(2N-1) \times N$ matrix of the form

$$
\mathbf{H} = \begin{pmatrix}
h_0 & 0 & 0 & \cdots & 0 \\
h_1 & h_0 & 0 & \cdots & 0 \\
h_2 & h_1 & h_0 & \cdots & 0 \\
\vdots & \vdots & \vdots & \ddots & \vdots \\
h_{N-1} & h_{N-2} & h_{N-3} & \cdots & h_0 \\
0 & h_{N-1} & h_{N-2} & \cdots & h_0 \\
0 & 0 & h_{N-1} & \cdots & h_0 \\
\vdots & \vdots & \vdots & \ddots & \vdots \\
0 & 0 & 0 & \cdots & h_{N-1}
\end{pmatrix}. \qquad (3.259)
$$

The convolution $\mathbf{H}c_k$ using this matrix results into a $2N - 1$ length vector, obtained when convolving two segments each of length N. Notice that in the covariance approach only the first N samples of the obtained convolution are considered and any samples beyond the subframe limit are not taken into consideration. Remembering that the impulse response $h(n)$ is a sharply decaying function (see Figure 3.16), it can be truncated at a value $R - 1 < N$ (say $R = 25$) without introducing any perceptually noticeable error. In this case the dimensions of the matrix in Equation (3.259) become $(2R - 1) \times N$ and the matrix of autocorrelations $\boldsymbol{\Phi}$ becomes a band matrix (when $R - 1 < N$) with $\phi(i) = 0$ for $i \geq R$. Henceforth, we will assume that the impulse response $h(n)$ is truncated at $R - 1$.

Using the autocorrelation approach the energy of the filtered codeword $c_k(n)$ in Equation (3.254) can be written as

$$\mathcal{E}_k = \sum_{i=0}^{N-1} c_k^2(i)\phi(0) + 2 \sum_{i=0}^{N-2} \sum_{j=i+1}^{N-1} c_k(i)c_k(j)\phi(j - i). \qquad (3.260)$$

Defining $\mu_k(i)$ to be the autocorrelation of the codeword $c_k(i)$ given by

$$\mu_k(i) = \sum_{n=i}^{N-1} c_k(n)c_k(n - i), \qquad (3.261)$$

Equation (3.260) can be written as

$$\mathcal{E}_k = \mu_k(0)\phi(0) + 2 \sum_{i=1}^{R-1} \mu_k(i)\phi(i). \qquad (3.262)$$

Evaluating the energy now requires R instructions, and the term T_k in Equation (3.257) requires $N + R + 3$ instructions. Figure 3.55 shows the optimum innovation sequence search procedure using the autocorrelation approach. For $N = 40$ and $R = 25$ less than 2000 multiplications per synthesized speech sample are needed when a 1024 sized codebook is used. The autocorrelations of the codewords $\mu_k(i)$ are precomputed and stored in another codebook. The autocorrelation approach has the disadvantage of needing a second codebook at the encoder to store the autocorrelations of the excitation codebook.

3.5.3.3   Using Structured Codebooks

The autocorrelation approach discussed in the previous section simplifies the representation of the mean squared weighted error in order to reduce the excessive computational load needed to search for the optimum innovation sequence. In this section, we look at methods to simplify the CELP system by utilizing structured codebooks where the codebook structure enables fast search procedures. We will discuss sparse, ternary, overlapping, and algebraic codebooks.

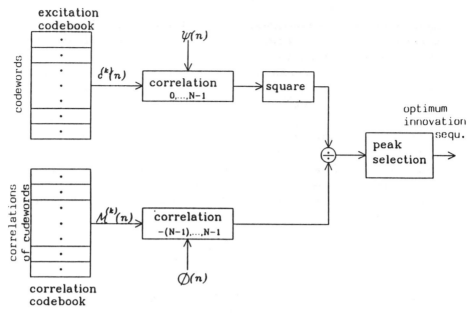

Figure 3.55: Searching for the optimum CELP innovation sequence using the autocorrelation approach

3.5.3.3.1   Sparse Excitation Codebooks   In the sparse excitation co-de-book, most of the excitation pulses in an excitation vector are set to zero. This is done by using center-clipping, where a zero-mean unit-variance Gaussian random process is used to populate the codebook, and the random variables are set to zero whenever their absolute value is below a specified threshold. The threshold value controls the codebook sparsity. Threshold values of 1.2 and 1.65 result in 77% and 90% sparsity, respectively. Sparse excitation (or center-clipping) was first proposed by Atal in his pioneer-ing work on Adaptive Predictive Coding (APC) [38], where he found that it is sufficient to quantise the high-amplitude portions of the prediction residual for achieving low perceptual distortion in the decoded speech. Us-ing center-clipping does not necessarily result in equal number of nonzero pulses in every excitation vector. To obtain w number of nonzero pulses per excitation vector, the largest w samples in the vector are retained and the remaining samples are set to zero. The use of sparse excitation vector codebooks was introduced by different authors independently [106, 108]. It was proposed by Davidson and Gersho [106] motivated by the multipulse LPC. It was shown that in MPE-LPC, about 8 pulses per pitch period are required to synthesize natural-sounding speech [89]. We have seen in Section 3.5.1 that it is sufficient to use 4 pulses in a 5 ms excitation frame (40 samples). In fact, when the speech is voiced, a few pulses are sufficient to represent the excitation signal, and setting most of the samples in the residual signal to zero does not affect the perceived speech quality. As the pulses in the excitation vector are not individually optimized, it is preferred to use sparse excitation vectors in case of voiced speech segments. On the other hand, in case of unvoiced speech segments, using nonsparse stochastic excitation vectors is more sensible. We found in our simulation that using 4 nonzero pulses in an excitation vector of 40 samples gives similar results to the original CELP where the whole excitation vector is populated from a Gaussian random process.

Using sparse excitation codebooks reduces the complexity of the CELP system by a factor around 10 when 4 nonzero pulses are used in an excitation vector of length 40. Since most of the excitation vector samples are equal to zero, most of the autocorrelations of the excitation vectors are also zero. The number of nonzero autocorrelations is not necessarily equal to the number of nonzero pulses. Using the autocorrelation approach with w nonzero pulses in the excitation vector, the cross-correlation term in Equation (3.253) can be expressed as

$$C_k = \sum_{i=0}^{w-1} \psi(m_i) g_k(i), \qquad (3.263)$$

where w is the number of nonzero pulses, $g(i)$ are the pulse amplitudes and m_i are the nonzero pulse positions. The energy term in Equation (3.254) is now given by

$$\mathcal{E}_k = \mu_k(0)\phi(0) + 2 \sum_{i=1}^{Q-1} \mu_k(n_i)\phi(n_i), \qquad (3.264)$$

where Q is the number of nonzero autocorrelations for the excitation vector at index k, and n_i are the indices of the nonzero autocorrelations μ. As the number of nonzero pulses w is much less than N, the computational effort needed to evaluate the term in Equation (3.257) is significantly reduced. Using sparse excitation vectors does not only simplify the search procedure but it also reduces the storage requirements of the codebooks. The excitation codebook will contain w pulses (usually 4) and their positions, and the autocorrelation codebook will contain the nonzero autocorrelations and their positions.

3.5.3.3.2 Ternary Codebooks The sparse excitation vector codebook search can be further simplified by using the ternary excitation approach [109], [108]. A ternary excitation vector is a sparse excitation vector in which the nonzero pulses are set to either -1 or 1. Similar to sparse excitation codebooks, ternary codebooks can be populated using a Gaussian random process with center-clipping where the random variables are set to zero if their values fall below a certain threshold, otherwise the variable is set equal to its sign (-1 or 1). This populating procedure, however, results in different number of nonzero pulses in every excitation vector. It is better to fix the number of nonzero pulses in every excitation vector in order to simplify the storage and search procedure. In this case the positions of the w pulses are chosen to be uniformly distributed between 0 and $N - 1$ and their amplitudes are randomly chosen to be either -1 or 1.

The ternary approach is more simple than using sparse excitation vectors, since $g(i)$ in Equation (3.263) is either -1 or 1 which means that multiplications in the numerator of the term are reduced to summations. Another advantage is the reduction in codebook storage requirement, since for every nonzero pulse, its position and sign can be stored in one byte with the most significant bit in the byte reserved for the pulse sign [109]. If the excitation vector contains 4 nonzero pulses, 4 bytes are needed to store each codeword. Simulation results have shown that ternary codebooks perform as well as Gaussian or sparse codebooks.

A geometric representation of the CELP excitation codebook was utilized in [109, 120] to show that sparse excitation vector codebooks and ternary codebooks are equivalent, in terms of coding performance, to the initially proposed Gaussian random codebook. We mentioned earlier that the CELP approach is based on representing the residual signal after short-term and long-term prediction by a slowly-varying power spectrum Gaussian random process. Due to the existence of the gain factor in the analysis-by-synthesis loop, all the codewords in the excitation codebook can be assumed to have unit energy, and the gain factor introduces the flexibility in changing the power spectrum of the excitation. Therefore, the excitation codebook of size L can be represented by L points on the surface of a unit sphere in N-dimensional space centered at the origin. In fact, the Gaussian process by which the excitation signal is modelled can be considered as all the points on the surface of the unit sphere. Now since the excitation codewords are chosen at random from this Gaussian process, and due to the spherical symmetry of the multi-dimensional Gaussian distribution, the L points representing the codewords are uniformly distributed over the surface of the sphere. In the case of sparse excitation or ternary excitation vectors, there are w nonzero pulses whose positions are chosen at random. The total number of position combinations is $^{N}C_{w}$. Provided that the codebook size L is much less than the number of position combinations, the sparse excitation or ternary excitation codebooks can still be considered as L points uniformly distributed over the surface of unit sphere. To explain this, let us take the ternary case with the typical values $N = 40$, $w = 4$ and $L = 1024$. The number of possible position combinations is 91390, and since we have 4 pulses with amplitudes -1 or 1, the total number of the possible codewords from which the codebook is chosen is 91390×16. Since these 1.5 million points are distributed all over the surface of the unit sphere, and since the excitation codebook is randomly chosen from this huge number of possibilities, the resulting ternary codebook can be considered as 1024 points uniformly distributed over the surface of the unit sphere in the 40-dimensional space. This illustrates the equivalence of the ternary codebook to the originally proposed one where all the N pulses in the codeword are Gaussian random variables.

Ternary codebooks can be properly structured to result into fast codebook search algorithms. Spherical lattice codebooks were proposed by Ireton and Xydeas [122] where large excitation codebooks can be used without requiring the CELP complexity. Regular pulse ternary codebooks were also proposed [123, 124] where the codebook can be efficiently exhaustively searched using Gray codes [113]. Adoul et al [115, 120, 121] used algebraic codes for populating the ternary codebook where the special structure of the codebook leads to fast search algorithms. A special structure of algebraic codes will be discussed in the next subsection.

3.5.3.3.3 Algebraic codebooks Algebraic codes can be used to populate the excitation codebooks. Efficient codebook search algorithms can be obtained using the highly structured algebraic codes. Initially, algebraic codebooks were obtained using binary error-correction codes [120]. We describe here an algebraic code whereby the excitation vectors are derived using interlaced permutation codes (IPC).

In the interlaced permutation codes, an excitation vector contains a few number of nonzero pulses with predefined interlaced sets of positions. The pulses have their amplitudes fixed to 1 or -1, and each pulse have a set of possible positions distinct from the positions of the other pulses. The sets of positions are interlaced. The excitation code is identified by the positions of its nonzero pulses. Thus, searching the codebook is in essence searching the optimum positions of the nonzero pulses. To further explain the codebook structure, we describe a 12 bit codebook used to encode 60 samples excitation vectors (utilized in 4.8 kb/s speech coding). The excitation vector contains 4 nonzero pulses having amplitudes of 1, -1, 1, and -1, respectively. Each pulse can take one of 8 possible positions, and each position is encoded with 3 bits resulting in a 12 bit codebook. If the sets of positions are denoted by $m_i^{(j)}$, $j = 0, \ldots, 3$ and $i = 0, \ldots, 7$ then

$$m_i^{(j)} = 2j + 8i, \qquad \begin{array}{l} j = 0, \ldots, 3, \\ i = 0, \ldots, 7. \end{array} \qquad (3.265)$$

The pulse amplitudes and sets of positions are given in Table 3.10.

This codebook structure has several advantages. Firstly, it does not require any storage. Secondly, it has inherent robustness against channel errors as the pulse positions are transmitted and one channel error will alter only the position of one pulse. The most important advantage, however, is that the codebook can be very efficiently searched. Denoting the pulse positions by m_i, $i = 0, \ldots, 3$, then the cross-correlation term of Equation (3.253) is given by

$$C = \psi(m_0) - \psi(m_1) + \psi(m_2) - \psi(m_3), \qquad (3.266)$$

amp.	potential positions
1	0, 8, 16, 24, 32, 40, 48, 56
−1	2, 10, 18, 26, 34, 42, 50, 58
1	4, 12, 20, 28, 36, 44, 52, (60)
−1	6, 14, 22, 30, 38, 46, 54, (62)

Table 3.10: Amplitudes and possible positions of the excitation pulses in the 12 bit algebraic code

and the energy term of Equation (3.254) is given by

$$
\begin{aligned}
\mathcal{E} \;=\; & \phi(m_0, m_0) \\
& +\phi(m_1, m_1) - 2\phi(m_1, m_0) \\
& +\phi(m_2, m_2) + 2\phi(m_2, m_0) - 2\phi(m_2, m_1) \\
& +\phi(m_3, m_3) - 2\phi(m_3, m_0) + 2\phi(m_3, m_1) - 2\phi(m_3, m_2)
\end{aligned}
$$

By changing only one pulse position at a time, the correlation and energy terms can be very easily updated. The search is accomplished in 4 nested loops where in the inner most loop, the correlation is updated with one addition and the energy with 4 additions and one multiplication. Despite the efficiency of the search procedure, the exhaustive search becomes rapidly complicated as the codebook size exceeds 2^{12}. For searching huge excitation codebooks, a focussed search strategy has been developed [131]. In this approach, a very small subset of the codebook is searched while guaranteeing a performance very close to that of full search.

3.5.3.3.4 Overlapping Codebooks Another efficient codebook structure is represented by overlapped excitation codebooks [108]. This overlapping concept can be combined with sparse or ternary excitation concepts yielding very efficient codebook search algorithms and minimal storage requirements. In an overlapping shift by k codebook, each codeword is obtained by shifting the previous codeword by k samples and adding k new samples. Therefore two adjacent codewords share all but k samples. The first advantage of overlapping codebooks is reducing the codebook storage requirement. For a codebook of size L with N dimension vectors, $N + k(L - 1)$ samples need to be stored. Using stochastically derived ternary excitation most of these samples (about 80%) are zero and the rest are −1 or 1. A shift by 2 was found very efficient and it gave identical results to those obtained using non-overlapping codebooks [110]. Besides reducing the codebook storage, the other advantage of overlapping codebooks is reducing the computational load needed to exhaustively search the codebook.

Consider the stochastic sequence $q(n)$, $n = 0, \ldots, k(L-1) + N - 1$. For a shift by k codebook the codewords are given by

$$c_j(n) = q(kj + n), \qquad \begin{aligned} n &= 0, \ldots, N-1, \\ j &= 0, \ldots, L-1, \end{aligned} \qquad (3.267)$$

where N is the excitation vector length and L is the codebook size. The first observation is the reduction in the codebook storage. For a shift by 2 codebook ($k = 2$), $2L + N - 2$ samples are needed to be stored instead of NL samples. The main advantage is the reduction in the search complexity, when evaluating Equation (3.242). The convolution

$$\rho_j(n) = c_j(n) * h(n)$$

requires $N(N+1)/2$ instructions and it has to be determined for every codeword. For a shift by k codebook

$$\begin{aligned}
\rho_j(n) &= \sum_{i=0}^{n} c_j(i)h(n-i) = \sum_{i=0}^{n} q(kj+i)h(n-i), \\
&= \sum_{i=0}^{k-1} q(kj+i)h(n-i) + \sum_{i=k}^{n} q(kj+i)h(n-i) \qquad (3.268) \\
&= \sum_{i=0}^{k-1} q(kj+i)h(n-i) + \sum_{m=0}^{n-k} q(k(j+1)+m)h(n-k-m).
\end{aligned}$$

From Equation (3.268)

$$\rho_j(n) = \sum_{i=0}^{k-1} q(kj+i)h(n-i) + \rho_{j+1}(n-k). \qquad (3.269)$$

For a shift by 1 codebook, as for the adaptive codebook, the following relation is obtained

$$\begin{aligned}
\rho_j(0) &= q(j)h(0) \\
\rho_j(n) &= q(j)h(n) + \rho_{j+1}(n-1), \qquad n = 1, \ldots, N-1. \,(3.270)
\end{aligned}$$

The convolution of the last codeword is first determined by

$$\rho_{L-1}(n) = \sum_{i=0}^{n} q(L-1+i)h(n-i), \qquad (3.271)$$

and then the relation in (3.270) is used to update the convolution from $j = L - 2$ down to 0. Updating the convolution $\rho_j(n)$ requires N instructions. The impulse response, $h(n)$, of the weighted synthesis filter can be truncated at $R - 1$ where R is usually 25 without any loss in accuracy [110] (see Figure 3.16). In this case R instructions are needed to update $\rho_j(n)$. In case of shift by 2 codebooks Equation (3.269) is reduced to

$$
\begin{aligned}
\rho_j(0) &= q(2j)h(0) \\
\rho_j(1) &= q(2j)h(1) + q(2j+1)h(0) \\
\rho_j(n) &= q(2j)h(n) + q(2j+1)h(n-1) + \rho_{j+1}(n-2), \quad n = 2 \ldots N-1.
\end{aligned}
$$

$$(3.272)$$

The value of $\rho_{L-1}(n)$ is initially computed as in Equation (3.271) then the relation in (3.272) is used to update $\rho_j(n)$ from $j = L-2$ down to 0. If $h(n)$ is truncated at $R-1$, then $2R-1$ instructions are needed to update $\rho_j(n)$.

In the case of the adaptive LTP codebook, the registration buffer $q(n)$ contains the excitation history at the input of the weighted synthesis filter $1/A(z/\gamma)$, that is $u(n)$ from $n = -L_a$ to -1 where L_a is the buffer length and its contents are updated in every new subframe by shifting the buffer contents to the left by N positions and introducing new N values. The term T_k to be maximized in this case requires about $2N + R$ instructions (R for the convolution ρ_j, N for the energy $\mathcal{E}_j$, and N for the correlation $\mathcal{C}_j$).

In case of the stochastic codebook, sparse or ternary sequences are usually used, and the ternary approach is preferred as it reduces the storage and complexity. In the ternary case the sequence $q(n)$ contains values -1, 1, or 0. The number of zeros (the sparsity) is usually 80-90%. The ternary sequence is derived by center-clipping a unit variance Gaussian sequence at a certain threshold (which determines the sparsity). In the DoD coder [82] a threshold of 1.2 is used which results in 77% sparsity. A shift by 2 sparse stochastic codebook has been found equivalent in performance to a non-overlapping codebook. Using sparse codebooks reduces the complexity dramatically. For a shift by 2 codebook, when either $c_j(0)$ or $c_j(1)$ is zero (that is $q(2j)$ and $q(2j+1)$) the number of instructions in the relation of (3.272) is reduced to R. Further, if both $c_j(0)$ and $c_j(1)$ are zero, no more instructions are needed to update $\rho_j(n)$.

The autocorrelation approach can also be used to update the energy term [110] which is given by

$$
\mathcal{E}_j = \mu_j(0)\phi(0) + 2 \sum_{n=1}^{N-1} \mu_j(n)\phi(n),
\tag{3.273}
$$

where $\mu_j(n)$ is the autocorrelation of the codeword $c_j(n)$ and for a shift by k codebook it is given by

$$
\mu_j(n) = \sum_{i=n}^{N-1} c_j(i)c_j(i-n) = \sum_{i=n}^{N-1} q(kj+i)q(kj+i-n).
\tag{3.274}
$$

It can be easily shown that the correlations $\mu_j(n)$ are updated by

$$\mu_{j+1}(n) = \mu_j(n) + \sum_{i=N}^{N+k-1} q(kj+i)q(kj+i-n) - \sum_{i=n}^{n+k-1} q(kj+i)q(kj+i-n).$$

$$(3.275)$$

Based on Equation (3.275) it can be easily shown that for a shift by 1 codebook the filtered codeword energy, using the autocorrelation approach, is updated by

$$\mathcal{E}_{j+1} = \mathcal{E}_j \quad - \quad q(j)\left[\phi(0) + 2\sum_{n=1}^{N-1} q(j+n)\phi(n)\right]$$

$$+ \quad q(j+N)\left[\phi(0) + 2\sum_{n=1}^{N-1} q(j+N-n)\phi(n)\right].(3.276)$$

When $\phi(n)$ is truncated at $R-1$, where R is typically 25, $2R+2$ instructions are needed to update the energy. Similarly, in case of shift by 2 codebooks the energy is updated by

$$\mathcal{E}_{j+1} = \mathcal{E}_j \quad - \quad q(2j)\left[\phi(0) + 2\sum_{n=1}^{N-1} q(2j+n)\phi(n)\right]$$

$$- \quad q(2j+1)\left[\phi(0) + 2\sum_{n=1}^{N-1} q(2j+1+n)\phi(n)\right]$$

$$+ \quad q(2j+N)\left[\phi(0) + 2\sum_{n=1}^{N-1} q(2j+N-n)\phi(n)\right]$$

$$+ \quad q(2j+N+1)\left[\phi(0) + 2\sum_{n=1}^{N-1} q(2j+N+1-n)\phi(n)\right].$$

$$(3.277)$$

In this case $4R + 4$ instructions are required to update the energy. Using the convolution approach $2R + N - 1$ operations are required. For 4.8 kb/s coding where $N = 60$ both approaches are similar in computational sense. For smaller frame sizes the convolution approach is preferred. Another great advantage of the convolution approach is its efficiency with sparse codebooks. In the convolution approach, when both $q(2j)$ and $q(2j + 1)$ are zero, no operations are needed to update the convolution, and the energy is updated by

$$\mathcal{E}_{j+1} = \mathcal{E}_j - [\rho_{j+1}(N-2)]^2 - [\rho_{j+1}(N-1)]^2 \qquad (3.278)$$

which requires 2 instructions. In case of the autocorrelation approach $q(2j)$, $q(2j+1)$, $q(2j+N)$ and $q(2j+N+1)$ have to be zero in order to update the energy without requiring any instructions, and this is less likely to happen. Thus for shift by 2 stochastic sparse codebooks the convolution approach is more attractive than the autocorrelation approach.

3.5.3.3.5 Self Excitation Another approach to define the excitation signal is the self-excitation concept. The self-excited LPC [117] can be seen as another variant of CELP in which the codebook is changing. The self-excitation structure offers some simplicity since the excitation can be viewed as a shift by 1 overlapping adaptive codebook (similar to the pitch codebook), but it has a serious disadvantage of propagating channel errors. In self-excitation systems, the excitation sequence is determined by searching through a buffer which contains the previous history of the excitation (or the past decoded speech [118]). Both the encoder and decoder use the same excitation buffer, and the buffers at the encoder and decoder are initially filled with the same Gaussian random sequence. The excitation sequence in the present frame is determined by searching through the excitation buffer for the sequence which minimizes the weighted error between the original and synthesized speech. The excitation is determined by a delay and a corresponding gain factor, and the excitation buffer is updated in every new frame using the excitation determined in the previous frame. The excitation buffers at both the encoder and decoder should have the same content in order to generate identical synthesized speech. This is true in the absence of channel errors. However, in practical applications, the encoded speech parameters could be perturbed due to the noise in the transmission channel, and a single bit error occurring to one of the excitation parameters will cause a mismatch between the excitation buffers at the encoder and decoder, and this will persist for the forthcoming frames.

The same algorithms used in overlapping codebooks and described in the previous section can be deployed in the self-excited (SE) coder as the excitation can be represented by an overlapping shift by 1 adaptive codebooks. However, overlapping fixed stochastic codebooks give similar performance with much less complexity (due to their sparsity), and they are more robust against channel errors than the self excited approach.

3.5.3.4 CELP performance

The CELP coder has been evaluated at the bit rates from 4.8 to 8 kb/s. The resulting speech quality ranged from communications quality at 4.8 kb/s to near-toll quality at 8 kb/s. The bit allocations at 4.4 and 8 kb/s are shown in Tables 3.11 and 3.12, respectively. In 4.4 kb/s coding the speech frame is 30 ms long divided into 4 subframes of 7.5 ms (60 samples) while in 8 kb/s coding a 16 ms speech frame is used and is divided in 4 subframes of 4 ms. The histogram of the magnitude of the excitation gain and its logarithm are shown in Figure 3.56 (a) and (b), respectively. The sign of the gain is quantised with one bit and the magnitude can be efficiently quantised with 4 bits using either logarithmic or non-uniform quantization.

Figures 3.57 and 3.58 show a speech segment, the CELP excitation, the synthesis filter excitation, and the reconstructed speech in cases of gaussian codebook and ternary codebook, respectively. A 5 ms excitation frame and a 512 sized codebook are used (6.6 kb/s coding). The variation of speech

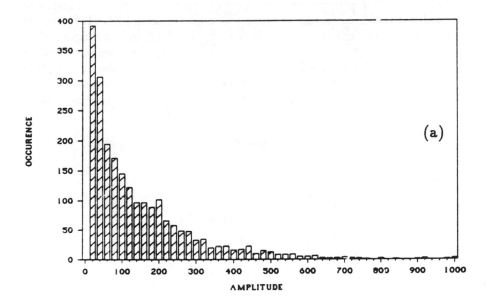

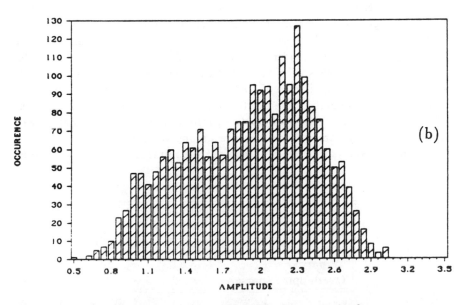

Figure 3.56: (a) Histogram of the codebook gain magnitude
(b) Histogram of the logarithmic codebook gain magnitude

Parameter	Number of Bits
LSF's	36 (3,3,4,4,4,4,4,4,3,3)
LTP delays	24 (7,5,7,5)
LTP gains	16 (4 × 4)
book indices	36 (4 × 9)
excitation gains	20 (4 × 5)
Total	132 bits per 30 ms

Table 3.11: Bit allocation for 4.4 kb/s CELP coding

Parameter	Number of Bits
LSF's	36 (3,3,4,4,4,4,4,4,3,3)
LTP delays	24 (7,5,7,5)
LTP gains	12 (4 × 3)
book indices	36 (4 × 9)
excitation gains	20 (4 × 5)
Total	128 bits per 16 ms

Table 3.12: Bit allocation for 8 kb/s CELP coding

power and SEGSNR vs. time for the sentence *"to reach the end he needs much courage"* uttered by a female speaker is seen in Figure 3.59.

The different CELP approaches described earlier were compared and they all showed similar performances. Table 3.13 shows the SEGSNR for the different approaches with 4 ms excitation vectors (32 samples) and 512 sized codebooks (at a bit rate of 7.8 kb/s). The equivalence of the different approaches for populating the excitation codebook becomes clear from the dB figures quoted. To our satisfaction, informal subjective listening tests did not show any perceivable difference among them either.

Figure 3.60 shows the SEGSNR against the number of codebook address

Codebook Population	SEGSNR (dB)
Gaussian	14.03
Sparse	14.06
Ternary	13.81
Overlapping sparse	14.09

Table 3.13: SEGSNR for different CELP approaches at 7.8 kb/s coding

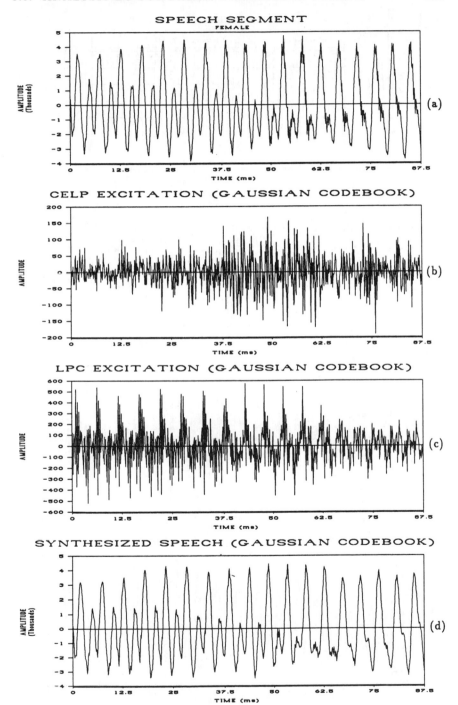

Figure 3.57: (a) A 87.5 ms speech segment,
(b) CELP excitation,
(c) LPC excitation,
(d) reconstructed speech, in case of a Gaussian codebook

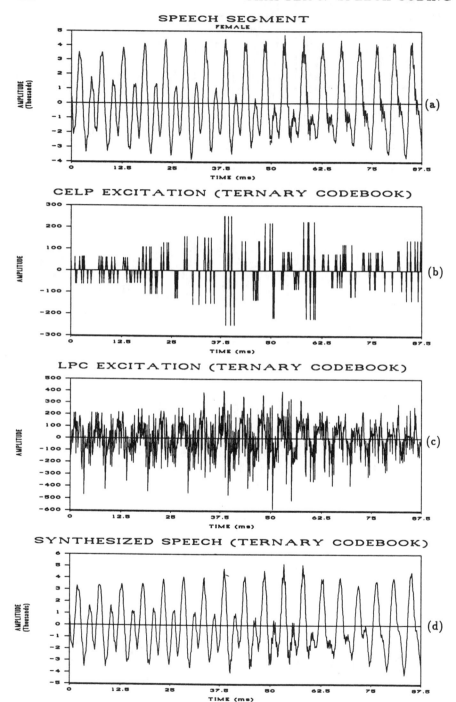

Figure 3.58: (a) A 87.5 ms speech segment,
                (b) CELP excitation,
                (c) LPC excitation,
                (d) reconstructed speech, in case of a ternary codebook

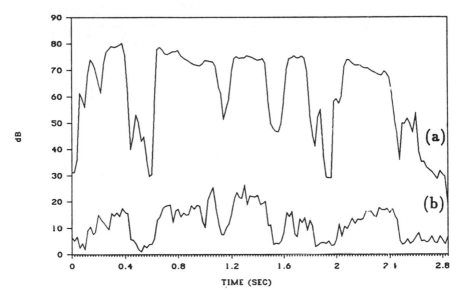

Figure 3.59: (a) Speech power and (b) SEGSNR variation vs. time for the CELP codec using ternary codebook

bits using a ternary excitation codebook with 5 ms excitation vectors. We found that at least a 9-bit codebook is needed to maintain high speech quality. Larger than 10-bit codebooks become impractical due to the exponential increase in the coder complexity.

3.5.4 Binary Pulse Excitation

We discussed in the previous section code-excited linear prediction coding and described several methods which reduce the coder complexity. These computationally efficient methods have reduced the excessive complexity of the original algorithm but the exhaustive search of the the excitation codebook has still to be performed. In this chapter, a novel approach for representing the excitation signal called *transformed binary pulse excitation (TBPE)* is described. In this approach the excitation signal consists of regularly spaced stochastically derived pulses, where very efficient algorithms for determining the excitation sequence can be obtained. We will describe the excitation definition and derive efficient algorithms for exhaustive and non-exhaustive search of the excitation sequence. A performance comparison between the TBPE and CELP will be given.

3.5.4.1 Transformed Binary Pulse Excitation

The block diagram of the TBPE coder is shown in Figure 3.61. The coder

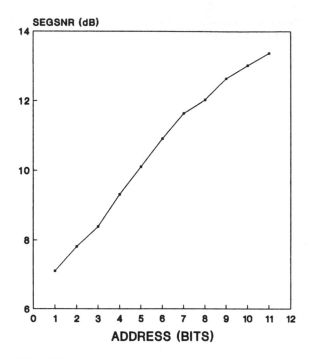

Figure 3.60: SEGSNR against codebook size in terms of number of address bits using ternary codebook with 5 ms excitation vectors

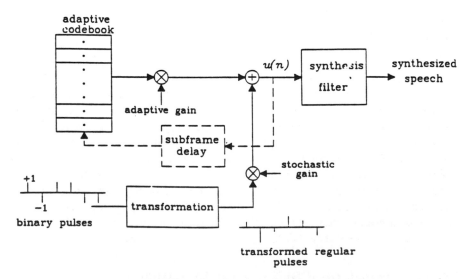

Figure 3.61: Block diagram of the transformed BPE coder

has common features with both the RPE and CELP. The excitation signal consists of a number of pseudo-stochastic pulses with predefined pulse positions. In an excitation frame of length N, we suppose that there are M nonzero pulses separated by $D - 1$ zeros, where $M = N$ DIV D, and DIV denotes integer division. The excitation vector is given by

$$v(n) = \beta \sum_{i=0}^{M-1} g_i \delta(n - m_i), \qquad n = 0, \ldots, N - 1 \qquad (3.279)$$

where $\delta(n)$ is the Kronecker delta, g_i are the pulse amplitudes, m_i are the pulse positions, and β is a scalar gain similar to that which appears in the CELP. As in the RPE approach [1], there are D sets of pulse positions given by

$$m_i^{(k)} = k + iD, \qquad \begin{array}{l} i = 0, \ldots, M - 1 \\ k = 0, \ldots, D - 1 \end{array} \qquad (3.280)$$

where D is the pulse spacing, and k is the position of the first pulse. In RPE coders, the optimum pulse amplitudes and first pulse position are determined by minimizing the mean-squared weighted error between the original and synthesized speech, and this requires solving a set of $M \times M$ equations D times. Further, the pulse amplitudes in RPE are each quantised with 3 bits after scaling by the maximum pulse, or the rms value of the pulses, which is quantised with 5 or 6 bits. The large number of bits needed to quantise the pulse amplitudes in RPE makes it difficult to achieve high quality speech below 9.6 kbit/s. In our TBPE approach, the pulses are pseudo-stochastic random variables, similar to the CELP concept, and they are quantised only with one bit per pulse, in addition to the scaling gain β.

Instead of obtaining the pulse amplitudes g_i, $i = 0, \ldots, M - 1$, from a large stochastic codebook as in the CELP approach, the pulses are determined by the transformation of a binary vector. That is

$$\mathbf{g} = \mathbf{Ab}, \qquad (3.281)$$

where $\mathbf{b}$ is an $M \times 1$ binary vector with elements -1 or 1, $\mathbf{A}$ is an $M \times M$ transformation matrix, and $\mathbf{g}$ is the excitation vector containing the pulse amplitudes. The vector $\mathbf{b}$ could be one of 2^M possible binary patterns, which means 2^M different excitation vectors can be obtained using the transformation in Equation (3.281). Thus, this transformation is equivalent to a 2^M sized codebook with the need to only store an $M \times M$ matrix. The equivalent of smaller codebook sizes can be obtained by setting some of the binary pulses to fixed values, or by omitting some of the columns of the matrix $\mathbf{A}$. If the hypothetic codebook size is to be reduced by a factor m, either m pulses in the binary vector are made fixed (say -1), or m columns are omitted from the matrix $\mathbf{A}$ resulting into an $M \times Q$ transformation matrix and $Q \times 1$ binary vector, where $Q = M - m$. On the other hand, the equivalent of larger codebooks is obtained by utilizing several transformation matrices. Using m different transformation matrices is equivalent to a book of size 2^{M+m}.

When the transformation matrix is of dimension $N \times M$ the resulting excitation vector is of dimension $N \times 1$. The excitation in this case is not sparse, and this transformation becomes equivalent to the VSELP approach [114, 113]. The M columns of the $N \times M$ transformation matrix are equivalent to the M basis vectors of the VSELP. If the columns of the $N \times M$ matrix are given by c_i, $i = 0, \ldots, M-1$, then the excitation vector v can be expressed as

$$v = \sum_{i=0}^{M-1} b_i c_i, \tag{3.282}$$

where b_i are the elements of the binary vector with values -1 or 1. The equivalence of an $N \times M$ transformation to the VSELP approach becomes plausible in Equation (3.282). However, using an $M \times M$ transformation reduces the complexity (as the excitation vectors become sparse with regularly spaced pulses) without affecting the speech quality.

For the special case where the transformation matrix is equal to the identity matrix I, the excitation pulses are binary with values -1 or 1. This regular pulse binary codebook has been suggested by many authors [123, 124, 127]. In fact, binary codebooks were first proposed by Leguyader et al. [128] where the excitation is non-sparse and the pulses are -1 or 1. Sparse codebooks yield better performance, and setting most of the binary pulses to zero results in ternary excitation vectors [108, 109]. Regular pulse binary vectors are ternary vectors where the nonzero pulse positions are predefined to be equally spaced. This eliminates the codebook storage and yields very efficient search algorithms as we will see later in this Chapter.

The regular binary pulse excitation vectors can be viewed as 2^M points regularly distributed over the surface of a sphere in N-dimensional space. When the transformation matrix A is orthogonal (i.e. $A^T A = I$), the transformation results into a vector containing Gaussian random variables. Generating binary pulses at random and examining the distribution of the pulses g_i resulting from the orthogonal transformation reveals that the variables g_i follow a Gaussian distribution with zero mean and unit variance. Applying an orthogonal transformation to the binary vectors rotates the vectors without changing their distribution in the N-dimensional space. Both identity and orthogonal transformations exhibited similar objective performances, however, using an orthogonal transformation resulted in slight improvement in the subjective speech quality.

A further speech quality improvement was achieved when the transformation matrix was derived from a training set of RPE vectors as the RPE approach gives the optimum amplitudes of the excitation pulses. Any other iterative algorithm which minimizes the expectation of the perceptually weighted error between the original and synthesized speech can be used to derive the transformation matrix. For example, the iterative method used to optimize the VSELP basis vectors [113] can be used here. In fact, the advantage of incorporating the transformation is that it introduces a general framework for defining the excitation characteristics, where the transformation matrix can be chosen in a way to obtain some desired codebook properties. Laflamme et al. [129] have recently proposed an elegant approach for defining the transformation, or the shaping matrix where the matrix is a function of the LPC filter $A(z)$, resulting in a codebook which is dynamically frequency-shaped. In the their implementation [131], the transformation is a lower triangular matrix containing the impulse response of the filter

$$F(z) = (1 - \mu z^{-1}) \frac{A(z/\gamma_1)}{A(z/\gamma_2)}. \tag{3.283}$$

This filter have a similar role to that of postfiltering (see Section 3.5.5) with the advantage of being implemented inside the analysis-by-synthesis loop.

3.5.4.2 Excitation Determination

The weighted error between the original and synthesized speech is given by

$$e_w(n) = s_w(n) - \hat{s}_w(n), \tag{3.284}$$

where $s_w(n)$ is the weighted input speech and $\hat{s}_w(n)$ is the weighted synthesized speech. The weighted synthesized speech can be written as

$$\hat{s}_w(n) = \sum_{i=0}^{n} v(i)h(n-i) + Gc_\alpha(n) * h(n) + \hat{s}_0(n), \tag{3.285}$$

where $h(n)$ is the impulse response of the weighted synthesis filter $1/A(z/\gamma)$, G is the adaptive codebook gain (or the LTP gain), $c_\alpha(n)$ is the codeword chosen from the adaptive codebook (or α is the LTP delay), and $\hat{s}_0(n)$ is the zero-input response of the weighted synthesis filter.

From Equations (3.284) and (3.285), the weighted error can be written as

$$e_w(n) = x(n) - \sum_{i=0}^{n} v(i)h(n-i), \tag{3.286}$$

where

$$x(n) = s_w(n) - Gc_\alpha(n) * h(n) - \hat{s}_0(n). \tag{3.287}$$

Now, substituting the excitation signal $v(n)$ from Equation (3.279) into Equation (3.286) gives

$$e_w(n) = x(n) - \sum_{i=0}^{n} \beta \sum_{k=0}^{M-1} g_k \delta(i - m_k) h(n - i),$$

$$= \quad x(n) - \beta \sum_{k=0}^{M-1} g_k h(n - m_k), \qquad n = 0, \ldots, N-1,$$

$$(3.288)$$

where $h(n-m_k) = 0$ for $n < m_k$. The excitation parameters are determined by minimizing the mean square of the weighted error $e_w(n)$ which is given by

$$E = \sum_{n=0}^{N-1} \left[x(n) - \beta \sum_{i=0}^{M-1} g_i h(n - m_i) \right]^2. \qquad (3.289)$$

Setting $\partial E / \partial \beta$ to zero leads to

$$\beta = \frac{\sum_{i=0}^{M-1} g_i \psi(m_i)}{\sum_{i=0}^{M-1} \sum_{j=0}^{M-1} g_i g_j \phi(m_i, m_j)}, \qquad (3.290)$$

where ψ is the correlation between $x(n)$ and the impulse response $h(n)$, given by

$$\psi(i) = \sum_{n=i}^{N-1} x(n) h(n - i), \qquad (3.291)$$

and ϕ is the autocorrelation of the impulse response $h(n)$ given by

$$\phi(i, j) = \sum_{n=\max(i,j)}^{N-1} h(n - i) h(n - j). \qquad (3.292)$$

By substituting Equation (3.290) into Equation (3.289), the minimum mean squared weighted error between the original and synthesized speech can be written as

$$E \quad = \quad \sum_{n=0}^{N-1} x^2(n) - \beta \sum_{i=0}^{M-1} g_i \psi(m_i),$$

$$= \quad \sum_{n=0}^{N-1} x^2(n) - \frac{\left[\sum_{i=0}^{M-1} g_i \psi(m_i) \right]^2}{\sum_{i=0}^{M-1} \sum_{j=0}^{M-1} g_i g_j \phi(m_i, m_j)}. \qquad (3.293)$$

Equation (3.293) can be written in matrix form as

$$E \quad = \quad \mathbf{x}^T \mathbf{x} - \frac{(\boldsymbol{\Psi}^T \mathbf{g})^2}{\mathbf{g}^T \boldsymbol{\Phi} \mathbf{g}} \quad = \quad \mathbf{x}^T \mathbf{x} - \frac{(\boldsymbol{\Psi}^T \mathbf{A} \mathbf{b})^2}{\mathbf{b}^T \mathbf{A}^T \boldsymbol{\Phi} \mathbf{A} \mathbf{b}}, \qquad (3.294)$$

where $\mathbf{x}$ is an $N \times 1$ vector, $\boldsymbol{\Psi}$ and $\mathbf{b}$ are $M \times 1$ vectors, and $\boldsymbol{\Phi}$ is an $M \times M$ matrix with elements $\phi(m_i, m_j)$, $i, j = 0, \ldots, M-1$. The autocorrelation approach can be used to express $\phi(m_i, m_j) = \phi(|m_i - m_j|)$. In this case, the matrix $\boldsymbol{\Phi}$ is reduced to a Toeplitz symmetric matrix with diagonal $\phi(0)$ and off-diagonals $\phi(D), \phi(2D), \ldots, \phi([M-1]D)$, respectively (see Equation (3.47)). Defining

$$z = A^T \Psi, \qquad (3.295)$$

and

$$\Theta = A^T \Phi A, \qquad (3.296)$$

Equation (3.294) becomes

$$E = x^T x - \frac{(z^T b)^2}{b^T \Theta b}. \qquad (3.297)$$

The excitation gain and binary code can be jointly optimized similar to the CELP case, and the mean squared weighted error becomes (see Equation (4.22))

$$E = x^T x - \hat{\beta}(2C - \hat{\beta}\mathcal{E}), \qquad (3.298)$$

where $\hat{\beta}$ is the quantised value of the gain $\beta = C/\mathcal{E}$, C is the cross-correlation between $x(n)$ and the filtered excitation (see Equation (4.19)) given by

$$C = z^T b, \qquad (3.299)$$

and $\mathcal{E}$ is the energy of the filtered excitation (see Equation (4.20)) given by

$$\mathcal{E} = b^T \Theta b. \qquad (3.300)$$

The optimum excitation vector is the one which maximizes the second term in Equation (3.298) given by

$$T = \hat{\beta}(2C - \hat{\beta}\mathcal{E}). \qquad (3.301)$$

3.5.4.2.1 Efficient Exhaustive Search: The Gray Code Approach

To determine the optimum innovation sequence, one could exhaustively search through all possible binary patterns and select the pattern which maximizes the term in Equation (3.301). This can be easily done using a Gray code counter [113, 126], where the Hamming distance between adjacent binary patterns is 1. As for every new pattern only one pulse is changed, C and $\mathcal{E}$ can be simply updated taking into account the pulse which has been toggled. Using a Gray code counter, the cross-correlation C is updated by

$$C_k = C_{k-1} + 2z_j b_j^{(k)}, \qquad (3.302)$$

where k is the index of the Gray code and j is the index of the pulse which has been toggled. The energy of the filtered excitation $\mathcal{E}$ can be similarly updated by

$$\mathcal{E}_k = \mathcal{E}_{k-1} + 4b_j^{(k)} \sum_{\substack{i=0 \\ i \neq j}}^{M-1} b_i^{(k)} \theta(i, j). \qquad (3.303)$$

To get rid of the multiplications by 2 and 4 in Equations (3.302) and (3.303) we define [113]

$$C' = C/2 \qquad \text{and} \qquad \mathcal{E}' = \mathcal{E}/4.$$

The term in (3.301) now becomes

$$T = 4\hat{\beta}(C' - \hat{\beta}\mathcal{E}'). \tag{3.304}$$

Equations (3.302) and (3.303) are reduced to

$$C'_k = C'_{k-1} + z_j b_j^{(k)}, \tag{3.305}$$

and

$$\mathcal{E}'_k = \mathcal{E}'_{k-1} + b_j^{(k)} \sum_{\substack{i=0 \\ i \neq j}}^{M-1} b_i^{(k)} \theta(i,j). \tag{3.306}$$

Equations (3.305) and (3.306) offer a very efficient method to exhaustively search for the best binary excitation pattern. For every new pattern, $M+1$ operations are needed to update both C' and $\mathcal{E}'$. This is much more efficient than the overlapping codebook approach described in Chapter 4. The Gray code approach is also used in the VSELP coder [113] which has been selected for the future American digital mobile radio system. The VSELP differs from the TBPE in defining the transformation matrix where a $N \times M$ dimension matrix is used (M basis vectors of length N), which results in nonsparse excitation vectors.

3.5.4.2.2 Non-exhaustive Search

Although the exhaustive search based Gray code approach is very efficient, the regular structure of the excitation pulses results into a much simpler excitation determination procedure in which the exhaustive search is ruled out [126, 123].

A closer look at the autocorrelation matrix Φ in Equation (3.47) suggests that it is strongly diagonal, because the magnitude of $\phi(nD)$ (D is usually 4) is much less than $\phi(0)$ (see Figure 3.16). As $\mathbf{A}^T\mathbf{A}$ is equal to the identity matrix $\mathbf{I}$ (when $\mathbf{A}$ is orthogonal), Θ of Equation (3.296) is also strongly diagonal. Therefore, as $\mathbf{b}^T\mathbf{b}$ is constant ($= M$), the denominator in Equation (3.297) (the energy of the filtered excitation $\mathcal{E}$) can be approximated by a constant equal to $M\phi(0)$. Figure 3.62 shows an example of the variation of the magnitude of both the numerator C^2 and the denominator $\mathcal{E}$ with varying the binary pattern ($M = 10$). It is clear that the change in the term to be maximized $C^2/\mathcal{E}$ is dominated by the value of the numerator. Thus, minimization of the error in Equation (3.297) can be performed by maximizing the numerator, i.e. maximizing the absolute value of the cross-correlation $C = \mathbf{z}^T\mathbf{b}$, and this can be simply done by choosing the pulses to be equal to the signs of $\mathbf{z}$, i.e.

$$b_i = \text{sign}\{z_i\}, \qquad\qquad i = 0,\ldots, M-1. \tag{3.307}$$

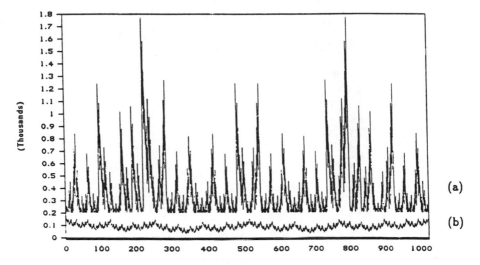

Figure 3.62: Variation of (a) the numerator $\mathcal{C}^2$, and (b) denominator $\mathcal{E}$, with varying the binary pattern from 0 to 1023 ($M = 10$)

Equation (3.307) offers an extremely simple excitation determination procedure in which no exhaustive search is needed.

Figure 3.63 shows the histogram of the Hamming distance between the binary vector b_0 determined using the simple relation of Equation (3.307) and the optimum binary vector b_{opt} determined by the exhaustive search through all the possible binary vectors for the one which minimizes the mean squared weighted error. The exhaustive search is performed with the joint optimization of the binary vector and excitation gain. We notice that the optimum vector is properly computed 72% of the time since the Hamming distance between b_0 and b_{opt} over that time is zero. For about 23% of the time the Hamming distance is one, which means that whenever Equation (3.307) fails to determine the optimum binary vector the computed vector differs from the optimum one by only one sign.

This observation has led us to the following efficient search procedure. An initial binary vector is first determined using Equation (3.307), then the second term of Equation (3.298) is evaluated using the initial vector and the other M vectors which have a Hamming distance of one from the initial vector. In this efficient procedure the search of a book of size 2^M is reduced to searching a local book of size $M + 1$, yet guaranteeing that 95% of the time the optimum binary vector is identified. Notice that for the $M + 1$ sized local codebook the efficient Gray code procedure is used to update $\mathcal{C}_k$ and $\mathcal{E}_k$ as in Equations (3.302) and (3.303).

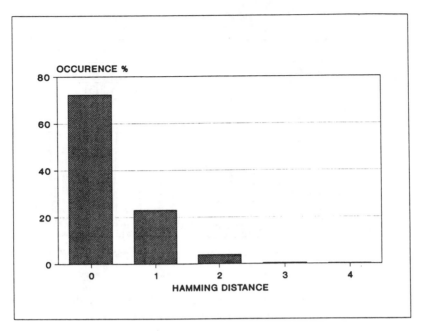

Figure 3.63: Frequency of occurrence vs. Hamming distance between the Binary code determined by Equation (5.30) and the one determined by exhaustive search

3.5.4.3   Evaluation of the BPE Coder

The BPE coder was evaluated at different bit rates in the range from 4.8 to 8 kbit/s. The subjective and objective speech quality was indistinguishable form that of the CELP coder at similar bit rates. Tables 3.14 and 3.15 show the bit allocation for BPE at 4.8 and 7.5 kb/s, respectively. In 4.8 kb/s coding a 30 ms speech frame is used and it is divided into 4 excitation frames of 7.5 ms (60 samples). In 7.5 kb/s coding a 24 ms speech frame is used and it is divided into 6 excitation frames of 4 ms (32 samples). Figure 3.64 shows SEGSNR against bit rate from 4.8 to 9.6 kb/s.

Figure 3.65 shows a speech segment, the binary pulse excitation, the synthesis filter excitation, and the reconstructed speech, where 8 pulses in a 5 ms excitation frame are used (6.6 kb/s coding). The variation of the speech energy and the SEGSNR with time for the sentence *"to reach the end he needs much courage"* uttered by a female speaker is shown in Figure 3.66.

The SNRs of the different search procedures described earlier are shown in Table 3.16. In a 5 ms excitation frame, 10 pulses are used. BP1 represents the simple approach using Equation (3.307), BP2 denotes the two stage approach in which an $M + 1$ sized local codebook is searched (M is the number of pulses), and BP3 denotes the exhaustive search. Using the

Parameter	Number of Bits
LSF's	36 (3,3,4,4,4,4,4,3,3)
LTP delays	24 (7,5,7,5)
LTP gains	12 (4 × 3)
binary pulses	48 (4 × 12)
pulse positions	8 (4 × 2)
excitation gains	20 (4 × 5)
Total	144 bits per 30 ms

Table 3.14: Bit allocation for 4.8 kb/s BPE coding

Parameter	Number of Bits
LSF's	36 (3,3,4,4,4,4,4,3,3)
LTP delays	36 (7,5,7,5,7,5)
LTP gains	18 (6 × 3)
excitation pulses	48 (6 × 8)
pulse positions	12 (6 × 2)
matrix identifier	6 (6 × 1)
excitation gains	24 (6 × 4)
Total	180 bits per 24 ms

Table 3.15: Bit allocation for 7.5 kb/s BPE coding

Search procedure	BP1	BP2	BP3
SEGSNR (dB)	13.130	13.350	13.373

Table 3.16: SNRs of different search procedures in BPE

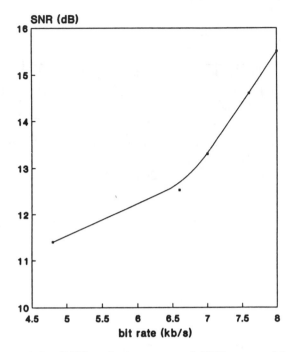

Figure 3.64: BPE codec's segmental SNR versus bit rate

	Binary regular pulses	Ternary excitation
5 ms excitation vectors	12.5208 dB	12.6356 dB
4 ms excitation vectors	13.85 dB	13.81 dB

Table 3.17: SEGSNR for the BPE and ternary CELP with 4 ms and 5 ms excitation vectors

simple search reduces the SNR by 0.24 dB as compared to the exhaustive search approach which is insignificant. Using the nonexhaustive two stage search brings the SNR closer to the exhaustive search case.

Table 3.17 shows the segmental SNRs of the BPE and ternary CELP coding using 4 ms and 5 ms excitation vectors. In case of the ternary codebook a 9-bit stochastic codebook was utilised with the gain quantised using 5 bits (4 bits for the magnitude and 1 for the sign). In case of BPE, 8 binary pulses were used with the first pulse position quantised using 2 bits and the gain using 4 bits (the BPE gain is always positive as the sign information is carried by the pulses themselves). It is clear from the segmental SNR figures in Table 3.17 that the objective quality of BPE is very close to that of the CELP. In fact, subjective listening tests did not show any difference in speech quality in either case.

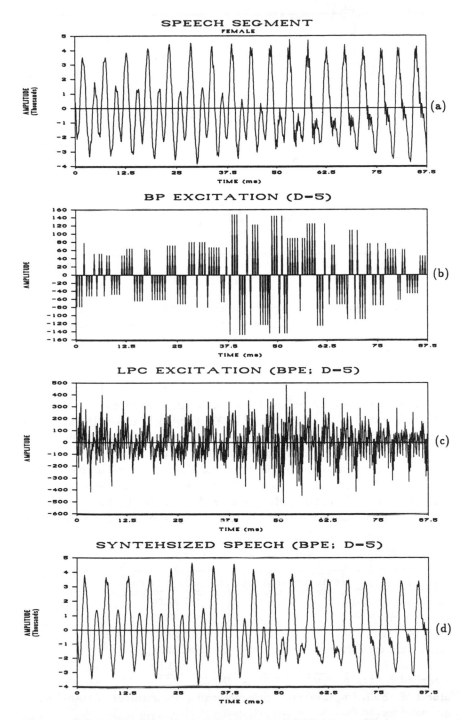

Figure 3.65: (a) A 87.5 ms speech segment, (b) binary excitation, (c) LPC excitation, (d) reconstructed speech

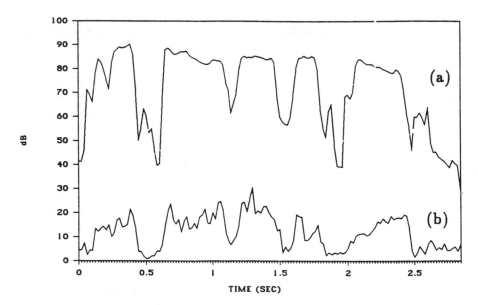

Figure 3.66: Variation of (a) speech power and (b) SEGSNR vs. time for the sentence *"to reach the end he needs much courage"* uttered by a female speaker using BPE

When the excitation vectors were not transformed, implying the utilisation of binary regular pulses, there was a slight degradation in speech quality compared to the transformed case. Table 3.18 shows the improvement in SEGSNR with increasing the number of transformation matrices, where 10 pulses are used in a 5 ms excitation vector. Zero number of matrices refers to untransformed binary pulses. Using 8 matrices gives 1.4 dB improvement in SNR at the expense of increasing the bit rate by 0.6 kb/s. Using untransformed binary pulses reduces the complexity since computing z and Θ in Equations (3.295) and (3.296) is not needed. In 8 kbit/s cod-

Number of matrices	SEGSNR (dB)
0	13.2734
1	13.3800
2	13.7100
4	14.3353
8	14.6198

Table 3.18: SNR improvement with increasing the number of transformation matrices

ing, better performances were obtained using two transformation matrices, where one of the matrices was set to the identity matrix (no transformation) to reduce the complexity and storage requirement.

The TBPE coder has several advantages over the CELP. The main advantage is the significant reduction in the computational complexity. As shown in Section 5.3.2, the search of an excitation codebook of size 2^M is reduced to searching a local book of size $M + 1$. In case of untransformed vectors, computing the second term of Equation (3.297) requires about $M^2 + M$ instructions and for the next M vectors in the local codebook, about $M(M + 3)$ instructions are needed to update the term. This is repeated D times (for all possible first pulse positions) which results in a total of $2M + 4$ instructions per speech sample. Using the transformation requires the computation of z and Θ in Equations (3.295) and (3.296), where z needs M^2 instructions and it is computed D times, which results in M instructions per speech sample, while Θ is computed only once since it is independent of the first pulse position, and it requires about $(3M^2 + M)/(2D)$ instructions per speech sample. For $M = 8$ and $D = 4$, and with no transformation, about 20 instructions per speech sample are required to jointly optimize the binary vector and the first pulse position, and this number rises to 50 when a transformation is used.

The second advantage of the TBPE is the ability to improve the speech quality by utilizing several transformation matrices, which is equivalent to using a very large excitation codebook; a task which becomes impractical with the CELP when the codebook address exceeds 10 bits. Another advantage is the reduction in the storage requirements of the excitation codebook. The equivalent of a 2^M sized codebook is obtained by storing an $M \times M$ matrix, and the storage is eliminated in case of untransformed pulses. Finally, the TBPE possesses an inherent robustness against transmission errors. As the excitation pulses are directly derived from the transmitted binary vector, a transmission error in the binary vector will cause little change in the transformed excitation vector, while in CELP coders a transmission error in the codebook address will cause the receiver to use an entirely different excitation vector.

3.5.4.4　Complexity Comparison between BPE and CELP

We will try here to give a comparison between the BPE and the different CELP approaches described in Chapter 4 in terms of complexity and storage requirement. Consider the case where 40 samples excitation frames (5 ms) are used. In case of CELP, a 9-bit codebook is used with the gain quantised using 5 bits. In case BPE, we use 8 pulses spaced by a distant $D = 5$ with 2 bits reserved for the first pulse position and 4 bits for the gain. We will assume that the impulse response $h(n)$ is truncated at $R - 1 = 24$, and the computation of $\psi(n)$ and $\phi(n)$ is not considered in the complexity assessment. Only the arithmetical operations are considered here (fetching stored data from memory is not considered).

Using the original CELP with the autocorrelation approach, $N + R = 65$ instructions (one instruction is one addition plus one multiplication) are needed to compute C and $\mathcal{E}$ in Equations (4.19) and (4.28), respectively. This results in a total of about 850 instructions per speech sample to search the 512 sized codebook. Assuming that every real-valued sample requires 4 bytes, then 160 kbytes are needed to store both the excitation and correlation codebooks.

In case of a ternary excitation codebook, with 4 nonzero pulses, and using the autocorrelation approach, 4 instructions are needed to compute C and about 6 to compute $\mathcal{E}$. The term $C^2/\mathcal{E}$ requires about 12 instructions, and the codebook search requires about 150 instructions per speech sample. Concerning the book storage, every excitation pulse position and sign are stored in one byte and the excitation codebook requires 2 kbytes. The problem here is the storage of the correlations of the codewords. For 4 nonzero excitation pulses, there is at most 6 nonzero correlations for every excitation vector [109]. The correlation $\mu(0)$ is always equal to 4 and needs not to be stored. The other correlations have integer values and can be stored with 2 bytes each (amplitude and position). A maximum of 5 kbytes is then required to store the 512 sized correlations codebook. Storage of the correlation codebook can be eliminated if the energies of the filtered codewords are computed on-line. The energy is expressed by

$$\mathcal{E} = 4\phi(0) + 2\sum_{i=0}^{2}\sum_{j=i+1}^{3} b_i b_j \phi(|m_i - m_j|),$$

where b_i is the amplitude $(-1,1)$ of the ith nonzero pulse and m_i is its position. Computing $\mathcal{E}/2$ requires 13 instructions and the term to be maximized now requires 19 instructions. A total of about 250 instructions are now needed with only 2 kbytes needed to store the excitation codebook.

In case of a ternary shift by 1 overlapping codebook, the excitation buffer contains $40 + 511 = 551$ samples with values $(-1,0,1)$. Assuming that every sample is stored in 2 bits, about 140 bytes are needed to store the excitation. The convolution $c(n) * h(n)$ is computed for the first codeword and it requires $N(N + 1)/2$ instructions. The correlation C needs N and the energy $\mathcal{E}$ needs N instructions. For the next codeword, the convolution is updated by $R = 25$ instructions, the correlation by N and the energy by N instructions. Remember that if the new pulse in the next codeword is zero then the convolution and energy need not to be updated. For a sparsity factor $1 - S$ (S is the ratio of nonzero pulses) and codebook size L, the convolution requires $N(N + 1)/2 + S(L - 1)R$, the energy $N + S(L - 1)N$, and the correlation NL instructions. For the typical values of $N = 40$, $L = 512$, $R = 25$, and $1 - S = 0.9$, over 600 instructions per speech sample are needed to search the codebook. Note that the shift by 2 codebook used by the DoD requires even more computational load in order to search the codebook. The only obvious advantage of overlapping codebooks is their storage efficiency. It is clear from the previous discussion that Xydeas ternary approach [109] is far more computationally efficient than DoD overlapping codebook approach [80], although both approaches utilize similar excitation definition (stochastic ternary).

In the VSELP approach [113] with 512 sized codebook, there is 9 basis vectors of dimension 40. The correlation term $\mathcal{C}$ is expressed by

$$\mathcal{C} = \mathbf{\Psi}^T \mathbf{A} \mathbf{b}$$

where $\mathbf{\Psi}$ is a 40×1 vector, $\mathbf{A}$ is a 40×9 matrix whose columns are the 9 basis vectors, and $\mathbf{b}$ is a 9×1 binary vector with elements -1 or 1. The filtered codeword energy is given by

$$\mathcal{E} = \mathbf{b}^T \mathbf{A}^T \mathbf{\Phi} \mathbf{A} \mathbf{b}$$

where $\mathbf{\Phi}$ is the 40×40 matrix of autocorrelations. For the first binary code, $\mathcal{C}$ requires $40 \times 9 + 9$ instructions and $\mathcal{E}$ requires about $40 \times 9 + 40 \times 40$ instructions. When the Gray code search is used, $\mathcal{C}$ is updated with 1 instruction and $\mathcal{E}$ with 9 instructions (see Equations (5.28) and (5.29)). Note that only half the codebook is searched since the complement of a binary code $\mathbf{b}$ yields the same value of $\mathcal{C}^2/\mathcal{E}$, and only the sign of the gain $\mathcal{C}/\mathcal{E}$ is changed. The total number of operations needed to search the codebook is about 130 instruction per speech sample. The codebook storage requires 40×9 real values which is about 1440 bytes.

In the BPE approach, when 8 binary pulses are used with no transformation, no codebook storage is required. The correlation $\mathcal{C}$ in Equation (5.22) requires 8 instructions and the energy $\mathcal{E}$ in Equation (5.23) requires about 8^2 instructions. This is repeated 4 times for all the possible positions. When the simple search approach is used, less than 10 instructions per speech sample are needed. When an 8×8 transformation matrix is used, the storage requirement is 64 real values which is about $0.25\,\mathrm{kbytes}$. The extra computational load is computing $\mathbf{z}$ in Equation (5.18) 4 times which requires $8^2 \times 4$ instructions, and computing Θ one time (independent of the pulse positions) which requires 800 instructions. The total is now less than 40 instructions per speech sample. Using the local codebook approach requires to search another M binary codes. As the Gray code approach is used, which needs $M + 1$ operations to update $\mathcal{C}$ and $\mathcal{E}$, the extra load is $8(8 + 1)/40$ which is less than two instructions per speech sample and this can be neglected. Table 3.19 shows the complexity and storage requirement figures described earlier. Objective and subjective performances of the approaches described in Table 3.19 were very close. However, the superiority of the BPE in both computational and storage efficiency is evident.

3.5.5 Postfiltering

Below $8\,\mathrm{kb/s}$ the quality of the reconstructed speech starts to degrade and speech enhancement techniques can be deployed to improve the speech

Approach	Complexity	Storage (kbytes)
original CELP	850	160
ternary with correlation codebook	250	7
ternary without correlation codebook	250	2
overlapping shift by 1 ternary	600	0.13
VSELP	130	1.4
binary (no transformation)	10	0
binary transformed pulses	40	0.25

Table 3.19: Comparison between the BPE and different CELP approaches in terms of codebook search complexity and storage requirements. The complexity is given by the number of instructions per speech sample. A 512 sized codebook is used with an excitation frame of 40 samples

quality at lower bit rates. Postfiltering has been efficiently used with AD-PCM [132] and the same concept can be utilized to enhance the quality of analysis-by-synthesis predictive coders. The postfilter emphasizes the speech formants and it has a spectral shape which lies between an all-pass filter and the synthesis filter. A commonly used postfilter is given by [133]

$$F(z) = \frac{A(z/\beta)}{A(z/\alpha)} = \frac{1 - \sum_{k=1}^{p} a_k \beta^k z^{-k}}{1 - \sum_{k=1}^{p} a_k \alpha^k z^{-k}}, \tag{3.308}$$

where $0 \le \alpha, \beta \le 1$. The filter spectrum is controlled by the values of α and β. The parameters α and β increase the bandwidth of the spectrum resonances. Figure 3.67 shows an example of the spectra of the synthesis filter $1/A(z)$ and the filters $1/A(z/0.8)$ ($\alpha = 0.8$ and $\beta = 0$) and $A(z/0.5)/A(z/0.8)$ ($\alpha = 0.8$ and $\beta = 0.5$). When $\beta = 0$ the filter $1/A(z/\alpha)$ is a bandwidth expanded synthesis filter with bandwidth expansion given by Equation (2.95). Due to the spectral tilt of the filter it acts as a low-pass filter and the postfiltered speech is somewhat muffled. Using the numerator reduces the spectral tilt while keeping the enhancement of the spectral resonances and this improves the postfilter performance. Choosing the parameters α and β depends on the bit rate used, and at 4.8 kb/s the values $\alpha = 0.8$ and $\beta = 0.5$ have been suggested [133]. To further remove the spectral tilt the following form can be used

$$F(z) = (1 - \mu z^{-1}) \frac{A(z/\beta)}{A(z/\alpha)}. \tag{3.309}$$

RELATIVE MAGNITUDE (dB)

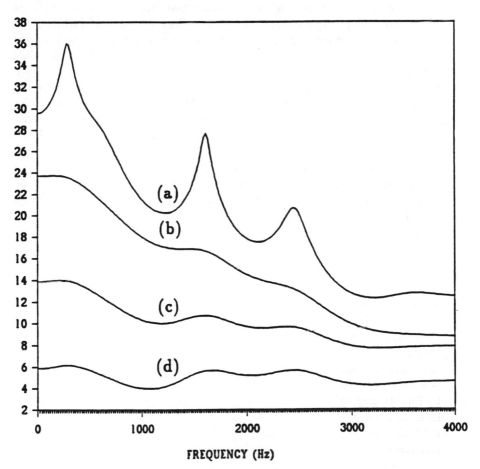

Figure 3.67: Spectra of (a) $1/A(z)$ (b) $1/A(z/0.8)$ (c) $A(z/0.5)/A(z/0.8)$ (d) $(1 - 0.5z^{-1})A(z/0.5)/A(z/0.8)$

The spectrum of this filter is shown in Figure 3.67-(d) for a value of $\mu = 0.5$. A value of $\mu = 0.5$ was found adequate [133], however, it is better to use adaptive postfiltering in which μ is changed according to the speech characteristics. The value of μ can be derived using the first reflection coefficient.

Pitch postfiltering has also been proposed to enhance the pitch periodicity in the reconstructed speech. The pitch postfilter has the form

$$\frac{1}{P'(z)} = \frac{1}{1 - \epsilon G z^{-M}}, \tag{3.310}$$

where G is the LTP gain, M is the LTP delay, and ϵ is a fraction which lies around 0.3 [102]. The pitch and spectral postfilters are cascaded where pitch postfiltering is applied first. Another enhancement configuration was proposed in [113] where the pitch enhancing filter is applied to the decoded LPC excitation rather than the reconstructed speech (the filter is called prefilter in this case). The pitch prefilter in this case enhances the periodicity in the excitation signal. The speech is then reconstructed using the prefiltered excitation and spectral postfiltering is applied afterwards. It is argued that this configuration reduces artifacts in the reconstructed speech due to waveform discontinuities which pitch postfiltering sometimes introduces.

In general, the postfilter causes an amplification of the speech signal and gain control techniques have to be used to compensate for the gain differences. This can be done either on sample-by-sample basis [133] or on a block basis.

Bibliography

[1] P. Kroon, E.F. Deprettere and R.J. Sluyter. "Regular-pulse excitation - A novel approach to efficient multipulse coding of speech". *IEEE Trans. ASSP, Vol. 34, No. 5*, pp. 1054–1063, October 1986.

[2] B.S. Atal and J.R. Remde. "A new model of LPC excitation for producing natural-sounding speech at low bit rates". *Proc. ICASSP'82*, pp. 614–617, 1982.

[3] R.E. Crochiere, S.A. Webber and J.L. Flanagan. "Digital coding of speech in sub-bands". *Bell System Tech. J.*, pp. 1069-1085, October 1976.

[4] M.R. Schroeder and B.S. Atal. "Code-Excited Linear Prediction (CELP): High quality speech at very low bit rates". *Proc. ICASSP'85, Tampa, Florida, USA* pp. 937–940, 26-29 March, 1985.

[5] R. Steele. "The cellular environment of light-weight hand-held portables". *IEEE Communications Magazine*, pp. 20-29, July 1989.

[6] R. Steele. "Delta modulation systems". *Pentech Press*, London, 1975.

[7] K.W. Cattermole. "Principles of pulse code modulation". *Iliffe*, London, 1969.

[8] B.S. Atal and M.R. Schroeder. "Predictive coding of speech signals". *Bell System Tech. J.*, pp. 1973-1986, October 1970.

[9] R.A. McDonald. "Signal-to-noise and idle channel performance of DPCM systems with particular application to voice signals". *Bell System Tech. J.*, pp. 1123-1151, September 1966.

[10] N.S. Jayant and P. Noll. "Digital coding of waveforms". *Prentice-Hall*, 1984.

[11] J.L. Flanagan. "Speech analysis synthesis and perception". *Springer-Verlag* Berlin, 1972.

[12] C.R. Rabiner and R.W. Schafer. "Digital Processing of Speech Signals". *Prentice Hall*, 1978.

[13] I.J. Wassell, D.J. Goodman and R. Steele. "Embedded delta modulation". *IEEE Trans. of Acoustics, Speech and Signal Processing, vol.36, no.8*, pp. 1236 - 1243, August 1988.

[14] **W.C. Wong, R. Steele, B. Glance** and **D. Horn.** "Time diversity with adaptive error detection to combat Rayleigh fading in digital mobile radio". *IEEE Trans. on Comms., COM-31, No.3,* pp. 378-387, March, 1983.

[15] **C-E. Sundberg, W.C. Wong** and **R. Steele.** "Weighting strategies for companded PCM transmitted over Rayleigh fading and Gaussian channels". *Bell Syst. Tech. J., Vol. 63, No. 6,* pp. 587-626, 1984.

[16] **R. Steele, C-E. Sundberg** and **W.C. Wong.** "Transmission errors in companded PCM over Gaussian and Rayleigh fading channels". *Bell Syst. Tech. J.,* pp. 955-990, July/Aug., 1984.

[17] **W.C. Wong, R. Steele** and **C-E. Sundberg.** "Soft decision demodulation to reduce the effect of transmission errors in logarithmic PCM transmitted over Rayleigh fading channels". *Bell Syst. Tech. J., Vol. 63, No. 10,* pp. 2193-2213, Dec. 1984.

[18] **R. Steele, C-E.W. Sundberg** and **W.C. Wong.** "Transmission of log-PCM via QAM over Gaussian and Rayleigh fading channels". *IEE Proc., vol.134, pt.F, Norb,* pp. 539-556, October 1987.

[19] **C-E.W. Sundberg, W.C. Wong, R. Steele.** "Logarithmic PCM weighted QAM transmission over Gaussian and Rayleigh fading channels". *IEE Proc., vol.134, pt.F, no.6,* pp. 557-570, October 1987.

[20] **K.H.J. Wong** and **R. Steele.** "ADPCM using vector predictor and forward adaptive quantiser with step-size prediction". *Electronic Letters, vol.21, no.2,* pp. 74-75, 17 January 1985.

[21] **C.S. Xydeas, C.C. Evci** and **R. Steele.** "Sequential adaptive predictors for ADPCM speech encoders". *IEEE Trans. on Communications, vol.COM-30, no.8,* pp. 1942-1954, August 1982.

[22] **J.L. Flanagan, M.R. Schroeder, B.S. Atal, R.E. Crochiere, N.S. Jayant** and **J.H. Tribolet.** "Speech coding". *IEE Trans. on Communications,* pp. 710-736, April 1979.

[23] **R. Zelinski** and **P. Noll.** "Adaptive transform coding of speech signals". *IEEE Trans. on Acoustics, Speech and Signal Processing,* pp. 299-309, August 1977.

[24] **D. Esteban** and **C. Galand.** "Application of quadrature mirror filters to split band voice coding scheme". *Proc. ICASSP'77,* pp. 191-195, May 1977.

[25] **D.J. Goodman.** "Embedded DPCM for variable bit rate transmission". *IEEE Trans. on Communication, vol.COM-28,* pp. 1040-1046, July 1980.

[26] **D.J. Goodman** and **C-E. Sundberg.** "Combined source and channel coding for variable-bit-rate speech transmission". *Bell System Tech. J., vol.62, no.7,* pp. 2017-2036, September 1983.

[27] **J.D. Johnston.** "A filter family designed for use in quadrature mirror filter banks". *ICASSP'80, Denver, Colorado, USA,* pp. 291-294, April 1980.

[28] **H.J. Nussbaumer.** "Complex quadrature mirror filters". *ICASSP'83, Boston, USA*, pp. 221-223, 1983.

[29] **C.R. Galand** and **H.J. Nussbaumer.** "New quadrature mirror filter structures". *IEEE Trans. on ASSP, vol.ASSP-32, no.3*, pp. 522-531, June 1984.

[30] **N.S. Jayant.** "Adaptive quantization with a one-word memory". *Bell System Tech. J., vol.52, no.7*, pp. 1119-1144, Sept 1973.

[31] **N.S. Jayant.** "Step-size transmitting differential coders for mobile telephony". *Bell System Tech. J.*, pp. 1557-1581, Nov 1975.

[32] **R.E. Crochiere.** "An analysis of 16 Kbit/s sub-band coder performance: dynamic range, tandem connections and channel errors". *BSTJ, vol.57, no.8*, pp. 2927-2952, October 1978.

[33] **M.R. Schroeder.** "Reference signal for signal quality studies". *The Journal of the Acoustical Society of America, vol.44, no.6*, pp. 1735-1736, 1968.

[34] **CCITT.** *Recommendation P.70, Red Book, Geneva*, pp. 111-114, 1985.

[35] **R.B. Hanes.** "A 16 kbit/s speech codec for four-channel 64 kbit/s transmission". *Br. Telecom. Technol. J., vol.3, no.1*, pp. 5-13, January 1985.

[36] **F.I. Itakura** and **S.I. Saito.** "Analysis-synthesis theory based on the maximum likelihood method". *Proc. of 6th Int. Congress on Acoustics, Tokyo*, pp. c17-20, 1968.

[37] **B.S. Atal.** "Speech analysis and synthesis by linear prediction of speech wave". *J. Acoust. Soc. Am., vol.47*, 1970.

[38] **B.S. Atal.** "Predictive coding of speech at low bit rates". *IEEE Trans. on Commun., vol.30*, pp. 600-614, April 1982.

[39] **C.K. Un** and **D.T. Magill.** "The residual-excited linear prediction vocoder with transmission rate below 9.6 Kbits/s". *IEEE Trans. on Commun., vol.23, no.12*, pp. 1466-1474, December 1975.

[40] **L.R. Rabiner** and **R.W. Shafer.** "Digital processing of speech signals". *Prentice-Hall Int.*, 1978.

[41] **J.D. Markel** and **A.H. Gray, Jr.** "Linear prediction of speech". *Springer-Verlag, New York*, 1976.

[42] **J. Makhoul.** "Stable and efficient lattice methods for linear prediction". *IEEE Trans. on ASSP, vol.25*, pp. 423-428, October 1977.

[43] **A.H. Gray, Jr.** and **D.Y. Wong.** "The Burg algorithm for LPC speech analysis/synthesis". *IEEE Trans. on ASSP, vol.28, no.6*, pp. 609-615, Dec. 1980.

[44] **A. Jennings.** "Matrix computation for engineers and scientists". *Wiley and Sons Ltd.*, 1977.

[45] **B.S. Atal** and **M.R. Schroeder.** "Predictive speech signal coding with reduced noise effects". *U.S. Patent, no. 4,133,976*, January 1979.

[46] **S. Singhal** and **B.S. Atal.** "Improving performance of multi-pulse LPC coders at low bit rates". *Proc. ICASSP'84*, pp. 1.3.1-1.3.4, 1984.

[47] **A.H. Gray, Jr.** and **J.D. Markel.** "Distance measures for speech processing". *IEEE Trans. ASSP, vol.24, no.5*, pp. 380-391, October 1976.

[48] **J. Makhoul, S. Roucos** and **H. Gish.** "Vector quantization in speech coding". *Proc. of IEEE, vol.73, no.11*, pp. 1551-1588, November 1985.

[49] **R.M. Gray.** "Vector quantization". *IEEE ASSP Magazine*, pp. 4-29, April 1984.

[50] **R.M. Gray, A. Buzo, A.H. Gray** and **Y. Matsuyama.** "Distortion measures for speech processing". *IEEE Trans. ASSP, vol.28, no.4*, pp. 367-376, August 1980.

[51] **Y. Linde, A. Buzo** and **R.M. Gray.** "An algorithm for vector quantiser design". *IEEE Trans. on Commun., vol.28, no.1*, pp. 84-95, January 1980.

[52] **L.R. Rabiner, M.M. Sondhi** and **S.E. Levinson.** "Note on the properties of a vector quantiser for LPC coefficients". *Bell Sys. Tech. J., vol.26, no.8*, pp. 2603-2616, October 1983.

[53] **D.Y. Wong, B.H. Juang** and **A.H. Gray, Jr.** "An 800 bit/s vector quantization LPC vocoder". *IEEE Trans. ASSP, vol.30, no.5*, pp.770-780, October 1982.

[54] **R. Viswanathan** and **J. Makhoul.** "Quantization properties of transmission parameters in linear predictive systems". *IEEE Trans. ASSP, vol.23*, pp. 309-321, June 1975.

[55] **N. Sugamura** and **N. Farvardin.** "Quantizer design in LSP analysis--synthesis". *IEEE J. on Selec. Areas in Commun., vol.6, no.2*, pp. 432-440, February 1988.

[56] **F. Itakura** "Line spectral representation of linear predictive coefficients of speech signals". *J. Acoust. Soc. Amer., vol.57, Supplement no.1, S35*, 1975.

[57] **G.S. Kang** and **L.J. Fransen.** "Low-bit rate speech encoders based on line-spectrum frequencies (LSFs)". *NRL Report 8857*, November 1984.

[58] **J. Makhoul.** "Linear prediction: a tutorial review". *Proc. of IEEE, vol.63, no.4*, pp. 561-580, April 1975.

[59] **F.K. Soong** and **B.H. Juang.** "Line spectrum pair (LSP) and speech data compression". *Proc. ICASSP'84*, pp. 1.10.1-1.10.4, 1984.

[60] **P. Kabal** and **R.P. Ramachandran.** "The computation of line spectral frequencies using Chebyshev polynomials". *IEEE Trans. ASSP, vol.34, no.6*, pp. 1419-1426, December 1986.

[61] M. Omologo. "The computation and some spectral considerations on line spectrum pairs (LSP)". *Proc. EUROSPEECH'89*, pp. 352-355, 1989.

[62] B.M.G. Cheetham. "Adaptive LSP filter". *Electronics Letters, vol.23, no.2*, pp. 89-90, 16th January 1987.

[63] F. Itakura and N. Sugamura. "LSP speech synthesizer". *Tech. Rept. 5, Speech Group, Acoustical Soc. of Japan*, November 1979.

[64] B.S. Atal, R.V. Cox and P. Kroon. "Spectral quantization and interpolation for CELP coders". *Proc. ICASSP'89, Glasgow, UK*, pp. 69-72, 23-26 May, 1989.

[65] N. Sugamura and F. Itakura. "Speech analysis and synthesis methods developed at ECL in NTT–From LPC to LSP". *Speech Commun., vol.5*, pp. 199-215, June 1986.

[66] A. Lepschy, G.A. Mian and U. Viaro. "A note on line spectral frequencies". *IEEE Trans. ASSP, vol.36, no.8*, pp. 1355-1357, August 1988.

[67] B.M.G. Cheetham and P.M. Huges. "Formant estimation from LSP coefficients". *Proc. IERE 5th Int. Conf. on Digital Processing of Signals in Commun., Univ. of Loughborough*, pp. 183-189, September 20-23, 1988.

[68] F.K. Soong and B.H. Juang. "Optimal quantization of LSP parameters". *Proc. ICASSP'88*, pp. 394-397, 1988.

[69] G.S. Kang and L.J. Fransen. "Application of line-spectrum pairs to low-bit-rate speech encoders". *Proc. ICASSP'85, Tampa, Florida, USA* pp. 244-247, 26-29 March, 1985.

[70] J.R. Crosmer and T.P. Barnwell, III. "A low bit rate segment vocoder based on line spectrum pairs". *Proc. ICASSP'85, Tampa, Florida, USA* pp. 240-243, 26-29 March, 1985.

[71] W.T.K. Wong and I. Boyd. "Optimal quantization performance of LPC parameters for speech coding". *Proc. EUROSPEECH'89*, pp. 344-347, 1989.

[72] C. Laflamme, J-P. Adoul and S. Morissette. "A real time 4.8 kbits/sec CELP on a single DSP chip (TMS320C25)". *IEEE Workshop on Speech Coding for Telecom., Vancouver, Canada*, September 5-8, 1989.

[73] D. Lin and B.M. McCarthy. "Efficient quantization and interpolation of LPC spectral parameters". *IEEE Workshop on Speech Coding for Telecom., Vancouver, Canada*, September 5-8, 1989.

[74] B.S. Atal and M.R. Schroeder. "Predictive coding of speech and subjective error criteria". *IEEE Trans. ASSP, vol.27, no.3*, pp. 247-254, June 1979.

[75] R.P. Ramachandran and P. Kabal. "Stability and performance analysis of pitch filters in speech coders". *IEEE Trans. ASSP, vol.35, no.7*, pp. 937-946, July 1987.

[76] **W.P. LeBlanc, S. Hanna** and **S.A. Mahmoud.** "Performance of a low complexity CELP speech coder under mobile channel fading conditions". *Proc. IEEE Vehicular Tech. Conf.,* pp. 647-651, May 1989.

[77] **P. Kroon** and **B.S. Atal.** "On improving the performance of pitch predictors in speech coding systems". *IEEE Workshop on Speech Coding for Telecom., Vancouver, Canada,* September 5-8, 1989.

[78] **P. Kroon** and **B.S. Atal.** "Pitch predictors with high temporal resolution". *Proc. ICASSP'90, Albuquerque, New Mexico, USA,* pp. 661-664, 3-6 Apr., 1990.

[79] **J.S. Marques, I.M. Trancoso, J.M. Tribolet** and **L.B. Almeida.** "Improved pitch prediction with fractional delays in CELP coding". *Proc. ICASSP'90, Albuquerque, New Mexico, USA,* pp. 665-668, 3-6 Apr., 1990.

[80] **Proposed Federal Standard 1016.** "Telecommunications: Analog to digital conversion of radio voice by 4,800 bit/second code excited linear prediction (CELP)". *First draft,* September 1, 1989.

[81] **C. Laflamme.** Unpublished work.

[82] **J.P. Campbell, Jr., T.E. Tremain** and **V.C. Welch.** "The DoD 4.8 kbps standard (proposed federal standard 1016)". in *Advances in Speech Coding,* Kluwer Academic Publishers, pp. 121-133, 1990.

[83] **R.P. Ramachandran** and **P. Kabal.** "Pitch prediction filters in speech coding". *IEEE Trans. ASSP, vol.37, no.4,* pp. 467-478, April 1989.

[84] **P. Kabal** and **R.P. Ramachandran.** "Joint optimization of linear predictors in speech coding". *IEEE Trans. ASSP, vol.37, no.5,* pp. 642-650, May 1989.

[85] **B.S. Atal** and **M.R. Schroeder.** "Optimizing predictive coders for minimum audible noise". *Proc. ICASSP'79,* pp. 453-455, 1979.

[86] **J. Makhoul** and **M. Berouti.** "Adaptive noise spectral shaping and entropy coding in predictive coding of speech". *IEEE Trans. ASSP, vol.27, no.1,* pp. 63-73, February 1979.

[87] **P. Kroon** and **E.F. Deprettere.** "Experimental evaluation of different approaches to the multipulse coder". *Proc. ICASSP'84,* pp. 10.4.1-10.4.4, 1984.

[88] **S. Singhal** and **B.S. Atal.** "Amplitude optimization and pitch prediction in multipulse coders". *IEEE Trans. ASSP, vol.37, no.3,* pp. 317-327, March 1989.

[89] **S. Singhal.** "Reducing computation in optimal amplitude multipulse coders". *Proc. ICASSP'86,* pp. 2364-2367, 1986.

[90] **T. Aresaki, K. Ozawa, S. Ono** and **K. Ochiai.** "Multi-pulse excited speech coder based on maximum cross correlation search algorithm". *Globecom 83,* pp. 794-798, 1983.

[91] M. Berouti, H. Garten, P. Kabal and P. Mermelstein. "Efficient computation and coding of the multipulse excitation for LPC". *Proc. ICASSP'84*, pp. 10.1.1-10.1.4, 1984.

[92] J-P. Lefevre and O. Passien. "Efficient algorithms for obtaining multipulse excitation for LPC coders". *Proc. ICASSP'85, Tampa, Florida, USA* pp. 957-960, 26-29 March, 1985.

[93] E.F. Deprettere and P. Kroon. "Regular excitation reduction for effective and efficient LP-coding of speech". *Proc. ICASSP'85, Tampa, Florida, USA* pp. 965-968, 26-29 March, 1985.

[94] J-P. Adoul, F. Didelot, P. Mabilleau and S. Morissette. "Generalization of the multipulse coding for low bit rate coding purposes: The generalized decimation". *Proc. ICASSP'85, Tampa, Florida, USA* pp. 256-259, 23-26 March, 1985.

[95] J.L. Flanagan, M.R. Schroeder, B.S. Atal, R.E. Crochiere, N.S. Jayant and J.M. Tribolet. "Speech coding". *IEEE Trans. on Commun., vol.27, no.4*, pp. 710-737, April 1979.

[96] R.D. Strum and D.E. Kirk. "First principles of discrete systems and digital signal processing". *Addison-Wesley Pub. Comp.*, 1988.

[97] K. Helwig, R. Hofman, P. Vary and R.J. Sluyter. "MATS-D speech codec: Regular pulse excitation LPC". *Proc. 2nd seminar on Land Mobile Digital Radio Communication, Stockholm*, 14-16 October 1986.

[98] R.J. Sluyter, G.J. Bosscha and H.M.P.T. Schmitz. "A 9.6 kb/s speech coder for mobile radio applications". *Proc. ICC'84*, pp. 1159-1162, 1984.

[99] P. Vary et al. "Speech codec for the European mobile radio system". *Proc. ICASSP'88*, pp. 227-230, 1988.

[100] S.P. Lloyd. "Least squares quantization in PCM". *IEEE Trans. Inf. Th., vol.28, no.2*, pp. 129-137, March 1982.

[101] M.J. Noah. "Optimal Lloyd-Max quantization of LPC speech parameters". *Proc. ICASSP'84*, pp. 1.8.1-1.8.4, 1984.

[102] P. Kroon and E.F. Deprettere. "A class of analysis-by-synthesis predictive coders for high quality speech coding at rates between 4.8 and 16 kbit/s". *IEEE J. on Selected. Areas in Commun., vol.6, no.2*, pp. 353-363, February 1988.

[103] R. Soheili, A.M. Kondoz and B.G. Evans. "New innovations in multipulse speech coding for bit rates below 8 kb/s". *Proc. EUROSPEECH*, pp. 298-301, 1989.

[104] B.S. Atal and M.R. Schroeder. "Stochastic coding of speech signals at very low bit rates". *IEEE Int. Conf. Commun.*, May 1984.

[105] M. Copperi and D. Sereno. "Vector quantization and perceptual criteria for low bit rate coding of speech". *Proc. ICASSP'85, Tampa, Florida, USA* pp. 252-255, 23-26 March, 1985.

[106] G. Davidson and A. Gersho. "Complexity reduction methods for vector excitation coding". *Proc. ICASSP'86,* pp. 3055-3058, 1986.

[107] G. Davidson, M. Yong and A. Gersho. "Real-time vector excitation coding of speech at 4800 bps". *Proc. ICASSP'87,* pp. 2189-2192, 1987.

[108] D. Lin. "New approaches to stochastic coding of speech sources at very low bit rates". *Signal Processing III: Theories and Applications (Proc. of EUSIPCO-86),* pp. 445-448, 1986.

[109] C.S. Xydeas, M.A. Ireton and D.K. Baghbadrani. "Theory and real time implementation of a CELP coder at 4.8 and 6.0 Kbits/sec using ternary code excitation". *Proc. of IERE 5th Int. Conf. on Digital Processing of Signals in Commun., Univ. of Loughborough,* pp. 167-174, 20-23 September 1988.

[110] W.B. Kleijn, D.J. Krasinsky and R.H. Ketchum. "An efficient stochastically excited linear predictive coding algorithm for high quality low bit rate transmission of speech". *Speech Commun., vol.7, no.3,* pp. 305-316, October 1988.

[111] R.A. Salami. "Binary pulse excitation: a novel approach to low complexity CELP coding" in *Advances in speech coding,* Kluwer Academic Publishers, pp. 145-156, 1991.

[112] M. Delprat, M. Lever and C. Gruet. "A 6 kbps regular pulse CELP coder for mobile radio communications". in *Advances in speech coding,* Kluwer Academic Publishers, pp. 179-188, 1991.

[113] I.A. Gerson and M.A. Jasiuk. "Vector sum excitation linear prediction (VSELP) speech coding at 8 kbps". *Proc. ICASSP'90, Albuquerque, New Mexico, USA,* pp. 461-464, 3-6 Apr., 1990.

[114] I.A. Gerson and M.A. Jasiuk. "Vector Sum Excited Linear Prediction (VSELP)". *IEEE Workshop on Speech Coding for Telecom., Vancouver, Canada,* September 5-8, 1989.

[115] J-P. Adoul et al. "Fast CELP coding based on algebraic codes". *Proc. ICASSP'87,* pp. 1957-1960, 1987.

[116] A.M. Kondoz and B.G. Evans. "CELP base-band coder for high quality speech coding at 9.6 to 2.4 kbps". *Proc. ICASSP'88,* pp. 159-162, 1988.

[117] R.C. Rose and T.P. Barnwell III. "Quality comparison of low complexity 4800 bps self excited and code excited vocoders". *Proc. ICASSP' 87,* pp. 1637-1640, 1987.

[118] N. Gouvianakis and C. Xydeas. "Advances in analysis by synthesis LPC speech coders". *J. IERE, vol.57, no.6 (supplement),* pp. S272-S286, November/December 1987.

[119] I.M. Trancoso and B.S. Atal. "Efficient procedures for finding the optimum innovation in stochastic coders". *Proc. ICASSP'86*, pp. 2375-2378, 1986.

[120] J-P. Adoul and C. Lamblin. "A comparison of some algebraic structures for CELP coding of speech". *Proc. ICASSP'87*, pp. 1953-1956, 1987.

[121] C. Lamblin, J-P. Adoul, D. Massaloux and S. Morissette. "Fast CELP coding based on the Barnes-Wall lattice in 16 dimensions". *Proc. ICASSP'89, Glasgow, UK*, pp. 61-64, 23-26 May, 1989.

[122] M.A. Ireton and C.S. Xydeas. "On improving vector excitation coders through the use of spherical lattice codebooks (SLC's)". *Proc. ICASSP'89, Glasgow, UK*, pp. 57-60, 23-26 May, 1989.

[123] M. Lever and M. Delprat. "RPCELP: A high quality and low complexity scheme for narrow band coding of speech". *Proc. EUROCON*, pp. 24-27, June 1988.

[124] R.A. Salami. "Binary code excited linear prediction (BCELP): new approach to CELP coding of speech without codebooks". *Electronics Letters*, *vol.25, no.6*, pp. 401-403, 16 March 1989.

[125] W.B. Kleijn. "Optimal codes to protect CELP against channel errors". *IEEE Workshop on Speech Coding for Telecom.*, *Vancouver, Canada*, September 5-8, 1989.

[126] R.A. Salami and D.G. Appleby. "A new approach to low bit rate speech coding with low complexity using binary pulse excitation (BPE)". *IEEE Workshop on Speech Coding for Telecomm.*, *Vancouver, Canada*, September 5-8, 1989.

[127] J.-M. Müller, H. Scheuermann and B. Wächter. "GSM half rate codec: a possible candidate". *IEEE Workshop on Speech Coding for Telecomm.*, *Vancouver, Canada*, September 5-8, 1989.

[128] A. Le Guyader, D. Massaloux and F. Zurcher. "A robust and fast CELP coder at 16 kbit/s". *Speech Communication, vol.7*, pp. 217-226, 1988.

[129] C. Laflamme, J-P. Adoul, H.Y. Su and S. Morissette. "On reducing computational complexity of codebook search in CELP coder through the use of algebraic codes". *Proc. ICASSP'90, Albuquerque, New Mexico, U.S.A.*, April 3-6, 1990.

[130] R.A. Salami, K.H.H. Wong, R. Steele and D.G. Appleby. "Performance of error protected binary pulse excitation coders at 11.4 kb/s over mobile radio channels". *Proc. ICASSP'90, Albuquerque, New Mexico, USA*, pp. 473-476, 3-6 Apr., 1990.

[131] C. Laflamme, J-P. Adoul, R. Salami, S. Morissette and P. Mabilleau. "16 kbps wideband speech coding technique based on algebraic CELP". *Proc. ICASSP'91, Toronto, Canada* pp. 13-16, 14-17 May, 1991.

[132] **V. Ramamoorthy and N.S. Jayant.** "Enhancement of ADPCM speech by adaptive postfiltering". *Bell Sys. Tech. J., vol.63*, pp. 1465-1475, October 1984.

[133] **J. Chen and A. Gersho.** "Real-time vector APC speech coding at 4800 bps with adaptive postfiltering". *Proc. ICASSP'87*, pp. 2185-2188, 1987.

Chapter 4

Channel Coding

K.H.H. Wong[1], and L. Hanzo[2]

4.1 Introduction

Sparked off by Shannon's pioneering discoveries back in 1948, the 1950s witnessed feverish activities in the field of forward error correction coding (FEC), which led to the introduction of the single error correcting Hamming block code [1] in 1950 and to the birth of the first convolutional code in 1955 [2]. The theory of FEC is lavishly published in classic references [3], [4], [5], [6], [7], [8], [9], [10], [11], however real-time implementations constituted serious limitations to their deployment in the past. Constrained by the error statistics of existing communications links, such as satellite channels or cables, almost exclusively random error statistics, i.e. memoryless channels have been studied. With the evolution of digital mobile radio communication it is important to provide a comprehensive overview of FEC techniques tailored for these hostile bursty channels.

The key to efficient FEC via bursty fading channels is the deployment of appropriately matched binary or non-binary channel interleavers to randomize the bursty error statistics and hence render the channel memoryless. If this condition is met, most of the memoryless theory applies. Whence we commence our discourse with a rudimentary discription of a number of interleaving schemes, followed by the encoding and decoding algorithms as well as theoretical and simulated performance of various convolutional codes using both soft and hard decisions. Block codes, in particular Bose-Chaudhuri-Hocquenghem (BCH) codes and Reed-Solomon(RS) codes are portrayed in contrast to convolutional codes in terms of encoding and de-

---

[1] University of Southampton
[2] University of Southampton and Multiple Access Communications Ltd.

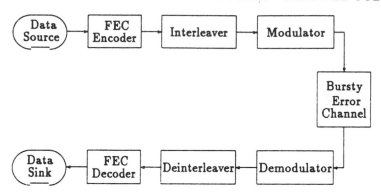

Figure 4.1: Typical communications system with FEC coding and interleaving

coding algorithms, complexity as well as theoretical and measured performance. A particularly powerful combination, concatenated codes, are also
highlighted and compared to convolutional and block codes in the context
of the interplay of complexity, coding gain as well as the ability to detect
and correct various channel error distributions for the tranmission of speech
and data signals.

4.2   Interleaving Techniques

Interleaving is a process of rearranging the ordering of a sequence of binary
or non-binary symbols in some unique one-to-one deterministic manner.
The reverse of this process is the deinterleaving which is to restore the
sequence to its original ordering.

In many applications in communication technology, interleaving is used
in conjunction with forward error correction (FEC) codes to enhance error
correction performance. The interleaver is inserted between the channel
encoder and the modulator as shown in Figure 4.1. Most block or convolutional codes are designed to combat random independent errors which
usually occur in a memoryless channel. For channels having memory, burst
errors are observed that are due to the mutually dependent signal transmission impairments. An example of such a channel is a fading channel.
The fading arises because the signals arriving at the antenna have trans-
versed different paths having various attenuations and delays. The effective
received signal is the vector sum of these multipath signals and exhibits
fades which depend on numerous factors, such as mobile speed, propagation delay spread and frequency. The received signal fading causes burst
errors during the signal fades.

Interleaving is deployed to disperse the burst errors when the received
signal level fades, and to reduce the concentration of the errors that must be
corrected by the channel code. Before a sequence of symbols is transmitted

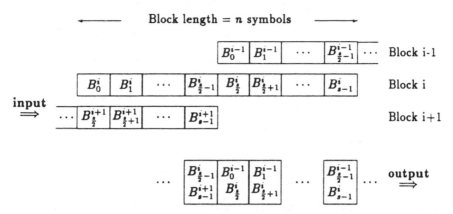

Figure 4.2: Diagonal interleaver

the symbols from several codewords are interleaved. Then when an error burst occurs the errors will be shared among the interleaved codewords and a less powerful code is required to correct them. Thus interleaving effectively makes the channel appear like a random error channel to the decoder. The idea behind interleaving is to separate the codeword symbols in time, thereby reducing the memory of the channel. As the interleaving period increases, the error performance can be expected to improve in the sense that noise bursts are more dispersed. On the other hand, the delays due to interleaving and deinterleaving increase. Consequently, there is always a tradeoff between error performance and interleaving delay.

The interleaver shuffles the code symbols, each consisting of m-bits, over a span of several codewords. If m is equal to one, bit interleaving is applied. The span required is often determined by the burst duration. Four types of interleavers are described in the Sections 4.2.1 to 4.2.4. The effect of the interleaving on the bit and symbol error distributions are presented in Sections 4.2.5-4.2.7.

4.2.1 Diagonal Interleaving

A diagonal interleaver [12] accepts coded symbols in blocks from the encoder, permutes these symbols, and feeds them to the modulator. The i-th input block of coded symbols is interleaved in a diagonal way with the previous $(i-1)$-th block and the following $(i+1)$-th blocks as shown in Figure 4.2. A block of n coded symbols, each symbol consisting of m-bits, is divided into s subblocks where $B_j^i, j = 1, 2, \ldots, s$ is the j-th subblock of the i-th block. The interleaved output is formulated by reading vertically two subblocks belonging to the $(i-1)$-th and i-th blocks or to the i-th and $(i+1)$-th blocks.

If the number of subblocks is equal to the number of symbols in a block, i.e., $s = n$, the subblock has only a single symbol and the interleaver shuffles

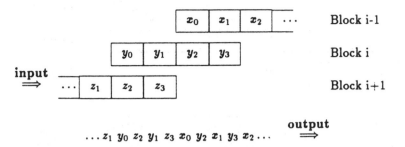

Figure 4.3: Example of Diagonal interleaving with single symbol subblock

the block on a symbol-by-symbol basis such that the output is a sequence of alternate symbols from two successive blocks. This is apparent from Figure 4.3, where the interleaved output is $\ldots z_1 y_0 z_2 y_1 z_3 x_0 y_2 x_1 y_3 x_2 \ldots$, and x, y and z are the symbols of the $(i-1)$-th, i-th and $(i+1)$-th block, respectively. If burst errors occurred such that the error symbol e in the received sequence is $\ldots z_1 e e y_1 z_3 x_0 y_2 e e x_2 \ldots$, then the deinterleaved sequence becomes $\ldots z_1 e z_3 e y_1 y_2 e x_0 e x_2 \ldots$, where the errors are distributed over two successive blocks.

The interleaver and deinterleaver end-to-end delay is three times the block length, i.e., 3nm bits. This is because the interleaving results in a delay of one and a half blocks, see Figure 4.2, and the deinterleaving results in a similar delay. This short delay is particularly important in some applications such as digital speech transmission. However, the dispersion of the burst errors is limited to the adjacent blocks which implies that the burst errors are halved and shared by only two successive blocks.

4.2.2 Block Interleaving

A block interleaver takes a codeword of n symbols and writes them by rows into a matrix with a depth of D rows and width of W columns as shown in Figure 4.4. Suppose W is equal to n, i.e. a row of symbols in the interleaver corresponds to a codeword. Each codeword is composed of k information symbols and $n-k$ parity symbols. After the matrix is completely filled, the symbols are fed to the modulator one column at a time and transmitted over the channel. At the receiver, the deinterleaver performs the inverse permutation of the matrix by feeding in one column at a time until the matrix is filled, and removing the symbols a row at a time. The important characteristics of this interleaving approach are that any burst of errors of length $b \leq D$ results in single errors in a codeword. Also any burst of length $b = rD, (r > 1)$ results in bursts of no more than r symbol errors in a codeword. However, a periodic sequence of single errors spaced by

$$\longleftarrow \quad\quad\quad W \text{ width} \quad\quad\quad \longrightarrow$$

	Information symbols $\longrightarrow$		$\longleftarrow$	Parity symbols $\longrightarrow$	

Input $\Rightarrow$

S_1^1	S_2^1	S_3^1	$\cdots$	S_k^1	P_1^1	$\cdots$	P_{n-k}^1
S_{k+1}^2	S_{k+2}^2	S_{k+3}^2	$\cdots$	S_{2k}^2	P_1^2	$\cdots$	P_{n-k}^2
S_{2k+1}^3	S_{2k+2}^3	S_{2k+3}^3	$\cdots$	S_{3k}^1	P_1^3	$\cdots$	P_{n-k}^3
$\vdots$	$\vdots$	$\vdots$		$\vdots$	$\vdots$		$\vdots$
$S_{(D-1)k+1}^D$	$S_{(D-1)k+2}^D$	$S_{(D-1)k+3}^D$	$\cdots$	S_{Dk}^D	P_1^D	$\cdots$	P_{n-k}^D

D depth

$\Downarrow$ Output

Figure 4.4: Block interleaver

D symbols results in a single codeword error, where every symbol in the codeword is incorrect. The interleaver and deinterleaver end-to-end delay is $2WD$ symbols. To be precise, only $W(D-1)+1$ storage needs to be filled before transmission can begin as soon as the first symbol of the last row of the $D*W$ matrix is filled. The same delay applies to the deinterleaver. Therefore, the end-to-end delay is $(2WD - 2W + 2)$ symbols. The memory requirement is WD symbols in both the interleaver and deinterleaver.

Another block interleaver [13], shown in Figure 4.5 is a derivation from the arrangement shown in Figure 4.4. Instead of writing rows of symbols into the matrix and appending the parity symbols to the k successive information symbols, the symbols are now written a column at a time. The parity symbols are encoded from a row of k information symbols each separated by D symbols in their natural order. As symbol $S_1^1, S_2^2, S_3^3, \ldots, S_{kD}^D$ occur they are placed into the matrix and also transmitted at the same time. Thus by the time S_{D+1}^1 arrives, symbols $S_1^1, S_2^2, \ldots, S_D^D$ have been transmitted. Armed with a knowledge of all the information symbols, the parity symbols P_1^1 to P_{n-k}^D are calculated immediately after symbol S_{kD}^D has entered the matrix. The advantage of this interleaving scheme is that information symbols are transmitted in their natural order. Hence, the interleaver delay is negligible, the end-to-end delay is WD symbols which is due to the deinterleaver.

The interleaver parameters D and W must be selected so that all expected burst lengths are less than D. However, this type of interleaver lacks robustness when a periodic sequence of single errors spaced by D symbols occurs. In this situation all the symbols in a row are erroneous and this overloads the channel codec. The interleaving scheme to be described in the next section exhibits the ability of dispersing bursty noise as well as periodic noise.

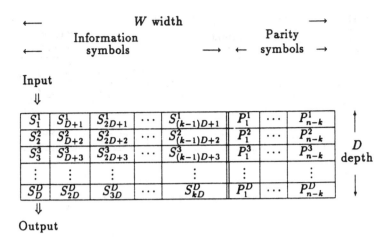

Figure 4.5: Modified block interleaver

4.2.3   Inter-Block Interleaving

The inter-block interleaver [12] takes an input block of NB symbols and disperses N symbols to each of the next B output blocks. Consider a coded symbol x from the encoder and an output symbol y from the interleaver. The mapping from the m-th symbol of the i-th coded input block to the $(j + Bt)$-th interleaved symbol of the $(i + j)$-th output block is given by

$$y(i + j, j + Bt) = x(i, m), \quad \text{for all } i \tag{4.1}$$
$$\text{with } j = m \bmod B$$
$$\text{and } t = m \bmod N.$$

An example of inter-block interleaving with $B = 3$ and $N = 2$ is illustrated in Figure 4.6. The symbols of the three successive i-th, $(i + 1)$-th and $(i + 2)$-th coded input blocks are denoted as a, b and c respectively. Here, $y(i + j, j + 3t) = x(i, m)$, for all i, with $j = m \bmod 3$, and $t = m \bmod 2$. For $m = 0, y(i, 0) = x(i, 0)$ and $m = 1, y(i + 1, 4) = x(i, 1)$ and so on. It is noted that the successive symbols of the i-th input block are mapped to the next B output blocks consecutively, but with the irregular offset position $(j + Bt)$ in the block. This irregular offset has the advantage of randomizing the periodic noise. In order to make sure that the mapping is one-to-one, B and N cannot have a common multiple. Usually this places a constraint on the block size of BN symbols. Another disadvantage of this interleaving scheme is the dispersive nature in that the output sequence is expanded by $(B - 1)$ blocks, the interleaving delay is $B^2 N$ symbols composed of the delay BN due to buffering the input block, plus the extra delay $(B-1)BN$ due to the dispersion of the symbols.

Input blocks:

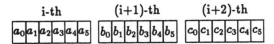

Output block:

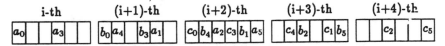

Figure 4.6: Example of inter-block interleaving with $B = 3$ and $N = 2$

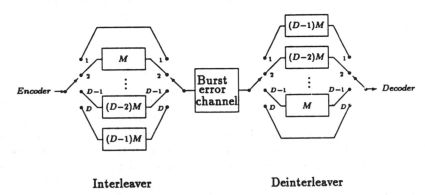

Figure 4.7: Convolutional interleaver and deinterleaver

4.2.4 Convolutional Interleaving

Convolutional interleavers have been proposed by Ramsey [14] and Forney [11]. The structure proposed by Forney is shown in Figure 4.7. The code symbols are sequentially shifted into the bank of D registers, where each successive register provides M symbols more memory than the preceding one. With the switch in position 1 the data is connected directly to the channel. With each new code symbol the commutator switches to the next register, and the oldest symbol in that register is shifted out. After the switch has reached position D, it returns to position 1 and the cycle of the switching continues. The deinterleaver performs the inverse operation, and the input and output commutators for both interleaver and deinterleaver must be synchronized.

The performance of a convolutional interleaver is very similar to that of a block interleaver. The important advantage of convolutional over block interleaving is that with convolutional interleaving the end-to-end delay is $W(D-1)$ symbols, where $W = DM$, and the memory required is $W(D -$

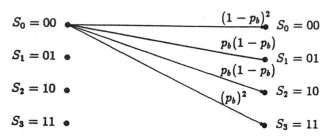

Figure 4.8: Non-binary and non-symmetric channel

1)/2 in both interleaver and deinterleaver. Therfore, there is a reduction of one-half in the delay and the memory of a convolutional interleaver compared to the block interleaver.

4.2.5   Discrete Memoryless Channel

The FEC encoded symbols consisting m-bits are serially transmitted. The receiver performs symbol regeneration prior to FEC decoding. If the channel is Gaussian, or if sufficient interleaving is employed in the case of a fading channel, the probability of any bit being in error is p_b. The probability of the regenerated symbol being erroneous depends on the value of m. For example, Figure 4.8 shows the probabilities of a transmitted two bits symbol 00 being regenerated as 00, 01, 10 and 11.The probability of receiving an error symbol with i number of error bits can be expressed as

$$p_{s,i} = \left(\begin{array}{c} m \\ i \end{array} \right) p_b^i (1 - p_b)^{m-i}. \tag{4.2}$$

The symbol error probability p_s is the sum of all the error symbols with i bit errors and is given by

$$p_s = \sum_{i=1}^{m} \left(\begin{array}{c} m \\ i \end{array} \right) p_b^i (1 - p_b)^{m-i}. \tag{4.3}$$

This Equation of symbol error probability is based on the assumption of the memoryless channel. However, the mobile channel has considerable memory and therefore Equation 4.3 is only valid when the interleaving period is sufficiently long. Unfortunately, the tolerable delay of some transmissions, such as digital speech communications restrict the interleaving period and prevent the establishment of a true memoryless channel. The next section illustrates the memoryless condition on the channel by using different interleaving methods with various delays.

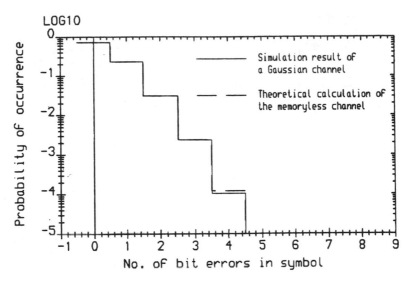

Figure 4.9: Histogram of the number of bit errors in 8-bit symbols via Gaussian channel with $SNR = 2dB$ and $BER = 3.74 \cdot 10^{-2}$

4.2.6 The Effect of Interleaving on Symbol Error Distribution

Under certain conditions, for example in some microcells [15], the mobile radio channel has minimal memory, i.e., it behaves as a Gaussian channel. In this channel the bit errors are random and there is no point in interleaving, as demonstrated by Figure 4.9, where the bit error distribution within eight bit symbols has been determined using computer simulation when the transmissions are over a Gaussian channel using MSK modulation. The bit error probability for a channel SNR of 2 dB is 3.74×10^{-2}. Inserting this p_b value into Equation 4.2 enabled the theoretical curves to be plotted. The simulation and theoretical results are in very close agreement, demonstrating that the Gaussian channel is indeed memoryless.

By constrast, the Rayleigh fading channel has memory as error bursts occur due to the presence of deep fades resulting from the movement of the mobile station. We conducted an experiment where data at 16 kbits/s was formulated into blocks, subjected to interleaving, and conveyed via MSK over a mobile radio channel that exhibited Rayleigh fading [16]. The mobile station travelled at 30 mph, and the propagation frequency was 900 MHz. The channel SNR at the mobile station was 5 dBs. In Figure 4.10 we display our findings. As bench markers we show the histogram of the number of bit errors in a symbol in the absence of interleaving, and also for the memoryless channel computed from Equation 4.2. The performance in the absence of interleaving indicates the bursty nature of the channel as there was a relatively high probability of 10^{-5} that all 8-bits of a symbol are

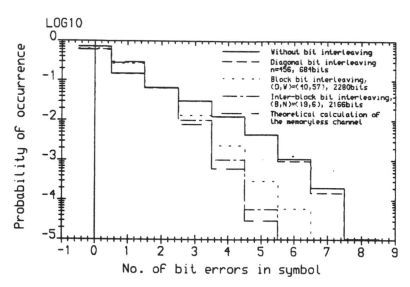

Figure 4.10: Histogram of the number of bit errors in 8-bit symbols via Rayleigh channel with $SNR = 2dB$ and $BER = 5.79 \cdot 10^{-2}$

erroneous. The randomising performance of the diagonal interleaver (see Figure 4.2) is poor because the block size was relatively small at 456 bits, and the interleaving only involves reallocating the bits over two adjacent blocks. This method of interleaving is not efficient at dispersing the errors, although it does have the advantage of a relatively small delay penalty. Clearly, when the diagonal bit interleaving is deployed, the histogram is similar to that of the no interleaving scenario and it is very different from that of the ideal memoryless channel.

When the block interleaver of Figure 4.4 was used having a long interleaving delay of 2280 bits, with $D = 40$ and $W = 57$, the probability of many bit errors per symbol significantly decreased compared to the diagonal interleaver case. The inter-block interleaver shown in Figure 4.6 was deployed with $B = 19$ and $N = 6$, numbers which produce an interleaving delay of 2166 bits, i.e., a number close to the 2280 bits used for the block interleaver. The inter-block interleaver had the best performance due to its ability to disperse the bits in the interleaving process in a near random fashion, rather than the periodic way of the block interleaver. Its histogram suggests a near-random in-symbol bit error distribution, when compared to that of the memoryless channel.

The dispersion of the burst errors by the bit interleaving results in random distribution of bit errors. As we will show in the forthcoming sections, this is desirable for the Viterbi decoding of convolutional codes and the trellis decoding of block codes since both decoding methods operate on bit-by-bit basis. However, the random distribution of the bit errors increases

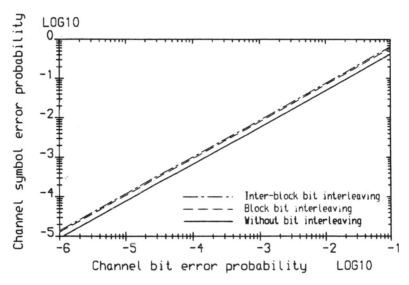

Figure 4.11: The effect of bit interleaving on 8-bit symbols via Rayleigh fading channels

the symbol error probability as shown in Figure 4.11. This figure implies that dispersion of the burst errors results into higher symbol error probability. This phenomenon is not desirable to those decoding methods operating on symbol basis such as the Berlekamp-Massey decoding of Reed-Solomon codes. In this case, symbol interleaving is applied in order to constrain the bit errors in the symbol to itself without spreading out to other symbols such that the symbol error probability after deinterleaving remains the same as on the channel. The channel symbol error probability of the fading channel is less than that of the memoryless channel. Hence, the symbol interleaving on the fading channel has a lower symbol error probability than the theoretical value calculated for the bit interleaved memoryless channel.

4.2.7 Effect of Symbol Size on Symbol Error Probability

From Equation 4.3 in Section 4.2.5, we notice that the symbol error probability p_s increases with the symbol size m. This increase is due to the summation of all the i, $i = 1, 2, \ldots, m$, bit errors in a symbol. Figure 4.12 shows the channel symbol error probability for the case of 1, 3, 4 and 8 bits symbols. This implies that large symbol sizes suffer from higher symbol error probability which then discourages the use of long codewords in block coding.

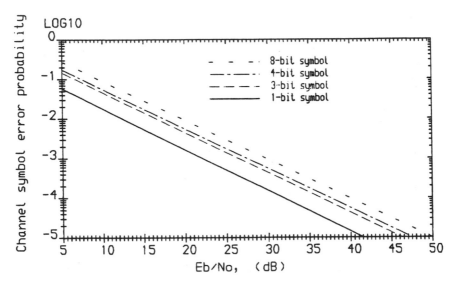

Figure 4.12: The effect of symbol-size on symbol error probability via Rayleigh fading channels

4.3   Convolutional Codes

Convolutional codes were first suggested by Elias [2] in 1955. Shortly afterwards, a sequential decoding algorithm for these codes was proposed by Wozencraft [17],[18], and its implementation was independently described by Fano [19] and by Massey [3] in 1963. Called threshold decoding, it has been successfully implemented in numerous communications systems. In 1967 Viterbi [20] proposed a maximum likelihood decoding algorithm that provided optimum error rate performance and achieved a shorter decoding delay compared to sequential decoding. The implementation of the Viterbi decoder became feasible with the advent of integrated circuit technology and was used in the deep-space and satellite communications of the early 1970s. It was also adopted by the GSM committee in the late 1980s for the Pan-European digital cellular mobile radio system.

4.3.1   Convolutional Encoding

A convolutional code (CC) is a sequence of encoded symbols which is generated by passing the information sequence through a binary shift register as shown in Figure 4.13. At each symbol instant, a k-bit information symbol is inserted into the input stages of the shift register. The register consists of Kk binary stages that constitute the present k-bit information symbol and the $(K - 1)$ previous k-bit input symbols. The parameter K is known

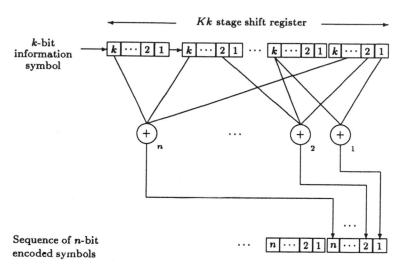

Figure 4.13: Convolutional encoder for the $CC(n, k, K)$ code

as the constraint length and it determines the memory length of the shift register. The n linear algebraic function generators, $g_1, g_2, \ldots, g_n$, defined by their connections to various register stages yield an n-bit convolutional coded symbol via the set of modulo-2 adders. At each symbol instant, a new information symbol enters the register, the other symbols move to the next symbol location and one symbol leaves the register. A new encoded symbol is then generated as previously described. The process continues, encoding the sequence of information symbols into a new sequence of output symbols.

The convolutional code is basically described by three parameters n, k and K and is denoted by $CC(n, k, K)$. The coding rate, defined by

$$R = k/n \tag{4.4}$$

represents the amount of information per encoded bit. If k is equal to one, namely one bit per symbol, a binary CC is generated, whereas for k greater than one, the CC code is referred to as, not surprisingly, a non-binary code.

In order to generate the $CC(n, k, K)$ code, a set of n generators, $g_1, g_2, \ldots, g_n$, is used to produce an n-bit symbol. Each generator is described as a vector with a dimension of full register length of Kk bits. The i-th element of the vector specifies the existence or otherwise the connection of the i-th bit position of the register to the modulo-2 adder. The value of the vector is defined by assigning a logical 1 to where the connections are made, and all other positions in the register are assigned to be logical zeros. We see in Figure 4.13 that various bit positions in the register are connected to modulo-2 adders. For example, we observe that the inputs to

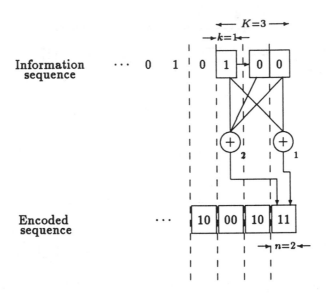

Figure 4.14: Convolutional encoder for the $CC(2, 1, 3)$ code

the modulo-2 adder defined by g_1 are in the last two symbol stages of the register. So generator g_1 is $[00 \ldots 0010000001]$ for $k = 4$, i.e., a sequence having only two logical ones. Notice that a generator defines the positions of the connections from the register to the modulo-2 adder. The output of the adder depends, of course, on the logical values of the information bits applied to its inputs.

For the sake of simplicity, we demonstrate the encoding process by considering the binary $(k = 1)$, half-rate $(R = k/n = 1/2)$, convolutional code, $CC(2, 1, 3)$ with constraint length of three binary stages $(K = 3)$. Once the basic concepts are established by this example, the same principle can be generalized to any convolutional code. The generators in Figure 4.14 are described by the vectors,

$$g_1 = [1\ 0\ 1] \quad \text{and} \quad g_2 = [1\ 1\ 1] \tag{4.5}$$

or are written in equivalent form of polynomials as

$$g_1(z) = 1 + z^2 \quad \text{and} \quad g_2(z) = 1 + z + z^2 \ . \tag{4.6}$$

The shift register has three binary stages, the first stage is for the present bit and the latter stages are for the two previous input bits. The status of the register is defined by the logical values of its bits, while the *state* of the code is represented by the two previous input bits. Initially, the shift register is set to the all-zero state and then an input bit of logical value 1 is applied. Generators g_1, associated with the first and third stage of the register, and g_2, associated with all three stages of the register, cause both the outputs from the two modulo-2 adders produce output bits of logical 1 that formulate the encoded symbol 11. At the next instant, data moves into the register from the right which is equivalent to the register being shifted to the left by $k = 1$ bit. The new input bit is a logical 0 and the state of the register is changed from 00 to 10. The new encoded symbol becomes 10. As the encoding process continues, the next input bits are ...,0,1, and the state transition becomes ..., 01, 10, 01. The string of the subsequent encoded symbols ..., 10, 00, is generated, and passed to the modulator for transmission.

Convolutional codes, as implied by the name, can be viewed as the convolution of the impulse response of the encoder with the information sequence. In our example, if an information sequence with a single logical 1 followed by all zeros is shifted into the register, the output sequence is referred to as the impulse response of the encoder and is given by $11, 10, 11, 00, 00, 00, \ldots$. For any arbitrary input sequence, the encoder output is the modulo-2 addition of the impulse responses arising from each logical 1 at the input, where each response is positioned in accordance with the location of the ones in the input sequence. The output sequence O corresponding to any arbitrary input sequence I can be found by multiplying the input vector by a generator matrix G, namely,

$$\mathbf{O} = \mathbf{I}^{\mathbf{T}}\mathbf{G} \qquad (4.7)$$

where G can be constructed as

$$\mathbf{G} = \begin{bmatrix} 11\ 10\ 11\ 00\ 00\ 00\ \ldots \\ 00\ 11\ 10\ 11\ 00\ 00\ 00\ \ldots \\ 00\ 00\ 11\ 10\ 11\ 00\ 00\ 00\ \ldots \\ \vdots \end{bmatrix}. \qquad (4.8)$$

As convolutional codes do not have a defined length, the generator matrix G is a semi-infinite matrix. For the example in Figure 4.14, $\mathbf{I}^{\mathbf{T}} = 101000\ldots$, the output sequence is obtained by adding the first and the third rows of G to produce $\mathbf{O}^{\mathbf{T}} = 11, 10, 00, 10, 11, 00, 00, \ldots$.

4.3.2   State and Trellis Diagrams

Figure 4.15 shows the action of the encoding process from an information sequence to a chain of state transitions and finally to a sequence of encoded symbols. As the information sequence is shifted into the register by $k = 1$ bit at a time, the encoder state obtains its value by observing the information sequence with a window size of $(K - 1)k = 2$ bits. A chain of state transitions of ..., 01, 10, 01, 10, 00, is obtained from an information sequence of ...010100. New encoded symbols are formed in response to

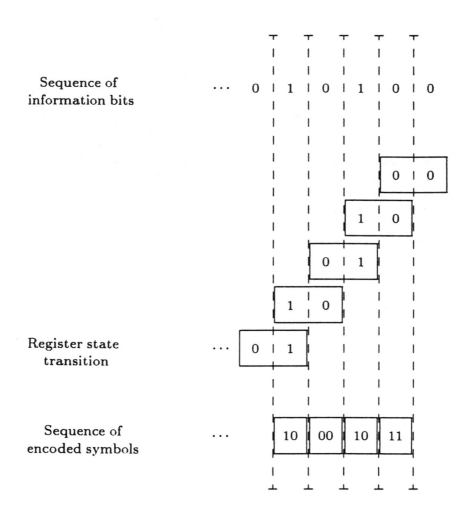

Figure 4.15: The action of the encoding process

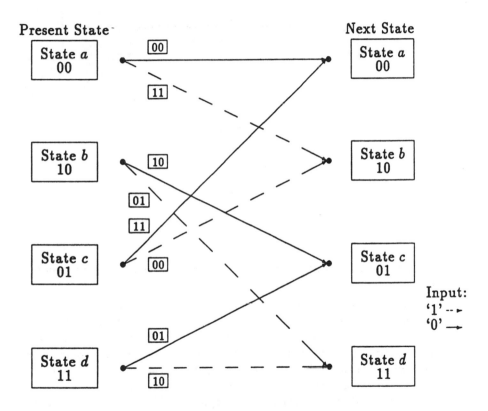

Figure 4.16: State transition diagram

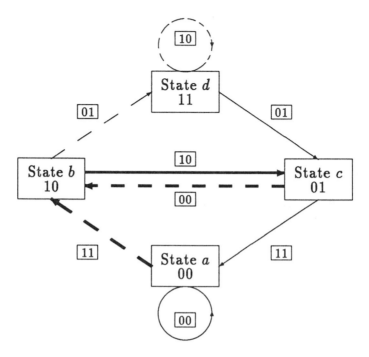

Figure 4.17: The state diagram

these state transitions and the sequence of the encoded symbols becomes ..., 10, 00, 10, 11.

The encoder state is given by the last $(K-1)k$ register stages, namely 2 in this example. The number of possible states is $2^{(K-1)k}$ ($=4$ here). The state at each instant can be either 00, 10, 01 or 11, which we label as state a, b, c or d, respectively. Figure 4.16 tabulates the possible state transitions at any symbol instant. The branch emanating from the present state to the next state indicates the state transition. The broken line branch is the transition initiated by input bit of logical 1, whereas the solid branch is due to the input bit being a logical 0. The symbol attached to each branch is an $n=2$ bit word and is the encoded symbol. The number of outgoing branches emanating from the present state is 2^k ($=2$ here), which corresponds to the number of the possible input patterns. Figure 4.16 represents the fundamental structure that can be used to construct the state diagram and trellis diagram of the convolutional code.

The state diagram corresponding to the state transition diagram is shown in Figure 4.17. It consists of a total number of $2^{(K-1)k}=4$ states connected by all the possible transitions shown in the state transition diagram. In the example being considered, the shift register is initially cleared setting the encoder to state a. The present input is a logical 1, the state changes

Symbol
Instant: $J = 0$ $J = 1$ $J = 2$ $J = 3$ $J = 4$

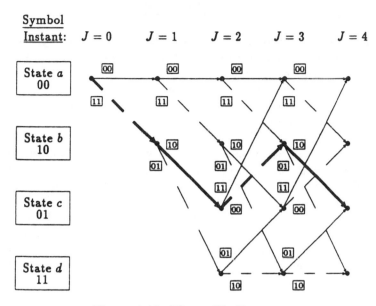

Figure 4.18: The trellis diagram

from a to b, as illustrated by the dotted branch emanating from state a in Figure 4.17. The encoder output is the symbol 11 that is assigned to the transition branch. At the next instant, the present state becomes state b and the encoder receives a new input of logical 0. This causes the state transition from b to c, and the encoded symbol becomes 10. The encoding cycle is repeated for subsequent inputs, which change the states and then generate the encoded symbols. By following the change of states throughout the encoding process, a particular path associated with states $a \rightarrow b \rightarrow c \rightarrow b \rightarrow c \rightarrow \ldots$, can be observed. This path is unique to the input sequence.

Another representation of the encoding process is the trellis diagram shown in Figure 4.18. This is formed by concatenating the consecutive instants of the state transition diagram of Figure 4.16 starting from the reset all-zero condition. The diagram illustrates all the possible paths, namely 16, that can occur for 4-bit information sequences. We observe in Figure 4.18 that after the initial transient i.e., after instant $J = 1$, the trellis contains four nodes at every symbol instant. The trellis is extended, for example to the $J = 5$ symbol instant, by appending the state transition diagram of Figure 4.16 every node, i.e., state. In our example, the path corresponding to the input sequence is drawn by the thick line in Figure 4.18 corresponding to the encoded sequence of 11, 10, 00, 10, $\ldots$.

4.3.3   Maximum Likelihood Decoding

We have seen in Section 4.3.1 that an information sequence changes the encoder states which in turn generates a sequence of encoded symbols. The encoded sequence is interleaved to combat the effects of signal fading in the channel, modulates a carrier and is transmitted over the mobile radio channel. The channel impairments will distort the received signal. After demodulation and deinterleaving, the convolutional decoder inverts the encoding process. The convolutional decoder operates on the input sequence by estimating the most likely path of state transitions in the trellis. Once identified, the corresponding information sequence is delivered as the decoded sequence. If the decoder employs the Viterbi algorithm [20], [21], *all* the possible paths in the trellis are searched, and their distances to the sequence at the decoder input are compared. The path with the smallest distance is then selected, and the information sequence regenerated. This method is known as maximum likelihood decoding in the sense that the most likely sequence from *all* the paths in the trellis is selected. It therefore results in the minimum bit error rate.

We have seen in Section 4.3.1 that a convolutional code is a long sequence of encoded symbols without a well-defined block length. In practice, convolutional codes can be truncated to a fixed length and concatenated, i.e., as if in a sequence of packets. The code is terminated by appending $(K-1)k$ logical zeros to the last information bit for the purpose of clearing the shift register of all information bits. By this means, the encoder returns to a known all-zero state. As the stuffing of logical zeros carries no information, the actual coding rate is now below R. In order to maintain the coding rate close to R, the encoded sequence (period of truncation) needs to be very long.

4.3.3.1   Hard-decision Decoding

The Viterbi algorithm is best explained using the trellis diagram of the simple binary $CC(2,1,3)$ code that was previously used as an example. We will assume that the demodulator provides only hard-decisions when regenerating the information sequence (soft-decision decoding is considered later). In this case the Hamming distances between the received symbols and the estimated transmitted symbols in the trellis are used as a metric, i.e., a confidence measure. Figure 4.19 records the history of the paths selected by the Viterbi decoder. Suppose there are no channel errors, the input sequence to the decoder is the same as the encoded sequence 11, 10, 00, 10, ... as illustrated in Figure 4.14. At the first instant $J = 1$, the received symbol is 11 which is compared with the possible transmitted symbols 00 and 11 of the branches from node a to a and from node a to b, respectively. The metrics of these two branches are their Hamming distances, namely the differences between the possible transmitted symbols 00 or 11 and between the received symbol 11. Their distances are 2 and 0,

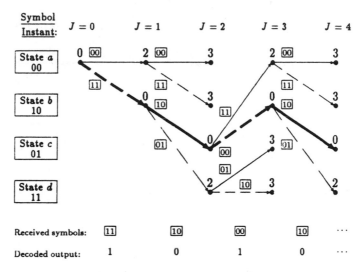

Figure 4.19: Example of Viterbi decoding

respectively.

Now we define that the branch metric is the Hamming distance of an individual branch, and the path metric at the J-th instant is the sum of the branch metrics at all its branches from $J = 0$ to the J-th instant. Hence the path metrics, printed on top of each node in Figure 4.19, at instant $J = 1$ are 2 and 0 for the paths $a \rightarrow a$ and $a \rightarrow b$, respectively. At the second instant $J = 2$, the received symbol is 10 and the branch metrics are 1, 1, 0 and 2 for the branches $a \rightarrow a$, $a \rightarrow b$, $b \rightarrow c$ and $b \rightarrow d$, respectively. The path metrics, or accumulated metrics, are 3, 3, 0 and 2 for the corresponding paths $a \rightarrow a \rightarrow a$, $a \rightarrow a \rightarrow b$, $a \rightarrow b \rightarrow c$ and $a \rightarrow b \rightarrow d$. At the third instant, the received symbol is 00. There are eight possible branches, as seen also in Figure 4.17 and their branch metrics are 0, 2, 2, 0, 1, 1, 1 and 1 for the branches $a \rightarrow a$, $c \rightarrow a$, $a \rightarrow b$, $c \rightarrow b$, $b \rightarrow c$, $d \rightarrow c$, $b \rightarrow d$ and $d \rightarrow d$, respectively. Let α_1 and α_2 denote the corresponding paths $a \rightarrow a \rightarrow a \rightarrow a$ and $a \rightarrow b \rightarrow c \rightarrow a$ that begin at the initial node a and remerge at node a at $J = 3$. Their respective path metrics are 3 and 2. Any further branches with $J > 3$ steming from the node a at $J = 3$ will add identical branch metrics to the path metric of both paths α_1 and α_2, and this means that the path metric of α_1 is larger at $J = 3$ and will continue to be larger at $J > 3$. The Viterbi decoder selects the path with the smallest metric and therefore decides to discard the α_1 path and retain the α_2 path. The α_2 path is called the *survivor*. This procedure is also applied at the other nodes b, c and d at $J = 3$. Notice that paths $a \rightarrow a \rightarrow a$ and $a \rightarrow a \rightarrow b$ cannot survive as their path metrics are larger than that of their counterparts of the merging pairs and

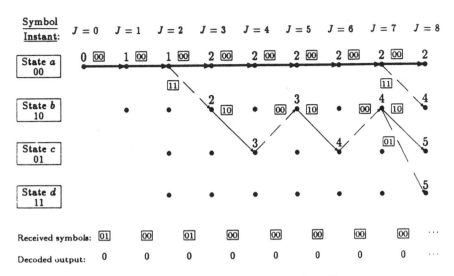

Figure 4.20: Example of correct decoding

they are therefore eliminated from the decoder memory. Thus there are four paths that survive at each instant and there are four path metrics, one for each of the surviving paths. Similarly, at the instant $J = 4$, there are again four survivors and the survivor with the smallest metric is the path $a \rightarrow b \rightarrow c \rightarrow b \rightarrow c$. This path corresponds to the transmitted sequence and hence the correct information sequence 1010... is delivered to the decoder output. This decoding method is optimum in the sense that it minimizes the probability that the entire sequence is in error.

4.3.3.1.1 Correct Decoding We will now introduce channel errors, and demonstrate how the Viterbi decoding corrects them. Figure 4.20 shows four survivor paths in the trellis at the instant $J = 8$. Suppose that the all-zero sequence is transmitted, and the received sequence is 01 00 01 00 00 ... , where the logical ones are the channel error bits. The first received symbol is 01 and the metric at $J = 1$ is 1 which reflects the number of error bits in the first received symbol. Similarly, the next error bit in the third received symbol 01 increments the metric of the all-zero path as well. At instant $J = 8$, four survivors are the estimated transmitted sequences which have Hamming distances of 2, 4, 5 and 5 for the received sequence shown in Figure 4.20. The Viterbi decoder favours the all-zero path, and its metric shows the number of channel error bits.

4.3.3.1.2 Incorrect Decoding When the number of channel errors exceeds the correcting capability of the code, incorrect decoding will occur as illustrated in Figure 4.21. Again, an all-zero sequence is transmitted and

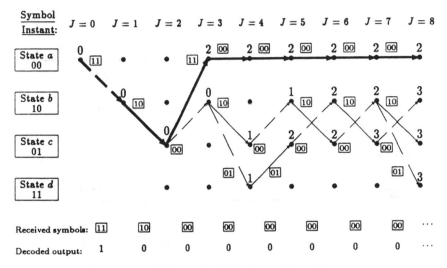

Figure 4.21: Example of incorrect decoding

the received sequence due to the channel impairments contains three channel error bits. The incorrect decoding occurs in the three initial branches which results in a single decoded information bit error determined at $J = 8$. The path metric of the selected path, drawn in thick line in Figure 4.21, is 2 and no longer corresponds to the actual number of the channel error bits, namely 3.

The last two examples of correct and incorrect decoding, are related to decisions that depend on whether the Hamming distance of the received sequence to the correct path is smaller than the distances to other paths in the trellis. If it is closer, we obtain correct decoding. We observe that the Hamming distance between the paths in the trellis plays an important role in the correcting capability of the code, and this will be discussed in Section 4.3.4.

4.3.3.2 Soft-decision Decoding

So far we have limited our discussion to hard-decision decoding. That is, the demodulated signal at the demodulator output is sampled and hard-limited to regenerate the binary signal for channel decoding. We now explore the techniques of soft-decision decoding. In this approach, the signal variations at the output of the demodulator are sampled and quantised. For an additive white Gaussian noise (AWGN) channel, hard quantization of the received signal results in a loss of about $2\,\mathrm{dB}$ in E_b/N_0 compared with infinitely fine quantization [22], while an 8-level quantization reduces the loss compared to infinite fine quantization to less than $0.25\,\mathrm{dB}$. This indicates that quantization with 8-levels is adequate for our purposes. Figure 4.22

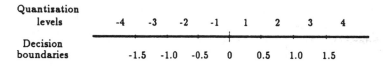

Figure 4.22: 8-level received signal quantization

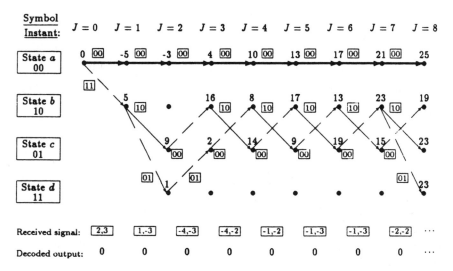

Figure 4.23: Example of Viterbi decoding with soft-decision

shows the range of the sampled output from a binary demodulator with decision boundaries spaced by 0.5 and eight quantization levels, where the transmitted signal level is unity. The magnitude of the quantized level {1, 2, 3, 4} represents the confidence we have in the received sample, where the higher the number the greater the confidence. For example, a level ±4 gives a high confidence that the bit is a logical ±1, whereas a level of ±2 gives us less confidence as to the polarity of the bit.

Figure 4.23 applies the Viterbi decoding with soft-decision to the same example as in Figure 4.21, which produced an erroneous output when hard-decision decoding was used. By using soft-decision, the demodulator passes a sequence of quantized levels to the FEC decoder instead of a sequence of data bits. Assuming that an all-zero sequence is transmitted, in the absence of channel noise and fading, the received signal at the output of the demodulator is quantized to a large negative level. Should the quantized output be positive, i.e., a transmission error occurs, the channel impairments have caused a stronger, but opposite polarity voltage against the transmitted signal. In Figure 4.23 we see that the first three quantized levels are positive, namely 2,3,1 are channel error bits in the absence of channel decoding. The first received symbol has confidence levels 2,3, and is compared with the transmitted symbol of the possible transitions of 00 and 11 in the trellis. We observe that levels 2 and 3 give us a measure of confidence that the transition should be 11. Comparing transition 00 with 2,3, the branch metric for path $a \rightarrow a$ at $J = 1$ is $-2 + (-3) = -5$. However, when we compare the possible transition 11 with levels 2,3, the branch metric is $2 + 3 = 5$, when a transition $a \rightarrow b$ occurred. Similar arguments at $J = 2$ result in accumulated confidences of -3, 9, 1 for paths $a \rightarrow a \rightarrow a$; $a \rightarrow b \rightarrow c$ and $a \rightarrow b \rightarrow d$. Let us denote α_1 and α_2 the paths $a \rightarrow a \rightarrow a \rightarrow a$ and $a \rightarrow b \rightarrow c \rightarrow a$, (not shown in Figure 4.23) respectively. The path metrics at $J = 3$ for paths α_1 and α_2 are 4 and $2 (= 9 - 7)$, respectively. The Viterbi decoder selects the path with the largest metric because of its stronger accumulated confidence. Hence, path α_1 is selected. At instant $J = 8$, the all-zero path has the largest metric among the four survivors. This example shows that the incorrect decoding, which occurred when hard-decision decoding was used can be rectified by using soft-decision decoding.

4.3.3.3 The Viterbi Algorithm

From the examples in Subsection 4.3.3.1 and 4.3.3.2, we found that the complexity of the trellis diagram directly reflects the computation and memory requirement of the Viterbi decoder. In general, for a $CC(n, k, K)$ code, there are $2^{(K-1)k}$ possible states in the encoder. In the Viterbi decoder all states are represented by a single column of nodes in the trellis at every symbol instant. At each node in the trellis, there are 2^k merging paths and the one with the minimum distance is selected as the survivor. As a consequence there are 2^k path comparisions at each merging node, and the same number of comparisions is repeated for all $2^{(K-1)k}$ nodes at a sampling instant. The computation increases exponentially with K and k, and this restricts them to relatively small values.

The Viterbi decoder selects and updates $2^{(K-1)k}$ surviving sequences and stores them in a memory at each instant. At the end of the encoded sequence or packet, the decoder selects the survivor with the minimum distance, i.e., the minimum Hamming distance with hard-decision decoding, or the maximum confidence measure with soft-decision decoding. In practice, the encoded packets are usually very long, and it is impractical to store the entire length of the surviving sequences before making a decision as to the information sequence because of the long decoding delay that would accrue. Instead, only the most recent L information bits in each of the surviving sequences are stored. Once the survivor with the minimum distance is identified the symbol associated with this path L periods ago is conveyed to the output as a decoded information symbol. In Section 4.3.1 we described how the present encoded symbol depends on the $(K - 1)$ previous symbols. The ramification of this is that the joint probabilities between symbols tends to be inversely proportional to the time separation between them. As a consequence we arrange for the parameter L to be made sufficiently large, normally $L \geq 5K$, for the present symbol of the surviving sequences to have a minimum effect on the decoding of the L previous symbols.

We have described the principles of hard-decision and soft-decision Viterbi decoding. The path metric in hard-decision decoding is a measure of the Hamming distance, and the path with the minimum metric is identified by the Viterbi decoder to yield the recovered sequence. With soft-decision decoding the path metric is a confidence measure and the path with the maximum metric provides the highest confidence and is therefore selected by the Viterbi decoder. Let us now summarize the essential features for the Viterbi algorithm that apply for hard-decision and for soft-decision decoding. For a $CC(n, k, K)$ code, there are $2^{(K-1)k}$ number of states and each state can change to 2^k possible states at the next instant in the trellis. The state transition in the trellis is represented by the branch, and there are 2^K possible branches at each instant. For a particular information sequence, a unique path is observed in the trellis. A path consists of a chain of branches and each branch in the trellis is associated with an encoded symbol. The symbol attached to the path forms the sequence of encoded symbols that are transmitted. At the receiver, the Viterbi decoder matches the received symbol to every possible encoded symbol in the trellis. Suppose c_{ji} denotes the bit at the instant J in the trellis, where j indicates the j-th branch and i indicates the i-th bit in the symbol. The value of c_{ji} in the trellis is an estimate of the corresponding bit r_{ji} at the output of the demodulator. The Viterbi decoding process can be described by the following steps:

1. **Branch metric calculation**
   The branch metric $m_j^{(\alpha)}$ at the J-th instant of the α path through the trellis is defined as the logarithm of the joint probability of the received n-bit symbol $r_{j1}r_{j2} \ldots r_{jn}$ conditioned on the estimated transmitted n-bit symbol $c_{j1}^{(\alpha)}c_{j2}^{(\alpha)} \ldots c_{jn}^{(\alpha)}$ for the α path. That is,

$$
\begin{aligned}
m_j^{(\alpha)} &= ln \prod_{i=1}^{n} P(r_{ji} \mid c_{ji}^{(\alpha)}) \\
&= \sum_{i=1}^{n} ln \, P(r_{ji} \mid c_{ji}^{(\alpha)}) .
\end{aligned}
\tag{4.9}
$$

2. **Path metric calculation**
   The path metric $M^{(\alpha)}$ for the α path at the J-th instant is the sum of the branch metrics belonging to the α path from the first instant

to the J-th instant. Therefore,

$$M^{(\alpha)} = \sum_{j=1}^{J} m_j^{(\alpha)} . \tag{4.10}$$

3. **Information sequence update**
There are 2^k merging paths at each node in the trellis and the decoder selects from the paths $\alpha_1, \alpha_2, \ldots, \alpha_{2^k}$, the one having the largest metric, namely,

$$\max \left(M^{(\alpha_1)}, M^{(\alpha_2)}, \ldots, M^{(\alpha_{2^k})} \right) \tag{4.11}$$

and this path is known as the survivor.

4. **Decoder output**
When all of the $2^{(K-1)k}$ survivors have been determined at the J-th instant, the decoder outputs the $(J - L)$-th information symbol from its memory of the survivor with the largest metric.

Let us apply the path metric calculation, defined in the above steps, for hard-decision decoding, where r_{ji} is either 0 or 1. We again use our previous example of Viterbi decoding, see Figure 4.19. The two paths α_1 and α_2 that begin at the initial node a at $J = 0$ and merge to node a after three branches in the trellis have Hamming distances of 2 and 3 with the received sequence 10 00 01 ..., respectively. According to Equation 4.9 and 4.10, the path metrics for α_1 and α_2 are

$$M^{(\alpha_1)} = \sum_{j=1}^{3} \sum_{i=1}^{2} \ln P(r_{ji} \mid c_{ji}^{\alpha^1}) = 4\ln(1 - p_b) + 2\ln(p_b)$$

$$M^{(\alpha_2)} = \sum_{j=1}^{3} \sum_{i=1}^{2} \ln P(r_{ji} \mid c_{ji}^{\alpha^2}) = 3\ln(1 - p_b) + 3\ln(p_b) \tag{4.12}$$

where p_b is the probability of a channel bit error. Assuming that $p_b < \frac{1}{2}$, then $M^{(\alpha_1)}$ is greater than $M^{(\alpha_2)}$ and therefore the α_1 path is selected as the survivor. We also note that, as expected, the path α_1 has the smallest Hamming distance.

If soft-decision is employed, each quantized sample at the output of the demodulator indicates a confidence measure of its associated data bit. This measure is the Euclidean distance of the received signal vector from the signal boundary in the constellation. For an M-ary modulation scheme, the signal points in the constellation are surrounded by more than one boundaries. The smaller the Euclidean distance of the received vector to a particular signal point, the stronger is our confidence in the value of the vector. However, for binary modulation, the two signal points are separated by a single boundary. The larger the Euclidean distance of the received signal from the boundary, the more confidence we have that the received signal has the correct polarity. As an example, if the modulation is minimum shift keying (MSK) [23], there is an in-phase (I) and quadrature phase (Q) signalling channel. The signal boundaries of the I and Q channels are spaced by a phase angle of 90° and they are therefore orthogonal. The confidence measure on one channel does not affect the measure on the other channel. Both I and Q can be then treated as independent binary signalling channels and our confidence is proportional to the Euclidean distance of the received vector from the signal boundary. Suppose the MSK transmitted signal is coherently detected to yield

$$r_{ji} = \frac{A_s T}{2} a_{ji}^{(\alpha)} + N(T) \tag{4.13}$$

where A_s is the transmitted signal amplitude, T is the bit duration, $a_{ji}^{(\alpha)}$ is -1 for $c_{ji}^{(\alpha)} = 0$; and 1 for $c_{ji}^{(\alpha)} = 1$, i.e., $a_{ji}^{(\alpha)} = 2c_{ji}^{(\alpha)} - 1$, and $N(T)$ is a Gaussian random noise signal with zero mean and variance $\sigma_N^2 = N_0 T/8$, and $N_0/2$ is the double-sided power spectral density of the receiver thermal noise. For the AWGN channel, the demodulator output signal has a probability density function (PDF) of

$$f(r_{ji} \mid c_{ji}^{(\alpha)}) = \frac{1}{\sqrt{2\pi}\sigma_N} \exp\left(-\frac{(r_{ji} - \frac{A_s T}{2} a_{ji}^{(\alpha)})^2}{2\sigma_N^2} \right) . \tag{4.14}$$

By substituting Equation 4.14 into Equation 4.9, it gives the branch metric which is then inserted into Equation 4.10 to obtain the path metric,

$$
\begin{aligned}
M^{(\alpha)} &= \sum_{j=1}^{J} \sum_{i=1}^{n} \ln P\left(r_{ji} \mid c_{ji}^{(\alpha)} \right) \\
&= \sum_{j=1}^{J} \sum_{i=1}^{n} \left[\ln\left(\frac{1}{\sqrt{2\pi}\sigma_N} \right) - \frac{(r_{ji} - \frac{A_s T}{2} a_{ji}^{(\alpha)})^2}{2\sigma_N^2} \right] \tag{4.15} \\
&= \sum_{j=1}^{J} \sum_{i=1}^{n} \left[\ln\left(\frac{1}{\sqrt{2\pi}\sigma_N} \right) - \frac{r_{ji}^2}{2\sigma_N^2} + \frac{\frac{A_s T}{2} r_{ji} a_{ji}^{(\alpha)}}{2\sigma_N^2} - \frac{(\frac{A_s T}{2} a_{ji}^{(\alpha)})^2}{2\sigma_N^2} \right]
\end{aligned}
$$

$$\tag{4.16}$$

The first and second terms of Equation 4.16 are common to all paths. Also $(a_{ji}^{(\alpha)})^2$ is always 1, and the fourth term is a constant and again is common to all paths. As the terms common to all path metrics do not change the path selection, they can be ignored from the calculation of path metrics. Futhermore, $(A_s T/4\sigma_N^2)$ in the third term can also be neglected. Therefore the path metric difference becomes

$$\Delta M^{(\alpha)} = \sum_{j=1}^{J} \sum_{i=1}^{n} r_{ji} a_{ji}^{(\alpha)} . \tag{4.17}$$

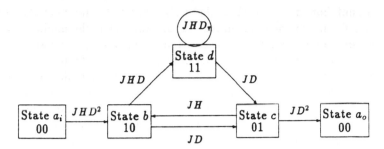

Figure 4.24: State diagram for the $CC(2,1,3)$ code

Equation 4.17 implies that the path metric at the J-th instant of the α path is the accumulated bit confidence measure. The path with the higher confidence reflects the larger metric and thus the Viterbi decoder selects the path with the larger metric as the survivor. This decision method has been demonstrated in our previous soft-decision decoding example in Figure 4.23.

4.3.4 Distance Properties of Convolutional Codes

From the examples in Figure 4.20 and Figure 4.21, we observe that the Hamming distance of the paths in the trellis determines the correcting capability of the code. The bit error rate (BER) performance of a convolutional code thus depends on its distance property. As convolutional encoding involves modulo-2 linear operations on the information sequence, the convolutional code is linear and therefore the distance separations of all the paths from a code sequence in the trellis is independent of which particular code sequence is considered.

For the sake of simplicity, we assume that the all-zeros sequence is transmitted. Consequently if any erroneous decoding occurs, the non-zero path in the trellis is favoured by the Viterbi decoder. For example, in Figure 4.21, we see that the trace of the incorrect path leaves the all-zero sequence before eventually merging back to the all-zero path. The trace can also be observed from the state diagram of the code shown in Figure 4.17. The initial state is at node a, and if there is no channel error, the received symbol is 00 and the next state is again at node a. Self-looping at node a occurs. Suppose the received sequence is no longer an all-zero sequence because of the channel noise and that the decoder fails to correct the errors. The state at node a changes to the other nodes for some instants before merging back to node a again. By splitting the node a of the state diagram in Figure 4.17 into an input stage a_i and an output stage a_o, the traces of all the incorrect paths are revealed by the possible connections from the input to the output as shown in Figure 4.24. The transition from node a_i to b represents the state transition of leaving the correct path, whereas that from node c to a_o is the transition of merging back to the all-zero path. The branches of the state diagram are labelled as either D^0, D^1 and D^2, where the exponent of D denotes the Hamming distance of the received symbol to the all-zero symbol in bits. The factor H is introduced into those branches activated by the information bit 1. Also, a factor of J is introduced into each branch such that the exponent of J will serve as a counter to indicate the number of instants in any given path before merging back to the correct sequence.

Let us illustrate the situation by considering an example of the factors J, H and D associated with the branch $a_i \rightarrow b$ in Figure 4.24. This branch consists of a single transition and therefore the exponent of factor J is unity. The state transition from 00 to 10 is due to a single information bit 1 to the input of the encoder, the exponent of H is thus 1. As the transition produces an encoded symbol of 11 having a distance of 2 bits from the all-zero sequence, the factor D^2 is attached to this branch. Consequently, the branch $a_i \rightarrow b$ is labelled by JHD^2. Let X_s be a variable representing the accumulated weight of each path that enters state s. The transfer function associated with all transitions from state a_i to a_o then provides the required enumeration of path weights. That is, we consider,

$$T(D, H, J) = X_{a_o}/X_{a_i} \qquad (4.18)$$

such that all the possible incorrect paths originating from the input stage and terminating at the output stage are illustrated. From Figure 4.24, the state equations provide the following recursive relationships,

$$
\begin{aligned}
X_b &= JHD^2 X_{a_i} + JHX_c \\
X_c &= JDX_b + JDX_d \\
X_d &= JHDX_b + JHDX_d \\
X_{a_o} &= JD^2 X_c \, .
\end{aligned}
\qquad (4.19)
$$

The transfer function of the $CC(2, 1, 3)$ code is obtained by determining X_{a_o}/X_{a_i} from Equation 4.19 and substituting the result into Equation 4.18, viz:-

$$T_{CC213}(D, H, J) = \frac{J^3 H D^5}{1 - JHD(1 + J)} \qquad (4.20)$$

and on dividing out,

$$
\begin{aligned}
T_{CC213}(D, H, J) &= J^3 H D^5 + J^4 H^2 D^6 + J^5 H^2 D^6 + J^5 H^3 D^7 \\
&\quad + 2J^6 H^3 D^7 + J^7 H^3 D^7 + \ldots
\end{aligned}
\qquad (4.21)
$$

The first term of the transfer function $T_{CC213}(D, H, J)$ indicates that there is an incorrect path having a Hamming distance of five bits (exponent of D) from the all-zero sequence that merges back to node a after three instants

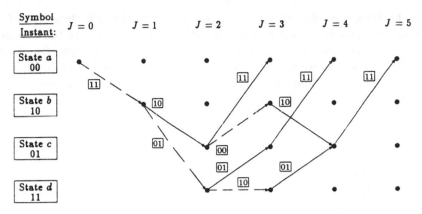

Figure 4.25: Pictorial description of $T_{CC(2,1,3)}(D, H, J)$

(exponent of J), and there is an erroneous information bit 1 (exponent of H). From the trellis diagram shown in Figure 4.25, the first term of $T_{CC213}(D, H, J)$ is the trace of the path which is observed from node a, through b, c and then merging back to node a again. Similarly, the second term of $T_{CC213}(D, H, J)$ is shown up as the path $a \rightarrow b \rightarrow d \rightarrow c \rightarrow a$ in Figure 4.25. This path of four branches, leaving at $J = 0$ and returning at $J = 4$, has a Hamming distance of six bits with the all-zero sequence and induces two erroneous information bits at the decoder output.

Each term of the transfer function represents an incorrect trellis path. The total number of incorrect paths increases exponentially with J and therefore the transfer function has infinite terms. An important property of the transfer function is that it provides the distance properties of all the paths of the convolutional code. The minimum distance between two paths of the code is called the *minimum free distance* and is denoted as d_{free}. The d_{free} of $CC(2, 1, 3)$ is equal to five, the exponent of D in the first summation term in Equation 4.21.

The factor J of the transfer function is to determine the number of branches spanned by the paths. If the convolutional code sequence is truncated after q instants, then the transfer function for the truncated code is obtained by truncating $T(D, H, J)$ at the term J^q. However, for a very long code sequence, the transfer function tends to have infinite number of terms and therefore J is no longer important in determining the truncation. The factor J can be suppressed by setting $J = 1$ in Equation 4.21 to yield,

$$T_{CC(2,1,3)}(D, H, 1) = HD^5 + 2H^2D^6 + 4H^3D^7 + \ldots \quad (4.22)$$

The transfer function in Equation 4.22 does not depend on the path length J. For instance, the second and third terms in Equation 4.21 indicate that either term has two non-zero, i.e. erroneous information bits and a Hamming distance of six bits with the all-zero path, but both terms are different in J such that their path lengths span over four and five instants, respectively. These two terms are combined together to form the second term in Equation 4.22 regardless of their path lengths. Furthermore, H is set to unity in Equation 4.22 in order to ignore the number of erroneous information bits associated with the path. The transfer function now depends only on D and the Hamming distance of all the incorrect paths to the all-zero sequence becomes:

$$T_{CC(2,1,3)}(D, 1, 1) = D^5 + 2D^6 + 4D^7 + \ldots . \qquad (4.23)$$

As the correct path is assumed to be an all-zero sequence, the Hamming distance between an incorrect and correct path is the weight (number of logical ones) of the incorrect path. In general, if d denotes the Hamming distance of a weight-d path, Equation 4.23 can be expressed as

$$T(D) = \sum_{d=d_{min}}^{\infty} A_d D^d \qquad (4.24)$$

where the coefficient A_d is the number of incorrect paths of weight-d regardless of the information bits on the path.

An important property of the code is the weight distribution, which determines the number of information bit errors at the decoder output if the weight-d path is incorrectly selected. The weight distribution $W_{CC213}(d)$ of the $CC(2, 1, 3)$ code is characterized by the total number of erroneous information bits of all weight-d trellis paths and is formally obtained by differentiating $T_{CC213}(D, H)$ with respect to H and setting $H = 1$:

$$W_{CC213}(d) = \left. \frac{d(T_{CC213}(D, H, 1))}{dH} \right|_{H=1} \qquad (4.25)$$

This is true, since differentiating $T(D, H)$ effectively yields the multiplication of the total number of incoming paths with the number of incorrect decoded bits per such path, giving the total number of incorrect decoded bits in all weigth-d paths. Substituting Equation 4.22 into 4.25, we have

$$W_{CC213}(d) = D^5 + 4D^6 + 12D^7 + \ldots . \qquad (4.26)$$

In general, the weight distribution of a code can be described by,

$$W(d) = \sum_{d=d_{min}}^{\infty} W_d D^d \qquad (4.27)$$

where the coefficient W_d is defined as the total number of erroneous information bits for all weight-d paths. The minimum distance among all these paths is d_{free} and therefore $W_d = 0$ for $d < d_{min}$.

Information sequence

Encoded sequence

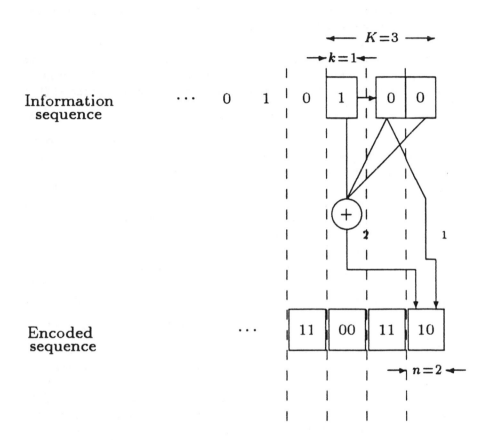

Figure 4.26: Convolutional systematic encoder for the $CC(2,1,3)$ code

The $CC(2,1,3)$ code of our example is a *nonsystematic* code because neither of the polynomial generators produce output bits that are identical to the information bits. Conversely, if either of the generator polynomials $g_1(z)$ or $g_2(z)$ have only a single connection to the register, the information sequence appears directly at the encoder output and the codes are described as *systematic* codes. Nonsystematic convolutional codes are usually preferred because of their higher error correcting capability compared to systematic codes, while for block codes the converse is true. The reasons become apparent if we compare the distance property of the systematic $CC(2,1,3)$ with our example of a nonsystematic $CC(2,1,3)$. The generators $g_1(z)$ and $g_2(z)$ of the systematic code are, as shown in Figure 4.26:

$$g_1(z) = z^2 \quad \text{and} \quad g_2(z) = 1 + z + z^2 \ . \tag{4.28}$$

The generator $g_1(z)$ is a direct hard-wire connection and therefore copies the information sequence to its output. The optimum performance code is formulated by computer search of optimum connections for the generator polynomials. As the $g_1(z)$ connection is fixed to be z^2, it reduces the degree of freedom for the search of the optimum code. The code cannot therefore achieve the maximum distances between encoded sequences as it can for nonsystematic codes. By eliminating one of the adders, there is a reduction in the minimum free distance.

If we calculate the transfer function of the systematic $CC(2,1,3)$ code with the generator polynomials given in Equation 4.28, then the weight distribution of this code is obtained with the aid of Equation 4.25 as

$$W_{CC213}(d) = 3D^4 + 15D^6 + 58D^8 + \dots \ . \tag{4.29}$$

The d_{free} of this code is 4, which is less than that of its nonsystematic counterpart ($d_{free} = 5$). We note that the d_{free} is the minimum separable distance of the code, i.e., any two encoded sequences must be separated by a Hamming distance of at least d_{free}. The performance of the code is proportional to d_{free}. The smaller d_{free} of the systematic code compared with the nonsystematic reflects the reduction of the relative distances between encoded sequences. In fact, Bucher and Heller [24] have shown that for large K, the performance of a systematic code of constraint length K is approximately the same as that of a nonsystematic code of constraint length $K(1 - R)$. Thus for $R = 1/2$ and very large K, systematic codes have the performance of nonsystematic codes of half the constraint length, while demanding the same decoder complexity.

So far, we have used the $CC(2,1,3)$ code as our simple example to illustrate the principle of convolutional coding. This code is weak because its contraint length $K = 3$ is short. The code advocated by the GSM recommendation [12] for speech communications over the Pan-European cellular mobile radio system is $CC(2,1,5)$. The generator polynomials of the full-rate speech channel [25] for this code are described by

$$g_1(z) = 1 + z^3 + z^4 \quad \text{and} \quad g_2(z) = 1 + z + z^3 + z^4 \ . \tag{4.30}$$

This is a binary half rate code with a constraint length of 5 bits. The state diagram has $2^{(K-1)k} = 16$ states and the transfer function of the code is found by solving 16 state equations. For the binary code where $k = 1$, the complexity of the computation grows exponentially with K. An alternative way of obtaining the transfer function for large values of K is to trace through every possible non-zero path in the trellis by an exhaustive computer search and record their path distances. The weight distribution of the code $CC(2, 1, 5)$ is obtained by recording the total weight of all information sequences which produce paths of distance d from the all-zero path. From our computer search we found the weight distribution to be

$$W_{CC215}(d) = 4D^7 + 12D^8 + 20D^9 + \dots . \qquad (4.31)$$

The coefficient W_7 of the first term in Equation 4.31 indicates that a total of 4 information bit errors are associated with all weight-7 paths. Similarly, the coefficients W_8 and W_9 of the second and third terms have 12 and 20 information bit errors associated with weight-8 and weight-9 paths, respectively. The minimum distance d_{free} among all of these paths is 7, the exponent of D in the first term in Equation 4.31.

The code used for satellite communications [25] is $CC(2, 1, 7)$, a binary half rate code with a longer constraint length of 7 bits. The generator polynomials [26] used in the encoder are given by

$$g_1(z) = 1 + z^2 + z^3 + z^5 + z^6 \quad \text{and} \quad g_2(z) = 1 + z + z^2 + z^3 + z^6 . \quad (4.32)$$

The weight distribution of this code is obtained by computer search in order to avoid the complexity of solving its $2^{(K-1)k} = 64$ state equations, and is given by

$$W_{CC217}(d) = 36D^{10} + 211D^{12} + 1404D^{14} + \dots . \qquad (4.33)$$

The minimum free distance of this code is $d_{free} = 10$. Although all of the $CC(2, 1, 3)$, $CC(2, 1, 5)$ and $CC(2, 1, 7)$ are half rate binary codes, the effect of increasing their contraint length is to increase the number of states in the code and their minimum free distance and thereby enhance their error correcting capability.

4.3.5   Punctured Convolutional Codes

For a convolutional code of rate k/n there are 2^k merging paths at each node in the trellis. The decoding of this code by the Viterbi algorithm selects the path with the highest metric out of the 2^k possibilities at each node. The number of calculations per selection at each node grows exponentially with k, rendering the implementation of the codec for operation at high speed a difficult task, particularly in case of high-rate codes. Yamada proposes a syndrome-former trellis [27] to decode high-rate codes. The method achieves the same performance [28] as the Viterbi decoding

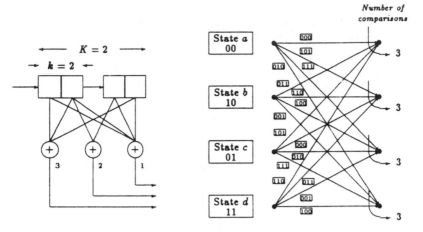

Figure 4.27: Encoder and trellis diagram for the $CC(3,2,2)$, $R = 2/3$ code

algorithm with reduced number of computations. However, the Viterbi decoding of high rate codes, where $k > 1$ can also be significantly simplified by employing punctured convolutional codes [29, 30]. Puncturing allows us to obtain a high-rate code by periodically deleting some of the coded bits from a low-rate encoder output. In addition, puncturing of the low-rate $1/n$ code results in the decoding trellis operating with k equal to unity.

Let us consider a high rate $R = 2/3$ code, which can be achieved by either the $CC(3,2,2)$ code or the punctured $PCC(2,1,3)$ code. Both examples ·are chosen with the same number of states i.e., $2^{(K-1)k} = 4$. The generator polynomials [31] of the $CC(3,2,2)$ code are

$$
\begin{aligned}
g_1(z) &= 1 + z + z^2 + z^3 \\
g_2(z) &= z + z^2 \\
g_3(z) &= 1 + z + z^3 \ .
\end{aligned}
\tag{4.34}
$$

At each instant, a 2-bit information symbol is inserted into the encoder and the three generators produce a 3-bit output symbol. The coding rate R is therefore equal to 2/3. Figure 4.27 illustrates all the possible state transitions at a symbol instant. The number of possible states of this code is $2^{(K-1)k} = 4$ and the symbol attached to each state transition is an $n = 3$ bit encoded symbol. Each state transition is activated by a new $k = 2$ bit information symbol at the encoder input. A 2-bit shift of the encoder state at each instant induces one of four possible state transitions. For each node in the trellis shown in Figure 4.27 there are four merging branches and thus three pairwise comparisons are required to select the survivor. The comparisons are repeated for the other nodes and the total number of comparisons for decoding every three received bits is 12.

The same coding rate can also be achieved by periodically deleting bits from the $CC(2,1,3)$ half rate code as demonstrated in Figure 4.28. A

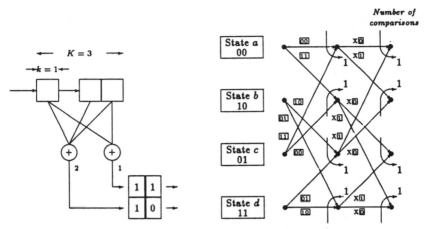

Figure 4.28: Encoder and trellis diagram for the $PCC(2,1,3)$, $R = 2/3$ code

punduring matrix is assigned at the output of a half rate encoder, where an element of 1 in the matrix allows its input bit to appear at the output, whereas an element of 0 in the matrix deletes the incoming bit at its input. The top row of the matrix consists of 11 which does not delete any bits from the output of adder 1. The second row of the matrix is 10 which deletes every alternate bit from the output of adder 2. The encoded sequence is formulated by sampling the adders output alternately. In this case, every fourth encoded bit is deleted, the encoder will produce three output bits for every two input bits resulting in a $R = 2/3$ code. The trellis of the punctured code shown in Figure 4.28 is basically constructed from Figure 4.16. It is equivalent to the trellis of the half rate code except that the X indicates the position of the deleted bits. There are only two merging paths to each node and therefore only a pairwise comparison is required at each node. For every three bits received over two symbol instants, the number of comparisons is 8, which is less than the 12 required by the $CC(3,2,2)$ code. Puncturing the code reduces its minimum free distance, in this case from 5 to 3. However, this is the largest minimum free distance [32] for any $CC(3,2,2)$ code, and therefore in this case there is no reduction in the minimum free distance due to puncturing.

Let us now illustrate an example of punctured coding by using the encoder described in Figure 4.28. It consists of a half-rate $CC(2,1,3)$ encoder as used in Figure 4.14, followed by a puncturing matrix at its output. Suppose we use the same input sequence to the encoder as in Figure 4.14, i.e., $\ldots, 0, 1, 0, 1$. The output sequence is punctured every fourth bit and becomes $\ldots, 0, 00, 0, 11$. On the basis of no channel errors, the input sequence to the decoder is the same as the encoded sequence. The received sequence can be either hard-decision or soft-decision decoded. Let us ap-

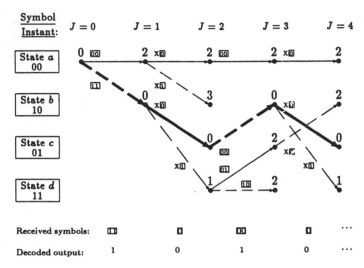

Figure 4.29: Example for the Viterbi decoding of a punctured code

ply hard-decision Viterbi decoding as shown in Figure 4.29. At the first instant, $J = 1$, the received symbol is 11 which is compared with the possible transmitted symbols 00 and 11 of the branches $a \rightarrow a$ and $a \rightarrow b$. Their Hamming distances are 2 and 0, respectively. At the second instant, $J = 2$, the received symbol consists of only a single bit 0, as the other bit of the symbol was punctured at the encoder. The corresponding bit being punctured in the trellis is marked by X in Figure 4.29. The received bit is then compared with the possible transmitted bits 0, 1, 0 or 1 for the corresponding branches $a \rightarrow a$, $a \rightarrow b$, $b \rightarrow c$ and $b \rightarrow d$. The path metrics are 2, 3, 0 and 1 for the branches $a \rightarrow a \rightarrow a$, $a \rightarrow a \rightarrow b$, $a \rightarrow b \rightarrow c$ and $a \rightarrow b \rightarrow d$, respectively. Similarly, Viterbi decoding algorithm is applied to the third and fourth instants. At the instant $J = 4$, there are four survivors and the survivor with the smallest metric is the path $a \rightarrow b \rightarrow c \rightarrow b \rightarrow c$. This path corresponds to the transmitted sequence and hence the correct information sequence 1010... is delivered to the decoder output.

Let us consider a punctured $R = 2/3$ code with a constraint length of 5 bits. This may be produced by puncturing the $CC(3, 1, 5)$ code where the generator polynomials are given by [30]

$$
\begin{aligned}
g_1(z) &= 1 + z + z^4 \\
g_2(z) &= 1 + z^2 + z^3 + z^4 \\
g_3(z) &= 1 + z^2 + z^4 .
\end{aligned}
\tag{4.35}
$$

This punctured code is designed to produce the optimum performance at BER= 10^{-5} in the presence of random bit errors. The puncturing pattern of the encoded bit sequence is to delete every alternate output bit, starting from the first bit generated by $g_1(z)$ and $g_2(z)$, and also every alternate output bit, starting from the second bit produced by $g_3(z)$. Hence, three output bit remained for every two input bits that results in a 2/3 rate code.

4.3.6 Hard-decision Decoding Theory

As the convolutional code is linear, we assume that the all-zero path is transmitted, knowning that our findings can be generalized for other non-zero paths. Convolutional codes, unlike block codes, do not necessarily have a fixed length. The Viterbi decoder selects the survivors in the trellis at every instant while the sequence is being received. Bit errors at the decoder output are due to selecting the incorrect path in the trellis. In order to derive the post-decoding bit error rate performance, we define the first-event error probability, P_{FE}, as the probability when for the first time an incorrect path merges to the correct path at a node with a metric that is smaller than the one for the correct path. We recall that for hard-decision decoding, the metrics in the Viterbi algorithm are the Hamming distances between the received sequence and the paths in the trellis. Let us assume that the decoder receives a sequence of demodulated bits having an average bit error rate of p_b over a memoryless channel. Suppose that the non-zero path merging at node a is separated by a Hamming distance of d from the all-zero path. If the number of error bits in the received sequence exceeds $d/2$, the received sequence is compared with both non-zero and all-zero paths and their Hamming distances are $< d$ and $> d$, respectively. The non-zero path has a smaller metric and is therefore favoured by the Viterbi decoder. Consequently an erroneous decoding occurs. If the Hamming distance between the non-zero and the all-zero paths is odd, i.e., d is odd, the probability of an incorrect decoding is the probability that the number of channel errors is $\geq (d+1)/2$,

$$P_{ICD}(d) = \sum_{i=(d+1)/2}^{d} \binom{d}{i} p_b^i (1-p_b)^{d-i} ; \qquad d \text{ is odd.} \qquad (4.36)$$

However, when the Hamming distance between the non-zero and the all-zero paths is even, and the number of channel errors equals $d/2$, the received sequence has equal distances from the non-zero and the all-zero paths. The metrics of both paths are therefore equal. In this case, the Viterbi decoder will randomly select one of the paths and therefore an erroneous decoding occurs for half of the time. Hence the probability of the incorrect decoding for even values of d is

$$P_{ICD}(d) = \sum_{i=d/2+1}^{d} \binom{d}{i} p_b^i (1-p_b)^{d-i} + \frac{1}{2} \binom{d}{d/2}$$

$$p_b^{d/2}(1-p_b)^{d/2} ; \qquad d \text{ is even.} \qquad (4.37)$$

From Equation 4.36, we may obtain an upper bound by noting that

$$
\begin{aligned}
P_{ICD}(d) \; &< \; \sum_{i=(d+1)/2}^{d} \binom{d}{i} p_b^{\frac{d}{2}} (1-p_b)^{\frac{d}{2}} \\
&= \; p_b^{\frac{d}{2}} (1-p_b)^{\frac{d}{2}} \sum_{i=(d+1)/2}^{d} \binom{d}{i} \\
&< \; p_b^{\frac{d}{2}} (1-p_b)^{\frac{d}{2}} \sum_{i=0}^{d} \binom{d}{i} \\
&= \; 2^d p_b^{\frac{d}{2}} (1-p_b)^{\frac{d}{2}} .
\end{aligned}
\tag{4.38}
$$

Similarly, it can be shown that Equation 4.38 is an upper bound on P_{ICD} for d even.

The transfer function of Equation 4.21 describes all the possible non-zero paths in the trellis that initially leave the all-zero path and eventually merge at node a after some symbol instants. The distance from the non-zero path increases with the number of instants of separation. Any non-zero paths terminating at node a with different distances have the possibility of being selected by the Viterbi decoder as the most likely path and will result in a first-event error. The union bound of the first-event error probability is obtained by summing the probability of incorrect decoding over all the possible non-zero paths merging at node a,

$$
P_{FE} < \sum_{d=d_{min}}^{\infty} A_d P_{ICD}(d)
\tag{4.39}
$$

where the coefficient of A_d is the number of non-zero paths with weight-d. Substituting the upper bound of P_{ICD} from Equation 4.38 into Equation 4.39 yields the upper bound on the first-event error probability,

$$
P_{FE} < \sum_{d=d_{min}}^{\infty} A_d 2^d p_b^{\frac{d}{2}} (1-p_b)^{\frac{d}{2}} .
\tag{4.40}
$$

Furthermore, by substituting $D = 2\sqrt{p_b(1-p_b)}$ into Equation 4.40 and using Equation 4.24, enables the upper bound of P_{FE} to be expressed as

$$
P_{FE} < \sum_{d=d_{min}}^{\infty} A_d D^d = T(D) \bigg|_{D = 2\sqrt{p_b(1-p_b)}} .
\tag{4.41}
$$

If a first-event error occurs, the non-zero path is selected and the decoder outputs the corresponding erroneous information sequence. As an all-zero sequence was transmitted, the number of post-decoding bit errors is equal to the number of non-zero bits in the information sequence. Hence, if each event error probability term is weighted by the total number of non-zero information bits on all the weight-d paths, W_d namely, the first-event error probability bound can be modified to provide a bound on the post-decoding bit error, p_{bp}. In addition, for a $R = k/n$ code, there is a k-bit symbol decoded at each instant. Thus, the hard-decision post-decoding biterror probability p_{bp} is union bounded by

$$p_{bp} < \frac{1}{k} \sum_{d=d_{min}}^{\infty} W_d P_{ICD}(d) \qquad (4.42)$$

where the coefficient W_d is the total number of information bit errors for all weight-d paths and P_{ICD} is obtained from Equation 4.36 and 4.37. However, if the upper bound of P_{ICD} is substituted into Equation 4.42, the upper bound of p_{bp} is expressed as

$$p_{bp} < \frac{1}{k} \sum_{d=d_{min}}^{\infty} W_d \left[4p_b(1 - p_b)\right]^{d/2} \qquad (4.43)$$

and from Equation 4.25 and Equation 4.27, we can write,

$$p_{bp} < \frac{1}{k} \frac{d\, T(D, H)}{dH} \bigg|\, H = 1, D = 2\sqrt{p_b(1 - p_b)} \qquad (4.44)$$

In our example of the $CC(2, 1, 3)$ code, the weight distribution is obtained by differentiating $T_{CC213}(D, H)$ with respect to H, see Equation 4.26. On substituting our result into Equation 4.43, the upper bound of p_{bp} is given by

$$p_{bp} < D^5 + 2D^6 + 4D^7 + \cdots \bigg|\, D = 2\sqrt{p_b(1 - p_b)} \qquad (4.45)$$

However, if the expressions of P_{ICD} in Equation 4.36 and 4.37 are used to substitute into Equation 4.42, a tighter bound of p_{bp} can be obtained.

4.3.7 Soft-decision Decoding Theory

In Section 4.3.6 we evaluated the performance of convolutional codes for hard-decision decoding. With this type of decoding the signal at the output of the demodulator is sampled and binary quantized and the resulting bits are passed to the input of the decoder. However, with soft-decision decoding the demodulated signal is quantized and conveyed to the decoder. In deriving the probability of bit error for convolutional codes we assume that the voltage levels at the demodulator output are statistically independent from each other. To simplify the analysis we transmit the all-zero sequence, knowing that our results are applicable to other transmitted sequences.

Suppose that the transmitted signal is MSK modulated, and that coherent demodulation is used at the receiver. The sampled demodulated signal is applied directly to the decoder without the samples being quantized. The path metric of the α path at J instant is given by Equation 4.17 and the Viterbi decoder selects the survivor with the largest metric among all the competing paths merging at node a. Let us denote α_0 and α_1 as the all-zero path and the non-zero path, respectively, merging to node a. If $M^{(\alpha_1)} > M^{(\alpha_0)}$, the non-zero path is selected that results in incorrect decoding, and the probability of an incorrect decoding is

$$
\begin{aligned}
P_{ICD}(d) &= P\left(M^{(\alpha_1)} > M^{(\alpha_0)}\right) \\
&= P\left(M^{(\alpha_1)} - M^{(\alpha_0)} \geq 0\right) \\
&= P\left(\sum_{j=1}^{J}\sum_{i=1}^{n} r_{ji}(a_{ji}^{(\alpha_1)} - a_{ji}^{(\alpha_0)}) \geq 0\right) .
\end{aligned}
\tag{4.46}
$$

Suppose the non-zero path α_1 that merges with the all-zero path α_0 differs in d number of encoded bits, i.e., there are d logical 1 bits in the non-zero sequence. We can therefore simplify Equation 4.46 by evaluating $P_{ICD}(d)$ only at the d positions to yield

$$
P_{ICD}(d) = P\left(\sum_{p=1}^{d} r_p > 0\right)
\tag{4.47}
$$

where the index p is the position of those d bits in which the two paths differ, and r_p represents the demodulator output corresponding to one of these d bits. For an AWGN channel, the signal r_p is a Gaussian random variable with a PDF described by Equation 4.14. A new random variable $r = \sum_{p=1}^{d} r_p = r_p d$ can be derived, where r is also gaussian with a mean and variance equal to μd and $\sigma_N^2 d$, respectively, and $\mu = A_s T/2$. The probability $P_{ICD}(d)$ of selecting the non-zero path is the probability of r (i.e., summation of r_p) being positive in Equation 4.47,

$$
\begin{aligned}
P_{ICD}(d) &= \frac{1}{2}\text{erfc}\left(\sqrt{\Gamma d}\right) \\
&= \frac{1}{2}\text{erfc}\left(\sqrt{E_b R d/N_0}\right)
\end{aligned}
\tag{4.48}
$$

where $\Gamma = A_s^2 T/\eta_0$ is the channel signal-to-noise ratio (SNR) and E_b is the energy per information bit.

Equation 4.48 gives the probability of incorrectly decoding from the pairwise comparisons into a path of distance d from the all-zero path. There are, of course, many other possible paths with different distances that merge with the all-zero path at a given node. However, we can upper bound the error probability by summing the incorrect decoding probabilities of the pairwise comparisons over all possible paths that merge with the all-zero path at a given node. By doing so we form the union bound of the first-event error probability as

$$P_{FE} < \sum_{d=d_{min}}^{\infty} A_d P_{ICD}(d)$$

$$= \frac{1}{2} \sum_{d=d_{min}}^{\infty} A_d \text{erfc}\left(\sqrt{E_b R d/N_0}\right) \qquad (4.49)$$

where A_d is the coefficient of the transfer function $T(D, H)$, representing the number of non-zero paths with weight-d. Furthermore, the expression of the union bound in Equation 4.48 can be simplified to obtain the upper bound by replacing the complementary error function by the exponential function because

$$\text{erfc}\left(\sqrt{E_b R d/N_0}\right) \leq \exp\left(-E_b R d/N_0\right) \qquad (4.50)$$

and therefore Equation 4.48 becomes

$$P_{ICD}(d) \leq \frac{1}{2} \exp\left(-E_b R d/N_0\right) . \qquad (4.51)$$

The probability of post-decoding bit errors p_{bp} is obtained by weighting the first-event error probability term with the weight, (i.e., number of bit 1 in the nonzero information sequence), of the incorrect path. Hence, similarly to hard decision, the union bound of the post-decoding bit error probability is given by

$$p_{bp} < \frac{1}{k} \sum_{d=d_{min}}^{\infty} W_d P_{ICD}(d)$$

$$< \frac{1}{2k} \sum_{d=d_{min}}^{\infty} W_d \text{erfc}\left(\sqrt{E_b R d/N_0}\right) . \qquad (4.52)$$

By substituting Equation 4.51 into 4.52, we obtain the upper bound of the post-decoding bit error probability,

$$p_{bp} < \frac{1}{2k} \sum_{d=d_{min}}^{\infty} W_d \exp\left(-E_b R d/N_0\right)$$

$$< \frac{1}{2k} \frac{d\,T(D, H)}{d\,H}\bigg|_{H=1,\,D=\exp(-E_b R d/N_0)} . \qquad (4.53)$$

4.3.8 Convolutional Code Performance

The Viterbi decoding of convolutional codes is analysed in references [20, 33], and their performance over additive white Gaussian noise (AWGN) channels is widely known [22, 30, 31] for various coding rates and constraint lengths. The AWGN channel results in every bit having an equal

probability of being erroneous. Convolutional codes rely on adjacent bits
to correct an error. As burst errors are infrequent on a random channel,
convolutional codes are appropriate. However, in the mobile radio environment where the narrowband transmission link is modelled by a Rayleigh
fading channel, burst errors occur due to the deep fades. The result is that
convolutional codes become occasionally overloaded and the BER performance deteriorates. In order to decrease the BER interleaving techniques
are introduced. Unfortunately interleaving introduces delay which may be
unacceptable for digital speech transimissions. This situation has motivated us to examine various interleaving methods having minimum delay,
while still providing an acceptable performance. We studied the effect of
code parameters, such as the constraint length, the coding rate, and the
performance of hard and soft decisions on the received signals. Our results
emphasized the gain in performance achieved by using the Viterbi decoding with soft-decisions when the transmissions were via Rayleigh fading
channels.

The system block diagram used in our simulations is shown in Figure 4.1.
The source data were protected by convolutional codes and scrambled by
diagonal interleaving, block interleaving or inter-block interleaving. The
interleaved data were MSK modulated and transmitted over AWGN or
Rayleigh fading channels. Previous experimental results [34] have showed
that the mobile radio channel in highway microcells is Rician, although
they can approach Gaussian or Rayleigh channels on occasions. In our
experiments the received signals were demodulated into symbols if hard-
decision decoding was used, or the demodulated signal was sampled and
quantized into values representing the confidence of the received signals if
soft-decision decoding was applied. The demodulated data was deinterleaved and convolutionally decoded to give the recovered data.

4.3.8.1 Convolutional Code Performance via Gaussian Channels

Figure 4.30 displays a set of theoretical and simulation results using Minimum Shift Keying (MSK) for the the half-rate $R = 1/2$ convolutional code
$CC(2, 1, 5)$ decoded by hard-decision Viterbi decoding [VD-HD]. The post-
decoding BER measured at the decoder output is shown as a function of
signal-to-noise ratio (E_b/N_0) for the AWGN channel. The union bound
and the upper bound in the Figure are obtained by substituting the weight
distribution $W_{CC215}(d)$ of the $CC(2, 1, 5)$ code, given by Equation 4.31,
into Equations 4.42 and 4.43, respectively. The union bound and the simulation results were in good agreement, especially at the high E_b/N_0 values.
At a BER of 10^{-6}, the difference between the upper bound and the simulation result was < 1dB, whereas the union bound and the simulation
agreed within a 0.1dB. We concluded that the simulation results for the
hard-decision decoding agreed with our theoretical calculations.

The minimum free distance d_{min} of the two-third rate punctured
$PCC(3, 2, 5)$ code generated by using Equation 4.35 was five, a value which

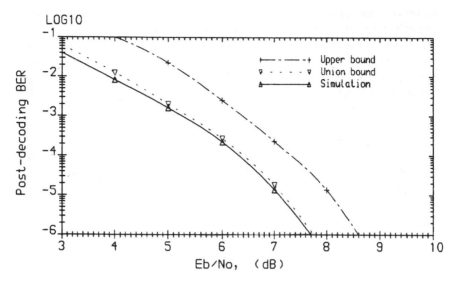

LOG10

Post-decoding BER

Eb/No, (dB)

Figure 4.30: Post-decoding BER of the $R = 1/2, CC(2, 1, 5)$ code with [VD-HD] via Gaussian channel

was the same as for the half rate $CC(2, 1, 3)$ code. The codes were recovered by hard-decision decoding and their BER performances against E_b/N_0 are displayed in Figure 4.31. It is interesting to note that the $CC(2, 1, 3)$ code had 1 dB loss over the $CC(2, 1, 5)$ code as shown in Figure 4.30 at a BER of 10^{-6}. Comparing the simulation results of the $PCC(3, 2, 5)$ code with the $CC(2, 1, 5)$ code again in Figure 4.30, we observe that it is only 0.1 dB and 0.6 dB inferior to the $CC(2, 1, 5)$ code at BERs of 10^{-6} and 10^{-2}, respectively. The performance was therefore improved by having codes with longer constraint length K, rather than by reducing the coding rate R. The constraint length of the code determined the coder complexity as the number of states of the binary code were 4 and 16, for $K = 3$ and $K = 5$, respectively. Although the $PCC(3, 2, 5)$ is more complex to implement than the $CC(2, 1, 3)$ code, it has a higher data throughput. In general we can exchange coder complexity for data throughput.

By using soft-decision Viterbi decoding [VD-SD], we obtained a set of theoretical and simulation results for the half rate convolutional codes $CC(2, 1, 5)$ and $CC(2, 1, 7)$, and for the two-third rate punctured convolutional codes $PCC(3, 2, 5)$ and $PCC(3, 2, 7)$ as shown in Figure 4.32. The soft-decision decoding was assumed to utilize infinite quantization levels from the demodulator. Again, the theoretical and simulation results were in close agreement. By comparing the decoding methods of soft and hard decision decoding, the $CC(2, 1, 5)$ code required an E_b/N_0 of 5.7 dB to achieve a BER of 10^{-6} using soft-decision, whereas it needed 7.7 dB (as

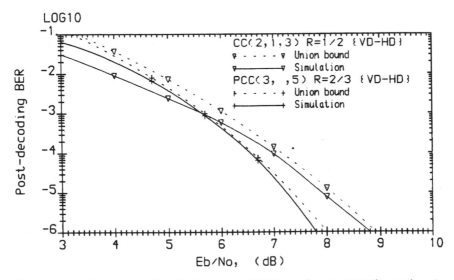

Figure 4.31: Post-decoding BER of the $CC(2, 1, 3)$ and $PCC(3, 2, 5)$ codes with [VD-HD] via Gaussian channel

indicated in Figure 4.30) to acquire the same BER employing hard-decision decoding. The performance gain of soft-decision decoding was 2 dB. In the case of the two-third rate punctured $PCC(3, 2, 5)$ code, it required E_b/N_0 values of 6.2 dB (Figure 4.32) and 7.8 dB (Figure 4.31) to yield a BER of 10^{-6} for soft and hard decision, respectively. There was 1.6 dB gain in E_b/N_0 by using soft-decision decoding. An interesting comparison in Figure 4.32 is that there is 1 dB gain in E_b/N_0 if the constraint length of either $CC(2, 1, 5)$ or $PCC(3, 2, 5)$ code is increased from five to seven binary stages. Also, only 0.5 dB is gained in E_b/N_0 by reducing the coding rate of either the $PCC(3, 2, 5)$ or the $PCC(3, 2, 7)$ code from a two-third to a half rate $CC(2, 1, 5)$ or $CC(2, 1, 7)$ code, respectively.

4.3.8.2   Convolutional Code Performance via Rayleigh Channels

When the transmissions were over a mobile radio channel to a MS in a vehicle travelling at 60 mph, the signal envelope was subjected to Rayleigh fading as shown in Figure 4.33. A deep fade of -20 dB relative to its root mean square value was common and led to the occurrence of burst errors. Interleaving techniques were employed to transform the bursty channel into a near memoryless channel. The theoretical union and upper bounds of the post-decoding BER of the $CC(2, 1, 5)$ code, decoded by using the hard-decision Viterbi algorithm, for the memoryless channel were evaluated by substituting the channel BER p_b into Equations 4.42 and 4.43, respec-

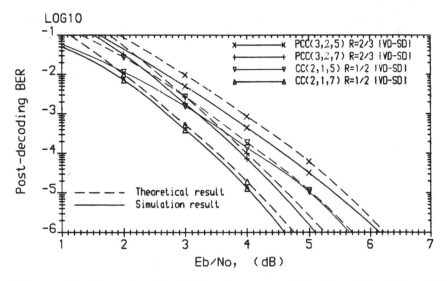

Figure 4.32: Post-decoding BER of various soft-decision Viterbi-decoded [VD-SD] convolutional codes via AWGN channel

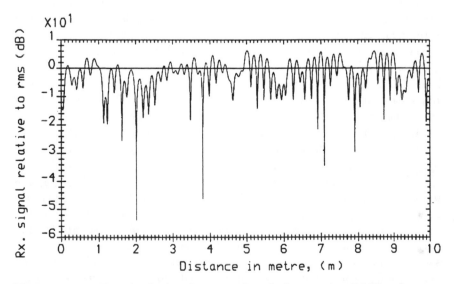

Figure 4.33: Received signal strength relative to its RMS value over Rayleigh-fading channel

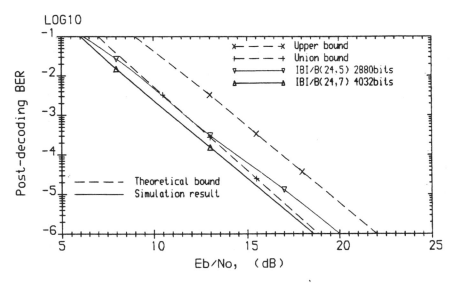

Figure 4.34: Theoretical bounds for the memoryless channel and simulation results using various interleavers for the post-decoding BER of the $CC(2, 1, 5)[VD - SD]$ code

tively. These bounds were displayed in Figure 4.34. The simulation results for inter-block bit interleaving with $B = 24$ and $N = 7$, i.e., IBI/B(24,7), showed a good approximation to the theoretical calculations. Inter-block bit interleaving with a delay of 4032 bits converted the channel into a memoryless one, and an $E_b/N_0 =18.5$ dB was required to achieve a BER of 10^{-6}. By reducing the interleaving delay to 2880 bits, the inter-block bit interleaving IBI/B(24,5) provided an approximately memoryless channel, but the performance was 1.5 dB inferior to that of IBI/B(24,7). The penalty of shorter delay was a reduction in the BER performance.

If the signal was not protected by coding and was not interleaved for transmission over the Rayleigh fading channel, then the simulations showed that an E_b/N_0 value of 52 dB was required to achieve a BER of 10^{-6}. After introducing the convolutional code $CC(2, 1, 5)$ without interleaving, the required E_b/N_0 value was reduced to 37.5 dB for a BER of 10^{-6}, as shown in Figure 4.35, a coding gain of 14.5 dB where hard-decision decoding was used. Then in addition to the error protection, diagonal bit interleaving over a delay of 456 bits was applied. The burst errors were divided into smaller segments and were dispersed into adjacent blocks. Smaller segments of errors had a better chance of being corrected and that reduced the E_b/N_0 value to 33 dB. However, the interleaving depth of this method was only two and the error segments in the adjacent block were still bursty as illustrated by the PDF in Figure 4.10. This E_b/N_0 value was also obtained using block

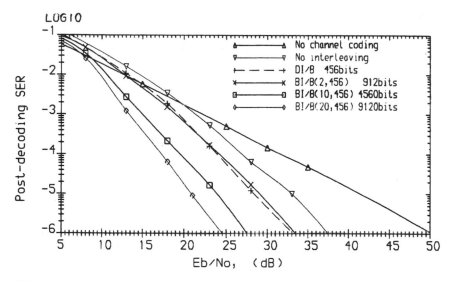

Figure 4.35: Effect of diagonal and block interleaving on the post-decoding BER of the $CC(2,1,5)[VD-SD]$ code over Rayleigh-fading channel

bit interleaving BI/B(2,456), but with the same depth. When the depth was increased to 10 and 20, the burst errors were more randomly distributed and the required E_b/N_0 values dropped to 27.5 and 24.5 dB, respectively. When comparing the E_b/N_0 at a BER of 10^{-6}, we found that for the $CC(2,1,5)$ code the BI/B(20,456) scheme with 9120 bits delay (displayed in Figure 4.35) required 6 dB more SNR than the IBI/B(24,7) interleaver with 4032 bits delay (see Figure 4.34). The inter-block bit interleaving dispersed the burst errors more randomly with a smaller delay penalty compared to block bit interleaving.

In Figure 4.36, the results for the $CC(2,1,5)$ code using soft-decision decoding and different inter-block bit interleaving delays is presented. With no interleaving the E_b/N_0 value of 32.5 dB gave a BER of 10^{-6} having a gain of 5 dB compared to the hard-decision version (see Figure 4.35). When IBI/B(8,7) with 448 bits delay was introduced, the E_b/N_0 value was reduced to 19.5 dB as illustrated in Figure 4.36. This delay was acceptable for speech transmissions and the performance of the code provided a guideline for speech codec design. If the speech quality of the codec requires it to operate with a BER> 10^{-3}, a minimum value of E_b/N_0 of 11.5 dB must be guaranteed. When the delay was increased to 1024 and 2880 bits, the required E_b/N_0 values were reduced to 15.5 and 11.5 dB at a BER of 10^{-6}. With the longer delay of 4032 bits, a further reduction of 1 dB was achieved. As the channel became random, the E_b/N_0 was reduced to the minimum of 10.5 dB for the BER of 10^{-6}.

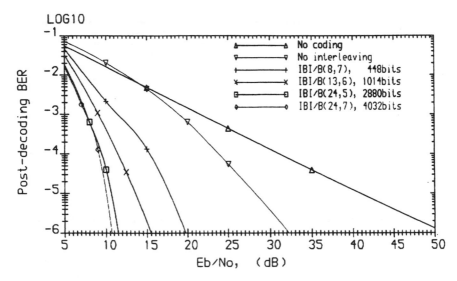

Figure 4.36: Effect of interblock interleaving on post-decoding BER for the $CC(2, 1, 5)[VD - SD]$ code over Rayleigh-fading channel

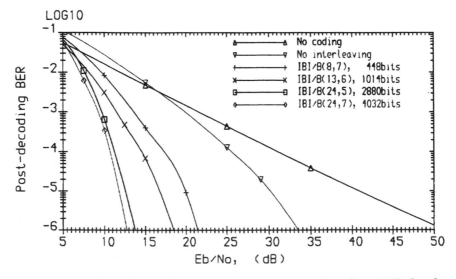

Figure 4.37: Effect of interblock interleaving on post-decoding BER for the $PCC(3, 2, 5)[VD - SD]$ code over Rayleigh-fading channel

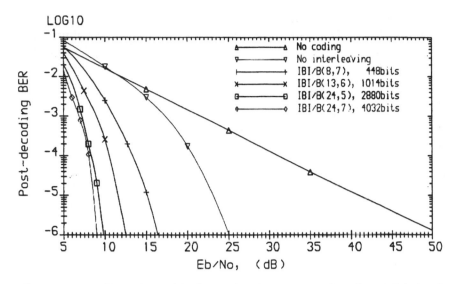

Figure 4.38: Effect of interblock interleaving on post-decoding BER for the $CC(2,1,7)[VD-SD]$ code over Rayleigh-fading channel

When the code was changed from half to two-third rate by keeping the same constraint length K of 5, a new set of results for the $PCC(3,2,5)$ code was obtained, see Figure 4.37. The E_b/N_0 values were 33.5, 21.5, 18.5, 13.7, and 12.6 dB for a BER of 10^{-6} for the delays of 0, 448, 1014, 2880, and 4032 bits, respectively. Despite the higher coding rate, the values were only about 2 to 3 dB inferior to that of the $CC(2,1,5)$ code. When speech was transmitted with an interleaving delay of 448 bits, a channel E_b/N_0 value of 13.5 dB was required for a BER of 10^{-3} that required 2 dB more compared with the half rate code with the same constraint length. It is interesting to note that at low E_b/N_0 values the BER with coding is even higher than that without coding. More bit errors were introduced at the decoder output because substantial burst errors could not be corrected. The Viterbi decoder selected the incorrect path and that precipitated more errors than if coding had not been used. The code had no error detection capability enabling the incorrect decoding to be identified. In addition, the non-systematic code was superior as the received bit sequence did not contain a copy of the information data. As a result, there was no remedy if incorrect decoding occurred.

When the constraint length of the codes was now extended to $K = 7$, the half rate $CC(2,1,7)$ code and the two-third rate $PCC(3,2,7)$ code yielded the results displayed in Figure 4.38 and Figure 4.39, respectively. The gains in E_b/N_0 for the $CC(2,1,7)$ code compared to the $CC(2,1,5)$ code were 7, 3, 3, 2, and 1.5 dB for delays of 0, 448, 1014, 2880, and 4032 bits,

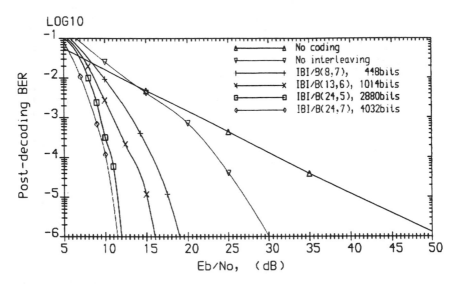

Figure 4.39: Effect of interblock interleaving on post-decoding BER for the $PCC(3,2,7)[VD - SD]$ code over Rayleigh-fading channel

respectively. The gradually decreasing gain in E_b/N_0 with increasing delay was because the behaviour of the channel became more random and approached a Gaussian channel. The gain of the $CC(2,1,7)$ code over the $CC(2,1,5)$ code for the case of the AWGN channel was as small as 1 dB as shown in Figure 4.32. Similarly, the gains of E_b/N_0 for the $PCC(3,1,7)$ code in Figure 4.39 over the $PCC(3,1,5)$ were 3.5, 2.5, 2.5, 1.7, and 1.3 dB for delays of 0, 448, 1014, 2880, and 4032 bits, respectively. The decreasing gain in E_b/N_0 with increasing delay was again observed. The gain in E_b/N_0 due to the increased constraint length for the two-third rate code was comparatively smaller than for the half rate code.

4.3.9   Conclusions on Convolutional Coding

For transmissions over the AWGN channel, the performance of convolutional codes is enhanced by 0.5 dB if the constraint length K is increased by 1. A degradation of at most 1 dB occurs when the coding rate is increased from half to two-thirds. As a result, if the data throughput is increased by increasing the coding rate from half to two-thirds, the constraint length of the code can also be increased in order to obtain the same performance. If soft-decision is applied, a further 1.5–2.0 dB gain in E_b/N_0 can be achieved. The coding gains of convolutional codes for transmissions over AWGN channel are summarized in Table 4.1.

In the Rayleigh fading channel, inter-block interleaving over 2880 bits can

	E_b/N_0 value		Coding gain	
	BER=10^{-3}	BER=10^{-6}	BER=10^{-3}	BER=10^{-6}
No coding	6.8dB	10.5dB	0dB	0dB
$CC(2,1,5)\ R=1/2\ [VD-HD]$	5.3dB	7.7dB	1.5dB	2.8dB
$PCC(3,1,5)\ R=2/3\ [VD-HD]$	5.6dB	7.8dB	1.2dB	2.7dB
$CC(2,1,5)\ R=1/2\ [VD-SD]$	3.2dB	5.6dB	3.6dB	4.9dB
$CC(2,1,7)\ R=1/2\ [VD-SD]$	2.7dB	4.6dB	4.1dB	5.9dB
$PCC(3,1,5)\ R=2/3\ [VD-SD]$	3.7dB	6.1dB	3.1dB	4.4dB
$PCC(3,1,7)\ R=2/3\ [VD-SD]$	3.2dB	5.1dB	3.6dB	5.4dB

Table 4.1: Coding gain of covolutional codes over AWGN channel

approximately render the fading channel into a memoryless one. For an acceptable delay of 448 bits introduced by the inter-block bit interleaving, digital speech transmission is possible with a BER of 10^{-3} or less, providing the E_b/N_0 value is above 11.5 dB. The soft-decision Viterbi decoding in the fading channel achieves a gain of 5 dB at BER of 10^{-6} compared to the hard-decision decoder. This gain is more than that for the AWGN channel (2 dB), and means that soft-decision decoding is more effective in the fading environment. A degradation of 2–3 dB occurs if the coding rate is increased from half to two-thirds. If the constraint length of the code is increased from 5 to 7, the gain in E_b/N_0 for the half rate code ranges from 7 to 1.5 dB for a delay ranging from 0 to 4032 bits, respectively. For the long delay, the channel tends to be random and therefore the gain is only 1.5 dB. This is equivalent to the gain (1.5–2 dB) achieved for transmission over the AWGN channel. The improvement for the two-third rate code ranges from 3.5 to 1.3 dB over the delay range from 0 to 4032 bits, respectively. The coding gains of convolutional codes in the Rayleigh fading channel are tabulated in Table 4.2. For the high rate codes, the BER is higher at the low E_b/N_0 values than when channel coding is not used. This is due to the lack of error detection capability, and to the use of non-systematic codes.

4.4 Block Codes

The history of block codes began in 1950 when a class of single error correcting block codes was introduced by Hamming [1]. The correcting capability of Hamming codes was, however, very weak and of limited practical value. A major breakthrough came when Hocquenghem [35] in 1959 and Bose-Chaudhuri [36, 37] in 1960 discovered a large class of multiple error correcting codes which are named after them as the Bose-Chaudhuri-Hocquenghem (BCH) binary codes. Soon after this pioneering work, the cyclic structure of these codes was discovered by Peterson [38]. The limitation of the theory to binary codes was removed by Gorenstein and Zierler in 1961 [39] providing coverage of both binary and nonbinary codes. An important subclass of BCH codes was discovered by Reed and Solomon [40],

| | E_b/N_0 value | | Coding gain | |
	BER=10^{-3}	BER=10^{-6}	BER=10^{-3}	BER=10^{-6}
No coding	21.5dB	52.0dB	0dB	0dB
$CC(2,1,5)$ $R = 1/2$ $[VD-HD]$				
No interleaving	21.3dB	37.5dB	0.2dB	14.5dB
DI/B 456bits	19.5dB	33.5dB	2.0dB	18.5dB
BI/B(2,456) 912bits	19.0dB	33.5dB	2.5dB	18.5dB
BI/B(10,456) 4560bits	15.0dB	27.5dB	6.5dB	24.5dB
BI/B(20,456) 9120bits	13.5dB	24.5dB	8.0dB	27.5dB
IBI/B(24,5) 2880bits	11.7dB	20.0dB	9.8dB	32.0dB
IBI/B(24,7) 4032bits	10.9dB	18.6dB	10.6dB	33.4dB
$CC(2,1,5)$ $R = 1/2$ $[VD-SD]$				
No interleaving	19.0dB	32.2dB	2.5dB	19.8dB
IBI/B(8,7) 448bits	11.5dB	19.7dB	10.0dB	32.3dB
IBI/B(13,6) 1014bits	9.0dB	15.5dB	12.5dB	36.5dB
IBI/B(24,5) 2880bits	7.6dB	11.5dB	13.9dB	40.5dB
IBI/B(24,7) 4032bits	7.5dB	10.6dB	14.0dB	41.4dB
$PCC(3,1,5)$ $R = 2/3$ $[VD-SD]$				
No interleaving	20.0dB	33.5dB	1.5dB	18.5dB
IBI/B(8,7) 448bits	13.7dB	21.5dB	7.8dB	30.5dB
IBI/B(13,6) 1014bits	11.5dB	18.5dB	10.0dB	33.5dB
IBI/B(24,5) 2880bits	9.7dB	13.8dB	11.8dB	38.2dB
IBI/B(24,7) 4032bits	9.3dB	12.7dB	12.2dB	39.3dB
$CC(2,1,7)$ $R = 1/2$ $[VD-SD]$				
No interleaving	17.3dB	25.0dB	4.2dB	27.0dB
IBI/B(8,7) 448bits	11.0dB	16.5dB	10.5dB	35.5dB
IBI/B(13,6) 1014bits	8.8dB	12.5dB	12.7dB	39.5dB
IBI/B(24,5) 2880bits	7.2dB	9.8dB	14.3dB	42.2dB
IBI/B(24,7) 4032bits	6.8dB	9.0dB	14.7dB	43.0dB
$PCC(3,1,7)$ $R = 2/3$ $[VD-SD]$				
No interleaving	19.2dB	30.0dB	2.3dB	22.0dB
IBI/B(8,7) 448bits	13.2dB	19.0dB	8.3dB	33.0dB
IBI/B(13,6) 1014bits	11.0dB	16.0dB	10.5dB	36.0dB
IBI/B(24,5) 2880bits	9.5dB	12.0dB	· 12.0dB	40.0dB
IBI/B(24,7) 4032bits	8.7dB	11.5dB	12.8dB	40.5dB

Table 4.2: Coding gain of covolutional codes over Rayleigh-fading channel

and these codes known as the Reed-Solomon (*RS*) codes achieve maximum separable distance between their codewords. The first decoding algorithm for binary *BCH* codes was suggested by Peterson in 1960 [38], followed by techniques [39, 41, 42] of how to practically implement the decoder. An efficient decoding algorithm proposed by Berlekamp [43, 4] and Massey [44, 45] became available for correcting a large number of errors. With the advance of digital integration circuit technology, the deployment of block codes became practical for a wide range of applications. Powerful Reed-Solomon (RS) block decoders [46, 47] have reportedly been built that operate at data rates above 120 Mbit/s. They have also become important as the outer layer code for use in concatenation with an inner layer convolutional code in both deep space communications [26] and in mobile satellite as well as radio applications [48].

Cyclic block codes are basically described by two parameters n and k, and a generator polynomial. A block of k information symbols at the input to the encoder is encoded into a block of n symbols. A characteristic of block codes is that each n symbol codeword is uniquely determined by a block of k input symbols. The ratio of k/n is the coding rate of the code and determines the amount of added redundancy.

4.4.1 The Structure of Block Codes

We begin to describe the algebraic structure that is fundamental in understanding the theory of block codes. Although both real and complex numbers are commonly employed in engineering applications, the algebraic theory for block coding requires the algebraic construction of fields. The operations for fields include addition, subtraction, multiplication and division, but their definitions are not the same as those of elementary arithmetic. The first step in understanding block codes is to grasp the concept of the fields with finite number of elements, and the arithmetic operations that can be performed. From this introduction, we will move to extension fields, polynomials and into coding algorithms.

4.4.1.1 Finite Fields

A finite field, also called a *Galois field*[1], is denoted by $GF(q)$. It describes a finite set of q elements with two defined operations, addition and multiplication. These operations performed on the inverse elements implicitly imply two further operations, that of subtraction and division. The rules of these operations are not significantly different from those employed in arithmetric operations with real and complex numbers. The rules of finite fields are illustrated as follows.

---

[1] Galois fields are named in honour of the French mathematician Évariste Galois (1811-1832) who was killed in a duel at the age of 20. On the eve of his death, he wrote a letter to a friend in which he gave the results of his theory of algebraic equations, already presented to the Paris Academy.

1. There are two operations, addition and multiplication, for operating on elements.

2. The field is closed. That is, the sum of the addition or the product of the multiplication results in a third element which is contained within the field.

3. The field always contains a unique additive identity element 0, and a unique multiplicative identity element 1, such that $u + 0 = u$ and $u \cdot 1 = u$ for any element u.

4. For every element u, there is a unique additive inverse element $-u$ such that $u + (-u) = 0$, and for $u \neq 0$, there is a unique multiplicative inverse element, denoted by u^{-1}, such that $u \cdot u^{-1} = 1$. The existence of the inverse elements implies the inverse operation, subtraction and division.

5. For operation on elements u, v and w, the following laws apply

$$
\begin{array}{lrcl}
\textit{associative:} & u + (v + w) & = & (u + v) + w \\
& u \cdot (v \cdot w) & = & (u \cdot v) \cdot w \\
\textit{commutative:} & u + v & = & v + u \\
& u \cdot v & = & v \cdot u \\
\textit{distributive:} & u \cdot (v + w) & = & u \cdot v + u \cdot w \ .
\end{array}
$$

In the ordinary arithmetic system, we observe examples of fields which conform to the above definitions of addition and multiplication. These fields are the set of all real numbers, the set of all complex numbers and the set of all rational numbers. Their elements obey the rules of ordinary addition and multiplication, and the fields are closed and contain an infinite number of elements. The additive identity element 0 and the multiplicative identity element 1 are among the elements in the fields, and every element has a unique additive and multiplicative inverse element. Furthermore, associative, commutative and distributive laws apply. In contrast, the set of all integers is not a field because integers other than unity have no multiplicative inverses in the set. These fields all have an infinite number of elements. The number of elements in a field is called the *order* of the field and it may be finite or infinite. A field with a finite number of elements is called a *finite field* and is denoted by $GF(q)$, where q is the number of elements in the field. For example, $GF(5)$ is a finite field containing a set of five integer elements $\{0,1,2,3,4\}$ operating under modulo-5 addition and multiplication. The addition and multiplication tables for the elements in $GF(5)$ are shown in Table 4.3. Notice that the $GF(5)$ contains the additive identity element 0 and the multiplicative identity element 1 in the field. Also, for each element in $GF(5)$, a unique additive inverse, and a unique multiplicative inverse, except for 0, always exist in the field. Thus, the addition to an inverse element implies the subtraction such that $2 - 3 =$

+	0	1	2	3	4
0	0	1	2	3	4
1	1	2	3	4	0
2	2	3	4	0	1
3	3	4	0	1	2
4	4	0	1	2	3

•	0	1	2	3	4
0	0	0	0	0	0
1	0	1	2	3	4
2	0	2	4	1	3
3	0	3	1	4	2
4	0	4	3	2	1

Table 4.3: Arithmetic tables for $GF(5)$ operations

$2 + (-3) = 2 + 2 = 4$, and similarly the multiplication by an inverse element implies the division such that $2/3 = 2 \cdot 3^{-1} = 2 \cdot 2 = 4$.

We have stated that a finite field $GF(q)$ consists of q integer elements. Suppose q is not a prime number, but a multiple of u and v. If $GF(q)$ is a finite field, u and v are the elements of the field, where u^{-1} and v^{-1} are their inverse elements, respectively. Hence,

$$u = R_q[u] = R_q[v^{-1}vu] = R_q[v^{-1}q] = 0 \qquad (4.54)$$

where $R_q[\bullet]$ is the modulo-q operation of $[\bullet]$. The proof shows a contradiction because $u \neq 0$. This means that $GF(q)$ consisting of integer elements is not a field if q is not a prime number. So, is it possible to have a $GF(q)$ with q not a prime number, such as $GF(8)$ or $GF(25)$? The answer is yes. In general, the finite field exists for $GF(q^m)$, where q is a prime number greater than 1 and m is an integer. The simplest field of $GF(q^m)$ with $m = 1$ is called the *prime field* $GF(q)$. The prime field consists of the set of all integer elements having values from zero to less than q. The operations on the integer elements are modulo-q addition and multiplication. For example, $GF(3)$ has a set of integer elements $\{0, 1, 2\}$ and similarly the set of integer elements of $GF(5)$ is $\{0, 1, 2, 3, 4\}$. The smallest prime field is $GF(2)$ which only consists of the additive and multiplicative identity elements 0 and 1, respectively. If m is greater than 1, $GF(q^m)$ is constructed as an extension of the prime field and is refered to as an *extension field*. Thus, $GF(8) = GF(2^3)$ is the extension field of $GF(2)$ and similarly $GF(25) = GF(5^2)$ is the extension field of $GF(5)$. The construction of the extension field is explained in Section 4.4.1.3 after the important algebraic concepts of vector spaces are described in Section 4.4.1.2.

4.4.1.2 Vector Spaces

The concept of vector space is closely related to the ideas of linear algebra and matrix theory in mathematics. The representation of an n-dimensional vector **v** is an enumeration of its coordinates $(v_1, v_2, \ldots, v_n)$. In two- or three-dimensional Euclidean space, the coordinates are simply the projections of **v** onto coordinate axes and the vector can be visualized geometrically as a directed line in a two- or three-dimensional plane. The properties of geometric vectors in ordinary coordinate systems provide an intuitive concept. The addition of two vectors is the addition of corresponding co-ordinates of the two vectors, and the multiplication of a vector by a real number is done by multiplying each coordinate by the number. These definitions can be extended mathematically to an n-dimensional vector space.

Having defined the concept of geometric vectors, we introduce the analogue of vector space over a *field F*. A set **V** is called a *vector space* and its elements are called *vectors*. The field elements of F are called *scalars*. A vector space **V** over a field F is structured by a set of vectors and a set of scalars under the operations of addition and multiplication in a mathematical system very much like a system of geometric vectors, real numbers and ordinary algebra. The defined operations are the addition of vectors called *vector addition*, and the multiplication of a vector by a scalar called *scalar multiplication*. The addition of any two vectors u and v in **V** i.e., u $+$ v, results in a vector that is also in **V**. The multiplication of a scalar a in F by a vector v in **V**, i.e., av, also gives a vector in **V**. The results of both vector addition and scalar multiplication are always vectors in **V** as the operations are subjected to the constraints imposed by the closure properties of fields.

For any vectors u and v in **V** and any scalars a and b in F, the following conditions must be satified:

1. A vector space **V** over a field F is a commutative group under vector addition.

2. The distributive laws apply, such that $a(u + v) = au + av$, and $(a + b)u = au + bu$.

3. The associative law applies, such that $(ab)u = a(bu)$.

4. Let 1 be the multiplicative identity element in F. Then for any u in **V**, $1u = u$.

For vector addition in **V**, an additive identity element, called the *origin* of **V** , exists and is denoted by 0 such that u $+ 0 = 0 + u = u$ for all u in **V**. Also, for scalar addition in F, an additive identity element exists in F called the zero scalar element and is denoted by 0. The two identity elements 0 and 0 are closely related. For all a in F and all u in **V**, we have $a0 = 0$ and $0u = 0$.

So far we have decribed the concept and the definition of vector space over a field. Let us now confine our interest to the application to the error control codes. A set of n elements $(u_1, u_2, \ldots, u_n)$, where u_i is a field element in F, is called an *n-tuple* over F. The addition of any two n-tuples is defined by an element-by-element addition, such as

$$(u_1, u_2, \ldots, u_n) + (v_1, v_2, \ldots, v_n) = (u_1 + v_1, u_2 + v_2, \ldots, u_n + v_n) \quad (4.55)$$

where the addition of $u_i + v_i$ is done in F and their sum is another element in F. The multiplication of an element from F by an n-tuple over F is the element-by-element scalar multiplication,

$$a(u_1, u_2, \ldots, u_n) = (au_1, au_2, \ldots, au_n) \qquad (4.56)$$

where each multiplication au_i is performed in F. Under the operations of elementwise addition and elementwise scalar multiplication, the distributive and associative laws apply, and the set of n-tuples constitutes a vector space over field F. A vector space can be constructed in this way with any field F, but the main feature in error control coding is that in a vector space over finite fields, the scalars represent code symbols and the n-tuples represent codewords.

If $v_1, v_2, \ldots, v_k$ are vectors in a vector space V over a field F, any sum of the form

$$\mathbf{u} = a_1 v_1 + a_2 v_2 + \cdots + a_k v_k \qquad (4.57)$$

where the scalars a_i are in the field F, is called a *linear combination* of $v_1, v_2, \ldots, v_k$. A set of k vectors $\{v_1, v_2, \ldots, v_k\}$ is said to be *linearly independent* if there is not a single set of scalars $\{a_1, a_2, \ldots, a_k\}$, except all a_i zero, such that

$$a_1 v_1 + a_2 v_2 + \cdots + a_k v_k = 0 \qquad (4.58)$$

where 0 is the zero vector. If only one set of scalars, not all equal to zero, is found to cause the linear combination equal to the zero vector, the set of vectors $\{v_1, v_2, \ldots, v_k\}$ are said to be *linearly dependent*. For example, the vectors $(0,0,1)$, $(0,1,0)$ and $(1,0,0)$ are linearly independent over any field, whereas the vectors $(1,1,1)$, $(0,1,1)$ and $(1,0,0)$ are linearly dependent over $GF(2)$ as their sum is equal to the zero vector.

In any vector space V, there is at least one set of linearly independent vectors that may generate any vector in V by means of linear combinations. The set of vectors $\{v_i\}$ is said to *span* the vector space V and any such set of vectors is called the *basis* of the vector space. For example, the three binary vectors $(0,0,1)$, $(0,1,0)$ and $(1,0,0)$ are linearly independent and constitute the basis of the vector space V_3 which consists of all eight binary vectors $(0,0,0)$, $(0,0,1)$, $\ldots$, $(1,1,1)$ formed by the linear combinations of the basis over $GF(2)$. If we use the two vectors $(0,0,1)$ and $(0,1,0)$ to form linear combinations over $GF(2)$, the resulting vectors $(0,0,0)$, $(0,0,1)$, $(0,1,0)$ and $(0,1,1)$ constitute a vector space V_2 which is a *subspace* of V_3. The vectors $(0,0,1)$ and $(0,1,0)$ are said to span V_2. The concept of the vector space is closely related to the familiar x, y and z axes of 3-dimensional coordinate systems in Euclidean space. The basis vectors in the vector space are represented by the unit vectors. However, the unit vectors are not the only basis vectors for V_3. The vectors $(1,0,0)$, $(0,1,0)$ and $(0,1,1)$ are the alternative basis for V_3. Similarly, the vectors $(0,0,1)$ and $(0,1,1)$ also form an alternative basis for subspace V_2. Furthermore, the vector $(1,1,1)$ is a basis for another subspace V_1, composed of only the vectors $(0,0,0)$ and $(1,1,1)$. The *dimension* of the vector space is the number of spaning

vectors that can be used to generate the space by linear combinations. In
the example, the vector spaces V_3, V_2, and V_1 have dimension three, two
and one respectively.

In summary, we have highlighted the important concept of vector spaces,
including the linear dependency among vectors, linear combinations of vectors, basis vectors of forming vector spaces, and dimensionality of a vector
space. These concepts lay the foundations for algebraic operations in a
finite field for error control coding.

4.4.1.3   Extension Fields

We now extend the concept of vectors, or m-tuples, in vector spaces to
polynomial representations in algebraic systems. In the last section, the m-
dimensional vector space over $GF(q)$ had q^m vectors, each constituted by
m-tuples of elements in field $GF(q)$. That is, for a two-dimensional vector
space over $GF(2)$, the vectors include a set of four 2-tuples (0,0), (0,1),
(1,0) and (1,1). The addition and subtraction operations on the vectors are
performed element-by-element over $GF(q)$ and the result of the operation
is another vector in the vector space. For example, $(0, 1) + (1, 0) = (1, 1)$,
where (1,1) is also a vector in the vector space. The multiplication and
division operations on the vectors are not obvious. Let us associate each
vector with a polynomial having coefficients corresponding to the elements
in the vector. The set of all 2-tuples defined on $GF(2)$ is replaced by the
set of all degree-1 polynomials defined on $GF(2)$. That is, the set of four
2-tuples over $GF(2)$ can be represented by 0, 1, z, $z + 1$, corresponding to
(0,0), (0,1), (1,0) and (1,1), respectively.

The addition and subtraction on polynomials are performed on their coefficients over $GF(q)$. In our example, the addition of $(0z + 1)$ and $(z + 0)$
results in another polynomial, $(z + 1)$, in the set. This shows the closure
property under addition. Similarly, the closure property under multiplication applies if the product of any two polynomials is another polynomial
in the set. We notice that all the polynomials in the set are of degree
$(m - 1)$ or less. The multiplication performed in a finite field can therefore
be defined by taking the remainder of the product with respect to a fixed
polynomial of degree m. By this definition, we always achieve a remainder
of degree $(m - 1)$ or less that must therefore be another polynomial in the
set.

The fixed polynomial is denoted by $p(z)$ and must be a *prime polynomial*
such that $p(z)$ is irreducible, that is, a degree-m polynomial that has no
factors of degree less than m and greater than 0. This requirement is
demonstrated by the following proof. Suppose that $p(z)$, whose degree is
at least 2, is not prime. Then $p(z) = u(z)v(z)$ for some $u(z)$ and $v(z)$ in
the set, each of degree at least 1, and their inverse polynomials are $u^{-1}(z)$

+	0	1	z	$z+1$
0	0	1	z	$z+1$
1	1	0	$z+1$	z
z	z	$z+1$	0	1
$z+1$	$z+1$	z	1	0

Table 4.4: Addition table for $GF(4)$

•	0	1	z	$z+1$
0	0	0	0	0
1	0	1	z	$z+1$
z	0	z	$z+1$	1
$z+1$	0	$z+1$	1	z

Table 4.5: Multiplication table for $GF(4)$

and $v^{-1}(z)$, respectively. Hence,

$$u(z) = R_{p(z)}[u(z)] = R_{p(z)}[v^{-1}(z)v(z)u(z)] = R_{p(z)}[v^{-1}(z)p(z)] = 0$$
$$(4.59)$$

where $R_{p(z)}[\bullet]$ represents the remainder of $[\bullet]$ upon division by $p(z)$. As $p(z)$ is a multiple of $u(z)$ and $v(z)$, $u(z)$ cannot be equal to 0. This contradicts to Equation 4.59, whence our initial assumption that $p(z)$ is not a prime was wrong. The proof then demonstrates that $p(z)$ must be a prime number in order to perform multiplication in a finite field.

Having defined the addition and multiplication on polynomials, we now illustrate the relationship between polynomials and fields. A set of polynomials of degree $(m-1)$ with coefficients defined over $GF(q)$ constitutes a finite field $GF(q^m)$ with a total of q^m polynomials. As an example, $GF(4)$ is constituted by four polynomials of degree-1 defined over $GF(2)$, i.e., $\{0, 1, z, z+1\}$. The addition of polynomials is performed on their coefficients over $GF(2)$. By using the prime polynomial, $p(z) = z^2 + z + 1$, the multiplication of polynomials is the remainder of their product divided by $p(z)$. The addition and multiplication tables for $GF(4)$ are tabulated in Table 4.4 and Table 4.5. The $GF(4)$ consists of four elements including the additive identity element 0 and the multiplicative identity element 1. We observe from the addition table that each polynomial is its own additive inverse. Also, from the multiplication table, each nonzero polynomial has a unique multiplicative inverse, z being the inverse of $z+1$ and vice versa, while 1 is its own inverse, as always. Thus, $GF(4)$ is a finite field constructed from $GF(2)$.

In general, a finite field $GF(q^m)$ exists for any number q^m, where q is a prime and m is a positive integer. The relationship between $GF(q)$ and $GF(q^m)$ is that $GF(q)$ is a *subfield* of $GF(q^m)$ such that the elements of $GF(q)$ are a subset of the elements in $GF(q^m)$. Equivalently, $GF(q^m)$ is

Exponential Notation	Polynomial Notation	Binary Notation
0	0	00
α^0	1	01
α^1	z	10
α^2	$z+1$	11

Table 4.6: Representations of $GF(4)$

called the *extension field* of $GF(q)$. For example, $GF(2)$ is a subfield of $GF(4)$ such that the elements, $\{0,1\}$, of $GF(2)$ are a subset of the elements, $\{0,1,z,z+1\}$, of $GF(4)$. Also $GF(4)$ is an extension field of $GF(2)$.

4.4.1.4   Primitive Polynomials

Every Galois field has at least one *primitive element*, denoted by α, which can represent every field element, except zero, as a power of α. For example, in the $GF(5)$, we have $2^1 = 2$, $2^2 = 4$, $2^3 = 3$ and $2^4 = 1$, where the results of 2^3 and 2^4 are their modulo-5 values. Thus $\alpha = 2$ is a primitive element of $GF(5)$. Similarly, $\alpha = 3$ is also a primitive element of $GF(5)$ such that $3^1 = 3$, $3^2 = 4$, $3^3 = 2$ and $3^4 = 1$. Consider the example of $GF(4)$. We try $\alpha = z$, the consecutive powers of α give $\alpha^1 = z$, $\alpha^2 = z+1$ and $\alpha^3 = 1$, where the results of α^2 and α^3 are their modulo $p(z) = z^2 + z + 1$ values. By employing $\alpha = z+1$, we again generate all the field elements by raising the power of α, i.e., $\alpha^1 = z+1$, $\alpha^2 = z$, and $\alpha^3 = 1$. In both examples of $GF(4)$ and $GF(5)$, we have found two primitive elements of which either one can generate a list of the nonzero field elements as powers of α. Once all the field elements have been found, we can adopt different notations to represent the elements. As an example of $GF(4)$, we can associate a binary 2-tuple with the polynomial, as shown in Table 4.6. For the binary notation, the addition of two elements is implemented by a bitwise exclusive-OR operation. We show in Table 4.6 that the addition of 1 and z is equal to $(z+1)$. This occurs because we represent two elements 1 and z by 01 and 10, and bitwise exclusive-OR of 01 and 10 is 11, corresponding to $(z+1)$. For the exponential notation, the field elements are represented by the successive powers of the primitive element. The advantage of this notation is that the multiplication of two elements is equivalent to the addition of their exponents. For example, the multiplication of z and $(z+1)$ gives $(z^2 + z)$ which is then taken modulo-$p(z)$ to yield the product, namely 1. Equivalently, the multiplication in exponential notation of these two elements α^1 and α^2 also results in α^3, namely 1.

We know that the field elements of prime field $GF(q)$ are a set of integer elements $\{0, 1, 2, \ldots, q-1\}$. However, for the extension field $GF(q^m)$, we would like to represent the polynomial elements as the successive powers of the primitive element, where multiplication of two polynomials can easily be done by the addition of the exponents of their corresponding exponential

Degree	Primitive Polynomials	Degree	Primitive Polynomials
2	$z^2 + z + 1$	14	$z^{14} + z^{10} + z^6 + z + 1$
3	$z^3 + z + 1$	15	$z^{15} + z + 1$
4	$z^4 + z + 1$	16	$z^{16} + z^{12} + z^3 + z + 1$
5	$z^5 + z^2 + 1$	17	$z^{17} + z^3 + 1$
6	$z^6 + z + 1$	18	$z^{18} + z^7 + 1$
7	$z^7 + z^3 + 1$	19	$z^{19} + z^5 + z^2 + z + 1$
8	$z^8 + z^4 + z^3 + z^2 + 1$	20	$z^{20} + z^3 + 1$
9	$z^9 + z^4 + 1$	21	$z^{21} + z^2 + 1$
10	$z^{10} + z^3 + 1$	22	$z^{22} + z + 1$
11	$z^{11} + z^2 + 1$	23	$z^{23} + z^5 + 1$
12	$z^{12} + z^6 + z^4 + z + 1$	24	$z^{24} + z^7 + z^2 + z + 1$
13	$z^{13} + z^4 + z^3 + z + 1$	25	$z^{25} + z^3 + 1$

Table 4.7: Primitive polynomials over $GF(2)$ (Blahut [6])

notation. This is usually convenient if the polynomial z corresponds to a primitive element of the field enabling the field elements to be found by computing the successive powers of z. A special prime polynomial, called a primitive polynomial, is selected to construct the field. A *primitive polynomial* $p(z)$ over $GF(q)$ is a prime polynomial over $GF(q)$ with z being a primitive element for constructing the field elements in the extension field. Table 4.6 lists the primitive polynomials of degree 2 to degree 25 over $GF(2)$ that enable us to construct a field from $GF(2^2)$ to $GF(2^{25})$.

Let us now summarize the construction of an extension field $GF(q^m)$, where q is a prime and m is an integer. We first generate all the field elements by using the primitive element and the primitive polynomial and then construct the addition and multiplication tables. As the $GF(q^m)$ is an extension field of $GF(q)$, the elements in the $GF(q^m)$ are represented by q^m polynomials of degree $(m-1)$ or less with coefficients in $GF(q)$. To generate the field elements, a degree-m primitive polynomial over $GF(q)$ is selected and the primitive element is $\alpha = z$. The power of α is raised successively, i.e., $\{\alpha^0, \alpha^1, \ldots, \alpha^{q^m-2}\}$ until all the elements except zero are generated. If the prime field is $GF(2)$, the coefficients of the polynomials are binary and therefore the degree-$(m-1)$ polynomial elements can be represented by $(m-1)$ binary digits notation. The addition of two polynomial elements is done by adding coefficients over $GF(q)$ of corresponding powers of z. The multiplication of two elements is the addition of the powers of their corresponding exponential notation.

We now illustrate the example of constructing $GF(16)$, which can be written as $GF(2^4)$ where $q = 2$ and $m = 4$. The primitive polynomial is therefore defined on $GF(2)$ of degree-4. From Table 4.7, $p(z) = z^4 + z + 1$ is used to generate the field elements. The power of the primitive element $\alpha = z$ is raised to represent all the non-zero field elements $\{\alpha^0, \alpha^1, \ldots, \alpha^{14}\}$ in $GF(16)$ and their modulo-$p(z)$ value is computed to give the polynomial representations as shown in Table 4.8. For instance,

$$\alpha^6 = z^6 \pmod{p(z)} = z^3 + z^2 .$$

Exponential Notation	Polynomials Notation	Binary Notation	Hexadecimal Notation
0	= 0	= 0000	= 0
$\alpha^0 \equiv \alpha^{15}$	= 1	= 0001	= 1
α^1	= z	= 0010	= 2
α^2	= z^2	= 0100	= 4
α^3	= z^3	= 1000	= 8
α^4	= $z + 1$	= 0011	= 3
α^5	= $z^2 + z$	= 0110	= 6
α^6	= $z^3 + z^2$	= 1100	= C
α^7	= $z^3 + z + 1$	= 1011	= B
α^8	= $z^2 + 1$	= 0101	= 5
α^9	= $z^3 + z$	= 1010	= A
α^{10}	= $z^2 + z + 1$	= 0111	= 7
α^{11}	= $z^3 + z^2 + z$	= 1110	= E
α^{12}	= $z^3 + z^2 + z + 1$	= 1111	= F
α^{13}	= $z^3 + z^2 + 1$	= 1101	= D
α^{14}	= $z^3 + 1$	= 1001	= 9

Table 4.8: Field elements of $GF(16)$ generated by $p(z) = z^4 + z + 1$

The addition of two elements requires us to add coefficients over $GF(2)$ of corresponding powers of z in each of the polynomials, and then to express the sum in its exponential representation. For example,

$$
\begin{aligned}
\alpha^6 + \alpha^7 &= (z^3 + z^2) + (z^3 + z + 1) \\
&= z^2 + z + 1 \\
&= \alpha^{10} .
\end{aligned}
$$

An alternative approach of directly adding two elements together in an exponential representation is done using the Zech logarithm, $Z(j)$, which is defined by

$$
\alpha^{Z(j)} = 1 + \alpha^j . \tag{4.60}
$$

Two elements α^i and α^j can be added by

$$
\begin{aligned}
\alpha^i + \alpha^j &= \alpha^i(1 + \alpha^{j-i}) \\
&= \alpha^{i+Z(j-i)} .
\end{aligned}
$$

Using this technique, we tabulate the Zech logarithms for $GF(16)$ in Table4.8. Let us illustrate the addition by using the following example,

$$
\alpha^3 + \alpha^7 = \alpha^{3+Z(4)} = \alpha^4 .
$$

By using Table 4.9, the exponential representations of the sums are tabulated in Table 4.10. The product of two elements is a new element with the power equal to the sum of the powers of the corresponding elements. For example,

$$
\alpha^6 + \alpha^{12} = \alpha^{6+12} = \alpha^{15} \cdot \alpha^3 = \alpha^3 .
$$

The multiplication table is shown in Table 4.11.

j	$Z(j)$
$-\infty$	0
0	$-\infty$
1	4
2	8
3	14
4	1
5	10
6	13
7	9
8	2
9	7
10	5
11	12
12	11
13	6
14	3

Table 4.9: Zech's Logarithms in $GF(16)$

$+$	0	α^0	α^1	α^2	α^3	α^4	α^5	α^6	α^7	α^8	α^9	α^{10}	α^{11}	α^{12}	α^{13}	α^{14}
0	0	α^0	α^1	α^2	α^3	α^4	α^5	α^6	α^7	α^8	α^9	α^{10}	α^{11}	α^{12}	α^{13}	α^{14}
α^0	α^0	0	α^4	α^8	α^{14}	α^1	α^{10}	α^{13}	α^9	α^2	α^7	α^5	α^{12}	α^{11}	α^6	α^3
α^1	α^1	α^4	0	α^5	α^9	α^0	α^2	α^{11}	α^{14}	α^{10}	α^3	α^8	α^6	α^{13}	α^{12}	α^7
α^2	α^2	α^8	α^5	0	α^6	α^{10}	α^1	α^3	α^{12}	α^0	α^{11}	α^4	α^9	α^7	α^{14}	α^{13}
α^3	α^3	α^{14}	α^9	α^6	0	α^7	α^{11}	α^2	α^4	α^{13}	α^1	α^{12}	α^5	α^{10}	α^8	α^0
α^4	α^4	α^1	α^0	α^{10}	α^7	0	α^8	α^{12}	α^3	α^5	α^{14}	α^2	α^{13}	α^6	α^{11}	α^9
α^5	α^5	α^{10}	α^2	α^1	α^{11}	α^8	0	α^9	α^{13}	α^4	α^6	α^0	α^3	α^{14}	α^7	α^{12}
α^6	α^6	α^{13}	α^{11}	α^3	α^2	α^{12}	α^9	0	α^{10}	α^{14}	α^5	α^7	α^1	α^4	α^0	α^8
α^7	α^7	α^9	α^{14}	α^{12}	α^4	α^3	α^{13}	α^{10}	0	α^{11}	α^0	α^6	α^8	α^2	α^5	α^1
α^8	α^8	α^2	α^{10}	α^0	α^{13}	α^5	α^4	α^{14}	α^{11}	0	α^{12}	α^1	α^7	α^9	α^3	α^6
α^9	α^9	α^7	α^3	α^{11}	α^1	α^{14}	α^6	α^5	α^0	α^{12}	0	α^{13}	α^2	α^8	α^{10}	α^4
α^{10}	α^{10}	α^5	α^8	α^4	α^{12}	α^2	α^0	α^7	α^6	α^1	α^{13}	0	α^{14}	α^3	α^9	α^{11}
α^{11}	α^{11}	α^{12}	α^6	α^9	α^5	α^{13}	α^3	α^1	α^8	α^7	α^2	α^{14}	0	α^0	α^4	α^{10}
α^{12}	α^{12}	α^{11}	α^{13}	α^7	α^{10}	α^6	α^{14}	α^4	α^2	α^9	α^8	α^3	α^0	0	α^1	α^5
α^{13}	α^{13}	α^6	α^{12}	α^{14}	α^8	α^{11}	α^7	α^0	α^5	α^3	α^{10}	α^9	α^4	α^1	0	α^2
α^{14}	α^{14}	α^3	α^7	α^{13}	α^0	α^9	α^{12}	α^8	α^1	α^6	α^4	α^{11}	α^{10}	α^5	α^2	0

Table 4.10: Addition table for $GF(16)$

•	0	α^0	α^1	α^2	α^3	α^4	α^5	α^6	α^7	α^8	α^9	α^{10}	α^{11}	α^{12}	α^{13}	α^{14}
0	0	0	0	0	0	0	0	0	0	0	0	0	0	0	0	0
α^0	0	α^0	α^1	α^2	α^3	α^4	α^5	α^6	α^7	α^8	α^9	α^{10}	α^{11}	α^{12}	α^{13}	α^{14}
α^1	0	α^1	α^2	α^3	α^4	α^5	α^6	α^7	α^8	α^9	α^{10}	α^{11}	α^{12}	α^{13}	α^{14}	α^0
α^2	0	α^2	α^3	α^4	α^5	α^6	α^7	α^8	α^9	α^{10}	α^{11}	α^{12}	α^{13}	α^{14}	α^0	α^1
α^3	0	α^3	α^4	α^5	α^6	α^7	α^8	α^9	α^{10}	α^{11}	α^{12}	α^{13}	α^{14}	α^0	α^1	α^2
α^4	0	α^4	α^5	α^6	α^7	α^8	α^9	α^{10}	α^{11}	α^{12}	α^{13}	α^{14}	α^0	α^1	α^2	α^3
α^5	0	α^5	α^6	α^7	α^8	α^9	α^{10}	α^{11}	α^{12}	α^{13}	α^{14}	α^0	α^1	α^2	α^3	α^4
α^6	0	α^6	α^7	α^8	α^9	α^{10}	α^{11}	α^{12}	α^{13}	α^{14}	α^0	α^1	α^2	α^3	α^4	α^5
α^7	0	α^7	α^8	α^9	α^{10}	α^{11}	α^{12}	α^{13}	α^{14}	α^0	α^1	α^2	α^3	α^4	α^5	α^6
α^8	0	α^8	α^9	α^{10}	α^{11}	α^{12}	α^{13}	α^{14}	α^0	α^1	α^2	α^3	α^4	α^5	α^6	α^7
α^9	0	α^9	α^{10}	α^{11}	α^{12}	α^{13}	α^{14}	α^0	α^1	α^2	α^3	α^4	α^5	α^6	α^7	α^8
α^{10}	0	α^{10}	α^{11}	α^{12}	α^{13}	α^{14}	α^0	α^1	α^2	α^3	α^4	α^5	α^6	α^7	α^8	α^9
α^{11}	0	α^{11}	α^{12}	α^{13}	α^{14}	α^0	α^1	α^2	α^3	α^4	α^5	α^6	α^7	α^8	α^9	α^{10}
α^{12}	0	α^{12}	α^{13}	α^{14}	α^0	α^1	α^2	α^3	α^4	α^5	α^6	α^7	α^8	α^9	α^{10}	α^{11}
α^{13}	0	α^{13}	α^{14}	α^0	α^1	α^2	α^3	α^4	α^5	α^6	α^7	α^8	α^9	α^{10}	α^{11}	α^{12}
α^{14}	0	α^{14}	α^0	α^1	α^2	α^3	α^4	α^5	α^6	α^7	α^8	α^9	α^{10}	α^{11}	α^{12}	α^{13}

Table 4.11: Multiplication table for $GF(16)$

4.4.1.5   Minimal Polynomials

In the last section the concept of an extension field was introduced. We now investigate the relationship between the extension field and the prime field, and this will lead to the introduction of the minimal polynomials. It is these polynomials that play a cardinal role in the formation of the generator polynomials for BCH codes to be described in Section 4.4.3. A special case of minimal polynomials is the primitive polynomial which we used as the prime polynomial in constructing the extension field in Section 4.4.1.4.

In ordinary algebraic arithmetic, a polynomial of degree-n with real coefficients has exactly n roots, some of which may be repeated. If the roots are not from the field of real numbers, they are from the field of complex numbers that contains the field of real numbers as a subfield. For example, the polynomial $f(z) = z^2 + 4z + 13$ defined in the field of real numbers is irreducible. It does not have real roots, but instead it has two complex conjugate roots, $-2 \pm 3\imath$, where $\imath = \sqrt{-1}$. Similarly, in finite field arithmetic, if the polynomial defined in the subfield is irreducible, it has no roots in the subfield, only in the extension field. Every polynomial $f(z)$ of degree-n has n roots, and if $f(z)$ is irreducible over the subfield then all n roots are in the extension field. For example, $f(z) = z^4 + z^3 + z^2 + z + 1$ is irreducible over $GF(2)$ and it has no roots from $GF(2)$. Instead, it has four roots, $\alpha^3, \alpha^6, \alpha^9$ and α^{12}, from the $GF(2^4)$, which is the extension field of $GF(2)$. By using the addition and the multiplication tables in Table 4.10 and Table 4.11, we can verify these roots by substituting into the polynomial. For α^3 we have,

$$f(\alpha^3) \;=\; (\alpha^3)^4 + (\alpha^3)^3 + (\alpha^3)^2 + \alpha^3 + 1$$

$$= \alpha^{12} + \alpha^9 + \alpha^6 + \alpha^3 + 1$$
$$= 0$$

and hence, α^3 is a root of $f(z)$. The other roots α^6, α^9 and α^{12} can also be verified by the same procedure. As $f(z)$ has a degree of four, with roots, $\alpha^3, \alpha^6, \alpha^9$ and α^{12}, then $(z + \alpha^3)(z + \alpha^6)(z + \alpha^9)(z + \alpha^{12})$ must be equal to $z^4 + z^3 + z^2 + z + 1$. Again, by using the addition and the multiplication tables in Table 4.10 and Table 4.11, we evaluate

$$
\begin{aligned}
&(z + \alpha^3)(z + \alpha^6)(z + \alpha^9)(z + \alpha^{12}) \\
&= (z^2 + \alpha^2 z + \alpha^9)(z^2 + \alpha^8 z + \alpha^6) \\
&= z^4 + (\alpha^2 + \alpha^8)z^3 + (\alpha^9 + \alpha^6 + \alpha^{10})z^2 + (\alpha^{17} + \alpha^8)z + \alpha^{15} \\
&= z^4 + z^3 + z^2 + z + 1 .
\end{aligned}
$$

The properties of these roots in extension fields are important in finite fields. Let $f(z)$ be an irreducible polynomial with coefficients from $GF(2)$, and β be a root of $f(z)$ such that $f(\beta) = 0$. As $f(z)$ is irreducible over $GF(2)$, it has no roots in $GF(2)$, and therefore β must be an element in some extension field $GF(2^m)$. The additions and multiplications required for the evaluation of the polynomial are performed in the extension field $GF(2^m)$, as $GF(2)$ is contained in any of its extension. Now let us describe the characteristics of these roots by the following key properties.

1. If $f(z)$ is an irreducible polynomial of degree-n over $GF(2)$ and has a root β from $GF(2^m)$, then for any $l \geq 0$, β^{2^l} is also a root of $f(z)$, i.e., $\beta, \beta^2, \beta^4, \beta^8, \ldots, \beta^{2^{n-1}}$ are all roots of $f(z)$. This property can be verified by the following proof. Let us consider

$$
\begin{aligned}
f^2(z) &= (f_0 + f_1 z + \cdots + f_n z^n)^2 \\
&= [f_0 + (f_1 z + f_2 z^2 + \cdots + f_n z^n)]^2 \\
&= f_0^2 + f_0 \cdot (f_1 z + f_2 z^2 + \cdots + f_n z^n) \\
&\quad + f_0 \cdot (f_1 z + f_2 z^2 + \cdots + f_n z^n) \\
&\quad + (f_1 z + f_2 z^2 + \cdots + f_n z^n)^2 \\
&= f_0^2 + (f_1 z + f_2 z^2 + \cdots + f_n z^n)^2 .
\end{aligned}
$$

Repeating the expansion of the above Equation, we obtain,

$$f^2(z) = f_0^2 + (f_1 z)^2 + (f_2 z^2)^2 + \cdots + (f_n z^n)^2 .$$

As $f(z)$ is defined on $GF(2)$, the coefficient f_i is either 0 or 1. Therefore $f_i^2 = f_i$ and the Equation becomes

$$
\begin{aligned}
f^2(z) &= f_0 + f_1 z^2 + f_2(z^2)^2 + \cdots + f_n(z^2)^n \\
&= f(z^2) . \tag{4.61}
\end{aligned}
$$

From Equation 4.61, we deduce that for any $l \geq 0$,

$$f^{2^l}(z) = f(z^{2^l}) \tag{4.62}$$

and for $z = \beta$,

$$f^{2^l}(\beta) = f(\beta^{2^l}) . \tag{4.63}$$

As β is a root of $f(z)$, it implies that $f(\beta) = 0$. The powers of $f(\beta)$, i.e., $f^{2^l}(\beta)$, are also equal to zero. From Equation 4.63 we see that $f^{2^l}(\beta) = 0$, $f(\beta^{2^l}) = 0$, and therefore β^{2^l} is also a root of $f(z)$. This shows that if β (an element from $GF(2^m)$) is a root of the polynomial $f(z)$ over $GF(2)$, then all the β^{2^l} (elements from $GF(2^m)$) for $l \geq 0$ are also roots of $f(z)$. The element β^{2^l} is called a *conjugate* of β. For example, the polynomial $f(z) = z^4 + z^3 + 1$ is irreducible over $GF(2)$, and has four roots. One of the roots is α^7 which is an element in $GF(2^4)$. This can be verified by substituting α^7 into $f(z)$,

$$
\begin{aligned}
f(\alpha^7) &= (\alpha^7)^4 + (\alpha^7)^3 + 1 \\
&= \alpha^{13} + \alpha^6 + 1 \\
&= 0 .
\end{aligned}
$$

The conjugates of α^7 are $(\alpha^7)^2 = \alpha^{14}$, $(\alpha^7)^{2^2} = \alpha^{28} = \alpha^{13}$ and $(\alpha^7)^{2^3} = \alpha^{56} = \alpha^{11}$. It should be noted that for $l > (m-1) = 3$, the conjugates repeat again such as $(\alpha^7)^{2^4} = \alpha^{112} = \alpha^7$, $(\alpha^7)^{2^5} = \alpha^{224} = \alpha^{14}$, and so on. We recall that the conjugates α^{14}, α^{13} and α^{11} are also the roots of $f(z)$.

2. If β is a nonzero element in $GF(2^m)$, then β^{2^m-1} is always equal to 1. Adding 1 to both sides of the equation $\beta^{2^m-1} = 1$ gives

$$\beta^{2^m-1} + 1 = 0 . \tag{4.64}$$

β is an element, and in the above Equation it is seen to be a root of the polynomial $(z^{2^m-1} + 1)$ over $GF(2)$. As the polynomial has degree of $2^m - 1$, it has $2^m - 1$ roots which are all the nonzero elements in $GF(2^m)$. As the zero element 0 of $GF(2^m)$ is the root of z, it then follows that the elements of $GF(2^m)$ form all the roots of $(z^{2^m} + z)$. As every element β in an extension field $GF(2^m)$ is a root of the polynomial $(z^{2^m} + z)$, there is a polynomial in $GF(2)$, called the *minimal polynomial* $\psi_\beta(z)$ of β. This polynomial is the smallest degree monic polynomial having β as a root, where a *monic* polynomial is defined as a polynomial with a leading coefficient of 1. In the case of $GF(2)$ the coefficient is either 0 or 1 and therefore all polynomials are monic. For example, a polynomial $(z^{2^4} + z)$ of degree-16 defined on $GF(2)$ has 16 roots which are all the elements in $GF(2^4)$.

Conjugate roots	Minimal polynomial
0	z
α^0	$z + 1$
$\alpha^1, \alpha^2, \alpha^4, \alpha^8$	$z^4 + z + 1$
$\alpha^3, \alpha^6, \alpha^9, \alpha^{12}$	$z^4 + z^3 + z^2 + z + 1$
α^5, α^{10}	$z^2 + z + 1$
$\alpha^7, \alpha^{11}, \alpha^{13}, \alpha^{14}$	$z^4 + z^3 + 1$

Table 4.12: Minimal polynomials of the elements in $GF(2^4)$

Let us express the polynomial $(z^{2^4} + z)$ over $GF(2)$ as the product of the smallest degree monic polynomials,

$$z^{2^4} + z = $$
$$z(z+1)(z^2+z+1)(z^4+z+1)(z^4+z^3+1)(z^4+z^3+z^2+z+1). \qquad (4.65)$$

Each factor of the polynomial $(z^{2^4} + z)$ represents a minimal polynomial $\psi_\beta(z)$ over $GF(2)$ of some element β in $GF(2^4)$. The minimal polynomial $\psi_0(z)$ of zero element 0 from $GF(2^4)$ is the factor z and the minimal polynomial $\psi_{\alpha^0}(z)$ of unit element $\alpha^0 = 1$ is the factor $(z+1)$. Also, the minimal polynomial of element α^3 is $(z^4 + z^3 + z^2 + z + 1)$. According to Property 1, the conjugates of α^3 are also the roots of the minimal polynomial. Thus, the elements α^3 and its conjugates, α^6, α^9 and α^{12}, have the same minimal polynomial. This can be verified as follows,

$$\psi_{\alpha^3}(\alpha^3) = (\alpha^3)^4 + (\alpha^3)^3 + (\alpha^3)^2 + \alpha^3 + 1 = 0$$
$$\psi_{\alpha^6}(\alpha^6) = (\alpha^6)^4 + (\alpha^6)^3 + (\alpha^6)^2 + \alpha^6 + 1 = 0$$
$$\psi_{\alpha^9}(\alpha^9) = (\alpha^9)^4 + (\alpha^9)^3 + (\alpha^9)^2 + \alpha^9 + 1 = 0$$
$$\psi_{\alpha^{12}}(\alpha^{12}) = (\alpha^{12})^4 + (\alpha^{12})^3 + (\alpha^{12})^2 + \alpha^{12} + 1 = 0 .$$

The minimal polynomials of all the elements in $GF(2^4)$ are tabulated in Table 4.12. Notice that the minimal polynomial of β is unique, that is, for every β there is one and only one minimal polynomial. However, different elements of $GF(2^4)$ can have the same minimal polynomial. Moreover, for every element in $GF(2^m)$, the degree of the minimal polynomial over $GF(2)$ is at most m.

3. From Property 2, we understand that the minimal polynomial of the element β from $GF(2^m)$ is defined as the smallest degree polynomial over $GF(2)$ with the root of β. The minimal polynomial is therefore irreducible. Also, Property 1 states that if the element β from $GF(2^m)$ is a root of an irreducible polynomial, then all the other roots of the polynomial are the conjugates of β. Hence, the element β and its conjugates form all the roots of the minimal polynomial,

and the total number of roots determines the degree of the minimal polynomial. Let e be the degree of the minimal polynomial of β from $GF(2^m)$, and e be defined as the smallest integer such that,

$$\beta^{2^e} = \beta .$$
(4.66)

As the element β and all its conjugates are all the roots of the minimal polynomial, the minimal polynomial of β is formed by

$$\psi_\beta(z) = \prod_{i=0}^{e-1}(z + \beta^{2^i}) .$$
(4.67)

For example, the conjugates of $\beta = \alpha^3$ in $GF(2^4)$ are

$$\beta^2 = \alpha^6, \quad \beta^{2^2} = \alpha^{12}, \quad \text{and} \quad \beta^{2^3} = \alpha^{24} = \alpha^9 .$$

The minimal polynomial of $\beta = \alpha^3$ is then formed as

$$
\begin{aligned}
\psi_{\alpha^3}(z) &= (z + \alpha^3)(z + \alpha^6)(z + \alpha^9)(z + \alpha^{12}) \\
&= (z^2 + \alpha^2 z + \alpha^9)(z^2 + \alpha^8 z + \alpha^6) \\
&= z^4 + z^3 + z^2 + z + 1 .
\end{aligned}
$$

The minimal polynomial of α^3 in $GF(2^4)$ can be verified with the aid of Table 4.12.

4. The minimal polynomial of a primitive element of $GF(2^m)$ has degree-m and is a primitive polynomial. In the construction of the Galois field $GF(2^m)$, we use a primitive polynomial $p(z)$ of degree-m and the primitive element which is a root of $p(z)$. The primitive element is α and the successive powers of α represent all the non-zero elements of $GF(2^m)$. They form a commutative group under multiplication and the group is closed as $\alpha^{2^m-1} = 1$. That is, if $l > (2^m - 1)$, the element $\alpha^l = \alpha^{(2^m-1)i+j} = \alpha^j$, $j \leq (2^m - 1)$ and therefore α^j is an element in the group. For example, in $GF(2^4)$, the minimal polynomial $z^4 + z + 1$ over $GF(2)$ of degree-4 of the primitive element α is used as the primitive polynomial to construct all the non-zero elements in $GF(2^4)$.

So far, we have studied the structure of the Galois field which introduces the finite field arithmetic in the cyclic codes. We now concentrate on the encoding and decoding algorithms of different error control codes.

4.4.2 Cyclic Codes

Cyclic codes were first introduced by Prange in 1957 [49]. They can be easily implemented by shift register circuits. For an (n, k) linear code C, k information symbols are encoded into an n-symbol codeword. This is a *cyclic code* if every cyclic shift of a vector in C is also a code vector in C. Thus if the elements of an n-tuple $\mathbf{v} = (v_{n-1}, \ldots, v_1, v_0)$ are cyclically shifted one place to the left, we obtain another n-tuple,

$$\mathbf{v}^{(1)} = (v_{n-2}, \ldots, v_0, v_{n-1}) .$$

This process is called a cyclic shift of $\mathbf{v}$. If the elements of $\mathbf{v}$ are cyclically shifted by i places to the left, the resultant n-tuple is

$$\mathbf{v}^{(i)} = (v_{n-i-1}, v_{n-i-2}, \ldots, v_1, v_0, v_{n-1}, \ldots, v_{n-i}) .$$

In order to explore the algebraic properties of the cyclic code, we express the code vector $\mathbf{v} = (v_{n-1}, \ldots, v_1, v_0)$ using the polynomial representation where the coefficients of the polynomial correspond to the elements of the vector, viz:-

$$\mathbf{v}(z) = v_{n-1}z^{n-1} + \cdots + v_2 z^2 + v_1 z + v_0 .$$

Hence, an n-tuple code vector is represented by a polynomial of degree $(n-1)$ or less. If $v_{n-1} \neq 0$, the degree of $\mathbf{v}(z)$ is $(n-1)$; if $v_{n-1} = 0$, the degree of $\mathbf{v}(z)$ is less than $(n-1)$. The corresponding polynomial representation of the cyclically shifted code vector $\mathbf{v}^{(i)}(z)$ can be written as

$$\mathbf{v}^{(i)}(z) = \underbrace{v_{n-i-1}z^{n-1} + \cdots + v_1 z^{i+1} + v_0 z^i}_{(n-i) \text{ terms}}$$
$$+ \underbrace{v_{n-1}z^{i-1} + \cdots + v_{n-i+1}z + v_{n-i}}_{i \text{ terms}} .$$

The equivalent operation of the cyclic shift in terms of polynomial representation can be achieved by the following manipulation. We first observe that the multiplication of $\mathbf{v}(z)$ by z^i is

$$z^i \mathbf{v}(z) = \underbrace{v_{n-1}z^{n+i-1} + \cdots + v_{n-i}z^n}_{i \text{ terms}} + \underbrace{v_{n-i-1}z^{n-1} + \cdots + v_1 z^{i+1} + v_0 z^i}_{(n-i) \text{ terms}}$$

and that the order of the first i terms in the above equation exceeds the degree $(n-1)$, while those terms with degree less than i are absent. Due to the cyclic property, those i terms with degree higher than $(n-1)$ are shifted to the lower order part of the polynomial. This cyclic arithmetic is done by

$$z^i \mathbf{v}(z) = \underbrace{v_{n-i-1}z^{n-1} + \cdots + v_0 z^i}_{(n-i) \text{ terms}} + \underbrace{v_{n-1}z^{i-1} + \cdots + v_{n-i+1}z + v_{n-i}}_{i \text{ terms}}$$
$$+ \underbrace{v_{n-1}z^{i-1}(z^n + 1) + \cdots + v_{n-i+1}z(z^n + 1) + v_{n-i}(z^n + 1)}_{i \text{ terms}}$$
$$= q(z)(z^n + 1) + \mathbf{v}^{(i)}(z) \qquad (4.68)$$

where $q(z) = v_{n-1}z^{i-1} + \cdots + v_{n-i+1}z + v_{n-i}$. This means that if $\mathbf{v}(z)$ is a code polynomial, $\mathbf{v}^{(i)}(z)$ is also a code polynomial for any cyclic shift i. From Equation 4.68, we note that the cyclically shifted code polynomial $\mathbf{v}^{(i)}(z)$ is the remainder resulting from dividing the polynomial $z^i\mathbf{v}(z)$ by $(z^n + 1)$. That is,

$$\mathbf{v}^{(i)}(z) = R_{z^n+1}\left[z^i\mathbf{v}(z)\right] \tag{4.69}$$

where $R_{f(z)}[\bullet]$ is the remainder of the modulo-$f(z)$ of $[\bullet]$.

So far we have defined cyclic codes, and now we will focus on their properties.

1. For an (n, k) linear code C, there are 2^k codeword polynomials $c(z)$. The codeword polynomials of degree $(n-1)$ are encoded by a generator polynomial $g(z)$ of degree $(n-k)$. As all the codeword polynomials of a cyclic code must be multiples of a generator polynomial $g(z)$, it then follows from Equation 4.69 that a codeword polynomial can be described by

$$c(z) = R_{z^n+1}\left[a(z)g(z)\right] \tag{4.70}$$

where $a(z)$ is an arbitrary polynomial. A code polynomial $c(z)$ is modulo $(z^n + 1)$ and this implies that the block length is n.

2. Another property of the cyclic code is that the generator polynomial $g(z)$ of an (n, k) code is a factor of $(z^n + 1)$. Let r be the degree of the generator polynomial, where $r = n - k$. Multiplying $g(z)$ by z^k results in a polynomial $z^k g(z)$ of degree n. Dividing $z^k g(z)$ by $(z^n + 1)$, we obtain

$$z^k g(z) = (z^n + 1) + g^{(k)}(z) \tag{4.71}$$

where $g^{(k)}(z)$ is the remainder. From Equation 4.69, we note that $\mathbf{v}^{(i)}(z)$ is a code polynomial given by cyclically shifting $\mathbf{v}(z)$ i times. Similarly, $g^{(k)}(z)$ is the code polynomial obtained by shifting $g(z)$ to the left cyclically k times. Hence, $g^{(k)}(z)$ is a multiple of $g(z)$, say $g^{(k)}(z) = a(z)g(z)$. Substituting $g^{(k)}(z)$ into Equation 4.71 yields

$$z^n + 1 = g(z)\left[z^k + a(z)\right] . \tag{4.72}$$

Hence, $g(z)$ is a factor of $(z^n + 1)$. Consequently, for any cyclic code having generator polynomial $g(z)$,

$$z^n + 1 = g(z)h(z) \tag{4.73}$$

where the polynomial $h(z)$ is the *parity-check polynomial*. Then for any codeword polynomial $c(z)$

$$
\begin{aligned}
R_{z^n+1}\left[c(z)h(z)\right] &= R_{z^n+1}\left[a(z)g(z)h(z)\right] \\
&= R_{z^n+1}\left[a(z)(z^n + 1)\right] \\
&= 0 .
\end{aligned}
\tag{4.74}
$$

Having presented the definition and the properties of cyclic codes, we now highlight their encoding. Suppose the information sequence is represented by a polynomial $i(z)$ of degree $(k-1)$. The set of information polynomials $i(z)$ is mapped into the set of codeword polynomials $c(z)$ using the generator polynomial $g(z)$. A simple encoding method is

$$c(z) = i(z)g(z) . \tag{4.75}$$

This method is called *nonsystematic* encoding because the codeword polynomial does not contain a copy of $i(z)$. Alternatively *systematic* encoding is where the information polynomial $i(z)$ is inserted into the high-order coefficients of the codeword $c(z)$, and the parities are appended to the low-order coefficients. The codeword polynomial for systematic codes is

$$c(z) = i(z)z^{n-k} + b(z) \tag{4.76}$$

where $b(z)$ is evaluated so that

$$R_{g(z)}\left[c(z)\right] = 0 .$$

It follows that,

$$R_{g(z)}\left[i(z)z^{n-k}\right] + R_{g(z)}\left[b(z)\right] = 0$$

and the degree of $b(z)$ is less than $(n-k)$, the degree of $g(z)$. Therefore,

$$b(z) = -R_{g(z)}\left[i(z)z^{n-k}\right] . \tag{4.77}$$

The systematic and nonsystematic encoding procedures are unique one-to-one mappings from a set of information polynomials to a set of of codeword polynomials, but the mappings are different for the two methods.

4.4.3   Bose-Chaudhuri-Hocquenghem Codes

The Bose-Chaudhuri-Hocquenghem (BCH) codes constitute a prominent class of cyclic block codes that have multiple-error detection and correction capabilities. In this section their theory and structure is studied. The class of binary and nonbinary BCH codes is considered in Section 4.4.3.1 and in Section 4.4.3.2, respectively. For the nonbinary BCH codes, an important subclass is that of the Reed-Solomon (RS) codes which achieve the maximum separable distance between codewords as it will be detailed in Section 4.4.3.2.1.

A BCH code accepts k information symbols and produces an n-symbol codeword. If a codeword is designed to correct t random errors, the code is called a *t-error-correcting* code and is denoted as a $BCH(n,k,t)$ code. A BCH code is a cyclic code and therefore can be constructed by its generator polynomial $g(z)$. According to the second property of cyclic codes given in Section 4.4.2 and Equation 4.72, the generator polynomial is a factor of $(z^n + 1)$. That is,

$$z^n + 1 = a(z)g(z)$$

where $a(z)$ is an arbitrary polynomial. Also, the second property of minimal polynomials in Section 4.4.1.5 states that the polynomial $(z^{q^m-1} + 1)$ is the least common multiple (LCM) of the minimal polynomials of all the non-zero elements in $GF(q^m)$. That is,

$$z^{q^m-1} + 1 = LCM[\psi_{\alpha^0}(z), \psi_{\alpha^1}(z), \ldots, \psi_{\alpha^{q^m-2}}(z)] .$$

On observing these two properties, we assign

$$n = q^m - 1 \tag{4.78}$$

such that

$$z^n + 1 = z^{q^m-1} + 1 .$$

The generator polynomial, which is a factor of $(z^n + 1)$, can be constructed by a product of the minimal polynomials over $GF(q)$ of elements from $GF(q^m)$. As the minimal polynomial $\psi_{\alpha^i}(z)$ over $GF(q)$ is defined to have roots of element α^i and its conjugates from $GF(q^m)$, the roots of the generator polynomial are also the element of α^i and its conjugates. For a t-error-correcting code, the generator polynomial $g(z)$ is defined by the least common multiple of the minimal polynomials over $GF(q)$ having $2t$ consecutive powers of α, i.e., $\alpha^{j_0}, \alpha^{j_0+1}, \ldots, \alpha^{j_0+2t-1}$, as their roots, where j_0 is an integer, and hence,

$$g(z) = LCM[\psi_{\alpha^{j_0}}(z), \psi_{\alpha^{j_0+1}}(z), \ldots, \psi_{\alpha^{j_0+2t-1}}(z)] . \tag{4.79}$$

The degree of the generator polynomial $g(z)$, regardless of nonsystematic encoding or systematic encoding described by Equations 4.75 and 4.77 respectively, determines the number of redundancy symbols in a codeword. The $2t$ roots of $g(z)$ allow to correct t error symbols. We require the degree of $g(z)$ to be as small as possible by keeping $2t$ roots so as to correct t errors with minimum redundancy. When j_0 is chosen to be 1, the first root is the primitive element α and this usually gives the $g(z)$ of smallest degree. These codes are described as the *primitive BCH codes* and their blocklength is $(q^m - 1)$. The distance between codewords required for a t-error-correcting code is

$$d = 2t + 1 \tag{4.80}$$

where d is called the *designed distance* of the code. But the minimum separable distance between codewords in some *BCH* codes may be greater, i.e., $d_{min} \geq 2t + 1$ and this is detailed in reference [7].

4.4.3.1 Binary BCH Codes

A BCH code is binary if the codeword symbols are binary, i.e., the symbol is defined on $GF(q)$, where $q = 2$. To represent a codeword as a polynomial, the symbols of a codeword are the coefficients of the polynomial. In the case of a binary code, the symbol is either 0 or 1, and the coefficients of the polynomial have binary values. The blocklength n of the code, as defined in Equation 4.78, is $(q^m - 1)$ symbols. The number of symbols in a codeword determines the extension field $GF(q^m)$ where the roots of the generator polynomial reside. As a consequence, the generator of the code is a binary polynomial, defined on $GF(2)$, having roots in the extension field $GF(q^m)$.

We now present examples of how to construct the generator polynomial of a binary BCH code. This generator polynomial is defined on $GF(2)$. As our first example, we let $m = 4$, giving a blocklength of 15 bits and a generator polynomial having roots from $GF(2^4) = GF(16)$. Table 4.8 gives $GF(16)$ as an extension field of $GF(2)$. The table is constructed using the primitive polynomial $p(z) = z^4 + z + 1$. The minimal polynomials over $GF(2)$ of all field elements in $GF(16)$ are listed in Table 4.12, where $\alpha = z$ is the primitive element. According to Equation 4.79, the generator polynomial for the double error correcting BCH code requires four consecutive roots in $GF(16)$ and can be constructed by

$$
\begin{aligned}
g(z) &= LCM[\psi_\alpha(z), \psi_{\alpha^2}(z), \psi_{\alpha^3}(z), \psi_{\alpha^4}(z)] \\
&= LCM[z^4 + z + 1, z^4 + z + 1, z^4 + z^3 + z^2 + z + 1, z^4 + z + 1] \\
&= (z^4 + z + 1)(z^4 + z^3 + z^2 + z + 1) \\
&= z^8 + z^7 + z^6 + z^4 + 1 \ .
\end{aligned}
\tag{4.81}
$$

The degree of $g(z)$ is eight, i.e., $n - k = 8$, and therefore $k = 7$. This means that 7 information bits are encoded into a codeword with blocklength of 15 bits. The code can correct up to 2 error bits occurring at any position in a codeword, and the code is denoted as $BCH(15, 7, 2)$.

Another example of constructing a generator polynomial for a triple error correcting binary code is by multiplying six minimal polynomials having consecutive roots in $GF(16)$,

$$
\begin{aligned}
g(z) &= LCM[\psi_\alpha(z), \psi_{\alpha^2}(z), \psi_{\alpha^3}(z), \psi_{\alpha^4}(z), \psi_{\alpha^5}(z), \psi_{\alpha^6}(z)] \\
&= (z^4 + z + 1)(z^4 + z^3 + z^2 + z + 1)(z^2 + z + 1) \\
&= z^{10} + z^8 + z^5 + z^4 + z^2 + z + 1 \ .
\end{aligned}
\tag{4.82}
$$

The generator polynomial is of degree 10, and therefore 10 parity bits are included in a codeword in order to correct 3 error bits. The code with blocklength of 15 bits thus carries 5 information bits and is therefore described as $BCH(15, 5, 3)$.

+	0	1	2	3
0	0	1	2	3
1	1	0	3	2
2	2	3	0	1
3	3	2	1	0

$\bullet$	0	1	2	3
0	0	0	0	0
1	0	1	2	3
2	0	2	3	1
3	0	3	1	2

Table 4.13: Arithmetic tables for $GF(4)$

4.4.3.2 Nonbinary BCH Codes

A nonbinary BCH code has symbols of more than one bit defined on $GF(q)$, where $q > 2$. The coefficients of the polynomial are also nonbinary and are elements of $GF(q)$. From Equation 4.78, the blocklength of the code is $(q^m - 1)$ symbols and the number of symbols in a codeword determines the extension field $GF(q^m)$. The generator of the code is defined as a polynomial over $GF(q)$ having roots in the extension field $GF(q^m)$.

Let us consider an example of a nonbinary BCH code. A symbol consisting of two bits is defined on $GF(4)$. If the blocklength of the code is selected to have $4^2 - 1$ symbols, i.e., $q = 4$ and $m = 2$, the generator polynomial over $GF(4)$ has roots in the extension field $GF(q^m) = GF(16)$. Table 4.13 is a decimal representation of the arithmetic tables of $GF(4)$ in Tables 4.4 and 4.5. We also show in Table 4.14 the minimal polynomials over $GF(4)$ of all the field elements in $GF(16)$, where α is primitive. For the double error correcting BCH code, the generator polynomial over $GF(4)$ requires four consecutive roots in $GF(16)$ and can be constructed as

$$
\begin{aligned}
g(z) &= LCM[\psi_\alpha(z), \psi_{\alpha^2}(z), \psi_{\alpha^3}(z), \psi_{\alpha^4}(z)] \\
&= LCM[z^2 + z + 2, z^2 + z + 3, z^2 + 3z + 1, z^2 + z + 2] \\
&= (z^2 + z + 2)(z^2 + z + 3)(z^2 + 3z + 1) \\
&= z^6 + 3z^5 + z^4 + z^3 + 2z^2 + 2z + 1 .
\end{aligned}
$$

The degree of the generator polynomial is 6, the remainder of the information polynomial after division by $g(z)$ results in the parity polynomial. The degree of the parity polynomial is 5 which therefore determines 6 parity symbols in a codeword. Therefore the code is described as a nonbinary $BCH(15, 9, 2)$ code over $GF(4)$ as it encodes 9 information symbols, each of 2 bits, into a codeword of 15 symbols. The code is able to correct 2 quaternary error symbols in a codeword with blocklength of 15 quaternary symbols (i.e., 30 bits). Furthermore, for a triple error correcting code, the generator polynomial can be constructed by

$$
\begin{aligned}
g(z) &= LCM[\psi_\alpha(z), \psi_{\alpha^2}(z), \psi_{\alpha^3}(z), \psi_{\alpha^4}(z), \psi_{\alpha^5}(z), \psi_{\alpha^6}(z)] \\
&= (z^2 + z + 2)(z^2 + z + 3)(z^2 + 3z + 1)(z + 2)(z^2 + 2z + 1) \\
&= z^9 + 3z^8 + 3^7 + 2z^6 + z^5 + 2z^4 + z + 2 .
\end{aligned}
$$

Conjugate roots	Minimal polynomial
0	z
α^0	$z + 1$
α^1, α^4	$z^2 + z + 2$
α^2, α^8	$z^2 + z + 3$
α^3, α^{12}	$z^2 + 3z + 1$
α^5	$z + 2$
α^6, α^9	$z^2 + 2z + 1$
α^7, α^{13}	$z^2 + 2z + 2$
α^{10}	$z + 3$
α^{11}, α^{14}	$z^2 + 3z + 3$

Table 4.14: Minimal polynomials of the elements in $GF(4^2)$

This code is described a $BCH(15, 6, 3)$ triple error correcting code over $GF(4)$.

4.4.3.2.1 Reed-Solomon Codes In Section 4.4.3.2, the nonbinary BCH codes define the symbols over $GF(q)$, where $q > 2$, and the roots of the generator polynomial are over $GF(q^m)$. The Reed-Solomon codes are a special case of nonbinary BCH codes with $m = 1$. The coefficients of the generator polynomial are from $GF(q)$ and the roots of the generator are also elements from $GF(q)$. Hence, the minimal polynomials for constructing the generator polynomial are defined on $GF(q)$ with roots from the same field. Notice that the minimal polynomial over $GF(q)$ of an element β in the same $GF(q)$ is

$$\psi_\beta(z) = z - \beta .$$

For a t-error-correcting RS code, the generator polynomial is

$$g(z) = (z - \alpha^{j_0})(z - \alpha^{j_0+1}) \cdots (z - \alpha^{j_0+2t-1}) . \qquad (4.83)$$

As all the minimal polynomials of any element in $GF(q)$ are of degree 1, the choice of j_0 will not optimize the degree of the generator polynomial. The convention is to set, $j_0 = 1$ to produce a generator polynomial of degree $2t$. Hence, for an RS code

$$n - k = 2t . \qquad (4.84)$$

The blocklength of the RS codes is determined by substituting $m = 1$ into Equation 4.78, whence

$$n = q - 1 .$$

An important characteristic of Reed-Solomon codes is their maximum minimum distance property. Observe that the designed distance d of the code is given by Equation 4.80 and the minimum separable distance d_{min}

between codewords may be actually greater than the designed distance. The lower bound of d_{min} is therefore,

$$d_{min} \geq d = 2t + 1 = n - k + 1 \qquad (4.85)$$

because $2t = n - k$. In addition, systematic codewords exist with only one nonzero information symbol and $(n - k)$ parity symbols. This codeword has a maximum symbol distance of $(n - k + 1)$ from the all-zero codeword. Hence, the minimum distance of the code cannot be greater than $(n - k + 1)$ and the upper bound of d_{min} is therefore

$$d_{min} \leq n - k + 1 . \qquad (4.86)$$

Equation 4.86 is known as the Singleton bound. Combining both Equation 4.85 and 4.86 yields d_{min} for the RS codes as

$$d_{min} = n - k + 1 . \qquad (4.87)$$

The Reed-Solomon code is a maximum-distance code, and the minimum distance is $(n - k + 1)$.

As an example, we will find the generator polynomial $g(z)$ for a double error correcting RS code, i.e., for $t = 2$. Suppose the symbols of the code are chosen to have four bits, defined on $GF(q)$, where $q = 16$. The roots of $g(z)$ are also defined on $GF(q)$, i.e, $GF(16)$. Consequently, for a double error correcting RS code, the generator polynomial $g(z)$ over $GF(16)$ has a set of four roots $\{\alpha, \alpha^2, \alpha^3, \alpha^4\}$ from $GF(16)$. From Equation 4.83, $g(z)$ can be constructed over $GF(16)$ as

$$\begin{aligned} g(z) &= (z - \alpha)(z - \alpha^2)(z - \alpha^3)(z - \alpha^4) \\ &= z^4 + \alpha^{13}z^3 + \alpha^6 z^2 + \alpha^3 z + \alpha^{10} . \end{aligned} \qquad (4.88)$$

The degree of $g(z)$ is four, $n - k = 4$. The blocklength of the code is given by Equation 4.85 such that $n = 15$ and so $k = 11$. An information sequence of 11 hexadecimal symbols (44 bits) is encoded into a codeword with blocklength of 15 hexadecimal symbols (60 bits). The Reed-Solomon code is denoted as $RS(n, k, t)$, and, in this example, it is an $RS(15, 11, 2)$ code.

4.4.4 Encoding of Block Codes

The structure of cyclic codes has the advantage that their encoders and decoders can be implemented using shift-register circuits. The cyclic codes can be encoded either nonsystematically or systematically. For the nonsystematic codes, the encoder multiplies an arbitrary information polynomial $i(z)$ with the generator polynomial $g(z)$ to obtain the codeword $c(z)$, as shown in Equation 4.75. Whereas, for the systematic codes, the encoder evaluates and then appends the redundancy symbols to the arbitrary information polynomial (see Equation 4.76). The systematic code thus contains

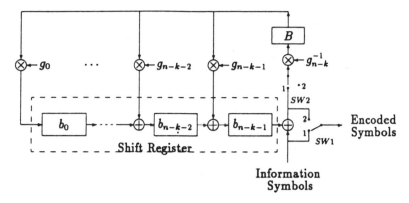

Figure 4.40: Systematic encoder for block codes

a copy of the information symbols. Suppose the codeword is transmitted over a noisy channel, and the number of error symbols in the received codeword exceed the correcting capability of the code. For the nonsystematic codes, the decoder is required to correct the received codeword and then divide the corrected codeword by the generator polynomial $g(z)$ to regenerate the information polynomial $i(z)$, viz:-

$$i(z) = c(z)/g(z) . \tag{4.89}$$

However, if the corrected codeword is erroneous, the regenerated information polynomial can have any arbitrary value which may result in disastrous symbol errors. By constrast to the non-systematic codes, the systematic *RS* decoder corrects the received codeword and then copies the information symbols from the corrected codeword to its output. As we are considering here the situation where the number of channel errors exceeds the correcting capability of the code, the information polynomial has error symbols due to both the channel noise and the correction error. However, the correctly received symbols in the information polynomial remain intact and are undisturbed by the decoder. The result is that systematic *RS* codes perform better than the nonsystematic ones.

For systematic codes the generator polynomial $g(z) = g_{n-k}z^{n-k} + \cdots + g_1 z + g_0$ formulates a codeword by appending $(n - k)$ parity symbols $b_{n-k-1}, \ldots, b_1, b_0$ to k information symbols. The encoder employs a shift register (SR) having $(n - k)$ stages as depicted in Figure 4.40. At the beginning of the encoding process, the SR is cleared and both switches $SW1$ and $SW2$ are placed in position 1. The first k number of information symbols are passed directly to the encoder output forming the information part of the codeword, and they are also multiplied over $GF(q^m)$ by the coefficient g_{n-k}^{-1}, where $q = 2$ and $m = 1$ for binary codes. This product is buffered in register B. The value of register B is multiplied by g_{n-k-1}, followed by $GF(q^m)$ addition with b_{n-k-2} to form a new parity symbol

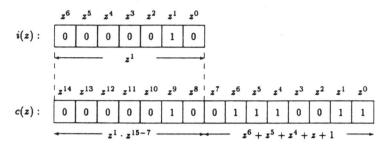

Figure 4.41: Binary representation of polynomials

of b_{n-k-1}. Again, the new b_{n-k-2} value is calculated by multiplying the content of register B with g_{n-k-2} and then adding b_{n-k-3} to this product. Similar multiplications and additions are performed to achieve new values from b_{n-k-3} to b_0. The second information symbol enters the encoder and the cycle of multiplications and additions is repeated to yield a new set of $(n-k)$ parity symbols. After the k-th information symbol has entered the encoder and has produced the parity symbols, the switches are turned to position 2, preventing data from entering the SR. The $(n-k)$ parity symbols are removed serially from the encoder.

4.4.4.1   Binary BCH Encoder

We now present an example of encoding systematically a binary BCH-$(15, 7, 2)$ double error correcting code. The numerical calculation of the encoding will first be demonstrated, followed by its implementation using a shift register circuit. The generator polynomial $g(z)$ of this code is given by Equation 4.81. Let us take an arbitrary information polynomial for our example as

$$i(z) = z^1 . \tag{4.90}$$

According to the systematic encoding described in Equation 4.76, the parity polynomial $b(z)$ is obtained using Equation 4.77 as,

$$b(z) = R_{(z^8+z^7+z^6+z^4+1)}[z^1 \cdot z^{15-7}] .$$

By performing long division over $GF(2)$ we have

$$
\begin{array}{r}
z+1 \\
\hline
z^8 + z^7 + z^6 + z^4 + 1 \overline{\smash{\big)}\, z^9} \\
\underline{z^9 + z^8 + z^7 + z^5 + z} \\
z^8 + z^7 + z^5 + z \\
\underline{z^8 + z^7 + z^6 + z^4 + 1} \\
z^6 + z^5 + z^4 + z + 1
\end{array}
\tag{4.91}
$$

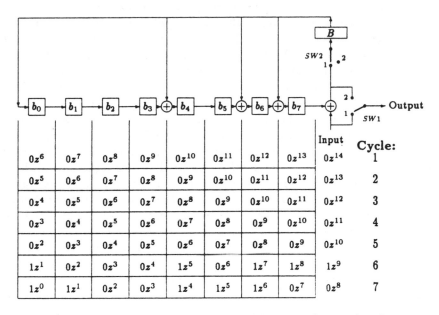

								Input	Cycle:
$0z^6$	$0z^7$	$0z^8$	$0z^9$	$0z^{10}$	$0z^{11}$	$0z^{12}$	$0z^{13}$	$0z^{14}$	1
$0z^5$	$0z^6$	$0z^7$	$0z^8$	$0z^9$	$0z^{10}$	$0z^{11}$	$0z^{12}$	$0z^{13}$	2
$0z^4$	$0z^5$	$0z^6$	$0z^7$	$0z^8$	$0z^9$	$0z^{10}$	$0z^{11}$	$0z^{12}$	3
$0z^3$	$0z^4$	$0z^5$	$0z^6$	$0z^7$	$0z^8$	$0z^9$	$0z^{10}$	$0z^{11}$	4
$0z^2$	$0z^3$	$0z^4$	$0z^5$	$0z^6$	$0z^7$	$0z^8$	$0z^9$	$0z^{10}$	5
$1z^1$	$0z^2$	$0z^3$	$0z^4$	$1z^5$	$0z^6$	$1z^7$	$1z^8$	$1z^9$	6
$1z^0$	$1z^1$	$0z^2$	$0z^3$	$1z^4$	$1z^5$	$1z^6$	$0z^7$	$0z^8$	7

Figure 4.42: Systematic encoder for the $BCH(15, 7, 2)$ code

where $b(z)$ is given by the remainder $(z^6 + z^5 + z^4 + z + 1)$. From Equation 4.76, the codeword $c(z)$ is

$$c(z) = z^9 + z^6 + z^5 + z^4 + z + 1 . \qquad (4.92)$$

Observe that the binary representation of the codeword $c(z)$ in Figure 4.41 has a blocklength of 15 bits, where each bit represents a coefficient of the codeword polynomial. By using systematic encoding, the information polynomial z^1 is multiplied by z^{15-7} to give the information part positioned at the high-order part $(z^{14}-z^8)$ of the codeword. The low-order part (z^7-z^0) of the codeword is constituted by the parity polynomial $z^6 + z^5 + z^4 + z + 1$.

The systematic encoder implemented by the shift register for the BCH-$(15, 7, 2)$ code is shown in Figure 4.42. It can be seen to be a derivative of Figure 4.40. The generator polynomial is $g(z) = z^8 + z^7 + z^6 + z^4 + 1$. The coefficient g_8 of the z^8 term is 1, the inverse of g_8, i.e., g_8^{-1} is also 1. In the case of binary codes, the coefficient of the generator polynomial is either 0 or 1 and the g_8^{-1} is always 1. Observe that all the multipliers illustrated in Figure 4.40 are absent in Figure 4.42. If the coefficient is 1, the multiplier is replaced by a direct hard-wire connection as shown in Figure 4.42, whereas if the coefficient is 0, no connection is made.

The shift register consists of 8 storage elements, $\{b_0, b_1, \ldots, b_7\}$, where each element buffers a parity bit. The feedback connections to these elements are determined by the coefficients of the generator polynomial. This circuit arrangement is actually performing a polynomial division similar to the division shown in Equation 4.91. The input to the shift register is the information polynomial $i(z)$ which is divided by the generator polynomial $g(z)$. The remainder of the division is buffered in the storage elements. In Figure 4.42, each row of the grid represents the contents stored in the elements for a particular cycle. Initially the shift register is cleared and all its elements are zeros. Both switches $SW1$ and $SW2$ are in position 1. At the first cycle, the zero coefficient of order z^6 in $i(z)$ is inserted into the shift register. According to the systematic code, this input bit corresponds to the order z^{14} in $c(z)$ and therefore passes directly to the output. The adder sums the input bit and b_7 over $GF(2)$ to produce the result of 0 which is buffered in register B. The value of register B is added to b_6 to form a new parity bit of b_7. Similarly, the new b_6 value is calculated by adding the content of register B to b_5. Furthermore, the value of b_4 is shifted to b_5 by forming its new value. Similar additions performed sequentially yield new values from b_4 to b_0. As a consequence, the storage elements contain the first partial remainder of the division of $i(z)$ by $g(z)$. The partial remainder polynomial has degree of z^{13} and its coefficients are all-zero. For the rest of the cycles, similar additions and shifts of the parity bits are made and the partial remainder polynomial in the storage elements is always one degree less than in the previous cycle. At the end of the seventh cycle, the storage devices contain the remainder of the division of $i(z)z^{n-k}$ by $g(z)$. The switches $SW1$ and $SW2$ are turned to position 2 and all the parity bits are removed serially from the encoder.

It is interesting to note the similarity of the long division calculation demonstrated in Equation 4.91 and the division implemented by the shift register circuit. At the sixth cycle, the nonzero coefficient of order z^9 input bit is inserted into the shift register. The partial remainder polynomial is $z^8 + z^7 + z^5 + z$ which corresponds to the partial remainder of the first step in the long division process. Similarly, at the next cycle, the content of the storage elements is the remainder of the long division.

4.4.4.2   Reed-Solomon Encoder

We will explain the operation of a RS encoder by considering the specific example of a double error correcting code with four bits per symbol. The arithmetic operations are over $GF(16)$ and the generator polynomial for the double error correcting code is described by Equation 4.88. The result is an $RS(15, 11, 2)$ code over $GF(16)$, encoding 11 information symbols into a codeword of 15 symbols. Given that the information polynomial over $GF(16)$ is

$$\begin{aligned}
i(z) \;=\;\; & 0z^{10} + 0z^9 + 0z^8 + \alpha^{12}z^7 + \alpha^{12}z^6 + \alpha^{12}z^5 + \alpha^{12}z^4 \\
& +\alpha^{12}z^3 + \alpha^{12}z^2 + \alpha^{12}z^1 + \alpha^{12}
\end{aligned} \tag{4.93}$$

consisting of 11 symbols, we will assume that the three high-order terms $(z^{10}-z^8)$ contain all-zero symbols and all the low-order terms (z^7-z^0) contain all-one (i.e., α^{12}) symbols. By using systematic encoding described in Equation 4.76, the parity polynomial $b(z)$ can be obtained with the aid of Equation 4.77 as,

$$b(z) = R_{(z^4+\alpha^{13}z^3+\alpha^6z^2+\alpha^3z+\alpha^{10})} \Big[(\alpha^{12}z^7 + \alpha^{12}z^6 + \alpha^{12}z^5 + \alpha^{12}z^4$$

$$+\alpha^{12}z^3 + \alpha^{12}z^2 + \alpha^{12}z^1 + \alpha^{12}) \cdot z^{15-11} \Big].$$

By performing long division over $GF(16)$, $b(z)$ is,

$$b(z) = \alpha^{14}z^3 + \alpha^2z^2 + \alpha^0z + \alpha^6 . \tag{4.94}$$

The polynomial $b(z)$ contains four parity symbols which are systematic encoded according to Equation 4.76. The codeword polynomial $c(z)$ becomes

$$\begin{aligned} c(z) = {} & 0z^{14} + 0z^{13} + 0z^{12} + \alpha^{12}z^{11} + \alpha^{12}z^{10} + \alpha^{12}z^9 + \alpha^{12}z^8 + \alpha^{12}z^7 \\ & +\alpha^{12}z^6 + \alpha^{12}z^5 + \alpha^{12}z^4 + \alpha^{14}z^3 + \alpha^2z^2 + \alpha^0z + \alpha^6 \end{aligned} \tag{4.95}$$

and is of degree 14 with 15 terms. The coefficient of each term represents a four bits symbol in the codeword. The blocklength of the codeword is 15 four-bit symbols, and therefore the codeword has 60 bits. The 11 information symbols reside in the high-order $(z^{14}-z^4)$ terms of $c(z)$, while the low-order (z^3-z^0) terms contains four parity symbols.

Reed-Solomon codes belong to the class of cyclic codes and their cyclic property enables the RS encoder to be implemented by shift register circuits. The systematic encoder for the $RS(15, 11, 2)$ code is shown in Figure 4.43. Its circuitry is similar to that for the binary BCH encoder shown in Figure 4.42. For Figure 4.43 we see that the shift register consists of four storage elements, $\{b_0, b_1, b_2, b_3\}$, each element buffers a parity symbol and each symbol has four bits. The feedback connections to these devices are determined by the coefficients of the generator polynomial. The generator polynomial of this code is $g(z) = z^4 + \alpha^{13}z^3 + \alpha^6z^2 + \alpha^3z + \alpha^{10}$, which is a monic polynomial such that the coefficient g_4 of the z^4 term is 1. The inverse of g_4 is also 1 and therefore the multiplier of g_4^{-1} (i.e., corresponding to the multiplier of g_{n-k}^{-1} in Figure 4.40) is a direct hard-wire connection. The multipliers of α^{13}, α^6, α^3 and α^{10} in the circuit correspond to the coefficients of z^3, z^2, z^1 and z^0, respectively. All the multiplications and additions are performed over $GF(16)$.

The input to the shift register is the information polynomial $i(z)$ which is divided by the generator polynomial $g(z)$. The partial remainder of the division is then buffered in the storage devices. As there are 11 information symbols in a codeword, the division is completed at the end of the eleventh cycle. The four parity symbols are the remainder of the division and are buffered in the shift register.

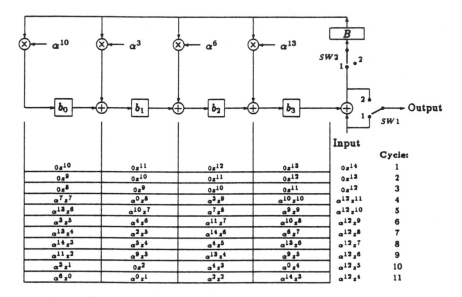

b_0	b_1	b_2	b_3	Input	Cycle:
$0z^{10}$	$0z^{11}$	$0z^{12}$	$0z^{13}$	$0z^{14}$	1
$0z^{9}$	$0z^{10}$	$0z^{11}$	$0z^{12}$	$0z^{13}$	2
$0z^{8}$	$0z^{9}$	$0z^{10}$	$0z^{11}$	$0z^{12}$	3
$\alpha^{7}z^{7}$	$\alpha^{0}z^{8}$	$\alpha^{3}z^{9}$	$\alpha^{10}z^{10}$	$\alpha^{12}z^{11}$	4
$\alpha^{13}z^{6}$	$\alpha^{10}z^{7}$	$\alpha^{7}z^{8}$	$\alpha^{9}z^{9}$	$\alpha^{12}z^{10}$	5
$\alpha^{3}z^{5}$	$\alpha^{4}z^{6}$	$\alpha^{11}z^{7}$	$\alpha^{10}z^{8}$	$\alpha^{12}z^{9}$	6
$\alpha^{13}z^{4}$	$\alpha^{2}z^{5}$	$\alpha^{14}z^{6}$	$\alpha^{6}z^{7}$	$\alpha^{12}z^{8}$	7
$\alpha^{14}z^{3}$	$\alpha^{5}z^{4}$	$\alpha^{4}z^{5}$	$\alpha^{13}z^{6}$	$\alpha^{12}z^{7}$	8
$\alpha^{11}z^{2}$	$\alpha^{9}z^{3}$	$\alpha^{13}z^{4}$	$\alpha^{9}z^{5}$	$\alpha^{12}z^{6}$	9
$\alpha^{3}z^{1}$	$0z^{2}$	$\alpha^{4}z^{3}$	$\alpha^{0}z^{4}$	$\alpha^{12}z^{5}$	10
$\alpha^{6}z^{0}$	$\alpha^{0}z^{1}$	$\alpha^{2}z^{2}$	$\alpha^{14}z^{3}$	$\alpha^{12}z^{4}$	11

Figure 4.43: Systematic encoder for the $RS(15, 11, 2)$ code

The cyclic code has a defined blocklength of n symbols of which k are information symbols, but systematic cyclic codes can be shortened by dropping s information symbols from each codeword. Now $(k - s)$ information symbols are combined with $(n - k)$ parity symbols to form a codeword with a blocklength of $(n - s)$ symbols. The code is converted from an (n, k) code to an $(n - s, k - s)$ code which is known as a *shortened cyclic code*. Those deleted symbols must be the ones in the high-order positions of the information sequence. As the deleted symbols in a shortened code are always set to zero and they are not transmitted, the receiver reinserts them and decodes just as if the code were not shortened. If the symbol of the code is q-ary, the original code consists of a set of q^k possible codewords and has minimum distance d_{min}. After the codeword has been shortened by s symbols, the set of possible codewords is reduced to q^{k-s} and therefore the minimum distance is d_{min} or larger. As the number of parity symbols of the codeword remains the same after shortening, the code is still able to correct t error symbols. For example, the $RS(15, 11, 2)$ code consists of 11 information symbols and 4 parity symbols. If the number of information symbols is reduced by three, i.e., $s = 3$, such that the three high-order terms $(z^{14} - z^{12})$ are always set to zeros, then the blocklength of the shortened code is reduced by three and the code becomes $RS(12, 8, 2)$. The generator polynomial $g(z)$ of the shortened code is given by Equation 4.88, and the four parity symbols in a codeword are able to correct two error symbols.

4.4.5 Decoding Algorithms for Block Codes

When a codeword is transmitted, errors can occur at any symbol position in a codeword. For binary codes, the magnitude of the error symbol is 1. For the nonbinary code defined on $GF(q)$, where $q > 2$, the error magnitude can be any value from 1 to $(q - 1)$. If the positions of the errors and their magnitudes are found, the original codeword can be recovered by correcting the magnitude in each error position of the received codeword. In Section 4.4.5.1, we derive the error magnitudes and positions as the solution of a set of non-linear syndrome equations. The set of equations is then turned into a matrix with its coordinate elements arranged in a special structure known as a Vandermonde matrix. The direct inversion of the Vandermonde matrix was first suggested by Peterson [38] to obtain the solution for binary codes and then extended for non-binary codes by Gorenstein and Zierler [39]. The decoding algorithm is known as Peterson-Gorenstein-Zierler method and is described in Section 4.4.5.2. Berlekamp [4] and Massey [44, 45] recognized that the best way to derive the decoding algorithm was as a solution to a problem in designing linear-feedback shift registers. Their method described in Section 4.4.5.3, is conceptually more complex than the direct matrix inversion, but is computationally much simpler. Finally, in Section 4.4.5.4, we describe the Forney algorithm for evaluating the error magnitudes for non-binary codes.

4.4.5.1 The Syndrome Equations

Consider a t error correcting $BCH(n, k, t)$ code constructed by a generator polynomial over $GF(q)$ having $2t$ consecutive roots in $GF(q^m)$. Notice that this is an RS code if $m = 1$. To simplify the equations we let $j_0 = 1$ in Equation 4.79 whereby the roots of the generator polynomial are $\alpha^1, \alpha^2, \ldots, \alpha^{2t}$. Suppose that the codeword $c(z)$ is expressed in polynomial form over $GF(q)$, viz:-

$$c(z) = c_{n-1}z^{n-1} + c_{n-2}z^{n-2} + \ldots + c_1z + c_0$$

where the coefficient c_i represents the i-th symbol in the codeword. As errors can occur at any symbol position in a codeword, the error polynomial

$$e(z) = e_{n-1}z^{n-1} + e_{n-2}z^{n-2} + \ldots + e_1z + e_0$$

is also of degree-$(n - 1)$, where the coefficients e_i with non-zero value indicate that the i-th symbol of the codeword is in error and that its error magnitude is e_i. The coefficient with zero value means that there is no error in the i-th position. For a t error correcting code, the maximum number of non-zero coefficient terms is t. If this condition is not met, the number of errors exceeds the correcting capability of the code and the received word cannot be corrected. The received word is therefore described as

$$r(z) = c(z) + e(z) \ . \tag{4.96}$$

Now suppose that there are v error symbols in a received codeword, where $0 \leq v \leq t$. These errors occur in unknown positions $[i_1, i_2, \ldots, i_v]$, with error magnitudes $[e_{i_1}, e_{i_2}, \ldots, e_{i_v}]$. By ignoring the zero coefficient terms in $e(z)$, we may express the error polynomial as

$$e(z) = e_{i_1} z^{i_1} + e_{i_2} z^{i_2} + \ldots + e_{i_v} z^{i_v} \qquad (4.97)$$

with its v number of unknown error magnitudes and v number of unknown positions. Moreover the value of v itself is also unknown as the number of error symbols v in a correctable codeword can be any value $\leq t$. If all these unknowns are found, the received word can be corrected.

We now define a syndrom S_l as the received polynomial $r(z)$ evaluated at $z = \alpha^l$, where α^l is a root of the generator polynomial. This implies that $c(\alpha^l) = 0$ as α^l is also a root of the codeword polynomial $c(z)$, and that the syndrome S_l depends only on the error polynomial $e(z)$, i.e.,

$$S_l = r(\alpha^l) = c(\alpha^l) + e(\alpha^l) = e(\alpha^l) \ .$$

Let us now evaluate the received polynomial $r(z)$ at α to obtain the syndrome S_1, where α is a root of the generator polynomial. Thus for $l = 1$, we have,

$$
\begin{aligned}
S_1 &= r(\alpha) = c(\alpha) + e(\alpha) = e(\alpha) \\
&= e_{i_1} \alpha^{i_1} + e_{i_2} \alpha^{i_2} + \cdots + e_{i_v} \alpha^{i_v} \ . \qquad (4.98)
\end{aligned}
$$

Notice that the position i_x of the error symbol can be any position in a codeword, namely $0 \leq i_x < (n - 1)$. From Equation 4.78, $n = q^m - 1$ for BCH codes, and the element α^{i_x} representing the error position is in the extension field $GF(q^m)$. To simplify the notation, we assign P_x to be the error position α^{i_x}, and M_x to be its error magnitude e_{i_x}. Hence, Equation 4.98 becomes

$$S_1 = M_1 P_1 + M_2 P_2 + \cdots + M_v P_v \ .$$

Similarly, by evaluating the received polynomial $r(z)$ at the set of $2t$ roots $\{\alpha, \alpha^2, \ldots, \alpha^{2t}\}$ of the generator polynomial $g(z)$, we obtain a set of syndromes,

$$
\begin{array}{lllll}
S_1 &= M_1 P_1 + M_2 P_2 + \cdots + M_v P_v & = \sum_{i=1}^{v} M_i P_i \\
S_2 &= M_1 P_1^2 + M_2 P_2^2 + \cdots + M_v P_v^2 & = \sum_{i=1}^{v} M_i P_i^2 \\
\vdots &= \qquad\qquad \vdots & = \quad \vdots \\
S_{2t} &= M_1 P_1^{2t} + M_2 P_2^{2t} + \cdots + M_v P_v^{2t} & = \sum_{i=1}^{v} M_i P_i^{2t}
\end{array}
\qquad (4.99)
$$

This set of syndrome equations can be expressed in the following matrix form,

$$
\begin{bmatrix} S_1 \\ S_2 \\ \vdots \\ S_{2t} \end{bmatrix}
=
\begin{bmatrix}
P_1 & P_2 & \ldots & P_v \\
P_1^2 & P_2^2 & \ldots & P_v^2 \\
\vdots & & & \vdots \\
P_1^{2t} & P_2^{2t} & \ldots & P_v^{2t}
\end{bmatrix}
\begin{bmatrix} M_1 \\ M_2 \\ \vdots \\ M_v \end{bmatrix} \ .
\qquad (4.100)
$$

The syndromes $S_1, S_2, \ldots, S_{2t}$ are computed from the received polynomial. They are constants in the set of simultaneous nonlinear equations in which the error positions $P_1, P_2, \ldots, P_v$ and their magnitudes $M_1, M_2, \ldots, M_v$ are unknowns.

We have to find $2 \cdot v$ unknowns using $2 \cdot t$ syndromes by evaluating a set of $2t$ non-linear equations which is difficult to solve. Instead, we now describe a method in Section 4.4.5.2 used by Peterson [38] for direct solution of these nonlinear equations.

4.4.5.2   Peterson-Gorenstein-Zierler Decoding

Here we use Peterson's method [38] for converting the syndrome equations into linear equations from which the error positions can be computed. Let us define an *error-locator polynomial* to be the polynomial with zeros at the inverse error positions P_i^{-1} for $i = 1, 2, \ldots, v$:

$$L(z) = (1 - zP_1)(1 - zP_2) \cdots (1 - zP_v) \tag{4.101}$$

and upon multiplying out the product terms we obtain,

$$L(z) = L_v z^v + \cdots + L_1 z + 1 . \tag{4.102}$$

If the coefficients of Equation 4.102 are found, we can find the zeros of $L(z)$ to obtain the error positions. Therefore we concentrate on finding the values of $L_1, L_2, \ldots, L_v$ from the given syndromes.

Let us now multiply both sides of Equation 4.102 by $M_i P_i^{j+v}$ and substitute $z = P_i^{-1}$ to give

$$M_i P_i^{j+v} L(P_i^{-1}) = M_i P_i^{j+v}(L_v P_i^{-v} + \cdots + L_1 P_i^{-1} + 1) . \tag{4.103}$$

As P_i^{-1} is a root of the error-locator polynomial, $L(P_i^{-1}) = 0$, and hence,

$$0 = M_i(L_v P_i^j + \cdots + L_1 P_i^{j+v-1} + P_i^{j+v}) . \tag{4.104}$$

The equation is valid for each i, and on summing over $i = 1, 2, \ldots, v$ for a given j, we have

$$\sum_{i=1}^{v} M_i(L_v P_i^j + \cdots + L_1 P_i^{j+v-1} + P_i^{j+v}) = 0$$

or

$$L_v \sum_{i=1}^{v} M_i P_i^j + \cdots + L_1 \sum_{i=1}^{v} M_i P_i^{j+v-1} + \sum_{i=1}^{v} M_i P_i^{j+v} = 0 \tag{4.105}$$

The summation of each term in Equation 4.105 is recognized as a syndrome in Equation 4.99, and therefore

$$L_v S_j + \cdots + L_1 S_{j+v-1} + S_{j+v} = 0 . \tag{4.106}$$

The syndromes in Equation 4.106 are all described in Equation 4.99 and they are written as $S_1, S_2, \ldots, S_{2t}$. As $v \le t$, the subscript of S_{j+v} suggests that the value of j is in the interval $1 \le j \le v$, and Equation 4.106 is for a particular value of j, whence,

$$L_v S_j + \cdots + L_2 S_{j+v-2} + L_1 S_{j+v-1} = -S_{j+v} \qquad j = 1, \ldots, v. \quad (4.107)$$

In matrix form the coefficients $L_1, L_2, \ldots, L_v$ of the error-locator polynomial $L(z)$ and the syndromes $S_1, S_2, \ldots, S_{2t}$ are related by

$$\begin{bmatrix} S_1 & S_2 & S_3 & \cdots & S_{v-1} & S_v \\ S_2 & S_3 & S_4 & \cdots & S_v & S_{v+1} \\ S_3 & S_4 & S_5 & \cdots & S_{v+1} & S_{v+2} \\ \vdots & & & & & \vdots \\ S_v & S_{v+1} & S_{v+2} & \cdots & S_{2v-2} & S_{2v-1} \end{bmatrix} \begin{bmatrix} L_v \\ L_{v-1} \\ L_{v-2} \\ \vdots \\ L_1 \end{bmatrix} = \begin{bmatrix} -S_{v+1} \\ -S_{v+2} \\ -S_{v+3} \\ \vdots \\ -S_{2v} \end{bmatrix}.$$

$$(4.108)$$

that can also be written as

$$\vec{\vec{S}} \vec{L} = \vec{S} . \quad (4.109)$$

The coefficients of the error-locator polynomial $L(z)$ can be computed by inverting the matrix $\vec{\vec{S}}$, namely,

$$\vec{L} = \vec{\vec{S}}^{-1} \vec{S} \quad (4.110)$$

provided that the matrix is nonsingular. The structure of the matrix $\vec{\vec{S}}$ is recognized as a Vandermonde matrix [8] which is nonsingular if its dimension corresponds to the number of unknowns. Hence, $\vec{\vec{S}}$ can be inverted if its dimension is $v \times v$, but it is singular and cannot be inverted if its dimension is greater than v [39, 9], where v is the actual number of errors occured. This theorem provides the basis of determining the actual number of errors v and $L_1, L_2, \ldots, L_v$.

Let us now summarize the method of obtaining the error positions and subsequently their error magnitudes as a series of steps.

1. Obtain a set of $2t$ syndromes $S_1, S_2, \ldots, S_{2t}$, by substituting the roots of the generator into the received polynomial $r(z)$.

2. We have to determine v in order to invert $\vec{\vec{S}}$ for the solution of Equation 4.110. We first assign $v = t$, as t is the maximum number of correctable errors.

3. The $(v \times v)$ matrix $\vec{\vec{S}}$ is formulated from the syndromes and its determinant, $\det(\vec{\vec{S}})$ is computed.

4. If $\det(\vec{\vec{S}}) = 0$, we cannot invert $\vec{\vec{S}}$. Therefore v is reduced by 1 and the procedure reverts to Step 3. However, if $\det(\vec{\vec{S}}) \neq 0$, $\vec{\vec{S}}$ can be inverted and we proceed with the matrix inversion to obtain $\vec{\vec{S}}^{-1}$.

5. The coefficients $L_1, L_2, \ldots, L_v$ of the error-locator polynomial $L(z)$ are evaluated from Equation 4.110, i.e., $\vec{L} = \vec{\vec{S}}^{-1}\vec{S}$.

6. As we define $L(z)$ to have zeros at the inverse error positions P_1^{-1}, $P_2^{-1}, \ldots, P_v^{-1}$, these are the elements of $GF(q^m)$. There is usually only a moderate number of field elements and the simplest way to find the zeros of $L(z)$ is by substituting every field element into $L(z)$ in turn. The element α^i is the inverse of the error position if $L(\alpha^i) = 0$, and hence the error symbol is at position $(\alpha^i)^{-1}$. This process of finding the zeroes by trial and error is known as a *Chien search*.

7. For binary BCH codes, the symbol of the code is a single bit and its value at the error positions is toggled for error correction. For non-binary codes, the error magnitudes $M_1, M_2, \ldots, M_v$ are determined from Equation 4.100. Although Equation 4.99 represents a set of $2t$ equations, the actual number of errors v is now known and the v unknown error magnitudes can be solved by a set of v equations. The error magnitudes $M_1, M_2, \ldots, M_v$ are given by

$$\begin{bmatrix} M_1 \\ M_2 \\ \vdots \\ M_v \end{bmatrix} = \begin{bmatrix} P_1 & P_2 & \cdots & P_v \\ P_1^2 & P_2^2 & \cdots & P_v^2 \\ \vdots & & & \vdots \\ P_1^v & P_2^v & \cdots & P_v^v \end{bmatrix}^{-1} \begin{bmatrix} S_1 \\ S_2 \\ \vdots \\ S_v \end{bmatrix}. \tag{4.111}$$

We have located the positions and magnitudes of the errors. This enables the error polynomial $e(z)$ to be formulated, and when it is added to the received polynomial $r(z)$ the original codeword $c(z)$ is recovered.

Let us now illustrate the decoding algorithm by considering a triple error correcting binary $BCH(15, 5, 3)$ code. Suppose the information polynomial

$$i(z) = z^1$$

is systematically encoded by the generator $g(z)$ given by Equation 4.82 to produce a codeword,

$$c(z) = z^{11} + z^9 + z^6 + z^5 + z^3 + z^2 + z.$$

If three or less errors occur, the received word $r(z)$ can be corrected. Suppose that double error symbols occur causing the received polynomial to be,

$$r(z) = z^{13} + z^{11} + z^9 + z^5 + z^3 + z^2 + z.$$

To rectify the errors, we proceed through the steps of the decoding algorithm. First we compute the syndromes over $GF(16)$ with the aid of Table 4.10 and Table 4.11. The six syndromes are

$$S_1 = r(\alpha^1) = \alpha^0 \qquad S_2 = r(\alpha^2) = \alpha^0 \qquad S_3 = r(\alpha^3) = \alpha^1$$
$$S_4 = r(\alpha^4) = \alpha^0 \qquad S_5 = r(\alpha^5) = \alpha^{10} \qquad S_6 = r(\alpha^6) = \alpha^2 \ .$$

$$(4.112)$$

Assigning $v = t = 3$, $\vec{S}$ is formulated with a dimension of (3×3),

$$\vec{S} = \begin{bmatrix} S_1 & S_2 & S_3 \\ S_2 & S_3 & S_4 \\ S_3 & S_4 & S_5 \end{bmatrix} = \begin{bmatrix} \alpha^0 & \alpha^0 & \alpha^1 \\ \alpha^0 & \alpha^1 & \alpha^0 \\ \alpha^1 & \alpha^0 & \alpha^{10} \end{bmatrix} \ .$$

The determinant of $\vec{S}$ is zero and therefore cannot be inverted. Next we assign $v = 2$, and formulate $\vec{S}$ again with a dimension of (2×2),

$$\vec{S} = \begin{bmatrix} S_1 & S_2 \\ S_2 & S_3 \end{bmatrix} = \begin{bmatrix} \alpha^0 & \alpha^0 \\ \alpha^0 & \alpha^1 \end{bmatrix} \ .$$

The $\det(\vec{S}) = \alpha^4$ and hence double errors are recognized. The inverse of $\vec{S}$ is calculated as

$$\vec{S}^{-1} = \begin{bmatrix} \alpha^{12} & \alpha^{11} \\ \alpha^{11} & \alpha^{11} \end{bmatrix} \ .$$

From Equation 4.110, we can evaluate the coefficients L_1, L_2 of $L(z)$,

$$\begin{bmatrix} L_2 \\ L_1 \end{bmatrix} = \begin{bmatrix} \alpha^{12} & \alpha^{11} \\ \alpha^{11} & \alpha^{11} \end{bmatrix} \begin{bmatrix} \alpha^1 \\ \alpha^0 \end{bmatrix} = \begin{bmatrix} \alpha^4 \\ \alpha^0 \end{bmatrix}$$

to give the error-locator polynomial,

$$L(z) = L_2 z^2 + L_1 z + 1 = \alpha^4 z^2 + \alpha^0 z + 1 \ .$$

By the Chien search, we find the zeros of the error-locator polynomial by substituting each field element of $GF(16)$ into $L(z)$,

$$\begin{array}{llll}
L(\alpha^0) = \alpha^4 & L(\alpha^1) = \alpha^{12} & L(\alpha^2) = 0 & L(\alpha^3) = \alpha^{11} \\
L(\alpha^4) = \alpha^{13} & L(\alpha^5) = \alpha^{11} & L(\alpha^6) = \alpha^{12} & L(\alpha^7) = \alpha^1 \\
L(\alpha^8) = \alpha^1 & L(\alpha^9) = 0 & L(\alpha^{10}) = \alpha^6 & L(\alpha^{11}) = \alpha^0 \\
L(\alpha^{12}) = \alpha^4 & L(\alpha^{13}) = \alpha^{13} & L(\alpha^{14}) = \alpha^6 &
\end{array} \ .$$

The elements α^2 and α^9 are the zeros of $L(z)$, and the inverses of these zeros are the error positions α^{13} and α^6, respectively. The error magnitudes of binary codes at these positions must be 1. Armed with the knowledge of the error positions and error magnitudes we formulate the error polynomial,

$$e(z) = z^{13} + z^6 \ .$$

The original codeword $c(z)$ is then recovered by adding the error polynomial $e(z)$ to the received word $r(z)$.

$$
\begin{aligned}
c(z) &= r(z) + e(z) \\
&= (z^{13} + z^{11} + z^9 + z^5 + z^3 + z^2 + z) + (z^{13} + z^6) \\
&= z^{11} + z^9 + z^6 + z^5 + z^3 + z^2 + z .
\end{aligned}
$$

As another example, consider a triple error correcting nonbinary $RS(15, 9, 3)$ code over $GF(16)$. Suppose the information polynomial $i(z)$ consisting of all-zero symbols,

$$i(z) = 0$$

is systematically encoded by the generator polynomial

$$
\begin{aligned}
g(z) &= (z - \alpha)(z - \alpha^2)(z - \alpha^3)(z - \alpha^4)(z - \alpha^5)(z - \alpha^6) \\
&= z^6 + \alpha^{10}z^5 + \alpha^{14}z^4 + \alpha^4z^3 + \alpha^6z^2 + \alpha^9z + \alpha^6 .
\end{aligned}
$$

The encoded polynomial $c(z)$ is also an all-zero codeword,

$$c(z) = 0 .$$

Suppose three error symbols occur such that the received polynomial $r(z)$ is

$$r(z) = \alpha^2 z^{14} + \alpha^{13}z^8 + \alpha^6 z^2 .$$

In the case of the all-zero encoded polynomial $c(z)$, all the non-zero symbols in the received polynomial $r(z)$ are the error symbols, and therefore $e(z) = r(z)$. To proceed with the decoding steps, we evaluate the six syndromes over $GF(16)$ by using Tables 4.10 and 4.11.

$$
\begin{array}{lll}
S_1 = r(\alpha^1) = \alpha^7 & S_2 = r(\alpha^2) = \alpha^{12} & S_3 = r(\alpha^3) = \alpha^{13} \\
S_4 = r(\alpha^4) = \alpha^8 & S_5 = r(\alpha^5) = \alpha^3 & S_6 = r(\alpha^6) = \alpha^2 .
\end{array}
$$

Next we assign $v = t = 3$, formulate $\vec{S}$ with a dimension of (3×3).

$$
\vec{S} = \begin{bmatrix} S_1 & S_2 & S_3 \\ S_2 & S_3 & S_4 \\ S_3 & S_4 & S_5 \end{bmatrix} = \begin{bmatrix} \alpha^7 & \alpha^{12} & \alpha^{13} \\ \alpha^{12} & \alpha^{13} & \alpha^8 \\ \alpha^{13} & \alpha^8 & \alpha^3 \end{bmatrix} .
$$

As $\det(\vec{S}) = \alpha^8$, triple errors are recognized. The inverse of $\vec{S}$ is

$$
\vec{S}^{-1} = \begin{bmatrix} 0 & \alpha^5 & \alpha^{10} \\ \alpha^5 & \alpha^6 & \alpha^{12} \\ \alpha^{10} & \alpha^{12} & \alpha^{13} \end{bmatrix} .
$$

The coefficients L_1, L_2 and L_3 of $L(z)$ can be found by using Equation 4.110,

$$\begin{bmatrix} L_3 \\ L_2 \\ L_1 \end{bmatrix} = \begin{bmatrix} 0 & \alpha^5 & \alpha^{10} \\ \alpha^5 & \alpha^6 & \alpha^{12} \\ \alpha^{10} & \alpha^{12} & \alpha^{13} \end{bmatrix} \begin{bmatrix} \alpha^8 \\ \alpha^3 \\ \alpha^2 \end{bmatrix} = \begin{bmatrix} \alpha^9 \\ \alpha^{11} \\ \alpha^3 \end{bmatrix}$$

to give the error-locator polynomial,

$$L(z) = L_3 z^3 + L_2 z^2 + L_1 z + 1 = \alpha^9 z^3 + \alpha^{11} z^2 + \alpha^3 z + 1 .$$

By means of the Chien search we find the zeros of the error-locator polynomial on substituting each field element of $GF(16)$ into $L(z)$,

$$\begin{array}{llll} L(\alpha^0) = \alpha^{13} & L(\alpha^1) = 0 & L(\alpha^2) = \alpha^3 & L(\alpha^3) = \alpha^0 \\ L(\alpha^4) = \alpha^8 & L(\alpha^5) = \alpha^1 & L(\alpha^6) = \alpha^0 & L(\alpha^7) = 0 \\ L(\alpha^8) = \alpha^3 & L(\alpha^9) = \alpha^7 & L(\alpha^{10}) = \alpha^2 & L(\alpha^{11}) = \alpha^{12} \\ L(\alpha^{12}) = \alpha^{10} & L(\alpha^{13}) = 0 & L(\alpha^{14}) = \alpha^4 \end{array}$$

The elements α^1, α^7 and α^{13} are the zeros of $L(z)$, and therefore the inverses of the zeros are the error positions α^{14}, α^8 and α^2, respectively. The error magnitudes of nonbinary codes at these positions have to be determined. From Equation 4.111, the error magnitudes M_1, M_2 and M_3 are

$$\begin{bmatrix} M_1 \\ M_2 \\ M_3 \end{bmatrix} = \begin{bmatrix} P_1 & P_2 & P_3 \\ P_1^2 & P_2^2 & P_3^2 \\ P_1^3 & P_2^3 & P_3^3 \end{bmatrix}^{-1} \begin{bmatrix} S_1 \\ S_2 \\ S_3 \end{bmatrix}$$

$$= \begin{bmatrix} \alpha^2 & \alpha^8 & \alpha^{14} \\ (\alpha^2)^2 & (\alpha^8)^2 & (\alpha^{14})^2 \\ (\alpha^2)^3 & (\alpha^8)^3 & (\alpha^{14})^3 \end{bmatrix}^{-1} \begin{bmatrix} \alpha^7 \\ \alpha^{12} \\ \alpha^{13} \end{bmatrix}$$

$$= \begin{bmatrix} \alpha^7 & \alpha^6 & \alpha^0 \\ \alpha^2 & \alpha^{14} & \alpha^1 \\ \alpha^7 & \alpha^{12} & \alpha^{12} \end{bmatrix} \begin{bmatrix} \alpha^7 \\ \alpha^{12} \\ \alpha^{13} \end{bmatrix} = \begin{bmatrix} \alpha^6 \\ \alpha^{13} \\ \alpha^2 \end{bmatrix}$$

The error positions and error magnitudes provide the error polynomial,

$$e(z) = M_3 P_3 + M_2 P_2 + M_1 P_1 = \alpha^2 z^{14} + \alpha^{13} z^8 + \alpha^6 z^2 .$$

As before, the original codeword $c(z)$ is then recovered by adding the error polynomial $e(z)$ to the received word $r(z)$.

$$\begin{aligned} c(z) &= r(z) + e(z) \\ &= (\alpha^2 z^{14} + \alpha^{13} z^8 + \alpha^6 z^2) + (\alpha^2 z^{14} + \alpha^{13} z^8 + \alpha^6 z^2) = 0 . \end{aligned}$$

Hence the received word is corrected and the all-zero codeword $c(z)$ is recovered.

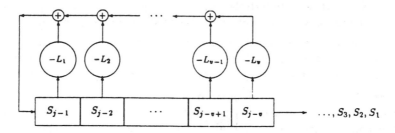

Figure 4.44: Autoregressive filter implemented by linear-feedback shift register

4.4.5.3   Berlekamp-Massey Algorithm

In Section 4.4.5.2, we showed that the error positions can be computed by direct matrix inversion. For v error symbols in a codeword, the number of computations required to invert a v by v matrix is proportional to v^3. Consequently the matrix inversion approach is only practical for moderate values of v. In practice, long codes with large t are usually preferable as they have the capability of correcting a large number of errors. For these codes, Berlekamp and Massey introduced an efficient method of decoding without recourse to matrix inversion. Their method of inverting the matrix is inspired by the analogy of designing a linear-feedback shift register.

The first row of the matrix equations in Equation 4.108 expresses S_{v+1} in terms of $S_1, S_2, \ldots, S_v$, the second row expresses S_{v+2} in terms of S_2, $S_3, \ldots, S_{v+1}$, and so on. If the vector $\vec{L}$ in the equation is known, the recursive expression of the syndromes in terms of themselves suggests the equation of an autoregressive filter,

$$S_j = -\sum_{i=1}^{v} L_i S_{j-i}, \qquad j = v+1, \ldots, 2v . \qquad (4.113)$$

This filter is implemented by a linear-feedback shift register as shown in Figure 4.44. The taps are represented by a polynomial $L(z)$ to give a sequence of syndromes. Now, the problem of finding $L_1, L_2, \ldots, L_v$ in Equation 4.108 is converted to a filter design problem by evaluating its tapping coefficients L_i, $i = 1, \ldots, v$ given in Equation 4.113 to produce the sequence of known syndromes S_j.

There are many linear-feedback shift registers with different lengths and different connection polynomials that are capable of generating the known sequence of syndromes. The length of the shift register may be greater than the degree of $L(z)$ such that some rightmost shift register stages may not be connected. However, there are two criteria for designing the shift register. Firstly, for the syndromes of a correctable error pattern, the solution of $\vec{L}$

expressed in Equation 4.108 has a dimension of v and is unique because only the v by v matrix is invertible. Hence, the polynomial $L(z)$ is of degree v. From Figure 4.44, we note that the register is designed to contain a row of the v by v matrix $\vec{\vec{S}}$, and therefore the register length is v. As a consequence, the connection polynomial $L(z)$ has the same length v as the linear-feedback shift register and the rightmost tap of the register is non-zero. Secondly, if the dimension of the matrix $\vec{\vec{S}}$ is smaller than v, $\vec{\vec{S}}$ cannot be inverted and $\vec{L}$ has no solution. This means that the smallest dimension of $\vec{L}$ is v. From the above, we choose the shift register having the shortest length such that the polynomial $L(z)$ has the smallest degree.

The problem becomes one of designing the filter with the shortest length to predict a given sequence of syndromes. If the register length corresponds to the degree of the connection polynomial $L(z)$, the coefficients of $L(z)$ are the solution of the matrix equations expressed in Equation 4.108. Let us now illustrate the design procedure by deriving a minimum length linear-feedback shift register to produce the binary sequence 11101011000110 shown in Figure 4.45. Our approach is inductive, progressively modifying the register at each step. Let $LFSR_i$ denote the linear-feedback shift register design at the i-th iteration. Also, l_i and $L^{(i)}(z)$ denote the register length and the connection polynomial, respectively.

Initially, at step $i = 1$, we use the simplest linear-feedback shift register $LFSR_1$ consisting of a single stage with a direct feedback from its output to its input giving $l_1 = 1$ and $L^{(1)}(z) = z + 1$. The register design $LFSR_1$ is initialized by the first bit of the sequence and generates the first three bits but fails on the fourth. This failure prompts us to modify $LFSR_1$ either in its length or in its tapping positions. It is obvious that a length-2 shift register cannot combine two ones (first and second bits) to produce a one (third bit) and then combine two ones (second and third bits) on the next shift to produce a zero (fourth bit). This suggests that the length of the register must be increased to 3 with an even number of feedback connections.

At step $i = 2$ in Figure 4.45, we use a length-3 shift register $LFSR_2$ (i.e., $l_2 = 3$) with connection polynomial $L^{(2)}(z) = z^3 + z + 1$. It can generate the 4-th, 5-th and 6-th bits correctly, but fails on the 7-th bit. At this point, we modify $LFSR_2$ such that the new register $LFSR_3$ continues to generate from the 7-th bit onward without reproducing the previous correct bits. Suppose l_3 is the length of the shift register at the next step $i = 3$, the register is modified in a way that the correction values added to the $(l_3 + 1)$-th, $(l_3 + 2)$-th, ..., (7-1)-th bits are zero, but the value added to the 7-th bit is one so as to change the generated bit from 0 to 1. In the example shown in Figure 4.45, $l_3 = 4$ and the correction values to the 5-th and 6-th bits are zero, but to the 7-th bit it is a one.

We observe that the linear combination (bitwise exclusive-OR for binary codes) of the generated bit from the shift register and the next bit to be

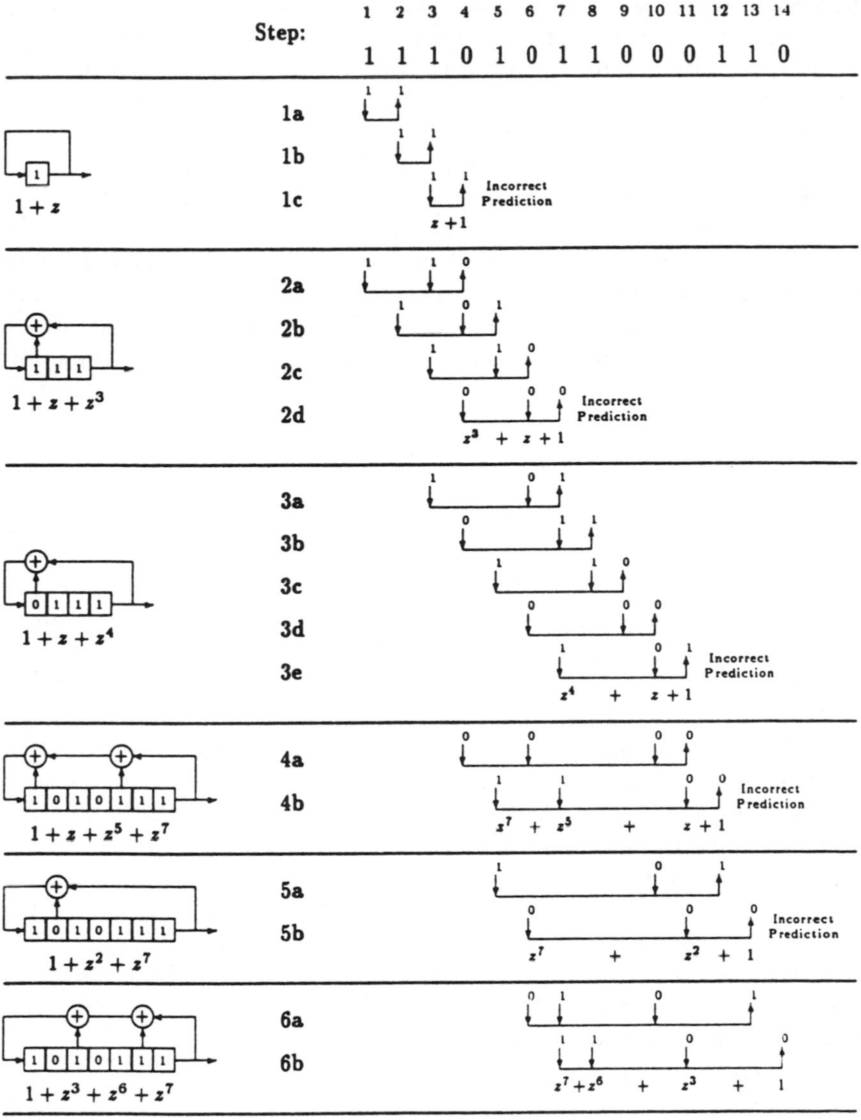

Figure 4.45: Constructing a minimum length shift register by the Berlekamp-Massey algorithm to generate the binary sequence

predicted in the sequence is always zero if the prediction is correct, and is one if the prediction is wrong. For example, at step 1a and 1b in Figure 4.45, the generated bit is 1 and the next bit in the sequence is also 1. The prediction of either steps is correct and therefore the linear combination of these two bits is zero. However, at step 1c, the generated bit is 1 and the fourth bit in the sequence is 0. The prediction is wrong and their linear combination is 1. The result of the non-zero linear combination is essential to generate the correction value about to be added to the output of the $LFSR_2$ design to compensate for its failure to generate the right prediction when a new $LFSR_3$ is formed.

As the bit generated from a shift register is the linear combination of its stages defined by the connection polynomial, the $LFSR_2$'s output can be modified by adding the appropriately shifted the linear combination defined by the connection polynomial $L^{(1)}(z)$ of the $LFSR_1$ to form a new $LFSR_3$. The length of the register is increased to $l_3 = 4$ and the connection polynomial becomes

$$
\begin{aligned}
L^{(3)}(z) &= L^{(1)}(z)z^3 + L^{(2)}(z) \\
&= (z+1)z^3 + (z^3 + z + 1) = z^4 + z + 1 \ .
\end{aligned}
$$

Now, at step 3, the linear combinations due to the tapping defined by the connection polynomial of the $LFSR_1$ at steps 1a and 1b are zero and the next symbol is predicted with zero error. Shifted by three positions according to the multiplication by z^3 this contributes a zero correction to the prediction by $LFSR_3$ at the fifth and the sixth bits. However, when the $LFSR_3$ predicts the seventh bit, the correction value (observed from the $LFSR_1$ at step 1c) is one which changes the previous prediction by the $LFSR_2$ to the current prediction by the $LFSR_3$, i.e., from 0 to 1. Hereafter, the $LFSR_3$ produces the 8-th, 9-th and 10-th bits in the sequence properly, but fails on the 11-th. At this point, the $LFSR_3$ is again modified for step 4.

To proceed with step 4, the $LFSR_3$ can be modified by adding the appropriately shifted non-zero output of either the $LFSR_1$ or the $LFSR_2$ to form a new $LFSR_4$. The length of the $LFSR_4$ will be either 8 or 7, according to the positioning shifts of z^7 or z^4, respectively, and their corresponding connection polynomials are

$$
\begin{aligned}
L^{(4)}(z) &= L^{(1)}(z)z^7 + L^{(3)}(z) \\
&= (z+1)z^7 + (z^4 + z + 1) = z^8 + z^7 + z^4 + z + 1
\end{aligned}
$$

or

$$
\begin{aligned}
L^{(4)}(z) &= L^{(2)}(z)z^4 + L^{(3)}(z) \\
&= (z^3 + z + 1)z^4 + (z^4 + z + 1) = z^7 + z^5 + z + 1 \ .
\end{aligned}
$$

As the criterion of designing the linear-feedback shift register is that it has a minimum length, we choose the $LFSR_2$ to modify the $LFSR_3$. The $LFSR_4$ generates the 11-th bit in the sequence, but fails on the 12-th.

At step 5 in Figure 4.45, the $LFSR_5$ is basically constructed from $LFSR_4$ with modification by $LFSR_3$. The length of $LFSR_5$ is still 7 and the connection polynomial is

$$
\begin{aligned}
L^{(5)}(z) &= L^{(3)}(z)z + L^{(4)}(z) \\
&= (z^4 + z + 1)z + (z^7 + z^5 + z + 1) = z^7 + z^2 + 1 .
\end{aligned}
$$

The register now predicts correctly until it reaches bit 13.

At this point, the $LFSR_5$ can be modified by one of the previous shift registers, $LFSR_1$, $LFSR_2$, $LFSR_3$ and $LFSR_4$. But we choose the one to form the minimum length shift register for step 6. That is $LFSR_3$ which forms a new $LFSR_6$ of length 7. The connection polynomial is

$$
\begin{aligned}
L^{(6)}(z) &= L^{(3)}(z)z^2 + L^{(5)}(z) \\
&= (z^4 + z + 1)z^2 + (z^7 + z^2 + 1) = z^7 + z^6 + z^3 + 1 .
\end{aligned}
$$

The new $LFSR_6$ is able to predict the 13-th bit and also the last bit 14. Finally, a length-7 $LFSR_6$ is the correct register. If it is initialized with the first 7 bits of the sequence, it will then generate the subsequent bits. If the bits in the sequence are the syndromes, the connection polynomial $LFSR_6$ is the error-locator polynomial, and the coefficients of the polynomial constitute the solution for $\vec{L}$ in Equation 4.108. We note that the length of the $LFSR_6$ corresponds to the degree of the connection polynomial $L^{(6)}(z)$ that satifies the criteria of filter design.

Let us now summarize the design procedure of the last example and also generalize it for the design of nonbinary codes. During the process of formulating the shift register, we always determine the register length l_i and the connection polynomial $L^{(i)}(z)$. A correction term $C^{(i)}(z)$ is updated for use at the next step whenever the linear-feedback shift register fails to predict the next symbol correctly. At each step, the shift register predicts the next symbol in the sequence of syndromes. If the prediction is correct, the connection polynomial remains intact and the correction term is multiplied by z. However, if the prediction is wrong, the connection polynomial is modified by adding the correction term. As all the previous connection polynomials are candidates for the correction term, we choose the one to give the minimum register length.

If the length of the register has increased after modification, the previous connection polynomial is multiplied by z and becomes the current correction term. For example, the prediction of the shift register at step 3e in Figure 4.45 is wrong. The connection polynomial $L^{(3)}(z)$ is therefore modified by adding the correction term $L^{(2)}(z)z^4$ to form $L^{(4)}(z)$ for the new register at the next step 4. The length of the register has extended from 4 to 7 stages, and the previous connection polynomial $L^{(3)}(z)$ is multiplied by z and is maintained as the correction term for the next step 4b. However, if the length of the register remains the same after the modification, the current correction term is multiplied by z and is kept for the next step.

For example, the correction term at step 4b in Figure 4.45 is $L^{(3)}(z)z$. The next prediction of the shift register is wrong, and the connection polynomial $L^{(4)}(z)$ is modified by adding the correction term $L^{(3)}(z)z$ to form the $L^{(5)}(z)$ for step 5. The length of the new register remains the same with 7 stages, and the correction term is maintained and updated by multiplying by z to become $L^{(3)}(z)z^2$ for step 5b.

In the example, the bit sequence is binary. The prediction of the shift register is either 0 or 1. If the prediction is wrong, the discrepancy is always one. However, for the sequence of syndromes defined on $GF(q)$, where $q > 2$, the discrepancy can be any value greater than zero and less than q. In this case, the connection polynomial of the shift register produces a wrong prediction and has a discrepancy d with the next symbol in the sequence. If the connection polynomial is selected as the correction term it is normalized such that the discrepancy is equal to one. When the shift register is modified at the subsequent step, this normalized term is then multiplied by the value of the discrepancy for the current prediction.

From our deliberations, we see that the Berlekamp-Massey algorithm is an iterative process of predicting $2t$ syndroms. Each iteration predicts one syndrom and the process lasts for $2t$ iterations. Let $L^{(i)}(z)$ and $C^{(i)}(z)$ be the connection polynomial and the correction polynomial, respectively, at i-th iteration, $l^{(i)}$ be the length of the linear-feedback shift register, $j^{(i)}$ be the displacement of the correction polynomial from the beginning of the syndrome sequence (i.e., the location of the oldest symbol of the polynomial in the sequence), and the discrepancy in the prediction is $d^{(i)}$.

1. Initially, the values are set as

$$i = 1, \quad j^{(0)} = 0, \quad l^{(0)} = 0, \quad L^{(0)}(z) = 1 \text{ and } C^{(0)}(z) = z .$$

2. The discrepancy of the prediction is evaluated as

$$d^{(i)} = S_i + \sum_{k=1}^{l^{(i)}} L_k^{(i)} S_{i-k} .$$

3. If the discrepancy is zero, the prediction at the current iteration i is correct and the correction polynomial is updated by

$$C^{(i+1)}(z) = zC^{(i)}(z) .$$

The register is left unchanged so that

$$L^{(i+1)}(z) = L^{(i)}(z), \quad l^{(i+1)} = l^{(i)} \text{ and } j^{(i+1)} = j^{(i)} .$$

The register then continues to predict the next symbol and goes back to step 2 to calculate the discrepancy.

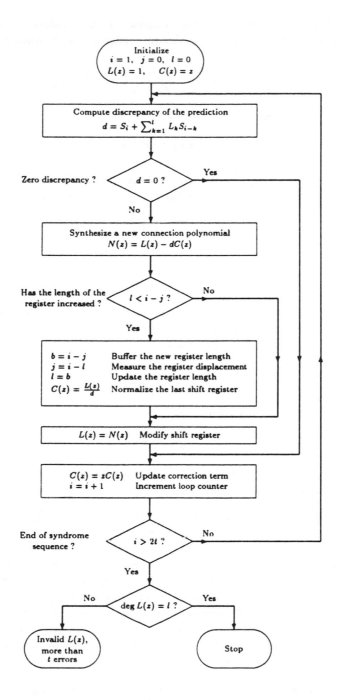

Figure 4.46: Computer flow diagram of the Berlekamp-Massey algorithm

4. However, if the discrepancy is non-zero, the prediction is wrong. The connection polynomial of the shift register is modified according to the following expression

$$L^{(i+1)}(z) = L^{(i)}(z) - d^{(i)}C^{(i)}(z)$$

and the length of the new register becomes

$$l^{(i+1)} = \max\left(l^{(i)}, i - j^{(i)}\right) .$$

(a) If the length of the new register has not increased, i.e., $l^{(i+1)} = l^{(i)}$, the last correction polynomial is retained and is updated as

$$C^{(i+1)}(z) = zC^{(i)}(z)$$

and its displacement remains the same,

$$j^{(i+1)} = j^{(i)} .$$

(b) If the length of the new register has increased, i.e., $l^{(i+1)} > l^{(i)}$, the last connection polynomial is normalized and becomes the new correction polynomial,

$$C^{(i+1)}(z) = z\frac{L^{(i)}(z)}{d^{(i)}}$$

and its displacement is,

$$j^{(i+1)} = i - l^{(i)} .$$

The new register then predicts the next symbol and repeats step 2 to calculate the discrepancy of the prediction.

The algorithm is depicted in the form of the computer flow diagram illustrated in Figure 4.46. Notice that the calculation of the discrepancy $d^{(i)}$ requires not more than t multiplications for each iteration, and also that the linear-feedback shift register and the correction polynomial update require at most $2t$ multiplications per iteration. The procedure iterates $2t$ times and therefore the algorithm requires at most $(t + 2t)2t = 6t^2$ multiplications. By constrast, the direct matrix inversion method derived from the Peterson-Gorenstein-Zierler method requires on the order of t^3 multiplications. The low complexity of the Berlekamp-Massey algorithm makes it perferable to the direct matrix inversion approach suggested by Peterson, if $t \geq 6$.

Let us now consider the previous examples in Section 4.4.5.2 to illustrate the procedure of the Berlekamp-Massey algorithm. The first example is the triple error correcting $BCH(15, 5, 3)$ code whose syndromes are given in Equation 4.112, while the procedures of determining the

i	$d^{(i)}$	$L^{(i)}(z)$	$C^{(i)}(z)$	$l^{(i)}$
1	α^0	$1+\alpha^0 z$	z	1
2	0	$1+\alpha^0 z$	z^2	1
3	α^4	$1+\alpha^0 z+\alpha^4 z^2$	$\alpha^{11}z+\alpha^{11}z^2$	2
4	0	$1+\alpha^0 z+\alpha^4 z^2$	$\alpha^{11}z^2+\alpha^{11}z^3$	2
5	0	$1+\alpha^0 z+\alpha^4 z^2$	$\alpha^{11}z^3+\alpha^{11}z^4$	2
6	0	$1+\alpha^0 z+\alpha^4 z^2$	$\alpha^{11}z^4+\alpha^{11}z^5$	2

Table 4.15: Computing the $L(z)$ of the $BCH(15,5,3)$ code by Berlekamp-Massey algorithm

i	$d^{(i)}$	$L^{(i)}(z)$	$C^{(i)}(z)$	$l^{(i)}$
1	α^7	$1+\alpha^7 z$	$\alpha^8 z$	1
2	α^5	$1+\alpha^5 z$	$\alpha^8 z^2$	1
3	α^{14}	$1+\alpha^5 z+\alpha^7 z^2$	$\alpha^1 z+\alpha^6 z^2$	2
4	α^{11}	$1+\alpha^{14} z+\alpha^{12} z^2$	$\alpha^1 z^2+\alpha^6 z^3$	2
5	α^2	$1+\alpha^{14} z+\alpha^{10} z^2+\alpha^8 z^3$	$\alpha^{13}z+\alpha^{12}z^2+\alpha^{10}z^3$	3
6	α^2	$1+\alpha^3 z+\alpha^{11} z^2+\alpha^9 z^3$	$\alpha^{13}z^2+\alpha^{12}z^3+\alpha^{10}z^4$	3

Table 4.16: Computing the $L(z)$ polynomial of the $RS(15,9,3)$ code by Berlekamp-Massey algorithm

error-locator polynomial by Berlekamp-Massey algorithm are shown in Table 4.15. The resulting $L(z)$ polynomial of the $BCH(15,5,3)$ code obtained by the Berlekamp-Massey algorithm can be verified by that of the Peterson method in Section 4.4.5.2. The second example is the nonbinary triple error correcting $RS(15,9,3)$ code over $GF(16)$. The procedure is illustrated in Table 4.16.

4.4.5.4 Forney Algorithm

The Peterson-Gorenstein-Zierler decoder described in Section 4.4.5.2, provides a method of decoding BCH codes. The method requires two v by v matrix inversions which consumes substantial computational power. The first matrix inversion evaluates the error-locator polynomial, and this can be circumvented by employing the computationally efficient Berlekamp-Massey algorithm. The second matrix inversion involves the evaluation of the error magnitudes, which can be carried out more efficiently using the Forney algorithm. In this Section, we commence by deriving the Forney algorithm and then illustrate the procedure with an example.

We recall from Equation 4.101 that the error-locator polynomial $L(z)$ is defined to have zeros at the inverse error positions P_i^{-1} for $i = 1, 2, \ldots, v$, namely,

$$L(z) = \prod_{i=1}^{v}(1 - zP_i) = (1 - zP_1)(1 - zP_2)\cdots(1 - zP_v) \ . \qquad (4.114)$$

We now define the *syndrome polynomial* $S(z)$ as

$$S(z) \triangleq \sum_{j=1}^{2t} S_j z^j = S_1 z + S_2 z^2 + \cdots + S_{2t} z^{2t} \ . \qquad (4.115)$$

The coefficient S_j of the polynomial is the syndrome, and from Equation 4.99 the syndrome polynomial can also be expressed as

$$S(z) = \sum_{j=1}^{2t} \sum_{i=1}^{v} M_i P_i^j z^j \ . \qquad (4.116)$$

We also define the *error-evaluator polynomial* $E(z)$ as the product of the syndrome polynomial $S(z)$ and the error-locator polynomial $L(z)$,

$$E(z) \triangleq S(z)L(z) \pmod{z^{2t}} \qquad (4.117)$$

and its expanded form is given by

$$E(z) = E_{2t-1} z^{2t-1} + \cdots + E_2 z^2 + E_1 z \ . \qquad (4.118)$$

Substituting Equation 4.116 and 4.114 into Equation 4.117, yields

$$
\begin{aligned}
E(z) &= \sum_{j=1}^{2t} \sum_{i=1}^{v} M_i P_i^j z^j \prod_{l=1}^{v}(1 - zP_l) \pmod{z^{2t}} \\
&= \sum_{i=1}^{v} M_i P_i z \sum_{j=1}^{2t}(P_i z)^{j-1} \prod_{l=1}^{v}(1 - zP_l) \pmod{z^{2t}} \\
&= \sum_{i=1}^{v} M_i P_i z \left[(1 - zP_i)\sum_{j=1}^{2t}(P_i z)^{j-1}\right] \prod_{l \neq i}(1 - zP_l) \pmod{z^{2t}}
\end{aligned}
$$

By extending and simplifying the square bracketed term we get:

$$
\begin{aligned}
E(z) &= \sum_{i=1}^{v} M_i P_i z \left[1 - (P_i z)^{2t}\right] \prod_{l \neq i}(1 - zP_l) \pmod{z^{2t}} \\
&= \sum_{i=1}^{v} M_i P_i z \prod_{l \neq i}(1 - zP_l) \ . \qquad (4.119)
\end{aligned}
$$

This error-evaluator polynomial has the following values at the inverse error positions P_l^{-1}:

$$
\begin{aligned}
E(P_l^{-1}) &= \sum_{i=1}^{v} M_i P_i P_l^{-1} \prod_{j \neq i} (1 - P_l^{-1} P_j) \\
&= M_1 P_1 P_l^{-1} \prod_{j \neq 1} (1 - P_l^{-1} P_j) + \cdots + M_l \prod_{j \neq l} (1 - P_l^{-1} P_j) \\
&\quad + \cdots + M_v P_v P_l^{-1} \prod_{j \neq v} (1 - P_l^{-1} P_j) \\
&= M_l \prod_{j \neq l} (1 - P_l^{-1} P_j) .
\end{aligned}
\tag{4.120}
$$

Using the product rule for differentiation, we evaluate the derivative of the error-locator polynomial $L(z)$ given by Equation 4.114 as

$$
L'(z) = \sum_{i=1}^{v} -P_i \prod_{j \neq i} (1 - z P_j)
\tag{4.121}
$$

and at the inverse error position $z = P_l^{-1}$,

$$
\begin{aligned}
L'(P_l^{-1}) &= -\sum_{i=1}^{v} P_i \prod_{j \neq i} (1 - P_i^{-1} P_j) \\
&= -P_1 \prod_{j \neq 1} (1 - P_1^{-1} P_j) - \cdots - P_l \prod_{j \neq l} (1 - P_l^{-1} P_j) \\
&\quad - \cdots - P_v \prod_{j \neq v} (1 - P_v^{-1} P_j) \\
&= -P_l \prod_{j \neq l} (1 - P_l^{-1} P_j) .
\end{aligned}
\tag{4.122}
$$

From Equations 4.120 and 4.122,

$$
\frac{E(P_l^{-1})}{L'(P_l^{-1})} = \frac{M_l}{-P_l} .
\tag{4.123}
$$

Rearranging Equation 4.123, we find the magnitude of the error symbol at position P_l to be,

$$
M_l = \frac{-E(P_l^{-1})}{P_l^{-1} L'(P_l^{-1})} = \frac{E(P_l^{-1})}{\prod_{j \neq l} (1 - P_l^{-1} P_j)} .
\tag{4.124}
$$

As the error-evaluator polynomial $E(z)$ is defined to be the product of the error-locator polynomial $L(z)$ and the syndrome polynomial $S(z)$,

$$
\begin{array}{r|lllll}
S_6 z^6 + & S_5 z^5 + & S_4 z^4 + & S_3 z^3 + & S_2 z^2 + & S_1 z \\
          &   & L_3 z^3 + & L_2 z^2 + & L_1 z + & 1 \\
\hline
S_6 z^6 + & S_5 z^5 + & S_4 z^4 + & S_3 z^3 + & S_2 z^2 + & S_1 z \\
L_1 S_6 z^7 + L_1 S_5 z^6 + & L_1 S_4 z^5 + L_1 S_3 z^4 + L_1 S_2 z^3 + L_1 S_1 z^2 \\
L_2 S_6 z^8 + L_2 S_5 z^7 + L_2 S_4 z^6 + & L_2 S_3 z^5 + L_2 S_2 z^4 + L_2 S_1 z^3 \\
L_3 S_6 z^9 + L_3 S_5 z^8 + L_3 S_4 z^7 + L_3 S_3 z^6 + & L_3 S_2 z^5 + L_3 S_1 z^4 \\
\hline
          & E_5 z^5 + & E_4 z^4 + & E_3 z^3 + & E_2 z^2 + & E_1 z \qquad (0.1)
\end{array}
$$

Figure 4.47: Long multiplication of $S(z)$ and $L(z)$

the coefficients of $E(z)$ can be expressed as the product of L_i and S_i. Consider the example of a triple error correcting code having a degree-5 error-evaluator polynomial whose coefficients are given in Figure 4.47:

$$E(z) = E_5 z^5 + E_4 z^4 + E_3 z^3 + E_2 z^2 + E_1 z \qquad (4.125)$$

Similarly, for a t error correcting code in general, the error-evaluator polynomial $E(z)$ is formulated with coefficients

$$E_j = S_j + \sum_{k=1}^{t} S_{j-k} L_k \qquad j = 1, \ldots, (2t-1) \qquad (4.126)$$

where $S_j = 0$ for $j \leq 0$.

Let us now illustrate the Forney algorithm by evaluating the error magnitudes of the nonbinary triple error correcting $RS(15, 9, 3)$ code over $GF(16)$, as previously discussed in Section 4.4.5.2. The syndrome polynomial $S(z)$ is written as

$$S(z) = \alpha^2 z^6 + \alpha^3 z^5 + \alpha^8 z^4 + \alpha^{13} z^3 + \alpha^{12} z^2 + \alpha^7 z$$

and the error-locator polynomial $L(z)$ is given by

$$L(z) = \alpha^9 z^3 + \alpha^{11} z^2 + \alpha^3 z + 1 .$$

By using the Equation 4.126, the coefficients of the error-evaluator polynomial $E(z)$ are evaluated as,

$$
\begin{aligned}
E_1 &= S_1 = \alpha^7 \\
E_2 &= S_2 + L_1 S_1 = \alpha^{12} + \alpha^3 \alpha^7 = \alpha^3 \\
E_3 &= S_3 + L_1 S_2 + L_2 S_1 = \alpha^{13} + \alpha^3 \alpha^{12} + \alpha^{11} \alpha^7 = \alpha^2 \\
E_4 &= S_4 + L_1 S_3 + L_2 S_2 + L_3 S_1 = \alpha^8 + \alpha^3 \alpha^{13} + \alpha^{11} \alpha^{12} + \alpha^9 \alpha^7 = 0 \\
E_5 &= S_5 + L_1 S_4 + L_2 S_3 + L_3 S_2 = \alpha^3 + \alpha^3 \alpha^8 + \alpha^{11} \alpha^{13} + \alpha^9 \alpha^{12} = 0
\end{aligned}
$$

giving

$$E(z) = \alpha^2 z^3 + \alpha^3 z^2 + \alpha^7 z .$$

The error positions are located by using either the Peterson-Gorenstein-Zierler method described in Section 4.4.5.2 or the Berlekamp-Massey algorithm presented in Section 4.4.5.3, and are

$$P_1 = \alpha^2, \qquad P_2 = \alpha^8 \text{ and } P_3 = \alpha^{14}$$

and their corresponding inverse values are

$$P_1^{-1} = \alpha^{13}, \qquad P_2^{-1} = \alpha^7 \text{ and } P_3^{-1} = \alpha^1 .$$

By using Equation 4.124, the error magnitudes are evaluated using the Forney algorithm as

$$M_1 = \frac{E(P_1^{-1})}{(1 - P_2 P_1^{-1})(1 - P_3 P_1^{-1})} = \frac{\alpha^2 \alpha^{39} + \alpha^3 \alpha^{26} + \alpha^7 \alpha^{13}}{(1 - \alpha^8 \alpha^{13})(1 - \alpha^{14} \alpha^{13})} = \alpha^6$$

$$M_2 = \frac{E(P_2^{-1})}{(1 - P_1 P_2^{-1})(1 - P_3 P_2^{-1})} = \frac{\alpha^2 \alpha^{21} + \alpha^3 \alpha^{14} + \alpha^7 \alpha^7}{(1 - \alpha^2 \alpha^7)(1 - \alpha^{14} \alpha^7)} = \alpha^{13}$$

$$M_3 = \frac{E(P_3^{-1})}{(1 - P_1 P_3^{-1})(1 - P_2 P_3^{-1})} = \frac{\alpha^2 \alpha^3 + \alpha^3 \alpha^2 + \alpha^7 \alpha^1}{(1 - \alpha^2 \alpha^1)(1 - \alpha^8 \alpha^1)} = \alpha^2 .$$

The error-evaluator polynomial $E(z)$ can be found from the error-locator polynomial $L(z)$ by multiplying it by the syndrome polynomial $S(z)$. Notice however, that a sequence of recursively updated polynomials $E^{(i)}(z)$ can be defined to obey the same recursive relationship as $L^{(i)}(z)$. In this way, one could use the Berlekamp-Massey algorithm to obtain $E(z)$ in parallel to $L(z)$. Let us therefore define ε by:

$$E(z) \triangleq z\varepsilon(z) . \tag{4.127}$$

The Berlekamp-Massey algorithm applied to $L(z)$ is also deployed to obtain $\varepsilon(z)$. The initial value of $\varepsilon^{(0)}(z) = 0$, and the correction term $D^{(0)}(z) = -1$. Then, $\varepsilon(z)$ is updated to

$$\varepsilon^{(i+1)}(z) = \varepsilon^{(i)}(z) - d^{(i)} D^{(i)}(z) .$$

If $l^{(i+1)} = l^{(i)}$, the correction polynomial is updated to

$$D^{(i+1)}(z) = zD^{(i)}(z)$$

and if $l^{(i+1)} > l^{(i)}$, it becomes

$$D^{(i+1)}(z) = z \frac{\varepsilon^{(i)}(z)}{d^{(i)}} .$$

An example of finding the error-evalutor polynomial by the Berlekamp-Massey algorithm is illustrated with reference to Table 4.17. This is again the example of the $RS(15, 9, 3)$ code over $GF(16)$. At the end of $2t = 6$ iterations, $\varepsilon(z)$ is found, and from Equation 4.127 the error-evaluator polynomial is obtained as

$$E(z) = \alpha^7 z + \alpha^3 z^2 + \alpha^2 z^3 .$$

i	$d^{(i)}$	$\varepsilon^{(i)}(z)$	$D^{(i)}(z)$	$l^{(i)}$
1	α^7	α^7	0	1
2	α^5	α^7	0	1
3	α^{14}	α^7	$\alpha^8 z$	2
4	α^{11}	$\alpha^7 + \alpha^4 z$	$\alpha^8 z^2$	2
5	α^2	$\alpha^7 + \alpha^4 z + \alpha^{10} z^2$	$\alpha^5 z + \alpha^2 z^2$	3
6	α^2	$\alpha^7 + \alpha^3 z + \alpha^2 z^2$	$\alpha^5 z^2 + \alpha^2 z^3$	3

Table 4.17: Computing the $\varepsilon(z)$ polynomial of the $RS(15, 9, 3)$ code by Berlekamp-Massey algorithm

4.4.6 Trellis Decoding for Block Codes

In our quest for decoding algorithms making use of the channel measurement information [50], we will now investigate the performance of trellis decoding [51, 52] of both binary and non-binary block codes. We commence by considering the trellis construction for binary BCH codes, and then applying the Viterbi algorithm to decode them. The trellis construction can easily be extended to non-binary codes. A t error correcting code, is represented by $BCH(n, k, t)$, where k information bits are encoded into n-bit codewords. The code is defined over $GF(2)$. In the case of systematic codes, the generator polynomial $g(z) = g_{n-k}z^{n-k} + \cdots + g_1 z + g_0$ formulates a codeword by appending $(n - k)$ parity bits $b_{n-k-1}, \ldots, b_1, b_0$, to k information bits.

4.4.6.1 Trellis Construction

To illustrate the trellis construction we commence by considering the example of the $BCH(15, 11, 1)$ code over $GF(2)$. The systematic encoder, displayed in Figure 4.48, is derived from Figure 4.40, and the trellis diagram of this code is shown in Figure 4.49. The generator polynomial, $g(z) = z^4 + z + 1$, produces parity bits b_0, b_1, b_2, and b_3 which are buffered in the shift register (SR). The values of the set $\{b_0, b_1, b_2, b_3\}$ represent the states of the register. The four parity bits result in 2^4 different states in the SR. The sequential change of states during the process of encoding a codeword can be cataloged as a particular path through the trellis. There are 2^{11} unique paths in the trellis and each path represents a particular codeword. The trellis has 2^{15-11} rows and $(15 + 1)$ columns. The nodes on the same row represent the same state, whereas the nodes on the same column illustrate all the possible states a, b, ..., p with their corresponding values 0000, 0001, ..., 1111. The state-changes between adjacent columns in the trellis are marked by the transition vector. In Figure 4.49, the vectors are drawn either by a solid line or by a dashed line according to whether the encoder input is a 0 or 1, respectively.

Initially all the parity bits in the SR are set to zero. The number of

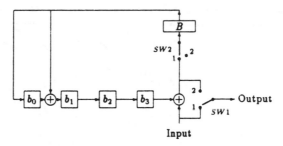

Figure 4.48: Systematic encoder for the $BCH(15, 7, 2)$ code

encoder states increases as each new information bit is inserted into the encoder. The symbol signalling instants corresponding to the column positions in the trellis, shown in Figure 4.49, are indexed by the integer J. On inserting the first information bit into the encoder, $J = 0$, and two different nodes are possible at the next instant. The arrival of the second information bit when $J = 1$ causes the number of possible nodes at the next instant to increase to 2^2. The number of possible nodes continues to increase with J until the maximum number 2^4 is reached. This maximum number of states is reached when $J = 4$, and from then on the number of possible states is constant. The state transitions pattern from instant $J = i$ to $J = i + 1$, shown in Figure 4.49, repeats for $J = 4, 5, \ldots, 10$, until the last information bit to be encoded has entered the SR at $J = 11$. At this moment, the switches $SW1$ and $SW2$ are turned to position 2, and the SR is cleared one parity bit at a time as the bits are removed from the encoder to leave behind the all-zero state in the SR. The number of possible states is thus divided by two at every column in the trellis merging towards the all-zeros state, which is reached after clocking the encoder 12 times.

The trellis for the $BCH(15, 11, 1)$ code is suitable for maximum likelihood decoding by the Viterbi algorithm [20, 33] in order to improve the BER performance. A further improvement in performance can be achieved by using soft-decision decoding.

4.4.6.2   Trellis Decoding

The method of trellis decoding of block codes is similar to Viterbi decoding of convolutional codes. The block decoder selects the path in the trellis having the smallest distance from the received word and thereby identifies the recovered codeword. The distance properties of the code determine its error correcting capability.

The distance properties of the $BCH(15, 7, 2)$ binary block code are derived in a similar way to those of convolutional codes, which suggests that this code can correct any combination of two bit errors. For convolutional codes the number of incorrect paths increases exponentially and indefinitely

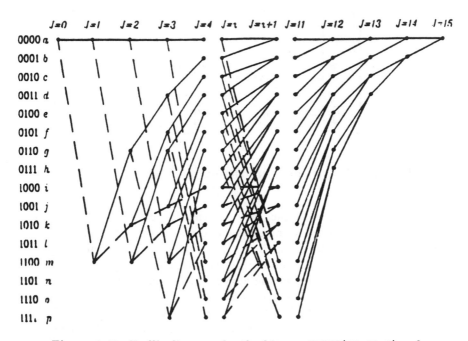

Figure 4.49: Trellis diagram for the binary $BCH(15, 11, 1)$ code

with the number of columns j in the trellis. However, block codes have a fixed codeword length, and consequently the trellis is truncated after n columns. The paths in the trellis initially diverge from, and finally converge to, the all-zero state at the codeword boundaries. All the paths have the same length of n bits. There is a total number of 2^k possible paths in the trellis and the weight distribution, $W_{BCH1572}(d)$, of the $BCH(15, 7, 2)$ code was found by computer search through all of these paths as

$$W_{BCH1572}(d) = 42d^5 + 84d^6 + 49d^7 + 56d^8 + 126d^9 + 84d^{10} + 7d^{15} , \quad (4.128)$$

giving a total of 42 error bits for all distance-5 paths, 84 for all distance-6 paths, etc. The minimum separable distance, d_{min}, between the codewords is 5 bits. We will represent the total number of information bit errors for all those paths having a distance d by the coefficient W_d. Following the procedure of Section 4.3.6, the union bound on the post-decoding bit error probability p_{bp} for binary block codes is

$$p_{bp} \leq \sum_{d=d_{min}}^{n} W_d P_{ICD}(d) \qquad (4.129)$$

where $P_{ICD}(d)$ is the probability of incorrect decoding, i.e., the probability that the decoder selects a path at distance d from the correct path. The

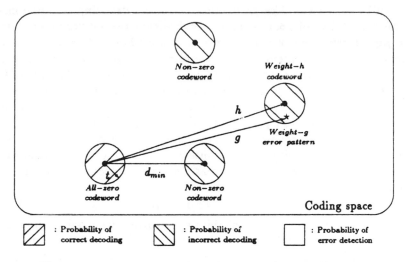

Figure 4.50: Representation of codewords in coding space

value of $P_{ICD}(d)$ is found from Equations 4.36 and 4.37 for hard-decision decoding, or from Equation 4.46 for soft-decision decoding. Reed-Solomon codes have a maximum separable distance of $(n - k + 1)$ symbols amongst codewords. Optimum symbol-by-symbol decoding methods [53, 54] that minimize the symbol error rate result in increased complexity compared to Viterbi decoding, while their performance is essentially the same. Consequently, we decode using a bit oriented Viterbi algorithm and describe the properties of the RS code in terms of bit-distance measures. For the $RS(4, 2, 1)$, $GF(16)$ code the weight distribution $W_{RS421}(d)$ is found by computer search to be

$$W_{RS421}(d) = 14d^4 + 58d^5 + 86d^6 + 134d^7 + 210d^8 + 218d^9 + 170d^{10}$$
$$+ 82d^{11} + 24d^{12} + 20d^{13} + 8d^{14}. \tag{4.130}$$

The union bound on the post-decoding bit error probability p_{bp} for this code is obtained by substituting $W_{RS421}(d)$ into Equation 4.129.

4.4.7 Block Decoding Theory

In this Section we consider the abilities of block codes to combat transmission errors. As the linear block code is considered here, we may analyse its behaviour by assuming that a particular codeword is transmitted, say the all-zero codeword, knowing that our findings are applicable for the transmission of any other codeword.

The concept of geometric coding space is particularly useful for visualizing the decoding situation. Figure 4.50 represents the coding space containing $(q^m)^n$ words of which $(q^m)^k$ are legitimate codewords. If the received word contains $\leq t$ errors, it lies within the all-zero codeword sphere and can be corrected. The probability of correct decoding is represented by the probability of codewords received within the all-zero codeword's decoding sphere. However, if the received word has $> t$ errors, it is uncorrectable and subsequently results in either incorrect decoding or error detection. In the former case [55], the received word falls within one of the non-zero codeword decoding spheres and is therefore erroneously decoded to be a legitimate non-zero codeword. The probability of incorrect decoding is constituted by the probability of codewords being received in the union of the non-zero codeword spheres. The probability of error detection is that of codewords being received outside the union of all the legitimate decoding spheres. Thus we may express the error detection probability of a codeword as

$$P_{ED} = 1 - P_{CD} - P_{ICD} \qquad (4.131)$$

where P_{CD} and P_{ICD} are the probabilities of correctly decoding and incorrectly decoding into another valid codeword, respectively. Considering that ratio of those uncorrectable error words which are detectable, we express this relative error detection probability as

$$P_{EDR} = \frac{P_{ED}}{P_{ED} + P_{ICD}} . \qquad (4.132)$$

Observe that for codes with high error correcting capability t the minimum coding space separation $(2t + 1)$ amongst legitimate codewords is high, consequently $P_{ICD} \ll 1$ and $P_{EDR} \sim 1$.

The block codes considered here are over $GF(q^m)$, where $q = 2$ and m is the number of bits in a symbol. The m-bit symbols are transmitted sequentially over a binary channel, which is modelled as an asymmetric memoryless channel as discussed in Section 4.2.5. The probability of receiving an error symbol for this channel is given by Equation 4.3 and the probability of receiving a symbol with i bit errors is expressed by Equation 4.2. We now derive analytical expressions for the probability of correct decoding, incorrect decoding, error detection, and subsequently the probability of post-decoding bit and symbol errors.

4.4.7.1 Probability of Correct Decoding

An (n, k, t) block code defined over $GF(q^m)$ with minimum distance $d_{min} = n - k + 1$ is able to correct t symbol errors. Hence, the probability of correct decoding P_{CD} is the probability of receiving an n-symbol word having t or fewer symbol errors and is given by

$$P_{CD} = \sum_{i=0}^{t} \binom{n}{i} [1 - (1 - p_b)^m]^i [(1 - p_b)^m]^{n-i} \qquad (4.133)$$

The index i is the number of error symbols in a codeword and ranges from zero to t. There are $\binom{n}{i}$ possible error patterns, and $[1 - (1 - p_b)^m]^i$ is the probability of i symbols being received in error, while $[(1 - p_b)^m]^{n-i}$ is the probability of $(n - i)$ symbols being received correctly. Notice that for channels having a high SNR, p_b is very small and P_{CD} approaches unity.

4.4.7.2 Probability of Incorrect Decoding

If the received word contains more than t error symbols, it is either decoded incorrectly, or an error detection flag is raised after identifying uncorrectable errors in the received word. The probability of incorrect decoding [10] is

$$P_{ICD} = \sum_{h=d_{min}}^{n} P_{ICD}(h) \qquad (4.134)$$

where $P_{ICD}(h)$ is the probability of incorrect decoding to a codeword having a symbol distance h from the all-zero codeword. In order to determine the probability of incorrectly decoding to a weight-h codeword, we define $N_{g,s}(h)$ as the number of weight-g error patterns that are at a distance s from a particular weight-h codeword. When $s \leq t$ and $h \geq d_{min}$, each such error pattern is decoded incorrectly into the weight-h codeword, as seen in Figure 4.50. Suppose $P(g)$ is the probability of occurence of a particular weight-g error pattern and A_h is the number of weight-h codewords, then the probability of incorrectly decoding as a weight-h codeword is the sum of all the probabilities $P(g)$ of weight-g error patterns which lie inside the decoding sphere of a particular weight-h codeword, that is,

$$P_{ICD}(h) = A_h \sum_{s=0}^{t} \sum_{g=h-s}^{h+s} N_{g,s}(h)P(g), \qquad 2t + 1 \leq h \leq n. \qquad (4.135)$$

Clearly, we have to determine A_h, $P(g)$ and $N_{g,s}(h)$ to be able to compute $P_{ICD}(h)$ in Equation 4.135. As we have seen in Equation 4.133, the probability p_g of having exactly g number of erroneous symbols out of the n symbols of a codeword is given by:

$$p_g = \begin{pmatrix} n \\ g \end{pmatrix} [1 - (1 - p_b)^m]^g [(1 - p_b)^m]^{n-g}. \qquad (4.136)$$

All the weight-g error polynomials are equiprobable in case of memoryless channels, since the errors can occur in any arbitrary position with the same probability. Also, as expected for any probability:

$$\sum_{g=0}^{n} p_g = 1.$$

The number of weight-g error patterns is found to be

$$N(g) = \begin{pmatrix} n \\ g \end{pmatrix} (2^m - 1)^g. \qquad (4.137)$$

Whence the probability of a specific weight-g error polynomial $P(g)$ is yielded as:

$$P(g) = \frac{p_g}{N(g)} = \frac{1}{(2^m - 1)^g}[1 - (1 - p_b)^m]^g[(1 - p_b)^m]^{n-g}. \qquad (4.138)$$

An alternative expression for $P(g)$ is derived as follows. Suppose an all-zero (n, k, t) codeword is transmitted. The probability of occurrence of a single error symbol, which we will refer to as producing a weight-1 error pattern, is

$$p_1 = \binom{n}{1} \cdot \sum_{i_1=1}^{m} \binom{m}{i_1} p_b^{i_1} (1 - p_b)^{mn-i_1} . \qquad (4.139)$$

The probability of a weight-2 error pattern is:

$$p_2 = \binom{n}{2} \sum_{i_1=1}^{m} \sum_{i_2=1}^{m} \binom{m}{i_1} \binom{m}{i_2} p_b^{i_1+i_2} (1 - p_b)^{mn-(i_1+i_2)} . \qquad (4.140)$$

Therefore the probability of having an arbitrary weight-g error pattern is given as:

$$p_g = \binom{n}{g} \sum_{i_1=1}^{m} \sum_{i_2=1}^{m} \cdots \sum_{i_g=1}^{m} \left[\binom{m}{i_1} \binom{m}{i_2} \cdots \binom{m}{i_g} \right.$$
$$\left. p_b^{i_1+i_2+\cdots+i_g} (1 - p_b)^{mn-(i_1+i_2+\cdots+i_g)} \right] . \qquad (4.141)$$

The probability of receiving a particular weight-g error pattern out of the possible $N(g) = \binom{h}{g} \cdot (2^m - 1)^g$ such patterns is found upon dividing p_g, by $N(g)$, yielding:

$$P(g) = \frac{1}{(2^m - 1)^g} \sum_{i_1=1}^{m} \sum_{i_2=1}^{m} \cdots \sum_{i_g=1}^{m}$$
$$\left[\binom{m}{i_1} \binom{m}{i_2} \cdots \binom{m}{i_g} p_b^{i_1+i_2+\cdots+i_g} \right.$$
$$\left. (1 - p_b)^{mn-(i_1+i_2+\cdots+i_g)} \right] . \qquad (4.142)$$

Having determined $P(g)$ to be used in Equation 4.135 we now evaluate the number of received words corrupted by a weight-g error pattern, having a distance s to the weight-h codeword, i.e. $N_{g,s}(h)$. First we define e_i and c_i to represent the symbol at the i-th position of the error pattern and the weight-h codeword, respectively. By observing the values of the symbols between the error pattern and the codeword, the symbols e_i and c_i can have the following relationships,

$$c_i \; = \; e_i = 0$$
$$c_i \; = \; e_i \neq 0$$
$$c_i \; = \; 0, \quad e_i \neq 0$$
$$c_i \; \neq \; 0, \quad e_i = 0$$
$$c_i \; \neq \; 0, \quad e_i \neq 0, \quad c_i \neq e_i \; .$$

We define [10, 56, 57] the following new variables.

1. w is the number of symbols when $c_i = e_i \neq 0$.

2. x is the number of symbols when $c_i \neq e_i$ and $e_i \neq 0$, $c_i \neq 0$.

3. y is the number of symbols when $c_i \neq 0$ and $e_i = 0$.

4. z is the number of symbols when $c_i = 0$ and $e_i \neq 0$.

Furthermore, we know that

1. g is the number of symbols when $e_i \neq 0$.

2. h is the number of symbols when $c_i \neq 0$.

From the above definitions, the variables obey the following equations,

$$g \; = \; w + x + z \tag{4.143}$$
$$h \; = \; w + x + y \tag{4.144}$$
$$s \; = \; x + y + z. \tag{4.145}$$

Consider now the total number of weight-g error words having a distance s from the weight-h codeword in terms of w, x, y and z. Observe from Figure 4.51 that there are $\dbinom{h}{w}$ ways of $c_i = e_i \neq 0$; $\dbinom{h-w}{x}$ ways of $c_i \neq e_i$, $e_i \neq 0$ and $c_i \neq 0$ where every e_i may take one of $(2^m - 2)$ possible values, excluding $e_i \neq 0$ and $e_i \neq c_i$; and finally $\dbinom{n-h}{z}$ ways of being $c_i = 0$ and $e_i \neq 0$, where every e_i may take one of $(2^m - 1)$ $e_i \neq 0$ possible values. As a result,

$$N_{g,s}(h \mid w, x, y, z) = \binom{h}{w} \binom{h-w}{x} (2^m - 2)^x \binom{n-h}{z} (2^m - 1)^z \; .$$
$$\tag{4.146}$$

To simplify the conditional probability at the left hand side of Equation 4.146 the expressions of w and x in Equations 4.143, 4.144 and 4.145 are written in terms of h, g, s and z as

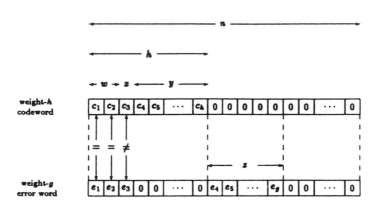

Figure 4.51: Representation of a weight-g error word and a weight-h code-word

$$w = h - s + z \tag{4.147}$$

$$x = g - h + s - 2z \tag{4.148}$$

and on substituting Equations 4.147 and 4.148 into Equation 4.146 we have

$$N_{g,s}(h \mid z) = \binom{h}{h-s+z}\binom{s-z}{g-h+s-2z}\binom{n-h}{z}$$
$$(2^m - 2)^{g-h+s-2z}(2^m - 1)^z . \tag{4.149}$$

The value of z is lower and upper limited by h and g. Consider the weight, h, of the codeword to be greater than g, such as $h = n$, then z can be as small as zero. When $h < g$, the smallest value of z is $g - h$. Therefore we define z_{min} to be the minimum value of z,

$$z_{min} \geq 0 \quad \text{if } h \geq g$$
$$z_{min} \geq g - h \quad \text{if } g > h , \tag{4.150}$$

that is,

$$z_{min} = \max\{0, g - h\} . \tag{4.151}$$

The expression of z can be found from Equations 4.143, 4.144 and 4.145 as,

$$z = \frac{g - h + s - x}{2} . \tag{4.152}$$

By setting $x = 0$, we define z_{max} to be the maximum value of z

$$z_{max} = \left\lfloor \frac{g - h + s}{2} \right\rfloor , \tag{4.153}$$

Weight-h	Number of weight-h codewords
h	A_h
5	18
6	30
7	15
8	15
9	30
10	18
15	1

Table 4.18: Number of weight-h codewords of $BCH(15, 7, 2)$ code

where $\lfloor (\bullet) \rfloor$ is the lower truncated integer value of $(\bullet)$. By summing over z, the number of weight-g error patterns at a distance s from the weight-h codeword is expressed by

$$
N_{g,s}(h) = \sum_{z=z_{min}}^{z_{max}} N_{g,s}(h \mid z)
$$

$$
= \sum_{z=z_{min}}^{z_{max}} \binom{h}{h-s+z} \binom{s-z}{g-h+s-2z} \binom{n-h}{z}
$$

$$
(2^m - 2)^{g-h+s-2z} (2^m - 1)^z . \tag{4.154}
$$

4.4.7.2.1 Number of Weight-h Codewords To evaluate $P_{ICD}(h)$ from Equation 4.135 we must compute A_h, the number of weight-h codewords. For example, the number of weight-h codewords of the $BCH(15, 7, 2)$ code is tabulated in Table 4.18.

For any maximum distance code, such as a Reed-Solomon code defined over $GF(q^m)$ with codeword length n and minimum distance d, the weight distribution A_h is given by [4]

$$
A_h = \binom{n}{h} (q^m - 1) \sum_{j=0}^{h-d} (-1)^j \binom{h-1}{j} (q^m)^{h-d-j} . \tag{4.155}
$$

By substituting $P(g)$, $N_{g,s}(h)$ and A_h from Equations 4.142, 4.154 and 4.155 into Equation 4.135 and then substituting $P_{ICD}(h)$ into Equation 4.134, we have,

$$
P_{ICD} = \sum_{h=d}^{n} \left[\binom{n}{h} (q^m - 1) \sum_{j=0}^{h-d} (-1)^j \binom{h-1}{j} (q^m)^{h-d-j} \right]
$$

$$
\sum_{s=0}^{t} \sum_{g=h-s}^{h+s} \left\{ \left[\sum_{z=z_{min}}^{z_{max}} \binom{h}{h-s+z} \binom{s-z}{g-h+s-2z} \right. \right.
$$

$$\left(\begin{array}{c} n-h \\ z \end{array} \right) (2^m - 2)^{g-h+s-2z} (2^m - 1)^z \left] \frac{1}{(2^m - 1)^g} \right.$$

$$[1 - (1 - p_b)^m]^g [(1 - p_b)^m]^{n-g}. \tag{4.156}$$

Observe that P_{ICD} tends to be zero when the SNR is high as p_b is approximately zero. Armed with P_{CD} and P_{ICD} the probabilty of error detection P_{ED} is computed using Equation 4.131, while the relative error detection probability, P_{EDR} is determined using Equation 4.132.

4.4.7.3 Post-decoding Bit and Symbol Error Probabilities

When the received codewords contain more than t symbol errors they are either known to be incorrect, or they are decoded into another codeword and the error is unknown. We will now determine the probability of the symbols and the bits in the regenerated codeword being in error for the cases when the errors are known and unknown.

When errors are detected, a systematic RS code conveys the information part of the codeword directly to the decoder output as decoded information. Consequently, the post-decoding error probability is equal to the pre-decoding error probability. Hence the contribution to the respective bit and symbol error probabilities p_{bp1} and p_{sp1} is

$$p_{bp1} = p_b P_{ED} \tag{4.157}$$

$$p_{sp1} = \sum_{i=1}^{m} \left(\begin{array}{c} m \\ i \end{array} \right) p_{bp1}^i (1 - p_{bp1})^{m-i}. \tag{4.158}$$

For error patterns that are undetectable the received word is decoded as a weight-h codeword, i.e., it contains h error symbols. The post-decoding symbol error probability p_{sp2} is given by

$$p_{sp2} = \frac{1}{n} \sum_{h=d}^{n} h P_{ICD}(h) \tag{4.159}$$

where $P_{ICD}(h)$ is given by Equation 4.156. The post-decoding bit error probability due to incorrect decoding can be evaluated from p_{sp2} as,

$$p_{sp2} = 1 - (1 - p_{bp2})^m \tag{4.160}$$

$$p_{bp2} = 1 - e^{\frac{1}{m} ln(1 - p_{sp2})}. \tag{4.161}$$

Then the total post-decoding symbol error probability p_{sp} and the total post-decoding bit error probability p_{bp} can be expressed in terms of the probability of pre-decoding bit errors p_b, the probability of error detection P_{ED} and the probability of incorrect decoding P_{ICD}, viz:-

$$p_{sp} = p_{sp1} + p_{sp2}$$

$$= \sum_{i=1}^{m} \left[\binom{m}{i} (p_b P_{ED})^i (1 - p_b P_{ED})^{m-i} \right]$$

$$+ \frac{1}{n} \sum_{h=d}^{n} h P_{ICD}(h) \tag{4.162}$$

and

$$p_{bp} = p_{bp1} + p_{bp2}$$

$$= p_b P_{ED} + \left[1 - e^{\frac{1}{m} ln \left(1 - \frac{1}{n} \sum_{h=d}^{n} h P_{ICD}(h) \right)} \right]. \tag{4.163}$$

4.4.8 Block Coding Performance

In this Section, we investigate the performance of BCH and RS codes operating at approximately half rate, but having different block lengths. The codes were transmitted via MSK modulation over AWGN or Rayleigh fading channels, and were decoded by either the hard-decision Berlekamp-Massey [BM-HD] algorithm, or by the soft-decision trellis decoding [TD-SD] method. For Rayleigh fading channels, interleaving was introduced at the transmitter in order to disperse burst errors at the receiver into the adjacent blocks resulting in an improvement in BER performance. The theoretical calculations of the probabilities of correct decoding, incorrect decoding, and relative error detection of each code were obtained from Equations 4.133, 4.134 and 4.132, respectively. These probabilities were compared with the results from simulations. Only systematic codes were used. When an incorrectable codeword was detected by syndromcheck, the information part of the codeword was passed to the decoder output. The probabilities of post-decoding symbol and bit errors were examined by simulation and compared with the theoretical results obtained from Equations 4.162 and 4.163.

4.4.8.1 Block Coding Performance via Gaussian Channels

Figure 4.52 and Figure 4.53 illustrate the probabilities of correct decoding and incorrect decoding, respectively, as a function of E_b/N_0 for the short binary $BCH(15, 7, 2)$ code decoded using the hard-decision Berlekamp-Massey algorithm [BM-HD]. Observe that for E_b/N_0 in excess of 7 dB $P_{CD} \sim 1$ and $P_{ICD} \sim 0$. The probabilities of detecting the incorrectable words calculated from Equation 4.132 were 0.57 for E_b/N_0 of 2 dB, and 0.60 for E_b/N_0 of 10 dB. Clearly, for practical E_b/N_0 values more than 57 % of the incorrectable words were detected, and 43 % were incorrectly decoded into other valid codewords. The comparatively low ability to detect code-overload is due to the densely packed, low correcting power ($t = 2$) code. The probability of post-decoding bit errors in Figure 4.54 was similar in its nature to P_{ICD} in Figure 4.53. The probability of incorrect decoding decreased to 10^{-6} when E_b/N_0 exceeded 9.7 dB, while the post-decoding bit

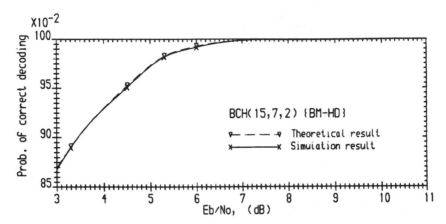

Figure 4.52: Probability of correct decoding for the $BCH(15,7,2)$ code using [BM-HD] decoding over AWGN channel

error probability was similarly reduced to 10^{-6} at E_b/N_0 of 9.6 dB. This demonstrated that the post-decoding bit errors were mainly contributed by incorrect decoding. Although the coding rate of this code was close to half, a block length of 15 bits allowed only 8 parity bits. The small number of parity bits meant that the separable distance between codewords was small, and as a result the transmitted codeword was easily corrupted into an incorrectable and undetectable word. Thus the small distance code gave a high percentage of incorrect decoding decisions. Figure 4.52 - Figure 4.54 also contain simulation results which were in close agreement to the theoretical ones.

We now compare the theoretical and simulation results for a number of non-binary Reed-Solomon codes having different codeword lengths, but all operating at a coding rate of 1/2. The Berlekamp-Massey hard-decision decoding method was used. Figure 4.55 shows the probability of correct decoding as a function of E_b/N_0 for the $RS(4,2,1)$ code over $GF(8)$, the $RS(12,6,3)$ code over $GF(16)$ and the $RS(57,29,14)$ code over $GF(256)$. The probability of correct decoding P_{CD} using shorter RS codes was found to be higher than for the longer codes for E_b/N_0 values below approximately 6dB. This was because the codes with larger number of bits per symbol suffered higher symbol error probability, as exemplified earlier in the interleaving section by Figure 4.12. This situation was reversed for higher E_b/N_0 values because the longer codes had larger t values, and therefore were able to correct more errors in a codeword. The simulation and theoretical results coincided.

Figure 4.56 shows the probability of incorrect decoding P_{ICD} as a function of E_b/N_0. The corresponding P_{ICD} values at E_b/N_0 of 3dB for $RS(4,2,1)$ and $RS(12,6,3)$ codes were 0.065 and 0.012, respectively. Although not displayed in Figure 4.56, the theoretical P_{ICD} value for the

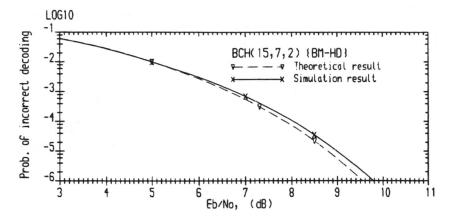

Figure 4.53: Probability of incorrect decoding for the $BCH(15, 7, 2)$ code using [BM-HD] decoding over AWGN channel

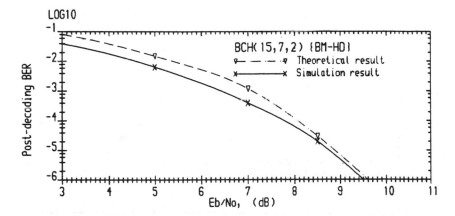

Figure 4.54: Post-decoding BER for the $BCH(15, 7, 2)$ code using [BM-HD] decoding over AWGN channel

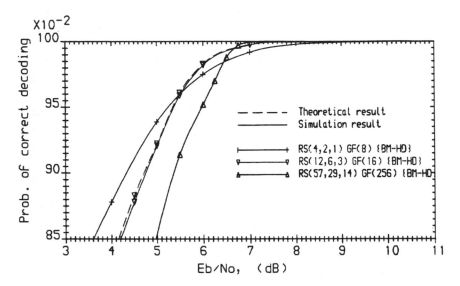

Figure 4.55: Probability of correct decoding for various RS codes over AWGN channel

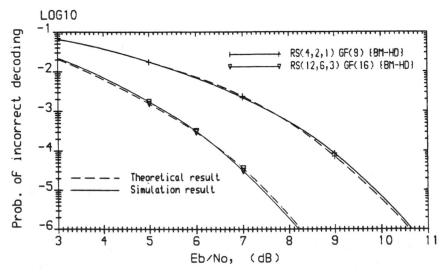

Figure 4.56: Probability of incorrect decoding of various RS codes over AWGN channel

$RS(57, 29, 14)$ code was 10^{-25}. This small P_{ICD} value was anticipated from the fact that the designed distances $(2t+1)$ of $RS(4, 2, 1)$, $RS(12, 6, 3)$, and $RS(57, 29, 14)$ codes, are 3, 7, and 29, respectively. The P_{ICD} value was significantly lower for longer codes because the separable distances between codewords were much larger. The probability of decoding an incorrectable word into another valid codeword was reduced as the larger distances restricted the occurrence of the incorrectable words. Again the simulation and theoretical values coincided.

Another property of RS codes is their capability to detect incorrectable error patterns, the probability of which is characterised in terms of the relative error detecting probability P_{EDR} as defined in Equation 4.132. As we experienced in the case of the $BCH(15, 7, 2)$ code, P_{EDR} was practically independent of E_b/N_0 for a particular code. For the short $RS(4, 2, 1)$ code, the average value of P_{EDR} was 0.73; for the $RS(12, 6, 3)$ code it was 0.98; whereas for the longest code the average P_{EDR} was $1 - 10^{-28}$. This suggested that the longer code protected by a larger number of parity symbols for the same coding rate, offered more reliable error detection. This is an important feature when transmitting computer data.

The post-decoding bit error probability p_{bp} was a function of both P_{ICD} and P_{ED} (see Equation 4.163). The simulation and theoretical results are displayed in Figure 4.57 for the three codes. The p_{bp} curves exhibited a cross-over region for E_b/N_0 between 5dB and 6dB. Below this region the bit error rate of the $RS(57, 29, 14)$ code was the highest as it had a higher channel symbol error rate than the other two codes, as shown in Figure 4.12. For E_b/N_0 in excess of 6dB the $RS(57, 29, 14)$ code had the best p_{bp} performance as the noise-induced corruption of the transmitted codeword was restricted for most of the time to the confines of the decoding sphere of that codeword. As a result, the slope of the curve is much sharper for the longer codes than for the shorter ones. The same tendency can be observed in Figure 4.58, where the post-decoding symbol error probability p_{sp} is displayed for the same conditions as those in Figure 4.57.

4.4.8.2 Block Coding Performance via Rayleigh Fading Channels

In this Section, we investigate the performance of block codes transmitted via MSK modulation over Rayleigh fading channels. Figure 4.59 and Figure 4.60 show the corresponding probabilities of correct and incorrect decoding as a function of E_b/N_0 for the binary $BCH(15, 7, 2)$ code. Without interleaving, burst errors occurred in the mobile channel, overloading the 2 bits per codeword correcting capability of the code, causing incorrect decoding. When inter-block bit interleaving IBI/B(24,5) with $B = 24$ and $N = 5$ was introduced, the channel became essentially memoryless. The burst errors were dispersed to the adjacent blocks, reducing the probability of incorrect decoding, and at the same time increasing the probability

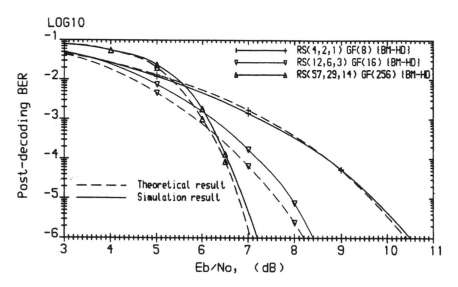

Figure 4.57: Post-decoding BER of various RS codes over AWGN channel

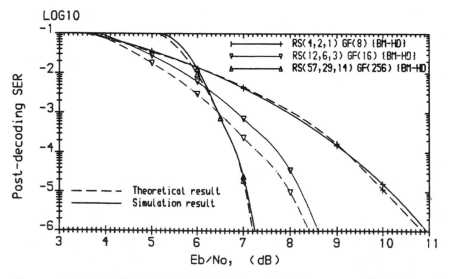

Figure 4.58: Post-decoding SER of various RS codes over AWGN channel

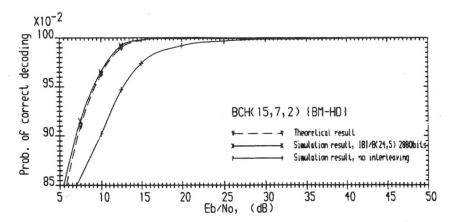

Figure 4.59: Probability of correct decoding of the $BCH(15, 7, 2)$ code over Rayleigh-fading channel

of correct decoding compared to the case without interleaving. The theoretical calculations were for the memoryless channel and were in a good agreement with the simulation results. The average probability of detecting the incorrectable words was calculated to be 0.58. This value was close to that obtained on the AWGN channel. This implies that the P_{EDR} depended on the distance separation between codewords, regardless of the channel error statistics. In Figure 4.61, the probability of post-decoding bit error p_{bp} is shown as a function of E_b/N_0. By using interleaving, the p_{bp} was reduced from 39 dB to 25 dB at a BER of 10^{-6}, this coding gain of 14 dB was achieved at a price of 2880 bits of delay.

We now consider the performance of the $RS(12, 6, 3)$ and $RS(57, 29, 14)$ codes. In Figure 4.62, we display the probability of correctly decoding a codeword as a function of E_b/N_0. When bit or symbol interleaving was deployed the P_{CD} was increased for both codes, and the longer the interleaving period the better was the performance. However, a bit interleaving period of 2880 bits appeared to be sufficient to randomize the error statistics of the channel, and yielded a performance very similar to the theoretical one. For an E_b/N_0 of 11 dB, the probability of correct decoding P_{CD} was approximately 0.86 for the $RS(57, 29, 14)$ code, and 0.92 for the $RS(12, 6, 3)$ code. The longer code performed better for high E_b/N_0 values than the shorter code, while for low E_b/N_0 values the situation was reversed.

The probability of incorrect decoding P_{ICD} is displayed in Figure 4.63 for the $RS(12, 6, 3)$ code with inter-block bit interleaving having a period of 2880 bits. The theoretical P_{ICD} curve is also displayed for that case when the interleaving period was infinitely long, representing the performance for the memoryless channel. For the $RS(57, 29, 14)$ code, the P_{ICD} was less than 10^{-23} for an E_b/N_0 of 10.5 dB, and therefore its performance curve cannot be shown in Figure 4.63.

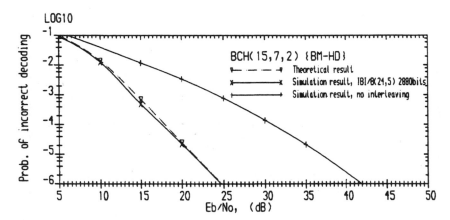

Figure 4.60: Probability of incorrect decoding of the $BCH(15,7,2)$ code over Rayleigh-fading channel

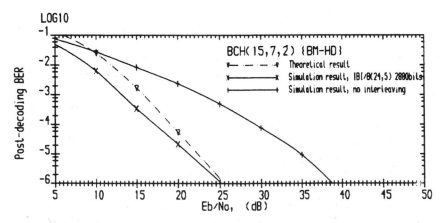

Figure 4.61: Post-decoding BER of the $BCH(15,7,2)$ code over Rayleigh-fading channel

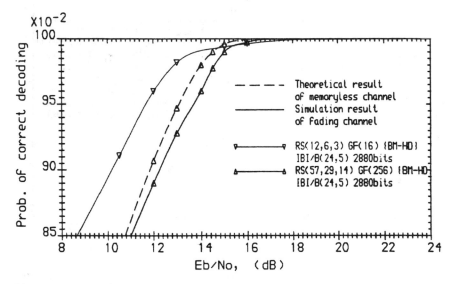

Figure 4.62: Probability of correct decoding of various RS codes with identical interleaving memory over Rayleigh-fading channel

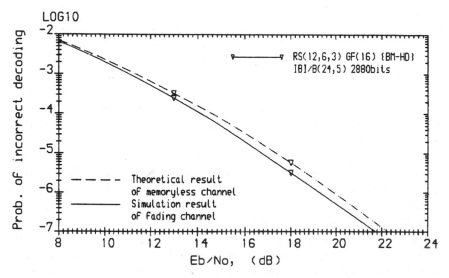

Figure 4.63: Probability of incorrect decoding of the $RS(12, 6, 3)$ code over Rayleigh-fading channel

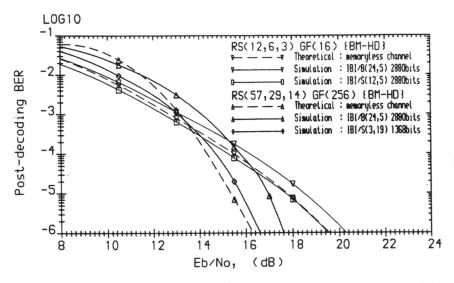

Figure 4.64: Post-decoding BER of various RS codes over Rayleigh-fading channel

The BER performance of the $RS(12, 6, 3)$ and $RS(57, 29, 14)$ code is compared in Figure 4.64 in terms of their post-decoding bit error probability with 2880 bits inter-block bit interleaving, as well as with two different symbol interleaving periods. Firstly, we focus our attention on the $RS(12, 6, 3)$ code, where we observe that even if the bit interleaving period was 2880 bits long, the BER performance was worse than that of the theoretical curve for the memoryless channel. However, if symbol interleaving was used with the same delay, practically the same BER performance was achieved as for the memoryless channel. The same tendency was noted in case of the longer $RS(57, 29, 14)$ code, where even a shorter symbol interleaving delay of 1368 bits resulted in a higher performance compared to when bit interleaving was used. This observation confirms that the symbol interleaving gives a better performance than bit interleaving with RS codes. Once again, the BER curves gave a sharper slope for the longer codes. Similar results were observed in terms of symbol error rates (SER) as shown in Figure 4.65.

In Figure 4.66 and Figure 4.67, the performance of the $RS(12, 6, 3)$, and the $RS(57, 29, 14)$ code with various interleaving delays is presented. The width of the block interleaver was determined by the blocklength. Consecutive error symbols on the channel were diverted into the adjacent blocks on deinterleaving. Observe that the performance of the shorter $RS(12, 6, 3)$ code hardly improved when the interleaving delay increased beyond 912 bits, while for the longer $RS(57, 29, 14)$ code 2280 bits was ade-

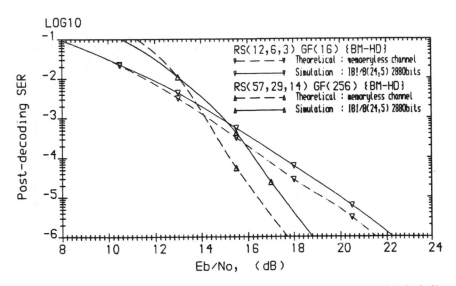

Figure 4.65: Post-decoding SER of various RS codes over Rayleigh-fading channel

quate. The longer delay for the longer code was incurred due to the width of the interleaver. However, as the delays were sufficiently long to randomize the channel to a memoryless one, the performances depended entirely on the correcting capabilities of the codes. This is evident by noting that the $RS(57, 29, 14)$ code achieved a BER of 10^{-6} at E_b/N_0 of 13 dB, while for the $RS(12, 6, 3)$ code 17.5 dB was necessary.

4.4.8.3 Soft/Hard Decisions via Gaussian Channels

In Figure 4.68, the post-decoding bit error probabilities of the $RS(4, 2, 1)$ code over $GF(16)$ and the $BCH(15, 7, 2)$ code are depicted both for hard-decision decoding and for soft-decision trellis decoding. The transmissions are over an AWGN channel. For both soft and hard decisions the $BCH(15, 7, 1)$ code performed better than the $RS(4, 2, 1)$ code as it corrected any combination of two bit errors. By constrast the RS code coped only with those bit errors which occurred in the same four bit symbols, as it can correct only one symbol error in a codeword. By using soft-decision, an improvement of about 2 dB was achieved over the hard-decision in both codes at a BER of 10^{-6}. The trellis decoding made use of the maximum likelihood method to select the path in the trellis that most resembled the received codeword. However, this probabilistic decoding did not have error detection capability, which is a serious disadvantage.

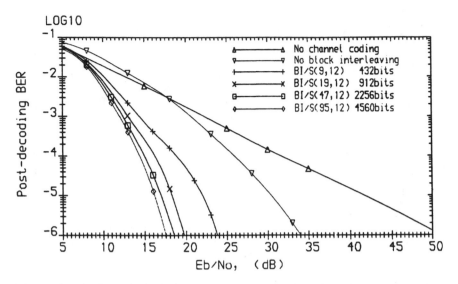

Figure 4.66: The effect of block interleaving on the post-decoding BER of the $RS(12,6,3)GF(16)$ code using [BM-HD] decoding over Rayleigh-fading channel

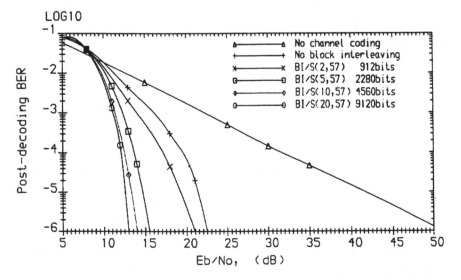

Figure 4.67: The effect of block interleaving on the post-decoding BER of the $RS(57,29,14)GF(256)$ code using [BM-HD] decoding over Rayleigh-fading channel

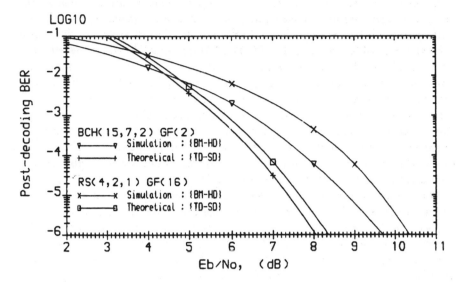

Figure 4.68: The effect of hard and soft decisions on the post-decoding BER of various block codes over AWGN channel

4.4.9 Conclusions on Block Coding

The performance of binary and non-binary block codes has been investigated theoretically, and simulation results have been presented. In the case of the Rayleigh fading channel, the interleaver selected for block codes is the block interleaver, with bit interleaving deployed for binary codes and symbol interleaving for non-binary codes. The block codes are decoded by either the hard-decision Berlekamp-Massey algorithm, or by the soft-decision trellis decoding method.

When the Berlekamp-Massey decoding algorithm is deployed, the data reliability increases with the amount of redundancy introduced into the

	E_b/N_0 value		Coding gain	
	BER=10^{-3}	BER=10^{-6}	BER=10^{-3}	BER=10^{-6}
No coding	6.8dB	10.5dB	0dB	0dB
$RS(4,2,1)\ GF(8)\ [BM-HD]$	7.2dB	10.5dB	-0.4dB	0dB
$RS(12,6,3)\ GF(16)\ [BM-HD]$	6.2dB	8.5dB	0.6dB	2.0dB
$RS(57,29,14)\ GF(256)\ [BM-HD]$	6.1dB	7.2dB	0.7dB	3.3dB
$RS(4,2,1)\ GF(16)\ [BM-HD]$	7.5dB	10.4dB	-0.7dB	0.1dB
$BCH(15,7,2)\ GF(2)\ [BM-HD]$	6.5dB	9.7dB	0.3dB	0.8dB
$RS(4,2,1)\ GF(16)\ [TD-SD]$	5.9dB	8.3dB	-0.9dB	2.2dB
$BCH(15,7,2)\ GF(2)\ [TD-SD]$	5.6dB	8.0dB	1.2dB	2.5dB

Table 4.19: Coding gain of block codes over AWGN channel

codeword. For the same coding rate, longer codes contain higher number of parity symbols than the shorter codes and the probability of detecting incorrectable words is very high. For example, the probability of detecting the incorrectable errors of the $RS(12, 6, 3)$ code is 0.98, as compared with the $RS(57, 29, 14)$ code whose is $1 - 10^{-28}$. This feature is valuable in automatic repeat request (ARQ) protocols, especially for computer data transfer. However, the penalty of using long codes is a large delay which may not be acceptable for speech transmissions. Also the complexity of the decoder increases with the amount of redundancy. The number of multiplications required by the Berlekamp-Massey algorithm (which is at most $6t^2$) increases exponentially with the number of the correctable errors. As a result, the complexity increases with t.

When the soft-decision trellis decoding method is used to decode an (n, k) block code over $GF(2^m)$, the number of bits representing the state of the encoder is equal to $(n - k) \times m$ used by the parities in a codeword. The number of states is therefore $2^{(n-k) \times m}$. The complexity of the decoder increases with the number of states. Low complexity codes operate at high coding rates or they are short codes. In either case, the separation between the codewords is small, making them weak error correcting codes. Nevertheless, it is the complexity of the decoder that limits the application of the trellis decoding. If trellis decoding is applied, the soft-decision decoding achieves a gain of 2 dB at BER of 10^{-6} for transmissions over an AWGN channel.

When the block symbol interleaving has a depth of 9 with a delay of 432 bits suitable for speech transmissions, the $RS(12, 6, 3)$ code over $GF(16)$ achieves an E_b/N_0 of 14.5 dB at a BER of 10^{-3}. Whereas the $RS(57, 29, 14)$ code over $GF(256)$ operating without interleaving encounters a similar delay of 456 bits, and needs an E_b/N_0 of 16 dB to acquire a BER of 10^{-3}. This suggest that the short codes require less decoder complexity, and perform slightly better at the target BER of 10^{-3} than the longer codes, and therefore might be preferable when speech signals are transmitted.

Table 4.19 and Table 4.20 tabulate the performance of the block codes for specific BERs when the transmissions are over an AWGN or a Rayleigh fading channel, respectively.

4.5   Concatenated Codes

Concatenated coding was first introduced by Forney [58] to utilize multiple levels of coding. Figure 4.69 illustrates two levels of coding and two levels of interleaving to combat the channel's burst errors. The level of the coding and interleaving closer to the channel is called the inner layer, whereas the level outside the inner layer is known as the outer layer. The inner and outer FEC codes can be convolutional codes or block codes. At the receiving end, the demodulator may produce either hard or soft decisions. In either case,

	E_b/N_0 value		Coding gain	
	BER=10^{-3}	BER=10^{-6}	BER=10^{-3}	BER=10^{-6}
No coding	23.0dB	52.0dB	0dB	0dB
$BCH(15,7,2)\ [BM-HD]$				
No interleaving 15bits	22.5dB	39.0dB	0.5dB	13.0dB
IBI/B(24,5) 2880bits	13.0dB	25.0dB	10.0dB	27.0dB
$RS(12,6)\ GF(16)\ [BM-HD]$				
No interleaving 48bits	20.5dB	34.0dB	2.5dB	18.0dB
BI/S(9,12) 432bits	14.5dB	24.0dB	8.5dB	28.0dB
BI/S(19,12) 912bits	13.0dB	20.0dB	10.0dB	32.0dB
BI/S(47,12) 2256bits	12.5dB	19.0dB	10.5dB	33.0dB
BI/S(95,12) 4560bits	12.0dB	17.0dB	11.0dB	35.0dB
$RS(57,29)\ GF(256)\ [BM-HD]$				
No interleaving 456bits	16.0dB	22.5dB	7.0dB	29.5dB
BI/S(2,57) 912bits	14.0dB	21.0dB	9.0dB	31.0dB
BI/S(5,57) 2280bits	12.5dB	15.5dB	10.5dB	36.5dB
BI/S(10,57) 4560bits	11.5dB	14.0dB	11.5dB	38.0dB
BI/S(20,57) 9120bits	11.0dB	13.0dB	12.0dB	39.0dB

Table 4.20: Coding gain of block codes over Rayleigh fading channel

these decisions are fed to the inner interleaver. The inner deinterleaver disperses the channel burst errors into random patterns. If the channel being considered is Gaussian, the inner interleaving is not required. The inner FEC decoder is designed to remove the random errors. If the inner FEC decoder cannot correct the word or erroneously decodes the word, the decoding errors are bursty in nature and the outer interleaver is used to disperse the errors into adjacent codewords of the outer code. The outer FEC decoder then attempts to correct the remaining errors. Essentially there are two types of concatenated codes, depending on the way the two layers are amalgamated: nested codes and product codes.

4.5.1   Nested Codes

Suppose the outer code and the inner code are denoted as (N, K, T) and (n, k, t) respectively. The outer coder encodes K outer symbols into N outer symbols as shown in Figure 4.70, where each outer symbol consists of k inner symbols. The inner coder then encodes each outer symbol of k inner symbols into n inner symbols. If the inner symbol is defined on $GF(q)$, the outer symbol consisting of k inner symbols is defined on $GF(q^k)$. Hence, the outer layer is described as the (N, K, T) code over $GF(q^k)$, and the inner layer is the (n, k, t) code over $GF(q)$. The combined layers of coding accept a sequence of Kk information symbols and produce a sequence of Nn symbols. An example of a nested code is the single error correcting shortened $RS(5, 3, 1)$ code over $GF(8) = GF(512)$ for the inner layer, and the double error correcting $RS(511, 507, 2)$ code over $GF(8^3)$ for the outer layer. Both layers of coding jointly form a nested code of (2555, 1521) over $GF(8)$.

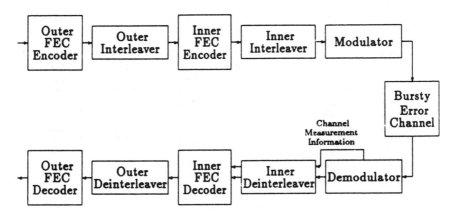

Figure 4.69: Concatenated coding system block diagram

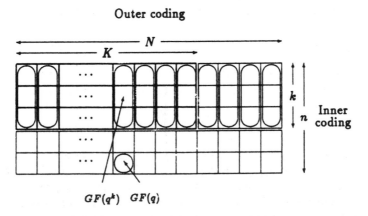

Figure 4.70: Structure of nested codes

Outer coding

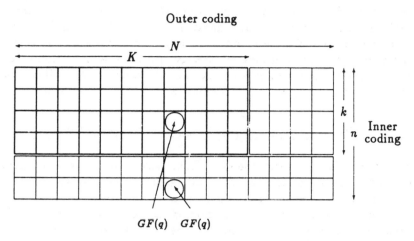

Figure 4.71: Structure of product codes

In Figure 4.70, the outer coder encodes a row at a time, and the inner coder encodes a column at a time. If the symbols are transmitted in rows to the channel, the burst errors corrupt consecutive symbols which are received and arranged in rows in the Figure. As a consequence, the bursts of error symbols are distributed over adjacent words of the inner codes. The dispersion of burst errors is equivalent to interleaving. The inner decoder corrects the burst errors and outputs N outer symbols to the outer decoder. The outer decoder corrects the decoding errors left by the inner layer.

A powerful nested code is formed by using convolutional codes as the inner layer and Reed-Solomon codes as the outer layer. This is because a convolutional code can achieve a high performance gain in correcting random errors. The errors at the convolutional decoder output due to erroneous decoding tend to be bursts and, the concatenated outer Reed-Solomon code is used to correct them.

4.5.2 Product Codes

Product codes are obtained by encoding a matrix of information symbols (defined on $GF(q)$) in two dimensions, namely, in rows and in columns. If each row is encoded by an (N, K, T) code, and each column is encoded by an (n, k, t) code, the information matrix with dimensions $(k \times K)$ is encoded to the dimension of $(n \times N)$. Figure 4.71 illustrates an example of a product code by using $RS(15, 11, 2)$ code over $GF(32)$ for the rows, and the shortened $RS(6, 4, 1)$ code over $GF(32)$ for the columns.

In a mobile radio channel, the signal fading occurs intermittently. Most of the codewords between fadings have unnecessary protection resulting in low data throughtput. A method proposed [59] is to use an adaptive product coding. The information protected with relatively low overhead error detection codes is transmitted via the channel. If the decoder detects errors, automatic repeat request (ARQ) is activated and the encoder transmits more and more redundancy until the decoder is able to correct the errors. By using this method, the amount of data protection adapts to the channel condition improving the data throughput. A similar adaptive coding scheme was proposed in References [60] and [61] as well.

4.6   Comparison of Error Control Codes

Every type of error control code such as convolutional codes, Reed-Solomon codes and concatenated codes have unique properties. Convolutional codes have an advantage in correcting random errors, whereas Reed-Solomon codes are good at correcting both random as well as burst errors and have reliable error detection capability. Concatenated codes possess high error correcting capability due to their long blocklengths. It is difficult to select 'the' best code for all the different channel conditions and system requirements, such as combined coding and interleaving delays, coding rate, data throughput, data integrity, coder complexity, types of channels, etc. In this Section, we investigate the performance of using different codes for speech and data signals for transmissions over mobile radio channels. In general, speech codecs are robust to moderate channel errors, but cannot tolerate long transmission delays. Hence, for speech signals short coding and interleaving delays, say about 500 bits are used, and moderate data protection is provided. For data transmission, data integrity is vital. The codes used for data signals must have a high error detection capability. When an incorrectable word is detected at the receiver, usually the retransmission of the codeword is necessary. As data transmission does not usually require realtime processing, coding and interleaving delays of about 2000 bits are long enough to randomize the Rayleigh fading channel's bursty error statistics even for low vehicular speeds. We have arranged for speech and data signals that the codecs operate with the same half rate in order to compare their bit error rate (BER) performance with the same amount of protection.

We experimented with both non-concatenated coding and concatenated coding of digital speech transmitted via MSK over Rayleigh fading channels. The BER results are shown in Figure 4.72 and Figure 4.73 respectively.

For non-concatenated codes, the $RS(57, 29, 14)$ code over $GF(256)$ deploying no interleaving had a coding delay of 456 bits. By stacking up nine $RS(12, 6, 3)$ codewords defined on $GF(16)$ in the interleaving matrix, the block symbol interleaver BI/S(9,12) achieved a delay of 432 bits. Both RS codes were decoded using the hard-decision Berlekamp-Massey algorithm. We observe from Figure 4.72 that the BER performance curves cross-over at an E_b/N_0 of 21 dB. Below 21 dB, the shorter $RS(12, 6, 3)$ code had a better performance than the longer $RS(57, 29, 14)$ code. The longer code

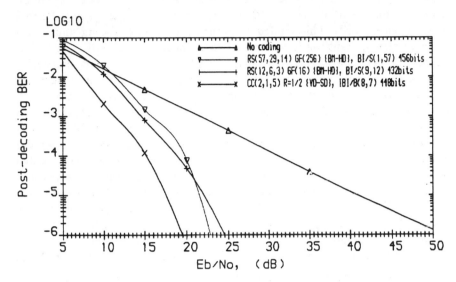

Figure 4.72: Post-decoding BER performance of half-rate non-concatenated block and convolutional codes for speech transmission over Rayleigh-fading channel

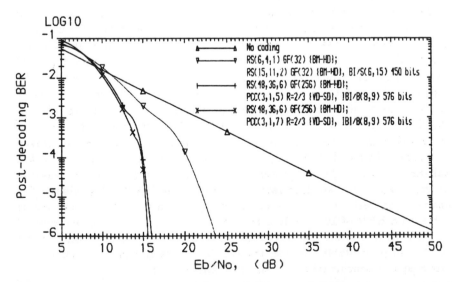

Figure 4.73: Post-decoding BER performance of half-rate concatenated codes for speech transmission over Rayleigh-fading channel

having a larger symbol field, i.e., 8 bits per symbol for the $RS(57, 29, 14)$
code compared with 4 bits per symbol for the $RS(12, 6, 3)$ code, suffered a
higher channel symbol error rate that was likely to cause more symbol errors
in a decoded word. This tended to induce incorrect decoding and incorrectable errors, resulting in a worse performance. Above 21 dB, the longer
$RS(57, 29, 14)$ code having more parity symbols in a codeword and better
burst-dispersing properties had a higher correcting capability, and therefore
had a better performance than the shorter $RS(12, 6, 3)$ code. Inter-block
bit interleaving IBI/B(8,7) of the convolutional code $CC(2, 1, 5)$ introduced
448 bits of delay. By using soft-decision Viterbi decoding, the BER performance of the $CC(2, 1, 5)$ code was 3 dB to 5 dB better than for both the
RS codes for BERs from 10^{-2} to 10^{-6}. The soft-decision decoding yielded
a better performance by making use of the channel information, especially
for transmissions over Rayleigh fading channels. For speech codecs able to
tolerate a BER of 10^{-3}, the $CC(2, 1, 5)$ code, the $RS(12, 6, 3)$ code, and the
$RS(57, 29, 14)$ code required E_b/N_0 values of 11.5 dB, 14.5 dB, and 15.5 dB,
respectively. The shorter RS code was better than the longer code for the
speech transmission. The CC code is the best among all three candidates.

Figure 4.73 compares the performance of the concatenated product code
and the nested code. The product code employed $RS(15, 11, 2)$ code over
$GF(32)$ for the outer coding and $RS(6, 4, 1)$ over $GF(32)$ for the inner
coding. The code was block interleaved BI/S(6,15) with a delay of 450 bits.
This code achieved a BER of 10^{-3} at 16.5 dB which was 1 dB and 2 dB
more than that required by the $RS(57, 29, 14)$ code and the $RS(12, 6, 3)$
code, respectively. The inferior performance was mainly due to the weak
correcting code in both dimensions, i.e., double error correcting code in
the rows, and single error correcting codes in the columns. The half-rate
nested code was constructed by concatenating the outer three-quarter rate
$RS(48, 36, 6)$ code over $GF(256)$ and the inner two-third rate $PCC(3, 1, 5)$
code, and the inter-block bit interleaving IBI/B(8,9) was over 576 bits. The
restricted delay did not allow us to use the outer interleaver. At a BER
of 10^{-3}, the required E_b/N_0 was 13.5 dB which was 1 dB less than that of
the $RS(12, 6, 3)$ code. If the constraint length of the inner CC code was
increased from five to seven binary stages, the performance was improved
by 0.5 dB. Notice that the improved nested code was still inferior to the
half-rate $CC(2, 1, 5)$ code by some 1.5 dB. The poor performance was due
to the weak inner two-third rate CC code at the low E_b/N_0 values. The
nested code performed better at higher E_b/N_0 values. As an example, it
achieved a BER of 10^{-6} at 15.5 dB, whereas the $CC(2, 1, 5)$ code required
19.5 dB. However, for speech codecs, any further reduction in BER below
10^{-3} is imperceptible. Hence, the $CC(2, 1, 5)$ is the most effective code
amongst our benchmarkers for speech transmission.

Figure 4.74 and Figure 4.75 illustrate the performance of non-
concatenated codes and concatenated codes, respectively, for the transmission of data. With a delay of over 2000 bits, the channel was randomized

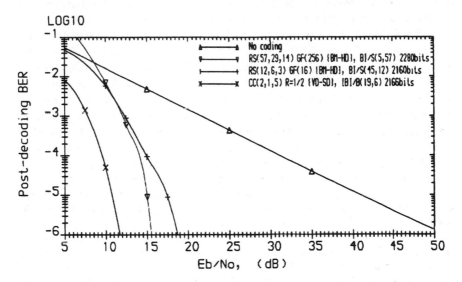

Figure 4.74: Post-decoding BER performance of half-rate non-concatenated block and convolutional codes for data transmission over Rayleigh-fading channel

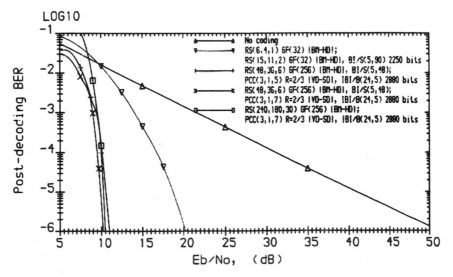

Figure 4.75: Post-decoding BER performance of half-rate concatenated codes for data transmission over Rayleigh-fading channel

to a near memoryless one, and the codes performed better than those used for the speech signals. For data transmission, we measure the performance of the codes at a BER of 10^{-6}. In Figure 4.74, the $RS(57, 29, 14)$ code and the $RS(12, 6, 3)$ code required 15.5 dB and 18.5 dB, respectively, yielding gains of 7 dB and 8 dB compared with their performance for speech data. However, the $CC(2, 1, 5)$ code decoded by soft-decision Viterbi algorithm required an E_b/N_0 of 11.5 dB at a BER of 10^{-6}. Although the $CC(2, 1, 5)$ achieved the best performance, it had no error detection capability, which is often a serious disadvantage. The $RS(57, 29, 14)$ code detected the incorrectable error words with a successful rate of $1 - 10^{-28}$, a value, sufficiently reliable to invoke retransmission in ARQ systems. Hence, the $RS(57, 29, 14)$ code was more appropriate for data protection.

In Figure 4.75, the product code constructed by the $RS(15, 11, 2)$ and $RS(6, 4, 1)$ codes again demonstrated lack of correcting power. For the nested code constructed by $RS(48, 36, 6)$ and $PCC(3, 1, 5)$ codes, the E_b/N_0 value was 10.5 dB at a BER of 10^{-6}. If the constraint length of the $PCC(3, 1, 5)$ code increased from 5 to 7, a gain of 0.5 dB was achieved. The $RS(48, 36, 6)$ code was used for the outer layer providing a successful error detection rate of $1 - 10^{-13}$. If the $RS(48, 36, 6)$ code was replaced by the $RS(240, 180, 30)$ code, the successful rate was virtually unity. However, this longer RS code had a higher probability of having an incorrectable error codeword, and therefore its performance deteriorated at low E_b/N_0 values.

Bibliography

[1] **R.W.Hamming**. "Error detecting and error correcting codes". *Bell Sys. Tech. J.,29*, pp.147–160, 1950.

[2] **P.Elias**. "Coding for noisy channels". *IRE Conv. Rec. pt.4*, pp.37–47, 1955.

[3] **J.L.Massey**. "Threshold decoding". *MIT Press, Cambridge, Mass.*, 1963.

[4] **E.R.Berlekamp**. "Algebraic Coding Theory". *McGraw-Hill, New York*, 1968.

[5] **S.Lin** and **D.J.Costello**. "Error Control Coding Fundamentals and Applications". *Prentice-Hall, Inc., New Jersey*, 1983.

[6] **R.E.Blahut**. "Theory and Practice of Error Control Codes". *Addison-Wesley*, 1983.

[7] **F.J.MacWilliams** and **J.A.Sloane**. "The Theory of Error-Correcting Codes". *North-Holland, Amsterdam*, 1977.

[8] **V.Pless**. "Introduction to the theory of error-correcting codes". *John Wiley and Sons*, 1982.

[9] **W.W.Peterson** and **E.J.Weldon**. "Error correcting codes". *MIT Press*, 1972.

[10] **A.M.Michelson** and **A.H.Levesque**. "Error-control techniques for digital communication". *John Wiley & Sons*, 1985.

[11] **G.D.Forney**. "Burst-correcting codes for the classic burst channel". *IEEE Trans. Commun. Technol., vol.COM-19*, pp.772–781, October 1971.

[12] **GSM Recommodation 05.03**. "Channel coding". *Draft Version 3.1.0*, February 1988.

[13] **K.Y.Liu** and **J.J.Lee**. "Recent results on the use of concatenated Reed-Solomon/Viterbi channel coding and data compression for space communications". *IEEE Trans. Commun., vol.COM-32, no.5*, pp.518–523, May 1984.

[14] **J.L.Ramsey**. "Realization of optimum interleavers". *IEEE Trans. Info. Theory, vol.IT-16, no.3*, pp.338–345, May 1970.

[15] **R.Steele**. "Towards a high-capacity digital cellular mobile radio system". *IEE Proc., Part F, 132, no.5*, pp.405–415, August 1985.

[16] **W.F.Bodtmann** and **H.W.Arnold**. "Fade-duration statistics of a Rayleigh-distributed wave". *IEEE Trans. Commun., vol.COM-30, no.3*, pp.549–553, March 1982.

[17] **J.M.Wozencraft.** "Sequential decoding for reliable communication". *IRE Natl. Conv. Rec., vol.5, pt.2*, pp.11–25, 1957.

[18] **J.M.Wozencraft** and **B.Reiffen.** *Sequential decoding.* MIT Press, Cambridge, Mass., 1961.

[19] **R.M.Fano.** "A heuristic discussion of probabilistic coding". *IEEE Trans. Info. Theory, vol.IT-9*, pp.64–74, April 1963.

[20] **A.J.Viterbi.** "Error bounds for convolutional codes and an asymphtotically optimum decoding algorithm". *IEEE Trans. Info. Theory, vol.IT-13*, pp.260–269, April 1967.

[21] **G.D.Forney.** "The Viterbi algorithm". *Proc. of the IEEE, vol.61, no.3*, pp.268–278, March 1973.

[22] **J.A.Heller** and **I.M.Jacobs.** "Viterbi decoding for satellite and space communication". *IEEE Trans. Commun. Technol., vol.COM-19, no.5*, pp.835–848, October 1971.

[23] **K.H.H.Wong** and **R.Steele.** "Transmission of digital speech in highway microcells". *Journal of Instn. of Electronic & Radio Engrs., vol.57, no.6 (supplement)*, pp.S246–S254, November-December 1987.

[24] **E.A.Bucher** and **J.A.Heller.** "Error probability bounds for systematic convolutional codes". *IEEE Trans. Inform. Theory, vol.IT-16*, pp.219–224, March 1970.

[25] **J.P.Odenwalder.** "Optimal decoding of convolutional codes". *PhD thesis, Dept. of Systems Sciences, School of Engineering and Applied Sciences, University of California, Los Angeles,* 1970.

[26] **Consultative Committee for Space Data Systems.** "Blue book". *Recommendations for Space Data System Standards: Telemetry Channel Coding,* May 1984.

[27] **T.Yamada, H.Harashima,** and **H.Miyakawa.** "A new maximum likelihood decoding of high rate convolutional codes using a trellis". *Trans. Inst. Electron. & Commun. Eng. Jpn. 66A*, pp.611–616, 1983.

[28] **L.H.C.Lee** and **P.G.Farrell.** "Error performance of maximum-likelihood trellis decoding of $(n, n-1)$ convolutional codes: a simulation study". *IEE Proc., vol.134, Pt.F, no.7*, pp.673–680, December 1987.

[29] **K.J.Larsen.** "Short convolutional codes with maximal free distance for rate 1/2, 1/3 and 1/4". *IEEE Trans. Info. Theory, vol.IT-19*, pp.371–372, May 1973.

[30] **J.B.Cain, G.C.Clark,** and **J.M.Geist.** "Punctured convolutional codes of rate $(n-1)/n$ and simplified maximum likelihood decoding". *IEEE Trans. Info. Theory, vol.IT-25, no.1*, pp.97–100, January 1979.

[31] **Y.Yasuda, K.Kashiki,** and **Y.Hirata.** "High-rate punctured convolutional codes for soft decision Viterbi decoding". *IEEE Trans. Commun., vol.COM-32, no.3*, pp.315–319, March 1984.

[32] **D.G.Daut, J.W.Modestino,** and **L.D.Wismer.** "New short constraint length convolutional code construction for selected rational rates". *IEEE Trans. Info. Theory, vol.IT-28*, pp.793–799, September 1982.

[33] **A.J.Viterbi.** "Convolutional codes and their performance in communication systems". *IEEE Trans. Commun. Technol., vol.COM-19, no.5,* pp.751–772, October 1971.

[34] **S.T.S.Chia, R.Steele, E.Green,** and **A.Baran.** "Propagation and bit error rate measurement for a microcellular system". *Journal of Instn of Electronic & Radio Engrs, vol.57, no.6 (supplement),* pp.S255–S266, November-December 1987.

[35] **A. Hocquenghem** "Codes correcteurs d'erreurs". *Chiffres (Paris), vol.2,* pp.147–156, September 1959.

[36] **R.C.Bose** and **D.K.Ray-Chaudhuri.** "On a class of error correcting binary group codes". *Information and Control, vol.3,* pp.68–79, March 1960.

[37] **R.C.Bose** and **D.K.Ray-Chaudhuri.** "Further results on error correcting binary group codes". *Information and Control, vol.3,* pp.279–290, September 1960.

[38] **W.W.Peterson.** "Encoding and error correction procedures for the Bose-Chaudhuri codes". *IRE Trans. Inform. Theory, vol.IT-6,* pp.459–470, September 1960.

[39] **D.Gorenstein** and **N.Zierler.** "A class of cyclic linear error-correcting codes in p^m synbols". *J. Soc. Ind. Appl. Math., 9,* pp.107–214, June 1961.

[40] **I.S.Reed** and **G.Solomon.** "Polynomial codes over certain finite fields". *J. Soc. Ind. Appl. Math., vol.8,* pp.300–304, June 1960.

[41] **R.T.Chien.** "Cyclic decoding procedure for the Bose-Chaudhuri-Hocquenghem codes". *IEEE Trans. Info. Theory, vol.10,* pp.357–363, 1964.

[42] **G.D.Forney.** "On decoding *BCH* codes". *IEEE Trans. Info. Theory, vol.11,* pp.549–557, 1965.

[43] **E.R.Berlekamp.** "On decoding binary Bose-Chaudhuri-Hocquenghem codes". *IEEE Trans. Info. Theory, vol.11,* pp.577–579, 1965.

[44] **J.L.Massey.** "Step-by-step decoding of the Bose-Chaudhuri-Hocquenghem codes". *IEEE Trans. Info. Theory, vol.11,* pp.580–585, 1965.

[45] **J.L.Massey.** "Shift-register synthesis and *BCH* decoding". *IEEE Trans. Info. Theory, IT-15,* pp.122–127, January 1969.

[46] **E.R.Berlekamp, R.E.Peile,** and **S.P.Pope.** "The application of error control to communications". *IEEE Commun. Magazine, vol.25, no.4,* pp.44–57, April 1987.

[47] **E.R.Berlekamp.** "The technology of error-correcting codes". *Proc. of the IEEE, vol.68, no.5,* pp.564–592, May 1980.

[48] **K.H.H.Wong, L.Hanzo,** and **R.Steele.** "Channel coding for satellite mobile channels". *International Journal on Satellite Comm., accepted for publication,* 1989.

[49] **E.Prange.** "Cyclic error-correcting codes in two symbols". *AFCRC-TN-57, 103, Air Force Cambridge Research Center, Cambridge, Mass.,* 1972.

[50] **D.Chase.** "A class of algorithms for decoding block codes with channel measurement information". *IEEE Trans. on Info. Theory, vol.IT-18, no.1,* pp.170–182, January 1972.

[51] **J.K.Wolf.** "Efficient maximum likelihood decoding of linear block codes using a trellis". *IEEE Trans. Info. Theory, vol.IT-24, no.1,,* pp.76–80, January 1978.

[52] **T.Matsumoto.** "Trellis decoding of linear block codes in digital mobile radio". *38 th IEEE Vehicular Technology Conf., Philadelphia, Pennsylvania,* pp.6–11, 15–17 June 1988.

[53] **C.R.P.Hartmann** and **L.D.Rudolph.** "An optimum symbol-by-symbol decoding rule for linear codes". *IEEE Trans. Info. Theory, vol.IT-22, no.5,,* pp.514–517, September 1976.

[54] **L.R.Bahl, J.Cocke, F.Jelinek,** and **J.Raviv.** "Optimum decoding of linear codes for minimizing symbol error rate". *IEEE Trans. Info. Theory, vol.IT-20,,* pp.284–287, March 1974.

[55] **T.Kasami** and **S.Lin.** "On the probability of undetected error for the maximum distance separable codes". *IEEE Trans. on Commun., vol.COM-32, no.9,* pp.998–1006, September 1984.

[56] **Z.McC.Huntoon** and **A.M.Michelson.** "On the computation of the probability of post-decoding error events for block codes". *IEEE Trans. on Info. Theory,* pp.399–403, May 1976.

[57] **A.M.Michelson.** "The calculation of post-decoding bit-error probabilities for binary block codes". *Nat. Telecomm. Conf. Rec.,* pp.24.3.1–24.3.4, 1976.

[58] **G.D.Forney.** *Concatenated codes.* MIT Press, Cambridge, Massachusetts, 1966.

[59] **S.D.Bate, B.Honary,** and **P.G.Farrell.** "Error control techniques applicable to HF channels". *IEE Proc., vol.136, Pt.I, no.1,* pp.57–63, February 1989.

[60] **U.H.-G.KreBel** and **P.A.M.Buné.** "Adapative forward error correction for fast data transmission over the mobile radio channel". *8 th European Conf. on Electrotechnics, Conf. Proc., on Area Commun., Sweden,* pp.170–173, June 1988.

[61] **P.A.M.Buné.** "A fast and secure data transmission scheme for the GSM system". *CEPT/GSM/WP2, Document 278, Stockholm, Sweden,* October 1987.

Chapter 5

Quaternary Frequency Shift Keying

I.J.Wassell[1] and R.Steele[2]

For the mobile radio system considered in this chapter the system signalling rates are sufficiently low for the mobile channel to exhibit flat fading. The modulated signal bandwidth is therefore less than the channel coherence bandwidth for a significant proportion of the time. Consequently there is no need to employ relatively expensive and power hungry adaptive equalisers to counteract the effects of time dispersive and frequency selective channels. The flat fading experienced by narrowband systems may be combatted by employing frequency hopping or space diversity. An effective way of decreasing the channel occupancy of a modulated signal, and to decrease the probability of intersymbol interference, is to use more than one bit per symbol. In this Chapter we consider quaternary frequency shift keying where two bits per symbol are transmitted.

5.1 An S900-D Like System

The deployment of multilevel modulation reduces the symbol rate compared to binary modulation and therefore the channel bandwidth can be decreased. However, a consequence of transmitting more than one bit per symbol is that the signal power must be commensurately increased for the same channel noise if the symbol error rate is not to increase. A particular

---

[1] Multiple Access Communications Ltd.
[2] University of Southampton and Multiple Access Communications Ltd.

advantage accrues if the TDMA multilevel signal has a symbol rate sufficiently low that the mobile radio channel exhibits flat fading rather than frequency selective fading. Thus by ensuring that the modulation bandwidth is less than the coherence bandwidth delay dispersion of the spectral components in the received signal is avoided and there is no need to employ equalisers to remove intersymbol interference (ISI). The flat fading can be combatted by means of diversity techniques.

Perhaps the most simple of all the multilevel modulation methods is quaternary frequency shift keying (QFSK), and a NB-TDMA system using QFSK has been studied by Ketterling, Pfitzmann and Tietgen [1, 2, 3]. In their QFSK/TDMA system they employ narrowband TDMA (NB-TDMA) and accommodate the channel induced ISI by restricting the number of TDMA channels per carrier to 10 and by employing 4-level FSK, i.e., QFSK. Their system is designed on the assumption that the time delays of the multipath signals do not exceed 10 μs and the fade depths are rarely more than 10 dB. The TDMA rate used is 128 kb/s or 64 k symbols/s, resulting in a symbol length of 15.6 μs, a duration not particularly long compared to the assumed excess delay spread of 10 μs. The system would have a better performance if the cellular clusters are reduced in size to ensure that the excess delay spread is significantly less than 10 μs, and if the symbol rate is decreased by arranging for fewer users per carrier. Alternatively higher level FSK can be used to reduce the symbol rate and thereby reduce the risk of ISI. However, the channel SNR would then have to be increased. Nevertheless the QFSK/NB-TDMA system [1, 2, 3] does have the virtue of simplicity. The systems described in Chapter 6, which are also designed to operate in large cells, have a better performance than QFSK/NB-TDMA but at the expense of greater complexity. We may speculate that eventually large cells will only exist in rural areas and sparsely populated countries. In the densely populated countries the large cell will be relatively rare, rendering this simple QFSK/NB-TDMA system adequate for many types of mobile communications.

Let us consider the QFSK/NB-TDMA system in more detail. The speech signals are residual excited linear prediction (RELP) encoded at 9.6 kb/s and converted to 11 kb/s by repeating essential and already protected parts of the information. The TDMA frame lasting 32 ms contains 10 voice band channels and a frame synchronisation word to yield a TDMA bit rate of 128 kb/s. Each consecutive two bits in the TDMA signal are formed into a symbol having four possible levels depending on the logical levels of the two-bits. This baseband waveform is applied to a voltage controlled oscillator to give the modulated signal. The dotted lines in Figure 5.1 show the waveforms of three sets of five consecutive arbitrary symbols. The sharp transitions at the symbol boundaries cause the eye pattern at a receiver with a noncoherent frequency discriminator to be wide open, but the spectral spillage of the RF modulated signal is unacceptably large. A

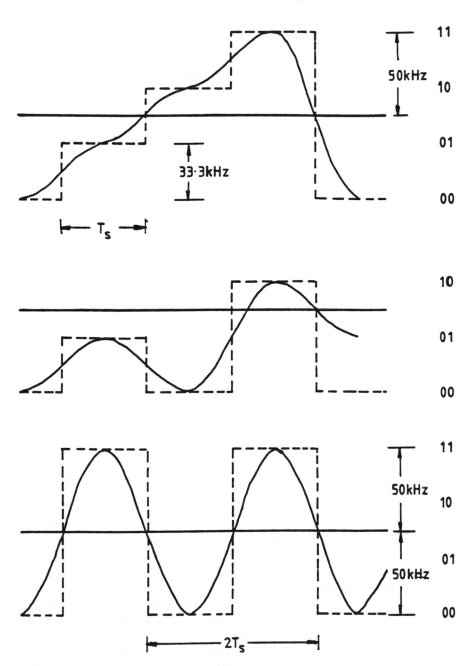

Figure 5.1: Baseband QFSK waveforms

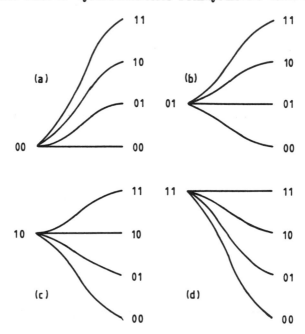

Figure 5.2: Baseband QFSK transition waveforms

compromise for containing the bandwidth of the modulated signal while providing an acceptable eye pattern at the receiver is to filter the symbols prior to modulation. By this approach the smooth waveforms shown in Figure 5.1 are obtained.

To generate the smooth transitions the current and previous symbols address a read-only memory (ROM) containing the 16 possible intersymbol transitions associated with the 4 possible symbols. These transitions are displayed in Figure 5.2 for a cosine square function, where 90° of the function is generated in a symbol period T_s. The sub-figures (a), (b), (c) and (d) show all the possible transitions from symbols 00, 01, 10 and 11, to the other symbol values, respectively. The filter action of the cosine squared shaping reduces the spectral spillage into adjacent RF channels, while providing an acceptable eye pattern. Figure 5.3 shows all the transitions from 00 (solid lines) and 11 (dotted lines) to all the possible symbols and then back again over a two symbol period. The corresponding transitions for 10 and 01 are shown in Figure 5.4. The eye pattern shown in Figure 5.5 is formulated by overlaying Figures 5.3 and 5.4. Notice that because of the raised cosine filtering the greatest width of the eye is 0.664 T_s, instead of T_s which would be available in the absence of filtering.

The smooth symbol waveforms of Figure 5.1 frequency modulate a RF carrier, and the frequency deviation between adjacent steps in the symbol

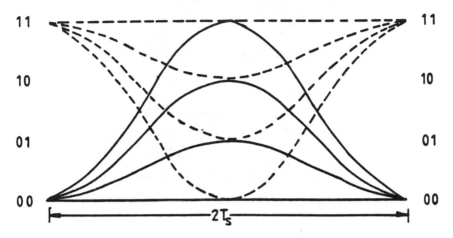

Figure 5.3: Baseband QFSK 00 to 11 transition waveforms

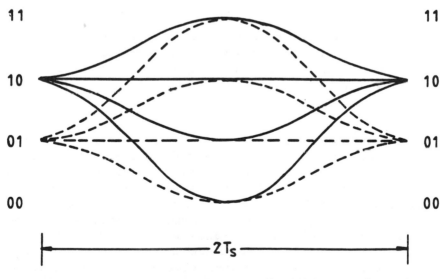

Figure 5.4: Baseband QFSK 10 to 01 transition waveforms

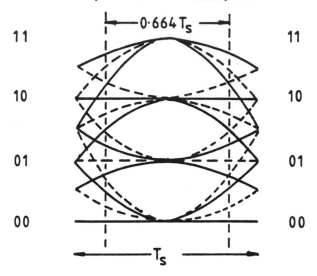

Figure 5.5: Baseband QFSK eye diagram

levels is set at 33.3 kHz. The highest modulation bandwidth occurs when the data are ... 0011001100 ... resulting in a peak deviation of ±50 kHz as shown in Figure 5.1, and a modulation signal that is nearly a pure tone of frequency $1/2T_s =$ as $1/T_s = 64$ kBd. The resulting line spectrum of the QFSK signal is displayed in Figure 5.6. Because it is considered that the adjacent RF carrier power should be attenuated by 70 dB relative to the unmodulated carrier, the bandwidth is seen to be 320 kHz. By acknowledging that the two zeros-two ones periodic sequence is statistically rare, the bandwidth of the transmitted signal is in reality of the order of 250 kHz. The modulation index is given by [4]

$$h_f = 2f_d T_s$$

where f_d is the frequency deviation from the carrier. The frequencies nearest the carrier are located at $f_c \pm f_d$, and the furthest ones at $f_c \pm 3f_d$ corresponding to $h_f = 0.52$. Applying a pseudo random binary sequence (PRBS) to the modulator results in the power spectrum of Figure 5.7. This Figure shows a 40 dB down bandwidth of only 240 kHz. To reduce spectral occupancy further one could consider reducing the modulation index, for example, with $h_f = 0.25$ the 40 dB bandwidth is now only 160 kHz. Unfortunately the power efficiency of the modulation scheme (in terms of bit error rate (BER)) has now been reduced. The carrier spacing recommended in References [1, 2, 3] is 320 kHz which gives good adjacent channel interference protection, while a receiver bandwidth of 250 kHz gives a reasonable compromise between signal distortion and noise band-limitation.

At the receiver, the frequency variations, i.e., the derivative of the phase

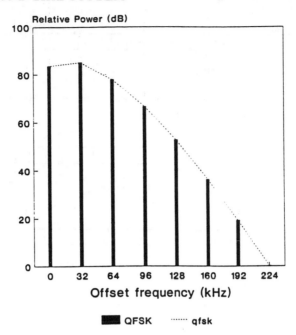

Figure 5.6: Line spectrum of QFSK signal for a ... 0011001100 ... data signal

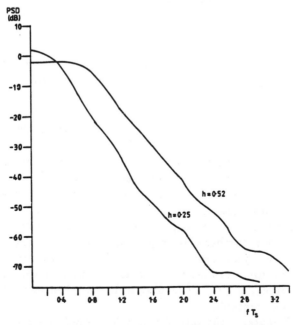

Figure 5.7: Power spectral density of the QFSK signal for a PRBS data signal

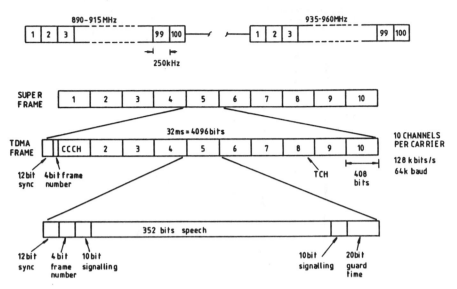

Figure 5.8: TDMA structure

of the received signal, are converted into voltage variations to yield the eye pattern of Figure 5.5. The decision levels to regenerate the symbols are set at the centre of the eye openings. This eye pattern is drawn for ideal equipment and transmission channels.

The TDMA channel burst power is 10 Watts, giving approximately 1 Watt mean power for the MS. This power level should accommodate cell sizes of 10 to 20 km radius. The BS antenna is approximately 150 m, and the receiver sensitivity is of the order 1.0 μV_{emf} at 50 Ω.

Figure 5.8 shows the basic TDMA structure. The MS to BS transmissions are in the frequency band 890–915 MHz, while the BS to MS are from 935–960 MHz. Each duplex band is divided into 100 sub-bands of 250 kHz, and as each carrier carries 10 traffic channels, the total number of duplex channels is 1000. For a seven-cell cluster there are 140 radio channels per cell. A ten frame super-frame is used, where each of its frames contains 10 TDMA traffic channels (TCHs) on a single carrier. A frame synchronisation word with a frame number is inserted at the beginning of each TDMA frame. The frame number is of importance as it identifies the position of the frame in the super-frame. Provided no call connection is established the MS is able to determine the reception quality of the other nine BSs in its super-frame, and it can request to be switched to another BS if the reception if unacceptable.

Each TCH contains 408 bits of which 352 constitute the digital speech. At the beginning of each time slot or packet, is a 12-bit synchronisation word, followed by a 4-bit frame number. Next comes a 10-bit signalling word made up of 8 bits of information and two bits for error detection. It is repeated as an error protection measure after the 352 bits of speech data have been processed. Because the MSs vary their distances from their BS as they travel the propagation times of the packets varies. The BS notes these variations, normalises them on a time basis and adjusts the synchronisation accordingly. It is able to achieve this by using the last 20 bits in a time-slot for transmission from MS to BS. Because each bit has a duration of 7.8 μs, the maximum propagation delay that can be accommodated is 20 × 7.8 = 156 μs. As the propagation velocity is 3.3 μs/km, a distance of $156/3.3 = 47.3$ km can be allowed. For BS to MS transmissions the last 20 bits in the time slot contain system information for the MS.

In each 10 channel frame, one channel is used as a common control channel (CCCH) to convey commands relating to MS registration, roaming, call set-up, and so forth. The CCCH is shown in Figure 5.8. The CCCH organisation is dependent on the direction of transmission. For BS to MS transmission the packet is divided into two identical sub-packets, each consisting of 12 synchronisation bits, followed by 172 information bits, and 20 BS identification bits. The use of the two sub-packets builds in redundancy to mitigate the effects of channel errors. Alternate packet structures are used for transmission from MS to BS. During odd-numbered frames four sub-packets, each having 102 bits are used. The first 12 bits in the sub-packet relate to synchronisation, the next 56 bits are message bits and the last 34 bits to equalise the propagation delay. These sub-packets are repeated as they are used in an ALOHA procedure [5]. The even numbered CCCH packets contain two sub-packets having 102 bits and one of 204 bits.

We have focused on a particular QFSK system that was a contender for the pan-European digital cellular mobile radio system. It was not a successful candidate because it had to compete with more complex systems that were more appropriate for large cell sizes. However, the QFSK/NB-TDMA system has the virtue of relative simplicity, and has applicability in other situations. We will, therefore, now embark on a closer inspection of QFSK, determining its theoretical performance in a variety of situations, leaving the reader to decide on its suitability for his or her application. We commence by considering the simplest channel, namely an additive white Gaussian noise (AWGN) channel. Later we will examine the performance of QFSK transmissions over Rayleigh fading channels.

5.2 QFSK Transmissions Over Gaussian Channels

The transmitted frequency in QFSK depends on the logical values of the two-bit symbols. A suitable symbol-to-carrier frequency mapping is specified in Table 5.1, where the signalling frequencies f_0, f_1, f_2 and f_3 are the

| Quaternary symbols | | Transmitted |
Natural binary	Gray code	carrier frequency
0 0	0 0	f_0
0 1	0 1	f_1
1 0	1 1	f_2
1 1	1 0	f_3

Table 5.1: Symbol to carrier frequency mapping for QFSK

values of the carrier frequency over the two-bit symbol interval. The frequencies are orthogonal and the spacing between adjacent tones is $2f_d$. These signalling frequencies, or elements, may be expressed as

$$s_i(t) = \sqrt{\frac{2E_s}{T_s}} \cos\left(2\pi f_i t + \phi_0\right) \qquad 0 \le t \le T_s \qquad (5.1)$$

where i is 0, 1, 2 or 3 depending on the logical values of the two bits being transmitted, E_s is the symbol energy, T_s is the symbol period and ϕ_0 is an arbitrary phase which can be set to zero with coherent demodulation. We will simplify our analysis by assuming square modulating waveforms; no bandlimiting by the channel; and the absence of intersymbol, adjacent channel and cochannel interference. Now because the channel contains additive white Gaussian noise (AWGN) $n(t)$, the signal at the input to the demodulator at the receiver during the symbol period from 0 to T_s is

$$r(t) = \alpha s_i(t) + n(t) \qquad (5.2)$$

where α is an attenuation factor. For a Gaussian channel, α is a constant for a MS at a given distance from its BS, and for simplicity we will set it to unity, unless otherwise stated. The AWGN signal may be represented as

$$n(t) = n_I(t) \cos 2\pi f_i t - n_Q(t) \sin 2\pi f_i t \qquad (5.3)$$

where $n(t)$, $n_I(t)$ and $n_Q(t)$ are all zero mean Gaussian random processes having the same average power level. The quadrature amplitude signals $n_I(t)$ and $n_Q(t)$ occupy the frequency band from -B/2 to B/2, where B is the bandwidth of $n(t)$ over positive frequencies. The two-sided PSD of $n_I(t)$ and $n_Q(t)$ is $N_o/2$.

5.2.1   Demodulation in the Absence of Cochannel Interference

We will now present expressions for the probability of symbol error as a function of channel SNR for QFSK transmissions over Gaussian channels. Coherent demodulation is considered first, followed by non-coherent demodulation. In Sections 5.2.2 and 5.2.3 we deal with the effects of cochannel interference.

5.2.1.1   Coherent Demodulation

The QFSK coherent demodulator is shown in Figure 5.9, and an analysis, following the approach of Clark [6], will now be given. The received signal $r(t)$ is multiplied by a set of i coherent RF tones, $\sqrt{2}\cos 2\pi f_i t$, to give $m_i(t)$, $i = 0, 1, 2$ and 3. The outputs from the four multipliers are

$$
\begin{aligned}
m_i(t) &= \left((2E_s/T_s)^{\frac{1}{2}}\cos 2\pi f_i t + n(t)\right)\sqrt{2}\cos 2\pi f_i t \\
&= (E_s/T_s)^{\frac{1}{2}} + (E_s/T_s)^{\frac{1}{2}}\cos 4\pi f_i t + \sqrt{2}n(t)\cos 2\pi f_i t. \quad (5.4)
\end{aligned}
$$

The final term in Equation 5.4 is the noise component, and from Equation 5.3 it can be expressed as

$$
(2)^{-\frac{1}{2}}\left\{n_I(t) + n_I(t)\cos 4\pi f_i t - n_Q(t)\sin 4\pi f_i t\right\}
$$

and consequently when $m_i(t)$ is low pass filtered it becomes

$$
y_i(t) = (E_s/T_s)^{\frac{1}{2}} + u(t) \qquad (5.5)
$$

where

$$
u(t) = n_I(t)/\sqrt{2}. \qquad (5.6)
$$

The signal $u(t)$ is a Gaussian random process with zero mean and a variance of $N_o/2$ over the frequency band of the baseband signal. Each $y_i(t)$ signal, $i = 0, 1, 2, 3$ is integrated to give

$$
\begin{aligned}
x_i(t) &= K\int_0^{T_s} y_i(t)dt \\
&= K(E_s/T_s)^{\frac{1}{2}} + v_i \qquad (5.7)
\end{aligned}
$$

where K is the gain of the integrator, and

$$
v_i = K\int_0^{T_s} u(t)dt \qquad (5.8)
$$

is the noise signal at the output of the integrator.

We digress at this stage to observe that the variance of a signal (here v_i) at the output of a linear filter (here an integrator) having a transfer function $G(f)$ and an input power spectral density (PSD) of $N_o/2$ is

$$
N = \frac{N_o}{2}\int_{-\infty}^{\infty}|G(f)|^2\,df \qquad (5.9)
$$

where

$$
G(f) = \int_{-\infty}^{\infty} g(t)e^{-j2\pi f t}dt \qquad (5.10)
$$

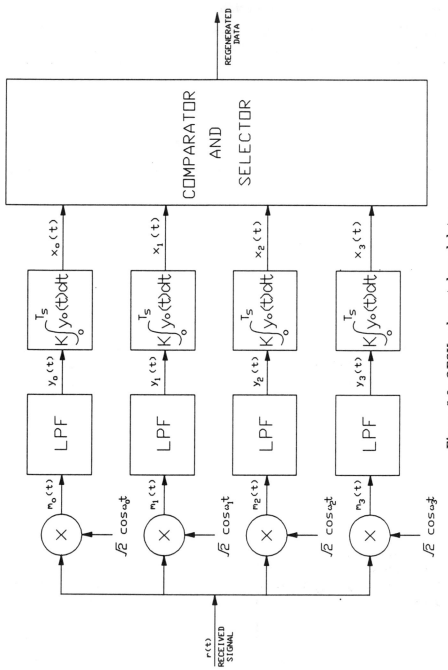

Figure 5.9: QFSK coherent demodulator

Integrator outputs x_i	Quaternary symbols			
	0	1	2	3
x_0	$K\sqrt{E_s T_s} + v_0$	v_0	v_0	v_0
x_1	v_1	$K\sqrt{E_s T_s} + v_1$	v_1	v_1
x_2	v_2	v_2	$K\sqrt{E_s T_s} + v_2$	v_2
x_3	v_3	v_3	v_3	$K\sqrt{E_s T_s} + v_3$

Table 5.2: Relationship between the quaternary symbols and the integrator outputs

and $g(t)$ is the impulse response of the filter. By Parsaval's theorem

$$\int_{-\infty}^{\infty} |G(f)|^2 \, df = \int_{-\infty}^{\infty} g^2(t) dt \tag{5.11}$$

and so

$$N = \frac{N_o}{2} \int_{-\infty}^{\infty} g^2(t) dt. \tag{5.12}$$

We now return to the main discourse, and observe that the integrators in the coherent demodulator of Figure 5.9 have an impulse response of

$$g(t) = \begin{cases} K; & 0 \le t \le T_s \\ 0; & \text{elsewhere} \end{cases} \tag{5.13}$$

and therefore from Equation 5.12

$$N = \frac{N_o}{2} K^2 T_s \tag{5.14}$$

is the variance of v_i.

The relationship between the quaternary symbols denoted by $i = 0, 1, 2, 3$ and the corresponding integrator outputs $x_i(t)$ are given in Table 5.2. Each of the quaternary symbols are orthogonal in that the transmitted carrier frequency f_i causes the i-th path in the demodulator to give a signal $K\sqrt{T_s E_s}$ plus noise, while all the other paths only pass the noise component. Let us consider the case when the symbol '00' was transmitted using the frequency f_o. The signals $x_o(t)$, $x_1(t)$, $x_2(t)$ and $x_3(t)$, are $K\sqrt{T_s E_s} + v_0$, v_1, v_2 and v_3, respectively. In order to decide which symbol was transmitted the $x_i(t)$ signals are compared, the largest $x_i(t)$ is selected, and the bit word associated with it is regenerated.

For an erroneous two-bit symbol to be regenerated one or more of the $x_i(t), i = 1, 2, 3$ must exceed $x_o(t)$. The probability that one or more of these three conditions are satisfied is the union

$$P_e \quad = \quad P\left[(x_1 > x_o) \cup (x_2 > x_o) \cup (x_3 > x_o)\right]$$

$$
\begin{aligned}
&= \ P\left(x_1 > x_o\right) + P\left(x_2 > x_o\right) + P\left(x_3 > x_o\right) \\
&- \ P\left[\left(x_1 > x_o\right) \cap \left(x_2 > x_o\right)\right] - P\left[\left(x_1 > x_o\right) \cap \left(x_3 > x_o\right)\right] \\
&- \ P\left[\left(x_2 > x_o\right) \cap \left(x_3 > x_o\right)\right] \\
&+ \ P\left[\left(x_1 > x_o\right) \cap \left(x_2 > x_o\right) \cap \left(x_3 > x_o\right)\right]
\end{aligned} \tag{5.15}
$$

and so

$$
P_e < P\left(x_1 > x_o\right) + P\left(x_2 > x_o\right) + P\left(x_3 > x_o\right) \tag{5.16}
$$

where $\cup$ and $\cap$ are the union and intersection of the events. The probability of a symbol error is, therefore, less than the sum of the individual probabilities of the three events. However, $x_i, i = 1, 2, 3$ are likely to exceed x_o with equal probability, and hence

$$
\begin{aligned}
P\left(x_i > x_o\right) &= \ P\left(v_i > K\sqrt{T_s E_s} + v_o\right) \\
&= \ P\left(v > K\sqrt{T_s E_s}\right)
\end{aligned} \tag{5.17}
$$

where

$$
v = v_o + v_i. \tag{5.18}
$$

The v_i are Gaussian distributed, and hence v is also a Gaussian variable with a variance equal to the sum of the variances of v_o and v_i, namely $2N$, and a mean of zero. Consequently the PDF of v is

$$
P(v) = \frac{1}{\sqrt{4\pi N}} exp\left(\frac{v^2}{4N}\right). \tag{5.19}
$$

Hence

$$
\begin{aligned}
P\left(x_i > x_o\right) &= \ \int_{K\sqrt{T_s E_s}}^{\infty} \frac{1}{\sqrt{4\pi N}} exp - \left(\frac{v^2}{4N}\right) dv \\
&= \ \int_{\frac{K\sqrt{T_s E_s}}{\sqrt{2N}}}^{\infty} \frac{1}{\sqrt{2\pi}} exp - \left(\frac{v^2}{2}\right) dv \\
&= \ Q\left(K\sqrt{\frac{T_s E_s}{2N}}\right)
\end{aligned} \tag{5.20}
$$

where the Q-function is defined by

$$
Q(\lambda) \triangleq \int_{\lambda}^{\infty} \frac{1}{\sqrt{2\pi}} exp\left(-\frac{x^2}{2}\right) dx. \tag{5.21}
$$

From Equations 5.15, 5.16, and 5.20,

$$
P_e < 3Q\left(K\sqrt{\frac{T_s E_s}{2N}}\right). \tag{5.22}
$$

Substituting K from Equation 5.14 into Equation 5.22 and replacing the less than sign by an equals sign on the assumption of high signal-to-noise ratios (SNR), yields

$$P_e = Q\left(K\sqrt{\frac{T_s E_s}{2N}}\right) = Q\left(\sqrt{\frac{E_s}{N_o}}\right) \qquad (5.23)$$

where E_s/N_o is the carrier-to-noise ratio C/N, assuming the receiver bandwidth is $1/T_s$. As we are transmitting two bits per symbol, the energy per bit is

$$E_b = E_s/2 \qquad (5.24)$$

giving a probability of symbol error in terms of bit energy as

$$P_e = Q\left(\sqrt{\frac{2E_b}{N_o}}\right). \qquad (5.25)$$

Observe that P_e is also the symbol error rate (SER). The variation of the probability of symbol error P_e as a function of E_b/N_o is shown in Figure 5.10. Now there are speech encoders that can operate with near toll quality performance with a bit error rate (BER) of 10^{-3}. As the BER is lower than the SER (often it is half), we observe from Figure 5.10 that toll quality speech is obtainable for E_b/N_o ratios exceeding 7 dB. If channel coding is employed, the SER before channel decoding can be of the order of 10^{-2}, necessitating a $E_b/N_o > 5 \; dB$. Notice that a SER of 10^{-8} can be achieved for an E_b/N_o of 12 dB.

5.2.1.2   Non-coherent Demodulation

We now turn our attention to a simpler type of demodulation known as non-coherent demodulation. Unlike coherent demodulation, there is no need to accurately acquire the transmitted carrier frequency in order to demodulate the received signal. Instead the received signal $r(t)$ is filtered by a bank of bandpass filters (BPFs) whose centre frequencies are those frequencies f_0, f_1, f_2 and f_3 that correspond to the frequencies used to convey the data symbols, see Table 5.1. As the received signal $r(t)$ is composed of a tone representing a data symbol plus Gaussian noise, the filtered signals are bandlimited noise signals, except for one filter whose output is the transmitted tone plus noise. Envelope detection of the filtered signals ensues to give signals x_0, x_1, x_2 and x_3. At a sampling instant these signals are sampled to give X_0, X_1, X_2 and X_3 respectively, and the largest X_i is noted. The value of i associated with X_i is used to regenerate the most probable transmitted symbol. The block diagram of the non-coherent QFSK demodulator is shown in Figure 5.11.

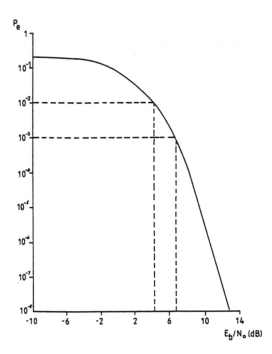

Figure 5.10: Coherent QFSK demodulation in the presence of AWGN

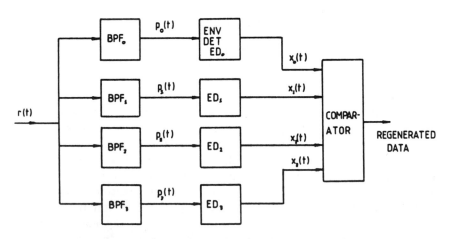

Figure 5.11: Non-coherent QFSK demodulator

In our analysis we will assume that the bandpass filters have a bandwidth B about a centre frequency f_i, and that B is the reciprocal of T_s. We will also assume that the signal components are filtered without introducing waveform distortion and that ideal envelope detection occurs. The probability of a correct symbol detection given symbol s_o was transmitted is

$$P_{co} = P\{X_1 < X_0, X_2 < X_0, X_3 < X_0 | s_o \text{ sent }\}$$

which we may express as

$$= \int_{-\infty}^{\infty} \int_{-\infty}^{x_o} \int_{-\infty}^{x_o} \int_{-\infty}^{x_o} [f(x_1, x_2, x_3|s_o) dx_1 dx_2 dx_3] f(x_o|s_o) dx_o \quad (5.26)$$

where the conditional PDF of x_o given the transmission of s_o is $f(x_o|s_o)$, and the joint PDF of X_1, X_2 and X_3 may be simplified to

$$f(x_1, x_2, x_3|s_o) = f(x_1)f(x_2)f(x_3). \quad (5.27)$$

As

$$\int_{-\infty}^{x_o} f(x_i) dx_i \quad (5.28)$$

is the same for $i = 1, 2, 3$, the probability of the QFSK non-coherent demodulator making a correct decision is

$$P_c = P_{co} = \int_{-\infty}^{\infty} \left[\int_{-\infty}^{x_o} f(x) dx \right]^3 f(x_o|s_o) dx_o \quad (5.29)$$

and consequently the probability of making an erroneous decision is

$$P_e = 1 - P_c. \quad (5.30)$$

To evaluate P_e, we must find expressions for the PDFs of $f(x)$ and $f(x_o|s_o)$. In order to do this we consider the effect of passing narrowband noise through an envelope detector. Now the quadrature representation of a narrowband noise signal $n(t)$ given in Equation 5.3 may be transformed into an equivalent envelope and phase form

$$n(t) = R(t) [\cos(2\pi f_c t + \theta(t)] \quad (5.31)$$

where the envelope $R(t)$ and the phase $\theta(t)$ are given by

$$R(t) = \sqrt{(n_I(t))^2 + (n_Q(t))^2} \quad (5.32)$$

and

$$\theta(t) = \tan^{-1}\left(\frac{n_Q(t)}{n_I(t)}\right). \quad (5.33)$$

To find the PDFs of $R(t)$ and $\theta(t)$, which we write for convenience as R and θ, we apply the PDF transform theorem, viz:-

$$f(R, \theta) = f(n_I, n_Q)|J_1| \qquad (5.34)$$

where J_1 is the Jacobian matrix. As n_I and n_Q are independent Gaussian random variables, the joint PDF of n_I and n_Q is

$$f(n_I, n_Q) = \frac{1}{2\pi N} \exp\left(-(n_I^2 + n_Q^2)/2N\right); -\infty < n_I, n_Q < \infty. \qquad (5.35)$$

Solving for n_I and n_Q in terms of R and θ gives a unique solution,

$$n_I = R\cos\theta \qquad (5.36)$$

and

$$n_Q = R\sin\theta. \qquad (5.37)$$

The Jacobian is

$$|J_1| = \begin{vmatrix} \frac{\partial n_I}{\partial R} & \frac{\partial n_I}{\partial \theta} \\ \frac{\partial n_Q}{\partial R} & \frac{\partial n_Q}{\partial \theta} \end{vmatrix} \qquad (5.38)$$

and from Equations 5.36 and 5.37,

$$|J_1| = \begin{vmatrix} \cos\theta & -R\sin\theta \\ \sin\theta & R\cos\theta \end{vmatrix} = R. \qquad (5.39)$$

Substituting $|J_1|$, n_I and n_Q from Equations 5.39, 5.36 and 5.37 into Equations 5.34 and 5.35 yields

$$f(R, \theta) = \frac{R}{2\pi N} \exp(-R^2/2N) \qquad (5.40)$$

where

$$0 \leq R \leq \infty \qquad (5.41)$$

and

$$-\pi \leq \theta < \pi. \qquad (5.42)$$

Integrating $f(R, \theta)$ over the range of θ and R gives the Rayleigh PDF

$$f(R) = \frac{R}{N} \exp(-R^2/2N) \qquad (5.43)$$

for the envelope R, and the uniform PDF

$$f(\theta) = \frac{1}{2\pi} \qquad (5.44)$$

for the phase θ.

Let us now return to Figure 5.11. If the message symbol transmitted is '00', say, the signal $x_0(t)$ is the envelope of the sum of a single tone f_0 and Gaussian noise. Signals x_1, x_2, and x_3 are the envelopes of band limited white Gaussian noise. However, we have just shown in Equation 5.43 that x_1, x_2 and x_3 have Rayleigh PDFs. Exchanging R in Equation 5.43 for x_i ; $i = 1, 2, 3$, we obtain the PDF of the envelope of the noise signals as

$$f(x_i) = \frac{x_i}{N} \exp\left[-\frac{x_i^2}{2N}\right].$$
(5.45)

We are now in a position to evaluate the cube term in Equation 5.29.

Our next objective is to determine $f(x_o|s_o)$. To do this we consider the receiver path that is carrying the data symbol '00' in the form of a carrier frequency f_o. Because of the receiver noise, the input to the envelope detector is a sinusoid plus the additive narrow band noise. From Equations 5.1 and 5.3 the signal applied to the envelope detector is

$$
\begin{aligned}
p_o(t) &= \sqrt{2S}\cos 2\pi f_o t + n_I(t)\cos 2\pi f_o t - n_Q(t)\sin 2\pi f_o t \\
&= \left(\sqrt{2S} + n_I(t)\right)\cos 2\pi f_o t - n_Q(t)\sin 2\pi f_o t
\end{aligned}
$$
(5.46)

where for convenience we have set E_s/T_s to S. The envelope of $p_o(t)$ is

$$R(t) = \sqrt{\left(\sqrt{2S} + n_I(t)\right)^2 + n_Q^2(t)}$$
(5.47)

and to determine its PDF we apply the following notation,

$$\chi_1 = \sqrt{2S} + n_I(t)$$
(5.48)

and

$$\chi_2 = n_Q(t)$$
(5.49)

where χ_1 and χ_2 are independent Gaussian random variables with

$$E[\chi_1] = \sqrt{2S}, \quad E[\chi_2] = 0$$
(5.50)

and

$$\sigma_{\chi_1}^2 = \sigma_{\chi_2}^2 = N.$$
(5.51)

Consequently the joint PDF of χ_1 and χ_2 can be written as

$$f(\chi_1, \chi_2) = \frac{1}{2\pi N} \exp\left(-\frac{(\chi_1 - \sqrt{2S})^2 + \chi_2^2}{2N}\right)$$
(5.52)

and the phase of $p_o(t)$ is

$$\theta(t) = \tan^{-1}(\chi_2/\chi_1).$$
(5.53)

Applying the PDF transformation of Equation 5.34 and the approach of determining polar PDFs used for receiver channels 1, 2, 3, we have

$$f(R, \theta) = \frac{R}{2\pi N} \exp\left(-\frac{(r^2 + 2S - 2\sqrt{2S}R\cos\theta)}{2N}\right). \tag{5.54}$$

The PDF of the envelope R is obtained by integrating over all values of θ,

$$
\begin{aligned}
f(R) &= \int_{-\pi}^{\pi} f(R, \theta)d\theta \\
&= \frac{R}{N} \exp\left(-\frac{2S + R^2}{2N}\right)\left[\frac{1}{2\pi}\int_{-\pi}^{\pi} \exp - \left(\frac{\sqrt{2S}R\cos\theta}{N}\right)d\theta\right]
\end{aligned}
\tag{5.55}
$$

The second term in Equation 5.55 is the modified Bessel function of the first kind and zero order defined by

$$I_o(a) \triangleq \frac{1}{2\pi}\int_{-\pi}^{\pi} \exp(a\cos u)du$$

enabling the PDF of R to be written as

$$f(R) = \frac{R}{N}I_o\left(\frac{\sqrt{2S}R}{N}\right)\exp\left(-\frac{R^2 + 2S}{2N}\right) \tag{5.56}$$

where $R \geq 0$. This is the Rician PDF.

As an aside, it is interesting to make the connection with the Rician PDF of fading signals in radio propagation channels described in Chapter 2. In the propagation case the Rician PDF of the fading envelope is due to a strong component (could be a line-of-sight) and a collection of scattered components whose combination have a Rayleigh envelope. In the QFSK demodulator, the envelope detector operates on the signal element tone $s_i(t)$ and an AWGN signal that has a Rayleigh envelope. The envelopes in both situations are the same, although the physical situations are radically different.

We are now able to formulate the conditional PDF $f(x_o|s_o)$ for the signal at the output of the envelope detector in the receiver path carrying the data symbol by merely replacing R in Equation 5.56 by the envelope signal $x_o(t)$, viz:-

$$f(x_o|s_o) = \frac{x_o}{N}I_o\left(\frac{\sqrt{2S}x_o}{N}\right)\exp\left(-\frac{x_o^2 + 2S}{2N}\right); 0 \leq x_o \leq \infty. \tag{5.57}$$

Having obtained expressions for the PDFs $f(x_i)$ in Equation 5.45 and $f(x_o|s_o)$ in Equation 5.57, we now continue with the evaluation of P_e. To determine the cube term in Equation 5.29 we first integrate $f(x_i)$ over the range 0 to x_o, namely

$$\int_0^{x_o} \frac{x}{N} \exp\left(-\frac{x^2}{2N}\right) dx = 1 - \exp\left(-\frac{x_o^2}{2N}\right) \tag{5.58}$$

to yield

$$\left[1 - \exp\left(-\frac{x_o^2}{2N}\right)\right]^3 = \sum_{j=0}^{3} \binom{3}{j} (-1)^j \exp\left(-j\frac{x_o^2}{2N}\right) \tag{5.59}$$

after applying the Binomial theorem. Substituting this result, and $f(x_o|s_o)$ from Equation 5.57 into Equation 5.29 yields the probability of making a correct symbol regeneration in the non-coherent demodulator as

$$P_c = \int_0^{\infty} \sum_{j=0}^{3} \binom{3}{j} (-1)^j \exp\left(-j\frac{x_o^2}{2N}\right) \frac{x_o}{N}$$

$$.I_o\left(\frac{\sqrt{2S}x_o}{N}\right) \exp\left(-\frac{x_o^2 + 2S}{2N}\right) dx_o. \tag{5.60}$$

Reversing the order of the summation and integration gives

$$P_c = \frac{1}{N} \sum_{j=0}^{3} \binom{3}{j} (-1)^j \exp(-2S/2N)$$

$$\cdot \int_0^{\infty} x_o \exp\left(-\frac{(1+j)x_o^2}{2N}\right) I_o\left(\frac{\sqrt{2S}x_o}{N}\right) dx_o. \tag{5.61}$$

As

$$\int_0^{\infty} x_o \exp\left(-\frac{(1+j)x_o^2}{2N}\right) I_o\left(\frac{\sqrt{2S}x_o}{N}\right) dx_o$$

$$= \frac{N}{1+j} \exp\left(\frac{2S}{2N(1+j)}\right) \tag{5.62}$$

we may express Equation 5.61 as

$$P_c = \exp(-2S/2N) \sum_{j=0}^{3} \binom{3}{j} \frac{(-1)^j}{1+j} \exp\left(\frac{S}{N(1+j)}\right). \tag{5.63}$$

Hence the probability of a symbol error becomes

$$P_e = 1 - P_c = \sum_{j=1}^{3} \binom{3}{j} \frac{(-1)^{j+1}}{1+j} \exp\left(-\frac{j}{j+1} \cdot \frac{S}{N}\right) \tag{5.64}$$

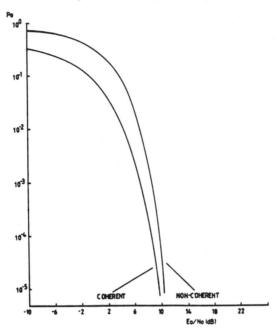

Figure 5.12: Non-coherent QFSK demodulation in the presence of AWGN

as the summation term is unity for $j = 0$.

Replacing S in Equation 5.64 by E_s/T_s, and as

$$N = N_o/T_s \tag{5.65}$$

then

$$P_e = \sum_{j=1}^{3} \binom{3}{j} \frac{(-1)^{j+1}}{1+j} \exp\left(-\frac{j}{j+1} \cdot \frac{E_s}{N_o}\right). \tag{5.66}$$

Because there are two bits in each symbol the energy per bit is given by Equation 5.24 enabling us to write the probability of symbol error in terms of energy per bit as

$$P_e = \frac{3}{2}\exp\left(-\frac{E_b}{N_o}\right) - \exp\left(-\frac{4}{3}\frac{E_b}{N_o}\right) + \frac{1}{4}\exp\left(-\frac{3}{2}\frac{E_b}{N_o}\right). \tag{5.67}$$

The variation of P_e as a function of E_b/N_o is shown in Figure 5.12, with the curve for coherent QFSK included as a bench mark.

5.2.2 Single Cochannel Interferer with Non-coherent Demodulation

An individual transmitted element of a QFSK signal is given by Equation 5.1, and therefore the signal from a single cochannel interferer has the form

$$c_i(t) = \sqrt{\frac{2E_I}{T_s}} \cos(2\pi f_i t + \phi) \tag{5.68}$$

where E_I is the symbol energy of the interferer, and ϕ is its phase whose fluctuations are uniformly distributed between zero and 2π. Suppose that each symbol is equally likely to be transmitted, and that the interference falls into only one demodulation channel at a time. As a consequence there are two mutually exclusive possibilities regarding the signal and its interference, namely, they can either fall into the same, or into different demodulation channels. An analysis utilising this approach is adopted in Reference [7].

When the interference and desired signal are both transmitting the symbol '00', say, the input to the non-coherent demodulator is

$$r_o(t) = s_o(t) + c_o(t) + n(t) \tag{5.69}$$

as the index i is zero. To ease the nomenclature we let

$$E_s/T_s = S \tag{5.70}$$

and

$$E_I/T_s = I \tag{5.71}$$

so the sum of the desired and interfering signal is

$$a(t) = \sqrt{2S} \cos 2\pi f_i t + \sqrt{2I} \cos(2\pi f_i t + \phi). \tag{5.72}$$

By re-writing Equation 5.72 as a phasor

$$a(t) = A(t) \cos(2\pi f_i t + \alpha(t)) \tag{5.73}$$

where

$$\alpha(t) = \tan^{-1} \left(\frac{-\sqrt{2T} \sin \phi}{\sqrt{2S} + \sqrt{2I} \cos \phi} \right) \tag{5.74}$$

is the phase angle and

$$A(t) = \sqrt{2S + 2I + 2\sqrt{2S}\sqrt{2I} \cos \phi} \tag{5.75}$$

is the envelope, the received signal is again composed of the sum of a sinusoidal input and additive noise. As a consequence the PDF of the envelope of the signal $x_o(t)$ at the output of the envelope detector in Figure 5.11, is Rician and is given by Equation 5.57 after exchanging $\sqrt{2S}$ for $A(t)$, viz:-

$$f(x_o|s_o, \phi) = \frac{x_o}{N} I_o \left(\frac{A(t)x_o}{N} \right) \exp \left(-\frac{x_o^2 + A(t)^2}{2N} \right). \tag{5.76}$$

The cochannel interference signal $c_o(t)$ is unable to pass through the BPFs having centre frequencies f_1, f_2, f_3. Only the noise signal $n(t)$ is filtered by these BPFs, to give signals whose envelopes have Rayleigh PDFs specified by Equation 5.45. As a consequence the conditional symbol error probability can be expressed as

$$P_{eo}|X_o, \phi = 1 - P(X_1 < X_o, X_2 < X_o, X_3 < X_o : X_o, \phi). \tag{5.77}$$

Following the procedure used in Section 5.2.1, we write Equation 5.77 in the form

$$P_{eo}|X_o, \phi = 1 - \int_0^\infty \left\{ \int_0^{x_o} \int_0^{x_o} \int_0^{x_o} \prod_{i=1}^3 f_{x_i}(x_i) dx_i \right\} f(x_o|s_o, \phi) dx_o \tag{5.78}$$

which can be expressed as

$$P_{eo}|X_o, \phi = 1 - \int_0^\infty \left[\int_0^{x_o} f_{x_j}(x_j) dx_j \right]^3 f(x_o|s_o, \phi) dx_o \tag{5.79}$$

to yield

$$P_{eo}|X_o, \phi = \sum_{k=1}^3 \binom{3}{k} \frac{(-1)^{k+1}}{1+k} \exp\left(-\frac{k}{k+1} \cdot \frac{A^2(t)}{2N} \right). \tag{5.80}$$

The average symbol error probability is determined by substituting for $A(t)$ and multiplying the above equation by the uniform PDF of ϕ and integrating over all possible values of ϕ to give

$$\begin{aligned}
P_{eo} = & \sum_{k=1}^3 (-1)^{k+1} \binom{3}{k} \frac{1}{1+k} \\
& I_o\left(\frac{2(SI)^{\frac{1}{2}}}{N} \left(\frac{k}{k+1} \right) \right) \exp\left(-\frac{(S+I)k}{N(1+k)} \right).
\end{aligned} \tag{5.81}$$

The probability of an error when the wanted signal and cochannel interference are passed by different bandpass filters in the demodulator will now be determined. The approach we use is similar to that employed in Section 5.2.1, and consequently many of the same equations are applicable. Assuming that the desired signal element has a carrier frequency f_o, while the interfering element has a carrier frequency f_1, then the envelopes of the signals at the outputs of envelope detectors ED_o and ED_1 in Figure 5.11 will have Rician PDFs. The envelope detectors ED_2 and ED_3 respond to Gaussian noise only and their outputs X_2 and X_3 have Rayleigh PDF's. The conditional probability of a symbol error is

$$P_{e1}|X_o = 1 - P\{X_1 < X_o, X_2 < X_o, X_3 < X_o|X_o\} \tag{5.82}$$

and so

$$P_{e1}|X_o = 1 - \int_0^\infty \left[\sum_{k=0}^{1} \binom{2}{k} (-1)^k \exp\left(-\frac{kx_o^2}{2N} \right) \right]$$

$$\left[\int_0^{x_o} \frac{x_1}{N} I_o \left(\frac{\sqrt{2I}x_1}{N} \right) \exp\left(-\frac{x_1^2 + 2I}{2N} \right) dx_1 \right] f(x_o|s_o) dx_o.$$
(5.83)

Substituting for $f(x_o|s_o)$ from Equation 5.57 and integrating gives

$$P_{e1} = 1 + \sum_{k=0}^{2} (-1)^{k+1} \binom{2}{k} \frac{1}{1+k} \exp(-k(2+k)b).$$

$$\left(1 - Q[(2a)^{\frac{1}{2}}, (2b)^{\frac{1}{2}}] + \frac{1}{2+k} \exp[-(a+b)] I_o \{2(ab)^{\frac{1}{2}}\} \right)$$
(5.84)

where

$$a = \frac{1}{N} \left(\frac{1+k}{2+k} \right) \quad \text{and} \quad b = \frac{S}{N} \frac{1}{(1+k)(2+k)}$$
(5.85)

and the Marcum Q function is defined as

$$Q(a,b) = \int_b^\infty I_o(at) \exp\left(-\frac{t^2 + a^2}{2} \right) dt.$$
(5.86)

On the assumption that the interfering element is equally likely to have any f_i ; $i = 0, 1, 2, 3$, the average probability of symbol error is

$$P_e = \frac{1}{4} P_{eo} + \frac{3}{4} P_{e1}.$$
(5.87)

The variation of P_e as a function of E_b/N_o for signal-to-interference ratios of 0, 2, 4, 6 and 10 dB are plotted in Figure 5.13. The curve for no cochannel interference is also displayed as a bench mark. With the exception of $E_b/N_o = 0$ dB, i.e., the interfering signal power is equal to the wanted signal power, all the curves exhibit a rapid fall in P_e with E_b/N_o. When the SIR exceeds 10 dB, E_b/N_o is always less than 2 dB from the curve for no cochannel interference for $P_e > 10^{-4}$.

5.2.3 Multiple Cochannel Interferers

Let us assume that there are a large number of interferers and that their effect is equivalent to an increase in the additive white Gaussian noise (AWGN) power at the receiver input. As a consequence P_e no longer tends towards zero as the SNR is increased. Instead it approaches an asymptotic

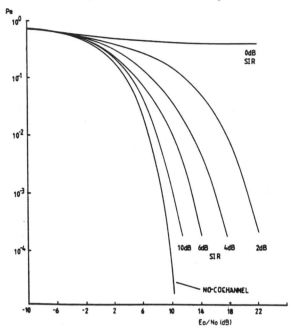

Figure 5.13: Non-coherent QFSK in the presence of one co-channel interferer

error probability determined by the level of the AWGN that corresponds to the multiple cochannel interference level. Results obtained by making this Gaussian noise assumption for the cochannel interference are pessimistic for small numbers of equal power interferers. However, if the number of interferers is greater than six, the assumption becomes more realistic.

5.2.3.1   Coherent Demodulation

The probability of symbol error P_e for a coherently detected QFSK system in the absence of cochannel interference is given by Equation 5.25. In order to express P_e in terms of carrier-to-noise power ratio C/N we note that

$$C = \frac{E_b}{T} \tag{5.88}$$

and

$$N = N_o B \tag{5.89}$$

where B is the receiver bandwidth. For the situation where

$$B = 1/T_s \tag{5.90}$$

we have the equivalence

$$\frac{C}{N} = \frac{E_b}{N_o}. \tag{5.91}$$

Consequently Equation 5.25 becomes,

$$P_e = Q\left(\sqrt{\frac{2C}{N}}\right). \tag{5.92}$$

We may model the effects of cochannel interference by adding an equivalent power C_I to the noise power N to give the probability of symbol error as

$$P_e = Q\left(\sqrt{\frac{2C}{N + C_I}}\right). \tag{5.93}$$

Defining the signal-to-interference power ratio as

$$\text{SIR} \triangleq \frac{C}{C_I} \tag{5.94}$$

and substituting into Equation 5.93 gives

$$
\begin{aligned}
P_e &= Q\left(\sqrt{\frac{2}{\frac{N}{C} + \frac{1}{\text{SIR}}}}\right) \\
&= Q\left(\sqrt{\frac{2E_b/N_o}{1 + \frac{1}{\text{SIR}}E_b/N_o}}\right). \tag{5.95}
\end{aligned}
$$

Figure 5.14 shows the graphical representation of Equation 5.102. Higher values of P_e occur at a given E_b/N_o compared to those in Figure 5.13, and the asymptotic nature of P_e at high E_b/N_o is apparent. The SIR must be above 7 dB to ensure that P_e can exceed 10^{-3} at high channel SNR values.

5.2.3.2 Non-Coherent Demodulation

The probability of symbol error P_e for a non-coherently detected QFSK system in the absence of cochannel interference is given by Equation 5.67. For convenience we define

$$\Upsilon \triangleq E_b/N_o \tag{5.96}$$

and so we can express the probability of error as

$$P_e = \frac{3}{2}\exp(-\Upsilon) - \exp\left(-\frac{4}{3}\Upsilon\right) + \frac{1}{4}\exp\left(-\frac{3}{2}\Upsilon\right). \tag{5.97}$$

As before we add an equivalent noise power C_I to account for the effect of cochannel interference. Noting the equivalence relation of Equation 5.91 we modify Υ to yield

$$\Upsilon = \frac{C}{N + C_I}. \tag{5.98}$$

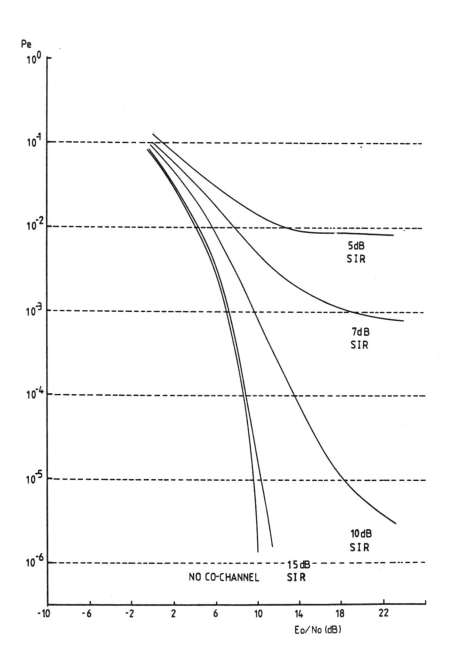

Figure 5.14: Coherent QFSK with multiple co-channel interferers

Using the definition for SIR of Equation 5.94 allows us to write

$$\Upsilon = \frac{E_b/N_o}{1 + \left(\frac{1}{SIR}\right)(E_b/N_o)}. \tag{5.99}$$

The probability of symbol error as function of channel SNR is displayed in Figure 5.15. The performance is seen to be worse than for coherent demodulation, amounting to an order of magnitude in P_e when the SIR is 7 dB.

5.3 QFSK Transmission Over Rayleigh Fading Channels

In this section we derive the symbol error rate performance when QFSK signals are transmitted over a non-frequency selective, slowly fading Rayleigh channel. This channel results in multiplicative distortion of the transmitted signal, and slow fading implies that the multiplicative process may be regarded as constant over at least one symbol period. The fading appears as a multiplicative factor α on the transmitted signal $s_i(t)$, [8, 9] and the received signal becomes

$$r(t) = \alpha s(t) + n(t). \tag{5.100}$$

We have already calculated the error probability for various modulation schemes with a time invariant channel i.e., a channel where α is a constant, which for convenience we set to unity. When α is no longer a constant we must include it in our equation for the probability of symbol error. In Section 5.2 the channel SNR, i.e., the carrier-to-noise ratio C/N, is E_s/N_o, but in fading conditions it becomes

$$\gamma = \alpha^2 \frac{E_s}{N_o} = \alpha^2 \frac{2E_b}{N_o}. \tag{5.101}$$

We compute the symbol error probability when α is a random variable by averaging $P_e(\gamma)$ over the PDF of γ, namely $f(\gamma)$, viz:-

$$P_e = \int_0^\infty P_e(\gamma)f(\gamma)d\gamma. \tag{5.102}$$

To obtain the PDF of γ we utilise the following transformation

$$f(\gamma) = f(\alpha_1)\left|\frac{d\alpha_1}{d\gamma}\right| + f(\alpha_2)\left|\frac{d\alpha_2}{d\gamma}\right| \tag{5.103}$$

where $f(\alpha)$ is a Rayleigh PDF,

$$f(\alpha) = \frac{\alpha}{\alpha_o^2}\exp(-\alpha^2/2\alpha_o^2) \tag{5.104}$$

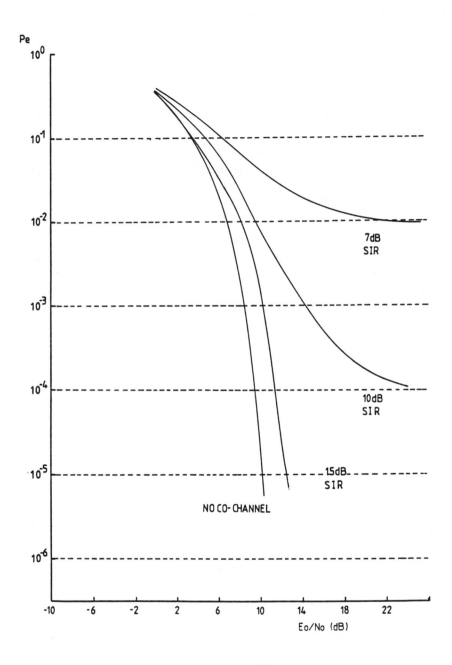

Figure 5.15: Non-coherent QFSK with multiple co-channel interferers

and the average value of α^2 is,

$$E[\alpha^2] = 2\alpha_o^2. \tag{5.105}$$

From Equation 5.101

$$\alpha_1 = \sqrt{\gamma N_o / E_s} \tag{5.106}$$

$$\alpha_2 = -\sqrt{\gamma N_o / E_s} \tag{5.107}$$

and furthermore,

$$\frac{d\alpha_1}{d\gamma} = \frac{1}{2}\sqrt{\frac{1}{\gamma}\frac{N_o}{E_s}} \tag{5.108}$$

$$\frac{d\alpha_2}{d\gamma} = -\frac{1}{2}\sqrt{\frac{1}{\gamma}\frac{N_o}{E_s}}. \tag{5.109}$$

However, the Rayleigh distribution does not exist for negative values and therefore from Equations 5.103 to 5.109,

$$f(\gamma) = f(\alpha_1)\left|\frac{d\alpha_1}{d\gamma}\right|. \tag{5.110}$$

Substituting $f(\alpha_1)$ and $d\alpha_1/d\gamma$ from Equations 5.104 and 5.108 into Equation 5.110 yields

$$f(\gamma) = \frac{N_o}{2\alpha_o^2 E_s} \exp\left\{-\left(\frac{N_o}{E_s}\frac{\gamma}{2\alpha_o^2}\right)\right\}. \tag{5.111}$$

Defining the average channel signal-to-noise ratio as

$$\Lambda \triangleq 2\alpha_o^2 \frac{E_s}{N_o} \tag{5.112}$$

gives the PDF of γ as

$$f(\gamma) = \frac{1}{\Lambda} \exp-\left(\frac{\gamma}{\Lambda}\right). \tag{5.113}$$

This function is known as the chi-square probability distribution.

5.3.1 Coherent Demodulation

The probability of symbol error for QFSK with coherent demodulation in a fading environment becomes with the aid of Equations 5.23 and 5.101

$$P_e(\gamma) = Q\left(\sqrt{\alpha^2 \frac{E_s}{N_o}}\right)$$

$$= Q\left(\sqrt{\gamma}\right) \tag{5.114}$$

and so from Equation 5.113 and 5.114,

$$P_e(\Lambda) = \int_0^\infty Q\left(\sqrt{\gamma}\right) \frac{1}{\Lambda} \exp(-\gamma/\Lambda) d\gamma. \tag{5.115}$$

Substituting for the Q function gives

$$P_e(\Lambda) = \frac{1}{\Lambda} \int_0^\infty \left\{ \frac{1}{\sqrt{2\pi}} \int_{\sqrt{\gamma}}^\infty \exp\left[-\frac{1}{2}\nu^2\right] d\nu \exp(-\gamma/\Lambda) \right\} d\gamma. \tag{5.116}$$

The inner integral can be evaluated using a series expansion, while $P_e(\Lambda)$ is obtained by numerical integration using Simpson's rule.

5.3.2   Non-Coherent Demodulation

The probability of symbol error for non-coherent demodulation as a function of the received SNR is given by Equation 5.67 for transmissions over Gaussian channels. We may rewrite this Equation for the Rayleigh fading channel as

$$P_e(\gamma) = \frac{3}{2} \exp\left(-\frac{\gamma}{2}\right) - \exp\left(-\frac{2}{3}\gamma\right) + \frac{1}{4} \exp\left(-\frac{3}{4}\gamma\right). \tag{5.117}$$

Substituting $P_e(\gamma)$ and $f(\gamma)$ into Equation 5.102 gives

$$\begin{aligned} P_e(\Lambda) &= \frac{1}{\Lambda} \int_0^\infty \exp\left(-\frac{\gamma}{\Lambda}\right) \left[\frac{3}{2} \exp(-\frac{\gamma}{2}) - \right. \\ &\quad \left. \exp\left(-\frac{2}{3}\gamma\right) + \frac{1}{4} \exp\left(-\frac{3}{4}\gamma\right) \right] d\gamma \end{aligned} \tag{5.118}$$

and upon integrating

$$P_e(\Lambda) = \frac{3}{2+\Lambda} - \frac{3}{2\Lambda+3} + \frac{1}{3\Lambda+4}. \tag{5.119}$$

Curves of $P_e(\Lambda)$ against average signal-to-noise ratio are given for coherent QFSK and non-coherent QFSK in Figure 5.16. When compared to the curves shown in Figure 5.12 for the Gaussian channel we see the devastating effect of fading on the performance of the QFSK modulation. This occurs because even at high *average* values of channel SNR, a deep fade means that the received signal is swamped by the channel noise giving a bit error rate of 0.5 for a short time. Nevertheless, P_e values of 10^{-3} can be achieved for channel SNR values above 30 dB. By deploying channel coding the reduction in channel SNR for this P_e will decrease by 10 dB or more.

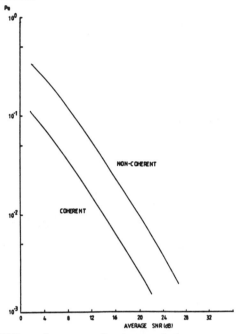

Figure 5.16: QFSK performance for transmissions over a flat Rayleigh fading channel

5.4 Conclusions

This chapter has examined QFSK for mobile radio systems where the signalling rates are sufficiently low to ensure that the received signal experiences flat fading rather than frequency selective fading for the majority of the time. The S900-D system conceived for the conventional large cell system, has its effective transmission rate (and hence bandwidth) reduced by the use of multilevel modulation. In this way the coherence bandwidth of the channel is not exceeded for a significant proportion of the time. The S900-D system is shown to be workable in static channels, flat fading channels and in the presence of co-channel interference when coherently demodulated. Non-coherent demodulation is not power efficient and its use with four-level FM modulation requires high signal-to-interference ratios at the receiver to achieve an acceptable performance. We note that the use of an S900-D like system may be more appropriate in a cordless telecommunications environment, where its low complexity would be an advantage.

Bibliography

[1] **K-H.Tietgen**: "Numerical Modulation Methods Applied in the FD/TDMA-System S900-D," *Proc. 2nd Nordic Seminar on Digital Land Mobile Radio Communications*, Stockholm, Paper No.33B, Oct 1986.

[2] **D.E.Pfitzmann and H-P.Ketterling**: "A New CP-4FSK Sampling Demodulator for S 900-D," *Proc. 2nd Nordic Seminar on Digital Land Mobile Radio Communications*, Stockholm, Paper No.33A, Oct 1986.

[3] **H-P.Ketterling**: "The Digital Mobile Radio Telephon System S 900-D. *Proc. 2nd Nordic Seminar on Digital Land Mobile Radio Communications*, Stockholm, Paper No.32, Oct 1986.

[4] **R.R.Anderson and J.Salz**: "Spectra of Digital FM," *BSTJ*,pp.1165-1189, Jul/Aug 1965.

[5] **G.L.Choudhury and S.S.Rappaport**: "Diversity ALOHA–A Random Access Scheme for Satellite Communications," *IEE Trans. Commun.*, Vol 31, No.3, pp.450–457, Mar 1983.

[6] **A.P.Clark**: *"Principles of Digital Data Transmission,"* Pentech Press, 1976.

[7] **M.J.Massaro**: "Error Performance of M-ary Non-coherent FSK in the Presence of CW Tone Interference," *IEEE Trans. Commun.*, pp.1363-1369, Nov 1975.

[8] **S.Stein** and **J.J.Jones**: *"Modern Communication Principles,"* McGraw-Hill, 1967.

[9] **J.G.Proakis**: *"Digital communications"*, MacGraw-Hill, New York, 1982.

Chapter 6

Wideband Systems

I.J.Wassell[1] and R.Steele[2]

6.1   Generalised Phase Modulation

Binary phase shift keying (BPSK) and quaternary phase shift keying (QPSK) were established by the mid-1960's. During the next decade new concepts in modulation resulted in the modulating data being shaped to yield smooth phase transitions in the carrier waveform at symbol boundaries. Further, the phase was constrained to change in a continuous fashion. An important example of this type of modulation is minimum shift keying (MSK) [1] in which the phase changes linearly over a symbol period. MSK is also called fast FSK [2] (FFSK) as well as continuous phase FSK (CPFSK) [3] with a modulation index of a 0.5. Its sidelobe spectral energy relative to non-continuous phase modulation methods, such as QPSK, is significantly reduced. MSK modulation belongs to the class of continuous phase modulation (CPM) [4] and as its phase response is shaped over a symbol period it is referred to a full response CPM.

As pressure mounted to increase the spectral efficiency of narrow band radio communications it became necessary to devise methods of reducing the spectral spillage of the CPM signal into adjacent channels. A basic technique evolved whereby the phase response to a symbol was spread over a number of symbol periods. By deliberately introducing intersymbol interference (ISI) the spectrum of the modulated signal became more compact. Although the CPM signal could be demodulated in the presence of ISI, an enhanced performance was achieved by removing ISI at the receiver by

[1] Multiple Access Communications Ltd.
[2] University of Southampton and Multiple Access Communications Ltd.

means of equalisation. The equalisation methods employed were essentially
derived from those used in high bit rate transmissions over telephone networks where ISI occurred.

CPM schemes that deliberately introduce ISI are generally known as
partial response modulations as only part of the symbol shaping is over a
symbol period. By increasing the spectral energy in the channel compared
to the out-of-channel energy, the power efficiency is increased. Thus for
a given BER the transmitted power can be reduced. Further, all CPM
systems have constant carrier envelopes enabling non-linear amplifiers to
be used thereby providing a high dc power-to-RF power conversion ratio.
Examples of partial response CPM are tamed frequency modulation (TFM),
generalised TFM (GTFM), Gaussian MSK (GMSK) and multi-h CPFSK.

6.1.1   Digital Phase Modulation

Digital phase modulation (DPM) [5, 6] is a form of CPM that is well suited
to VLSI implementation. The main distinction between DPM and most
other forms of CPM is that it is essentially a phase modulation technique in
which the shaped symbol pulses are applied directly to a phase modulator.
The demodulation of DPM is easier to implement than with other forms
of CPM which integrate the data signal prior to phase modulation, see
Section 6.1.2.

In DPM the data signal $\alpha(t)$ is applied to the modulator which houses a
digital phase shaping filter having an impulse response $q(t)$. For simplicity
we will consider the data to be binary, although CPM can accommodate
M-ary data symbols. The convolution of $\alpha(t)$ with $q(t)$ yields the phase
signal $\phi(t, \alpha)$, a phase that is dependent on both the data α and time t.
This $\phi(t, \alpha)$ signal addresses two ROMs to yield $\cos \phi(t, \alpha)$ and $\sin \phi(t, \alpha)$
as shown in Figure 6.1. The front-end of the DPM modulator is therefore
completely digital. To produce the modulated signal for radio transmission,
digital-to-analogue conversion (DAC) of both $\cos \phi(t, \alpha)$ and $\sin \phi(t, \alpha)$ ensues, and after passing through anti-aliasing filters, the resulting analogue
signals modulate the quadrature carriers $\cos 2\pi f_o t$ and $\sin 2\pi f_o t$, where f_o
is the carrier frequency. The radiated signal is formulated as

$$\begin{aligned}
\tilde{s}(t, \alpha) &= A \cos \phi(t, \alpha) \cos 2\pi f_o t - A \sin \phi(t, \alpha) \sin 2\pi f_o t \\
&= A \cos \left(2\pi f_o t + \phi(t, \alpha) \right)
\end{aligned} \tag{6.1}$$

where the amplitude A of $\tilde{s}(t, \alpha)$ is constant, independent of the data. This
constant envelope feature is an advantage when efficient class-C amplifiers
are used. The power in $\tilde{s}(t, \alpha)$ is

$$\frac{A^2}{2} = \frac{E_b}{T} \tag{6.2}$$

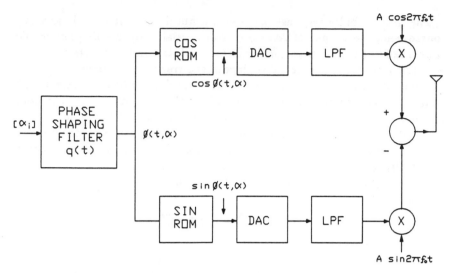

Figure 6.1: Block diagram of the bandpass DPM modulator

where E_b is the energy per bit and T is the duration of a bit. Hence the bandpass RF signal becomes

$$\tilde{s}(t, \alpha) = \sqrt{\frac{2E_b}{T}} \cos(2\pi f_o t + \phi(t, \alpha)) \tag{6.3}$$

which we may write as

$$\begin{aligned} \tilde{s}(t, \alpha) &= Re\left[\sqrt{\frac{2E_b}{T}} \exp\left\{j(2\pi f_o t + \phi(t, \alpha))\right\}\right] \\ &= Re\left[s(t, \alpha) \exp(j2\pi f_o t)\right] \end{aligned} \tag{6.4}$$

where $s(t, \alpha)$ is the complex baseband signal

$$s(t, \alpha) = \sqrt{\frac{2E_b}{T}} \exp j\phi(t, \alpha). \tag{6.5}$$

The inphase (I) and quadrature (Q) components of $s(t, \alpha)$ are

$$s_I(t, \alpha) = \sqrt{\frac{2E_b}{T}} \cos \phi(t, \alpha) \tag{6.6}$$

and

$$s_Q(t, \alpha) = \sqrt{\frac{2E_b}{T}} \sin \phi(t, \alpha) \tag{6.7}$$

respectively.

The filter impulse response $q(t)$ spans a number of bit periods and as a consequence the output $\phi(t, \alpha)$ consists of partially overlapping pulses. We allow the filter $q(t)$ to intentionally introduce intersymbol interference (ISI) in order to contain the spectral spillage of the transmitted signal $\tilde{s}(t, \alpha)$ into adjacent channels. As a consequence of the modulator filter an equaliser is employed at the receiver to remove these ISI effects, even if the transmission channel is ideal. For a Gaussian channel the design of the equaliser is a relatively straightforward procedure as we know exactly how the ISI was introduced. We will see that the task is more daunting in the presence of the multipath effects experienced in mobile radio channels.

The information carrying phase in Equation 6.3 is

$$\phi(t, \alpha) = \sum_{i=-\infty}^{\infty} \alpha_i q(t - iT) \tag{6.8}$$

where $\alpha_i = \ldots \alpha_{-2}, \ \alpha_{-1}, \ \alpha_0, \ \alpha_1, \ \ldots$ is an infinitely long sequence of uncorrelated data bits. The phase shaping filter of Figure 6.1 has the discrete time FIR filter arrangement shown in Figure 6.2. The data is applied to the filter at a rate $1/T$, and the delay D in each stage of the filter is less than T. Thus D is the sample period of the filter, and

$$\eta = T/D \tag{6.9}$$

is the oversampling ratio. Accordingly $1/D$ is the sampling rate of the filter. We may, therefore, express $\phi(t, \alpha)$ as a sequence of samples at instants $t = nD$, where n is the sampling instant number, viz:-

$$\phi(n, \alpha) = \sum_{i=-\infty}^{\infty} \alpha_i q_{n-i\eta} \tag{6.10}$$

and $\{q_n\}$ is the weighting sequence of the filter. The duration of the impulse response of the phase shaping filter is L symbol periods, and hence the number of filter coefficients is

$$K = \eta L. \tag{6.11}$$

The restriction on $(n - i\eta)$ is clearly 0 to $K - 1$.

During each bit period the phase shaping filter is excited by the sampled value of the bit and $\eta - 1$ consecutive zeros. The modulation index is defined [5] in terms of the maximum possible phase change during one bit interval T, namely,

$$h_p \triangleq \max\{\phi(t + T, \alpha) - \phi(t, \alpha)\}/\pi \ . \tag{6.12}$$

If an impulse sequence $\ldots 0, 0, 1, 0, 0, 0, 0, \ldots$ is applied to the filter its output is the filter weighting sequence $\{q_n\}$ whose maximum value is $\max\{q_n\}$.

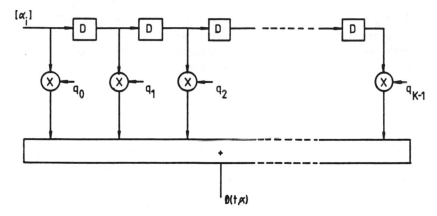

Figure 6.2: Phase shaping FIR filter

When the input sequence ... $0, 0, -1, 0, 0, 0, 0,$ ... the minimum value of the filter output is $-\max\{q_n\}$. However, when the DPM modulator is transmitting data there are a number of symbols passing through the FIR filter at any one time, and as α can be either $+1$ or -1, the value of h_p becomes

$$h_p = 2\max\{q_n\}/\pi \; ; \; n = 0, \; 1, \ldots, \; K - 1 \qquad (6.13)$$

The selection of the filter coefficients is critical as it determines both the power spectrum of the transmitted signal and the BER performance of the modem. It has been found [5] that raised cosine and similarly shaped pulses are suitable. Of particular interest is the raised cosine (RC) impulse response that spans L symbol periods, abbreviated to "L-RC", and defined by

$$q(t) = \begin{cases} \frac{\beta}{LT} \left(1 - \cos \frac{2\pi}{LT} t\right) & ; \quad 0 < t < LT \\ 0 & ; \quad \text{elsewhere} \end{cases} \qquad (6.14)$$

where β is a system parameter. The phase shaping filter is generally implemented in a digital form, and the weighting sequence $\{q_n\}$ of this FIR filter has coefficients [7] that are the samples of $q(t)$ weighted by the sampling period D, viz:-

$$q_n = Dq(nD) \qquad (6.15)$$

By replacing t by nD in Equation 6.14, and with the aid of Equation 6.9 we may express Equation 6.15 as

$$q_n = \frac{\beta}{L\eta} \left[1 - \cos \frac{2\pi}{L\eta} n\right] \qquad (6.16)$$

where the parameter β is adjusted to achieve the desired modulation index.

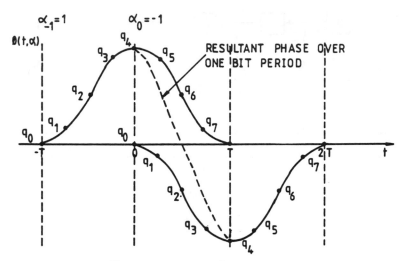

Figure 6.3: DPM phase response

The maximum value of q_n is $2\beta/L\eta$ and hence

$$\max\{q_n\} = \frac{2\beta}{L\eta} \tag{6.17}$$

From Equations 6.13 and 6.17

$$\beta = \frac{L\eta\pi h_p}{4} \tag{6.18}$$

and on substituting β into Equation 6.9 gives the weighting sequence

$$q_n = \frac{\pi h_p}{4}\left(1 - \cos\frac{2\pi}{L\eta}n\right) \tag{6.19}$$

Example

Consider an FIR impulse response extending over two symbol periods, with four samples per data bit, i.e., $L = 2$, $\eta = 4$. Figure 6.3 shows the separate contributions of α_{-1} and α_o to $\phi(t,\alpha)$, where α_{-1} and α_o are +1 and -1, respectively. We see that α_{-1} and α_o generate phase waveforms having amplitudes q_0 to q_7 originating at the instants when α_{-1} and α_o are applied. The resulting phase waveform over the interval 0 to 4 D, i.e., over one bit period is shown by the dotted line, where it is assumed that no other bits have activated the filter. From Equation 6.5,

$$\phi(n,\alpha) = \alpha_{-1}q_{n+4} + \alpha_o q_n \tag{6.20}$$

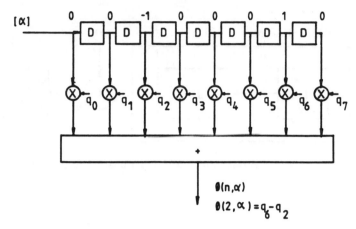

Figure 6.4: DPM FIR filter with the data sequence that yields the response in Figure 6.3.

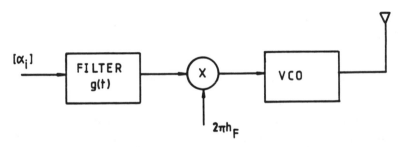

Figure 6.5: Direct FM modulator

and for a particular instant, say when $n = 2$,

$$\phi(2, \alpha) = q_6 - q_2 \qquad (6.21)$$

and Figure 6.4 shows the condition of the FIR filter that yields this $\phi(2, \alpha)$.

6.1.2 Digital Frequency Modulation

We designate forms of CPM where the integration of the phase occurs before phase modulation as digital frequency modulation (DFM) [8, 9]. A DFM signal can be produced by applying the data sequence $\{\alpha_i\}$ to a filter having an impulse response $g(t)$ that spreads each data bit over a number of bit intervals. The resulting signal is multiplied by $2\pi h_F$, where h_F is the DFM modulation index, and applied to a voltage control oscillator (VCO), see Figure 6.5. The filtered data sequence changes the frequency of the VCO directly thereby producing DFM.

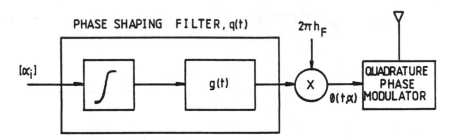

Figure 6.6: Production of FM via phase modulation

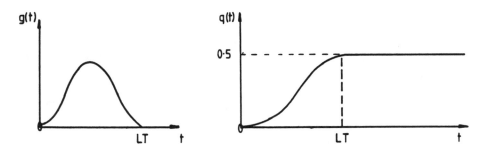

Figure 6.7: Frequency and phase shaping filter responses

An alternative approach that is easier to implement is to integrate the data sequence $\{\alpha_i\}$ prior to filtering, and apply the resultant signal to a phase modulator. The arrangement is shown in Figure 6.6, where the phase signal is

$$\phi(t, \alpha) = 2\pi h_F \int_{-\infty}^{t} \sum_{i=-\infty}^{\infty} \alpha_i g(\tau - iT) d\tau. \qquad (6.22)$$

The impulse response $q(t)$ is given by

$$q(t) = \int_{-\infty}^{t} g(\tau) d\tau \qquad (6.23)$$

where for a causal system

$$g(t) = 0 \; ; \; LT \; \leq \; t \; \leq \; 0 \qquad (6.24)$$

and the impulse response $g(t)$ is normalised such that

$$q(t) = 0.5 \; ; \; t \; \geq \; LT \; . \qquad (6.25)$$

Stylised responses of $g(t)$ and $q(t)$ are shown in Figure 6.7.

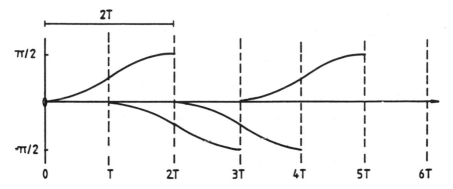

Figure 6.8: Output of filter q(t)in response to a sequence of data, $L = 2$

The phase of the DFM signal in the n-th symbol interval can be represented with the aid of Equations 6.22 and 6.23 as

$$\phi(t, \alpha) = 2\pi h_F \sum_{i=-\infty}^{n} \alpha_i q(t - iT) \; ; \quad nT \leq t \leq (n+1)T \qquad (6.26)$$

and upon rearranging and employing Equation 6.25,

$$\begin{aligned} \phi(t, \alpha) &= 2\pi h_F \sum_{i=n-L+1}^{n} \alpha_i q(t - iT) + \pi h_F \sum_{i=-\infty}^{n-L} \alpha_i \\ &= \theta(t, \alpha) + \theta_n \end{aligned} \qquad (6.27)$$

where $\theta(t, \alpha)$ is the correlative state vector that depends on the L most recent symbols currently in the filter.

In Figure 6.8 we display the $q(t)$ response for successive bits applied to the filter when $L = 2$. The second term θ_n in Equation 6.27 is the accumulated phase of all the previous symbols that have passed through the filter and it is referred to as the phase state. It is the elimination of this second term that gives DPM an implementation advantage over DFM, or over any similar modulation technique such as GMSK [10] or TFM [11].

Minimum Shift Keying Minimum shift keying (MSK) can be produced in a similar way to off-set quadrature phase shift keying (OQPSK) except that sinusoidal rather than rectangular shaping of the RF quadrature signal envelopes is performed. A good description of this approach is given by Pasupathy [1]. An alternative method, and the one that fits appropriately in our discourse, is to simply select $g(t)$ in Figure 6.6 to be

$$g(t) = \left\{ \begin{array}{ll} 1/(2T) & ; \quad 0 \leq t \leq T \\ 0 & ; \quad \text{elsewhere} \end{array} \right. . \qquad (6.28)$$

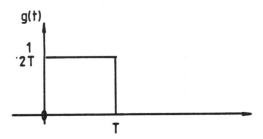

Figure 6.9: Impulse response $g(t)$ of filter for MSK modulation

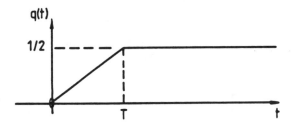

Figure 6.10: Impulse response q(t) of filter for MSK modulation

The impulse response $g(t)$ is shown in Figure 6.9, and the corresponding $q(t)$ in Figure 6.10. From Equation 6.26 the linear variation of $q(t)$ from 0 to T yields a linear phase transition of $\pi/2$ because h_F for MSK is defined to be a half, i.e., $2\pi h_F \alpha_i$ can take the values $\pm\pi$ and q_{t-iT} changes linearly from 0 to 0.5 during a symbol period.

The phase tree for MSK is displayed in Figure 6.11, where logical ones and logical zeros are represented by 1 and -1 respectively. When the input data bit is a logical one the phase increases by $\pi/2$, while it decreases by $\pi/2$ when the data is a logical zero. Notice that after two bit periods the phase tree is fully developed as any phases in excess of $\pm\pi$ wrap around to $\mp\pi/2$. Only four possible phases exist enabling us to represent the tree structure by the trellis arrangement shown in Figure 6.12. If in the tree diagram a logical 1 increases the phase from π to $3\pi/2$, the change in the trellis is from $-\pi$ (as this is identical to $+\pi$) to $-\pi/2$, and so forth.

When viewed as a form of frequency modulation the filter $g(t)$ in Figure 6.5 with the impulse response shown in Figure 6.9 implies a step change in frequency upon receipt of a data bit of opposite polarity. Thus a logical "1" ($\alpha_i = 1$) at the input of the filter in Figure 6.5 will cause the modulated signal to have a frequency of $\omega_c + \omega_d$, where ω_c is the nominal centre frequency and ω_d is the frequency deviation. From Figure 6.5 it can be seen

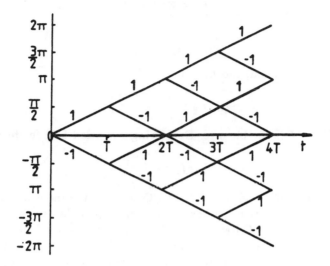

Figure 6.11: MSK phase tree

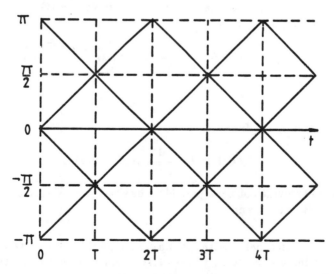

Figure 6.12: MSK phase trellis

that the frequency deviation of the VCO is given by the expression

$$\omega_d = 2\pi h_f g(t). \tag{6.29}$$

For MSK

$$g(t) = \frac{1}{2T} \tag{6.30}$$

over a symbol interval, and $h_f = 0.5$, giving

$$\omega_d = 2\pi \cdot \frac{1}{2} \cdot \frac{1}{2T} = \frac{\pi}{2} f_b \tag{6.31}$$

where $f_b = 1/T$ is the bit rate. Expressing Equation 6.31 in Hertz gives

$$f_d = \frac{f_b}{4}. \tag{6.32}$$

So upon receipt of a logical "1" the VCO frequency is increased by $f_b/4$ Hz from the nominal centre frequency. Similarly upon the receipt of a logical "0" ($\alpha_i = -1$) the VCO frequency is decreased by $f_b/4$ Hz from the nominal centre frequency. Because we have a simple frequency shift keyed system (albeit without phase discontinuities at bit intervals) it is possible to employ non-coherent frequency discriminator demodulation, as well as coherent phase demodulation. The loss in performance incurred by using a frequency discriminator is balanced by somewhat easier implementation.

Gaussian Minimum Shift Keying

A particular form of DFM where the impulse response $g(t)$ is Gaussian shaped is known as Gaussian minimum shift keying (GMSK) [10] having an impulse response $g(t)$ of

$$g(t) = \frac{1}{2T} \left[Q\left(2\pi B_b \frac{t - T/2}{\sqrt{\ell n 2}}\right) - Q\left(2\pi B_b \frac{t + T/2}{\sqrt{\ell n 2}}\right) \right] \tag{6.33}$$

for

$$0 \leq B_b T \leq \infty$$

where $Q(t)$ is the Q-function

$$Q(t) = \int_t^\infty \frac{1}{\sqrt{2\pi}} \exp(-\tau^2/2) d\tau \tag{6.34}$$

B_b is the bandwidth of a low pass filter having a Gaussian shaped spectrum, T is the bit period, and

$$B_N = B_b T \tag{6.35}$$

is the normalised bandwidth. It may be shown [12] that $g(t)$ is the result of convolving a non return to zero (NRZ) data stream of unity amplitude with a Gaussian low pass filter whose impulse response is

$$h_t(t) = \sqrt{\frac{2\pi}{\ell n 2}} B_b \exp\left(-\frac{2\pi^2 B_b^2}{\ell n 2} t^2\right). \tag{6.36}$$

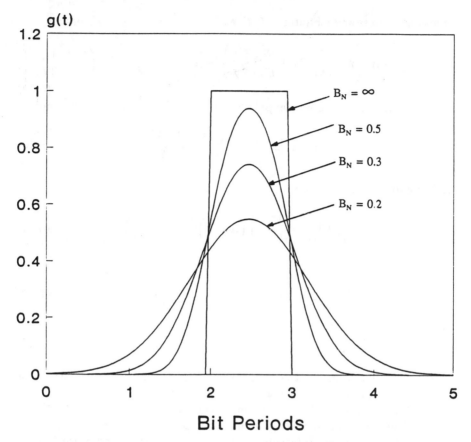

Figure 6.13: Impulse response $g(t)$ for GMSK for different values of B_N

To give the desired discrete time values of the Gaussian impulse response Equation 6.33 is multiplied by D and the time variable t is replaced by nD, where n is an integer. Upon using Equations 6.9 and 6.35, Equation 6.33 becomes

$$g_n = \frac{1}{2\eta} \left[Q \left(\frac{2\pi}{\sqrt{\ell n 2}} B_N \left(\frac{n}{\eta} - \frac{1}{2} \right) \right) - Q \left(\frac{2\pi}{\sqrt{\ell n 2}} B_N \left(\frac{n}{\eta} + \frac{1}{2} \right) \right) \right] \quad (6.37)$$

The response $g(t)$ is shown in Figure 6.13 for different B_N values.

To enable g_n to be represented in terms of a power series, the Q-functions are first replaced by error functions, i.e.,

$$Q(\theta) = \frac{1}{2} - \frac{1}{2} \operatorname{erf} \left(\theta / \sqrt{2} \right) \quad (6.38)$$

enabling us to rewrite Equation 6.37 as

$$g_n = \frac{1}{4\eta} \left[\text{erf}\left(\pi\sqrt{\frac{2}{\ell n2}}B_N\left(\frac{n}{\eta}+\frac{1}{2}\right)\right) - \text{erf}\left(\pi\sqrt{\frac{2}{\ell n2}}B_N\left(\frac{n}{\eta}-\frac{1}{2}\right)\right) \right]$$

(6.39)

The error functions are now expressed by the following series expansion

$$\text{erf}(z) = \frac{2}{\sqrt{\pi}}\sum_{k=0}^{\infty}\frac{(-1)^k z^{2k+1}}{k!(2k+1)}$$

(6.40)

yielding after simplification

$$g_n = c\left[\sum_{k=0}^{\infty}\frac{(-1)^k b^a}{ak!}\left(\left(n+\frac{\eta}{2}\right)^a - \left(n-\frac{\eta}{2}\right)^a\right)\right]$$

(6.41)

where

$$a = 2k+1, \; b = \pi\sqrt{\frac{2}{\ell n2}}\frac{B_N}{\eta}$$

(6.42)

and

$$c = \frac{1}{2\sqrt{\pi\eta}}.$$

(6.43)

The use of 100 terms in the expansion of Equation 6.41 and quadruple precision arithmetic are satisfactory for generating g_n.

In implementing a GMSK modulator the time impulse response $g(t)$ of the Gaussian filter must clearly be limited. Specifically, it is symmetrically truncated to L symbol intervals. When the modulation index for GMSK is 0.5, the phase state θ_n in Equation 6.27 can, like MSK, only assume four values 0, $\pi/2$, π and $3\pi/2$. By using $h_F = 0.5$, a GMSK signal can be demodulated by a parallel coherent minimum shift keying (MSK) demodulator, and other sub-optimal receivers.

When viewed as a form of frequency modulation the effect of the Gaussian filter with impulse response $g(t)$ is to prevent the instantaneous changes of frequency inherent in MSK. Consequently, a modulated signal power spectrum with much lower levels of side-lobe energy than that of MSK results. The instantaneous frequency changes due to spreading each bit over more than one bit interval is similar to that of the phase signal in the DPM modulator. Thus if $g(t)$ has a duration of 2 bits and a Gaussian shape, then Figure 6.3 describes the instantaneous frequency of the GMSK modulator output signal. As the modulator filter bandwidth decreases, the duration of its impulse response increases. Normalised bandwidths, B_N, of 0.5 and 0.3 correspond to impulse response durations of approximately 2 and 3 bits, respectively. Examples of instantaneous frequency variations are shown in Figure 6.14. It can be seen that as B_N is decreased, the

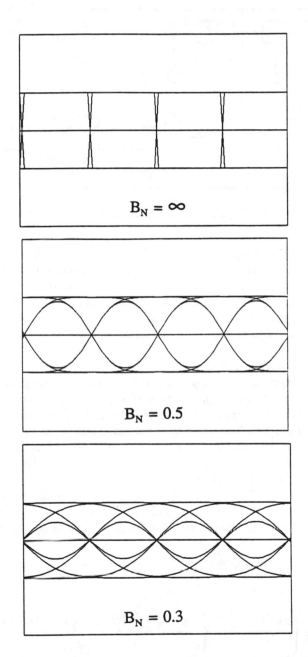

Figure 6.14: Instantaneous frequency variations of GMSK

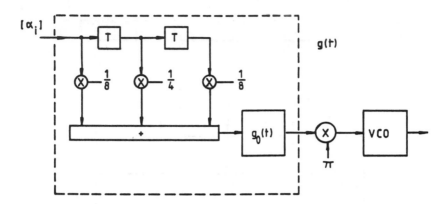

Figure 6.15: TFM modulator

number of possible frequency trajectories increases. This means that a frequency discriminator with simple threshold decisions has significantly degraded performance in demodulating GMSK when B_N is decreased to 0.3 [13]. For $B_N < 0.3$, discriminator detection can be employed provided additional post-discriminator processing is used. The advantage of smaller normalised filter bandwidths is the reduction in spectral occupancy of the modulated signal.

Tamed Frequency Modulation

In 1978 de Jager and C B Dekker [11] proposed a method of frequency modulation that resulted in the modulated signal having a compact power spectrum without sidelobes. They called the technique tamed frequency modulation (TFM). In TFM, the spectral compactness of the transmitted signal is achieved by careful control of its phase transitions. The premodulation filter $g(t)$ consists of a cascaded 3-tap transversal filter, and a low pass filter $g_o(t)$, see Figure 6.15.

The phase shifts of the modulated carrier over one bit period are restricted to either 0, $\pm\pi/4$ or $\pm\pi/2$, and are determined by the three latest consecutive input binary data bits. The frequency shaping function $g(t)$ is

$$g(t) = \frac{1}{8}[g_o(t - T) + 2g_o(t) + g_o(t + T)] \qquad (6.44)$$

where

$$g_o(t) \approx \frac{1}{T}\left[\frac{\sin\left(\frac{\pi t}{T}\right)}{\frac{\pi t}{T}} - \frac{\pi^2}{24}\left\{\frac{2\sin\left(\frac{\pi t}{T}\right) - \frac{2\pi t}{T}\cos\left(\frac{\pi}{T}t\right) - \left(\frac{\pi t}{T}\right)^2\sin\left(\frac{\pi T}{2}\right)}{\left(\frac{\pi t}{T}\right)^3}\right\}\right]$$

$$\approx \sin\left(\frac{\pi t}{T}\right)\left[\frac{1}{\pi t} - \frac{2 - \frac{2\pi t}{T}\cos\left(\frac{\pi t}{2}\right) - \frac{\pi^2 t^2}{T^2}}{\frac{24\pi t^3}{T^2}}\right]$$

$$(6.45)$$

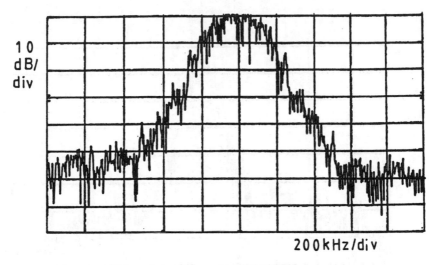

Figure 6.16: Measured DPM spectrum

An extension [14] of TFM, called generalised tamed frequency modulation (GTFM), provides flexibility in selecting $g(t)$. The tap coefficients of the transversal filter and the roll-off of the low pass filter response can be chosen to trade increase spectrum spillage into neighbouring bands for lower bit error rate (BER), and vice versa. Both the GTFM and TFM modulators can generate their signals by means of look-up tables.

6.1.3 Power Spectra

The spectral spillage of a modulated signal into adjacent channels is of prime importance in digital cellular mobile radio. Figures 6.16, 6.17 and 6.18 show the spectra of DPM, GMSK and MSK, respectively, for pseudo random modulating data at a rate of 250 kbits/s and a carrier frequency of 910 MHz [15]. The DAC in the modulator has 256 levels. The DPM spectrum of Figure 6.16 is for raised cosine pulse shaping over 3-bit periods (3RC), a modulation index h_p of 1.08, and 8 samples of the phase per bit period. The GMSK spectrum of Figure 6.17 is for $B_b T$ 0.25, a modulation index h_F of 0.5 and an oversampling ratio of 8. The spectrum of Figure 6.18 is for MSK with a modulation index h_F of 0.5. Stylised spectra for these modulations, in addition to those of TFM and generalised TFM (GTFM), are displayed in Figure 6.19. The rectangular shaped impulse response $g(t)$ used in MSK manifests itself as the sinc-like function with its deep narrow troughs and broad peaks. The narrow main-lobe and the low side-lobe levels of GMSK and GTFM are obtained at the expense of considerable ISI, and this increases the complexity of the equaliser located at the receiver.

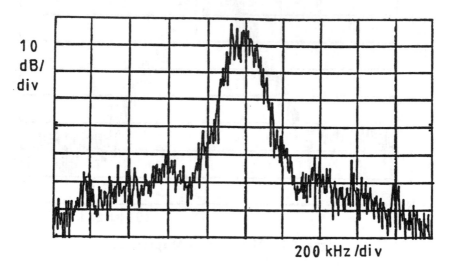

Figure 6.17: Measured GMSK spectrum

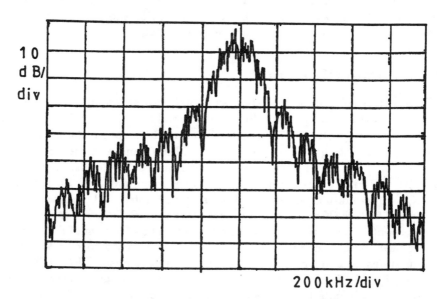

Figure 6.18: Measured MSK spectrum

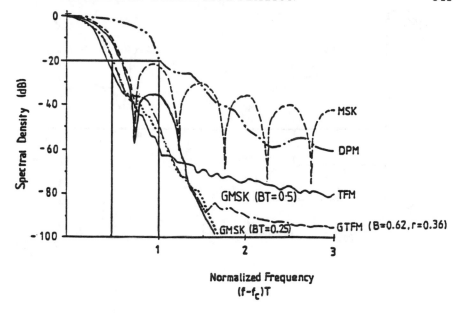

Figure 6.19: Stylised power spectral density curves for different types of modulation signals

DPM introduces less ISI than the other DFM methods, and for a given excess delay in the mobile radio channel it is able to operate with fewer states in its Viterbi equaliser compared to GMSK. However, DPM does have considerable out-of-band energy. Nevertheless when DPM operates in the presence of co-channel interference this out-of-band energy is not the limiting factor and is acceptable.

6.1.3.1 Modulated Signal Power Spectral Density Estimation

The PSDs shown in the previous section were obtained from prototype hardware modulators. Instead of building new hardware each time we wish to investigate the PSD of the modulated signal, spectral estimation [16] can be performed instead. To achieve this pseudo-random data is used and the baseband in-phase and quadrature signals analyzed on the computer to yield the PSD of the modulated signal. One way of computing the PSD is to decompose the complex modulator output sequence into k subsequences, each of M samples. These subsequences are spaced B samples apart, where $B = M/2$, as shown in Figure 6.20. Each subsequence is multiplied by a Hanning window and its FFT computed. Next the periodogram (normalised magnitude squared) of each FFT is calculated, and the k periodograms are averaged to give the PSD estimate. To perform the spectral estimate, an FFT block size of $M = 256$ is appropriate, and

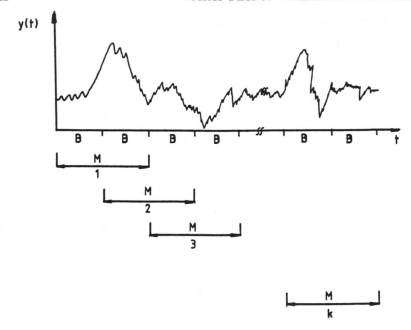

Figure 6.20: Spectral estimation process

$k = 100$ periodograms should be averaged. Theoretical treatments for the evaluation of the PSD of FM signals are given in references [17, 18, 19].

6.1.4 TDMA Format for DPM and DFM Transmissions

Having described the basic principles of DPM and DFM, we now consider how these modulation methods are arranged for TDMA transmissions. The speech signal is digitally encoded, and the resulting data stream is channel coded followed by interleaving. Because the channel data is to be transmitted via TDMA in a time slot, the channel data is conveyed to a packetiser. Other vital information, such as a propagation sounding sequence and system control information, is also inserted into the packetiser to yield packets that are forwarded at a constant continuous rate into the TDMA burst buffer. The arrangement is shown in Figure 6.21. For a mobile station the packet is removed from the buffer at the TDMA rate during an appropriate time slot. No data is removed from the buffer until a further TDMA frame period has elapsed. By contrast a base station may continuously provide packets for its numerous mobile stations at the TDMA rate in every frame slot.

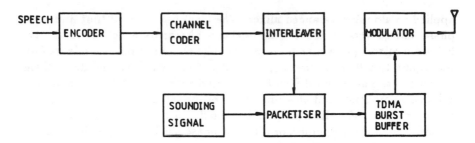

Figure 6.21: Basic TDMA transmitter for mobile radio

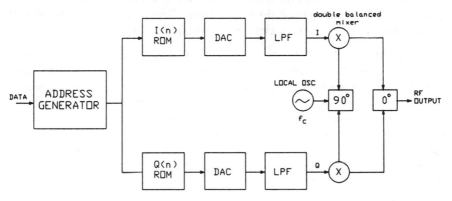

Figure 6.22: CPM modulator structure

6.1.5   Hardware Aspects

There are two basic ways to perform digital frequency modulation. Either direct frequency modulation as shown in Figure 6.5, or via phase modulation shown in Figure 6.1. The disadvantage of direct frequency modulation is that it is difficult to keep the centre frequency within the allowable bounds, while maintaining linearity and constant deviation sensitivity. Consequently a modulator employing this approach must adopt extra measures, such as closed loop control of the phase, to overcome these problems [11].

We will now consider DFM modulators employing the phase modulation approach shown in Figure 6.1. A practical realisation is shown in Figure 6.22. The phase shaping filter having impulse response $q(t)$, and the cos and sin read only memories (ROMS) are combined into two ROMS, I(n) and Q(n). Binary data applied to the modulator is used to generate appropriate addresses for the I(n) and Q(n) ROMS whose outputs are 8 bit sample values of $\cos \phi(t, \alpha)$ and $\sin \phi(t, \alpha)$ respectively. After D/A conversion and anti-alias low-pass filtering, the resulting I and Q signals are each applied to a double balanced mixer. The local oscillator output and a 90° phase shifted version provide the orthogonal signals which drive the two balanced mixers (shown as multipliers). The outputs from the mixers form the inputs to a 0° combiner i.e., an adder, from which the modulated signal emerges. The major difficulty associated with this type of modulator is achieving amplitude and phase matching of the signals in the I and Q paths. Deviation from 90° phase shift in the quadrature splitter, deviation from 0° addition in the combiner, or different phase shifts through the balanced mixers cause the modulated signal to have unwanted amplitude and phase variations. These distortions in the modulated signal are compounded if the modulator is followed by stages of non-linear amplification. The consequence is that at the receiver the eye pattern of the demodulated signal will be severely impaired making clock recovery and bit regeneration difficult.

Obtaining propriety RF components with the required close tolerances becomes increasingly difficult as the carrier frequency increases, rendering operation over a wide frequency band (greater than 200 MHz) in the vicinity of 2 GHz difficult to achieve at the time of writing. For wide-band operation it may be necessary to perform the modulation at a fixed intermediate frequency and up-convert the modulated signal to the required operating frequency. This approach, however, involves considerable extra complexity, including additional filtering, mixers and local oscillators, and may be inappropriate for a hand portable.

Another form of modulator structure is based on a simple PSK modulator followed by a phase locked loop. However, there are limits on the usable modulation index and the duration of the filter impulse response $g(t)$. A description of this type of modulator is provided in Reference [4].

6.2   CPM Receivers

The mobile radio communications considered in this section are concerned with CPM transmissions over channels that may exhibit frequency selective fading i.e., the modulated signal bandwidth will be greater than the channel coherence bandwidth for a significant proportion of the time. However, we commence our deliberations by considering what form the optimal receiver takes when the transmissions are over Gaussian channels. This will lead us to the notion of maximum likelihood sequence detection and then to the use of the Viterbi algorithm. The use of the Viterbi algorithm to equalise the effects of inter-symbol interference due to frequency selective fading will then be addressed.

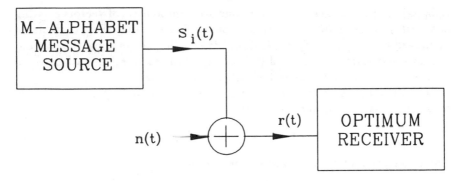

Figure 6.23: Maximum likelihood reception

6.2.1 Optimal Receiver

Let us consider a transmitter having an alphabet of M unique messages $m_i, i = 1, 2, \ldots, M$ which are transmitted as a signal $s_i(t); i = 1, 2, \ldots, M$ with exact correspondence, over an additive white Gaussian noise (AWGN) channel to a maximum likelihood (ML) receiver [20, 4]. Figure 6.23 shows the arrangement. The receiver selects from the corrupted received signal $r(t)$ what it deems to be the most likely transmitted signal $s_i(t)$, given the channel noise is $n(t)$. For convenience it will be assumed all signals in this subsection are bandpass, however later sections will draw distinctions between bandpass and low-pass signals. The receiver achieves this by maximising the *a posteriori* probability $P[s_i(t)|r(t)]$. This procedure gives rise to the name, maximum a posteriori (MAP) receiver. If no prior knowledge of the value of $s_i(t)$ is known to the receiver, the MAP process is equivalent to the ML detection. Such a process provides the minimum probability of message error when all the transmitted messages are equally probable. As $P[s_i(t)|r(t)]$ is unknown, we express it using the mixed form of Bayes rule as

$$P[s_i(t)|r(t)] = \frac{f_R[r(t)|s_i(t)]P[s_i(t)]}{f[r(t)]} \qquad (6.46)$$

where $f[r(t)]$ is the probability density function (PDF) of $r(t)$ and $P[s_i(t)]$ is the probability that $s_i(t)$ was transmitted. As $f[r(t)]$ is independent of the signal transmitted, the optimum receiver attempts to maximise the numerator in Equation 6.46. The probabilities $P[s_i(t)]$ are often unknown to the receiver, and hence the term $f_R[r(t)|s_i(t)]$ is maximised.

The reason $r(t)$ differs from $s_i(t)$ is due to the channel noise

$$n(t) = r(t) - s_i(t) \qquad (6.47)$$

which is statistically independent of $s_i(t)$. We wish to express these signals as a convergent series of so called basis functions [21]. The coefficients (or weights) applied to these basis functions are components of vectors existing in a vector space known as the signal space. To show that these coefficients do indeed lie in a vector space, and that these coefficients allow the calculation of the maximum likelihood probability requires detailed mathematical treatment which we do not feel is justified in this text.

In setting up the vector space, we have intimated that a set of basis functions must be specified, for example

$$\Psi_i(t) \; ; \; i = 1, 2, ..., n. \tag{6.48}$$

The inner product of two vectors is defined as

$$< h(t), g(t) >= \int h(t)g^*(t)dt \tag{6.49}$$

and an orthonormal basis set is one for which

$$< \Psi_i(t)\Psi_j(t) >= \left\{ \begin{array}{ll} 1 & ; \quad i = j \\ 0 & ; \quad i \neq j \end{array} \right. . \tag{6.50}$$

The range of integration for the inner product is over the range defined for that vector space. The components in vector space for a function $h(t)$ are given by

$$h_j =< h(t), \Psi_j(t) > . \tag{6.51}$$

To express $h(t)$ in terms of these components we write

$$h(t) = \sum_{j=1}^{n} h_j \Psi_j(t). \tag{6.52}$$

We now define a basis set that forms a suitable basis for all the transmitted messages $s_i(t)$. Thus

$$s_i(t) = \sum_{j=1}^{N} s_{ij} \Psi_j(t) \tag{6.53}$$

where N is the dimension of the signal space which has a maximum value equal to the number of transmitted messages. The coefficients are given by

$$s_{ij} =< s_i(t), \Psi_j(t) > . \tag{6.54}$$

We may also express the noise signal $n(t)$ using the same basis set, i.e.,

$$n(t) = \sum_{j=1}^{N} n_j \Psi_j(t). \tag{6.55}$$

Similarly the coefficients are

$$n_j =< n(t), \Psi_j(t) > . \tag{6.56}$$

Finally, the received signal $r(t)$ may be expressed as

$$r(t) = \sum_{j=1}^{N} r_j \Psi_j(t). \tag{6.57}$$

We note that in terms of signal space components

$$r_j = s_{ij} + n_j. \tag{6.58}$$

The coefficients $\{r_j\}$ are Gaussian random variables because they are produced by linear operations on other Gaussian random variables. The mean of r_j for the i'th transmitted message is

$$\begin{aligned}
E[R_j | s_i(t)] &= E[s_{ij} + N_j] \\
&= s_{ij} + E[N_j] \\
&= s_{ij} \\
&= \eta \quad . \tag{6.59}
\end{aligned}$$

Note that the use of capital letters denote random variables. We now wish to evaluate the conditional variance of r_j,

$$\begin{aligned}
E[(R_j - \eta)^2 | s_i(t)] &= E[(s_{ij} + N_j - s_{ij})^2] \\
&= E[N_j^2] \quad . \tag{6.60}
\end{aligned}$$

Substituting for N_j yields

$$\begin{aligned}
E[N_j^2] &= E[< n(t), \Psi_j(t) >< n(\tau)\Psi_j(\tau) >] \\
&= E\left[\int n(t)\Psi_j(t)dt \int n(\tau)\Psi_j(\tau)d\tau \right] \\
&= E\left[\int \int n(t)n(\tau)\Psi_j(t)\Psi_j(\tau)dtd\tau \right] \\
&= \int \int E[n(t)n(\tau)]\Psi_j(t)\Psi_j(\tau)dtd\tau \\
&= \frac{N_o}{2} \int \int \delta(t-\tau)\Psi_j(t)\Psi_j(\tau)dtd\tau \\
&= \frac{N_o}{2} \int \Psi_j^2(t)dt \\
&= \frac{N_o}{2} \tag{6.61}
\end{aligned}$$

where $N_o/2$ is the power spectral density of the Gaussian noise $n(t)$. The multivariate density function of $r_1, r_2, \ldots, r_N$ conditional on $s_i(t)$ is therefore

$$f[r_1, r_2, \ldots, r_N | s_i(t)] = \prod_{j=1}^{N} \frac{\exp[-(r_j - s_{ij})^2/N_o]}{\sqrt{\pi N_o}}$$

$$= \frac{\exp\left[-\sum_{j=1}^{N}(r_j - s_{ij})^2/N_o\right]}{(\pi N_o)^{N/2}} \qquad (6.62)$$

From Equation 6.46 we noted that maximising the term $f_R[r(t)|s_i(t)]$ or equivalently the term $f[r_1, r_2, \ldots, r_N|s_i(t)]$ over all possible messages $s_i(t)$ yields maximum likelihood reception. The conditional density of Equation 6.62 depends only upon the term

$$\sum_{j=1}^{N}(r_j - s_{ij})^2. \qquad (6.63)$$

It is possible to use Parcevals identity to show that

$$\sum_{j=1}^{N}(r_j - s_{ij})^2 = \int (r(t) - s_i(t))^2 dt , \qquad (6.64)$$

which is the square Euclidean distance between the signals $r(t)$ and $s_i(t)$. Noting the monotonicity of the exponential function, the density function f_R is maximised by choosing $s_i(t)$ that is closest to $r(t)$ in terms of Euclidean distance. The maximum likelihood receiver may then be implemented by calculating

$$\int (r(t) - s_i(t))^2 dt = \int r^2(t)dt + \int s_i^2(t)dt - 2\int r(t)s_i(t)dt \qquad (6.65)$$

for each i. We note that the first term is constant with respect to i, hence the receiver needs to perform only the correlation $\int r(t)s_i(t)dt$ and subtract it from the second term. The second term is the energy of $s_i(t)$ and if all the transmitted messages contain equal energy, then only the correlation need be formed. The integrals required to evaluate the expression may be implemented using linear filters. This technique is known as matched filtering.

For DFM and DPM systems, the phase of the transmitted signal in any particular symbol interval usually depends on previous symbols as well as on the latest symbol. Consequently an optimum maximum likelihood (ML) receiver must observe many symbol intervals before reaching a decision on the value of a specific symbol. In theory one must find the most likely sequence at the receiver corresponding to the whole of the transmitted sequence. This procedure is known as maximum likelihood sequence detection (MLSD) [4, 3] and is complex in terms of implementation and analysis. However, the Viterbi Algorithm (VA) [20, 22] provides a recursive optimal solution to the problem of estimating the sequence of states in a phase modulated signal. As the transition between phase states corresponds to a unique data sequence, the VA also provides MLSD. It will be shown later that it is the recursive application of the VA that gives the reduction in complexity required to produce an optimal receiver.

6.2.2 Probability of Symbol Error

The probability of symbol error is directly determined by the distance properties of the signal space. We will commence by examining the relationship between signal space and the probability of symbol error. We will then discuss factors affecting modulation distance properties. Suppose the signal $s_i(t)$ is transmitted and a maximum likelihood receiver recovers the signal $s_k(t)$ with a probability $P_e(k; i)$. This event will only occur if the received signal $r(t)$ is closer in *distance* to $s_k(t)$ than to $s_i(t)$. Consider the signal space representation of Figure 6.24 where one of the orthogonal signal space axes is seen to join $s_i(t)$ with $s_k(t)$. As stated previously, white, zero mean Gaussian noise will appear on all of the orthogonal axes. The signal space distance D between the two signals is given by

$$
\begin{aligned}
D &= \|s_i(t) - s_k(t)\| \\
&= \sqrt{\int (s_i(t) - s_k(t))^2 \, dt}.
\end{aligned}
\tag{6.66}
$$

The probability that $r(t)$ lies nearer to $s_k(t)$ than $s_i(t)$ is the probability that the first vector space component $r_1 - s_{i1}$ exceeds $D/2$. Thus the probability of wrongly identifying $s_i(t)$ is

$$
P_e(k; i) = \int_{\frac{D}{2}}^{\infty} \frac{1}{\sqrt{\pi N_o}} \exp\left(\frac{-u^2}{N_o}\right) du
\tag{6.67}
$$

as $r_1 - s_{i1}$ is Gaussian noise with zero mean and variance $N_o/2$.

To find the probability of detecting any other 'incorrect' signal we utilise the "union bound". This states that the probability of one or more events occurring is overbounded by the summation of the individual probabilities. Consequently the probability of error, given signal $s_i(t)$ was sent is

$$
\begin{aligned}
P_e(i) &\leq \sum_{k \neq i} P_e(k, i) \\
&\leq \sum_{k \neq i} \int_{\frac{D}{2}}^{\infty} \frac{1}{\sqrt{\pi N_o}} \exp\left(-\frac{u^2}{N_o}\right) du
\end{aligned}
\tag{6.68}
$$

and the total probability of error is

$$
P_e = \sum_i P_e(i) P\{s_i(t) \text{ sent }\}.
\tag{6.69}
$$

If all messages are equally likely and if $P_e(i)$ is identical for every i then P_e and $P_e(i)$ are identical. Equation 6.68 can also be expressed using Q-functions to yield

$$
P_e(i) \leq \sum_{k \neq i} Q\left(\frac{\|s_i(t) - s_k(t)\|}{\sqrt{2N_o}}\right).
\tag{6.70}
$$

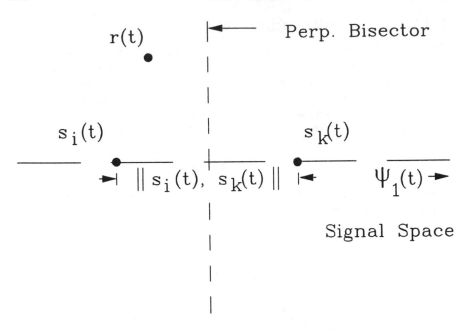

Figure 6.24: Signal space probability of error

The concept of signal space will now be applied to phase modulated signals. Suppose that two signals $s_i(t)$ and $s_k(t)$ are different over a duration of N intervals, where each interval has a duration of T and energy E. The square Euclidean distance between the signals is from Equation 6.66

$$D^2 = \int_0^{NT} s_i^2(t)dt + \int_0^{NT} s_k^2(t)dt - 2\int_0^{NT} s_i(t)s_k(t)dt. \qquad (6.71)$$

The first term is,

$$\int_0^{NT} s_i^2(t)dt = \int_0^{NT} \left[\sqrt{\frac{2E}{T}}\cos(w_c t + \phi_i(t))\right]^2 dt$$

$$= \frac{E}{T}\int_0^{NT} [1 + \cos 2(w_c t + \phi_i(t))]\,dt$$

$$= NE + \text{term depending on } \left(\frac{1}{w_c}\right). \qquad (6.72)$$

The second term in Equation 6.72 can be neglected for systems of interest here. The energy contributed by the second squared term in Equation 6.71 will also be NE. The cross signal term in Equation 6.71 is,

$$-2\int_0^{NT} s_i(t)s_k(t)dt = \frac{4E}{T}\int_0^{NT} \cos(w_c t + \phi_i(t))\cos(w_c t + \phi_k(t))dt$$

$$= \frac{2E}{T} \int_0^{NT} \cos(2w_c t + \phi_i(t) + \phi_k(t))$$

$$+ \frac{2E}{T} \int_0^{NT} \cos(\phi_i(t) - \phi_k(t))dt \qquad (6.73)$$

after expressing this equation in terms of complex exponentials, re-arranging and changing back to cosine terms. The first term depends on $1/w_c$ and will be neglected and so the cross-correlation term can be expressed as

$$\frac{2E}{T} \int_0^{NT} \cos \Delta\phi(t)dt \qquad (6.74)$$

where $\Delta\phi(t) = \phi_i(t) - \phi_k(t)$. Consequently the square Euclidean distance of Equation 6.71 can be expressed as

$$D^2 = 2NE - \frac{2E}{T} \int_0^{NT} \cos \Delta\phi(t)dt \qquad (6.75)$$

or equivalently

$$D^2 = \frac{2E}{T} \int_0^{NT} [1 - \cos \Delta\phi(t)]dt. \qquad (6.76)$$

For binary systems, $E = E_b$, where E_b is the energy per bit. With an M-ary system then,

$$E_b = \frac{E}{\log_2 M}. \qquad (6.77)$$

Thus the square Euclidean distance is now

$$D^2 = \frac{2E_b \log_2 M}{T} \int_0^{NT} [1 - \cos \Delta\phi(t)]dt. \qquad (6.78)$$

We now define the normalised Euclidean distance function as

$$d^2(s_i(t), s_k(t)) \triangleq \frac{\log_2 M}{T} \int_0^{NT} [1 - \cos \Delta\phi(t)]dt. \qquad (6.79)$$

Thus the square Euclidean distance may be written as

$$D^2 = 2E_b d^2(s_i(t), s_k(t)). \qquad (6.80)$$

Equation 6.70 enables us to express the probability of error $P_e(i)$ in terms of the signal space distance D as follows

$$P_e(i) \leq \sum_{k \neq i} Q\left(\frac{D}{\sqrt{2N_o}}\right). \qquad (6.81)$$

Substituting for D from Equation 6.80 yields

$$P_e(i) \leq \sum_{k \neq i} Q\left(\sqrt{\frac{E_b}{N_o}} d(s_i(t), s_k(t))\right). \tag{6.82}$$

When the ratio of signal energy to noise energy is reasonably high, say in excess of 10 dB, then one inter-signal distance completely dominates the expression for $P_e(i)$. Indeed we can go further and say that the worst case combination of $s_i(t)$ and $s_k(t)$ will eventually dominate the total error probability P_e as the SNR increases. The worst case distance for any observation interval N is called the minimum distance d_{min}. Consequently the probability of error may be expressed as

$$P_e \cong Q\left(\sqrt{\frac{E_b}{N_o} d_{min}^2}\right). \tag{6.83}$$

In practice an upper bound to d_{min}^2 called d_B^2, is easier to evaluate. This term can be evaluated for both full and partial response systems with arbitrary pulse shapes and modulation indices.

We now consider a practical implementation of the optimal receiver using the recursive solution provided by the Viterbi algorithm.

6.2.3   Principle of Viterbi Equalisation

In the presence of AWGN, the bandpass signal at the receiver is given by

$$\tilde{r}(t) = \tilde{s}(t, \alpha) + \tilde{n}(t) \tag{6.84}$$

where $\tilde{s}(t, \alpha)$ is the bandpass (signified by a raised tilda $\sim$) transmitted signal at time t and for data α, and $\tilde{n}(t)$ is the bandpass AWGN signal. As demonstrated in Section 6.2.1, the maximum likelihood ML receiver minimises Equation 6.65. The possible received signals $\tilde{s}(t, \bar{\alpha})$ now depend on the infinitely long estimated sequence $\{\bar{\alpha}\}$, and accordingly the ML receiver minimises the function

$$\int (\tilde{r}(t) - \tilde{s}(t, \bar{\alpha}))^2 dt \tag{6.85}$$

with respect to the estimated data sequence $\{\bar{\alpha}\}$, i.e., it minimises the Euclidean distance.

From Equations 6.65 minimising Equation 6.85 is equivalent to maximising the correlation

$$C(\bar{\alpha}) = \int_{-\infty}^{\infty} \tilde{r}(t)\tilde{s}(t, \bar{\alpha}) dt. \tag{6.86}$$

State	$\alpha_{n-1},\ \alpha_{n-2}$
Γ_o	-1, -1
Γ_1	-1, 1
Γ_2	1, -1
Γ_3	1, 1

Table 6.1: State Table

It would be possible to construct a receiver based on Equation 6.86 in which all the possible transmitted sequences are correlated with the received signal. The sequence $\bar{\alpha}$ chosen would be that which maximised $C(\bar{\alpha})$. However, even with short bursts this structure becomes unmanageable as the number of comparisons increases exponentially with sequence length. To overcome this problem we define [4, 20, 22]

$$C_n(\bar{\alpha}) \triangleq \int_{-\infty}^{(n+1)T} \tilde{r}(t)\tilde{s}(t,\bar{\alpha})dt \qquad (6.87)$$

and so

$$C_n(\bar{\alpha}) = C_{n-1}(\bar{\alpha}) + Z_n(\bar{\alpha}) \qquad (6.88)$$

where $C_n(\bar{\alpha})$ and $Z_n(\bar{\alpha})$ are referred to as a metric and an incremental metric, respectively, for a particular $\bar{\alpha}$. The incremental metric is given by

$$Z_n(\bar{\alpha}) = \int_{nT}^{(n+1)T} \tilde{r}(t)\tilde{s}(t,\bar{\alpha})dt. \qquad (6.89)$$

Thus $C_n(\bar{\alpha})$ can be evaluated recursively using Equation 6.88, where $Z_n(\bar{\alpha})$ is an incremental metric generated by correlating the received signal with an estimated signal over the nth symbol interval.

Example

In order to clarify how the optimal receiver works in terms of Equation 6.88, we will consider a modulator whose output waveform over a bit interval is dependent on the current and previous two bits, namely α_n, α_{n-1}, and α_{n-2}, yielding $2^3 = 8$ possible waveform segments. We may represent this situation by the state transition diagram shown in Figure 6.25.

Table 6.1 displays the four states that are associated with the values of α_{n-1}, α_{n-2}. In both the Figure and the Table, logical 0 and logical 1 data bits are represented by -1 and +1, and by solid and dotted lines, respectively.

Assuming the system is initially in state Γ_o, and $\alpha_n = -1$, the state is unchanged, and the modulator generates a wave form segment $\tilde{S}_o$. However, if $\alpha_n = 1$, the state changes to Γ_2 as the new values of α_{n-1} and α_{n-2} are 1 and -1. The modulator output is waveform segment $\tilde{S}_2$. Should the

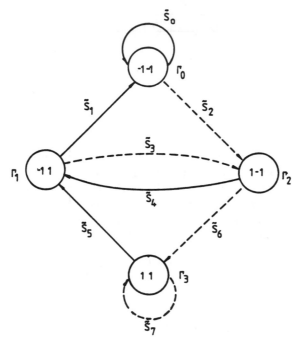

Figure 6.25: State transition diagram

next data bit be another logical 1 the system moves to Γ_3 and the output waveform is $\tilde{S}_6$. Any further logical ones applied to the modulator generate waveform segment $\tilde{S}_7$ as the transitions are from Γ_3 back to Γ_3. A logical 0 will cause a transition to Γ_1 and $\tilde{S}_5$ to occur, and we note that it is impossible to move directly from Γ_3 to Γ_2.

An alternative representation of the state transition diagram is the trellis diagram of Figure 6.26. The two columns of circles represent the four states Γ_0, Γ_1, Γ_2, Γ_3, at instants $n-1$ and n, while $\tilde{S}_i$; $i = 0, 1, \ldots, 7$, are the waveform segments generated by the modulator during a bit period. Observe that each state is connected to two other states as we are dealing with binary modulation, and one connection is due to the presence of a logical 1 and the other with a logical 0 data bit.

Knowing how the transmitter generates the waveform segments $\tilde{S}_i$ based on the present and previous two input data bits, we now consider how the optimal receiver regenerates the bit sequence in the presence of channel noise. As we know, the receiver will correlate the received signal waveform $\tilde{r}$ over one bit interval with all the possible known transmitted waveforms $\tilde{S}_i$; $i = 0, 1, \ldots, 7$. Now this cross-correlation process yields the eight incremental metrics $Z_n(\tilde{\alpha}) = Z_{ni}$; $i = 0, 1, \ldots, 7$, as shown in Figure 6.27. Although the largest Z_{ni} implies that $\tilde{S}_i$ was the most likely transmitted

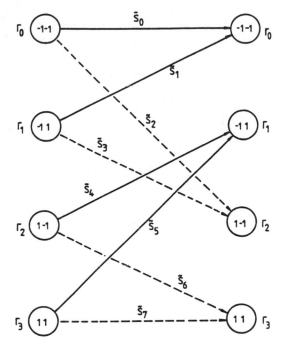

Figure 6.26: Trellis diagram, where _____ -1 and - - - 1

waveform, we refrain from making a decision on a single transmitted bit. Instead we compute each of the eight values of $C_n(\bar{\alpha})$ given by Equation 6.88
from a knowledge of the four values of $C_{n-1}(\bar{\alpha})$ and the eight values of $Z_{n,i}$.
For example, the two metrics for state Γ_o at instant n are represented by
$C_n(\alpha_n, \Gamma_j)$ where $\alpha_n = \pm 1$, and Γ_j are the states at $n-1$, namely

$$C_n(-1, \Gamma_o) = C_{n-1}(\Gamma_o) + Z_{no}$$

and

$$C_n(-1, \Gamma_1) = C_{n-1}(\Gamma_1) + Z_{n1}.$$

Only the larger value of these two metrics is retained and designated
$C_n(\Gamma_o)$, along with the logical value of α_n, namely -1. This procedure
is repeated for each of the four states to yield four metrics that will be
employed in the next bit period.

This recursive process continues for each successive bit interval, namely,
the path in the trellis associated with the larger C_n at each state is retained,
and the logical value of the bit is stored along with the other bits associated
with the path leading to that state. At the end of a data sequence we inspect
the final metrics for each of the four states. The largest metric identifies
the optimum path through the trellis, and the sequence of bits associated
with this path is deemed to be the most probable transmitted one.

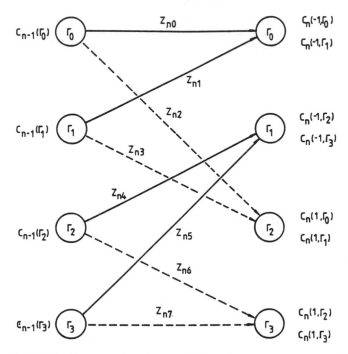

Figure 6.27: Trellis diagram showing metrics, where _____ -1 and --- 1

Figure 6.28 shows the development of the trellis for the 4-state modem described above over a ten bit period. The reader should refrain from calculating the numbers shown on the Figure as much data has been omitted to avoid obfuscation. Starting with an all zero sequence, the sub-figure for the first bit period, i.e., when $k = 1$ shows the effect of a logical 0 and a logical 1. By $k = 3$ all the states are in use. The numbers assigned to each state are in terms of the path having the *lowest* cumulative metric, as the computer program that generated these metrics operated on the basis of minimum Euclidean distance rather than the equivalent maximum cross-correlation, see Equation 6.65. Thus at each node, or state, the lowest Euclidean distance metric and the bit sequence from $k = 1$ to the current k are stored. For a 10 bit sequence we examine the accumulated Euclidean distance metrics for each of the four states, and select the lowest, namely zero in Figure 6.28. Thus we trace the path back from state Γ_3 to the beginning of the sequence. When the path is a dotted line a logical 1 occurs, and when it is solid line a logical 0 is formed. The regenerated sequences is seen to be 0000111011, as -1 signifies a logical 0.

Discussion

We have seen that maximising the cross-correlation (of Equation 6.87) is equivalent to minimising the Euclidean distance between the received and

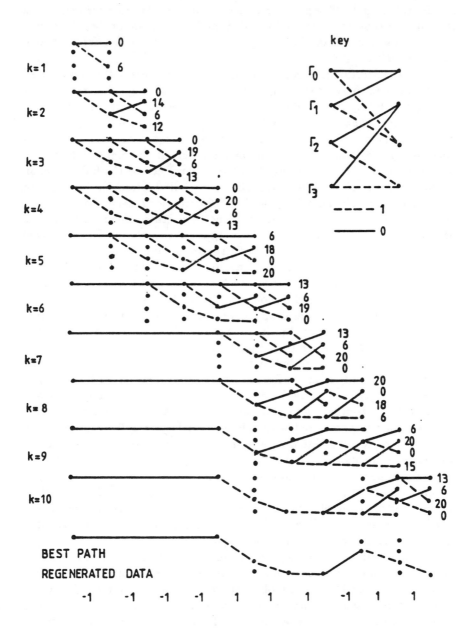

Figure 6.28: Trellis development of a four state VE

transmitted signals. The procedures described in this Section for recovering the data are known as the Viterbi algorithm (VA) [22]. Later we will employ it to equalise the effects of a fading channel, when we will refer to it as a Viterbi equaliser (VE).

In order to decrease the probability of generating errors in the regenerated bit stream, i.e., to avoid selecting the wrong path through the trellis, $v - 1$ dummy zero bits are inserted prior to the data in order to initialise the trellis of the Viterbi equaliser to the all-zero state. This is done in Figure 6.28. As a consequence all the possible paths to be traced through the trellis diagram by the Viterbi algorithm can be assumed to originate from the zero state node. After $v - 1$ data bits have been received the full recursion of the algorithm is reached. After processing all the information data in a manner previously described, the Viterbi processor receives v postcursor dummy zero bits, which ensures that the final state can be assumed to be the zero state. Subsequently the output data sequence for the whole burst, corresponding to the path with the minimum accumulated metric and which passes through the zero state terminal mode, is read out.

We have confined our discussions to binary data as that is our concern in this text. Multilevel modulation can be handled using the VA but at a considerable increase in complexity. Even with binary data the implementation of the Viterbi algorithm is not trivial. However, the power of the algorithm resides in that only 2^{v-1} paths need to be stored, rather than the vast number of paths associated with a tree structure, and consequently the hardware complexity required for the VA does not increase with the length of the data sequence, except for a linear increase in storage requirements. The number of computations is proportional to the data sequence. These remarks can be appreciated with reference to Figure 6.28. However, the situation is radically different if each bit's influence is associated with more waveform segments, i.e., if the ISI is deliberately increased to improved spectral compactness of the modulated signal. For example, if the output waveform segment is dependent on α_n, α_{n-1}, α_{n-2}, α_{n-3}, α_{n-4} ; $v = 5$, the number of states becomes $2^{v-1} = 16$, and the number of incremental metrics increases to 32. The hardware complexity of the Viterbi algorithm to remove the ISI therefore increases exponentially with v.

Having described the principle of optimum reception of bursts of TDMA data, we will now commence our description of the regeneration of CPM data signals transmitted over mobile radio channels. A pre-requisite to the complex baseband signal processing required in the recovery of the data is the demodulation of the received RF signal, and this topic is now addressed.

6.2.4   RF to Baseband Conversion

The NB-TDMA radio frequency (RF) signal conveys the data in bursts occupying one TDMA slot. The receiver at either a MS or BS tunes to

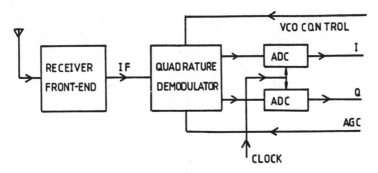

Figure 6.29: RF to baseband conversion

the appropriate carrier and using a conventional receiver front-end down-converts the RF signal to the intermediate frequency (IF). Quadrature demodulation follows, see Figure 6.29, to yield baseband analogue in-phase and quadrature signals.

Figure 6.30 shows the quadrature demodulator in more detail. The level of the IF signal is adjusted by the automatic gain control (AGC) signal via a variable gain amplifier. If the phase off-set between the voltage controlled oscillator (VCO) signal and the IF signal is zero the demodulation is coherent. Generally it is not necessary to force the phase off-set to zero because it can be accommodated as a channel imperfection when estimating the channel impulse response. This estimation procedure is described in Section 6.2.5. The low pass filters in Figure 6.30 remove the second harmonic of the IF at the outputs of the multipliers to leave the quadrature baseband signals. Thus the action of the quadrature demodulator is to accept the high frequency bandpass IF signal whose magnitude spectrum is of the form displayed in Figure 6.31 and yield baseband quadrature spectrum of the type shown in Figure 6.29.

Returning to Figure 6.29, we see that the baseband analogue quadrature signals are analogue-to-digitally converted (ADC). This process can be viewed as sampling the input signal at a rate

$$f_s = \eta_R/T \tag{6.90}$$

where η_R is the receiver oversampling ratio, and T is the bit duration. Typically, $\eta_R \geq 2$, thereby ensuring that the degree of spectral aliasing is tolerable. Each sample is encoded into n-bits in the ADC. The value of n is selected to provide sufficient waveform integrity without making excessive demands on the subsequent digital signal processing. Experiments [15] have shown that η_R values of 2 and 4 are satisfactory for GMSK and DPM demodulation, respectively.

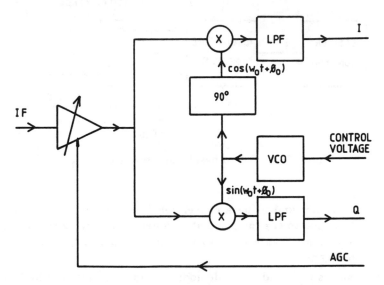

Figure 6.30: Quadrature demodulator

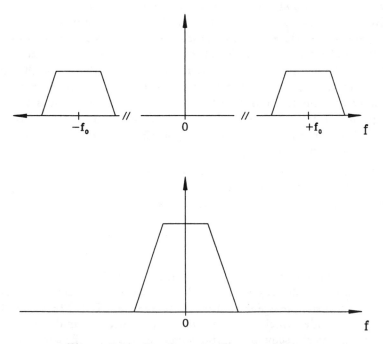

Figure 6.31: Bandpass and lowpass spectra

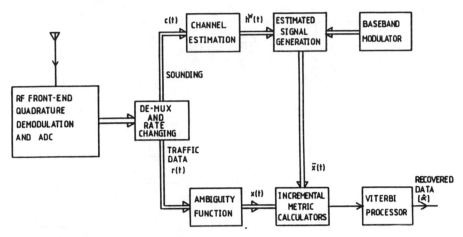

Figure 6.32: Block diagram of the receiver having a Viterbi equaliser

6.2.5 Baseband Processing

After generating the digital I and Q signals, and assuming slot synchronisation has been achieved, the data in the slot, i.e., the packet, is rate converted down to one that allows digital signal processing to be performed within a TDMA frame duration. The data in the packet is separated into the received baseband sounding signal $\hat{c}(t)$, and the received baseband traffic data $r(t)$. The arrangement is shown in Figure 6.32, where the double connecting lines represent complex baseband signals, i.e., inphase and quadrature, and the single lines represent real signals.

The principle of data regeneration is that a channel estimate $h^w(t)$ is formed from $\hat{c}(t)$, and all the possible baseband signals $\bar{s}(t)$ over one bit period are convolved with $h^w(t)$ to yield the signal estimates $\bar{x}(t)$. The waveforms of $\bar{s}(t)$ over a bit period are all those generated at the transmitter, plus additional ones to allow for multipath propagation, as described later in this section. The traffic data $r(t)$ is convolved with the ambiguity function to allow for imperfect channel estimation and the resulting signal $x(t)$, along with $\bar{x}(t)$, enable the incremental metrics required by the Viterbi algorithm (VA) to be calculated. Viterbi processing is performed, according to the description given in Section 6.2.3, and the traffic data $\{\hat{\alpha}\}$ is recovered.

As we are now concerned with baseband processing we will redraw Figure 6.32 entirely at baseband, removing the RF part. Further we will do the same for the transmitter, and convert the real mobile radio channel to its complex baseband equivalent as described in Chapter 2. The resulting diagram is displayed in Figure 6.33, although system controls are not shown. We will now commence a detailed description of how the regenerated sequence $\{\hat{\alpha}\}$ is produced.

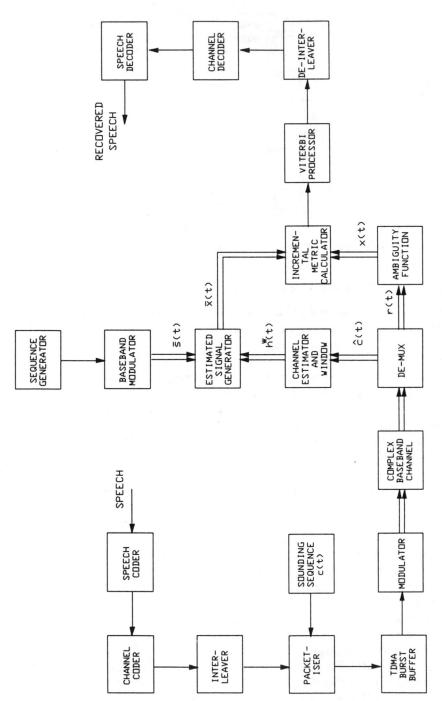

Figure 6.33: Baseband system

Channel Estimation

In order to obtain an estimate of the channel impulse response the data corresponding to the sounding sequence is processed. Let us momentarily digress to consider the case where the baseband sounding sequence is conveyed over an ideal channel and applied to a matched filter in the receiver. The resulting filter output signal $\rho(t)$ is an approximation to an impulse function. Now suppose we return to reality and replace the ideal channel with a mobile radio channel. In a baseband representation, and assuming linearity, we may remove the baseband radio channel to after the matched filter, and as a consequence the impulse-like signal at the output of the matched filter is convolved with the channel impulse response $h(t)$ of the mobile radio channel to give an estimate of the baseband channel impulse response $h'(t)$. In other words the convolution of the sounding sequence, the impulse response of the channel and the matched filter impulse response provides us with a good estimate of $h(t)$.

Let us consider the situation in more detail. The matched filtering of the received complex baseband sounding signal $\hat{c}(t)$, a corrupted version of the transmitted sounding signal $c(t)$, provides an estimate of the channel baseband impulse response, namely

$$h'(t) = \hat{c}(t) * p(t). \tag{6.91}$$

The equivalent low-pass impulse response of the matched filter, $p(t)$ is a time reversed version of the complex conjugate of the sounding signal, delayed to make it causal, i.e.,

$$p(t) = c^\circ(T_c - t). \tag{6.92}$$

The duration of $c(t)$ is T_c, and the raised zero denotes the complex conjugate operation. To clarify the operation of the channel estimator, we initially consider that the channel is noiseless. The baseband sounding signal at the output of the baseband mobile channel is

$$\hat{c}(t) = c(t) * h(t) \tag{6.93}$$

and after matched filtering the resulting signal is

$$h'(t) = c^\circ(T_c - t) * \hat{c}(t) * h(t) \tag{6.94}$$

or

$$h'(t) = \rho(t) * h(t) \tag{6.95}$$

where $\rho(t)$ is known as the ambiguity function [6]. Thus the estimated channel response $h'(t)$ is the actual response $h(t)$ convolved with the ambiguity function

$$\rho(t) = c^\circ(T_c - t) * c(t). \tag{6.96}$$

It should be noted that due to the even symmetry of $c(t)$ about its midpoint

$$c^o(T_c - t) = c^o(t) \tag{6.97}$$

so that we can put

$$
\begin{aligned}
\rho(t) &= c^o(t) * c(t) \\
&= [c_I(t) - jc_Q(t)] * [c_I(t) + jc_Q(t)] \\
&= c_I(t) * c_I(t) + c_Q(t) * c_Q(t). \tag{6.98}
\end{aligned}
$$

Consequently the ambiguity function is wholly real, which enables simplified processing to be performed in the receiver.

Channel estimation for both DPM and DFM can be performed using either a swept frequency signal (a chirp), or by transmitting a pseudo random binary sequence (PRBS) via phase (or frequency) modulation.

Chirp Sounding

The instantaneous frequency of the transmitted chirp signal is given by

$$f(t) = f_c + \frac{\lambda}{\pi} t \tag{6.99}$$

where f_c is the nominal centre frequency, λ is the sweep parameter and T_c is the chirp duration. The bandpass transmitted chirp signal is

$$\tilde{c}(t) = \cos 2\pi \left(f_c + \frac{\lambda t}{\pi} \right) t \tag{6.100}$$

which can also be expressed in the form

$$
\begin{aligned}
\tilde{c}(t) &= \Re \left[\exp j2\pi (f_c + \frac{\lambda t}{\pi}) t \right] \\
&= \Re \left[\exp(j2\lambda t^2) \exp(j2\pi f_c t) \right] \\
&= \Re[c(t) \exp(j2\pi f_c t)] \tag{6.101}
\end{aligned}
$$

where

$$
\begin{aligned}
c(t) &= \exp(j2\lambda t^2) \\
&= \cos 2\lambda t^2 + j \sin 2\lambda t^2 \tag{6.102}
\end{aligned}
$$

is the complex baseband representation of the bandpass chirp signal $\tilde{c}(t)$. If we define the maximum allowable frequency sweep in terms of the bit rate T, then from Equation 6.99

$$\frac{\lambda T_c}{\pi} = a \cdot \frac{1}{T} \tag{6.103}$$

where a is the new sweep parameter. Letting

$$T_c = nT \qquad (6.104)$$

i.e., the chirp duration is an integer number of bit periods, we can write

$$\lambda = \frac{a\pi}{nT^2}. \qquad (6.105)$$

Substituting for λ in Equation6.102 yields

$$c(t) = \cos\left(\frac{2a\pi}{nT^2}\right) t^2 + j \sin\left(\frac{2a\pi}{nT^2}\right) t^2. \qquad (6.106)$$

To express $c(t)$ in discrete time we have

$$T = \eta D \qquad (6.107)$$

where η is the oversampling ratio, and we express time as

$$t = kD \quad ; \quad k = -\infty \text{ to } \infty, \qquad (6.108)$$

to yield

$$c(k) = \cos\left(\frac{a2\pi}{n\eta^2}\right) k^2 + j \sin\left(\frac{a2\pi}{n\eta^2}\right) k^2. \qquad (6.109)$$

Due to the symmetry of $c(t)$ about its midpoint the ambiguity function is entirely real as shown in Figure 6.34, where $a = 0.5$. In order to ensure that the matched filtering (channel estimation) at the receiver is performed correctly, a period of zero carrier follows the frequency chirp. This period should be long enough, e.g., a duration of 6 bits, to accommodate the largest expected channel delay spread. The channel estimate is produced by performing matched filtering with a filter whose impulse response is given by $c^o(T_c - t)$.

Sequence Sounding

To reduce hardware complexity the channel estimation process can be facilitated by transmitting a signal which has undergone phase (or frequency) modulation by a pseudo random binary code. This code is selected to exhibit good autocorrelation properties, i.e., a peaked response with low sidelobes and preferably no imaginary components, by means of a computer search. As an example, for both DPM (3RC) and DFM (GMSK, $B_N = 0.3$), 16-bit codewords can be used, and to ensure accurate channel estimation at the receiver, the first 6 bits of a codeword can be appended to the end of the codeword and the final 6 bits to the start of the codeword. Thus the final 28-bit channel sounding preamble can tolerate 6 bits of delay spread before the channel estimate is corrupted by the following section of the TDMA burst. Suitable codewords for DPM (3RC) and GMSK ($B_N = 0.3$)

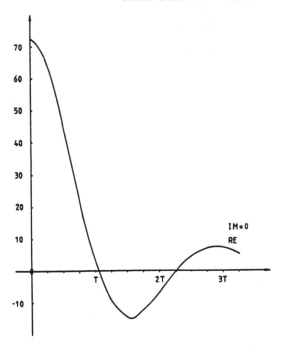

Figure 6.34: Chirp ambiguity function, a=0.5

are 1001101011001000 and 1110000111010011, respectively. Their ambiguity functions are shown in Figures 6.35 and 6.36, respectively. The DPM ambiguity function can be seen to possess a much narrower main lobe than that of the GMSK ambiguity function (1 bit duration as opposed to 2 bits). As a consequence the resolution of the channel estimation will be higher in the DPM case. This is expected because the bandwidth occupied by the DPM signal is considerably greater than that of the GMSK signal. The GMSK ambiguity function also possess a small imaginary component. However, the imaginary component can be neglected without significantly affecting the link performance.

The ambiguity function $\rho(t)$, i.e., the autocorrelation function of the sounding signal $c(t)$, can be made more impulse-like. This can be achieved in the case of a chirp signal by increasing its frequency sweep, an approach that is usually unacceptable because of the need to occupy a greater channel bandwidth. Figure 6.37 displays the frequency-time characteristics and the autocorrelation function of a chirp signal for two different frequency sweeps. By increasing the range of the swept frequency the autocorrelation signal has a narrower mainlobe and therefore the ability to resolve multipaths is enhanced. Typically in narrowband TDMA (NB-TDMA) applications, the width of the mainlobe of $\rho(t)$ is approximately of two bits duration. In the

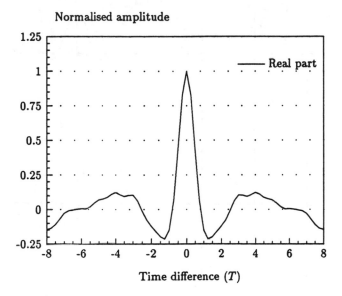

Figure 6.35: DPM sequence ambiguity function

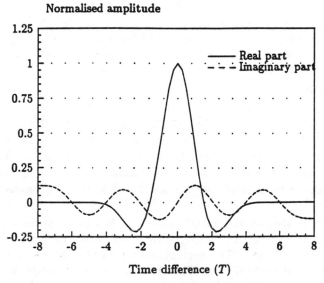

Figure 6.36: GMSK sequence ambiguity function

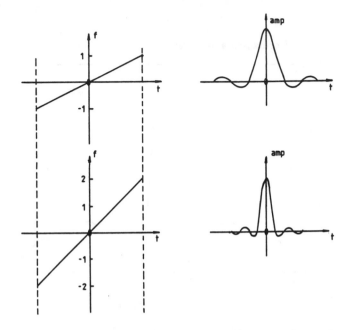

Figure 6.37: Chirp sounding

case of sequence sounding, increasing the length of the sequence will also narrow the main lobe of the autocorrelation function

In general terms, chirp soundings have autocorrelation functions with narrower main lobes than those produced by sequence sounding. However, the small performance difference between chirp and sequence sounding does not justify the extra complexity involved in chirp sounding.

Channel Windowing

A Viterbi channel equaliser of a given complexity can only cope with a certain delay spread of signal paths, and therefore the full $h'(t)$ cannot be used in the decoding process. Consequently $h'(t)$ is windowed to give a shortened $h'(t)$, say $h^w(t)$, and it is $h^w(t)$ that is generated in the channel estimator shown in Figure 6.33. In addition to truncation, the window function may involve amplitude weighting of $h'(t)$ near the edges of the window.

The need to shorten the duration of $h'(t)$ can be illustrated as follows. Suppose that in the modulator, the duration of one data pulse is spread over $L = 3$ bit periods. The corresponding number of states required in the Viterbi equaliser to remove ISI (when no channel multipath is present) is 2^{L-1} and the number of incremental metrics to be calculated is 2^L, i.e., there are eight possible received waveforms in one bit period. When multipath is present each bit is effectively spread over additional bit periods,

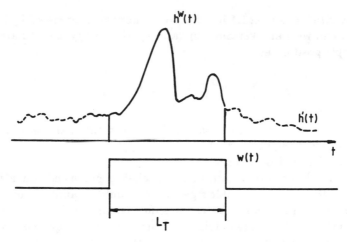

Figure 6.38: Channel windowing of $h'(t)$ to give $h^W(t)$

and consequently the number of possible received waveforms per bit period increases beyond eight. To remove the dispersive effects of the multipath channel as well as the ISI deliberately introduced in the modulator, the number of states in the Viterbi processor must be increased. Let the number of states in the Viterbi processor be 2^{v-1}, where $v > L$. Correspondingly the number of incremental metrics is now 2^v, as there are two incremental metrics associated with each state. When the sum of the duration L of the modulator filter impulse response and the duration of the estimated channel response L_p is greater than v, full equalisation is not possible. The receiver is now obliged to operate on a segment of the channel impulse response which must be selected to maximise the BER performance. This arrangement is the so-called reduced-state Viterbi equaliser, in which none of the estimated waveforms will match exactly the received waveforms; nevertheless good BER performance can still be achieved.

As we have said, a reduced-state equaliser selects an appropriate segment of the estimated channel impulse response. The duration of the response is limited to L_T bits, where

$$L_T = v - L \tag{6.110}$$

because we can only accommodate L_T bits of $h'(t)$ before we exceed the number of states in the equaliser. We will call this selection process rectangular windowing. A reasonable basis on which to select the appropriate segment of the estimated channel impulse response $h'(t)$ is to slide the rectangular window of length L_T bits over the whole of the estimated response, calculating the energy contained within the window at each point and then to identify the window position where the energy is maximum. Hopefully, the window resides on that part of $h'(t)$ which enables the VE to regenerate the data with the lowest BER. Thus for a channel estimate of L_p bits total duration and with an oversampling ratio η_R the energy is calculated at the i-th sample position as

$$E_i = \sum_{k=0}^{\eta L_T} |h'_{i+k}|^2 \; ; \; i = 0 \text{ to } \eta(L_p - L_T). \qquad (6.111)$$

Figure 6.38 shows an arbitrary $h'(t)$, the movement of the sliding window and the selected segment $h^w(t)$.

Estimated Signal Generation

Before the information data in the packet is processed, an attempt is made to determine the possible signals that would emanate from a channel having an impulse response $h^w(t)$. Specifically, all possible v-bit sequences are generated by the local modulator within the Viterbi equaliser (VE). The length, v, of the sequence is determined by the number of states (2^{v-1}) that can be accommodated in terms of the acceptable complexity in the Viterbi channel equaliser. Thus a baseband local modulator is supplied with all the possible combinations of the v-bit sequence to yield phase estimates, $\bar{\phi}(t)$. The quadrature estimates $(2E/T)^{\frac{1}{2}} \cos \bar{\phi}(t)$ and $(2E/T)^{\frac{1}{2}} \sin \bar{\phi}(t)$ are computed and it is these terms that are convolved with the estimated wideband baseband channel impulse response $h^w(t)$ to give the estimates of the possible received waveforms over one bit interval. However, in reality all the $\cos \bar{\phi}(t)$ and $\sin \bar{\phi}(t)$ estimates are stored in a ROM, i.e., the filtering and the trigonometric functions are reduced to a set of stored numbers, addressed by the possible v-bit data sequences. Notice in Figure 6.33 that

$$\bar{s}(t) = (2E/T)^{\frac{1}{2}} [\cos \bar{\phi}(t) + j \sin \bar{\phi}(t)]. \qquad (6.112)$$

The role of the Ambiguity Function

From Equation 6.95 the estimate of the channel impulse response $h'(t)$ is the actual channel impulse response $h(t)$ convolved with $\rho(t)$. Further, $h'(t)$ is windowed to give $h^w(t)$. Consequently the estimated signals in the receiver will be based on $h^w(t)$, not on the unknown $h(t)$. During the reception of information, as distinct from the sounding sequence $c(t)$, the received signal is

$$r(t) = s(t) * h(t), \qquad (6.113)$$

where $s(t)$ is the information signal. The receiver estimates that the received signals are

$$
\begin{aligned}
\bar{x}(t) &= \bar{s}(t) * h^w(t) \\
&= \bar{s}(t) * (w(t)h'(t)) \\
&= \bar{s}(t) * w(t)(h(t) * \rho(t)) \\
&= w(t)(h(t) * \rho(t)) * \bar{s}(t) \\
&= w(t)(\rho(t) * h(t) * \bar{s}(t)) \qquad (6.114)
\end{aligned}
$$

where $\bar{s}(t)$ is the local modulator output, and $w(t)$ is the windowing function applied to the estimated channel $h'(t)$, see Figure 6.38. In order to obtain the closest match between the received and estimated signals, $r(t)$ in Equation 6.113 is convolved with $w(t)\rho(t)$. Consequently we can write the baseband received signal as

$$\begin{aligned} x(t) &= r(t) * (w(t)\rho(t)) \\ &= h(t) * s(t) * w(t)\rho(t) \\ &= w(t)\rho(t) * h(t) * s(t) \end{aligned} \qquad (6.115)$$

The use of the weighted ambiguity function allows better gain matching between the received signal $x(t)$ and the estimated signals $\bar{x}(t)$ compared to that achieved using $\rho(t)$ directly. We will call the weighted ambiguity function $w(t)\rho(t) = \rho_w(t)$ and because $\rho(t)$ is entirely real we may write

$$\begin{aligned} x(t) &= [r_I(t) + jr_Q(t)] * \rho_w(t) \\ &= r_I(t)\rho_w(t) + jr_Q(t)\rho_w(t) \\ &= x_I(t) + jx_Q(t) \end{aligned} \qquad (6.116)$$

where $x_I(t)$ and $x_Q(t)$ are the inphase and quadrature components of the baseband signal. We may write the quadrature received baseband components explicitly in the presence of additive channel noise as

$$x_I(t) = \sqrt{\frac{2E}{T}}[h_I(t) * s_I(t) - h_Q(t) * s_Q(t)] * \rho_w(t) + n_I(t) * \rho_w(t) \quad (6.117)$$

and

$$x_Q(t) = \sqrt{\frac{2E}{T}}[h_I(t) * s_Q(t) + h_Q(t) * s_I(t)] * \rho_w(t) + n_Q(t) * \rho_w(t) \quad (6.118)$$

where $h_I(t)$ and $h_Q(t)$ and $n_I(t)$ and $n_Q(t)$ are the complex components of $h(t)$ and the additive baseband channel noise $n(t)$, respectively.

Notice that although we have ignored the effect of channel noise in producing the channel estimate, we have included noise in the above equations. This may be justified by considering the high energy of the sounding sequence compared with the energy per bit. The difference means that the effective SNR of the channel estimate is much higher (approximately 12 dB greater for a 14-bit preamble) than that experienced during data reception. Consequently when the SNR becomes too low to ensure reasonable channel estimation, the data is unusable anyway.

Incremental Metric Calculator

Initially, as an aid to understanding we will dispense with quadrature representation, and consider real partial response baseband signals enabling us to express the received signal as

$$x(t) = \alpha(t) * q(t) * h(t) * \rho_w(t) \qquad (6.119)$$

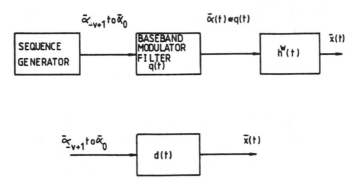

Figure 6.39: Locally generated signal estimates

where $\alpha(t)$ is a sequence of impulses representing the data vector at the transmitter, $q(t)$ is the impulse response of the modulator filter, and $h(t)$ is the channel impulse response. We will also ignore the effects of channel noise. Similarly, the estimated signal sequences from the local modulator in the receiver after convolution with the estimated windowed channel response are given by

$$\bar{x}(t) = \bar{\alpha}(t) * q(t) * h^w(t) \qquad (6.120)$$

where $h^w(t)$ in this case is a truncated channel estimate obtained by applying a rectangular window to $h'(t)$.

To generate these signal sequences over each bit period we feed all the possible v-bit patterns $\bar{\alpha}_{-v+1} \ldots, \bar{\alpha}_o$, into a local baseband modulator filter and convolve its output with the truncated channel estimate $h^w(t)$, see Figure 6.39. Let us represent the combined effect of the modulator filter and the estimated windowed channel by a network having an impulse response

$$d(t) = q(t) * h^w(t) \qquad (6.121)$$

such that $\bar{x}(t)$ is the convolution of the data sequence $\bar{\alpha}(t)$ with $d(t)$ as shown in Figure 6.39.

The estimated, locally generated baseband waveforms $\bar{x}(t)$ are seen to be dependent on the number of data bits v, the modulator filter impulse response $q(t)$, and the windowed estimated channel impulse response $h^w(t)$. However, in generating the digital representation of the received signal $x(t)$ it is necessary to sample at a rate η_R times the bit rate in order to prevent excessive aliasing.

Consider the case where $\eta_R = 2$ and $v = 5$. The estimates $\bar{x}(t)$ are formulated using five bits, namely $\bar{\alpha}_{-4}, \bar{\alpha}_{-3}, \bar{\alpha}_{-2}, \bar{\alpha}_{-1}$ and $\bar{\alpha}_0$. As $\eta_R = 2$, there are two samples per bit, achieved by introducing a zero between the $\bar{\alpha}$ values. Figure 6.40 shows an arbitrary response $d(t)$ together with its sampled values $d_0, d_1, \ldots, d_9$. Also displayed is the transversal arrangement of the network having the weighting sequence $\{d_n\}$ showing the data during the first sampling interval. The output of this transversal filter is seen to be

$$\bar{x}_{u,0} = \bar{\alpha}_0 d_0 + \bar{\alpha}_{-1} d_2 + \bar{\alpha}_{-2} d_4 + \bar{\alpha}_{-3} d_6 + \bar{\alpha}_{-4} d_8. \tag{6.122}$$

In the next sampling interval the data is shifted one delay stage to the right, and a zero is arranged to follow $\bar{\alpha}_0$ to yield

$$\bar{x}_{u,1} = \bar{\alpha}_0 d_1 + \bar{\alpha}_{-1} d_3 + \bar{\alpha}_{-2} d_5 + \bar{\alpha}_{-3} d_7 + \bar{\alpha}_{-4} d_9. \tag{6.123}$$

As $v = 5$, there are 32 possible received values of $\bar{x}_{u,0}$ and $\bar{x}_{u,1}$, i.e., $u = 0, 1, ..31$ and these are used in the computation of the incremental metrics employed in the Viterbi Algorithm (VA).

At the first sampling instant of the k-th bit the estimated signals $\bar{x}_{uo}$ are subtracted from the received signal x_{ko} and the results squared to give the Euclidean distances $(x_{ko} - \bar{x}_{uo})^2$. At the second sampling instant the distances $(x_{k1} - x_{u1})^2$ are formed. Notice that the second subscript of x and $\bar{x}$ signifies the first or second sampling instant during a bit period. The incremental metrics for one bit interval are therefore

$$m_{uk} = (x_{ko} - \bar{x}_{uo})^2 + (x_{k1} - \bar{x}_{u1})^2; u = 0 \text{ to } 31. \tag{6.124}$$

For general values of η_R and v this Equation becomes

$$m_{uk} = \sum_{i=0}^{\eta_R - 1} (x_{ki} - \bar{x}_{ui})^2; u = 0, 1, \ldots 2^v - 1. \tag{6.125}$$

Having formed the 32 incremental metrics they are applied to the Viterbi processor in Figure 6.41, and the new metrics for each state are established according to the description given in Section 6.2.3. At the next bit interval, $k + 1$, the inputs are $x_{k+1,0}$ and $x_{k+1,1}$. However, the $\bar{x}_{u,o}$ and $\bar{x}_{u,1}; u = 0, 1, \ldots, 31$, do not change as they are constant for the TDMA burst. They will only change when an updated windowed estimate of the channel impulse response is formulated, and that does not occur until the next packet is received. Observe that the samples applied to the incremental metric calculators are essentially sampled values of the analogue waveform segments $x(t)$ and $\bar{x}(t)$, and that the Viterbi processor favours the minimum values of the incremental metrics, as the Euclidean distance rather than the correlation criterion is used.

Having described how the Viterbi equalisation operates with real signals, we now address our actual problem where the I/Q network furnishes us with inphase and quadrature signals. If x_{Iki} and x_{Qki} represent the inphase and

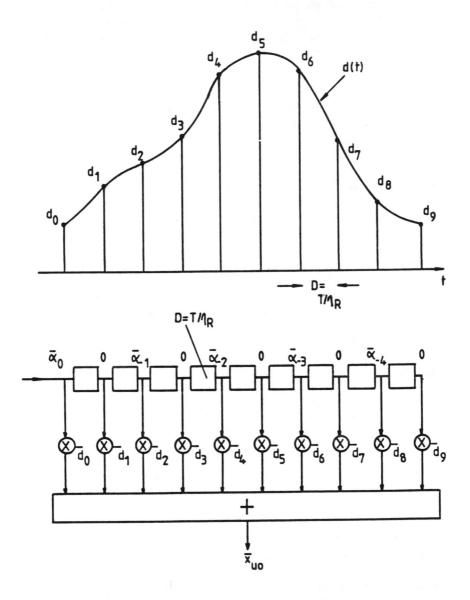

Figure 6.40: An example of how the estimated signals are formed

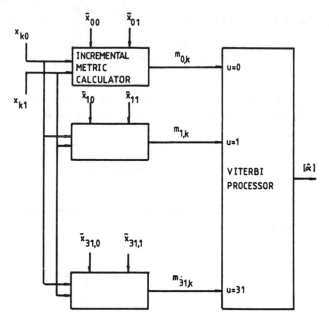

Figure 6.41: Incremental metric formation

quadrature components of the received signal at the i-th sample of the k-th bit interval, and $\bar{x}_{Iui}$ and $\bar{x}_{Qui}$ are the corresponding values for the locally generated signal estimates, $u = 0, 1, \ldots, 2^v - 1$, then the square of the Euclidean distance is

$$(x_{Iki} - \bar{x}_{Iui})^2 + (x_{Qki} - \bar{x}_{Qui})^2. \tag{6.126}$$

As i ranges from 0 to $\eta_R - 1$, the incremental metric becomes the sum of the square of the Euclidean distances at each sampling instant during a bit period; viz:-

$$m_{uk} = \sum_{i=0}^{\eta_R-1} \{(x_{Iki} - \bar{x}_{Iui})^2 + (x_{Qki} - \bar{x}_{Qui})^2\}; u = 0, 1, \ldots 2^v - 1. \tag{6.127}$$

6.2.6 Viterbi Equalisation of Digital Phase Modulation

Having described the basic principles of baseband processing we now consider the specific case of Viterbi equalisation of DPM signals [5, 6]. From Chapter 2 it is shown that for DPM the local modulator complex baseband output signal at the receiver is

$$\bar{s}(t) = \exp j \left[\sum_p \bar{\alpha}_p q(t - pT) \right] \qquad (6.128)$$

where to avoid cluttering the text we have set the amplitude $(2E/T)^{\frac{1}{2}}$ to unity and ϕ_0 to zero. The discrete time version of $\bar{s}(t)$ derived from Equation 6.128 is

$$\bar{s}(n, \bar{\alpha}) = \exp j \left[\sum_p \bar{\alpha}_p q_{n - p\eta_R} \right] \qquad (6.129)$$

and as $\bar{x}(n, \bar{\alpha})$ is the convolution of $\bar{s}(n, \bar{\alpha})$ with $h^w(n)$, see Equation 6.114, we will write $\bar{x}(n)$ as

$$\bar{x}_{ui} = \sum_{n=i-\eta_R L_T}^{i} h_{i-n}^w \cdot \exp j \left[\sum_{p=\left\lceil \frac{n+1}{\eta_R} \right\rceil - L}^{\left\lfloor \frac{n}{\eta_R} \right\rfloor} \bar{\alpha}_p q_{n - p\eta_R} \right] ; u = 0, 1, \ldots, 2^v - 1$$

$$(6.130)$$

where $\lceil \cdot \rceil$ and $\lfloor \cdot \rfloor$ denote the nearest integers above and below $\cdot$, respectively, i is the index of the sample during a per bit period and u is the incremental metric index. The limits of the first summation ensure that h^w is limited to the range 0 to $\eta_R L_T$, where L_T is given by Equation 6.110. The coefficients in the modulator filter span the range from q_o to $q_{\eta_R L - 1}$, as shown in Chapter 3, and hence the limits on the second summation.

We are interested in all the possible samples $\bar{x}_{u,i}$ of the estimated received waveforms in one bit interval. Firstly we will reconsider the earlier example given in Section 6.2.5, but with $\eta_R = 1$. In this example the estimated channel response is truncated to a duration equivalent to $L_T = 2$ bits. With $v = 5$, $L = 3$, the valid modulator filter coefficients are q_0, q_1, q_2 and the truncated channel response is h_o^w, h_1^w, h_2^w. The estimated received signals for a 5-bit sequence $\bar{\alpha}_{-4}$ to $\bar{\alpha}_0$ are, upon substituting into Equation 6.130,

$$\bar{x}_{ui} = \sum_{n=i-L_T}^{i} h_{i-n}^w \exp j \left[\sum_{p=n-2}^{n} \bar{\alpha}_p q_{n-p} \right]$$

$$= \sum_{n=i-2}^{i} h_{i-n}^w \exp j[\bar{\alpha}_{n-2} q_2 + \bar{\alpha}_{n-1} q_1 + \bar{\alpha}_n q_0]. \qquad (6.131)$$

Now i is zero as $\eta_R = 1$, and thus we can write

$$\begin{aligned} \bar{x}_{uo} = \ & h_2^w \exp j[\bar{\alpha}_{-4} q_2 + \bar{\alpha}_{-3} q_1 + \bar{\alpha}_{-2} q_0] \\ + \ & h_1^w \exp j[\bar{\alpha}_{-3} q_2 + \bar{\alpha}_{-2} q_1 + \bar{\alpha}_{-1} q_0] \\ + \ & h_0^w \exp j[\bar{\alpha}_{-2} q_2 + \bar{\alpha}_{-1} q_1 + \bar{\alpha}_0 q_0]. \end{aligned} \qquad (6.132)$$

Conceptually the process of producing samples of the estimated received waveforms can be viewed as the local modulator and channel model being cycled through all of the 32 possible 5-bit sequences every bit period.

To ensure a good performance the oversampling ratio η_R should be at least 2, which means that the receiver must produce $\bar{x}_{u,i}, i = 0, 1$; and $u = 0, 1, \ldots 31$ each bit period. For this case we can write the i-th sample as

$$\bar{x}_{ui} = \sum_{n=i-4}^{i} h_{i-n}^w \exp j \left[\sum_{p=\lceil \frac{n+1}{2} \rceil - 3}^{\lfloor n/2 \rfloor} \bar{\alpha}_p q_{n-2p} \right] \qquad (6.133)$$

where valid modulator filter coefficients are q_o to $q_{\eta_R L - 1}$, i.e., q_o to q_5; the truncated channel response is h_o^w to h_4^w, and i can be 0 or 1. The two samples are therefore given by

$$
\begin{aligned}
\bar{x}_{uo} = \ & h_4^w \exp j[\bar{\alpha}_{-4}q_4 + \bar{\alpha}_{-3}q_2 + \bar{\alpha}_{-2}q_0] \\
+ \ & h_3^w \exp j[\bar{\alpha}_{-4}q_5 + \bar{\alpha}_{-3}q_3 + \bar{\alpha}_{-2}q_1] \\
+ \ & h_2^w \exp j[\bar{\alpha}_{-3}q_4 + \bar{\alpha}_{-2}q_2 + \bar{\alpha}_{-1}q_0] \\
+ \ & h_1^w \exp j[\bar{\alpha}_{-3}q_5 + \bar{\alpha}_{-2}q_3 + \bar{\alpha}_{-1}q_1] \\
+ \ & h_0^w \exp j[\bar{\alpha}_{-2}q_4 + \bar{\alpha}_{-1}q_2 + \bar{\alpha}_0 q_0]
\end{aligned}
\qquad (6.134)
$$

and

$$
\begin{aligned}
\bar{x}_{u1} = \ & h_4^w \exp j[\bar{\alpha}_{-4}q_5 + \bar{\alpha}_{-3}q_3 + \bar{\alpha}_{-2}q_1] \\
+ \ & h_3^w \exp j[\bar{\alpha}_{-3}q_4 + \bar{\alpha}_{-2}q_2 + \bar{\alpha}_{-1}q_0] \\
+ \ & h_2^w \exp j[\bar{\alpha}_{-3}q_5 + \bar{\alpha}_{-2}q_3 + \bar{\alpha}_{-1}q_1] \\
+ \ & h_1^w \exp j[\bar{\alpha}_{-2}q_4 + \bar{\alpha}_{-1}q_2 + \bar{\alpha}_0 q_0] \\
+ \ & h_0^w \exp j[\bar{\alpha}_{-2}q_5 + \bar{\alpha}_{-1}q_3 + \bar{\alpha}_0 q_1].
\end{aligned}
\qquad (6.135)
$$

The set of signal estimates are complex quantities so we can write

$$\bar{x}_{ui} = x_{Iui} + jx_{Qui} \quad \text{for } u = 0 \text{ to } 31; \text{ and } i = 0, 1. \qquad (6.136)$$

The incremental metrics for the Viterbi demodulator are given by the Euclidian distances between the received signal and the estimated signals over one bit period are from Equation 6.127.

$$
\begin{aligned}
m_{uk} = \ & [(x_{Iko} - \bar{x}_{Iuo})^2 + (x_{Qko} - \bar{x}_{Quo})^2] \\
+ \ & [(x_{Ik1} - \bar{x}_{Iu1})^2 + (x_{Qk1} - \bar{x}_{Qu1})^2]
\end{aligned}
\qquad (6.137)
$$

where x_{Iki} and x_{Qki} are the inphase and quadrature components of the complex baseband received signal during the k-th bit. The set of signal estimates $\bar{x}_{ui}$ is only updated once per burst, although it is used once per bit period in order to evaluate the set of incremental metrics.

Non-rectangular Weighting Functions

In order for the receiver to handle long multipath delays, the truncation length (L_T) of the estimated channel impulse response must be of the order of the maximum expected excess delay. From Equation 6.130, increasing L_T requires more bits in $\{\bar{\alpha}_k\}$ to produce each signal estimate $\bar{x}_{ui}$, and each additional bit in $\{\bar{\alpha}_k\}$ doubles the number of states in the Viterbi equaliser and hence the hardware complexity. An alternative is to employ a non-rectangular weighted truncation of the channel estimate.

To illustrate the concept, consider an example where $\eta_R = 1$, $L = 3$, $v = 5$, and where the estimated channel impulse response is truncated to a length of $(v + 1)$ bits. As stated earlier, 2^{v-1} is the number of states in the demodulator. Modifying Equation 6.130 by setting $i = 0$ as $\eta_R = 1$, and n ranges from 0 to $v + 1$ as there are $v + 2$ samples in the estimated channel impulse response, the signal estimates are

$$\bar{x}_{uo} = \sum_{n=-(v+1)}^{0} h_{-n}^w \exp j \left[\sum_{p=n+1-L}^{n} \bar{\alpha}_p q_{n-p} \right] \qquad (6.138)$$

and on expanding,

$$\begin{aligned}
\bar{x}_{uo} = \ & h_6^w \exp j[\bar{\alpha}_{-8} q_2 + \bar{\alpha}_{-7} q_1 + \bar{\alpha}_{-6} q_0] \\
+ \ & h_5^w \exp j[\bar{\alpha}_{-7} q_2 + \bar{\alpha}_{-6} q_1 + \bar{\alpha}_{-5} q_0] \\
+ \ & h_4^w \exp j[\bar{\alpha}_{-6} q_2 + \bar{\alpha}_{-5} q_1 + \bar{\alpha}_{-4} q_0] \\
+ \ & h_3^w \exp j[\bar{\alpha}_{-5} q_2 + \bar{\alpha}_{-4} q_1 + \bar{\alpha}_{-3} q_0] \\
+ \ & h_2^w \exp j[\bar{\alpha}_{-4} q_2 + \bar{\alpha}_{-3} q_1 + \bar{\alpha}_{-2} q_0] \\
+ \ & h_1^w \exp j[\bar{\alpha}_{-3} q_2 + \bar{\alpha}_{-2} q_1 + \bar{\alpha}_{-1} q_0] \\
+ \ & h_0^w \exp j[\bar{\alpha}_{-2} q_2 + \bar{\alpha}_{-1} q_1 + \bar{\alpha}_0 q_0].
\end{aligned} \qquad (6.139)$$

It can be seen that the sampled value of the signal estimates depend on $\bar{\alpha}_{-8}$ to $\bar{\alpha}_0$. A rectangular window would require the Viterbi processor to have 2^8 states. However, as we are limited to 16 states we can only choose a subset of five consecutive $\bar{\alpha}$'s to generate the sampled values of the signal estimates. It seems reasonable from consideration of symmetry, and also to use the longest possible impulse response, that $\bar{\alpha}_{-2}$ to $\bar{\alpha}_{-6}$ be selected as the 'deterministic' state subset and $\bar{\alpha}_0$, $\bar{\alpha}_{-1}$, $\bar{\alpha}_{-7}$, $\bar{\alpha}_{-8}$ be designated as the 'stochastic' subset, as shown in Figure 6.42. The elements in this 'stochastic' subset may have the values of ± 1 with equal probability. One way to generate the sampled value of the signal estimate $\bar{x}_{uo}$ for a given deterministic set is to produce sampled values of the signal estimates for all possible combinations of the stochastic sub-set and take their average value. Thus to compute $\bar{x}_{uo}$ in Equation 6.139 for a particular deterministic set of $\bar{\alpha}_{-2}$, $\bar{\alpha}_{-3} \ldots$, $\bar{\alpha}_{-6}$, we consider all 16 combinations of $\bar{\alpha}_0$, $\bar{\alpha}_{-1}$, $\bar{\alpha}_{-7}$, $\bar{\alpha}_{-8}$,

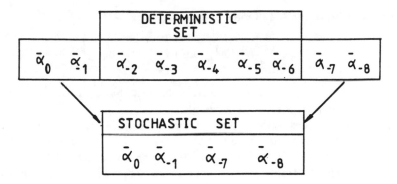

Figure 6.42: Deterministic and stochastic sets

to give 16 values of $\bar{x}_{uo}$. The value of $\bar{x}_{uo}$ adopted for the metric is the average of these 16 values of $\bar{x}_{uo}$, namely $\bar{X}_{uo}$.

This procedure is time consuming and it is possible to produce the same effect by weighting the estimated channel response. For example, consider evaluating the average value of the final term in Equation 6.139, namely,

$$h_0^w \exp j[\bar{\alpha}_{-2}q_2 + \bar{\alpha}_{-1}q_1 + \bar{\alpha}_0 q_0]. \tag{6.140}$$

This term has two stochastic elements $\bar{\alpha}_0$ and $\bar{\alpha}_{-1}$, and its average value on expanding the four possible combinations is,

$$
\begin{aligned}
&(h_0^w/4) \exp j(\bar{\alpha}_2 q_2)[\exp j(q_0 + q_1) + \exp j(q_1 - q_0) + \\
&\quad \exp j(-q_1 + q_0) + \exp j(-q_1 - q_0)] \\
=\ &(h_0^w/4) \exp j(\bar{\alpha}_2 q_2)(\exp j q_0 + \exp j(-q_0)) \\
&\quad (\exp j q_1 + \exp j(-q_1)) \\
=\ &(h_0^w/4) \exp j(\bar{\alpha}_2 q_2) 4 \cos q_0 \cos q_1 \\
=\ &h_0^w w_0^0 \exp j \bar{\alpha}_{-2} q_2
\end{aligned}
\tag{6.141}
$$

where the superscript zero implies $i = 0$. Thus it can be appreciated that multiplying h_o^w by a weighting factor

$$w_o^o = \cos q_1 \cos q_o \tag{6.142}$$

is equivalent to averaging when $\bar{\alpha}_0$ and $\bar{\alpha}_{-1}$ are allowed their possible ± 1 values.

The above procedure is repeated for the other terms in Equation 6.139 having elements in the stochastic set. Notice that the middle four terms in Equation 6.139 will produce weights of unity as they contain none of the stochastic set $\bar{\alpha}_0$, $\bar{\alpha}_{-1}$, $\bar{\alpha}_{-7}$, $\bar{\alpha}_{-8}$. The sampled values of the signal estimates in Equation 6.139 may be expressed as

$$\bar{X}_{uo} = \sum_{n=-6}^{0} h_{-n}^{w} w_{-n}^{0} \exp j \left[\sum_{p=n-2}^{n} \bar{\alpha}_p q_{n-p} \right] \tag{6.143}$$

where p is from -2 to -6 as the stochastic elements have already been taken into consideration in terms of the weights.

The sampled values of the signal estimates produced in this manner will now no longer exactly match any of the possible received signals and as a result, static BER performance will suffer. The advantage of this scheme is that it enables tolerable performance to be achieved over a much larger range of excess delay for a given receiver complexity.

When the oversampling ratio η_R is increased from unity a dif ferent weighting function is required for each sampling instant i. However, this approach is impractical, and so an average of the weighting factors over one bit interval,

$$W_n = \frac{1}{\eta_R} \sum_{i=0}^{\eta-1} w_n^i \tag{6.144}$$

is performed and W_n replaces w_n^o in Equation 6.143.

For explanatory purposes we consider a slightly different situation where the system parameters are $L = 3$, $v = 3$ and $\eta_R = 4$. The Viterbi equaliser has 4 states, and the estimated channel response has $\eta_R(v + 1) + 1 = 17$ samples. The deterministic sub-set has elements $\bar{\alpha}_{-2}$, $\bar{\alpha}_{-3}$ and $\bar{\alpha}_{-4}$ and the stochastic sub-set elements are $\bar{\alpha}_0$, $\bar{\alpha}_{-1}$, $\bar{\alpha}_{-5}$ and $\bar{\alpha}_{-6}$. From Equations 6.130 and 6.138 the signal estimates are given by

$$\bar{x}_{ui} = \sum_{n=i-(v+1)\eta_R}^{i} h_{i-n}^{w} \exp j \left[\sum_{p=\lceil \frac{n+1}{\eta_R} \rceil - L}^{\lfloor n/\eta_R \rfloor} \bar{\alpha}_p q_{n-p\eta_R} \right] \tag{6.145}$$

upon changing the summation limit $n = i - \eta_R L_T$ to $i - (v + 1)\eta_R$. The index i goes from 0 to $\eta_R - 1$. When $i = 0$,

$$\bar{x}_{ui} = h_0^w \exp j[\bar{\alpha}_{-2}q_8 + \bar{\alpha}_{-1}q_4 + \bar{\alpha}_0 q_0] + \ldots$$

$$.$$
$$.$$
$$.$$

$$h_{16}^w \exp j[\bar{\alpha}_{-6}q_8 + \bar{\alpha}_{-5}q_4 + \bar{\alpha}_{-4}q_0] \tag{6.146}$$

and similar expressions are obtained for sampling instants $i = 1$, 2 and 3.

The weighting factor w_o^o associated with the first term in Equation 6.143 is due to the stochastic sub-set elements $\bar{\alpha}_{-1}$, $\bar{\alpha}_0$, and can be shown to be

$$w_o^o = \cos q_o \cos q_4. \tag{6.147}$$

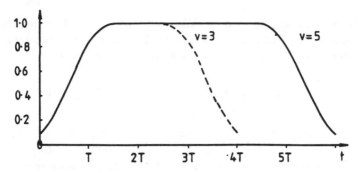

Figure 6.43: Weight functions

Similarly for the weighting factor for the final term in Equation 6.143 is

$$w^o_{16} = \cos q_4 \cos q_8. \tag{6.148}$$

When there are no stochastic elements within a term in Equation 6.146 the weight function is unity. Sets of weights $\{w^i_n\}$ are also produced at the other sampling instants of $i = 1, 2$ and 3. By employing Equation 6.146 the weighting factors for each of the estimated channel samples are generated. These weights are given in Table 6.2, where for convenience $Q(\cdot)$ means $\cos q(\cdot)$. Clearly to evaluate the average weight, W_n, the sum of the terms in each row of Table 6.2 must be divided by η_R as in Equation 6.144. Thus when $\eta_R = 4$, then

$$W_n = \frac{1}{4} \sum_{i=0}^{3} w^i_n. \tag{6.149}$$

The weight function for this example of 3RC DPM having $h_p = 1$ is plotted in Figure 6.43. Computer simulations indicate that when $v = 3$ the receiver can handle up to two bit periods excess path delay when weighting is used. With rectangular weighting, all the available states are required to remove ISI introduced by the modulator, and negligible excess delay can be accommodated. As a 16-state Viterbi equaliser is often used in practice, we also present in Figure 6.43 the weighting window for $v = 5$.

The optimum position of the weighted window is found by an approach similar to that described in Section 6.2.5 for the rectangular window. The energy values used to locate the position of the window are computed according to

$$E_i = \sum_{k=0}^{\eta_R(v+1)} W^2_k |h^w_{i+k}|^2 \text{ for } i = 0, \text{ to } \eta(L_p - (v+1)) \tag{6.150}$$

where L_p is the duration of the estimated channel impulse response in bit periods prior to windowing. The positioning of the centre of the window is where the value of i that results in the largest value of E_i resides.

This method of producing a so called reduced state VE is not the only possible approach. References [23, 24, 25, 26] show other schemes which attempt to approximate the received signal set using fewer states.

6.2.7   Viterbi Equalisation of GMSK Signals

The use of a frequency modulation scheme (in this case GMSK) [9, 15] requires the Viterbi processor to have extra states compared to DPM. These extra states are a consequence of the p possible values of the phase state term θ_n. Considering the situation at the modulator, we note that DPM requires 2^{L-1} states while $p2^L$ states are required when frequency modulation is used. The integer p is related to the modulation index by

$$h_f = 2k/p \qquad (6.151)$$

where k and p are integers and h is rational [8]. The phase states θ_n have values

$$\theta_n \in \{0,\ 2\pi/p,\ 2.2\pi/p \ldots (p-1)2\pi/p\} \qquad (6.152)$$

and each trellis state is now defined by a phase state θ_n and one of 2^{L-1} possible α sequences $\{\alpha_{n-1}, \alpha_{n-2} \ldots \alpha_{n-v+1}\}$. For GMSK we have $h_f = 0.5$, $k = 1$, $p = 4$ and so from Equation 6.152 there are four phase states with values 0, $\pi/2$, π and $3\pi/2$.

Consider the case of GMSK, $h_f = 0.5$ and $L = v = 3$ (i.e., an ideal channel) whose trellis diagram is shown in Figure 6.44. The four phase states, $0, \pi/2, \pi$ and $3\pi/2$ are shown, as are the bits defining the correlative state vector at time instants $n-1$ and n. Figure 6.45 shows the relationship between the correlative state vector and the phase state at time instants n and $n+1$. With the arrival of the latest bit, α_{n+1}, it may be observed that α_{n-2} is no longer part of the correlative state vector and contributes a value of either $\frac{\pi}{2}$ or $\frac{-\pi}{2}$ to the previous value of the phase state θ_n. Some the transitions within the trellis are shown for explanatory purposes. Consider phase state $\pi/2$ and correlative state $\alpha_{n-1} = -1, \alpha_{n-2} = -1$. If the latest bit $\alpha_n = 1$, then the new correlative state becomes $\alpha_n = 1, \alpha_{n-1} = -1$, alternatively if $\alpha_n = -1$, the correlative state is $\alpha_n = -1, \alpha_{n-1} = -1$. The new phase state is determined by bit α_{n-2}. It is the oldest correlative state bit and so with the arrival of the latest bit α_n, it will now contribute a constant value of phase $(\pm\pi/2)$ to the phase state vector, depending whether its value is ± 1. Returning to our example transitions, both of them will end up in phase state 0, because $\alpha_{n-2} = -1$, i.e., a subtraction of $\pi/2$ from the current phase state. These transitions are labelled 1 in Figure 6.44. Further example transitions, this time with $\alpha_{n-2} = 1$ are labelled 2 in Figure 6.44. In this case the transitions end in phase state π.

It can be appreciated from Figure 6.44 that the trellis structure for GMSK is four times more complicated than that required for DPM with an equivalent value of L.

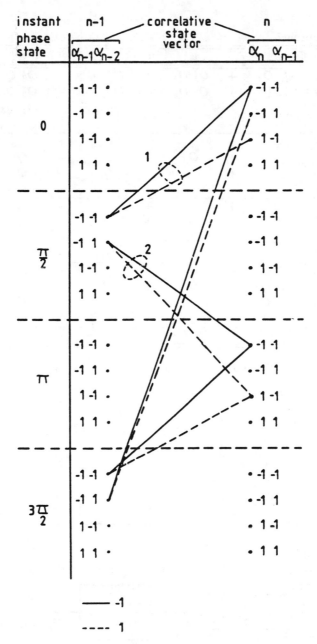

Figure 6.44: GMSK trellis diagram, where ——— -1 and - - - 1

n	w_n^0		w_n^1		w_n^2		w_n^3
0	$Q_0\,Q_4$	+	$Q_1\,Q_5$	+	$Q_2\,Q_6$	+	$Q_3\,Q_7$
1	Q_3	+	$Q_0\,Q_4$	+	$Q_1\,Q_5$	+	$Q_2\,Q_6$
2	Q_2	+	Q_3	+	$Q_0\,Q_4$	+	$Q_1\,Q_5$
3	Q_1	+	Q_2	+	Q_3	+	$Q_0\,Q_4$
4	Q_0	+	Q_1	+	Q_2	+	Q_3
5	1	+	Q_0	+	Q_1	+	Q_2
6	1	+	1	+	Q_0	+	Q_1
7	1	+	1	+	1	+	Q_0
8	1	+	1	+	1	+	1
9	Q_{11}	+	1	+	1	+	1
10	Q_{10}	+	Q_{11}	+	1	+	1
11	Q_9	+	Q_{10}	+	Q_{11}	+	1
12	Q_8	+	Q_9	+	Q_{10}	+	Q_{11}
13	$Q_{11}\,Q_7$	+	Q_8	+	Q_9	+	Q_{10}
14	$Q_{10}\,Q_6$	+	$Q_{11}\,Q_7$	+	Q_8	+	Q_9
15	$Q_9\,Q_5$	+	$Q_{10}\,Q_6$	+	$Q_{11}\,Q_7$	+	Q_8
16	$Q_8\,Q_4$	+	$Q_9\,Q_5$	+	$Q_{10}\,Q_6$	+	$Q_{11}\,Q_7$

Table 6.2: Values of w_n for $n = 0$ to 16

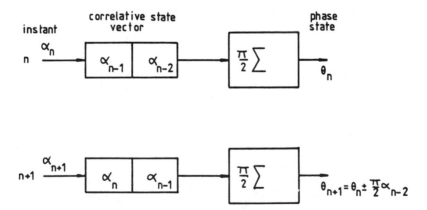

Figure 6.45: Relationship between the correlative state vector and phase state

The Viterbi Equaliser structure for GMSK is very similar to that proposed for DPM. To generate the estimated received signals at the receiver, we must convolve all the possible local modulator outputs with the estimated channel impulse response $h^w(t)$. The signal estimates will be used to generate $p2^v$ incremental metrics per received bit by calculating the Euclidean distances in a similar manner to that employed for DPM. Once the Viterbi algorithm has been applied recursively at each bit interval over the whole of the received TDMA burst, the path through the trellis with the smallest total metric is chosen as the most likely path, and the data sequence associated with it is regenerated.

One way to simplify the Viterbi equalisation of GMSK is to remove the effects of the additional phase states by tracking the phase states at each trellis node. The equaliser stores the present phase state at each trellis node to allow the signal estimates corresponding to the appropriate phase state to be chosen during the next bit interval. After updating each trellis node by storing the new cumulative metric and the updated information sequence, the latest phase state is also recorded for use in the following recursion. With only the additional complexity needed to store the phase state at each node in the trellis, the number of equaliser states for GMSK can be reduced to that required for a similar DPM scheme.

A characteristic of the Viterbi equalisation (VE) of GMSK is that single bit errors do not occur. Either double or multiples of double errors occur. The production of double errors can be demonstrated by considering the simple trellis of Figure 6.46 which has one correlative state and four phase states. The path corresponding to the transmission of an all logical zero (-1) sequence is shown, along with a diverging erroneous path occuring at instant i. It can be seen that two logical ones will be generated before the erroneous path rejoins the correct path at instant $i + 3$. Also shown in Figure 6.46 is another erroneous path, again diverging from the correct path at instant i. We will now investigate the effect of allowing only one error to occur. Consequently a path giving a logical 1 output is chosen for the transition from instant i to $i + 1$, and a path giving logical 0 outputs for the remainder of the burst. It can be seen that the correct and erroneous paths will never converge and so it is not possible for single burst errors to occur in GMSK.

An improvement in BER can be achieved if the double errors are converted to single errors. This can be achieved by differential encoding the data stream prior to the GMSK modulator, and differential decoding after the VE. The effect is demonstrated in Table 6.3, which shows various serial bit streams associated with a differential encoder/decoder pair of Figure 6.47. The first column shows the input bit stream, and the delayed bit stream is shown in the second column. The delay element is initialised to a logical '0'. The differential encoder output is given in the third column. When this bit stream is not corrupted, then the differentially decoded bit

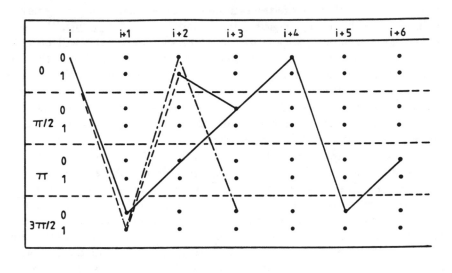

Figure 6.46: Demonstration of double errors in GMSK

Input	Delayed Input	TX	Error TX	No Errors		Two Errors	
				Delay Output	Output	Delay Output	Output
1	0	1	1	0	1	0	1
1	1	0	0	1	1	1	1
1	1	0	0	1	1	1	1
0	1	1	1	1	0	1	0
1	0	1	0*	0	1	0	0*
1	1	0	1*	1	1	0	1
1	1	0	0	1	1	1	1
1	1	0	0	1	1	1	1
0	1	1	1	1	0	1	0
0	0	0	0	0	0	0	0
0	0	0	0	0	0	0	0

Table 6.3: Differential coding and decoding

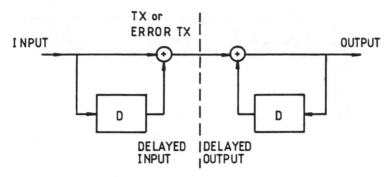

Figure 6.47: Differential coding and decoding

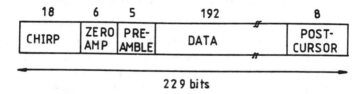

229 bits

Figure 6.48: Format of the data packet using chirp sounding

stream shown in column six is identical to the input bit stream. If we now introduce two errors into the transmitted bit stream TX as shown in column 4, and proceed to decode it as before, then there is only one bit in error when we compare the decoder output shown in column eight, with the encoder input. Note that bits in error are denoted * in Table 6.3. Consequently double bit errors are converted into single bit errors by adopting differential data encoding.

6.2.8   Simulations of DPM Transmissions

We now present simulation results for the transmission of data over radio channels via DPM. In our simulations the bandpass system elements were replaced by their complex baseband equivalents. Our experiments were conducted using 13448 data bits, and as a consequence the experiments are valid for BERs down to the order of 10^{-3}. The TDMA burst format is shown in Figure 6.48 for the chirp sounding, and in Figure 6.49 for the sequence sounding. The DPM had a 3-RC phase pulse shape and $h_p = 1.0$. Four times over sampling was used in both the modulator and the VE. A reduced 16-state VE was employed during all the simulations. No channel coding was used.

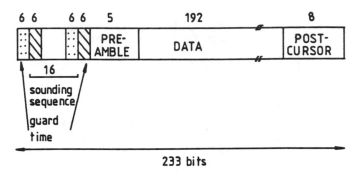

Figure 6.49: Format of the data packet using sequence sounding

6.2.8.1   DPM Transmissions over an AWGN Channel

If we assume there is a single propagation path between a static transmitter and static receiver, then the received signal does not experience fading or Doppler phenomena. The only sources of impairment are the transmitter and receiver filters (which for the purpose of the simulation we assumed introduced no band limiting of the signal) and the system noise, which we modelled as additive white Gaussian noise (AWGN). To simulate the AWGN channel we made use of the complex baseband representation of narrowband noise, adding sample values of zero-mean independent Gaussian processes to the in-phase and quadrature complex baseband signals. By manipulating the variance of the Gaussian sources we determined the BER as a function of carrier to noise (C/N) ratio, or alternatively E_b/N_o, where E_b is the energy per bit and N_o the one sided noise power spectral density (PSD). The relationship between C/N and E_b/N_o is simply

$$\frac{C}{N} = \frac{E_b}{T} \cdot \frac{1}{BN_o} \tag{6.153}$$

$$= \frac{E_b}{N_o} \cdot \frac{1}{BT}. \tag{6.154}$$

where B is the receiver bandwidth and T is the bit duration. It was convenient to introduce a parameter a that specified the relationship between B and T, namely

$$B = a \left(\frac{1}{T} \right) \tag{6.155}$$

giving

$$\frac{E_b}{N_o} = a \frac{C}{N}. \tag{6.156}$$

In order to calibrate the system, the carrier power was measured. Its value remained constant for DPM because of its constant envelope property. Next

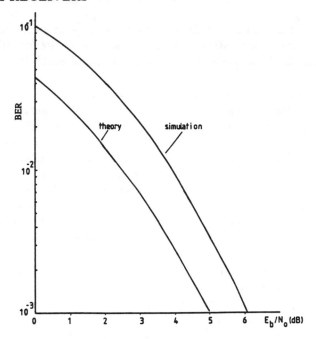

Figure 6.50: BER Performance of DPM in an AWGN channel

the Gaussian noise generators were set to give a PDF with variance N. Finally, knowledge of the transmitted DPM signal bandwidth gave the parameter $a = 3$. Figure 6.50 shows the simulated system's BER as a function of E_b/N_o.

The theoretical determination of BER as a function of E_b/N_o is a difficult problem. One approach [4] is to utilise a bound based on the minimum free distance of the transmitted signal set. For 3RC DPM, $h_p = 1$, the value of $d_{min} = 3$ [5] and at high values of E_b/N_o the probability of bit error may be approximated by

$$P_e \approx Q \left(\sqrt{d_{min}^2 \frac{E_b}{N_o}} \right). \tag{6.157}$$

This theoretical result is also plotted in Figure 6.50 for comparison. The simulation curve is roughly 1 dB poorer in performance than that given by theory.

6.2.8.2 DPM Transmissions over NonFrequency Selective Rayleigh and Rician Channels

The frequency non-selective Rayleigh fading channel is modelled by multiplying the transmitted signal envelope by an attenuation factor α, chosen from a Rayleigh PDF, and shifting the phase angle of the received signal by an angle θ that can have values in the range $(0 - 2\pi)$ with equal probability. For ease of simulation, the fading parameters were held constant for the duration of one TDMA burst, thereby providing slightly optimistic results.

The BER as a function of E_b/N_o is shown in Figure 6.51 for different values of the Rician parameter K in dBs, along with the results for the AWGN situation which we use as a bench marker. The BER performance was very much worse for the Rayleigh fading channel ($K = -\infty dB$) compared to that for the AWGN channel ($K = +\infty dB$). This was because even when E_b/N_o had a high value, the received signal was on occasions in a deep fade, enabling the noise to induce errors. The Rayleigh channel is the most severe type of non-frequency selective fading channel.

A study of propagation in microcells [27] revealed that in the majority of cases a dominant path exists between the transmitter and the receiver. The PDF of the fading envelope is Rician in this situation. To model a Rician channel a direct path, concurrent in time with the Rayleigh fading path, was added to the channel model. The Rician parameter K is the ratio of power in the dominant path to the power in the Rayleigh fading path. Figure 6.51 presents a family of Rician BER curves for which the AWGN channel and Rayleigh channel appear as special cases.

Inclusion of a direct path, even for $K = 0$ dB, caused a substantial improvement in performance, some 10 dB reduction in channel SNR compared to a Rayleigh fading channel at a BER of 10^{-3}. The improvement continued with increasing K, and at $k=14$ dB the performance approached that of the AWGN channel. Reference [28] suggests that a Rician channel with $K > 13$ dB behaves predominantly like an AWGN channel, an observation substantiated by this simulation. A Rayleigh fading channel was produced when the direct path was overwhelmed by the Rayleigh fading component [29]. It cannot be over stressed that by opting for microcellular structures, rather than large conventional cells, the PDFs of the received signal envelopes become Rician and may have high values of K, yielding significant gains in system performance.

6.2.8.3   DPM Transmissions over Frequency Selective Two-Ray Static Channels

This channel gives an insight into VE performance over a non-time varying frequency selective channel. The channel was modelled as a complex baseband finite impulse response (FIR) filter, having complex coefficients at delay intervals given by the reciprocal of the system sampling frequency. Only two of the tap coefficients were non-zero, giving rise to the two-ray channel. A typical complex impulse response where the delay between the two equal amplitude rays was eight sampling intervals (i.e., 2 bits with four times oversampling) is shown in Figure 6.52. This complex response was

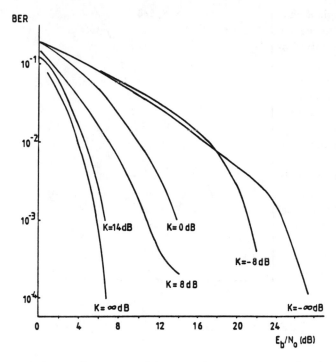

Figure 6.51: BER performance of DPM in frequency non-selective Rayleigh and Rician channels

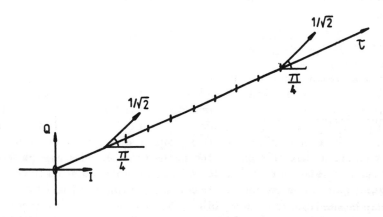

Figure 6.52: Typical complex impulse response of a frequency selective two-ray static channel

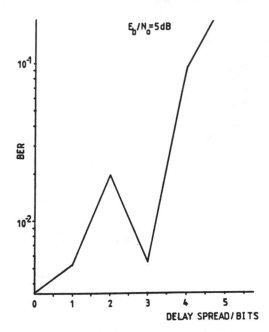

Figure 6.53: BER performance of DPM in a frequency selective two-ray static channel

resolved into its inphase and quadrature components and complex baseband convolution performed. Figure 6.53 shows the BER as a function of delay between the paths at an E_b/N_o of 5 dB. The degradation in BER as a result of the frequency selective channel are evident.

6.2.8.4   DPM Transmissions over Frequency Selective Two-Ray Fading Channels

The mobile radio channel is dynamic in nature and as a consequence its impulse response varies with time. When there is no direct path between the transmitter and receiver, the amplitude distributions on each channel path (or ray) have Rayleigh statistics. Where a direct path existed, as may be the case in a microcell, or in air-to-ground radio communications, then it was assumed that only one of the paths experienced Rayleigh fading; produced as a result of many scatterers. For ease of simulation the channel impulse response was assumed to be constant during a TDMA burst.

Consider the situation where both paths possessed equal average power and experienced independent Rayleigh fading. Figure 6.54 shows the BER as a function of path delay separation when $E_b/N_o = 16.5$ dB. The value of E_b/N_o was chosen to ensure that with zero path separation (i.e., the two paths coalesced into one path), the BER was of the order of 10^{-2}. For path

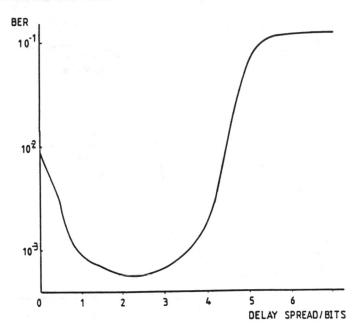

Figure 6.54: BER performance of DPM in a frequency selective two-ray Rayleigh fading channel

separations in the range from one to four bit periods, the BER improved by an order of magnitude compared with that achieved with zero path separation. This indicates that the VE uses the delayed path to provide a form of diversity. When the delay spread was less than one bit period the VE was unable to resolve the paths and so no improvement resulted. From Figure 6.54 it is clear that the reduced 16-state VE can cope with delay spreads of up to four bit periods duration. Delay spreads in excess of four bit periods caused the equalizer performance to degrade seriously and the saturated BER was much worse than that achieved with zero path separation. With four bit periods of path separation, the Viterbi equaliser could not accommodate both paths within its impulse response window. In the situation where one path dominated and the impulse response window was selected to accommodate it, then the interference caused by the second path lying outside the window was not removed. The result was a higher error rate than if both paths had been accommodated within the impulse response window.

The second part of this experiment involved only one of the two paths experiencing Rayleigh amplitude fluctuations. Figure 6.55, shows the BER as a function of path separation for values of $K = 0$ dB, 8 dB and 14 dB respectively. Again the values of E_b/N_o in each case were chosen to ensure

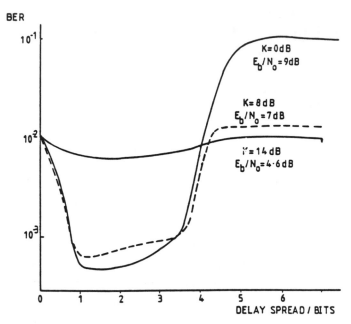

Figure 6.55: BER Performance of DPM in a frequency selective two-ray Rician fading channel

a BER of 10^{-2} for zero path separation. When $K = 0$ dB, the results were similar to those produced when both paths experienced Rayleigh variations (see Figure 6.54), except for a small improvement in the saturated BER. The results for $K = 8$ dB were more interesting because the saturated BER was now only marginally inferior to that measured at zero path separation. It appears that the diffuse path was sufficiently large to provide diversity gain in the one to four bit duration path separation region, but not large enough to cause significant problems at greater path separations. With $K = 14$ dB the channel was virtually AWGN and in this situation the VE could not improve the BER. Notice that as the channels go from Rayleigh, to Rician to virtually Gaussian, the required E_b/N_o to achieve a BER of 10^{-2} at zero separation decreased from 16.5 dB to 4.6 dB. In practice, the channel impulse response will not remain constant over the duration of a burst and consequently the channel estimates used in the VE will not be as accurate towards the end of a burst. However, the simulation results will not be changed dramatically, except in the rare cases where the channel is changing rapidly (i.e., significant Doppler phenomena).

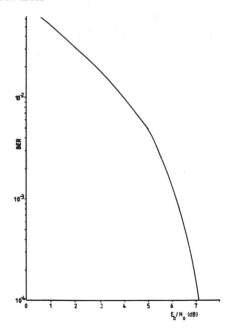

Figure 6.56: BER performance of GMSK in an AWGN channel

6.2.9   Simulations of GMSK Transmissions

Initially we simulated the VE for GMSK ($B_N=0.3$, $h_f=0.5$) with a trellis
where all four phase states were explicitly represented. This meant the VE
possessed $4 \times 16 = 64$ states. The BER performance of the VE was evaluated
over both static AWGN channels and Rayleigh fading channels. Next, we
incorporated phase state memory into the simulation of the VE, reducing
the number of states in the VE by a quarter of the previous value. Simulations showed that the BER performances of the equalisers were identical.
This result was significant because a factor of four reduction in complexity
gave no performance penalty.

6.2.9.1   GMSK Transmissions over an AWGN Channel

The simulations were similar to those with DPM, except that the bandwidth parameter a was reduced to 1.5 in order to account for the reduced
spectral occupancy of GMSK. The BER as a function of E_b/N_o is shown
in Figure 6.56. At a BER of 10^{-3} there was a degradation of about 1.5 dB
in AWGN performance using GMSK ($B_N = 0.3$) as compared to DPM
(3RC). This was expected because GMSK ($B_N = 0.3$, $h_f = 0.5$) had a free
distance d_f of two compared with three for DPM (3RC, $h_p = 1.0$)[4].

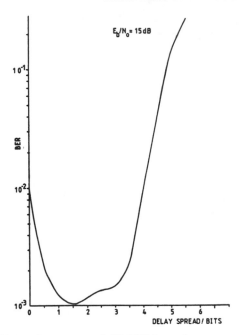

Figure 6.57: BER performance of GMSK in a frequency selective two-ray Rayleigh fading channel

6.2.9.2 GMSK Transmissions over Frequency Selective Rayleigh Fading Channels

The BER performance of the GMSK Viterbi equaliser for frequency selective Rayleigh fading channels is shown in Figure 6.57 for $E_b/N_o = 15dB$. The frequency selective channel had two independent equal power paths each with Rayleigh distributed amplitudes and phases uniformly distributed between 0 and 2π. The curve in Figure 6.57 is similar for DPM, see Figure 6.54. Again in the delay range from half a bit to four bits, there was a substantial improvement in BER performance due to multipath diversity. For delays in excess of four bits, the windowed estimated channel impulse response formed within the VE could no longer accommodate both paths and consequently BER performance was seriously degraded.

6.2.9.3 Comment

We have described how moderately wideband systems known as NB-TDMA employing Viterbi equalisation and either DPM or DFM modulation are suitable for use with dispersive mobile radio channels. There is scope for employing sophisticated reduced state equalisers which can cope with greater excess path delays, without needing additional states. Further refinements include adaption of the equaliser within the TDMA burst, and the use of non-linear decision feedback equalisers (DFE). The DFE potentially has a performance similar to that achieved by Viterbi equalisation, but with a considerable reduction in the processing load. The reduction in complexity and power consumption gained in using a DFE would be valuable for hand held portables.

Bibliography

[1] **S.Pasupathy**: "Minimum shift keying: A spectrally efficient modulation," *IEEE Trans. Commun. Mag.*, pp.14-22, Jun 1979.

[2] **R.de Buda**: "Coherent demodulation of frequency shift keying with low deviation ratio," *IEEE Trans. Commun.*, pp.196-209, Jun 1972.

[3] **W.P.Osborne and M.B.Luntz**: "Coherent and non-coherent detection of CPFSK," *IEEE Trans. Commun.*, Vol. 22, pp.1023-1036, Aug 1974.

[4] **J.B.Anderson, T.Aulin and C-E. Sundberg**: "Digital phase modulation," *Plenum Press*, New York, 1985.

[5] **T.Maseng**: "Digitally phase modulated (DPM) signals," *IEEE Trans. Commun.*, Vol. 33, pp.911-918, Sept 1985.

[6] **T.Maseng and O.Trandem**: "Adaptive digital phase modulation," *Nordic Seminar on Digital Land Mobile Communications,"* Stockholm, Paper 19, Oct 1986.

[7] **A.Papoulis**: "Circuits and Systems — A modern approach," *Holt, Rinehart, Winston Inc.*, 1980.

[8] **T.Aulin and C-E.Sundberg**: "Continuous phase modulation — Part 1: Full response signalling," *IEEE Trans. Commun.*, Vol. 29, pp.196-209, Mar 1981.

[9] **T.Aulin and C-E.Sundberg**: "Continuous phase modulation — Part 2: Partial response signalling," *IEEE Trans. Commun.*, Vol. 29, pp.210-225, Mar 1981.

[10] **K.Murota and K.Hirade**: "GMSK modulation for digital mobile radio telephony," *IEEE Trans. Commun.*, Vol. 29, pp.1044-1050, Jul 1981.

[11] **F.de Jager and C.B.Dekker**: "Tamed frequency modulation — A novel method to achieve spectrum economy in digital transmission," *IEEE Trans. Commun.*, Vol. 26, pp.534-542, May 1978.

[12] **M.K.Simon and C.C.Wang**: "Differential detection of Gaussian MSK in a mobile radio environment," *IEEE Trans. Vehicular Technology*, Vol. 33, No. 4, pp.307-320, Nov 1984.

[13] M.Hirono, T.Miki and K.Murota: "Multilevel decision method for bandlimited digital FM with limiter-discriminator detection," *IEEE J. Sel. Areas Commun.*, Vol. 2, No. 4, pp.498-506, Jul 1984.

[14] K-S.Chung: "Generalised tamed frequency modulation and its application for mobile radio communications," *IEEE J. Sel. Areas Commun.*, Vol. 2, No. 4, pp.487-497, Jul 1984.

[15] S.W.Wales, P.H.Waters and M.L.Streeton: "Experimental program phase II final report," Report No: 72/87/R/515/U, *Plessey Research Roke Manor*, Dec 1987.

[16] P.D.Welch: "The use of fast Fourier transform for the estimation of power spectra: A method based on time averaging over short modified periodograms," *IEEE Trans. Audio and Elect.*, Vol. 15, No. 2, pp.70-73, Jun 1967.

[17] R.R.Anderson and J.Salz: "Spectra of digital FM," *BSTJ*, pp.1165-1189, Jul/Aug 1965.

[18] H.E. Rowe and V.K.Prabhu: "Power spectrum of a digital frequency modulation signal," *BSTJ*, pp.1095-1125, Jul/Aug 1975.

[19] T.Maseng: "The power spectrum of digital FM as produced by digital circuits," *Signal Processing*, Vol. 9, No. 4, pp.253-261, Dec 1985.

[20] J.G.Proakis: "Digital communications," *MacGraw-Hill*, New York, 1982.

[21] J.M.Wozencraft and I.M.Jacobs: "Principles of Communication Engineering," *Wiley*, 1965.

[22] G.D.Forney, Jr: "The Viterbi algorithm," *Proc. IEEE*, Vol. 61, No. 3, pp.268-278, March 1973.

[23] A.P.Clark, J.D.Harvey and J.P.Driscoll: "Near maximum likelihood detection processes for distorted radio signals," *The Radio and Electronic Engineer*, Vol. 48, No. 6, pp.301-307, Jun 1978.

[24] A.Svensson, C-E.Sundberg and T.Aulin: "A class of reduced complexity Viterbi detectors for partial response CPM," *IEEE Trans. on Commun.*, Vol. 32, No. 10, pp.1079-1087, Oct 1984.

[25] A.D.Fagan and F.D.O'Keane: "Performance comparison of detection methods derived from maximum-likelihood sequence estimation," *IEE Proc.*, Vol. 133, Pt. F, No. 6, pp.535-542, Oct 1986.

[26] J.C.S.Cheung and R.Steele: "Modified Viterbi equaliser for mobile radio channel having large multipath delays," *Electronics Letters*, Vol. 25, No. 19, pp.1309-1311, Sept 1989.

[27] S.T.S.Chia, R.Steele, E.Green and A.Baran: " Propagation and BER measurements for a microcellular system," *J. IERE*, Vol. 57, No. 6 (supplement), pp.S255-S266, Nov/Dec 1987.

[28] **F.Davarian:** "Fade margin calculations for channels impaired by Ricean fading," *IEEE Trans. Vehic. Tech.*, pp.41-44, Feb 1985.

[29] **M.Schwartz:** "Communication systems and techniques," *McGraw-Hill*, 1966.

Chapter 7

Frequency Hopping

D.G. Appleby[1], and Y.F. Ko[2]

7.1   Introduction

In this chapter we will consider frequency hopping in the context of narrowband mixed time and frequency division multiple access (TD/FDMA) for digital cellular mobile radio systems. In the type of system under consideration, such as for example the GSM Pan European system, traffic channels are designated as combinations of time slot position and carrier frequency. However, the cellular multiple access protocol is based primarily on frequency division multiplexing (FDM), since the orthogonal sets of radio channels, needed to avoid cochannel interference between neighbouring cells, are obtained by assigning distinct sets of frequencies to all the cells in a reuse cluster, as in a pure FDMA system. Employing time division multiplexing (TDM) as well as FDM results in increased transmission symbol rates and fewer more widely spaced carrier frequencies are required in a given allocated band, which leads to less stringent transceiver design specifications on specifications on selectivity and frequency drift. In addition, higher spectral efficiency is possible with TDM, because of the greater precision achievable with digital switching technology. Narrowband operation is broadly defined by the provision that the transmitted symbol duration in most urban situations should be much greater than the delay spread caused by multipath propagation. This is widely regarded as a desirable condition, since it avoids the necessity for complex equalization techniques to remove the severe intersymbol interference which would otherwise occur.

---

[1]University of Southampton
[2]University of Southampton

Firstly we must differentiate between "slow" and "fast" frequency hopping techniques. In the former case, which is of primary interest here, the hop rate is much less than the information symbol rate and thus many symbols are sent on the same carrier frequency during each hop, maintaining narrowband transmission conditions within each hop, provided of course that the symbol modulation bandwidth does not exceed the coherence bandwidth. On the other hand fast frequency hopping, in which the hop rate is equal to, or often greater than, the symbol rate, has been advocated for mobile radio [1] because of its spread spectrum properties. It can thus be classified as a wideband code division multiple access (CDMA) technique and so suffers many of the disadvantages of such techniques, in particular, implementation would be more difficult because of the requirement for very fast frequency synthesizers.

Slow frequency hopping multiple access (SFHMA), using code division multiplexing as the main multiple access mechanism has been the subject of considerable research effort in recent years [2, 3, 4] leading to a viable system design, the SFH900 [5], which was a very strong contender for the GSM pan-European digital mobile radio network. Not only does SFHMA provide inherent frequency diversity, but it has the property of randomising cochannel interference, referred to as "interferer diversity" [5], which allows error correction coding to be applied effectively to correct errors caused by interference as well as signal fading.

Slow frequency hopping without CDMA has also been proposed for use in various narrowband TDMA systems [6, 7], including the GSM Pan-European system, often as an optional 'add-on' feature. In all these applications it is only the frequency diversity advantage of frequency hopping which is being exploited to avoid the problem of stationary or slowly moving mobile stations being subjected to prolonged deep fades.

In the following section we discuss the principles and the characteristics of SFHMA and then in Section 7.3 an exemplary SFHMA system based on the SFH900 proposal [5, 8] is described in detail. Sections 7.4 and 7.5 are devoted to analyses of the error rate performance of the SFHMA system in AWGN and in cochannel interference respectively. Estimates of spectral efficiency of the system are presented in Section 7.6 followed by a summary of the main conclusions drawn.

7.2 Principles of Slow Frequency Hopping Multiple Access

In this section we shall describe three multiple access techniques based on SFH. Possible reuse cellular structures which may be employed for SFHMA systems are discussed in Section 7.2.2, followed by a study on the factors that affect propagation in Section 7.2.3.

7.2.1   SFHMA Protocols

Three code division multiple access techniques employing SFH with pseudo random sequences have been defined in [2], they are namely, **orthogonal, random, and mixed;**

ORTHOGONAL: orthogonal hopping sequences of length N can be assigned to active users within a cell to which N hop frequencies have been allocated to ensure that during any hop only one user can transmit on a particular frequency. In neighbouring cells orthogonality of transmissions is attained by allocating distinct sets of frequencies, as in FDMA. A given set of hop frequencies and sequences can be reused in cells which are at or beyond the reuse distance. This protocol assumes complete synchronization of sequences so that any users with the same sequence of hop frequencies will experience continuous cochannel interference on every hop and thus the situation is the same as for FDMA, except for the inherent frequency diversity.

RANDOM: each active user is assigned a unique hop sequence which is uncorrelated with, but not necessarily orthogonal to, all other sequences. The constraint of low cross correlation of hopping sequences is less stringent than that of orthogonality and thus allows a much larger set of sequences to be selected. Cochannel interference will occur in this case, but it will be caused by a different subset of the other active users on each hop and in consequence the interference intensity will be subjected to random hop-to-hop variations. This behaviour, which is known as "interferer diversity", is an important feature of frequency hopping, since it spreads the interference effects evenly, on average, over all the available frequencies. It also enables the effects to be counteracted by powerful error correction coding. Reuse cellular structures are not necessary.

MIXED: as in the orthogonal protocol, a set of N orthogonal sequences of the N available hop frequencies is assigned to each cell, but in this case the sets of sequences assigned to reuse cells operating with the same frequencies are distinct and are selected to ensure that any two sequences from different cells are uncorrelated. Thus cochannel interference is caused only by transmissions in reuse cells, and in addition, its effects are mitigated by interferer diversity, as in the random protocol. It is assumed that a pair of uncorrelated sequences of length N hop to the same frequency in only one of the N hops.

Frequency diversity is the only advantage of the orthogonal protocol, whereas the random protocol has the additional important advantage of interferer diversity and was the preferred SFHMA protocol in early investigations [2]. However, later studies [5, 8] of SFHMA system design have opted for the mixed protocol, because when used in conjunction with fractional reuse structures, as described in the following sub-section, it offers more freedom for system design optimisation and hence may be expected to

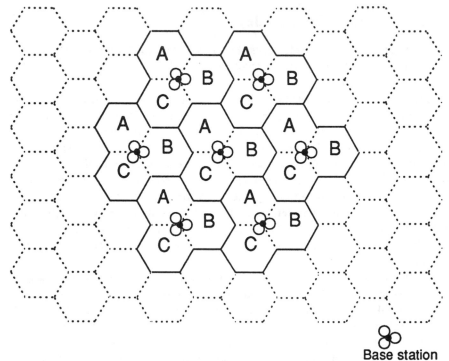

Figure 7.1: Three colour reuse structure.

yield higher spectral efficiency. Therefore in the remainder of this chapter we shall consider only the mixed protocol.

7.2.2   Reuse Cellular Structures

The basic form of reuse structure proposed for SFHMA systems [5, 8] is the three colour cluster, as shown in Figure 7.1, where the colours, indicated by the letters A, B and C, represent distinct sets of N frequencies assigned to the three cells. Also shown in the diagram is the collocated corner base station configuration, with directional antennas pointing into the respective cells, which is favoured primarily because of the obvious economies in the number of base station sites required and in the connections to the fixed cable network. Some authors refer to this configuration as a single sectored cell, but we prefer to describe it as a cluster of three separate cells.

Choosing a reference cell with colour A, then all other cells marked A in Figure 7.1 are full reuse cells using the same frequencies. It is assumed that it is possible to assign uncorrelated sets of hopping sequences to all the reuse cells near enough to produce significant levels of cochannel interference. An active user in the reference cell will experience interference from an active

Frequency Group	Shade						
	A1	*A2*	*A3*	*A4*	*A5*	*A6*	*A7*
f1	*	*	*				
f2	*			*	*		
f3	*					*	*
f4		*		*		*	
f5		*			*		*
f6			*	*			*
f7			*		*	*	

Table 7.1: Shade allocation scheme for a 21/3 reuse structure

user in a reuse cell on only one of each sequence of N hops, when both transceivers are tuned to the same frequency. Each additional active user in the same or another reuse cell will cause similar frequency collisions, but on a different frequency in the reference user's sequence. The dependence on the number of active users in reuse cells can be expressed in terms of the proportion of the hop sequence occupied by frequency collisions per active user. This quantity is known as the frequency collision rate and for the basic three colour cluster under consideration it has the value of $1/N$.

Fractional reuse structures, which offer greater flexibility for design trade-offs, are obtained by dividing each colour into M overlapping subsets of L groups of frequencies, which henceforth we will refer to as "shades" of that colour (Verhulst uses the term "pseudo-colour" in [5]). The resulting reuse cluster contains $3M$ cells arranged as M sub-clusters of size 3, each using a different shade of the 3 colours and centred on a common base station site. This is termed a $3M/L$ fractional structure, where L/M is the fraction of the N frequencies making up a colour, which is contained in each shade. As an example take $M=7$ and $L=3$, giving a 21/3 structure as shown in Figure 7.2. A shade allocation scheme for colour A is presented in Table 7.1 which indicates that the fractional overlap of hop frequencies is 1/3. Other schemes can be obtained from the one shown by interchanging pairs of rows in the allocation matrix.

Figure 7.3 shows the positions of both full and partial reuse cells for colour A, ie, the potential sources of cochannel interference, out to the first ring of full reuse cells. For full reuse cells, as in the case of the three colour cluster, the frequency collision rate is simply the inverse of the sequence length, ie $M/LN = 7/3N$. However, for partial reuse cells we must allow for the fact that, because of the incomplete frequency overlap, certain pairs of sequences will produce no collisions, with a probability equal to the fractional overlap, k. Thus a general expression for the frequency collision

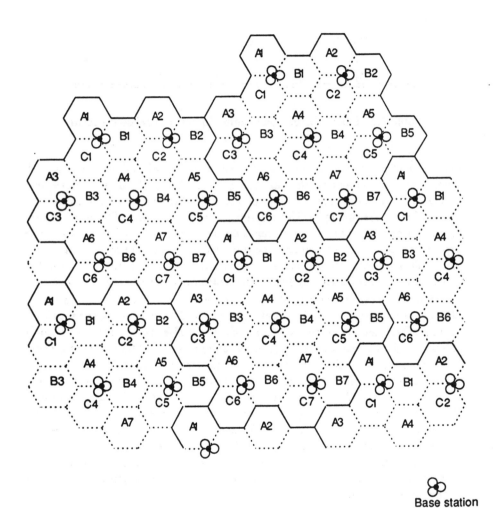

Figure 7.2: 21/3 reuse structure.

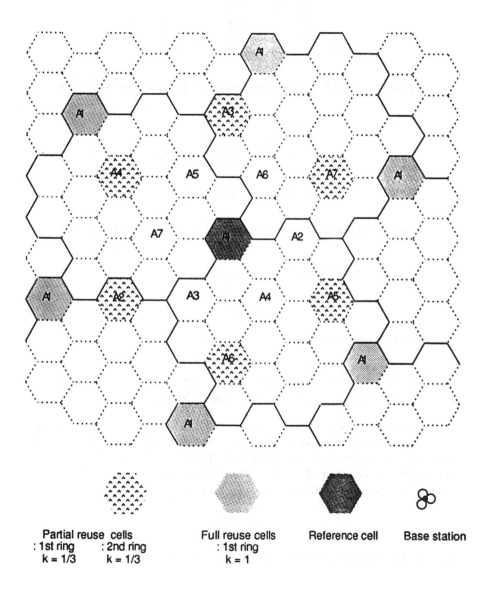

Figure 7.3: Reuse cells in 21/3 structure.

	Shade			
Frequency Group	A1	A2	A3	A4
f1	*	*	*	
f2	*	*		*
f3	*		*	*
f4		*	*	*

Table 7.2: Shade allocation scheme for a 12/3 reuse structure

Type of	y	
reuse cell	21/3	12/3
partial	7/9N	8/9N
full	7/3N	4/3N

Table 7.3: Frequency collision rates of partial and full reuse cells

rate, y, in this case is given by:

$$y = \frac{7k}{3N} \tag{7.1}$$

where k can have the following values:

$k = 0$ - for cells of different colour
$k = 1/3$ - for cells of same colour but different shade, ie partial reuse
$k = 1$ - for cells of same colour and shade, ie full reuse

For comparison, consider a second example in which we make $M = 4$ and $L = 3$, giving a 12/3 reuse structure as shown in Figure 7.4. One possible shade allocation matrix is presented in Table 7.2. Figure 7.5 shows the reuse cells for this case, for which $k = 2/3$ in cells with partial frequency reuse. The frequency collision rate now becomes

$$y = \frac{4k}{3N} \tag{7.2}$$

Values of y for both structures are given in Table 7.3.

7.2.3   Propagation Factors

In this section we will discuss briefly a set of factors which can be used to characterise the radio channel for studies of the SFHMA system [5,

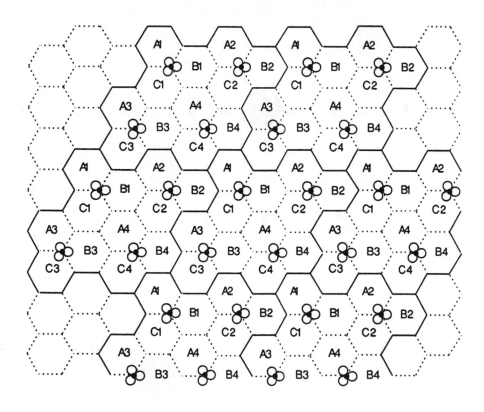

Base station

Figure 7.4: 12/3 reuse structure.

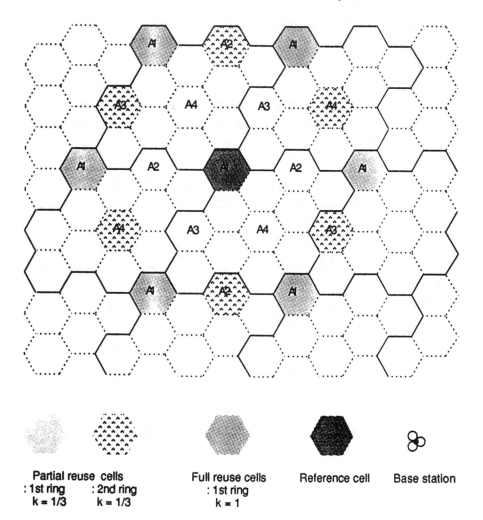

Figure 7.5: Reuse cells in 12/3 structure.

8]. More detailed treatments of propagation characteristics are presented in Chapters 1 and 2. For our purpose here it is convenient to assume a narrowband model for multipath distortion, even though in the system to be considered the normal narrowband criterion that the symbol duration should be much greater than the delay spread is not always satisfied.

Using the narrowband model, the power levels of both the wanted signal and interfering signals can be estimated as the product of three principal factors;

- mean received level A_i

- shadowing attenuation B_i

- multipath fading F_i.

A_i is deterministic and may include factors such as antenna radiation patterns, transmitter power control and receiver adjacent channel rejection, in addition to the mean path loss, which, based on the usual simplified model, can be assumed to be proportional to $d^{-\alpha}$, where d is the distance from the relevant transmitter and α is the propagation exponent, which depends on the environment. For all the calculations reported here α has been taken as 3.5.

B_i depends on large scale variations in the locations of the transmitter and receiver relative to the respective local topographies, and in practice it is unpredictable. Thus this factor is usually taken to be a random variable with lognormal distribution having a mean value of 0 dB and a standard deviation in the range 6 to 8 dB for typical urban environments. For normal vehicle speeds B_i may be assumed to be constant in frequency over a few MHz and in time over a few hundred ms.

F_i represents the rapid and severe fluctuations in signal level experienced on a radio link with a moving vehicle, often referred to as fast fading. These fluctuations are caused by the changes in relative phases between the multipath components due to small scale variations in vehicle location (of the order of half the wavelength). It can usually be assumed, especially in medium to large cells, that the resulting signal envelope has a Rayleigh amplitude distribution and a negative exponential pdf of power. Thus a negative exponential pdf with mean of unity is appropriate for this factor. If the spacing of consecutive hop frequencies is greater than the coherence bandwidth (of the order of a few hundred kHz) and if the transmission duration during each hop is less than the coherence time (of the order of a few ms) then the F_i can be assumed to be constant during each hop and to be statistically independent from hop-to-hop.

Interfering signals are characterised by a fourth factor E_i, which is a discrete random variable having two values 0 and 1 with a Bernoulli distribution, representing the effective on/off status of the interference source.

This factor accounts for the random nature of frequency collisions inherent in the mixed SFHMA protocol.

An important aspect in evaluating the effects of these propagation factors is the rate at which they are likely to change. A factor can be broadly categorized as "fast" if it changes independently from hop-to-hop, as noted in the case of the fading factor, F_i. On the other hand a factor which remains highly correlated over many hop periods can be categorized as "slow". For the wanted signal the factors A_o and B_o are slow, but for interfering signals all the A_i and B_i are fast in the mobile-base direction, ie uplink, because of the changes in the locations of sources of interference from hop-to-hop, although they are slow in the other direction. F_i and E_i are always fast.

The value zero for the index i refers to conditions relevant to the wanted signal within the reference cell and non-zero values refer to interfering signals from other cells. Hence the received carrier-to-interference ratio (CIR) is given by:

$$\lambda = \frac{A_o B_o F_o}{\sum_{i=1}^{M} A_i B_i F_i E_i} \tag{7.3}$$

7.3   Description of an SFHMA System

The system described in this section is closely related to the SFH900 system proposed by the French collaboration LCT and Sagatel [5, 8], as a candidate for the GSM Pan-European digital mobile radio system.

7.3.1   Multiple Access Protocol

A mixed SFHMA cellular protocol is assumed, in which, as explained in Sub-section 7.2.1, strictly orthogonal hopping sequences are used for CDMA within each cell, but in neighbouring reuse cells different uncorrelated sets of sequences are deployed. Thus there will be no interference between users within a cell since in any time slot only one user should be transmitting on any given carrier frequency. Contention with immediate neighbour cells is prevented by using orthogonal sets of hopping frequencies. Interference from further cells within the main reuse cluster, in which partial overlap of frequency sets is allowed, is reduced because of interferer diversity, ie, on each hop there is a new subset of possible interfers with statistically independent propagation characteristics.

7.3.2 Time Division Multiplexing

The hop duration of 4 ms is divided into 3 time slots, each 1.23 ms long, separated by guard times of 100 μs. These slots are used in sequence for transmission, reception and frequency switching. The users in a cell are divided into 3 approximately equal subsets which use the 3 possible phases of this sequence. Apart from other advantages in simplifying the design of the transceivers, this feature allows the number of FH channels in a cell to be trebled for a given set of FH sequences and also results in further randomisation of the interference generated by transmissions from the mobile stations.

We consider that the slot duration is short enough to justify the assumption that the transmission channel impulse response is static during each hop, even for fast vehicles.

7.3.3 Modulation and Equalization

Binary GMSK modulation with a normalised premodulation filter bandwidth $B_t T = 0.3$ is assumed, together with quasi-coherent demodulation using a Viterbi Algorithm equalization technique to counteract intersymbol interference caused by the pre-modulation Gaussian LPF and multipath distortion. Reference should be made to Chapter 6, for further details of GMSK modulation and Viterbi equalization.

Transmission of an 8-bit training pattern during each slot enables the demodulator to derive the complex baseband impulse response of the transmission channel and also to determine the carrier phase. Hence there is no need for a phase-locked loop, but the carrier frequency error must be limited to less than about 30 Hz to ensure that there is negligible degradation due to the consequent phase drift during the slot. Adequate frequency control is achieved by measurements on the unmodulated carrier transmissions on certain hops assigned to the Master Channel (see Section 7.3.5 below) or alternatively by sending two training patterns during each data hop (at 1/4 and 3/4 of the slot duration) so that the phase difference can be estimated.

7.3.4 Speech and Channel Coding

A 16 kbits/s sub-band speech coder, similar to that described in Chapter 3, is used with three levels of protection against digital errors. An important feature of the system under consideration is the highly redundant forward error correction (FEC) channel coding to enable correction of long error bursts extending over complete hops caused by fading or frequency collisions. The overall average code rate is 1/3 on speech traffic channels, which is achieved by concatenating two shortened Reed-Solomon (RS) codes. Figure 7.6 shows the cascaded arrangement of the two coders and two decoders.

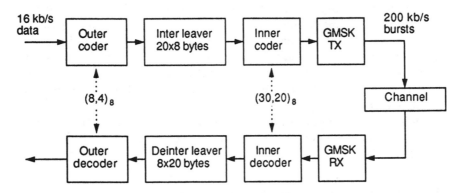

Figure 7.6: Concatenated RS channel coding.

The inner code is a (30,20) RS code with 8-bit symbols which codes a data block of length 160 bits to be transmitted in one hop. This code is applied to all traffic including control signals. The outer RS code has three levels of redundancy to give different levels of protection to various parts of the sub-band speech coder frame. These codes are (8,3), (8,4) and (8,5), also with 8-bit symbol length, but the net rate is maintained at 1/2. Symbols from a given codeword are transmitted on separate hops so that errors due to fading and interference are decorrelated. This is achieved by interleaving symbols from blocks of 20 codewords (bytes) at the input to the inner coder, as shown in Figure 7.6.

It is possible to use different cosets of the inner code in cells assigned the same hop frequencies so that captures by strong interfering cochannel signals can be identified and hence rejected. In the analyses of error probability described in Sections 7.4 and 7.5, a single (8,4) outer code has been assumed.

7.3.5   Transmitted Signal Structure

A hop frame comprises 60 hops, which are divided into 3 sub-sets as follows:

H1 - Common Control Channels including   4 hops
       the Master Channel
H2 - Associated Control Channels   8 hops
H3 - Traffic Channels   48 hops

The hop frame duration is $60 \times 4 = 240$ ms. Allowing for the proportion of the frame allocated to speech traffic, the time division multiplexing (including guard times), the channel coding redundancy and the dual training patterns the transmitted bit-rate required becomes very close to 200 kbits/s. For this bit rate the optimum hop frequency spacing is about 150 kHz, which provides 17 dB adjacent channel rejection.

7.3.6 Frequency Reuse

Base stations are assumed to be located at the junctions of three cells into which they would radiate via 120° sectorial antennas. Following [5], we represent the base station antenna radiation pattern with a back-to-front ratio of -20 dB by the expression

$$g(\theta) = \max[\cos(1.1\theta), 0.01] \qquad (7.4)$$

A reuse cluster size of 3 is appropriate for the primary 6 MHz band in which the Master Channel is located. The Master Channel is designed to have a simple and regular time-frequency structure to enable rapid acquisition of network control information by a mobile entering the system. In the remainder of the allocated system bandwidth the available frequencies will be divided into groups to be allocated as "shades" in fractional reuse structures, as described in Section 7.2.2. In practice some terrain dependent variations of reuse structures would be required to cope with the consequent propagation problems.

7.4 BER Performance in the Absence of Co-channel Interference

In this section the performance of the SFHMA system described above is investigated in AWGN channels. A pure AWGN channel is a highly idealised channel condition in mobile radio environments, in which the system is expected to perform well. In order to achieve spectral efficiency cellular mobile radio systems are designed to be limited by cochannel interference rather than noise. However, it is often useful to be able to compare systems on the basis of performance in the ideal AWGN channel, which is further characterised here according to whether the received signal level is assumed to be either static, or varying due to Rayleigh fading.

GMSK belongs to the class of modulation schemes called partial response Continuous Phase Modulation (CPM). BER performance is of course dependent on the particular type of demodulator. A comprehensive discussion on various demodulator structures for CPM signals can be found in [9]. The two classes of demodulator considered in this section are the sub-optimal MSK-type orthogonal coherent detector as suggested by Murota in his original paper on GMSK [10], and the optimum Maximum-Likelihood Sequence

Estimation (MLSE) detector, which is closely approximated by the quasi-coherent demodulator and equalizer based on the Viterbi Algorithm assumed for the SFHMA system. The MSK-type receiver is of interest here partly because of its simplicity in practical implementation with only a minor degradation in power efficiency compared to the MLSE detector, but primarily because its performance in more realistic mobile radio conditions is easier to analyse than the MLSE performance. Furthermore the SFHMA receiver is expected to perform similarly in these channel conditions, while the MLSE detector upper bounds the performance of GMSK.

As many published modulation scheme performances are evaluated at a typical coded speech data rate of 16 kbits/s, we start our analysis by considering the two types of GMSK receiver structure operating at this data rate with no FEC coding in Sections 7.4.1 and 7.4.2. In Section 7.4.3 some of the key assumptions on the mobile radio channels and system operations are discussed. The basis for comparison between the SFHMA and uncoded systems is also addressed in this section. The effects on BER performance resulting from the FEC coding and TDM specified for the SFHMA system are investigated in Section 7.4.4 for the case of the static AWGN channel. Finally we evaluate and compare the BER performances in the more realistic Rayleigh fading AWGN channel in Section 7.4.5.

7.4.1   BER Performance of the MLSE Detector

General discussions on the principle of MLSE detectors can be found in Chapter 6 and also in [11], while MLSE receiver structures suitable for GMSK are reviewed in [12] and [13]. Although the structure of the optimal MLSE is known, it is difficult to evaluate its BER performance analytically. However, it is generally accepted [9, 10] that at fairly high SNR the BER performance is asymptotically dominated by the minimum Euclidean distance between the symbols in the signal space. For the type of modulation being considered here, the probability of bit errors for the ideal coherent MLSE detector is bounded in the high SNR condition as [10]:

$$P_e = Q\left(d_{\min}\sqrt{\frac{E_b}{N_o}}\right) \qquad (7.5)$$

where $d_{\min}$ is the normalised minimum Euclidean distance, E_b is the mean energy per data bit, N_o is the single-sided power spectral density of the AWGN and $Q()$ is the normalised Gaussian integral, defined as:

$$Q(x) = \frac{1}{\sqrt{2\pi}}\int_x^\infty \exp(-u^2/2)du \qquad (7.6)$$

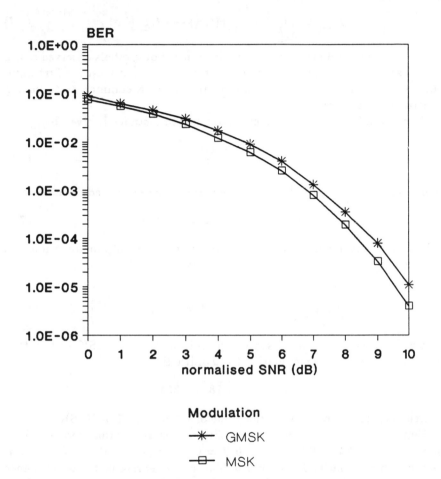

Figure 7.7: BER vs normalised SNR for GMSK ($B_t T = 0.3$) and for MSK using an ideal MLSE detector, in a static AWGN channel with continuous uncoded transmission at 16 kbits/s.

In general the minimum squared Euclidean distance in the signal space is
defined as:

$$D^2_{min} = \min_{\underline{\alpha}, \underline{\beta}} \left\{ \int_0^{LT} [v(t, \underline{\alpha}) - v(t, \underline{\beta})]^2 \, dt \right\} \tag{7.7}$$

where T is the symbol duration, L is the number of symbols observed and $\underline{\alpha}$
and $\underline{\beta}$ are digital sequences of length L, that differ in at least the first digit,
and which ensure that the modulator finishes in a common state having
started from a common state.

The normalised minimum squared Euclidean distance is given by:

$$d^2_{min} = \frac{D^2_{min}}{2E_b} \tag{7.8}$$

For constant envelope CPM signals, which can be expressed in the form:

$$v(t, \underline{\alpha}) = \sqrt{2S} \cos[\omega_0 t + \theta(t, \underline{\alpha})] \tag{7.9}$$

where S is the mean signal power, the normalised minimum Euclidean
distance squared can be shown [9] to be given by:

$$d^2_{min} = \min_{\underline{\alpha}, \underline{\beta}} \left\{ \frac{1}{T} \int_0^{LT} [1 - \cos[\theta(t, \underline{\alpha}) - \theta(t, \underline{\beta})]] \, dt \right\} \tag{7.10}$$

From the calculated data presented by Murota [10] for GMSK we find that
for the normalised premodulation bandwidth $B_t T = 0.3$

$$d_{min} = \sqrt{1.78} = 1.334 \tag{7.11}$$

which may be compared with the value of $\sqrt{2}$ for MSK or BPSK.

Figure 7.7 shows bit error probability versus the normalised signal-to-
noise ratio E_b/N_o for GMSK with $B_t T = 0.3$ using an ideal MLSE detector,
obtained by evaluating Equation 7.5. This curve represents the performance
of continuous uncoded transmission of data at 16 kbits/s. Also included is a
graph of error probability for MSK, which when compared with the GMSK
graph, shows that there is only a small penalty in performance in using
GMSK, which is more than outweighed by the superior spectral efficiency
of GMSK.

7.4.2   BER Performance of the MSK-Type Detector

In the original paper on GMSK [10] Murota and Hirade described an orthogonal coherent MSK-type detector and presented experimental results
showing that a BER performance close to that of an ideal MLSE detector
can be obtained. Although the MSK-type detector is theoretically sub-
optimum for partial response CPM signals and is suitable only for schemes

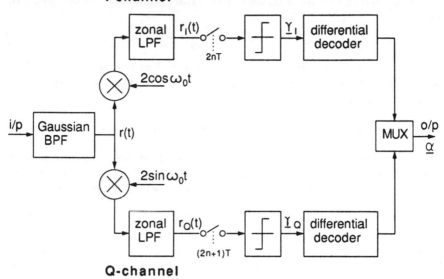

Figure 7.8: Conceptual structure of a parallel MSK-type receiver for GMSK signals.

with modulation index of 0.5, it is of special interest because of the relative ease of implementation compared with an MLSE detector. MSK-type receivers can be implemented in a serial form as well as the more familiar parallel form, but since it has been shown [14] that, assuming ideal phase and symbol timing recovery, both forms have equal performances, we shall analyse in this section the performance in a static AWGN channel of a parallel MSK-type detector only.

Figure 7.8 shows the conceptual structure of a parallel MSK-type receiver, in which the main demodulation noise filtering is provided by a Gaussian band-pass filter (BPF) at the receiver input, as in the experimental circuit described in [10]. Perfect synchronisation of both carrier phase and symbol timing is assumed in the following analysis. The impulse response of the Gaussian BPF can be expressed in the form:

$$h_r(t) = 2Re\{h_{rb}(t)\exp(j\omega_o t)\} \qquad (7.12)$$

where ω_o is the centre frequency and $h_{rb}(t)$ is the complex envelope of the impulse response of the equivalent low-pass filter, which assuming zero delay is given by the real Gaussian function:

$$h_{rb}(t) = \frac{\mu_r}{\sqrt{\pi}}\exp(-\mu_r^2 t^2) \qquad (7.13)$$

where the parameter μ_r is related to the 3 dB bandwidth of the BPF, B_r as:

$$\mu_r = \frac{\pi B_r}{\sqrt{2\ell n2}} \tag{7.14}$$

Thus the filter output waveform can be obtained as the sum of the GMSK input signal convolved with $h_r(t)$ and the filtered AWGN, which can be expressed in the form:

$$r(t) = Re\{[s(t) * h_{rb}(t)]\exp(j\omega_o t)\} + n_I(t)\cos\omega_o t - n_Q(t)\sin\omega_o t \tag{7.15}$$

where $n_I(t)$ and $n_Q(t)$ are the zero mean low-pass Gaussian processes of the inphase and quadrature components of the band-pass noise output and $s(t)$ is the complex envelope of the input signal which is given by:

$$s(t) = \sqrt{2S}\exp[j\theta(t,\underline{\alpha})] \tag{7.16}$$

After multiplying by the local inphase and quadrature carrier waveforms, $2\cos\omega_o t$ and $-2\sin\omega_o t$, and then zonal filtering to remove the sum frequency components, the baseband signals in the I and Q channels can be expressed as:

$$r_I(t) = \sqrt{2S}\cos\theta(t,\underline{\alpha}) * h_{rb}(t) + n_I(t) \tag{7.17}$$

and

$$r_Q(t) = \sqrt{2S}\sin\theta(t,\underline{\alpha}) * h_{rb}(t) + n_Q(t) \tag{7.18}$$

The output binary data sequence $\underline{\alpha}$ is determined by in effect differentially decoding the offset binary quadrature symbol sequences γ_I and γ_Q, which are obtained symbol by symbol after simple threshold detection of samples of $r_I(t)$ and $r_Q(t)$ taken at even and odd multiples of T, respectively. Since the processes of filtering and symbol detection are identical for the two quadrature channels, apart from a time shift of T, it is sufficient to consider only the I channel in evaluating the probability of quadrature symbol errors.

For a given data sequence $\underline{\alpha}$ the amplitude of the sample of $r_I(t)$ at the optimum sampling instant $(t = 0)$ is a Gaussian distributed random variable, y, which, putting $S = E_b/T$, has a mean value given by:

$$M_y(\underline{\alpha}) = \sqrt{\frac{2E_b}{T}}\int_{-\infty}^{\infty}\cos\theta(\tau,\underline{\alpha})h_{rb}(-\tau)d\tau \tag{7.19}$$

and a variance given by:

$$\begin{aligned}\sigma_y^2 &= <n_I^2(t)> \\ &= N_o B_{rn} \tag{7.20}\end{aligned}$$

where B_{rn} is the noise bandwidth of the BPF. Noting that the BPF has unity gain at the centre frequency ω_o and using Parseval's theorems B_{rn}

can be determined as:

$$B_{rn} = \int_{-\infty}^{\infty} h_{rb}^2(t)dt$$

$$= \frac{B_r}{2}\sqrt{\frac{\pi}{\ell n2}} \tag{7.21}$$

As a function of the data sequence, the probability of a quadrature symbol error is given by:

$$P_{qe}(\underline{\alpha}) = Q\left\{\frac{M_y(\underline{\alpha})}{\sigma_y}\right\} \tag{7.22}$$

It is of course the ISI in the baseband signal waveforms, produced by both the premodulation filter and the predetection filter, which causes the above error probability to vary with the data sequence. The overall probability can be obtained by averaging over the set of possible relevant sequences, having a length equal to the effective duration of the ISI, which is dependent on the normalised bandwidths of the filters. However, at high values of SNR, as for the MLSE detector, the overall error probability will be dominated by the contribution due to the minimum value of M_y. By analogy with the treatment of the MLSE detector we introduce a squared distance parameter due to Sundberg [14] defined as:

$$\delta^2 = \left[\frac{M_y(\alpha)}{\sigma_y}\right]^2 \cdot \frac{N_o}{E_b} \tag{7.23}$$

Hence the bound on the quadrature symbol error probability can be expressed in the form:

$$P_{qe} = Q\{\delta_{\min}\sqrt{E_b/N_o}\} \tag{7.24}$$

where using the convolution integral $\delta_{\min}$ is obtained from:

$$\delta_{\min}^2 = \min_{\underline{\alpha}}\left\{\frac{2}{TB_{rn}}\left[\int_{-\infty}^{\infty}\cos\theta(\tau,\underline{\alpha})h_{rb}(-\tau)d\tau\right]^2\right\} \tag{7.25}$$

Because of the differential decoding of the detected quadrature symbols each isolated symbol error will result in two output bit errors, but a group of consecutive quadrature symbol errors, occurring alternately on both I and Q channels, will also only give rise to two bit errors at the beginning and end of the group. Consequently it can be shown that the probability of bit errors is given by:

$$P_b = 2P_{qe}(1 - P_{qe})$$

$$\approx 2P_{qe} \quad \text{if } P_{qe} << 1 \tag{7.26}$$

Thus for high SNR we can approximate the bit error probability as:

$$P_b = 2Q\{\delta_{\min}\sqrt{E_b/N_o}\} \tag{7.27}$$

It has been found by Murota [15], and confirmed during computations made by the authors, that the maximum ISI is caused by the sequence $\underline{\alpha} = \cdots 0, 0, 0, 1, 1, 1 \cdots$ which can therefore be used to evaluate $\delta_{\min}$. This sequence produces the following phase waveform [15]:

$$\theta(t) = \frac{\pi}{2T} \int_0^t erf(\mu_t \tau) d\tau + \frac{\sqrt{\pi}}{2T \mu_t} \tag{7.28}$$

where μ_t is a parameter of the premodulation Gaussian low pass filter related to the 3 dB bandwidth, B_t, as:

$$\mu_t = \pi B_t \sqrt{\frac{2}{ln2}} \tag{7.29}$$

and $erf()$ is the Error Function defined as:

$$erf(x) = \frac{2}{\sqrt{\pi}} \int_0^x \exp(-u^2) du \tag{7.30}$$

Equation 7.27 was evaluated for normalised bandwidths of $B_t T = 0.3$ and $B_r T = 0.63$, the latter value being the empirical optimum value found by Murota and Hirade [10] and which has been also confirmed during these computations by the authors. The convolution integral in Equation 7.25 was determined numerically, with the limits reduced to $\pm T$, since outside this range $h_{rb}(t)$ becomes negligible.

Figure 7.9 presents graphs of bit error rate (BER) probability versus E_b/N_o for the MSK-type detector and for the MLSE detector (from Section 7.4.1) for comparison. Again in the context of the analysis of the SFHMA system this performance may be taken as being representative of continuous transmission of uncoded data at 16 kbits/s. It will be seen that the degradation in performance compared to the optimum MLSE detector is only about 1 dB in SNR. This computed performance for the MSK-type detector is also commensurate with the experimental results reported in [10].

7.4.3   Channel Models and System Assumptions

In the previous sub-sections we have analysed bit error rates in a static AWGN channel to provide an indication of the performance of the GMSK modulation scheme achievable under idealized conditions. In the following sub-sections we shall determine the BER performances of the SFHMA system, described in Section 7.3, and of an equivalent uncoded FDMA system in both static and fading AWGN channels. However, prior to starting the system analysis we wish to present the dynamic channel model and the principal assumptions affecting system performance.

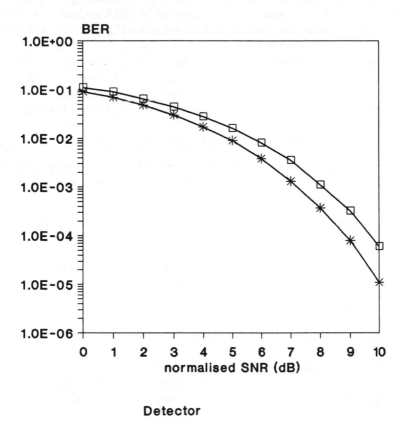

Figure 7.9: BER vs normalised SNR for GMSK $(B_t T = 0.3)$ using an ideal MSK-type detector $(B_r T = 0.63)$ and also using an ideal MLSE detector, in a static AWGN channel with continuous uncoded transmission at 16 kbits/s.

As indicated previously in Section 7.2.3 we assume a narrowband flat Rayleigh fading channel model to account for multipath propagation distortion, although we realise that the specified SFHMA system is not strictly narrowband because the transmitted bit duration ($5\mu s$) is comparable with the maximum expected delay spread. In [8], the use of the Rayleigh channel model is justified on the grounds that propagation tests at 900 MHz in urban areas showed that most of the multipath distortion energy occurs at short delay values and hence flat fading is the dominant channel impairment. Thus the complex envelope of the received GMSK signal can be expressed as:

$$s(t) = \rho(t) \exp\{j\phi(t)\} \exp\{j\theta(t, \underline{\alpha})\} \qquad (7.31)$$

where both $\rho(t)$, which is the Rayleigh distributed real envelope, and $\phi(t)$, which is a uniformly distributed phase perturbation, vary slowly enough to justify the following assumptions:

(1) the receiver can estimate the phase perturbations without error and hence can fully maintain the signal phase coherence;

(2) the magnitude of the signal envelope remains constant over the receiver observation interval of L_r bits.

The baseband signal in the I channel now becomes:

$$r_I(t) = \rho(t) \cos \theta(t, \underline{\alpha}) * h_{rb}(t) + n_I(t) \qquad (7.32)$$

where the Rayleigh distributed envelope $\rho(t)$ has a long term mean squared value of $2S$. It can be shown that the normalised short term signal-to-noise ratio $\gamma_b = E_b/N_o$ averaged over L_r bits has an exponential pdf given by:

$$
\begin{aligned}
p(\gamma_b) &= \tfrac{1}{\Gamma_b} \exp\left(-\tfrac{\gamma_b}{\Gamma_b}\right) & \text{for } \gamma_b \geq 0 \\
&= 0 & \text{for } \gamma_b < 0
\end{aligned}
\qquad (7.33)
$$

where Γ_b is the long term mean value.

In the SFHMA system coded speech data signals are transmitted in bursts at a rate of 200 kbits/s. In our analysis, we shall compare the SFHMA system performance with the performance of a conventional FDMA system with no error control coding and with continuous transmission at a rate of 16 kbits/s, which is the information bit rate of the SFHMA system. The principal performance measure we shall use is the probability of output data bit errors. Furthermore, for fair comparisons of the coded and uncoded systems, we assume that the same amount of energy is used to transmit the same number of information bits in the same fixed duration. As illustrated in Figure 7.10, the increase in bit rate due to FEC coding implies that the transmitted bit period in the coded system, T_c, is less than the data bit period, T, of the uncoded continuous transmission case and the

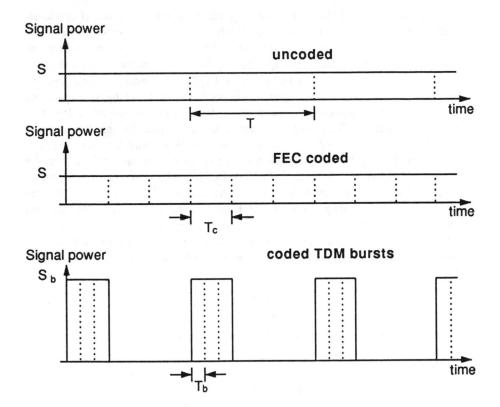

Figure 7.10: Basis of comparison between SFHMA and uncoded FDMA systems.

transmitted power S is assumed constant. This means that the energy per bit is decreased by the use of coding because $E_c = ST_c$, and therefore the channel bit error probability is increased. Note that the further increase of transmitted bit rate due to the burst mode of operation does not lead to a reduction of energy per channel bit because there is a compensating increase of signal power during the burst to S_b.

In all the following analysis sub-sections we have assumed MSK-type detectors in both systems, because this greatly reduces the computational complexity, especially for the analysis of cochannel interference. Furthermore, we make the assumption that hard decision symbol detection is used. In the SFHMA system no allowance has been made for signal validity information being supplied by the demodulator to the inner channel decoder.

7.4.4   BER Analysis of SFHMA System in a Static AWGN Channel

We shall use the MSK-type detector performance as the basis of our analysis in this section. In order to establish the overall bit error probability performance of the SFHMA system in AWGN channels, we must first determine the channel BER by applying Equation 7.27 modified to take into account the redundancy due to the overall 1/3 rate channel coding and the inclusion of an 8-bit training pattern in each transmitted burst. Then the probability of uncorrected errors in the output from the cascaded decoders can be evaluated. Note that to simplify the notation the inner and outer decoders will be referred to henceforth as decoders 1 and 2 respectively.

In each 60-hop frame the speech coder of an active user will produce a total of 16 kbits/s × 240 ms = 3840 data bits, which are carried as equal length blocks by the 48 data hops in the frame. Thus, allowing for the coding and the training patterns, the total number of bits transmitted over the channel per frame to convey this speech data will be $3 \times 3840 + 48 \times 8 = 11904$. If we apply the constraint of equal total energies per frame, i.e., equal average signal powers, then the energies per channel bit and per data bit are related as:

$$E_b \;=\; \frac{11904}{3840} E_c \;=\; 3.100 \, E_c \eqno(7.34)$$

From Equation 7.27 the channel BER at the input to decoder 1 is given

by:

$$
\begin{aligned}
P_{ib1} \;&=\; 2Q\{\delta_{\min}\sqrt{E_b/(3.1\,N_o)}\} \\
&=\; 2Q\{0.692\,\sqrt{E_b/N_o}\}
\end{aligned}
\eqno(7.35)
$$

where $\delta_{\min}$ has been evaluated for $B_t T = 0.3$ and $B_r T = 0.63$ as described in Section 7.4.2.

The next step of this analysis is to derive the BER performance improvement resulting from the channel coding. Consider initially the analysis of a single RS decoder capable of correcting up to t symbol errors in an (n,k) RS coded signal, often denoted as a (n,k,t) RS decoder. We will assume that:

(1) the RS code used is systematic, ie the input data symbols appear unchanged in the codeword

(2) the decoder is capable of detecting all decoding failures, ie where more than t symbol errors occur in a codeword

(3) in such events the decoder will output the received data symbols containing the original channel errors without attempting error correction.

In practice of course some severely corrupted codewords may lie close to a completely different codeword and thus be apparently successfully, but in fact incorrectly, decoded. A rigorous analysis of the post-decoding symbol and bit error probabilities for RS codes is presented in Section 4.4.7. However, for the codes used in the SFHMA system it has been found that the approximate method described below, based on the above assumptions, gives adequate accuracy in the output bit error probability calculations with much less computational complexity.

Neglecting incorrect decoding events, decoder output symbol errors can be considered to be due solely to input data symbol errors in those codewords containing more than t symbol errors in total. Thus the mean number of data symbol errors per received codeword is obtained by averaging the total number of symbol errors in such cases of decoding failure and multiplying by the proportion of data symbols in each codeword, ie:

$$N_{ds} = \frac{k}{n} \sum_{j=t+1}^{n} j \binom{n}{j} P_{is}^j (1 - P_{is})^{n-j} \tag{7.36}$$

where P_{is} is the probability of input symbol errors. Expressing the probability of output symbol errors as the mean number per data symbol gives:

$$P_{os} = \frac{N_{ds}}{k}$$
$$= \frac{1}{n} \sum_{j=t+1}^{n} j \binom{n}{j} P_{is}^j (1 - P_{is})^{n-j} \tag{7.37}$$

To determine the output bit error probability, P_{ob}, we postulate that the output symbol errors are caused by a sequence of independent bit errors with probability P_{ob}, giving the relationship:

$$P_{os} = 1 - (1 - P_{ob})^m \tag{7.38}$$

where m is the number of bits per symbol. Thus by a simple rearrangement of the above expression we obtain:

$$P_{ob} = 1 - (1 - P_{os})^{1/m} \tag{7.39}$$

A check was made on the justification of applying the approximate method to the performance analysis of the (30,20,5) and (8,4,2) decoders in the SFHMA system by using the exact expressions given in Section 4.4.7 to compute the relative probability of incorrect decoding P_{IDR}, given by the ratio;

$$P_{IDR} = \frac{P_{ICD}}{1 - P_{CD}} \tag{7.40}$$

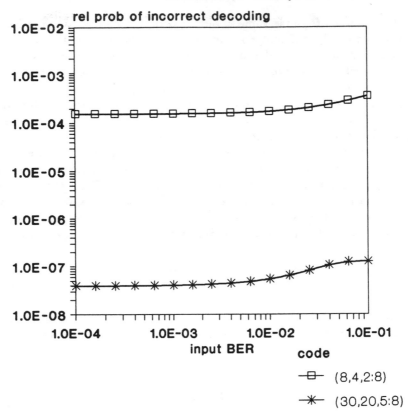

Figure 7.11: Relative probability of incorrect decoding vs input BER for the RS codes used in the SFHMA system.

where P_{ICD} is the probability of incorrect decoding, P_{CD} is the probability of correct decoding and $1 - P_{CD}$ is therefore the probability of decoding failure. Graphs of P_{IDR} vs P_{ib} for the two SFHMA decoders are presented in Figure 7.11. The low values of P_{IDR} obtained fully justify the use of the approximate method in all the evaluations of the bit error rate performance of the SFHMA system described in the remainder of this chapter. Note that the maximum value of t is assumed for each decoder, which for an RS code is given by;

$$t = \lfloor (n - k)/2 \rfloor \qquad (7.41)$$

where $\lfloor \; \rfloor$ denotes the next lower integer. Choosing a value of t less than the maximum will increase the reliability of detecting decoding failures and decrease the probability of incorrect decoding, but this does not seem to be required in these cases. Consider now the cascaded decoders in the SFHMA receiver. At the input to the first (inner) decoder one or more channel bit errors in one 8-bit symbol will constitute an input symbol error with probability given by:

$$
\begin{aligned}
P_{i s 1} &= 1 - \mathcal{P}\{\text{no bit error}\} \\
&= 1 - (1 - P_{i b 1})^8
\end{aligned}
\tag{7.42}
$$

Before the second decoding stage each output block of 20 data symbols from the first decoder are deinterleaved into 20 different 8-symbol codewords, each of which will contain symbols from 7 other hops. Hence it is reasonable to assume that symbol errors at the input to the second (outer) decoder are independent, even when the first decoder is dealing with bursts of errors due to fading.

The input symbol error rate (SER) of the second decoder is equal to the output SER of the (30,20,5) decoder, which from Equation 7.37 can be expressed as:

$$
\begin{aligned}
P_{i s 2} &= P_{o s 1} \\
&= \frac{1}{30} \sum_{j=6}^{30} j \binom{30}{j} P_{i s 1}^{j} (1 - P_{i s 1})^{30-j}
\end{aligned}
\tag{7.43}
$$

Applying Equation 7.37 again we obtain the output SER of the (8,4,2) (second) decoder as:

$$
P_{o s 2} = \frac{1}{8} \sum_{j=3}^{8} j \binom{8}{j} P_{i s 2}^{j} (1 - P_{i s 2})^{8-j}
\tag{7.44}
$$

Finally the output BER is obtained from Equation 7.39 as:

$$
P_{o b 2} = 1 - (1 - P_{o s 2})^{1/8}
\tag{7.45}
$$

The SFHMA system BER performance is obtained by evaluating Equations 7.35 and 7.42 to 7.45. The results are presented in Figure 7.12 as a graph of BER versus E_b/N_o, where they are compared with the performance of an uncoded system (from the analysis in Section 7.4.2). Somewhat surprisingly, this comparison shows that the powerful FEC coding gives no improvement in performance of the SFHMA system, in fact just the opposite, in the critical BER range 10^{-3} to 10^{-2}, where the speech decoder input BER threshold is most likely to be located.

7.4.5 BER Analysis in a Rayleigh Fading AWGN Channel

In this section, we shall first develop the BER performance for continuous transmission of uncoded speech data with an MSK-type detector in

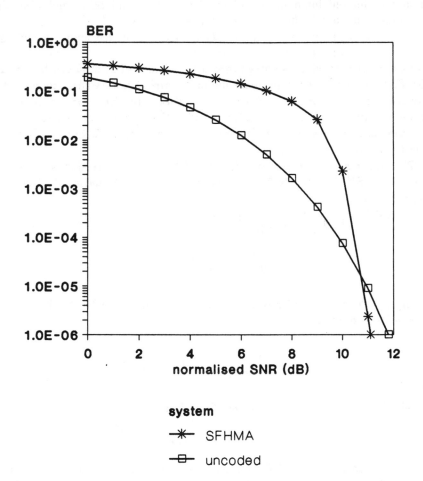

Figure 7.12: BER vs normalised SNR for both SFHMA and uncoded FDMA systems in a static AWGN channel [MSK-type detector: $B_t T = 0.3$ and $B_r T = 0.63$].

a Rayleigh fading channel with AWGN. Next the SFHMA system performance is derived for a similar channel condition and compared with the continuous uncoded case.

For the analysis of the continuous transmission uncoded reference system, the effective receiver observation interval, $L_r T$, may be considered to be the duration of the total ISI caused by both the premodulation filters. Thus the value of L_r is inversely related to the normalised bandwidths of the filters. For $B_t T = 0.3$ and $B_r T = 0.63$ then $L_r = 6$ would be an appropriate value.

The dynamic bit error probability in the Rayleigh fading channel, as a function of the long term mean normalised signal-to-noise ratio, Γ_b, is found by averaging the error probability with respect to the short term normalised signal-to-noise ratio γ_b as:

$$P_b(\Gamma_b) = \int_0^\infty P_b(\gamma_b) p(\gamma_b) d\gamma_b \qquad (7.46)$$

where $P_b(\gamma_b)$ is determined by evaluating Equations 7.25 to 7.28 and $p(\gamma_b)$ is the pdf given by Equation 7.33.

For the analysis of the SFHMA system performance the observation interval is taken as one hop slot duration (1.23 ms), during which time one complete codeword of the inner RS code is received. The corresponding value of L_r is approximately 250. The block of 20 symbols at the output of the inner decoder is deinterleaved into 20 consecutive codewords at the input to the outer decoder. Thus the codewords at this point can be assumed to contain independent symbol errors with a probability which is obtained by using Equations 7.35, 7.42 and 7.43 to determine the input SER of the outer (second) decoder P_{is2} as a function of γ_b and then averaging with respect to γ_b as:

$$P_{is2}(\Gamma_b) = \int_0^\infty P_{is2}(\gamma_b) p(\gamma_b) d\gamma_b. \qquad (7.47)$$

As before, the integral is evaluated numerically. The dynamic output bit error probability can then be determined as a function of the long term mean signal-to-noise ratio by means of Equations 7.44 and 7.45.

Figure 7.13 presents the results of the analysis of both the SFHMA and the uncoded systems, as graphs of BER versus mean signal-to-noise ratio Γ_b. For this case of the fading channel the benefits of the FEC coding incorporated in the SFHMA system are clearly demonstrated by the superior performance of the system in the critical BER range of 10^{-3} to 10^{-2}. At the mid-range BER value of 3×10^{-3} the coding gain is approximately 4.5 dB. It should be noted that the FEC coding scheme, comprising concatenated RS codes with block interleaving, is specifically designed to combat bursts of errors in a fading channel rather than the random errors arising in a static AWGN channel. Thus some difference in the efficiency of the FEC coding in the two channel conditions should be expected.

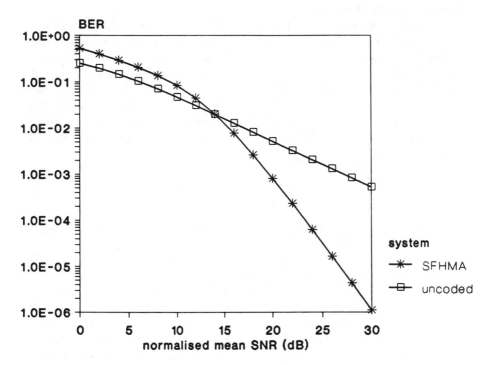

Figure 7.13: BER vs normalised mean SNR for both SFHMA and uncoded
FDMA systems in a Rayleigh fading AWGN channel [MSK-
type detector: $B_t T = 0.3$ and $B_r T = 0.63$].

7.5   BER Performance in the Presence of Co-channel Interference

Cochannel interference arises from frequency reuse which is a feature of any
efficient mobile radio systems, [16, 17, 18]. It is probably the most important parameter in the design of cellular mobile radio systems, because it
ultimately determines the system capacity. A unique advantage of SFHMA
systems is that they provide interferer diversity, ie, the set of interfering
sources changes from hop to hop, however this feature makes the analysis
of performance much more difficult.

In this section we will examine cochannel interference in the classical
sense without taking account of interferer diversity. Two approaches to the
analysis are presented, the first is a simplified analysis based on Murota's
treatment, [15], which assumes noise free reception, whereas in the sec-

ond approach by Ko [19] a more elaborate analysis is entailed which also accounts for receiver noise. Both treatments are based on the MSK-type receiver because of the greatly reduced computational complexity. Its BER performance is believed to be only slightly inferior to the MLSE detector.

We shall first consider the cochannel interference in non-fading channels for both the uncoded continuous transmission system and for the SFHMA system. Then the analysis is extended to cover frequency non-selective, Rayleigh fading channels.

7.5.1 BER Analysis in a Noiseless Static Channel

In this analysis, following [15], the presence of cochannel interference from a single source in a static noiseless channel is considered for the uncoded continuous system. In Verhulst's original analysis of SFH digital cellular radio systems [2], and in his subsequent analysis of the SFH900 system [3, 5, 8], this noiseless channel condition is also assumed. No account has been taken of the additional ISI caused by pre-demodulation filtering of the received GMSK signals. As in the previous analysis of the performance of the MSK-type detector in a static AWGN channel (see Section 7.4.2), we assume that the probability of decision errors in the receiver's quadrature channels is dominated by the worst case ISI condition which occurs for the data sequence $\cdots 0, 0, 0, 1, 1, 1 \cdots$. Therefore, we restrict our attention to the special case for which the signal phase waveform $\theta(t)$ is given by Equation 7.28.

The interference signal waveform can be expressed in a similar form to that of the desired signal except that the signal power, S, is replaced by the interference power, I. The interferer's phase is modelled by a random process, $\psi(t)$, uniformly distributed between $-\pi$ and π, resulting from the assumption that the signal and interference are not phase coherent and are modulated by independent data sequences. The total received signal $r_s(t)$ in this case is:

$$r_s(t) = \sqrt{2S} \cos[\omega_o t + \theta(t)] + \sqrt{2I} \cos[\omega_o t + \psi(t)] \qquad (7.48)$$

for which the equivalent low pass complex envelope is given by:

$$z(t) = \sqrt{2S} \exp[j\theta(t)] + \sqrt{2I} \exp[j\psi(t)] \qquad (7.49)$$

It is convenient to regard $z(t)$ as the complex sum of the desired signal and the interference signal phasors. Figure 7.14 shows a phasor diagram of the resultant complex envelope at the decision instant $t = 0$, the correct decision being obtained in the I channel when the resultant complex envelope remains in the right half plane. Furthermore, to allow for the inherent ISI in the GMSK signal the reference phase is defined as the modulation phase

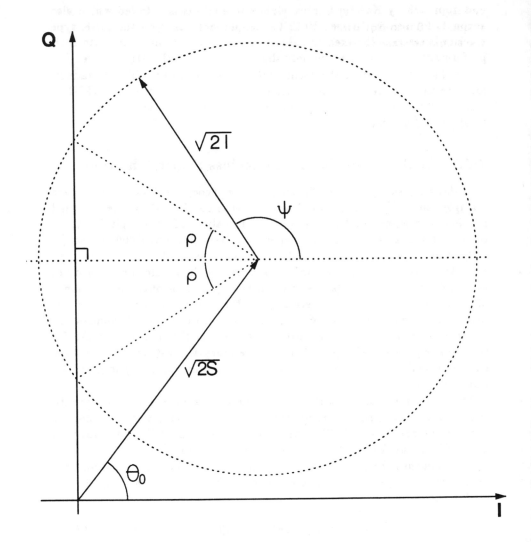

Figure 7.14: Phasor diagram for GMSK signal with cochannel interference signal at decision instant $(t = 0)$.

change at $t = 0$ of a classical MSK signal, ie, $B_t T = \infty$. Therefore, substituting $t = 0$ into Equation 7.28 and substituting for μ_t from Equation 7.29, the signal reference phase is obtained as:

$$\theta_o = \frac{\sqrt{\ell n 2}}{2 B_t T \sqrt{2\pi}} \qquad (7.50)$$

Due to the symmetry of the two quadrature channels and to the uniform distribution of $\psi(t)$ it can be argued, with reference to Figure 7.14, that the probability of quadrature symbol decision errors is given by the ratio of the angle subtended by the intersection of the Q axis with the circular locus of the resultant envelope, 2ρ, to the overall range of the phase angle, ie 2π. Thus the quadrature symbol error probability can readily be determined in terms of the signal reference phase, θ_o and the carrier signal-to-interference ratio (CIR), $\lambda = S/I$, as:

$$
\begin{aligned}
P_{qe}(\lambda) &= \frac{\rho}{\pi} = \frac{1}{\pi} \cdot \cos^{-1} \sqrt{\lambda \cos \theta_o} \quad \text{for } 0 \le \sqrt{\lambda} \le 1/\cos \theta_o \\
&= 0 \qquad\qquad\qquad\qquad\quad \text{for } \sqrt{\lambda} > 1/\cos \theta_o
\end{aligned}
\qquad (7.51)
$$

From Equation 7.26 the output bit error probability is given by:

$$P_b(\lambda) = 2 P_{qe}(\lambda)[1 - P_{qe}(\lambda)] \qquad (7.52)$$

In Figure 7.15 we display the performance curve of BER against CIR, which is obtained by evaluating Equations 7.50 to 7.52 for $B_t T = 0.3$.

For the SFHMA system we used Equations 7.50 to 7.52 to calculate the channel bit error probability at the input to the inner decoder, P_{ib1}, then the effect of channel coding is evaluated by means of Equations 7.42 to 7.44, as described above in Section 7.4.4. The results obtained are also displayed in the graph shown in Figure 7.15, for comparison with results for the uncoded system. It can be seen that the two curves are very close with a sharp knee at a CIR value of 1.4 dB, and then they fall extremely steeply. This behaviour is entirely consistent with the assumed model, which involves constant amplitudes of both the signal and the interference.

7.5.2 BER Analysis in a Static AWGN Channel

For the situation considered here, where the GMSK signal is accompanied by both cochannel interference and AWGN, we have employed an approximate analysis developed by Ko [19], which was shown to provide a very large saving in computation time, with only moderate degradation of accuracy, compared to more exact methods of analysis. Intersymbol interference due

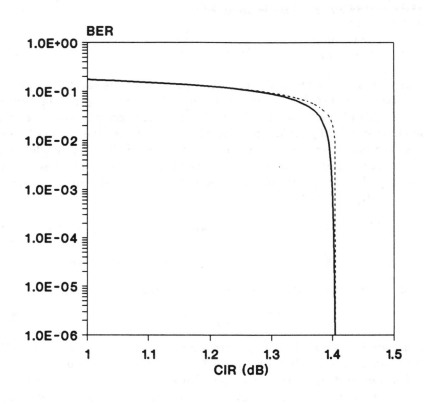

system

——— SFHMA

----- uncoded

Figure 7.15: BER vs CIR for both SFHMA and uncoded FDMA systems with cochannel interference in static noiseless channel [MSK-type detector: $B_t T = 0.3$].

to the receiver filter is taken into account, but the treatment is simplified by considering only the data sequence $\underline{\alpha}_{min}$ giving the minimum Euclidean distance and by approximating multiple interfering signals as arising from a single source.

With some minor changes to be compatible with the previous analysis of the MSK-type detector in section 7.4.2, Ko's approximate expression (see Equation 2.67 in [19]) for the probability of quadrature symbol error becomes:

$$P_{qe} = \frac{1}{2\pi} \int_0^{2\pi} Q \left\{ \frac{\sqrt{2S}}{\sigma_y} \int_{-\infty}^{\infty} h_{rb}(-\tau) \cos \theta(\tau, \underline{\alpha}_{min}) d\tau \right.$$

$$\left. + \frac{\sqrt{2I}}{\sigma_y} \int_{-\infty}^{\infty} h_{rb}(-\tau) \cos \psi d\tau \right\} d\psi \qquad (7.53)$$

where S is the input signal power, I is the total cochannel interference power, $h_{rb}(t)$ is the impulse response of the low-pass equivalent of the Gaussian band-pass filter in the MSK detector given by Equations 7.13 and 7.14, $\theta(t, \underline{\alpha}_{min})$ is the signal phase waveform with the worst case intersymbol interference (ISI) due to the transmitter Gaussian filter (for the data sequence $\cdots 0, 0, 0, 1, 1, 1 \cdots$) given by Equations 7.20 and 7.21. and 7.29, ψ is the interference phase assumed to be random and uniformly distributed over 0 to 2π and σ_y^2 is the variance of the noise in the baseband quadrature channel given by Equations 7.20 and 7.21.

The signal power can be expressed in the form:

$$S = E_b/T \qquad (7.54)$$

where E_b is the mean per energy/bit. In terms of S and the CIR, λ, the interference power is given by:

$$I = S/\lambda = E_b/\lambda T \qquad (7.55)$$

Assuming that;

(1) the interference phase ψ is constant over the effective duration of $h_{rb}(t)$,

(2) the zero frequency gain of the equivalent low-pass filter is unity, which implies that:

$$\int_{-\infty}^{\infty} h_{rb}(-\tau) d\tau = \int_{-\infty}^{\infty} h_{rb}(\tau) d\tau = 1 \qquad (7.56)$$

(3) the noise bandwidth of the receiver band-pass filter is equal to its -3 dB bandwidth B_r,

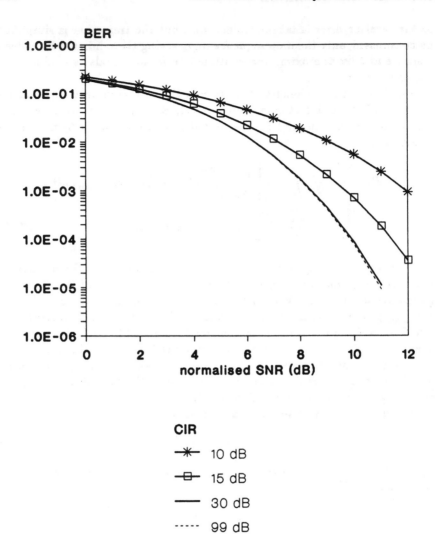

Figure 7.16: BER vs normalised SNR for uncoded FDMA system with cochannel interference in a static AWGN channel [MSK-type detector: $B_t T = 0.3$ $B_r T = 0.63$].

and also noting that $h_{rb}(-\tau) = h_{rb}(\tau)$, Equation 7.53 can be simplified and expressed as a function of the normalised signal-to-noise ratio (SNR) E_b/N_o and λ, in the form:

$$P_{qe} = \frac{1}{2\pi} \int_0^{2\pi} Q \left\{ \left(\Phi_s + \frac{\cos \psi}{\sqrt{\lambda}} \right) \sqrt{\frac{2}{B_r T}} \sqrt{\frac{E_b}{N_o}} \right\} d\psi \qquad (7.57)$$

where Φ_s is the signal phase integral, which is given by:

$$\Phi_s = \int_{-\infty}^{\infty} h_{rb}(\tau) \cos \theta(\tau, \underline{\alpha}_{min}) d\tau \qquad (7.58)$$

which from Equations 7.13 and 7.28 can be expressed as:

$$\Phi_s = \frac{\mu_r}{\sqrt{\pi}} \int_{-\infty}^{\infty} exp(-\mu_r^2 t^2) \cos \left[\frac{\pi}{2T} \int_0^t erf(\mu_t \tau) d\tau + \frac{\sqrt{\pi}}{2\mu_t T} \right] dt \qquad (7.59)$$

where the receiver and transmitter filter parameters, μ_r and μ_t, are related to the normalised filter bandwidths, $B_r T$ and $B_t T$ by Equations 7.14 and 7.29. In all the calculations described here a fixed value of ϕ_s was employed, which was obtained by putting $B_t T = 0.3$ and $B_r T = 0.63$.

Calculations of bit error rate using Equations 7.57 and 7.52 have been made for the uncoded system with $B_t = 0.3$ and $B_r T = 0.63$. Graphs of BER vs SNR and of BER vs CIR are presented in Figures 7.16 and 7.17, respectively. As would be expected, the curve for the highest value of CIR (99 dB) in Figure 7.16 is very similar to the curve for the uncoded FDMA system in Figure 7.12 for the static AWGN channel. As CIR is decreased below 30 dB significant deterioration in BER performance becomes apparent. Plotting the results against CIR, as in Figure 7.17, is of more interest when considering system performance under interference limited conditions, which will be our prime concern in the remainder of this chapter.

In the SFHMA system case the value of SNR used in the Equation 7.57 has been adjusted to allow for the overall FEC coding rate as in section 7.4.4. Then having used Equation 7.52 to obtain the input BER at the inner decoder, Equations 7.42 to 7.45 were employed to obtain the output BER. A graph of BER vs CIR is shown in Figure 7.18. As in Figure 7.17, it will be seen that for very high values of SNR the BER falls extremely steeply for CIR greater than 3 dB.

The reason that this steep cut-off occurs at a higher value of CIR than the value of 1.4 dB obtained for the noiseless channel analysis, which can be seen in Figure 7.15, is that in this case there is additional ISI due to the receiver filter, which was not accounted for in the previous analysis. It can be shown that in the limit as the SNR $\rightarrow \infty$ the cut-off will occur at $\lambda = \phi_s^{-2}$ giving CIR values very close to 3 dB for $B_r T = 0.63$ and 1.4 dB for $B_r T \gg 1$.

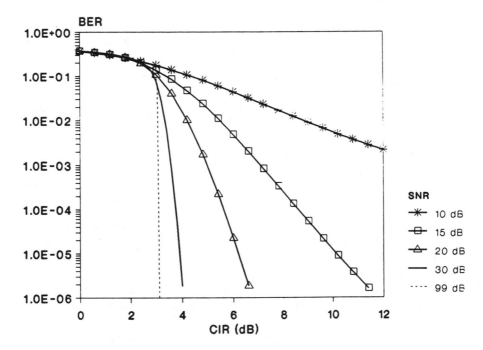

Figure 7.17: BER vs CIR for uncoded FDMA system with cochannel interference in a static AWGN channel [MSK-type detector: $B_t T = 0.3$ and $B_r T = 0.63$].

Further comparison of Figures 7.17 and 7.18 shows once more that the FEC coding employed in the SFHMA system is ineffective against the random channel errors produced in a static channel situation, the coding gain varying from low to negative values as the SNR changes from high to low.

7.5.3 BER Analysis in a Rayleigh Fading AWGN Channel

Again we base the analysis of system performance in a noiseless Rayleigh fading channel on [15]. Therefore we treat the multiple interferers as an equivalent single source producing a received power equal to the total power of the component signals and we can apply the expressions obtained in Section 7.5.1 to determine the channel bit error probability. It is assumed that the desired and interfering signals both fade independently, that the signal-to-interference ratio, λ, is constant over each hop burst duration and

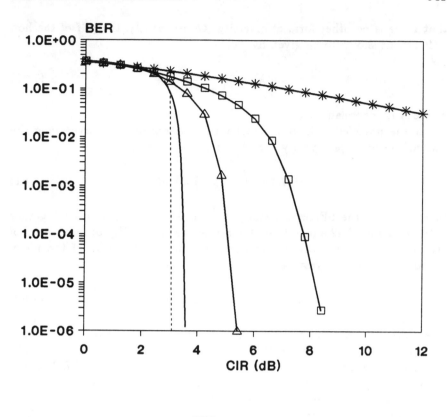

Figure 7.18: BER vs CIR for SFHMA system with cochannel interference in a static AWGN channel [MSK-type detector: $B_t T = 0.3$ and $B_r T = 0.63$].

that there is no other form of diversity. Consequently, the pdf of the hop CIR can be shown to be given by [20]:

$$p(\lambda) = \frac{\Lambda}{(\lambda + \Lambda)^2} \qquad (7.60)$$

where Λ is the mean CIR.

For the uncoded system the mean BER is obtained by averaging the channel bit error probability over λ, using the above pdf as:

$$P_b(\Lambda) = \int_0^\infty P_b(\lambda)p(\lambda)d\lambda \qquad (7.61)$$

In the case of the SFHMA system the above expression is used together with Equations 7.42 and 7.43 to determine the probability of input symbol errors at the outer decoder, averaged over λ, as a function of the mean signal to interference ratio as:

$$P_{is2}(\Lambda) = \int_0^\infty P_{is2}(\lambda)p(\lambda)d\lambda \qquad (7.62)$$

Thereafter, the output bit error probability is evaluated by using Equations 7.44 and 7.45. The results for both systems are presented as graphs of BER versus mean CIR in Figure 7.19. They show that the FEC coding of the SFHMA system is even more effective against cochannel interference in a fading channel condition than it was against noise, as can be seen by comparison with Figure 7.13. In this case the coding gain at a BER of 3×10^{-3} is approximately 11 dB.

7.5.4 BER Analysis of a Noiseless Rayleigh Fading Channel

This is a severe channel condition in which the wanted and the interfering signals are all subjected to independent fading and perturbed by AWGN. It has been shown in [19] that in a Rayleigh fading channel, the interference can be regarded as arising from an equivalent single source. Moreover, as in the previous analysis of fading, it is assumed that the instantaneous SNR and CIR are constant over the duration of a hop, but from hop to hop the wanted signal and the interference are subjected to mutually independent Rayleigh fading. Therefore before averaging it is necessary to modify Equation 7.57 to express the quadrature error probability as a function of the signal-to-noise voltage ratio $\rho_s = \sqrt{E_b/N_o}$ and of the interference-to-noise voltage ratio $\rho_i = \sqrt{IT/N_o} = \sqrt{E_b/N_o\lambda}$ in the form;

$$P_{qe}(\rho_s, \rho_i) = \frac{1}{2\pi} \int_0^{2\pi} Q\left\{ (\rho_s \Phi_s + \rho_i \cos\psi)\sqrt{\frac{2}{B_r T}} \right\} d\psi \qquad (7.63)$$

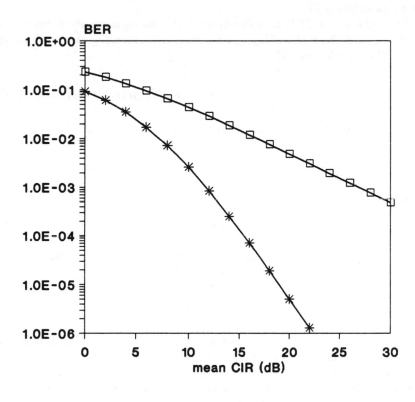

Figure 7.19: BER vs mean CIR for SFHMA and uncoded FDMA systems with cochannel interference in noiseless Rayleigh fading channel [MSK-type detector: $B_t T = 0.3$].

From which the bit error probability as a function of ρ_s and ρ_i is obtained using Equation 7.52.

Then the mean BER as a function of the mean normalised normalised SNR, Γ_b, and the mean CIR, Λ, is obtained by averaging over both ρ_s and ρ_i via the double integral:

$$P_b(\Gamma_b, \Lambda) = \int_0^\infty \int_0^\infty P_{be}(\rho_s, \rho_i) p(\rho_s) p(\rho_i) d\rho_s d\rho_i \qquad (7.64)$$

The two Rayleigh pdfs can be expressed as;

$$p(\rho_s) = \frac{2\rho_s}{\Gamma_b} exp\left(-\frac{\rho_s^2}{\Gamma_b}\right) \qquad (7.65)$$

and

$$p(\rho_i) = \frac{2\rho_i}{\Gamma_i} exp\left(-\frac{\rho_i^2}{\Gamma_i}\right) \qquad (7.66)$$

where Γ_i is the mean normalised interference-to-noise power ratio, which is related to the mean CIR and the mean SNR by:

$$\Gamma_i = \frac{\Gamma_b}{\Lambda} \qquad (7.67)$$

Simpsons's rule was used to evaluate the double integral in Equation 7.64. It was found that great care had to be taken with setting the upper limits in order to obtain sensible results without excessive computation time.

Computation of the SFHMA system performance proceeded in a similar way, using Equations 7.63 and 7.52 to give the channel error rate and then Equations 7.42 and 7.43 to determine the input symbol error rate of the outer decoder as a function of ρ_s and ρ_i. It was this probability which was averaged as indicated in Equation 7.64 and then used to compute the output BER by means of Equations 7.44 and 7.45. This procedure is essentially the same as used in previous calculations described in Section 7.4.5 and is justified by the same assumption, ie all the symbols in each codeword at the input to the outer decoder have been transmitted on different hops and so contain independent errors. Graphs of BER vs mean CIR for various values of mean SNR for the uncoded and SFHMA systems are shown in Figures 7.20 and 7.21. The curves for the highest SNR (99 dB) in both cases are very similar to those in Figure 7.19 for a fading channel without noise.

7.6 Estimation of Spectral Efficiency

Efficient utilisation of the limited radio spectrum available for mobile radio is one of the most important considerations in the design of digital cellular systems. In this section we will describe two methods for estimating the

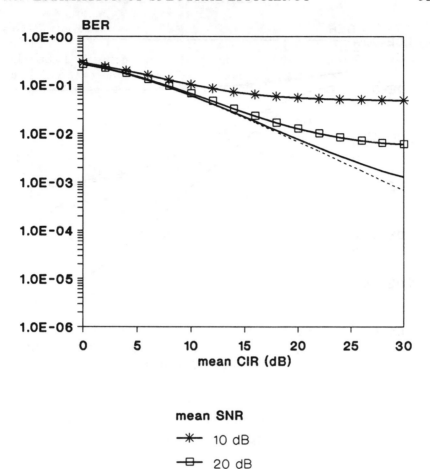

Figure 7.20: BER vs mean CIR for uncoded FDMA system with cochannel interference in Rayleigh fading AWGN channel [MSK-type detector: $B_t T = 0.3$ and $B_r T = 0.63$].

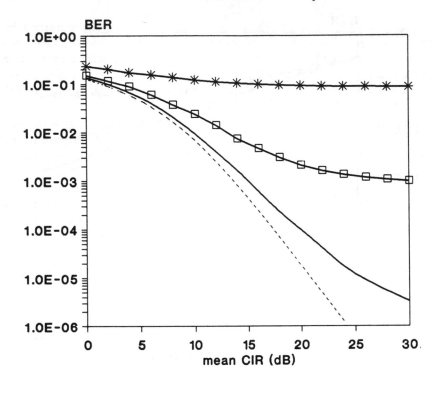

Figure 7.21: BER vs mean CIR for SFHMA system with cochannel interference in Rayleigh fading AWGN channel [MSK-type detector: $B_t T = 0.3$ and $B_r T = 0.63$].

spectral efficiency of a SFHMA system and present results obtained by the application of both methods for the base-to-mobile (down) link. For comparison, spectral efficiency estimates are also determined by a more conventional method for a TDM/FDMA system, which is assumed to use the same digital modulation and demodulation schemes, but with no FEC coding.

There are a number of definitions of spectral efficiency in general use, but the one which seems most appropriate for our present purpose is the maximum traffic, in Erlangs, carried per cell per unit of spectral width, which is conveniently expressed in units of erl/cell/MHz.

Interferer diversity, which is claimed to be one of the main advantages of SFHMA, requires a different approach in determining spectral efficiency since cochannel interference from a reuse cell is spread statistically over all hop frequencies as random frequency collisions with a probability dependent on the system traffic loading. Whereas in conventional FDMA (or TDMA) system analysis it is necessary to assume a worst case condition in which there is continuous interference from an active user on the same channel in each reuse cell, irrespective of the traffic loading.

7.6.1 Spectral Efficiency of SFHMA System : Method A

This method is the one described by Dornsetter and Verhulst in their paper [8] on the analysis of the SFH900 system, which differs only in small details from that presented in an earlier paper by Verhulst [5]. It is assumed that the system performance is interference limited. In the interest of simplicity only the simpler down-link (base-to-mobile) situation will be considered here, but both directions are dealt with in [8].

The combined characteristics of the demodulation and inner (intra hop) FEC decoding are represented by a simple threshold comparison, ie a transmitted hop is assumed valid if the hop CIR $\lambda > \gamma$ and invalid if $\lambda < \gamma$, where γ is the value of CIR such that the probability of correct reception of a hop is 0.5. A value of 7 dB is quoted for γ in [8], which is based on the measured BER vs SNR characteristics of an SFH system with no FEC coding using a quasi coherent GMSK receiver with Viterbi Algorithm equalizer. The measurements were made in a simulated dispersive channel having static multipath distortion with AWGN, but in the absence of cochannel interference. Although a satisfactory explanation is not provided in [8] for the choice of this value of SNR, which yielded a BER close to 5×10^{-2}, as a suitable CIR threshold value, it can be argued that this choice is justified on the grounds that the amplitude distribution of the sum of several interfering signals with independent phases and subject to independent dynamic multipath distortion will be approximately Gaussian by application of the Central Limit theorem.

Stipulation of a value for the probability of correct reception of a hop, q, is used to account for the effectiveness of the outer FEC coding and the speech coding to give an adequate decoded speech quality. A minimum value of 0.7 for q is suggested in [8] to ensure acceptable quality and in the case of the RS (8,4) outer code it is claimed that the corresponding output BER would be approximately 6×10^{-3}. Our analysis of the outer decoder indicates that 0.8 might be a better choice for the minimum q to ensure a similar value of output BER. Spectral efficiency is found by determining q as a function of the system traffic load, allowing for fast fading, interference diversity and shadowing, and then finding the load which ensures $q > 0.7$ with a high probability over the whole cell.

Using Equation 7.3 from Section 7.2.3, we can write:

$$
\begin{aligned}
q &= \mathcal{P}\{\lambda \geq \gamma\} \\
&= \mathcal{P}\left\{\frac{A_o B_o F_o}{\sum_{i=1}^{M} A_i B_i F_i E_i} \geq \gamma\right\}
\end{aligned}
\tag{7.68}
$$

where the probability is with respect to the "fast" variables, ie the fading factors F_i and the interference indicators E_i, while the "slow" variables, ie mean power levels A_i and shadowing factors B_i, are treated as constants. Assuming that the F_i values are independent and exponentially distributed (Rayleigh fading) and that the binary E_i values are independent with a Bernoulli distribution characterised by the frequency collision probability $p_i = \mathcal{P}\{E_i = 1\}$ and the probability of no collision $1 - P_i = \mathcal{P}\{E_i = 0\}$, it is shown in Appendix A that q is given by:

$$
q = \prod_{i=1}^{M} \left(1 - \frac{p_i \gamma}{\Lambda_i + \gamma}\right)
\tag{7.69}
$$

where $\Lambda_i = A_o B_o / A_i B_i$ is the short term mean CIR with respect to the i_{th} interferer in the worst case for $p_i = 1$. The simple on/off model used above to represent hop frequency collisions implies a further assumption that the cellular system is fully synchronised. The collision probability p_i is given by the product of the frequency collision rate per active user in the i_{th} cell, y_i, and the mean number of active users per TDM slot in that cell, where 'active' means that the mobile is engaged in a call and is currently receiving transmissions from its own BS. Assuming that all cells are carrying the same traffic load of a_{cell} Erlangs and that silence detection with no transmission during silent periods is implemented to reduce cochannel interference, the collision probability can be expressed as:

$$
p_i = \frac{y_i a_{cell} r_a}{n_t}
\tag{7.70}
$$

where n_t is the number of slots per TDM frame (= 3 for SFHMA system under consideration) and r_a is the speech activity ratio (for the down link

r_a is taken to be 0.5 in [5]). As described in Section 7.2.2, the collision rate y_i depends on the reuse cell structure, having a value of $1/N$ for the basic 3-cell full reuse cluster, where N is the number of hop frequencies assigned to each of the three basic colours. In general for a full reuse cluster of size C the collision rate y_i becomes $C/3N$. In the case of a $3M/L$ fractional reuse structure the appropriate value of y_i is $k_i M/LN$, where k_i is the fractional overlap between the sets of frequencies, ie shades, assigned to the i_{th} and to the reference cells. In terms of the total bandwidth W allocated to the downlink and the frequency spacing f_s, N is simply:

$$N = \frac{W}{3f_s} \tag{7.71}$$

Thus substituting in Equation 7.70, firstly for the case of the general full reuse structure, we can write:

$$p_i = \frac{C f_s a_{cell} r_a}{n_t W} \tag{7.72}$$

and secondly for fractional reuse structures:

$$p_i = \frac{3M k_i f_s a_{cell} r_a}{L n_t W} \tag{7.73}$$

Denoting the spectral efficiency by $\eta = a_{cell}/W$ and rewriting Equations 7.72 and 7.73 gives η in terms of p_i:

$$\eta = \frac{n_t p_i}{C f_s r_a} \tag{7.74}$$

for full reuse and:

$$\eta = \frac{L n_t p_i}{3M k_i f_s r_a} \tag{7.75}$$

for fractional reuse.

Dornsetter and Verhulst applied Equation 7.69 as an essential part of a simulation of a 19-cell network, the results of which are presented in [8]. The network comprised the central reference cell with the nearest ring of 6 reuse cells of the same colour plus two closer rings of different colours producing adjacent channel interference. Mobile locations within the reference cell were selected at random and for each location, having calculated the set of products of the deterministic A_i factors and the random B_i factors (with log-normal pdf), q was computed for various values of traffic load expressed as the mean number of users engaged in calls per MHz per cell, X. For each value of X the values of q for the various locations were analysed to determine the "90% worst case value", q_{90}, defined as that value of q which is exceeded with a probability of 90%. The spectral efficiency can then be found from a graph of q_{90} vs X at the point where $q_{90} = 0.7$. X is related

size	3	9	12	21	27
Λ_{90} (dB)	5.2	13.0	14.5	18.2	20.6

Table 7.4: Values of Λ_{90} vs reuse cluster size [5]

to the collision probability and spectral efficiency as follows. The traffic per cell is given by:

$$a_{cell} = WX \tag{7.76}$$

Hence the spectral efficiency becomes simply:

$$\eta = a_{cell}/W = X \tag{7.77}$$

and from Equations 7.70 and 7.76 we have:

$$p_i = \frac{y_i r_a W X}{n_t} \tag{7.78}$$

Also described by Dornsetter and Verhulst in [8] is a simplified model based on their experiences with the simulation. They noted that there was a close association between sample locations with values of q close to q_{90} and those with values of mean CIR close to the "90% worst case value", Λ_{90}, ie $\mathcal{P}\{\Lambda > \Lambda_{90}\} = 90\%$. In addition, because of the directional BS antenna involved, only 3 cells out of the ring of 6 can produce significant interference levels in the reference cell and it was noticed that in most instances only 2 of these were simultaneously giving significant levels. Thus for a network having K rings, the proposed model gives a simple analytical expression for q_{90} in terms of the sets of values of p_i and Λ_{90i}:

$$q_{90} = \prod_{i=1}^{K} \left(1 - \frac{p_i \gamma}{\gamma + 2\Lambda_{90i}}\right)^2 \tag{7.79}$$

Curves of q_{90} vs X plotted in [8] show very close agreement with the simulation results. The crucial values of Λ_{90} used in these calculations are presented in Table 7.4. They were taken from the paper by Stjernvall [21] (NB. referenced in [5] but not in [8]), which presents computed distributions of CIR in the down link direction due to the closest ring of 6 reuse cells, for various sizes of reuse clusters having corner BS sites and assuming correlated lognormal shadowing, with sigma of 6 dB and a correlation coefficient of 0.7.

Our own calculations of q_{90} for the basic 3-cell structure, using the above values of γ and Λ_{90} and taking a frequency spacing of $f_s = 1/7$ MHz (also as in [5]), are presented in Figures 7.22 to 7.24 as a graphs of q_{90} vs X for $r_a = 0.5$, q_{90} vs p_i (for any r_a) and q_{90} vs the spectral efficiency η

for $r_a = 1$, respectively. In each case curves are shown for the following conditions;

(1) nearest reuse ring (size 3) only

(2) nearest reuse ring plus the two nearer rings of cells of the other two colours which produce adjacent-channel interference, for which the values of Λ_{90} quoted in [5] are $R_{AC} + 2dB$ and $R_{AC} + 5dB$, where R_{AC} is the adjacent-channel interference rejection, for which the value appropriate to $f_s \simeq 150$ kHz was taken as 17 dB.

(3) the second reuse ring (size 9) only.

Condition (2) was the one used for the computations reported in [5], however, comparing the curves for conditions (1) and (2) it will be seen that adjacent-channel interference has only a small effect on spectral efficiency and can safely be neglected, as was done in all our other calculations.

In Figure 7.24 a fourth curve is plotted for the more accurate approximation obtained by using the first 5 reuse rings (up to size 27). The difference between this curve and that for condition (1) is of greater significance and thus we have used multiple reuse rings out to the largest size possible in all subsequent computations to determine spectral efficiency.

For each reuse structure there is an upper limit on the value of η which can be attained, the theoretical value of which is determined as follows. The probability of frequency collisions, p_i, can be considered to be the product of three factors, the fractional overlap between the sets of hop frequencies in the reuse and reference cells, the speech activity ratio and the mean channel utilisation, ie the probability of a channel being occupied by a call, which may also be regarded as the traffic carried per channel. Thus p_i can be expressed as:

$$p_i = k_i U r_a \qquad (7.80)$$

where U is the channel utilisation. The collision probability will be maximum when all channels are occupied, ie $U = 1.0$ and so we can write:

$$p_{max} = r_a \qquad (7.81)$$

for full reuse structures, and;

$$p_{max} = k_i r_a \qquad (7.82)$$

for fractional reuse structures. Substituting in Equations 7.74 and 7.75 we have for full reuse structures:

$$\eta_{max} = \frac{n_t}{C f_s} \qquad (7.83)$$

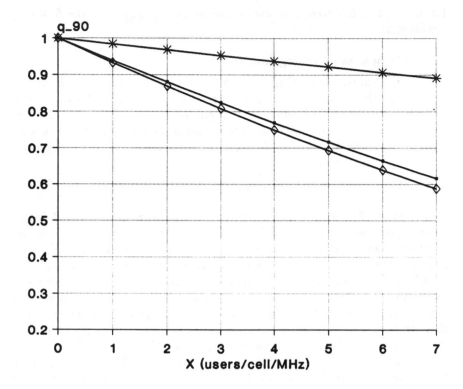

Figure 7.22: q_{90} vs X for $\gamma = 7dB$, full reuse cluster size 3 and $r_a = 0.5$

    (1) first reuse ring (size 3) only

    (2) first reuse ring plus 2 nearest rings of other colours giving adjacent-channel interference

    (3) second reuse ring (size 9) only.

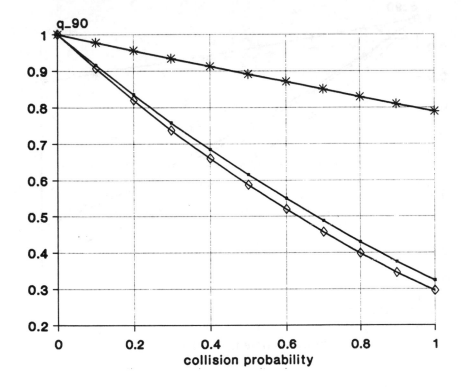

Figure 7.23: q_{90} vs collision probability p_i for $\gamma = 7dB$ and full reuse cluster size 3

(1) first reuse ring (size 3) only

(2) first reuse ring plus 2 nearest rings of other colours giving adjacent-channel interference

(3) second reuse ring (size 9) only.

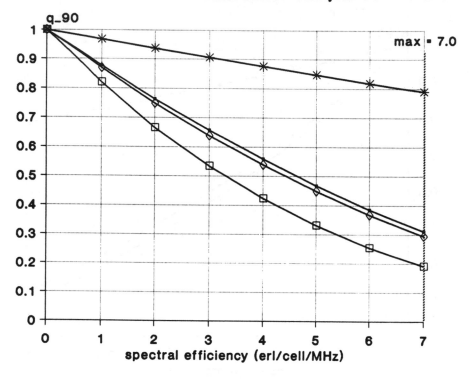

Figure 7.24: q_{90} vs η for $\gamma = 7dB$, full reuse cluster size 3 and $r_a = 1.0$

(1) first reuse ring (size 3) only

(2) first reuse ring plus 2 nearest rings of other colours giving adjacent-channel interference

(3) second reuse ring (size 9) only

(4) first 5 reuse rings (sizes 3,9,12,21 and 27).

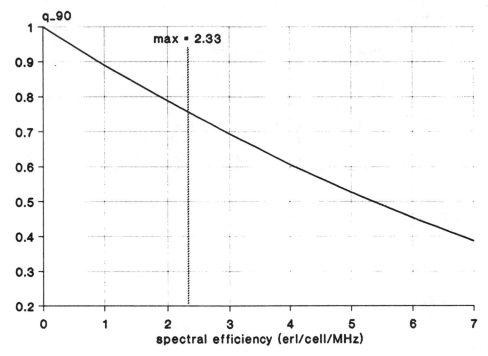

Figure 7.25: q_{90} vs η for $\gamma = 7dB$, full reuse cluster size 9 and $r_a = 1.0$.

and for fractional reuse structures:

$$\eta_{max} = \frac{Ln_t}{3Mf_s} \qquad (7.84)$$

A graph of q_{90} vs η for a full reuse structure of size 9 with $r_a = 1$ is shown in Figure 7.25. With the limited data available on Λ_{90} it was only possible to account for two reuse rings of sizes 9 and 27. However because of the increased distance between rings in this case we believe that two rings are sufficient to give an adequate accuracy.

Calculations of q_{90} as a function of η for various fractional reuse structures, all with $r_a = 1$, have been made using Equations 7.75 and 7.79 together with the data in Table 7.5. The graphs of Λ_{90} vs η shown in Figures 7.26 and 7.27 are examples of the results of the above calculations.

Note that although in principle the maximum value of the channel utilisation, U, is 1.0, it is not possible to obtain this ideal value in practical channel assignment sub-systems. Achievable maximum values can be determined using the Erlang B formula and will depend on the number of channels per cell, n_{ch}, and the permissible blocking probability, P_B. Taking the system bandwidth (for downlink) to be $W = 24$ MHz, hop frequency spacing to

structure	M	L	k	y	
				fractional	full
9/2	3	2	1/2	9/4N	3/2N
12/2	4	2	1/2	1/N	2/N
12/3	4	3	2/3	8/9N	4/3N
21/3	7	3	1/3	7/9N	7/3N
21/4	7	4	1/2	7/8N	7/4N

Table 7.5: Fractional reuse structure characteristics.

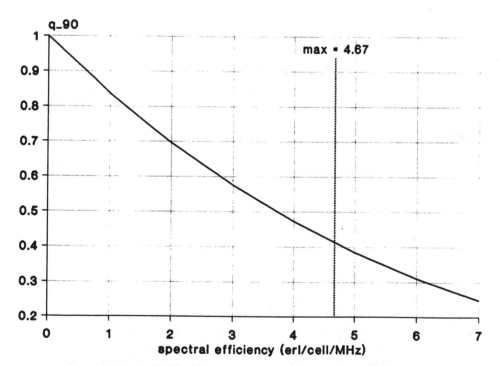

Figure 7.26: q_{90} vs η for $\gamma = 7dB$, fractional reuse cluster size 9/2 and $r_a = 1.0$.

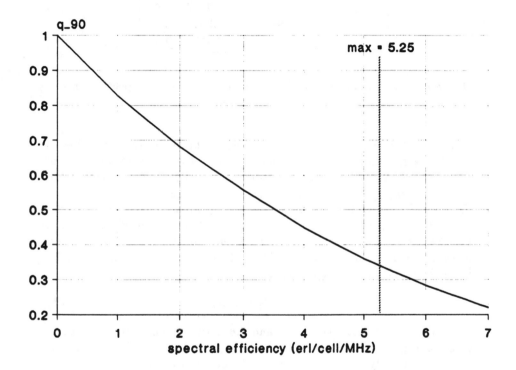

Figure 7.27: q_{90} vs η for $\gamma = 7dB$, fractional reuse cluster size 12/3 and $r_a = 1.0$.

structure	η_{max} ($U{=}1$)	n_{ch}	U_{max}	η_{max}
3	7.00	168	0.901	6.31
9	2.33	56	0.803	1.87
9/2	4.67	112	0.872	4.07
12/2	3.50	84	0.846	2.96
12/3	5.25	126	0.881	4.63
21/3	3.00	72	0.831	2.49
21/4	4.00	96	0.859	3.44

Table 7.6: Maximum values of channel utilisation and spectral efficiency (erl/cell/MHz) : $P_B = 2\%$

be f_s = 1/7 MHz and putting $P_B = 2\%$, we have determined values of U_{max} for each reuse structure and hence obtained more practical values of η_{max}. The results of the computations are listed in Table 7.6, where it will be seen that η_{max} is reduced by a factor in the range 0.8 to 0.9.

Threshold values of spectral efficiency, η_{th}, for $r_a = 1$ have been determined, for which q_{90} becomes equal to threshold values of 0.7 and 0.8. These values are listed below in Tables 7.7 and 7.8, together with η_{max} and the values of attainable system spectral efficiency, η_{syst}, given simply by taking the smaller of η_{th} and η_{max}, ie:

$$\eta_{syst} = \min\{\eta_{th}, \eta_{max}\} \tag{7.85}$$

The last column of both Tables 7.7 and 7.8 is the channel utilisation computed from η_{syst}. To obtain values of η_{syst} for the case of $r_a = 0.5$ and $q_{90} = 0.7$ it is only necessary to multiply the values of η_{th} given in Table 7.7 by $1/r_a$, ie by 2, and then apply Equation 7.85. The results of doing this are presented in Table 7.9. The results for η_{syst} summarised in Tables 7.7 and 7.8 suggest that for both values of q_{90} the optimum structure is the 9/2 cluster, with the 12/3 structure in second place, but having lower requirements on channel utilisation, which may provide greater network operational flexibility.

As might be expected, the effect of employing silence detection is shown by the results for $r_a = 0.5$, given in Table 7.9, to be a large increase in spectral efficiency and channel utilisation in most cases. Comparison of the values with these results for structures of size 3, 9, 12/2, 21/3 and 21/4 reported in [5] and [8] shows reasonable agreement, taking into account that Verhulst and Dornsetter do not allow for channel utilisation being limited to values less than 100%. In [8] it is suggested that trunking efficiency

cluster size	η_{max}	η_{th}	η_{syst}	$U\%$
3	6.31	1.76	1.76	25
9	1.87	2.93	1.87	80
9/2	4.07	1.98	1.98	42
12/2	2.96	1.58	1.58	45
12/3	4.63	1.87	1.87	36
21/3	2.49	2.05	2.05	68
21/4	3.44	1.91	1.91	48

Table 7.7: Spectral efficiency (erl/cell/MHz) for various reuse structures: $r_a = 1 : P_B = 2\% : q_{90} = 0.7$

cluster size	η_{max}	η_{th}	η_{syst}	$U\%$
3	6.31	1.12	1.12	16
9	1.87	1.88	1.87	80
9/2	4.07	1.25	1.25	27
12/2	2.96	1.00	1.00	29
12/3	4.63	1.19	1.19	23
21/3	2.49	1.30	1.30	43
21/4	3.44	1.21	1.21	30

Table 7.8: Spectral efficiency (erl/cell/MHz) for various reuse structures: $r_a = 1 : P_B = 2\% : q_{90} = 0.8$.

cluster size	η_{max}	η_{th}	η_{syst}	$U\%$
3	6.31	3.52	3.52	50
9	1.87	5.86	1.87	80
9/2	4.07	3.96	3.96	85
12/2	2.96	3.16	2.96	85
12/3	4.63	3.74	3.74	71
21/3	2.49	4.10	2.49	83
21/4	3.44	3.82	3.44	86

Table 7.9: Spectral efficiency (erl/cell/MHz) for various reuse structures : $r_a = 0.5 : P_B = 2\% : q_{90} = 0.7$

factors should not normally be applied to SFHMA systems, but this is only
true if situations involving high channel utilisation can be avoided, which
may involve some sacrifice of spectral efficiency.

7.6.2 Spectral Efficiency of SFHMA System : Method B

An alternative approach has been developed by the authors, which has
the merit of allowing the inclusion of the full analysis of the concatenated
FEC coding in a channel degraded by both noise and cochannel interference,
as described in Section 7.5. The key to this method is the derivation of an
expression for the pdf of the hop CIR, λ, as a function of the mean CIR
and the probability of frequency collisions, starting from the expression for
the probability of correct reception of a hop given in [8] and also derived
in Appendix A. Armed with this pdf the variation of output BER with
mean CIR can be determined, as described previously in Section 7.5.4 for
a fading channel, but with collision probability as a parameter. By these
means it is possible to find the value of the collision probability, p_i, and
hence the spectral efficiency, which causes the BER to equal or exceed a
threshold value (specified to ensure adequate speech decoder output) at a
mean CIR equal to the 90% worst case value for the reuse structure under
consideration. From Equation 7.69 the cumulative distribution of the hop

CIR, λ, is given by:

$$
\begin{aligned}
F(\gamma) &= \mathcal{P}\{\lambda \le \gamma\} = 1 - \mathcal{P}\{\lambda > \gamma\} = 1 - q \\
&= 1 - \prod_{i=1}^{M}\left(1 - \frac{p_i\gamma}{\Lambda_i + \gamma}\right)
\end{aligned}
\tag{7.86}
$$

where Λ_i is the short term mean CIR due to the ith interferer. The pdf is then obtained by differentiating $F(\gamma)$ wrt γ and putting $\lambda = \gamma$ to give:

$$
f(\lambda) = \sum_{i=1}^{M} \frac{p_i\Lambda_i}{(\Lambda_i + \lambda)^2} \prod_{\substack{j=1 \\ j \ne i}}^{M}\left(1 - \frac{p_j\lambda}{\Lambda_j + \lambda}\right).
\tag{7.87}
$$

Because of the directionality of the BS antennas assumed for the corner BS configuration, not all of the 6 cells in a given reuse ring will produce a significant level of interference at a mobile location in the reference cell. Generally the number of significant interferers alternates between 3 and either 2 or 4 in successive rings, as shown diagrammatically in Figure 7.28. Rings giving 4 interferers, for example of size 21, have 12 cells in total and may be regarded as two interlaced rings of 6 reuse cells. To evaluate Equation 7.87 for a reuse cluster of size 3 we used only the first 3 rings with 8 significant interferers. To simplify the computations required it is assumed that Λ_i remains constant over the surface of the reference cell at the value estimated at the BS location. This value is given by;

$$
\Lambda_i = \frac{\Lambda_M \sum_{j=1}^{8} I_{rj}}{I_{ri}}
\tag{7.88}
$$

where Λ_M is the overall mean CIR and I_{ri} is the relative interference power received from the ith source, which is obtained as the product of the ith BS antenna power gain, G_{Ai}, given by Equation 7.4 and the path gain, G_{Pi}, relative to the path from the central source in the nearest ring, ie;

$$
I_{ri} = G_{Ai}G_{Pi}
\tag{7.89}
$$

G_{Pi} is given by;

$$
G_{Pi} = \left[\frac{(D/R)_i}{3}\right]^{-3.5}
\tag{7.90}
$$

where D/R is the reuse distance to cell radius ratio which is related to the ring size C by the well known expression;

$$
D/R = \sqrt{3C}
\tag{7.91}
$$

Numerical values for the above quantities are presented in Table 7.10. As

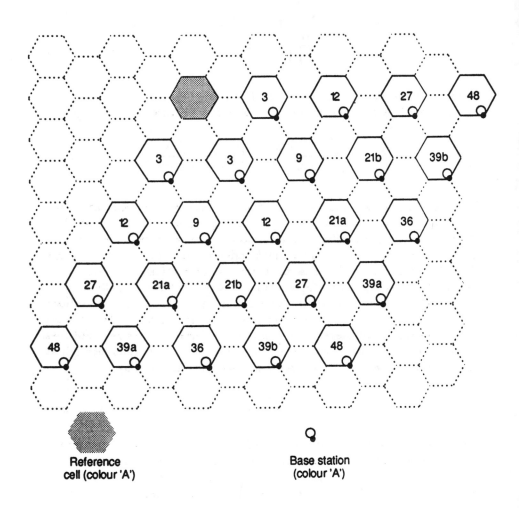

Figure 7.28: Positions of reuse cells producing significant cochannel interference for cluster size 3.

ring size	D/R	G_P	source No.	G_A	I_r	Λ/Λ_M
3	3.00	1.000	1	0.407	0.407	5.45
			2	1.000	1.000	2.22
			3	0.407	0.407	5.45
9	5.20	0.146	4	0.839	0.122	18.2
			5	0.839	0.122	18.2
12	6.00	0.0884	6	0.407	0.0360	61.6
			7	1.000	0.0884	25.1
			8	0.407	0.0360	61.6
					====	
					2.218	
21	7.94	0.0332	9	0.738	0.0245	
			10	0.917	0.0304	
			11	0.917	0.0304	
			12	0.738	0.0245	
27	9.00	0.0214	13	0.407	0.0087	
			14	1.000	0.0214	
			15	0.407	0.0087	

Table 7.10: Data for calculation of Λ_i for reuse cluster size 3.

in the previous analyses of performance the SFHMA system in a fading channel, it is necessary to determine the SER at the input to the second (outer) RS decoder as a function of the hop SNR normalised wrt the coded bit rate, γ_{cb}, and the hop CIR, λ, and then to evaluate a double integral to obtain the mean SER in terms of the mean CIR, Λ, and the mean SNR, Γ_{cb}, ie;

$$P_{is2}(\Lambda, \Gamma_{cb}) = \int_0^\infty \int_0^\infty P_{is2}(\lambda, \gamma_{cb}) f(\lambda) p(\gamma_{cb}) d\lambda d\gamma_{cb} \qquad (7.92)$$

where $f(\lambda)$ is the pdf of λ given by Equation 7.86 as above and $p(\gamma_{cb})$ is the pdf of γ_{cb}, which is assumed to have the negative exponential form associated with the Rayleigh fading model, as in Equation 7.33. Equations 7.57 (with SNR $= \gamma_{cb}$), 7.52, 7.42 and 7.43 are used to determine the first factor in the integrand. Finally the output BER from the second decoder is found by application of Equations 7.44 and 7.45. The mean SNR is normalised wrt the data bit rate by scaling Γ_{cb} as described in Section 7.4.4.

For all the calculations described here we put $r_a = 1$ and the SNR was kept constant at the relatively high, but arbitrarily chosen value of 30 dB, so that the system performance is primarily limited by cochannel interference rather than by noise.

A graph of output BER vs mean CIR, for cluster size 3 and $r_a = 1$, is presented in Figure 7.29 with η as parameter. From this data, for each value of η, the threshold value of mean CIR, denoted by Λ_{th}, was found at which the BER crosses a threshold value of 3×10^{-3}, which is a representative input BER threshold for typical 16 kbits/s toll quality speech decoders. Figure 7.30 presents a graph of Λ_{th} vs η, from which the value of η can be determined, denoted by η_{th}, at which $\Lambda_{th} = \Lambda_{90}$ for cluster size 3 (5.2 dB).

Similar calculations were made for reuse cluster size 9, for which case 5 significant interferers in the two nearest reuse rings of size 9 and 27 were allowed for in evaluating the hop CIR pdf via Equation 7.87. The results are also plotted in Figure 7.30. In this case it will be seen that the curve does not extend to the point where $\Lambda_{th} = \Lambda_{90}$ for cluster size 9 (13.0 dB) before η reaches its maximum value of 2.33 (for $U = 1$).

The computations required for the fractional reuse structures are complicated by the need to allow for two different values of collision probability depending on whether the interferer being considered is in either a fractional or a full reuse cell. For a $3M/L$ structure the full reuse rings will be those appropriate for a full reuse structure of size $3M$, while the fractional reuse rings will be those appropriate to the basic 3-cell structure. In order to limit the computational complexity involved in the calculations reported here we used only 3 rings of size 3, 9 and 12 with 8 significant interferers, which allowed for the determination of the performance of the 9/2, 12/2

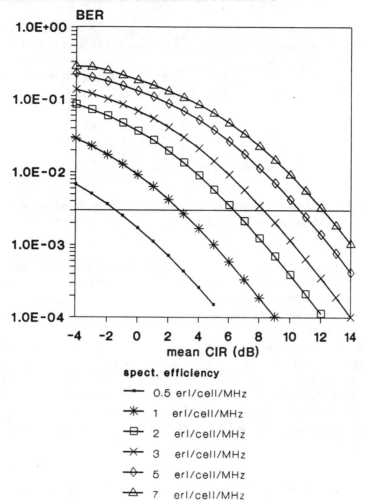

Figure 7.29: BER vs mean CIR for SFHMA system with cochannel interference in Rayleigh fading channel allowing for interferer diversity in full reuse cluster size 3 : normalised SNR = 30 dB and r_a = 1.0 [MSK-type detector: $B_t T$ = 0.3 and $B_r T$ = 0.63].

cluster size	η_{max}	η_{th}	η_{syst}	$U\%$
3	6.31	1.60	1.60	23
9	1.87	>2.33	1.87	80
9/2	4.07	1.92	1.92	41
12/2	2.96	1.47	1.47	42
12/3	4.63	1.74	1.74	33

Table 7.11: Spectral efficiency (erl/cell/MHz) for various reuse structures: $r_a = 1 : P_B = 2\%$

and 12/3 structures. In other respects the calculations were carried out in the same way as those for the full reuse structures described above. A graph of Λ_{th} vs η for the three structures is presented in Figure 7.31 and it will be seen that all three curves cross the relevant $\Lambda_{th} = \Lambda_{90}$ ($= 5.2$ dB) line for the basic 3-cell structure.

Table 7.11 summarises the values of η_{th} derived from Figures 7.30 and 7.31, together with the values of system spectral efficiency, η_{syst}, found by applying Equation 7.85 and the corresponding values of channel utilisation. The values of η_{max} are taken from Table 7.6. Apart from the cluster size 9 case all the values of η_{syst} and U lie between the corresponding values obtained using method A for $q_{90} = 0.7$ and $q_{90} = 0.8$, but closer to the former values.

7.6.3 Spectral efficiency of TD/FDMA system.

In this section we shall determine the spectral efficiency of an equivalent TD/FDMA system employing the same basic features such as, modulation, type of receiver, 16 kbits/s speech coding and corner base station cell configuration, as in the SFHMA system, but with constant carrier frequencies and no FEC coding. The transmitted bit rate is set at 172 kbits/s, which allows for 8 TDM slots/frame, together with a guard time of almost 8% of the slot duration and a 25% overhead for network control signalling, as in the SFHMA system. For this bit rate the minimum frequency spacing is taken as 125 kHz.

To determine the threshold value of mean CIR, Λ_{th}, which gives a BER threshold of 3×10^{-3} it was necessary to repeat the calculations described in Section 7.5.4 using Equations 7.63 to 7.67, but with the normalised SNR adjusted to allow for the total overhead of 34% in the transmitted bit rate. The results for a normalised SNR of 30 dB are shown in Figure 7.32 as

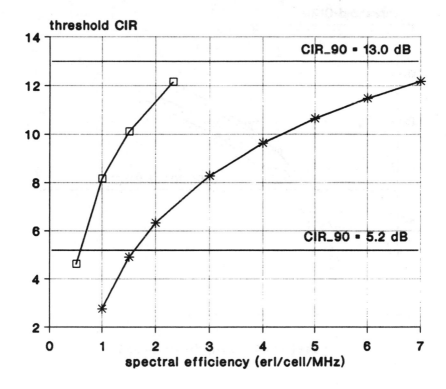

Figure 7.30: Λ_{th} vs η for full reuse clusters sizes 3 and 9 : $BER_{th} = 3 \times 10^{-3}$, SNR = 30 dB and $r_a = 1.0$.

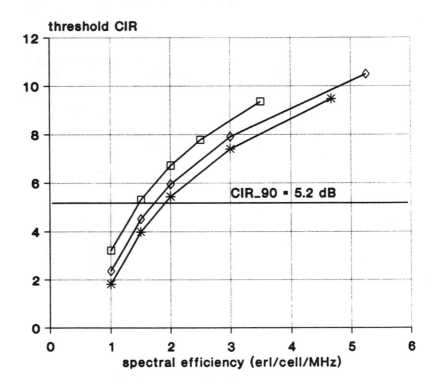

cluster size

✳	:	9/2
⊟	:	12/2
◇	:	12/3

Figure 7.31: Λ_{th} vs η for fractional reuse clusters sizes 9/2, 12/2 and 12/3 : $BER_{th} = 3 \times 10^{-3}$, SNR = 30 dB and $r_a = 1.0$.

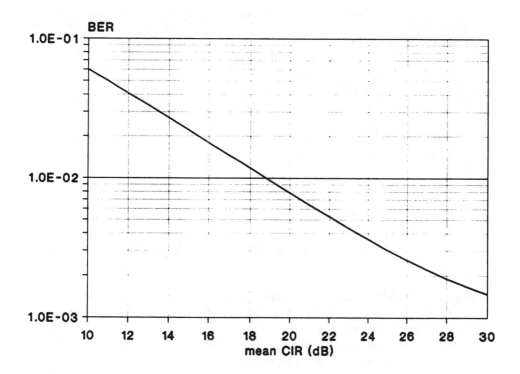

Figure 7.32: BER vs mean CIR for uncoded TD/FDMA system with cochannel interference in Rayleigh fading AWGN channel at normalised mean SNR = 30 dB [MSK-type detector : $B_t T = 0.3$ and $B_r T = 0.63$].

size	36	39	48
Λ_{90} (dB)	22.9	23.5	25.1

Table 7.12: Extrapolated values of Λ_{90} vs reuse cluster size.

quantity		TD/FDMA	SFHMA
system bandw.(MHz)	W	24	24
cluster size	C	48	9
No. TDM slots	n_t	8	3
frequency spacing(MHz)	f_s	0.125	0.143
No. freq./system	$n_{fs} = W/f_s$	192	168
No. channels/cell	$n_{ch} = n_{fs}n_t/C$	32	56
traffic/channel (erl)	a_{ch}	0.727	0.803
traffic/cell (erl)	a_{cell}	23.3	45.0
spectral efficiency (erl/cell/MHz)	$\eta_{syst} = a_{cell}/W$	0.971	1.88

Table 7.13: Calculation of spectral efficiency for SNR $= 30$ dB $: P_B = 2\%$

a graph of BER vs mean CIR from which the value of λ_{th} is found to be 25 dB. The minimum cluster size is determined from the requirement that the corresponding $\Lambda_{90} \geq \Lambda_{th}$. However the available data for Λ_{90} (see Table 7.4) only extends to a maximum value of 20.6 dB and so we have extrapolated the data to cover three additional cluster sizes, based on the assumption that Λ_{90} is proportional to $(D/R)^{-3.5}$ for larger clusters. The values obtained are presented in Table 7.12 and it will be seen that the minimum cluster size is 48. To determine the spectral efficiency we proceed to calculate the number of channels available per cell, the traffic carried per channel from the Erlang B formula for a blocking probability of 2% and finally the traffic carried per cell divided by the system bandwidth as indicated in Table 7.13. For comparison we also show the calculations for the SFHMA system, assuming the same worst case operation with no account taken of interferer diversity and for full reuse structures only. From the results of the analysis described in Section 7.5.4 for the SFHMA system in a fading AWGN channel, shown in Figure 7.21, at 30 dB SNR the CIR giving BER $= 3 \times 10^{-3}$ is 12.5 dB, which implies a minimum cluster size of 9.

These results demonstrate the superiority of the SFHMA system even with interferer diversity and silence detection discounted. Comparison with results for the SFHMA system obtained in the previous sections, which do allow for these effects, show good agreement for the cluster size 9. This is not as surprising as it might seem at first sight, since in this case the spectral efficiency is limited by the maximum channel utilisation, which implies that interferer diversity is not significant.

7.7 Conclusions

In this chapter we have described the principles and essential characteristics of SFHMA, concentrating on the mixed protocol for the basic code division multiplexing. A specification of a complete exemplary SFHMA system is provided as a basis for the performance analyses.

Details and results of BER performance analyses, mostly developed by the authors, are presented for the following conditions;

(1) a comparison of the ideal MLSE receiver for GMSK signals with a more practical MSK-type detector, which showed that the performance was only degraded by about 1 dB,

(2) system performance in static and Rayleigh fading channels,

(3) system performance with cochannel interference, both with and without noise and also with and without fading.

In the studies under the headings (2) and (3) comparative results are presented for an equivalent basic FDMA system which has no FEC coding.

No allowance was made initially in the cochannel interference analyses of the SFHMA system, described in Section 7.5, for the effects of interferer diversity, the main emphasis being placed on the performance of the GMSK receiver with concatenated FEC coding. However interferer diversity is accounted for in the two methods presented in the following section for estimating spectral efficiency. Method A is essentially the simplified model reported by Verhulst [5], with some refinement to allow for non-ideal channel utilisation, while method B is a development of the authors' more rigorous analysis of SFHMA performance in a channel degraded by both cochannel interference and noise, but utilising an expression for the pdf of hop CIR derived from Verhulst's analysis [5]. The results produced by the two methods agree surprisingly well, considering the different gross approximations involved.

Finally, in order to demonstrate the advantage of SFHMA we have estimated the spectral efficiency of a comparable TD/FDMA system. Comparing the results presented in Tables 7.7, 7.9 and 7.13, it can be seen that for the case of no silence detection ($r_a = 1$) the SFHMA system can provide a

spectral efficiency just over twice that of the TD/FDMA system, while if
silence detection is implemented the potential improvement is by a factor
. of almost four. Clearly there is a considerable advantage in spectral efficiency to be gained by employing SFHMA, under the assumptions used in
the analyses, but too much reliance should not be placed on the absolute
values of the improvement factors indicated above. Many crude approximations have been made in the underlying analyses and, moreover, the use
of the mean input BER to the unspecified speech decoder as the basis of
comparison of the systems can be criticised, because it does not allow for
the temporal characteristics of the stream of errors.

SFHMA also offers the advantage of greater operational flexibility because of the variety of reuse structures available and because of the potential ability to cope with non-uniform traffic loading, which is implied by
not needing to operate at high levels of the channel utilisation to achieve
good spectral efficiency.

7.8   Appendix A:

Derivation of the probability of correct reception of a hop

A hop will be correctly received if the hop CIR, $\lambda = S/I$, is not less than
the threshold value, γ, with a probability given by:

$$q = \mathcal{P}\{S/I \geq \gamma\} \tag{7.93}$$

where $S = A_o B_o F_o$ is the received hop signal power and I is the total
received hop interference power, obtained by summing the powers received
from the sources in the reuse cells, ie:

$$I = \sum_{i=1}^{M} A_i B_i E_i F_i. \tag{7.94}$$

Substituting for S and rearranging, Equation 7.93 can be written as:

$$
\begin{aligned}
q &= \mathcal{P}\{F_o \geq \gamma I / A_o B_o\} \\
&= \int_0^\infty p(I) \left[\int_{\gamma I / A_o B_o}^\infty p(F_o) dF_o \right] dI
\end{aligned}
\tag{7.95}
$$

where $p(I)$ is the pdf of the total interference power and $p(F_o)$ is the pdf of
the signal fading factor, which has an exponential pdf with a mean value
of unity, ie:

$$p(F_o) = \exp(-F_o) \tag{7.96}$$

Thus we can determine the inner integral in the RHS of Equation 7.95 giving:

$$q = \int_0^\infty p(I) \exp(-\gamma I / A_o B_o) dI \tag{7.97}$$

This expression is of the form of a Laplace transform of $p(I)$ for $s = \gamma / A_o B_o$ so that we can write:

$$q = \mathcal{L}\{p(I)\} \mid_{s=\gamma/A_o B_o} \tag{7.98}$$

Since I is the sum of a set of independent random variables I_i, we can apply the known relationship that the transform, ie characteristic function, of the pdf of the sum is the product of the transforms of the pdfs of the components:

$$\mathcal{L}\{p(I)\} = \prod_i \mathcal{L}\{p(I_i)\} \tag{7.99}$$

The discontinuous nature of the power received from the i_{th} interferer may be modelled as random on/off switching of a continuous exponentially distributed process, which has a mean value of $A_i B_i$. hence the pdf $p(I_i)$ can be determined as the sum of two conditional pdfs for the "on" ($E_i = 1$) and "off" ($E_i = 0$) states, weighted by the state probabilities p_i and $(1 - p_i)$ respectively, where p_i is the frequency collision probability for the i_{th} reuse cell. Conditional on $E_i = 1$ the pdf of I_i is the same as that of the continuous process, ie:

$$p(I_i | E_i = 1) = \frac{1}{A_i B_i} \exp\left(\frac{-I_i}{A_i B_i}\right) \tag{7.100}$$

The alternative condition of $E_i = 0$ implies that I_i can only take the single discrete value of zero and so the pdf in this case becomes a unit impulse at $I_i = 0$, ie:

$$p(I_i | E_i = 0) = \delta(I_i) \tag{7.101}$$

Combining these conditional pdfs gives the overall pdf of I_i as:

$$p(I_i) = (1 - p_i)\delta(I_i) + \frac{p_i}{A_i B_i} \exp\left(\frac{-I_i}{A_i B_i}\right) \tag{7.102}$$

Taking Laplace transforms gives:

$$\mathcal{L}\{p(I_i)\} = 1 - p_i + \frac{p_i}{1 + s A_i B_i} \tag{7.103}$$

Substituting into Equation 7.99 and putting $s = \gamma / A_o B_o$ we obtain:

$$q = \prod_{i=1}^{M} \left(1 - p_i + \frac{p_i}{1 + \gamma A_i B_i / A_o B_o}\right) \tag{7.104}$$

A somewhat different derivation of this expression is briefly described in [8]. We will define the ratio $A_o B_o / A_i B_i$ as the short term mean CIR with the respect to the i_{th} interference source, denoted by Λ_i. Note that this quantity relates the mean received signal power to the maximum mean interference power which could be received from the i_{th} source if it operated continuously. Finally, substituting in the above equation and rearranging gives:

$$
\begin{aligned}
q &= \prod_{i=1}^{M} \left(1 - p_i + \frac{p_i \Lambda_i}{\Lambda_i + \gamma} \right) \\
&= \prod_{i=1}^{M} \left(1 - \frac{p_i \gamma}{\Lambda_i + \gamma} \right)
\end{aligned}
\tag{7.105}
$$

Bibliography

[1] **G.R.Cooper**, and **R.W.Nettleton**, "A spread spectrum technique for high capacity mobile communications", *IEEE Trans. Veh. Technol.*, Vol. VT-27, No. 4, pp. 264-75, November 1978.

[2] **D.Verhulst, M.Mouly,** and **J.Szpirglas,** "Slow frequency hopping multiple access for digital cellular radiotelephone", *IEEE Trans. Veh. Technol.*, Vol. VT-33, No. 3, pp. 179–190, August 1984.

[3] **D.Verhulst,** and **M.Mouly,** "Spectrum efficiency evaluation techniques for future digital mobile systems", *Proc. 1st Nordic Seminar Digital Land Mobile Radio*, Espoo, Finland, 5–7 February 1985.

[4] **E. A.Geraniotis,** and **M. B.Pursley,** "Error probabilities for slow frequency-hopped spread-spectrum multiple-access communications over fading channels", *IEEE Trans. Commun.*, pp. 996–1009, May 1982.

[5] **D.Verhulst,** "Spectrum efficiency analysis of the digital system SFH900", *Proc. 2nd Nordic Seminar Digital Land Mobile Radio*, Stockholm, Sweden, 14–16 October 1986.

[6] **ANT/Bosch,** "Description of the experimental system S-900D for digital radiotelephone", GSM doc No 85/85, 1985.

[7] **J.Udenfeldt,** "DMS90—an experimental TDMA digital mobile telephone system", Ericsson Radio System AB, Stockholm, Sweden.

[8] **J-L.Dornstetter** and **D.Verhulst,** "Cellular efficiency with slow frequency hopping: Analysis of the digital SFH900 mobile system", *IEEE J. Select. Areas Commun.*, Vol. SAC-5, No. 5, pp. 835–848, June 1987.

[9] **J.B.Anderson, T.Aulin** and **C-E.Sundberg,** "Digital Phase Modulation", *Plenum Press*, New York 1986.

[10] **K.Murota** and **K.Hirade,** "GMSK modulation for digital mobile radio telephony", *IEEE Trans. Commun.*, Vol. COM-29, pp. 1044–1050, July 1981.

[11] **J. G.Proakis,** "Digital communications", *McGraw-Hill Book Company*, 1983.

[12] **T.Aulin,** **N.Rydbeck,** and **C-E.Sundberg,** "Continuous phase modulation—Part II: Partial response signaling", *IEEE Trans. Commun.,* Vol. COM-29, No.3, pp. 210–225, March 1981.

[13] **S. G.Wilson** and **M.G.Mulligan,** "An improved algorithm for evaluating trellis phase codes", *IEEE Trans. Inform. Theory,* Vol. IT-30, No. 6, pp. 846–851, November 1984.

[14] **A.Svensson** and **C-E.Sundberg,** "Serial MSK-type detection of partial response continuous phase modulation", *IEEE Trans. Commun.,* Vol. COM-33, No. 1, pp. 44–52, January 1985.

[15] **K.Murota,** "Spectrum efficiency of GMSK land mobile radio", *IEEE Trans. Veh. Technol.,* Vol. VT-34, No. 2, pp. 69-75, May 1985.

[16] **J.Oetting,** "Cellular mobile radio—an emerging technology", *IEEE Trans. Commun. Mag.,* pp. 10–15, November 1983.

[17] **V.H.MacDonald,** "Advanced mobile phone service: The cellular concept", *Bell Syst. Tech. J.,* Vol. 58, No. 1, pp. 15–41, January 1979.

[18] **R.Steele** and **V.K.Prabhu,** "High-user-density digital cellular mobile radio systems", *IEE Proc. on Communications, Radar and Signal Processing,* Pt. F, Vol. 132, No. 5, pp. 396–404, August 1985.

[19] **Y.F.Ko,** "Digital cellular mobile radio links and networks", *PhD Thesis,* Dept of Electronics and Computer Science, University of Southampton, Section 2.4.2., December 1989.

[20] **W.C.Jr.Jakes,** "Microwave mobile communications", *Wiley,* New York, pp. 367.

[21] **J.E.Stjernvall,** "Calculation of capacity and co-channel interference in a cellular system", *Proc. Nordic Seminar on Digital Land Mobile Radio,* Espoo, Finland, pp. 209-217, 5-7, February 1985.

Chapter 8

The Pan-European Digital Cellular Mobile Radio System—known as GSM

L. Hanzo[1] and J. Stefanov[2]

8.1 Introduction

The first cellular radio system in Europe was installed in Scandinavia in 1981 and it served initially only a few thousand subscribers. At the time of writing there are six different cellular systems operating in 16 European countries and serving more than 1.2 million subscribers. There is however, a general incompatibility of systems and user equipment. A mobile station (MS) designed for one system cannot be used in another which makes it impossible for mobiles to roam across international borders while making calls. The low scale of equipment production also results in higher equipment cost and call charges. In 1982 CEPT (Conference Europeene des Pastes et Telecommunication), the main governing body of the European PTT's, created the Groupe Speciale Mobile (GSM) Committee and tasked it with specifying a cellular pan-European public mobile communication system to operate in the 900 MHz band.

The first important decision made by the GSM Committee was the selection of a digital system. This was followed by the launch of experimental

---

[1] University of Southampton and Multiple Access Communications Ltd.
[2] University of Southampton and Multiple Access Communications Ltd.

	Access Type	Transm. Bit Rate KBIT/S	Carrier Spacing kHz	Mod. Type	Channels Per Carrier	Chan. Cod. Reuse Factor
CD-900	CDMA/TDMA	7980	4500	4-PSK	63	6.67
MATS-D/W	CDMA/TDMA	2496	1250	QAM	32	1.13
ADPM	TDMA	512	600	ADPM	12	2.0
DMS-90	TDMA	340	300	GMSK	10	1.85
MOBIRA	TDMA	252	250	GMSK	9	1.6
SFH-900	TDMA	200	150	GMSK	3	2.6
S900-D	TDMA	128	250	4-FSK	10	1.15
MAX II	TDMA	104.7	50	8-PSK	4	1.5
MATS-D/N	FDMA	19.5	25	GTFM	1	1.13

Table 8.1: Systems tested by GSM in Paris, 1986

programmes of different types of digital cellular radio systems in a number of European countries. By the middle of 1986 nine proposals were received for the future pan-European system, and GSM organised a trial in Paris to identify the one having the best performance. The technical details of the candidate systems are described in references [1], [2], [3], [4] and [9] . A short summary [6] of their salient features is listed in Table 8.1.

The first generation cellular radio systems use analogue frequency modulation. Single-channel-per-carrier (SCPC) Frequency Division Multiple Access (FDMA) is utilised, where the channel bandwidth is either 25 or 30 kHz. The proposed digital systems for GSM employed a variety of access methods, transmission rates and modulation schemes. Six systems adopted Time Division Multiple Access (TDMA), another two employed Code Division Multiple Access (CDMA) combined with TDMA, and one used FDMA on its mobile station to base station (MS to BS) uplink. The transmission bit rates for the various systems spanned the range from 20 kb/s to 8 Mbit/s. Altogether seven different modulation schemes were used in the nine systems.

In addition to the field trials in Paris, laboratory testing of the equipment was carried out using a propagation simulator. The test arrangements allowed for measurements with or without interference under static or dynamic conditions. The channel simulator had two independent Rayleigh fading paths, enabling five different propagation profiles to be used corresponding to various rural, suburban and urban propagation environments.

Based on the field trials and laboratory tests the candidate systems were assessed in order of importance against the following criteria [8]: spectrum efficiency, subjective voice quality, mobile cost, hand-portable feasibility, base station cost, ability to support new services and co-existance with current systems.

The test results demonstrated that the spectrum efficiency of all the candidate systems was equal or better than that of the first generation analogue cellular systems. However, not all the systems provided an acceptable transmission quality over the range of propagation conditions. Furthermore, there were significant implementational and operational risk factors involved with some of the systems. The two "hybrid" CDMA/TDMA systems proposed were wideband systems requiring broadband radio subsystems and very complex baseband signal processing. These systems had the highest degree of risk in terms of being implementable within the GSM time scale for the system to be operational in 1992 and that their costs might be commercially too high. The predicted performance of broadband receivers was considered to be difficult to achieve in practice [6] and broadband systems can be susceptible to spurious interference. The implementation of the complex VLSI's needed for baseband processing would be difficult and costly, although the resulting microcircuits were expected to be of low cost because of the large size of the market.

On the whole, the TDMA technique was preferred over FDMA and TDMA-CDMA. Partly, because for good transmission quality FDMA systems need antenna diversity not only at the base station but also at the mobile station, which in most cases is unacceptable. However, the high bitrate associated with TDMA implies dispersive wideband channel models and requires the use of channel coding, interleaving, as well as channel equalisation. In TDMA systems the lack of a duplexer in the mobile station and the multiplexing of several channels on one RF carrier in the base station leads to simpler, more compact and more cost-efficient design of both the MS and BS. TDMA systems are also more flexible than FDMA systems in accommodating new services in the future. The hand-over in TDMA systems can be performed more efficiently than in FDMA systems which is particularly important for high traffic density microcells.

Based upon the results of the tests in Paris, GSM decided at the beginning of 1987 to adopt a narrowband TDMA system with the basic system features specified as follows: 8 TDMA channels per carrier; regular pulse excited linear predictive (RPE-LPC) speech codec operating at 13 kb/s; half-rate (R=1/2), constraint length five (K=5) convolutional codecs CC(2,1,5); carrier spacing of 200 kHz; constant envelope Gaussian minimum shift keying (GMSK) modulation. By the end of 1988 the working parties of GSM and their supporting expert groups had substantially completed the specifications of the pan-European system. The specifications were released by GSM as 13 sets of recommendations [9] covering various aspects of the system. The following description of the GSM system, its major elements and their functions, is based almost entirely on the GSM recommendations [9]. Since the system is still evolving at the time of writting, fine detailes of the material presented in this chapter are subject to changes.

The GSM system operates in two paired bands: 890–915 MHz for uplink transmission, where the mobile transmits and the base station receives, and 935–960 MHz for downlink transmission, where the base station transmits and the mobile receives. A guard band of 200 kHz is provided at the lower end of each duplex band and the remaining spectrum is divided into 124 paired duplex channels with 200 kHz channel spacing in each band. The spacing between the duplex bands is 45 MHz.

The access scheme is TDMA with 8 timeslots per radio carrier detailed later by referring to Figure 8.2. The duration of each timeslot is approximately 0.58 ms, which leads to a TDMA frame duration of approximately $8 \cdot 0.58 = 4.6$ *ms*. The information is transmitted in bursts at a rate of approximately 271 kbit/s using Gaussian Minimum Shift Keying (GMSK) with a bandwidth- bitinterval product of BT=0.3. For the channel spacing of 200 kHz the use of this type of modulation allows the carrier separation to be 18 dB for the first adjacent channel and 50 dB for the next one. A rudimentary description of GMSK is provided in Section 8.6, while a detailed treatment is given in Chapter 6.

At the data rate of 271 kbit/s the multipath propagation leads to deep fades and to uncontrolled intersymbol interference in addition to that introduced in a controlled manner by the GMSK modulator. Transmission errors are combated by using channel coding and channel equalisation. The channel coding method employs two concatenated codes. A block code provides error detection for the most significant 50 speech bits, followed by a half-rate convolutional code, while the 78 least significant speech bits are not protected at all. The number of coded bits per block and the convolutional code rate depend on the type of information transmitted (speech, data, signalling). Aspects of the channel coding and interleaving are provided in Section 8.5, while the equalisation of GMSK signals is summarised in Section 8.7 and described in detail in Chapter 6. Suffice to state here that the GSM equaliser is expected to handle excess path delays of up to 16 μs. Slow frequency hopping with 217 hops/s is used to provide diversity effect and to increase the efficiency of coding and interleaving for slow moving mobile stations, as explained in Section 8.3. It helps also to decrease the effects of the co-channel interference.

A general objective of the GSM system is to provide a wide range of services and facilities, both voice and data, that are compatible with those offered by the existing fixed Public Services Telephone Networks (PSTN), Public Data Networks (PDN) and Integrated Services Digital Networks (ISDN). Another objective is to give compatibility of access to the GSM network for any mobile subscriber in any country which operates the system, and these countries must provide facilities for automatic roaming, locating and updating the mobile subscriber's status.

8.2   Overview of the GSM System

Mobile radio communications in a GSM Public Land Mobile Network (GSM-PLMN) is facilitated by a series of network functions and procedures. Figure 8.1 shows the simplified structure of a typical GSM PLMN with the functional entities of the system and their logical interconnections. The Mobile Station (MS) is the equipment used by a subscriber to access the services offered by the system. Functionally the MS includes a Mobile Termination (MT), and Terminal Equipment (TE) which may consist of more

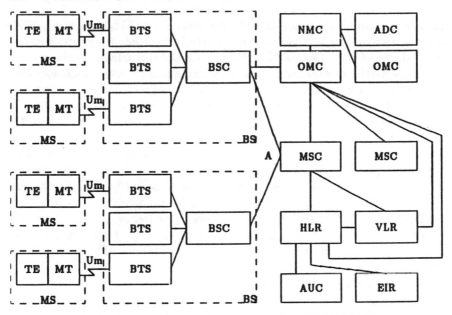

Figure 8.1: Simplified structure of a GSM PLMN

than one piece of equipment such as a telephone set and Data Terminal Equipment (DTE). If necessary, one or more Terminal Adapters (TA) may also be included. The MT performs functions needed to support the physical channel between the MS and the base station, such as radio transmissions, radio channel management, channel coding/decoding, speech encoding/decoding, and so forth, where the MS-BS radio interface is designated by U_m. There are various types of mobile stations such as vehicle mounted stations, portable stations and hand-held stations.

The Base Station System (BSS), defined also as a Base Station (BS), is divided functionally into a Base Transceiver Station (BTS) and Base Station Controller (BSC) and they are interconnected by the A-bis interface. The BS is associated with the radio channel management including channel allocation, link quality supervision, transmission of associated signalling information and broadcast messages, as well as controlling transmitted power levels and frequency hopping. Its further functions entail error correction encoding and decoding, digital speech transcoding or data rate adaptation, intracell handover initiation to a 'better' RF channel, as well as data and signalling encryption. The BTS is the transmission equipment used to give radio coverage for a traffic cell. All control functions in the base station are performed by the BSC. The radio equipment in a BS may serve more than one cell, in which case the BS will consist of several BTS's under the control of one BSC.

The Mobile Switching Centre (MSC) is linked to the BS via the A interface and performs all the switching functions needed for the operation of the mobile stations in the group of cells it services. The functions of an MSC include call routing and call control; procedures needed for interworking with other networks (e.g., PSTN, ISDN); procedures related to the mobile station's mobility management such as paging to receive a call, location updating while roaming and authentication to prevent unathorised access; as well as procedures required to implement handovers. Handover(HO) is the process of re-assigning the mobile station's communications to a different base station, when the mobile moves outside the range of the serving base station. The GSM system supports also another type of handover, the intra-cell handover, which is a call transfer from one channel to another one within the same cell, when a channel cannot be used any longer due to interference disturbance or maintenance problems. Due to the high traffic demand expected, RF channels have to be frequently reused, which results in small microcells and increased probability of high cochannel interferences. To counteract the interference limitation, efficient handover algorithms based on intelligent and hence complex received signal quality evaluations are absolutely essential.

The Home Location Register (HLR) is a data base unit for the management of mobile subscribers. Part of the mobile location information is stored in the HLR, which allows the incoming calls to be routed to the MSC in command of the area where the MS roames. The MS has to periodically inform the PLMN about its geographic location by updating the contents of the HLR. To assist this process the PLMNs are divided into disjoint geographic areas characterised by unique identifiers broadcast regularly to all MSs via the so-called Broadcast Control Channels (BCCHs) conveyed over reserved RF carriers. Should the MS observe a change of identifier, it issues a location update request. The HLR contains the International Mobile Subscriber Identity (IMSI) number which is used for the authentication of the subscriber by his AUthentication Centre (AUC). This enables the system to confirm that the subscriber is allowed to access it. Every subscriber belongs to a home network and the specific services which the subscriber is allowed to use are entered into his HLR. The Equipment Identity Register (EIR) allows for stolen, fraudulent or faulty mobile stations to be identified by the network operators.

The Visitor Location Register (VLR) is the functional unit that attends to a MS operating outside the area of its HLR. The visiting MS is automatically registered at the nearest MSC and the VLR is informed of the MSs arrival. A roaming number is then assigned to the MS and this enables calls to be routed to it. The Operations and Maintenance Centre (OMC), Network Management Centre (NMC) and ADministration Centre (ADC) are the functional entities through which the system is monitored, controlled, maintained and managed.

When a mobile user initiates a call the MS searches for a BS providing a sufficiently high received signal level on the BCCH carrier, it will synchronise to it, then the BS allocates a bidirectional signalling channel and also sets up a link with the MSC serving the area. The MSC uses the IMSI received from the mobile station to interrogate the subscriber's HLR. The subscriber's data obtained from the HLR is then sent to the local VLR. After the user is accepted by the network, the MS defines the type of service it requires and provides the destination number of its call. The BS serving the cell allocates a traffic channel and the MSC routes the call to its destination. If the MS moves to another cell, it is re-assigned to another BS and a handover occurs. If both BSs in the handover process are controlled by the same BSC, the handover takes place under the control of the BSC. If the base stations are controlled by different BSCs, then the handover is performed by the MSC.

The procedure of setting up a call connection from a fixed network to a MS is similar to the procedure described above. The main difference is that the MS must be paged by the BSC. A paging signal is transmitted on a paging channel (PCH) monitored continuously by all MSs and covers the location area in which the MS has registered. When the MS receives the paging signal, it starts an access procedure identical to that employed when the MS initiates a call.

8.3 Mapping Logical Channels onto Physical Channels

8.3.1 Logical Channels

The elaborate design of the MS-BS radio interface (U_m interface in GSM terminology) is motivated by the needs of providing appropriate signalling and traffic channels in the system. The traffic channels might take the form of circuit and packet switched bearer services at various synchronous and asynchronous speeds as well as teleservices, such as speech, short message, teletext or facsimile communications. Appropriate signalling and traffic communications are provided by the first three layers of the seven-layer Open Systems Interconnection (OSI) model. Accordingly, the physical layer (L1) interfaces with the data link layer (L2) via a number of logical channels constituted by speech and data traffic channels (TCHs) as well as signalling or control channels (CCHs). Our prime objective is to transmit the traffic channel's speech or data information, however their transmission via the network requires a variety of control channels. A feasible solution to efficient networking is provided by the set of logical control channels defined by GSM. The need for each specific control channel provided arises from the system architecture discussed previously , although further alternative concepts can readily be contrived.

As seen in Table 8.2, there are two general forms of speech and data traffic channels: the full rate traffic channels (TCH/F), which carry information at a gross rate of 22.8 kbit/s, and the half rate traffic channels (TCH/H),

Logical Channels					
Duplex Traf. Chan.: TCH		Control Channels: CCH			
FEC-coded Speech	FEC-coded Data	Broadc. CCH BCCH $BS \rightarrow MS$	Common CCH CCCH	Stand-alone Dedicated CCH SDCCH $BS \leftrightarrow MS$	Associated CCH ACCH $BS \leftrightarrow MS$
TCH/F 22.8 kbit/s	TCH/F9.6 TCH/F4.8 TCH/F2.4 22.8 kbit/s	Freq.Corr.Ch: FCCH	Paging Ch: PCH $BS \rightarrow MS$	SDCCH/4	Fast ACCH: FACCH/F, FACCH/H
TCH/H 11.4 kbit/s	TCH/H4.8 TCH/H2.4 11.4 kbit/s	Synchron. Ch: SCH	Rand. Access Ch: RACH $MS \rightarrow BS$	SDCCH/8	Slow ACCH: SACCH/TF, SACCH/TH SACCH/C4, SACCH/C8
		General Inf.	Access Grant Ch: AGCH $BS \rightarrow MS$		

Table 8.2: GSM logical channels

which communicate at a gross rate of 11.4 kbit/s. A physical channel carries either a full rate traffic channel, or two half rate traffic channels. In the former the traffic channel occupies one timeslot, while in the latter the two half-rate traffic channels are mapped onto the same timeslot, but in alternate frames.   Encoded speech and user data can be conveyed on a variety of full or half rate traffic channels. These may be notated as TCH/$\alpha\beta$, where α is either F or H signifying full rate or half rate, respectively, and β is either S when speech is being carried, or 9.6, 4.8 or 2.4 representing the data rate in kbit/s. The bit rate in the full rate traffic channel for speech (TCH/FS) is 13 kbit/s, becoming 22.8 kbit/s after embedded channel coding. The channel coded data rate in the half rate speech channel TCH/HS is 11.4 kbit/s. This channel is envisaged for future evolution of the system and allows the traffic capacity to be doubled, when a low bit rate toll quality voice codec becomes available. The traffic channels may carry a wide variety of user information, but not signalling information. However, when user data is transmitted, they may carry protocols such as, e.g., in the case of packet switching services according to CCITT Recommendation X25.

The control channels carry signalling or synchronisation data. As summarised in Table 8.2, four categories of control channels are used, known as the broadcast control channel (BCCH), the common control channel (CCCH), the stand-alone dedicated control channel (SDCCH) and the associated control channel (ACCH). The broadcast control channels are used only in downlink communications from the base station to the mobile stations in its vicinity. There are three types of BCCHs. The frequency correction channel (FCCH) is provided to facilitate frequency synchronisation of the mobile station to the master radio frequency source in the base station. The transmitted information on the FCCH is equivalent to an unmodulated carrier with a fixed frequency offset from the nominal carrier frequency. The function of the synchronisation channel (SCH) is to enable frame synchronisation of the mobile station and the identification of the serving base station. Accordingly, the data transmitted on the SCH contains the TDMA frame number (FN) and the base station identity code (BSIC). When not acting as FCCH or SCH, the BCCH carries general information, such as the number of common control channels, whether these common control channels are combined with stand-alone dedicated control channels and associated control channels on the same physical channels.

There are three types of CCCHs. The paging channel (PCH) contains paging signals from the BS to the MSs in case of a network originated call. There is a random access channel (RACH), which is used only in the uplink communications whereby the MSs request the allocation of a bidirectional stand-alone dedicated control channel for BS-MS signalling. An access grant channel (AGCH) is provided for downlink communications for the purpose of allocating a stand-alone dedicated control channel or a traffic channel to MSs, which was previously requested via the RACH.

There are two types of SDCCHs. The stand-alone dedicated control channel having 4 sub-channels and notated by SDCCH/4, and SDCCH/8 which has 8 sub-channels. These SDCCH channels are used for setting up the services required by the user. This involves interrogation of the mobile station as to the services required, the availability response of the base station and the allocation of a free traffic channel.

The ACCHs, like the SDCCHs, are also bidirectional channels. In the downlink they carry, for example, control commands from the base station to the MS to set its transmitted power level, while in the uplink they convey the status of the mobile station, such as the received signal levels from various adjacent BSs, etc.. An ACCH is always allocated in conjunction with either a traffic channel, or with a stand-alone dedicated control channel, as it will be explained later. There are two types of ACCH. The fast associated control channel (FACCH) facilitates urgent actions, such as handover commands and channel reassignment in intra-cell handovers. This type of channel can be associated either with a full rate traffic channel (FACCH/F), or with a half rate traffic channel (FACCH/H), and it is provided by stealing bits from its traffic channels, hence degrading their performance. The slow associated control channel (SACCH) is sub-divided into 4 types, depending on what type of channel it is associated with. Thus SACCH/TF is associated with a full rate traffic channel; SACCH/TH is associated with a half rate traffic channel; SACCH/C4 is associated with SDCCH/4; while SACCH/C8 is associated with SDCCH/8. In downlink transmissions the SACCH carries commands from the base station to the mobile station for setting its output power level. The mobile station responds by informing the BS of the set level of its output power, the measured received RF signal strength and the quality of the signals from adjoining cells.

After this rudimentary characterisation of the logical channels used in the GSM system we are ready to describe the TDMA physical channels carrying the information of the logical channels and the way logical channels are mapped onto physical ones.

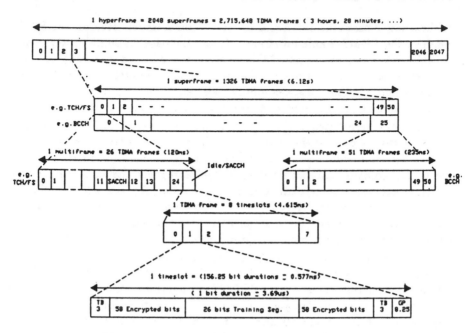

Figure 8.2: The GSM TDMA frame structure

8.3.2   Physical Channels

A physical channel in a TDMA system is defined as a timeslot with a timeslot number TN in a sequence of TDMA frames. However, the GSM system deploys TDMA combined with frequency hopping and hence the physical channel is partitioned in both time and frequency. Consequently the physical channel is defined as a sequence of radio frequency channels and timeslots. Each carrier frequency supports 8 physical channels mapped onto 8 timeslots within a TDMA frame. A given physical channel uses always the same timeslot number TN in every TDMA frame. Therefore, a timeslot sequence is defined by a timeslot number TN and a TDMA frame number FN sequence.

8.3.2.1   Mapping the TCH/FS and its SACCH as well as FACCH onto Physical Channels

In our deductive approach we use the example of the full rate speech traffic channel (TCH/FS) to explain how this logical channel is mapped onto the physical channel constituted by a so-called Normal Burst (NB) of the TDMA frame structure. This mapping is explained in macroscopic terms by referring to Figures 8.2 and 8.3. Bit-level fine details of the individual mapping steps will be provided in later sections, as details of the speech coding, error correction coding, etc., become available. Then this example will be extended to other physical bursts such as the Frequency Correction

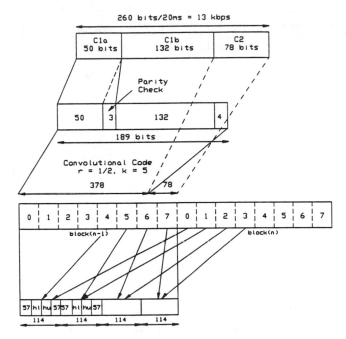

Figure 8.3: Mapping the TCH/FS logical channel onto physical channels

(FB), Synchronisation (SB), Access (AB) and Dummy Burst (DB) carrying logical control channels, as well as to their TDMA frame structures, as seen in Figures 8.2 and 8.7.

The Regular Pulse Excited (RPE) speech encoder delivers 260 bits/20 ms at a bitrate of 13 kbit/s, which are divided into three significance classes: Class 1a (50 bits), Class 1b (132 bits) and Class 2 (78 bits). The Class 1a bits are encoded by a systematic (53,50) cyclic error detection code by adding three parity bits. Then the bits are reordered and four zero tailing bits are added to perodically reset the subsequent half rate, constraint length five convolutional codec CC(2,1,5), as protrayed in Figure 8.3. Now the unprotected 78 Class 2 bits are concatenated to yield a block of 456 bits/20 ms, which implies an encoded bitrate of 22.8 kbit/s. This frame is partitioned into eight 57-bit subblocks that are blockdiagonally interleaved before undergoing interburst interleaving to be detailed later in Section 8.5. The flag bits hl and hu are included to classify, whether the burst being transmitted is really a TCH/FS burst or it has been 'stolen' by an urgent FACCH message. Now the bits are encrypted and positioned in a Normal Burst(NB), as depicted at the bottom of Figure 8.2, where three tailing bits (TB) are added at both ends of the burst to reset the memory of the Viterbi equaliser(VE). This mapping process is also summarised in

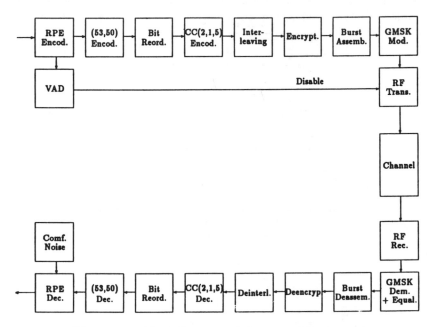

Figure 8.4: Blockdiagram of the TCH/FS channel

form of a hardware oriented block diagram in Figure 8.4, where the Voice Activity Detector (VAD) is included to enable or disable transmissions depending on whether speech is deemed to be present at the input of the RPE speech encoder. This allows namely a substantial reduction of the power consumption as well as that of the interferences imposed on other users. The effects of subjectively annoying silent periods are mitigated at the receiver by adding comfort noise during these intervals.

The 8.25 bit-interval duration guard space (GP) at the bottom of Figure 8.2 is provided to prevent burst overlapping due to delay fluctuations. Finally, a 26-bit equaliser training segment is included in the centre of the normal traffic burst. This segment is constructed by a 16-bit Viterbi channel equaliser training pattern surrounded by five quasi-periodically repeated bits on both sides. These bits provide a sufficiently long quasi-periodic extension of the training sequence before useful data is entered to the modulator, to keep the side-lobes of its autocorrelation function, as well as those of its spectra sufficiently low. The 16-bit pattern was determined by evaluating the autocorrelation of the modulated signals due to all 2^{16} candidate training sequences and selecting the one with the highest autocorrelation peak, while maintaining low main lobe to side-lobes ratio. For GMSK with BT=0.3 and modulation index of 0.5 several good sequences can be found. Since the MS has to be informed about which BS it communicates with, for neighbouring BSs different training patterns are used. Therefore, in the GSM system the eight best training patterns are used to be associated with eight different BS colour codes.

This 156.25 bit duration TCH/FS normal burst(NB) constitutes the basic timeslot of the TDMA frame structure, which is input to the GMSK modulator at a bitrate of approximately 271 kbit/s. Since the bit interval is 3.69 μs, the timeslot duration is $156.25 \cdot 3.69 \approx 0.577$ ms. Eight such normal bursts of eight appropriately staggered users are multiplexed onto one RF carrier giving a TDMA frame of $8 \cdot 0.577 \approx 4.615$ ms duration, as we see in Figure 8.2. The physical channel as characterised above provides a physical timeslot with an effective information throughput of 114 bits/4.615 ms=24.7 kbit/s, which is sufficiently high to transmit the 22.8 kbit/s TCH/FS. It even has a 'reserved' capacity of 24.7-22.8=1.9 kbit/s, which can be exploited to transmit slow control information associated with this specific traffic channel, i.e., to construct a so-called Slow Associated Control Channel (SACCH). The TCH/FS has a repetition delay of 20 ms and an interleaving delay of $8 \cdot 4.615 = 37$ ms, yielding a total speech delay of 20+37=57 ms.

To understand how we accommodate and access the SACCH we have to proceed with the construction of the TDMA frame hierarchy, as highlighted in Figure 8.2. The TCH/FS TDMA frames of the eight users are multiplexed into multiframes of 24 TDMA frames, but the 13th frame will carry a SACCH message, rather than the 13th TCH/FS frame, while the 26th frame will be an idle or dummy frame, as seen at the left hand side of Figure 8.2 representing the traffic channel hierarchy. The general control channel frame structure shown at the right of Figure 8.2 is discussed later. This way 24 TCH/FS frames are sent in a 26-frame multiframe during $26 \cdot 4.615 = 120$ ms. This reduces the traffic throughput to $\frac{24}{26} \cdot 24.7 = 22.8$ kbps required by TCH/FS, allocates $\frac{1}{26} \cdot 24.7 = 950$ bps to the SACCH and 'wastes' 950 bps in the idle frame. Observe that the SACCH frame has eight timeslots to transmit the eight 950 bps SACCHs of the eight users on the same carrier. The 950 bps idle capacity will be used in case of half rate channels, where 16 users will be multiplexed onto alternate frames of the TDMA structure to increase system capacity, when a half rate speech codec becomes available. Then sixteen 11.4 kbps encoded TCH/HSs will be transmitted in a 120 ms multiframe, where also sixteen SACCHs are available.

The construction of SACCH bursts is slightly different from TCH/FS bursts in that only 184 control bits are transmitted during 20 ms, in contrast to 260 speech bits, as portrayed in Figure 8.5. The additional channel capacity is exploited to accommodate a (224,184) external block code, extended by four zero tailing bits to reset the subsequent internal half rate, constraint length five convolutional codec. The total number of bits is now $(224+4) \cdot 2 = 456$ transmitted via four consecutive bursts, each carrying 114 bits. Each of these bursts are accommodated in a new 120 ms multiframe,

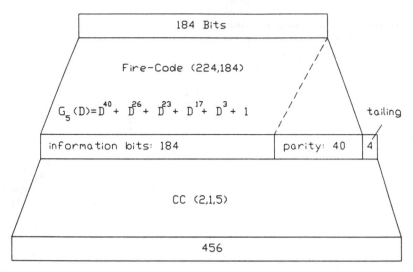

Figure 8.5: Mapping of SACCH, FACCH, BCCH, SDCCH, PCH and AGCH logical channels

yielding a repetition delay of $4 \cdot 120 = 480$ ms. The 456 error protected bits are transmitted at an average rate of 950 bps, which provides an unprotected information rate of $\frac{184}{456} \cdot 950 = 382$ bps for the SACCH, when associated with a traffic channel.

As opposed to independent SACCH frames, Fast Associated Control Channel (FACCH) messages are transmitted via the physical channels provided by bits 'stolen' from their own host traffic channels. The construction of the FACCH bursts from 184 control bits is identical to that of the SACCH, as also shown in Figure 8.5, but its 456-bit frame is mapped onto eight consecutive 114-bit TDMA traffic bursts, exactly, as specified for TCH/FS. This is carried out by stealing the even bits of the first four and the odd bits of the last four bursts, which is signalled by setting $hu = 1$, $hl = 0$ and $hu = 0$, $hl = 1$ in the first and last bursts, respectively. The unprotected FACCH information rate is 184 bits/20 ms=9.2 kbps, which is transmitted after concatenated error protection at a rate of 22.8 kbps. The repetition delay is 20 ms and the interleaving delay is $8 \cdot 4.615 = 37$ ms, resulting in a total of 57 ms delay.

In a subsequent stage of Figure 8.2 51 TCH/FS multiframes are amalgamated into one superframe lasting $51 \cdot 120 \; ms = 6.12$ s, which contains $26 \cdot 51 = 1326$ TDMA frames. There would be no need for any further levels of TDMA hierarchy, if it was not for the encryption, which uses the TDMA frame number (FN) as a parameter in its algorithm. However, with 1326 FNs only the encryption rule is not sufficiently secure. Therefore 2048 superframes are concatenated to form a hyperframe of $1326 \cdot 2048 = 2\,715\,648$ TDMA frames lasting $2048 \cdot 6.12$ s ≈ 3 h 28 min, using a satisfactorily high number of FNs in the encryption algorithm. This step now concludes our example of mapping the TCH/FS and its SACCH logical channel onto an appropriate physical channel constituted by a specific timeslot dedicated to a specific user of a specific RF channel carrying the messages of eight TDMA users. To reduce the complexity of MSs they do not have to receive and transmit simultaneously, their receive and transmit timeslots carrying the same TN are shifted by three in the TDMA frame with respect to each other.

8.3.2.2 Mapping Broadcast and Common Control Channels onto Physical Channels

In our TCH/FS and SACCH example in the previous subsection the RF channel was shared amongst 8 or 16 TDMA users and a specific timeslot was dedicated to one TCH/FS or two TCH/HS users. In contrast, the BCCH and CCCH logical channels of all MSs roaming in a specific cell share the physical channel provided by timeslot zero of the so-called BCCH carriers available in the cell. Furthermore, all BCCHs and CCCHs are simplex channels operating in uplink or downlink directions, as opposed to traffic channels, stand-alone dedicated control channels and fast or slow associated control channels, which are full duplex channels. Another difference is that in case of BCCHs and CCCHs 51 TDMA frames are mapped onto a $51 \cdot 4.615 = 235$ ms duration multiframe, rather than on a 26-frame, 120 ms duration multiframe. To compensate for the extended multiframe length of 235 ms, 26 multiframes constitute a 1326-frame superframe of 6.12 s duration, as demonstrated by Figure 8.2. Also the allocation of the uplink and downlink frames is different, since these control channels exist only in one direction, as seen in Figure 8.6.

Specifically, the random access channel (RACH) is only used by the MSs in uplink direction if they request, for example, a bidirectional stand-alone dedicated control channel (SDCCH) to be mapped onto an RF channel to register with the network and set up a call. The uplink RACH carries messages of eight bits per 235 ms multiframe, which is equivalent to an unprotected control information rate of 34 bps. These messages are concatenated FEC coded to a rate of 36 bits/235 ms=153 bps. They are not transmitted by the Normal Bursts (NB) derived for TCH/FS, SACCH or FACCH logical channels, but by the so-called Access Bursts (AB), depicted in Figure 8.7 in comparison to a NB and other types of bursts to be described when introduced. The tailing and synchronisation bits are given in Recommendation 05.02. The FEC coded, encrypted 36-bit messages, containing amongst other parameters also the encoded 6-bit BS identifier code (BSIC) constituted by the 3-bit PLMN colour code and 3-bit BS colour code for unique BS identification, are positioned after the 41-bit synchronisation sequence, which is extended to ensure reliable access burst recognition. These messages have no interleaving delay, while they are

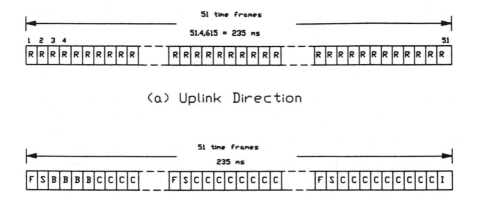

(a) Uplink Direction

(b) Downlink Direction

R: Random Access Channel

F: Frequency Correction Channel

S: Synchronisation Channel

B: Broadcast Control Channel

C: Access Grant/Paging Channel

I: Idle Frame

Figure 8.6: The control multiframe

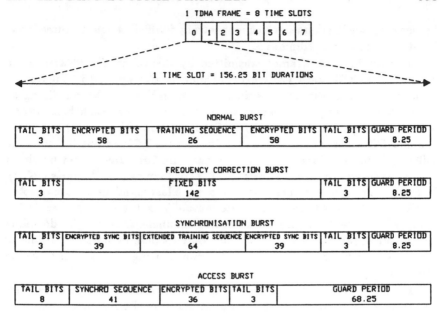

Figure 8.7: GSM burst structures

transmitted with a repetition delay of one control multiframe length, i.e. 235 ms.

The GSM system is specified to allow operation of mobile stations in cells when they are up to 35 km from their base station. The time a radio signal takes to travel the 70 km from the base station to the mobile station and back again is 233.3 μs. As signals from all the mobiles in the cell must reach the base station without overlapping each other, a guard period of 68.25 bits (252 μs) is provided in the access burst. This long guard period in the access burst is needed wₐen the mobile station attempts its first access to the base station, or after handover has occurred.

To provide the same long guard period in the other bursts would be spectrally inefficient. The GSM system overcomes this problem by using adaptive frame alignment. When the base station detects a 41-bit random access synchronisation sequence with a long guard period, it measures the received signal delay relative to the expected signal from a mobile station of zero range. This delay, called the timing advance, is signalled using a 6-bit number to the mobile station, which advances its timebase over the range of 0 to 63 bits, i.e., in units of 3.69 μs. By this process the TDMA bursts arrive at the BS in their correct timeslots and do not overlap with adjacent ones. This process allows the guard period in all other bursts to be reduced to $8.25 \cdot 3.69 \ \mu s \approx 30.46 \ \mu s$ (8.25 bits) only. In normal operation the BS continously monitors the signal delay from the MS and will instruct the MS to update its time advance parameter. In very large traffic cells there is an option to actively utilise every second timeslot only to cope with higher propagation delays, which is spectrally inefficient, but in these large, low-traffic rural cells admissible.

The downlink multiframe transmitted by the BS is shared amongst a number of BCCH and CCCH logical channels, as depicted in Figure 8.6. In particular, the last frame is an idle frame (I), while the remaining 50 frames are divided in five blocks of ten frames, where each block starts with a frequency correction channel (FCCH) followed by a synchronisation channel (SCH). In the first block of ten frames the FCH and SCH frames are followed by four broadcast control channel (BCCH) frames and by either four access grant control channels (AGCH) or four paging channels (PCH). In the remaining four blocks of ten frames the last eight frames are devoted to either PCHs or AGCHs, which are mutually exclusive for a specific MS being either paged or granted a control channel. Clearly, the downlink control multiframe hosts a total of four BCCHs, five FCCHs and SCHs, as well as $4 \cdot 8 + 4 = 36$ AGCHs or PCHs, constituting nine so-called paging blocks of four frames.

Each MS communicating via the downlink control multiframe is associated with one paging block of four AGCHs or PCHs out of the nine existing such blocks, although each block of four can be shared by several MSs. Therefore the bitrate of the physical channels provided for the concatenated error protected BCCH and AGCH or PCH logical channels is $4 \cdot 114 = 456$ bits per 235 ms, which is 1.94 kbps. The 456-bit error protected messages are derived from the 184-bit unprotected control messages exactly in the same way, as explained by referring to Figure 8.5 for the SACCHs and FACCHs in the previous subsection and have a repetition delay of 235 ms. The unprotected transmission rate available for BCCH, AGCH and PCH logical channels is $\frac{184}{456} \cdot 1.94$ kbps = 782 bps.

The FCCH uses the frequency correction burst (FCB) shown in Figure 8.7. The three tailing bits are used to reset the Viterbi equaliser's memory, while the 8.25 bit length guard space prevents eventual burst overlapping. The 142 fixed bits are chosen to yield a modulated signal which is equivalent to an unmodulated carrier with a fixed frequency offset above the nominal carrier frequency.

The SCH logical channel's control information is conveyed by the synchronisation burst (SB), seen also in Figure 8.7. The 25 synchronisation bits are concatenated FEC coded as described in Section 8.5 and encrypted to yield 78 protected bits, which are allocated both sides of the 64-bit extended training sequence specified in Recommendation 05.02. The role of the remaining bits is identical to those in other bursts.

There is a fifth burst type not shown in Figure 8.7, called Dummy Burst (DB), which has an identical structure to that of a NB, except for the fact that it carries no useful data. The 116 fixed encrypted bits have an equal probability of logical ones and zeros and are transmitted when no useful data is available, but the link has to be maintained to monitor the powers of the BCCH carriers of adjacent cells by the MSs.

It is important to note that a physical channel is specified by defining both the RF carrier and a TDMA timeslot number (TN). This is equivalent to saying that an RF channel carries eight physical channels. We emphasize that different timeslots of a specific RF carrier can carry both traffic and control logical channels, and therefore timeslots of the same RF channel can be assigned to different multiframe structures. It is plausible that multiframe types can only be altered at superframe boundaries, i.e., every 6.12 s, but with this proviso they are allowed to change arbitrarily in hyperframes. If the physical channels have to accommodate heavy control logical channel traffic, the allocation of timeslot zero only on the BCCH RF carrier would result in excessive access delays. In this case timeslots two, four and six can also be assigned to BCCH/CCCH logical channels. The 51-frame control multiframe structure is then slightly modified to transmit dummy bursts (DB) in the SCH and FCCH to maintain carrier transmissions, the power of which has to be monitored by the MSs. If there is only very little control information to be sent, timeslot zero of the BCCH carrier can be shared by BCCH/CCCH as well as by four SDCCH logical channels or by eight SDCCHs.

8.3.2.3 Broadcast Control Channel Messages

The BCCH transmits a variety of parameters to the MSs, of which those used to determine the legitimate combinations of control logical channels per physical channel are to be highlighted here. The number of basic physical channels supporting CCCHs is transmitted in the form of the 2-bit parameter BS_CC_CHANS. All CCCHs have to use timeslots on the BCCH carrier often referred to as C0 in GSM jargon. The first CCCH uses timeslot 0, the second CCCH will utilise timeslot 2 of C0 and for high control traffic demands timeslots 4 and 6 of C0 are allocated to further CCCHs. This means that $1 \leq BS\_CC\_CHANS \leq 4$.

A 1-bit flag indicating whether the CCCHs are combined with four SDCCHs and four SACCHs onto the same physical channel called BS_CCCH_SDCCH_COMB is also broadcast on C0. Namely, if they are combined, the number of 'available' AGCH/PCH paging blocks must be reduced from nine. The AGCH and PCH share the physical channel on a 'block of ten frames' basis, which enables the MS to determine after deinterleaving and FEC decoding whether the block contains an AGCH or PCH message. As mentioned earlier, these two messages are mutually exclusive, since the MS is either being granted a channel or paged. The physical channel is assigned to AGCH or PCH messages on a demand basis, where PCH messages have higher priority than AGCH messages, since the former already represent partially established calls occupying parts of the system. However, to prevent long call set up delays, a number of the available blocks in each 51-frame multiframe can be reserved for AGCHs. The number of such reserved blocks (BS_AG_BLKS_RES) is also encoded using three bits and broadcast via the BCCH, while the number of 'available' PCH blocks has to be reduced by BS_AG_BLKS_RES.

Another parameter broadcast in the BCCH is the number of 51-frame multiframes between transmissions of paging messages to MSs of the same paging group, which is abbreviated as BS_PA_MFRMS. Since $2 \leq BS\_PA\_MFRMS \leq 9$, it is encoded by three bits. The total number of paging blocks 'available' on a CCCH logical channel is denoted by N and is computed as the product of 'available' blocks (ABL) in a multiframe and the number of multiframes between paging MSs of the same paging group, i.e.: N=ABL · BS_PA_MFRMS. Once N is known, the MSs are only required to monitor every N^{th} paging block of their CCCH. All MSs listening to a particular paging block out of the available total of N belong to a specific paging group (PAGING_GROUP). Hence, there are N paging groups and the MS computes which one it belongs to from the International Mobile Subscriber Identity (IMSI), N and from the assigned number of CCCH timeslots ($1 \leq BS\_CC\_CHANS \leq 4$) as given below:

$$PAGING\_GROUP(0 \ldots N - 1) = ((IMSI \bmod 1000) \bmod (BS\_CC\_CHANS \cdot N)) \bmod N. \quad (8.1)$$

The knowledge of the paging group is only required to pinpoint, which paging block of which 51-frame multiframe has to be monitored by any MS. This determines for the MS, when it is supposed to be in active and dormant mode, thereby contributing towards the power-efficient MS operation. The required multiframe is encountered when:

$$PAGING\_GROUP \operatorname{div} (N \operatorname{div} BS\_PA\_MFRMS) = ((FN \operatorname{div} 51) \bmod (BS\_PA\_MFRMS)) \quad (8.2)$$

where *div* represents integer division, *mod* is short for modulo and the paging block index (PBI) to be monitored is given by:

$$PBI = (PAGING\_GROUP \bmod (N \operatorname{div} BS\_PA\_MFRMS). \quad (8.3)$$

8.3.3   Carrier and Burst Synchronisation

The GSM Recommendations do not specify the BS-MS synchronisation algorithms to be used, these are left to the equipment manufacturers. However, a unique set of timebase counters is defined to ensure perfect BS-MS synchronism. The BS sends frequency correction(FCB) and synchronisation bursts(SB) on specific timeslots of the BCCH carrier to the MS to ensure that the MS's frequency standard is perfectly aligned with that of the BS, as well as to inform the MS about the required initial state of its internal counters. The MS sends its uniquely numbered traffic and control bursts staggered by three timeslots with respect to those of the BS to prevent simultaneous MS transmission and reception, and also takes into account the required timing advance (TA) to cater for different BS-MS-BS round-trip delays.

The timebase counters used to uniquely describe the internal timing states of BSs and MSs are the Quarter bit Number (QN=0...624) counting

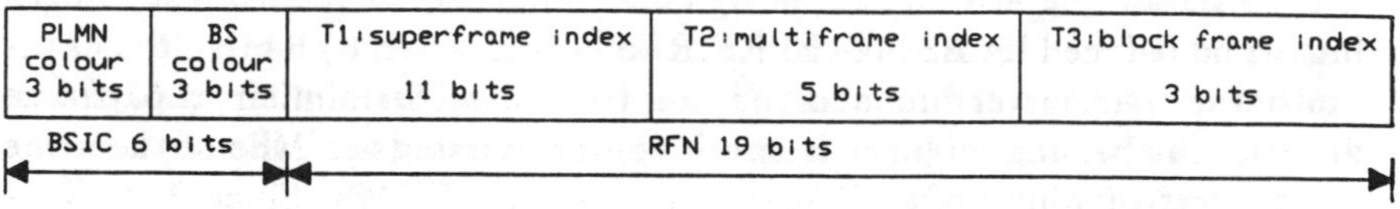

PLMN colour 3 bits	BS colour 3 bits	T1:superframe index 11 bits	T2:multiframe index 5 bits	T3:block frame index 3 bits

BSIC 6 bits   RFN 19 bits

Figure 8.8: Synchronisation channel(SCH) message format

the quarter bit intervals in bursts, Bit Number (BN=0...156), Timeslot Number (TN=0...7) and TDMA Frame Number (FN=0...26 · 51 · 2048), given in the order of increasing interval duration. The MS sets up its timebase counters after receiving a SB by determining QN from the 64-bit extended training sequence in the centre of the SB, setting TN=0 and decoding the 78 encrypted, protected bits carrying the 25 SCH control bits. The SCH carries frame synchronisation information as well as BS identification information to the MS, as seen in Figure 8.8, and it is provided solely to support the operation of the radio subsystem. The first six bits of the 25-bit segment consist of three PLMN colour code bits and three BS colour code bits supplying a unique BS Identifier Code (BSIC) to inform the MS, which BS it is communicating with. The second 19-bit segment is the so-called Reduced TDMA Frame Number (RFN) derived from the full TDMA Frame Number (FN), constrained to the range of $[0...(26 · 51 · 2048) − 1] = [0...2, 715, 647]$ in terms of three subsegments T1, T2 and T3. These subsegments are computed as:

$$T1(11bits) = (FN \ div \ (26 · 51)) \tag{8.4}$$

$$T2(5bits) = (FN \ mod \ 26) \tag{8.5}$$

$$T3'(3bits) = ((T3 − 1) \ div \ 10), \ where \ T3 = (FN \ mod \ 5). \tag{8.6}$$

Here T1 determines the superframe index in a hyperframe, T2 the multiframe index in a superframe, T3 the frame index in a multiframe, while T3' the block index of a frame in a specific control multiframe. Their role is best understood by referring to Figure 8.2. Once the MS has received the Synchronisation Burst (SB), readily computes the FN required in various control algorithms, such as encryption, handover, etc., as shown below:

$$FN = 51 · ((T3 − T2) \ mod \ 26) + T3 + 51 · 26 · T1, \ where \ T3 = 10 · T3' + 1. \tag{8.7}$$

It is desirable to have perfect synchronism of all the channels under the control of a BS, and hence all its RF carrier frequencies and timebase counter frequencies are derived from the same reference frequency. It is possible but not mandatory to synchronise different BSs together. When the BS detects a RACH message with its 41-bit synchronisation sequence, its unique BSIC and 68.25-bit guard period, it will notice that the MS is asking for random access to it. Using the 41-bit synchronisation sequence in this decoded AB the BS can now evaluate the propagation delay, which will be the timing advance to be signalled, rounded to the nearest integer bit period, to the MS. As the timing advance is encoded by 6 bits, it is hard-limited to 64·3.69=236 μs and it is kept constant for higher propagation delays. This timing advance is continuously updated and the adjustment error is less than half of a bit period.

8.3.4   Frequency Hopping

Frequency hopping combined with interleaving is known to be very efficient in combating channel fading, and it results in near-Gaussian performance even over hostile Rayleigh-fading channels. The principle of Frequency Hopping (FH) is that each TDMA burst is transmitted via a different RF CHannel (RFCH). If the present TDMA burst happened to be in a deep fade, then the next burst most probably will not be, as long as hopping is carried out to a frequency sufficiently different from the present one, having a differently fading envelope. However, this is not easily ensured due to the limited bandwidth available for GSM, since, for example, uplink transmissions are carried out in the 890-915 MHz band, where the maximum relative hopping frequency is $\sim$25 MHz/900 MHz$\approx$2.8%. Nevertheless, FH reduces the amount of time spent by the MS in a fade to 4.615 ms, the duration of a TDMA burst, which brings substantial gains in case of slowly moving MSs, such as pedestrians. The GSM frequency hopping algorithm is shown in Figure 8.9. The algorithm's input parameters include the TDMA Frame Number (FN) specified in terms of the indices FN(T1), FN(T2) and FN(T3), as received in the synchronisation burst (SB) via the Synchronisation Channel (SCH). A further parameter is the set of RF channels called mobile allocation(MA) assigned for use in the MS hopping sequence, which is limited to 1$\leq$N$\leq$64 channels out of the legitimate 124 GSM channels. The Mobile Allocation Index Offset (MAIO) determines the minimum value of the Mobile Allocation Index (MAI), which is the output variable of the FH algorithm determining the next RF channel to which frequency hopping is required. Lastly, the Hopping Sequence generator Number 0$\leq$HSN$\leq$63 is a further control parameter, which results in cyclic hopping if HSN=0, as seen in Figure 8.9, and in pseudo-random hopping patterns if $1 \leq HSN \leq 63$. This is, because for HSN=0 the mobile allocation index is computed as:

$$MAI = ((FN + MAIO) \; mod \; N), \qquad (8.8)$$

where (mod N) is taken to ensure that MAI remains an element of the set MA.

For $1 \leq HSN \leq 63$ somewhat more complex operations have to be computed, using a number of intermediate internal variables, as demonstrated by Figure 8.9. The only undefined variable in the figure is NB, representing the number of bits required for the binary encoding of N, the number of RF channels in the set MA. The function RNTABLE simply assigns

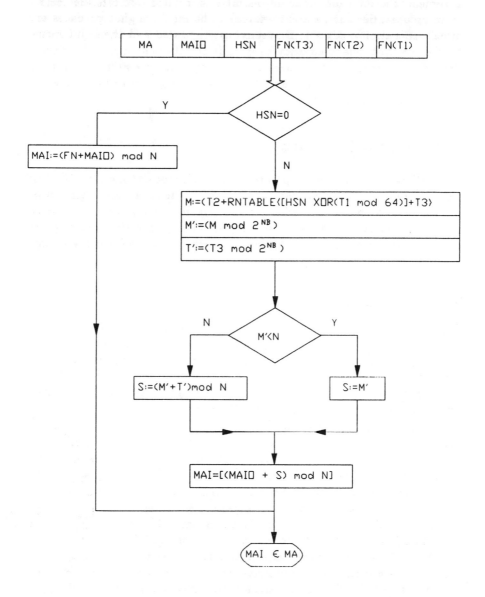

Figure 8.9: The GSM frequency hopping algorithm

one out of 114 pseudo-random numbers specified by GSM according to its argument, the XOR operator means bit-wise exclusive OR, while the remaining operations are self-explanatory. The result of the process is the mobile allocation index (MAI) specifying the next RF channel to be used by the MS. Note that frequency hopping is not allowed on timeslot zero of the BCCH carrier, which is ensured by using a single RFCH, i.e., setting N=1 and MAIO=0. In this case the FH sequence generation is uneffected by the value of HSN.

8.4   Speech Coding

The selection of the most appropriate speech codec for the GSM system from the set of candidate codecs was based on extensive comparative tests among various operating conditions. The rigorous comparisons published in [7] are interesting and offer deep insights for system designers as regards to the pertinent trade-offs in terms of speech quality, robustness against channel errors, complexity, system delay, etc.

8.4.1   Candidate codecs

Originally the participating countries have proposed six different codecs with an overall channel coding and speech coding rate of $16kbps$ for comparison. At a preliminary test the codecs were compared to the presently used companded FM system, and then two of the codecs were withdrawn. The remaining codecs were two different sub-band codecs and two pulse-excited codecs, which are detailed in Chapter 3.

SBC-APCM: Subband codec with block adaptive PCM. This codec used quadrature mirror filters (QMF) to split the input signal into 16 subbands of 250 Hz bandwidth, out of which the two highest bands were not transmitted. Adaptive bit allocation was used in the subbands on the basis of the power ratios of the various subbands, which constituted the side-information to be transmitted. The gross transmission rate of the subband signals was 10 kbit/s, the side-information was 3 kbit/s, which was protected by 3 kbit/s forward error correction coding (FEC) redundancy.

SBC-ADPCM: Subband codec with adaptive delta PCM. In this scheme the speech input signal was split into 8 subbands, out of which only 6 were transmitted. The subband signals were encoded by differential coding with backward estimation and adaptation, as opposed to the SBC-APCM candidate, where forward estimation and adaptation were used. The bit allocation of the subbands was fixed, hence no side-information was transmitted, which made the scheme more noise resilient, hence no FEC protection was required, and the bitrate was 15 kbit/s only.

MPE-LTP: Multi-pulse excited LPC codec with long term predictor. The particular speech codec implementation used in the comparisons required 13.2 kbit/s transmission bitrate, and 2.8 kbit/s embedded FEC coding was deployed to protect the most important bits of the speech codec.

RPE-LTP: Regular pulse excited LPC codec. A thorough theoretical analysis of this method was given in Chapter 3 and the forthcoming subsection was devoted to implementational details of this scheme, since it was selected for standardisation on the basis of the overall comparison tests.

These four codecs were compared in terms of speech quality, robustness, processing delays and computational complexity. From the experience with the companded FM reference system two benchmarker bit error rates (BER) were supposed, at which performance comparisons were carried out. The pessimistic case was a BER of $10^{-2} = 1\%$, which would require a carrier to noise ratio (CNR) of approx. 18 dB, exceeded probably in 90% of the reception area in the FM system. The optimistic channel exhibits a BER of $10^{-3} = 0.1\%$, requiring approx. CNR=26 dB, guaranteed for at least 50% of the coverage area for the reference FM system. The average mean opinion scores (MOS) on a five point scale over the various test conditions were found to be [7]: $MOS(RPE - LPC) = 3.54$, $MOS(MPE - LTP) = 3.27$, $MOS(SBA - APCM) = 3.14$, $MOS(SBC - ADPCM) = 2.92$, $MOS(FM) = 1.95$. These results have emphasized the superiority of the pulse-excited codecs and the importance of the long-term predictor (LTP). The RPE codec exhibiting the most favourable properties was further improved by deploying a LTP, and the RPE-LTP codec guarantees an $MOS \approx 4.0$ over a wide range of operating conditions.

8.4.2 The RPE-LTP Speech encoder

The schematic diagram of the RPE-LTP encoder is shown in Figure 8.10, where the following functional parts can be recognised [9], [10] and [11]: 1. Pre-processing, 2. STP analysis filtering, 3. LTP analysis filtering, 4.RPE computation.

1. Pre-processing: Pre-emphasis can be deployed to increase the numerical precision in computations by emphasizing the high-frequency, low-power part of the speech spectrum. This can be carried out by the help of a one-pole filter with the transfer function of:

$$H(z) = 1 - c_1 z^{-1}, \tag{8.9}$$

where $c_1 \approx 0.9$ is a practical value. The pre-emphasized speech $s_p(n)$ is segmented into blocks of 160 samples in a buffer, where they are windowed by a Hamming-window to counteract the spectral domain Gibbs oscillation, caused by truncating the speech signal outside the analysis frame. The Hamming-window has a tapering effect towards the edges of a block, while it has no influence in its middle ranges:

$$s_{psw}(n) = s_{ps}(n) \cdot c_2 \cdot (0.54 - 0.46 \cos 2\pi \frac{n}{L}). \tag{8.10}$$

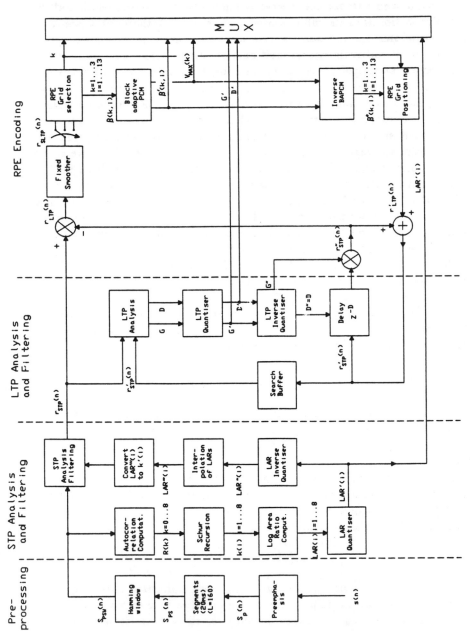

Figure 8.10: Blockdiagram of the RPE-LTP Encoder

where $s_{ps}(n)$ represents the pre-emphasized, segmented speech, $s_{psw}(n)$ is its windowed version and the constant $c_2 = 1.5863$ is determined from the condition that the windowed speech must have the same power as the non-windowed.

2. STP analysis filtering: For each segment of L=160 samples nine autocorrelation coefficients R(k) are computed from $s_{psw}(n)$ by:

$$R(k) = \sum_{n=0}^{L-1-k} s_{psw}(k)s_{psw}(n+k) \qquad k = 0\ldots 8. \qquad (8.11)$$

From the speech autocorrelation coefficients R(k) eight reflection coefficients k_i are computed according to the Schur-recursion [12], which is an equivalent method to the Durbin algorithm used for solving the LPC key-equations to derive the reflection coefficients k_i, as well as the STP filter coefficients a_i. However, the Schur-recursion delivers the reflection coefficients k_i only. The reflection coefficients k_i are converted to logarithmic area ratios (LAR(i)), because the logaritmically companded LARs have better quantisation properties than the coefficients k_i:

$$LAR(i) = \log_{10}(\frac{1+k(i)}{1-k(i)}), \qquad (8.12)$$

where a piecewise linear approximation with five segments is used to simplify the real-time implementation:

$$LAR'(i) = \begin{cases} k(i), & if \quad |k(i)| < 0.675 \\ sign\ [k(i)][2|k(i)| - 0.675], & if \quad 0.675 < |k(i)| < 0.95 \\ sign\ [k(i)][8|k(i)| - 6.375], & if \quad 0.975 < |k(i)| < 1.0 \end{cases}$$
$$(8.13)$$

The various LAR(i) $i = 1\ldots 8$ filter parameters have different dynamic ranges and differently shaped probability density functions (PDFs) as seen in Figure 3.6. This justifies the allocation of 6, 5, 4 and 3 bits to the first, second, third and fourth pairs of LARs, respectively. The quantised LAR(i) coefficients LAR'(i) are locally decoded into the set LAR''(i), as well as transmitted to the speech decoder. So as to mitigate the abrupt changes in the nature of the speech signal envelope around the STP analysis frame edges, the LAR parameters are linearly interpolated, and towards the edges of an analysis frame the interpolated LAR'''(i) parameters are used. Now the locally decoded reflection coefficients k'(i) are computed by converting LAR'''(i) back into k'(i), which are used to compute the STP residual $r_{STP}(n)$ in a socalled PARCOR (partial correlation) structure. The PARCOR scheme directly uses the reflection coefficients k(i) to compute the STP residual $r_{STP}(n)$, and it constitutes the natural analogy to the acoustic tube model of the human speech production.

3. LTP analysis filtering: As we have seen in Chapter 3, the LTP prediction error is minimised by that LTP delay D, which maximises the crosscorrelation between the current STP residual $r_{STP}(n)$ and its previously received and buffered history at delay D, i.e., $r_{STP}(n - D)$. To be more specific, the L=160 samples long STP residual $r_{STP}(n)$ is divided into four N=40 samples long subsegments, and for each of them one LTP is determined by computing the crosscorrelation between the presently processed subsegment and a continuously sliding N=40 samples long segment of the previously received 128 samples long STP residual segment $r_{STP}(n)$. The maximum of the correlation is found at a delay D, where the currently processed subsegment is the most similar to its previous history. This is most probably true at the pitch periodicity or at a multiple of the pitch periodicity. Hence the most redundancy can be extracted from the STP residual, if this highly correlated segment is substracted from it, multiplied by a gain factor G, which is the normalised crosscorrelation found at delay D. Once the LTP filter parameters G and D have been found, they are quantised to give G' and D', where G is quantised only by two bits, while to quantise D' seven bits are sufficient.

The quantised LTP parameters (G',D') are locally decoded into the pair (G",D") so as to produce the locally decoded STP residual $r'_{STP}(n)$ for use in the forthcoming subsegments to provide the previous history of the STP residual for the search buffer, as shown in Figure 8.10. Observe that since D is integer, we have D=D'=D". With the LTP parameters just computed the LTP residual $r_{LTP}(n)$ is calculated as the difference of the STP residual $r_{STP}(n)$ and its estimate $r''_{STP}(n)$, which has been computed by the help of the locally decoded LTP parameters (G",D) as shown below:

$$r_{LTP}(n) \quad = \quad r_{STP}(n) - r''_{STP}(n) \qquad (8.14)$$

$$r''_{STP}(n) \quad = \quad G''r'_{STP}(n - D). \qquad (8.15)$$

Here $r'_{STP}(n - D)$ represents an already known segment of the past history of $r'_{STP}(n)$, stored in the search buffer. Finally, the content of the search buffer is updated by using the locally decoded LTP residual $r'_{LTP}(n)$ and the estimated STP residual $r''_{STP}(n)$ to form $r'_{STP}(n)$, as shown below:

$$r'_{STP}(n) = r'_{LTP}(n) + r''_{STP}(n). \qquad (8.16)$$

4. RPE computation: The LTP residual $r_{LTP}(n)$ is weighted with the fixed smoother, which is essentially a gracefully decaying band limiting low-pass filter with a cut-off frequency of 4 kHz/3=1.33 kHz according to a decimation by three about to be deployed, as detailed in Chapter 3. The impulse response of this filter is also given in Chapter 3. The smoothed LTP residual $r_{SLTP}(n)$ is decomposed into three excitation candidates, by actually discarding the 40th sample of each subsegment, since the three candidate sequences can host 39 samples only. Then the energies E1, E2, E3 of the three decimated sequences are computed, and the candidate with the highest energy is chosen to be the best representation of the LTP residual. The excitation pulses are afterwards normalised to the highest amplitude

$v_{max}(k)$ in the sequence of the 13 samples, and they are quantised by a three bit uniform quantiser, whereas the logarithm of the block maximum $v_{max}(k)$ is quantised with six bits. According to three possible initial grid positions k, two bits are needed to encode the initial offset of the grid for each subsegment. The pulse amplitudes $\beta(k,i)$, the grid positions k and the block maxima $v_{max}(k)$ are locally decoded to give the LTP residual $r'_{LTP}(n)$, where the 'missing pulses' in the sequence are filled with zeros.

8.4.3 The RPE-LTP Speech Decoder

The block diagram of the RPE-LTP decoder is shown in Figure 8.11, which exhibits an inverse structure, consituted by the functional parts of: 1. RPE decoding, 2. LTP synthesis filtering, 3. STP synthesis filtering, 4. Post-processing.

1. RPE decoding: In the decoder the grid position k, the subsegment excitation maxima $v_{max}(k)$ and the excitation pulse amplitudes $\beta'(k,i)$ are inverse quantised, and the actual pulse amplitudes are computed by multiplying the decoded amplitudes with their corresponding block maxima. The LTP residual model $r'_{LTP}(n)$ is recovered by properly positioning the pulse amplitudes $\beta(k,i)$ according to the initial offset k.

2. LTP synthesis filtering: Firstly the LTP filter parameters (G',D') are inverse quantised to derive the LTP synthesis filter. Then the recovered LTP excitation model $r'_{LTP}(n)$ is used to excite this LTP synthesis filter (G',D') to recover a new subsegment of length N=40 of the estimated STP residual $r'_{STP}(n)$. To do so, the past history of the recovered STP residual $r'_{STP}(n)$ is used, properly delayed by D' samples and multiplied by G' to deliver the estimated STP residual $r''_{STP}(n)$, according to:

$$r''_{STP}(n) = G'.r'_{STP}(n - D'), \qquad (8.17)$$

and then $r''_{STP}(n)$ is used to compute the most recent subsegment of the recovered STP residual, as given below:

$$r'_{STP}(n) = r''_{STP}(n) + r'_{LTP}(n). \qquad (8.18)$$

3. STP synthesis filtering: To compute the synthesized speech $\hat{s}(n)$ the PARCOR synthesis is used, where similarly to the STP analysis filtering the reflection coefficients k(i) i=1...8 are required. The LAR'(i) parameters are decoded by using the LAR inverse quantiser to give LAR"(i), which are again linearly interpolated towards the analysis frame edges between parameters of the adjacent frames to prevent abrupt changes in the character of the speech spectral envelope. Finally, the interpolated parameter set is transformed back into reflection coefficients, where filter stability is guaranteed, if recovered reflection coefficients, which fell outside the unit circle are reflected back into it, by taking their reciprocal values. The inverse

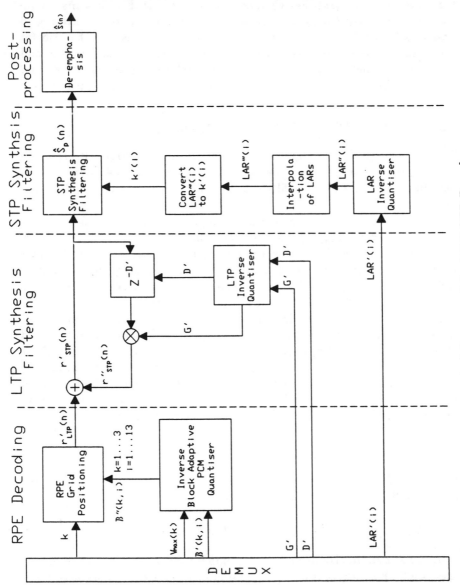

Figure 8.11: Blockdiagram of the RPE-LTP Decoder

Parameter to be encoded	No. of bits
8 STP LAR coefficients	36
4 LTP Gains G	4 x 2 = 8
4 LTP Delays D	4 x 7 = 28
4 RPE Grid-positions	4 x 2 = 8
4 RPE Block maxima	4 x 6 = 24
4x13=52 Pulse amplitudes	52 x 3 = 156
Total number of bits per 20 ms	260
Transmission bitrate	13 kbit/s

Table 8.3: Summary of the RPE-LTP bit-allocation scheme

formula to convert LAR(i) back into k(i) is given by:

$$k(i) = \frac{10^{LAR(i)} - 1}{10^{LAR(i)} + 1}. \tag{8.19}$$

4. Post-processing: The post-processing is constituted by the de--emphasis, using the inverse of the filter $H(z)$ in Equation 8.9.

The summarised RPE-LTP bit allocation scheme is tabulated in Table 8.3 for a period of 20 ms, which is equivalent to the encoding of L=160 samples, while the detailed bit-by-bit allocation is given in [9].

The 260 bits derived have to be reordered according to their subjective importances before error correction coding, as proposed by GSM, and classified into categories of Class 1a, Class 1b and Class 2 in descending order of prominence to facilitate a three-level error protection scheme, as it will be highlighted in Section 8.5. Note that this sensitivity order is based on subjective tests. Objective bit-sensitivity analysis based on a combination of segmental signal-to-noise ratios and cepstrum distances results in a similar significance order [13].

8.5 Channel Coding and Interleaving

The TDMA frame-hierarchy used in the GSM system has been highlighted by the help of Figure 8.2 in Section 8.3, where we have seen how the timeslots are multiplexed into TDMA frames, multiframes, superframes and hyperframes. The major unknown is now, how speech and data bits are mapped onto timeslots of the TDMA frames. First we focus our attention on the embedded forward error correction (FEC) and interleaving scheme used for speech transmissions.

8.5.1 FEC for Speech Channels

As explained in Section 8.4, the RPE-LTP codec delivers 260 bits/20 ms at a bitrate of 13 kbit/s according to the bit-allocation summarised in

Table 8.3. The detailed mapping of the sensitivity-ordered Class 1a (C1a), Class 1b (C1b) and Class 2 (C2) speech bits onto a normal burst(NB) is given in Figures 8.12 and 8.13.

As seen in Figure 8.12b [14], the first 50 C1 bits $(d_0 \ldots d_{49})$ are protected by a weak error detecting block code. The code implemented is a (53,50) shortened, systematic, cyclic code with the generator polynomial $G_4(D) = D^3 + D + 1$. The corresponding encoder is displayed in Figure 8.14. Observe that the taps of the linear shift register encoder are allocated at the positions specified by the generator polynomial $G_4(D)$. Since a systematic encoding rule has been adopted, the switch SW is closed for the duration of the first fifty clock pulses, and the information bits enter the encoder, as well as are passed on to the "Reordering & Tail" block in Figure 8.4. After fifty clock pulses the switch SW is opened and the parity bits P_0, P_1 and P_2 exit the encoder, while the rest of the bits, i.e., the C1b and C2 bits are unaltered.

At this stage a first interleaving step depicted in Figure 8.12c is carried out. Namely, the C1a and C1b bits with even indices, i.e., $d_0, d_2 \ldots d_{180}$, are collected in the first part of the data word, followed by the three parity bits P_0, P_1 and P_2. Then the C1a and C1b bits with odd indices, i.e., $d_1, d_3 \ldots d_{179}$, are stored in the buffer, followed by the 78 uncoded C2 bits.

Observe also that four zero tailing bits have been added at the end of the C1 section, which are necessary to reset the constraint-length K=5, rate=1/2 convolutional encoder CC(2,1,5) to be deployed to protect the C1 bits. Now the 189 C1 bits $(u_0, u_1 \ldots u_{188})$ are encoded by the powerful half-rate convolutional code in the block designated by CC(2,1,5) in Figure 8.4. The CC(2,1,5) codec uses the generator polynomials $G_0 = 1 + D^3 + D^4$ and $G_1 = 1 + D + D^3 + D^4$ and the encoder structure displayed in Figure 8.15 in the full-rate speech channels. The total frame length amounts to $2 \cdot 189 + 78 = 456$ bits, as demonstrated by Figure 8.12d.

Thereafter the 456 bits long encoded frame is partitioned into eight 57 bits long sub-blocks $(B_0 \ldots B_7)$, as shown in Figure 8.13a. The interleaving seen in Figure 8.4 is carried out by assigning the coded bits $c(n, k)$ into the interleaved sequence i(B,j) as stated below:

$$i(B, j) = c(n, k), \qquad (8.20)$$

where $k = 0, 1 \ldots 455$ is the bit index in the nth coded frame, displayed in Figure 8.12d, $n = 0, 1, 2 \ldots$ is the frame index, $B = b + 4n + [k \bmod 8]$ is the subblock index with an initial value of b and

$$j = \begin{cases} 2[(49k) \bmod (57)] & \text{if } [k \bmod 8] \leq 3 \\ 2[(49k) \bmod (57)] + 1 & \text{if } [k \bmod 8] > 3 \end{cases} \qquad (8.21)$$

is the bit index in the interleaved 57-bit subblocks. Each of the $N = 57$ bits long subblocks disperses its bits over eight consecutive such subblocks. Each of the 57 bits being spread will be followed or preceeded by the

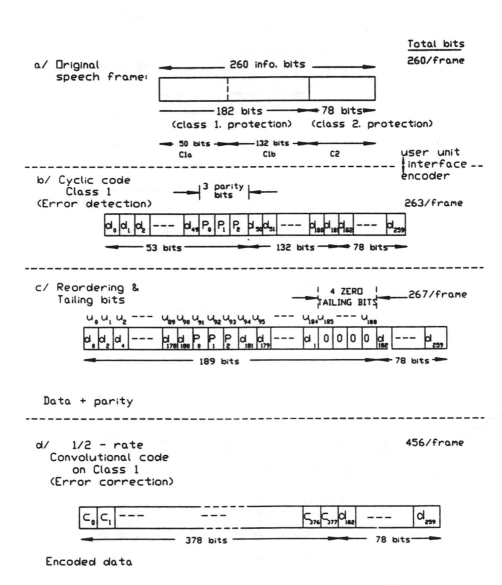

Figure 8.12: Forward error correction coding in TCH/FS

a/ Partitioning

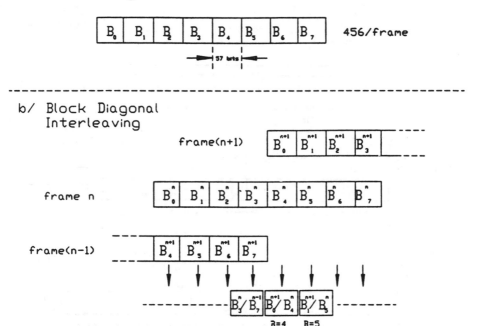

b/ Block Diagonal
   Interleaving

c/ Inter-burst Interleaving

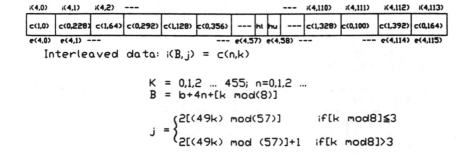

Interleaved data: $i(B,j) = c(n,k)$

$$K = 0,1,2 \ldots 455;\ n=0,1,2 \ldots$$
$$B = b+4n+[k \bmod(8)]$$

$$j = \begin{cases} 2[(49k) \bmod(57)] & \text{if}\,[k \bmod 8] \leq 3 \\ 2[(49k) \bmod (57)]+1 & \text{if}\,[k \bmod 8] > 3 \end{cases}$$

Figure 8.13: Partitioning and interleaving in TCH/FS

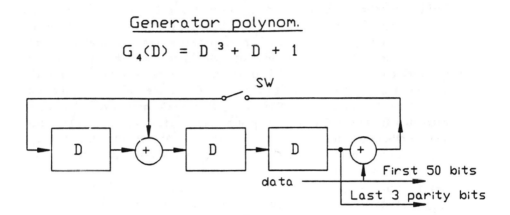

Generator polynom.

$$G_4(D) = D^3 + D + 1$$

Operation: 1 ... 50 CKL: SW closed
51 ... 53 CKL: SW open

Figure 8.14: The C1a (53,50) systematic, cyclic block encoder

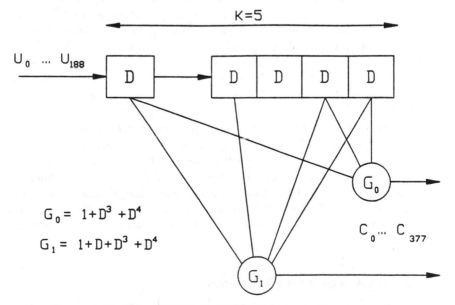

$$G_0 = 1 + D^3 + D^4$$
$$G_1 = 1 + D + D^3 + D^4$$

Figure 8.15: The CC(2,1,5) TCH/FS convolutional encoder

'position-equivalent' bit of the subblock half-a-frame, i.e., four 57-bit sub-blocks apart. This is represented by the block diagonal interleaving shown in Figure 8.13b and facilitated by the shift $4n$ in $B = b+4n+[k \bmod 8]$. Two originally adjacent $c(n, k)$ bits will be reordered so that their separation in the reordered frame becomes $2[(49k) \bmod (57)]$ or $2[(49k) \bmod (57)] + 1$ depending on the value of $[k \bmod 8]$. Since 49 and 57 are relative primes, this separation distance becomes pseudo-random, which is advantageous in terms of combating periodic error bursts. Observe also that for fixed b and n values B has a periodicity of eight in terms of k, which means that every eighth bit of $c(n, k)$ is directed to the same subblock B with a pseudo-random offset. Due to the term $4n$ in the definition equation $B = b + 4n + [k \bmod 8]$ four 114 bits long inter-block interleaved blocks are constructed from a 456 bits long speech frame. However, the 114-bit interleaved blocks are derived from two consecutive 456-bit speech frames by using additional block-diagonal interleaving, as displayed in Figure 8.13b and explained as follows.

Let us assume that frames $n = 0$ and $n = 1$ are being interleaved and $b = 0$. Then, for example, $c(0, 4), c(0, 12), c(0, 20), \ldots c(0, 452)$ will be mapped onto $i(4, 51), i(4, 37), i(4, 23), \ldots i(4, 65)$, respectively. Observe that even bits of block $n = 0$ are mapped onto odd positions of the fourth 114 bits long interleaved burst. It is also easily seen that $i(4, 1) = c(0, 228), i(4, 3) = c(0, 228+64), i(4, 5) = c(0, 228+128)$, etc. On the other hand, the even bits of the fourth interleaved burst $(B = 4)$ are supplied by the even bits of the speech block $n = 1$ as follows: $i(4, 0) = c(1, 0), i(4, 2) = c(1, 64), i(4, 4) = c(1, 128)$, etc. In other words, bits originally 228 positions apart follow each other in the interleaved bursts, as evidenced by Figure 8.13c, where the final step of including the stealing flags $e(B, 57) = hl(B)$ and $e(B, 58) = hu(B)$ is portrayed as well. These bits are used in fast associated control channel (FACCH) signalling. Specifically, $hl = 0$, $hu = 0$ indicates that the current frame carries traffic channel (TCH) information, $hu = 1$ means that every even-numbered bit is stolen by the FACCH for signalling information, while $hl = 1$ represents that the odd bits are signalling information.

The inner working of the reordering and interleaving process is best understood by generating three consecutive coded frames and programming the interleaving formula in a spreadsheet. The 116-bit bursts are then encrypted, split into two 58 bits long sequences and amalgamated in a normal burst, as shown in Figure 8.2 to create a $577\mu s$, i.e., 156.25-bit duration long TDMA timeslot, which is transmitted at a burstrate of approximately 271 kHz.

8.5.2 FEC for Data Channels

As mentioned earlier, the full- and half-rate data traffic channels standardised in the GSM system are: $TCH/F9.6$, $TCH/F4.8$, $TCH/F2.4$, as well as $TCH/H4.8$, $TCH/H2.4$. As a representative example, here we only consider the format of the $TCH/F9.6$.

The user unit's interface defined in R.04.21 delivers blocks of 60 bits every $5ms$ in accordance with the modified CCITT V.110 standard derived from the so-called $80bits/5ms$ frame or $16kbit/s$ V.110 standard. The 80 bits of the original V.110 Recommendation entail a combination of 48 data bits (D1 ... D48), eight fixed zeros and nine fixed ones, i.e., a total of 17 synchronisation bits, three bits (E1, E2, E3) to transmit the code of the user data-rate, four bits (E4 ... E7) for network independent clocking and multiframe synchronisation and 11 channel control bits. The $60bits/5ms$ modified V.110 standard frame is yielded by discarding the 17 fixed synchronisation bits (ones and zeros), as well as the data-rate code specified by the bits (E1, E2, E3), while still keeping a channel capacity of $(12-9.6)kbps = 2.4kbps$ for control channel information, such as $RS-232$ or $V.24$ standard signalling.

The burst structure of the $TCH/F9.6$ 9.6 kbit/s full-rate traffic channel is explained by referring to Figure 8.16. Four consecutive 60-bit data blocks are arranged to constitute a 240-bit information block, which is followed by four zero tailing bits for resetting the subsequent punctured convolutional codec after the transmission of a frame. The 240-bit data blocks are encoded by the half-rate, constraint length K=5 punctured convolutional code $PCC(2,1,5)$ using the same generator polynomials $G_0 = 1+D^3+D^4$ and $G_1 = 1+D+D^3+D^4$, as in the speech channel. This $PCC(2,1,5)$ convolutional code produces 488 encoded bits from the 244 data bits, but the following 32 coded bits are consistently punctured, i.e., not transmitted: $b(11+15j)$; $j = 0, 1, \ldots 31$. (On details of puncturing see Section 4.6.) The 456 encoded bits $b(0)\ldots b(455)$ are then mapped onto four consecutive 114-bit TDMA bursts (K, K+1, K+2 and K+3) using the mapping rule:

$$c(K,k) = b(k) \tag{8.22}$$

$$c(K+1,k) = b(k+114) \tag{8.23}$$

$$c(K+2,k) = b(k+228) \tag{8.24}$$

$$c(K+3,k) = b(k+342) \quad k = 0, 1 \ldots 113. \tag{8.25}$$

The stealing flags hl and hu used in the full-rate speech channel TCH/FS are included also here in the centres of the bursts, as seen in Figure 8.16b and have the same interpretation as in TCH/FS. The encoded bits are now reordered according to the following inter-burst interleaving rule:

$$i(K,j) = c(n,k) \tag{8.26}$$

$$K = K_0 + n + (k \bmod 19) \tag{8.27}$$

$$j = (k \bmod 19) + 19(k \bmod 6) \quad k = 0 \ldots 113, \tag{8.28}$$

where k is the bit-index in the 114-bit encoded bursts, n is the encoded burst-index, K is the interleaved burst-index and j is the bit-index in the interleaved burst. Inter-burst interleaving is described in general in Section 4.9, and is known to possess good randomising properties in dispersing

Full-rate data channel, 9.6 kb/s

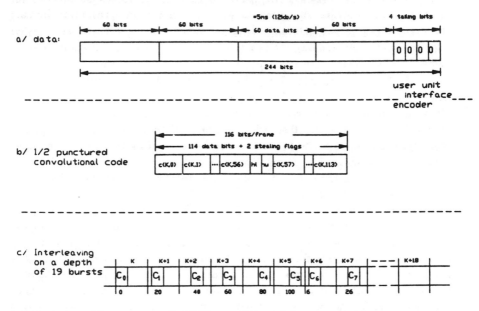

Figure 8.16: TCH/F9.6 FEC and burst structure

bursty channel errors, particularly, if the interleaving memory is sufficiently long.

In the specific scheme highlighted above $N \cdot B = 19 \cdot 6 = 114$ bits of an encoded burst are dispersed over $N = 19$ consecutive interleaved bursts, including the current one, while donating $B = 6$ bits to each one of them. A representative example of mapping the encoded bits of the Kth encoded burst $c(K, k), k = 0 \ldots 113$ onto bits of the subsequent 19 interleaved bursts is given in Figure 8.16c. Viewing the mapping from a different angle, bits $0, 19, 38, 57$ etc. of the Kth encoded burst are mapped onto itself, then bits $1, 20, 39, 58$ etc. onto burst $(K + 1)$, etc. This arrangement disperses 114 bits of a burst over $19 \cdot 114 = 2166$ bits, which appears sufficiently long, when combined with frequency hopping to randomise the bursty error statistics even for the slowly fading received signal envelopes of pedestrians as well.

8.5.2.1   Low-Rate Data Transmission

The 4.8 kbit/s full-rate data channel (TCH/F4.8) assumes 60 data bits from the user unit every 10 ms, as opposed to 5 ms in the previously considered TCH/F9.6. Furthermore, an R=1/3 rate convolutional encoder with the generator polynomials $G_1 = 1 + D + D^3 + D^4$, $G_2 = 1 + D + D^2 + D^4$, $G_3 = 1 + D + D^2 + D^3 + D^4$ is deployed combined with an inter-burst interleaver similar to that described for the TCH/F9.6 full-rate channel. The half-rate 4.8 kbit/s TCH/H4.8 channel also assumes 60 bits/10 ms from the user interface, but uses the generator polynomials and interleaving scheme of TCH/F9.6. Details of these channels along with those of other low-rate channels not discussed here are readily found in Recommendation 05.03.

8.5.3 FEC in Control Channels

As seen in Figure 8.5 of Section 8.3, the 184-bit control channel messages are delivered to the FEC encoder, where systematic shortened binary cyclic FIRE coding, using the generator polynomial $G_5(D) = (D^{23} + 1)(D^{17} + D^3 + 1) = D^{40} + D^{26} + D^{23} + D^{17} + D^3 + 1$ is carried out. Hence the 184 information bits are followed by 40 parity bits and 4 zero tailing bits to periodically clear the memory of the subsequent convolutional codec, yielding a 228-bit sequence. Then the R=1/2, K=5 convolutional code CC(2,1,5) is deployed using the generator polynomials $G_0 = 1 + D^3 + D^4$ and $G_1 = 1 + D + D^3 + D^4$ identical to those of TCH/FS, delivering 456 encoded bits. These blocks are then reordered and interleaved exactly as specified for TCH/FS, and the stealing flags are set to $hl = 1$, $hu = 1$ for SACCHs to indicate that no frame stealing is taking place. The block and convolutional encoding procedures, as well as the interleaving schemes used in BCCH, PCH, AGCH and SDCCH are perfectly identical to those described for the SACCH and are more powerful than those of speech channels due to the additional outer (224, 184) block coding deployed.

The 456 encoded bits of a FACCH are mapped onto 8 consecutive 114-bit bursts, as explained for TCH/FS, stealing even and odd bits in the first and last four bursts, respectively. This clearly implies that the present 20 ms, 456-bit speech frame is wiped out by an urgent FACCH message, such as a hand-over command. Hence, in contrast to SACCHs, for FACCHs $hl = 0$, $hu = 1$ is set in the first four 114-bit bursts, where even-indexed bits are stolen from the traffic channel, and $hl = 1$, $hu = 0$ is set in the last four 114-bit bursts of a traffic channel to signal that odd bits are stolen. This way an FEC-protected 456-bit link is created for the transmission of 184 FACCH bits. Again, since the control information requires higher integrity than speech channels, this link is enhanced by the deployment of the (224,184) FIRE-code, which provides an additional layer of error correction when compared to speech and data channels.

Random Access Channels (RACHs) have different message and coding formats, as portrayed in Figure 8.17. The eight RACH information bits $d(0) \ldots d(7)$ are input to a simple systematic cyclic shiftregister encoder characterised by the feedback generator polynomial $G_6(D) = D^6 + D^5 + D^3 + D^2 + D + 1$, yielding six parity bits $p(0) \ldots p(5)$. The six bits of the Base Station Identifier Code (BSIC) $b(0) \ldots b(5)$, to which random access is intended are bitwise modulo 2 added to $p(0) \ldots p(5)$ to deliver six so-called

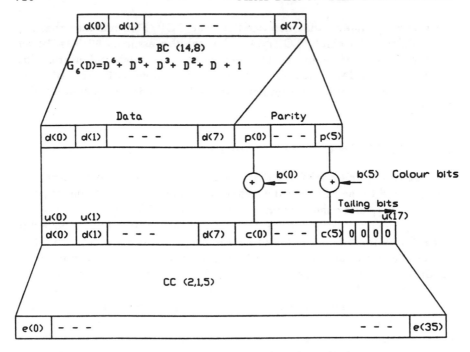

Figure 8.17: FEC in random access control channels

colour bits $c(0)\ldots c(5)$:

$$c(k) = b(k) \oplus p(k), \qquad k = 0\ldots 5 \qquad (8.29)$$

where $b(0)$ is the most significant bit (MSB) of the PLMN colour code and $b(5)$ is the least significant bit (LSB) of the BS colour code. The sequence $d(0)\ldots d(7), c(0)\ldots c(5)$ is then followed by four zero tailing bits to periodically reset the subsequent convolutional codec, and this 18-bit codeword is then convolutionally encoded using the R=1/2, K=5 CC(2,1,5) code by means of the generator polynomials $G_0 = 1 + D^3 + D^4$ and $G_1 = 1 + D + D^3 + D^4$, as explained for TCH/FS and shown in Figure 8.17.

The burst carrying the Synchronisation Channel (SCH) information on the downlink BCCH is constituted by the information bits $d(0)\ldots d(24)$, parity bits $p(0)\ldots p(9)$ and four resetting zero tailing bits for the subsequent CC(2,1,5) code, as demonstrated by Figure 8.18. The systematic cyclic shiftregister encoder defined by the generator polynomial $G_7 = D^{10} + D^8 + D^6 + D^5 + D^4 + D^2 + 1$ outputs the parity bits $p(0)\ldots p(9)$ and the resulting 38 bits are encoded using the CC(2,1,5) code known from the TCH/FS. For a comprehensive catalog of various FEC generator polynomials used in speech, data and control channels see Table 8.4.

$G_0 = D^4 + D^3 + 1$	TCH/FS, TCH/F9.6, TCH/H4.8, SACCH, FACCH, SDCCH, BCCH, PCH, AGCH, RACH, SCH
$G_1 = D^4 + D^3 + D + 1$	TCH/FS, TCH/F9.6, TCH/H4.8, SACCH, FACCH, SDCCH, BCCH, PCH, AGCH, RACH, SCH, TCH/F4.8, TCH/F2.4, TCH/H2.4
$G_2 = D^4 + D^2 + 1$	TCH/F4.8, TCH/F2.4, TCH/H2.4
$G_3 = D^4 + D^3 + D^2 + D + 1$	TCH/F4.8, TCH/F2.4, TCH/H2.4
$G_4 = D^3 + D + 1$	TCH/FS
$G_5 = D^{40} + D^{26} + D^{23} + D^{17} + D^3 + 1$	SACCH, FACCH, BCCH, PCH, AGCH, SDCCH
$G_6 = D^6 + D^5 + D^3 + D^2 + D + 1$	RACH (uplink)
$G_7 = D^{10} + D^8 + D^6 + D^5 + D^4 + D^2 + 1$	SCH (downlink BCCH)

Table 8.4: Summary of Generator Polynomials Used for Data, Speech and Control Channels

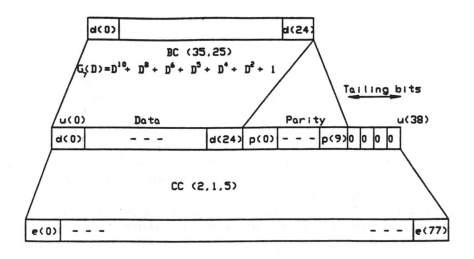

Figure 8.18: FEC in synchronisation channels

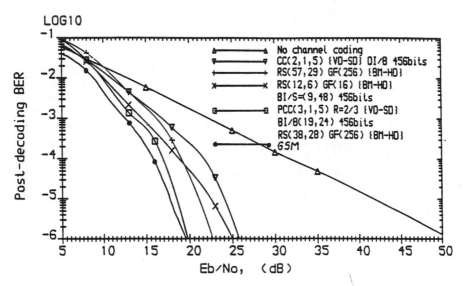

Figure 8.19: BER vs. E_b/N_0 performance of benchmark schemes for TCH/FS

8.5.4   FEC performance

To conclude the GSM error correction coding section, we conducted a number of experiments via narrowband flat-fading Rayleigh channels to assess the BER performance of the full-rate speech and data channels in contrast to a wide variety of bench-mark schemes [14], [15]. The simulations were carried out using minimum shift keying (MSK) and a Rayleigh-fading envelope with a propagation frequency of 900 MHz and vehicular speed of 30 mph. Our findings are summarised in Figures 8.19 and 8.20 for the speech and data channels, respectively. The overall coding and interleaving delay was close to those standardised for the speech and data channels, i.e., 456 bits and 2-3000 bits, respectively, to provide a fair basis for comparisons. The overall coding rate was R=0.5, but some of the bench-markers had two layers of coding and interleaving to efficiently randomise the bursty error statistics and provide error detection as well, as explained in Section 4.23.

For the speech channel we have used the following bench-mark systems:

1. The single layer convolutional code CC(2,1,5) with parameters recommended in the GSM system, but the interleaving scheme used is a simple diagonal bit interleaver (DI/B) over the specified 456-bit interval. In the decoder Viterbi decoding with soft decisions (VD-SD) has been used.

2. Another single layer arrangement deployed is the Reed-Solomon RS-

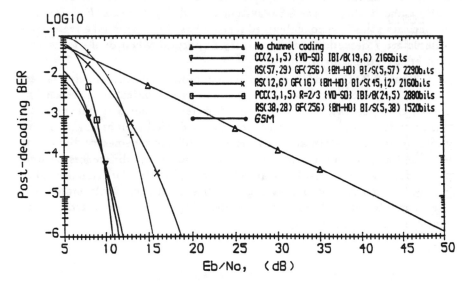

Figure 8.20: BER vs. E_b/N_0 performance of benchmark schemes for TCH/F9.6

(57,29) codec over Galois Field GF(256) decoded by the Berlekamp-Massey hard decision (BM-HD) decoding method.

3. Also a short Reed-Solomon codec, the RS(12,6) codec over GF(16) associated with a symbol block interleaver (BI/S) has been investigated.

4. A powerful concatenated coding arrangement is constituted by the combination of the 2/3-rate punctured convolutional inner codec PCC(3,1,5) with constraint length K=5 and the outer layer RS(38,28) codec over GF(256). The convolutional code has been decoded by using VD-SD and the block code by BM-HD.

5. Finally the GSM 05.03 codec has been simulated and compared to our bench-markers.

For speech channels the following observations can be made. It is clear that the GSM scheme is the most powerful one in terms of its BER versus E_b/N_0 performance, where E_b represents the bit-energy and N_0 the one-sided noise spectral density. Observe the dramatic performance increase due to the sophisticated interleaving scheme deployed, when compared to the curve representing our first bench-marker using the same codec with less powerful interleaving. The RS(12,6) code performs better than the RS(57,29) code above $E_b/N_0 = 20\,dB$, while below this cross-over point the situation is reversed. The best contender among our bench-mark systems is the concatenated $PCC(3,1,5)/RS(38,28)$ scheme giving very similar performance to that of the GSM system around $E_b/N_0 = 20\,dB$. It is worth emphasizing that our concatenated scheme has the additional advantage of reliable error detections, which is very important when deploying speech post-enhancement algorithms [23].

In Figure 8.20 the performance of data communication channels has been compared. The coding schemes implemented are identical to those used for speech channels having merely longer interleaving periods. Since for the duration of 2500 bits the channel can be considered memoryless, a significantly increased performance is achieved for all the schemes studied. However, at this long interleaving period the performance differences between the two single layer convolutional codes are negligible. Furthermore, above $E_b/N_0 = 10\,dB$ the concatenated scheme has the highest performance with the favourable error detection capability, which can be exploited in automatic repeat request (ARQ) systems.

8.6   Transmission and Reception

The family of constant envelope, continous phase modulation schemes is widely used over fading mobile radio channels due to their robustness against signal fading and interference, while maintaining good spectral efficiency. High interference resistance is achieved, if high modulation index and moderate filtering are used, which keeps the phase changes engendered by interference relative to those due to modulation low. Unfortunately, a high modulation index requires higher bandwidth, which means that a compromise has to be found. In other words, the slower and smoother are the phase changes, the better is the spectral efficiency. Spreading phase changes with zero initial and final slopes to three or four modulation intervals yields a partial response system. A representative of this family, called Gaussian Minimum Shift Keying (GMSK) is widely deployed via fading channels. It is derived from the full response Minimum Shift Keying (MSK) scheme, where phase changes between adjacent bit periods are piecewise linear, which results in discontinous phase derivative, i.e., instantaneous frequency. This clearly widens the spectrum, but by smoothing the phase using a Gaussian filter this problem is circumvented, as seen in Figure 8.21, where the GMSK signal is generated by modulating and adding two quadrature carriers.

The key parameter of GMSK in controlling both bandwidth and interference resistance is the 3 dB-bandwidth · bitinterval product (B·T) referred to as normalised bandwidth. It was found by GSM that as the B·T product is increased from 0.2 to 0.5, the interference resistance is improved by approximately 2 dB at the cost of increased bandwidth occupancy, and best compromise was achieved for B·T=0.3. The surprising fact is that the spectral efficiency gain due to higher interference tolerance and hence

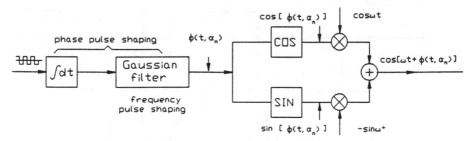

Figure 8.21: GMSK modulator schematic diagram

smaller microcells was deemed to be more significant than the spectral loss caused by wider GMSK spectral lobes.

The GSM system will initially operate in the 890–915 MHz band for uplink transmission and 935–960 MHz band for downlink transmission. There are 124 paired duplex radio channels with a carrier spacing of 200 kHz and duplex spacing of 45 MHz between uplink and downlink directions. A guard band of 200 kHz is left between the bottom edge of each band and the first RF carrier. Thus, the carrier frequencies in the two bands for the n-th duplex radio channel will be (in MHz):

$$F_{nI} = 890.2 + 0.2(n - 1) \text{ MHz} \tag{8.30}$$
$$F_{nII} = F_{nI} + 45 \text{ MHz}. \tag{8.31}$$

A TDMA system with 8 timeslots per RF carrier and 200 kHz channel spacing has the same spectral occupancy, as an SCPC (Single Channel Per Carrier) system with 25 kHz channel spacing. When taking into account the 22.8 kbit/s channel coded data rate, the bandwidth of 25 kHz implies an approximate spectral efficiency of 1 bit/Hz. The actual RF output spectrum of the transmitted signals in a TDMA system, however, is determined by the modulation process and the switching transients occuring, when bursts of RF signals are transmitted.

The recommendations for the output RF spectrum mask due to the GMSK modulation are given in graphic form in Figure 8.22. At the nominal bandwidth of 200 kHz the spectrum must have decayed by $30dB$ with respect to the carrier frequency component. The specified relative power levels for frequency offsets from the carrier equal to or greater than 400 kHz depend on the output power level at which the transmitter is operating and the type of station. For transmitter output power levels below 43 dBm (20W) the specifications allow slightly higher levels in the spectrum at points 600 kHz to 1.8 MHz away from the carrier. The example of curve *Con* : 2W representing a transmitted power of 2W at the antenna connector in Figure 8.22 demonstrates that for $10dB$ lower transmitted power a $10dB$ higher modulation spectral mask is acceptable, when compared with a 20W transmitter. Higher levels at points 400 kHz to 1.8 MHz away

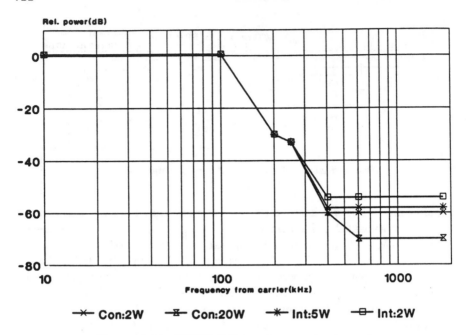

Figure 8.22: GMSK RF spectrum output mask

from the carrier are allowed also for equipment with integral antennas (e.g., portable sets) operating at power levels below 37 dBm (5W), as represented by the curves *Int* : 5*W* and *Int* : 2*W*, respectively.

Switching transients and power ramping The switching transients caused by the transmission of bursts of RF energy widen the output RF spectrum. The RF spectrum due to the switching transients is required to be 23, 26, 32 and 36 dBm down relative to the level specified by the modulation mask at frequencies of 400, 600, 1200 and 1800 kHz measured from the carrier frequency, respectively. The switching transients can be reduced by ramping the output power up and down when transmitting a burst, instead of just keying the transmitter on and off. The information transmitted in the burst must not be affected by the process of power ramping, which is performed at the beginning and end of the timeslot using the mask illustrated in Figure 8.23. The timeslot in the figure corresponds to a duration of 156.25 bits, that of the burst length. In a normal burst, frequency correction burst or a synchronisation burst a guard period of 8.25 bit periods is inserted between adjacent ones. The remaining 148 bit periods form the 'active part' of the bursts. In an access burst the guard period after the burst is 68.25 bits long leaving an active part of 88 bit periods.

The 'useful part' of a burst in all cases is one bit period shorter than the active part and it begins half way through the first bit period as shown in Figure 8.24 for a burst with a 148 bit periods long active part. During

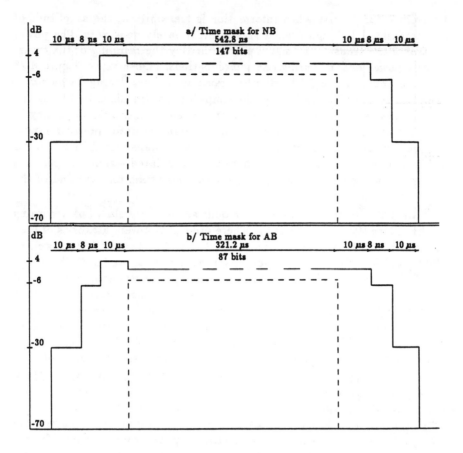

Figure 8.23: Power ramping time-masks for a./Normal bursts b./Access bursts

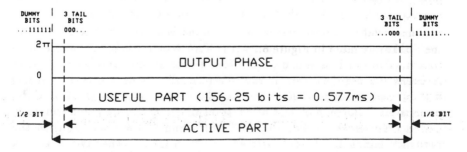

Figure 8.24: Active and useful parts of a normal burst

that part of the burst when information is transmitted, the amplitude of the modulated RF signal must stay approximately constant. The power control of the transmitted signal exemplified by the ramping of the transmitted power occurs during the guard periods. Observe in Figure 8.23 that the approximately 70 dB power ramp-up occurs during 28 μs corresponding to 7.6 bit intervals, while ramp-down takes place in 18 μs, i.e., 4.9 bit intervals. When bursts are transmitted at the same frequency in consecutive timeslots, i.e., no frequency hopping is used, power ramping between the slots is not required and the signal transmitted in the guard times between the active slots may be any modulated signal. In this case the recommended time masks apply for the beginning and the end of the series of consecutive bursts.

In the GSM system the base and mobile stations are classified according to the transmitter output power. There are eight power categories of base station transmitted power, namely 2.5, 5, 10, 20, 40, 80, 160 as well as 320W, and four for mobile stations given by 2, 5, 8, as well as 20W. Adaptive RF power control is mandatory for the mobile stations, but optional for the base stations. This feature reduces cochannel interference whilst maintaining the quality of the radio channel. It also decreases the power consumption, which is important for hand-held mobile stations. Provisions are made for 16 different power control levels with 2 dB spacing between adjacent steps. The lowest power level for all mobile stations, regardless of their power class, is 13 dBm (20 mW) and the highest power level is equal to the maximum peak power corresponding to the class of the particular mobile station, as listed above. In the base stations the same 16 steps of 2 dB-spaced power levels are provided to achieve adaptive RF power control. Level "0" corresponds to the maximum peak power specified for the transmitter. For example, for a transmitter power class "3" the power level "0" is 80 W.

The output power of a base station transmitter can be reduced from its maximum level in at least six steps of 2 dB (with each step accurate to ±0.5 dB) to adjust the radio coverage by the network operator. This RF output power adjustment is provided in every base station, and it is not connected in any way to the adaptive RF power control procedure.

The allowable spurious emissions from the base and mobile stations in the GSM system are higher than for the TACS analog cellular mobile radio system. The spurious signal power from a base station transmission must be below 0.5 pW (-93 dBm), 250 nW (-36 dBm) and 1 μW (-30 dBm) in the frequency bands 890–915 MHz, 100 kHz–1 GHz and 1–12.75 GHz, respectively. For a mobile station transmitter the permissible levels of spurious emissions are 250 nW (-36 dBm) and 1 μW (-30 dBm) in the frequency bands 100 kHz–1 GHz and 1–12.75 GHz, respectively, when a channel is allocated to the mobile station. These permitted emission levels decrease to 2 nW (-57 dBm) and 20 nW (-47 dBm) in the corresponding bands when the MS is in its idle mode. In a base station two or more transmitters are often combined onto a single antenna. An undesirable consequence of the combining is the generation of intermodulation signals by the nonlinearities in the transmitters. This effect occurs also when each transmitter feeds a separate antenna, but the transmitters are in close vicinity of each other. The peak power of any intermodulation product must not exceed 0.5 pW (-93 dBm) in the frequency band 890–915 MHz, nor 250 nW (-36 dBm) in the band 935–960 MHz. These limits apply not only to base station transmitters, but also to mobile PBX transmitters when operating in close vicinity of each other.

The receiver's performance is assessed in terms of a reference sensitivity level, which is the RF signal level at the receiver input assuming a specific propagation channel, for which the required error rate performance is achieved. For the hand portable receivers the reference sensitivity level is -102 dBm, and for all other types of mobile and base station receivers this level is -104 dBm. The receiver's blocking and intermodulation characteristics are specified assuming static propagation conditions for both wanted and unwanted signals, and a wanted RF signal level at the receiver input equals to the reference sensitivity level. It is required that a mobile station receiver performance does not degrade by more than 3 dB when an interfering continuous wave signal is applied to the receiver at a level of -23 dBm or less, and that the frequency spacing between the wanted and unwanted signals are not less than 800 kHz and not more than 45 MHz. The requirements for the base station receivers are 10 dB higher, e.g., the level at which the 3 dB degradation in performance occurs is set at -13 dBm.

The intermodulation characteristics are measured by applying a signal with a frequency f_0 to the receiver input at the reference sensitivity level, plus two interfering signals of frequencies f_1 and f_2 such that $2f_1 - f_2 = f_0$ and $|f_2 - f_1| = 800$ kHz. The signal at frequency f_1 is a sinusoid, while the other interfering signal is a GMSK modulated carrier f_2 that has been modulated by any 148 bits subsequence of a 511 bits pseudorandom sequence. The receiver performance under these conditions must not degrade by more than 3 dB when the peak levels of both interfering signals are -43 dBm or less.

The error rate performance of the RF subsystem is specified in the GSM system for various propagation conditions, referred to as $NAMEx$, where $NAME$ is the name of the propagation model and x is the vehicle speed in km/h. The models for rural area, hilly terrain, typical urban area and the profile for equalisation testing are referred to as RAx, HTx, TUx and EQx channels, respectively. An additive white Gaussian noise channel, i.e., a static channel, is also considered. Depending on the type of traffic or control channel, the performance is described in terms of frame erasure rate (FER), bit error rate (BER), or residual bit error rate (RBER). The RBER is defined as the ratio of the number of errors detected due to unprotected Class 2 bits over the frames defined as "good", to the number of transmitted bits in the "good" frames, where a frame is deemed to be good for example,

Type of Channel		Propagation Conditions			
		Static	TU 50	RA 250	HT 100
SDCCH	(FER)	0.1 %	4 %	4 %	6 %
RACH	(FER)	0.1 %	10 %	10 %	10 %
SCH	(FER)	1 %	15 %	15 %	15 %
TCH/F9.6, H4.8	(BER)	10^{-5}	0.3 %	0.1 %	0.8 %
TCH/F4.8	(BER)		10^{-4}	10^{-4}	10^{-4}
TCH/F2.4	(BER)		10^{-5}	10^{-5}	10^{-5}
TCH/H2.4	(BER)		10^{-4}	10^{-4}	10^{-4}
TCH/FS	(FER)	10^{-3}	3 %	2 %	7 %
C1b	(RBER)	0.4 %	0.2 %	0.2 %	0.5 %
C2	(RBER)	2 %	8 %	7 %	8 %

Table 8.5: Reference sensitivity performance

if the (53,50) cyclic error detecting block code protecting the Class 1a speech bits does not indicate code overload.

In GSM parlance there are three different types of error rates. One is concerned with the conditions of operation at reasonable signal levels and in the absence of interference. Another applies when the receiver is operating with signal levels close to its noise floor, and the third category applies for operation in the presence of interference. The nominal error rates (NER) apply to propagation conditions, when there is no interference and the received RF signal level is equal to -85 dBm. Under these conditions the chip error rate (channel BER), which is equivalent to the bit error rate of the non-protected C2 bits of a full rate traffic channel for speech (TCH/FS), is specified as $\leq 10^{-4}$ for the static channel, $\leq 4 \times 10^{-3}$ for the rural channel RA250, $\leq 4 \times 10^{-4}$ for the typical urban channel TU3 and $\leq 1\%$ for the channel used for the equaliser testing.

The reference sensitivity performance is the error rate performance when a received RF signal level is equal to the reference sensitivity level of -102 dBm for hand-portables and -104 dBm for all other mobile and base stations and when no interference is present. The reference sensitivity performance specifications are shown in Table 8.5, while the reference interference figures are summarised in Table 8.6 for the different types of channels and propagation conditions.

The reference interference performance is the error rate limit when the wanted input RF signal level is -85 dBm and a random GSM modulated interfering signal is present. The signal-to-interference ratio is ≤ 9 dB, while the signal-to-adjacent channel ratios at 200 and 400 kHz from the carrier are $\leq$ -9 dB and -41 dB, respectively. This interference ratio is called the "reference interference ratio" and is identical for all types of base and mobile stations. In the tests, the wanted and interfering signals are subject to the same propagation profiles, and when frequency hopping is used, they have the same hopping sequence. Under these conditions the error rates for the various types of channels and propagation conditions satisfy the limits shown in Table 8.6.

Type of Channel		Propagation Conditions			
		TU 3 (No FH)	TU 3 (FH)	TU 50	RA 250
SDCCH	(FER)	8 %	4 %	4 %	4 %
RACH	(FER)	12 %	12 %	12 %	10 %
SCH	(FER)	15 %	15 %	15 %	15 %
TCH/F9.6/H4.8	(BER)	1.5 %	0.3 %	0.3 %	0.2 %
TCH/F4.8	(BER)		10^{-4}	10^{-4}	10^{-4}
TCH/F2.4	(BER)		10^{-5}	10^{-5}	10^{-5}
TCH/H2.4	(BER)		10^{-4}	10^{-4}	10^{-4}
TCH/FS	(FER)	7 %	2.5 %	3.5 %	3 %
C1b	(RBER)	0.4 %	0.2 %	0.2 %	0.2 %
C2	(RBER)	8 %	8 %	8 %	8 %

Table 8.6: Reference interference performance

8.7 Wideband Channels and Viterbi Equalisation

8.7.1 Channel Models

The understanding of the GSM wideband channel models and Viterbi equalisation assumes a sound appreciation of the mobile radio propagation phenomena, as detailed in Chapter 2. If the transmitted signal's bandwidth is narrow compared to the channel's coherence bandwidth (B_c), all transmitted frequency components encounter nearly identical propagation delays, i.e., the so-called narrow band condition is met and the signal is subjected to non-frequency-selective or flat envelope fading. When the signal bandwidth is increased, for example, to accommodate several TDMA timeslots as in the GSM system, the channel becomes more dispersive which results in intersymbol interference. The channel's coherence bandwidth (B_c) is defined as the frequency, where the correlation of two received signal components' attenuation becomes less than 0.5 and (B_c) is inversely proportional to the delay-spread (d), i.e., $B_c=1/2\pi d$. Clearly, the wideband propagation channel is the superposition of a number of dispersive fading paths, suffering from various attenuations and delays, aggrevated by the phenomenon of Doppler shift caused by the MS's movement. The maximum Doppler shift $(f_D max)$ is given by $f_D max = v/\lambda_c = v \cdot f_c/c$, where v is the vehicular speed, λ_c is the wavelength of the carrier frequency f_c and c is the velocity of light. The momentary Doppler shift f_D depends on the angle of incidence α, which is uniformly distributed, i.e., $f_D = f_D max \cdot cos\alpha$, which has hence a random cosine distribution with a Doppler spectrum limited to $-f_D max < f < f_D max$. Due to time-frequency duality, this 'frequency dispersive' phenomenon results in 'time-selective' behaviour and the wider the Doppler spread, i.e., the higher the vehicular speed, the faster is the time-domain impulse response fluctuation.

In order to provide exactly specified, identical test conditions for different implementations of the GSM system, in particular for various Viterbi

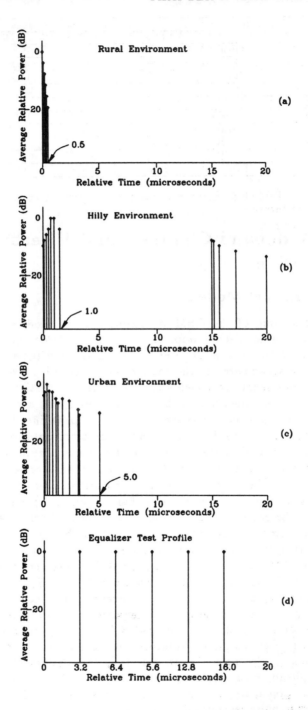

Figure 8.25: Typical GSM channel impulse responses

equalisers, a set of 12-tap and 6-tap typical channel impulse responses were defined, some of which are depicted in Figure 8.25. The Rural Area (RA) response is the least hostile amongst all standardised responses, decaying fast within one bit interval and in terms of bit error rate performance it behaves as a single-path non-dispersive channel, where no Viterbi Equaliser (VE) is required. The Hilly Terrain (HT) model has a short-delay section due to local reflections and a long-delay part around 15 μs due to distant reflections, therefore in practical terms it can be considered a two- or three-path model, providing useful diversity gain, when using a VE. The Typical Urban (TU) impulse response spreads over a delay interval of 5 μs, which is almost two 3.69 μs bit intervals duration and therefore results in serious InterSymbol Interference (ISI). Whence in simple terms it can be treated as a two-path model. The last standardised impulse response is artificially contrived to test the VE's performance and is constituted by six equidistant unit-amplitude impulses representing six equal-powered independent Rayleigh-fading paths with a delay-spread over 16 μs. With these impulse responses in mind the required channel is simulated by summing the appropriately delayed and weighted received signal components. In all but one cases the individual components are assumed to have Rayleigh amplitude distribution. In the RA model the main tap at zero delay is supposed to have Rician distribution with the presence of a dominant path.

In summary, we highlighted four dispersive channel models, three of which represent realistic propagation environments, while the most hostile equaliser test response is worse than any practical channel. If reliable communications is expected, the uncontrolled ISI introduced by the mobile channel, as well as the controlled ISI introduced in the partial response modulator have to be removed, which requires a channel equaliser. The Bit Error Rate (BER) is minimised if a 'Maximum Likelihood Sequence Estimator' (MLSE) is deployed to decide upon the most likely transmitted sequence, rather than deciding on a 'maximum likelihood decoded symbol' basis. The Viterbi Algorithm (VA) detailed in Chapter 4 is a well-suited efficient method for MLSE and is deployed in all proposed implementations of the GSM system.

8.7.2   Viterbi Equaliser

Both during and after the definition of the GSM standards a number of VE implementations have been proposed in the literature, which have different complexities and performances [16], [17], [18], [19], [21]. A simple general VE block diagram is shown in Figure 8.26. Once a call is set up, communications is maintained using Normal Bursts (NB) incorporating the 26-bit midamble in the centre of the burst, of which 16 bits constitute the frame synchronisation word and 5 bits are quasi-periodically repeated at both ends to keep autocorrelation function and frequency domain oscillations low. As mentioned before, there are eight different, specially selected

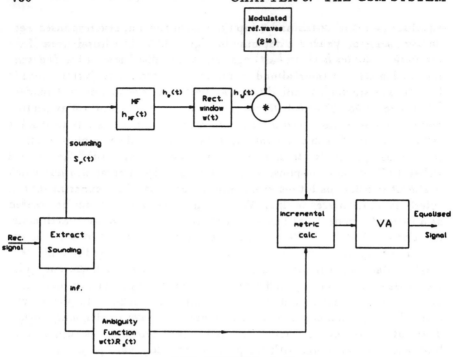

Figure 8.26: MLSE block diagram

synchronisation words associated with eight different adjacent BS colour codes. These special synchronisation sequences have been found by computer search, evaluating the autocorrelation functions of all possible 2^{16} sequences. Favourable are those sequences, which have the highest autocorrelation function main- to side-lobe ratio with near-zero values around the sampling instants $\pm T, \pm 2T, \pm 3T, \pm 4T$, etc., when quasi-periodically extended at both ends. It is highly desirable that both MSs and BSs use the same VE to keep development and production costs low, which additionally requires the recognition of synchronisation (SB) and access bursts (AB) as well, where 64- and 41-bit long synchronisation words are used, respectively. However, for the sake of simplicity, we only consider NBs with 26-bit midambles.

The modulated NB with the channel sounding sequence $s(t)$ in its centre is convolved with the channel's impulse response $h_c(t)$ and corrupted by noise. Neglecting the noise for simplicity, the received sounding sequence becomes:

$$s_r(t) = s(t) * h_c(t), \qquad (8.32)$$

which is then matched-filtered using the impulse response $h_{MF}(t)$ to derive an estimate of the channel's impulse response:

$$h_e(t) = s_r(t) * h_{MF}(t) = s(t) * h_c(t) * h_{MF}(t) = R_s(t) * h_c(t), \qquad (8.33)$$

where $R_s(t)$ is the sounding sequence's autocorrelation function. Clearly, if $R_s(t)$ is a highly peaked Dirac impulse-like function, then its convolution with $h_c(t)$ becomes $h_e(t) \approx h_c(t)$. With $s(t)$ in the middle of the NB the estimated $h_c(t)$ is quasi-stationary for the 0.577 ms burst duration, and can be used to equalise the 114 bits of useful information on both sides of it, although the time-variant channel precipitates higher error rates towards the burst edges. Since the complexity of the VE grows exponentially with the number of signalling intervals in the legitimate modulated reference sequences generated from all possible transmitted sequences for metric comparisons, the estimated channel response $h_c(t)$ has to be windowed to a computationally affordable length using the rectangular function $w(t)$, while having sufficiently long memory to compensate for the typical GSM impulse responses of the previous section.

Specifically, in addition to the duration L_{CISI} of the controlled ISI, also the channel's delay-spread L_c has to be considered in calculating the required observation interval $L_o = L_{CISI} + L_c$ of the 2^{L_o-1}-state VE. In practical terms, using a bit interval of 3.69 μs and maximum channel impulse response durations of around 15–20 μs, a VE with a memory of 4-6 bit intervals is a good compromise, where $h_c(t)$ is retained over that 4-6 bit interval of its total time domain support length, where it is exhibiting the highest energy. L_o consecutive transmitted bits give rise to 2^{L_o} possible transmitted sequences, which are first input to a local modulator to generate the modulated waveforms, and then convolved with the windowed estimated channel response $h_w(t)$ to derive the legitimate reference waveforms for metric calculation, as portrayed in Figure 8.26.

Recall that the condition $h_e(t) = h_c(t)$ is met only, i.e. the estimated impulse response is identical to the true channel impulse response only, if $R_s(t)$ is the Dirac delta function, which is not fulfilled when finite-length sounding sequences are used. The true channel response $h_c(t)$ could only be computed by deconvolution from Equation 8.33 upon neglecting the rectangular window $w(t)$. Alternatively, the received signal can be convolved for the sake of metric calculation with the known windowed autocorrelation function $w(t) \cdot R_s(t)$ often referred to as ambiguity function, as seen in the lower branch of Figure 8.26 after extracting the sounding sequence from the received normal burst. Clearly, this way the received signal is 'predistorted' using the ambiguity function, identically to the estimated impulse response in Equation 8.33. This filtered signal is then compared to all possible reference signals and the incremental metrics m_i, $i = 0 \ldots (2^{L_o-1})$ are computed, which are utilised by the Viterbi algorithm (VA) to determine the maximum likelihood transmitted sequence, as explained in Chapter 6 and [20], [21].

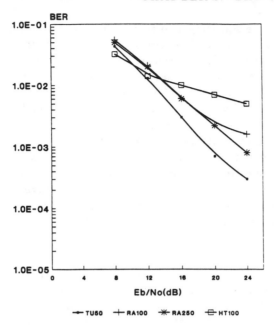

Figure 8.27: Viterbi equaliser BER vs. E_b/N_0 performance

8.7.3   Performance

The various implementations referenced have different complexities and
performances. As a representative example we quote the BER vs. SNR
performance published in [19], as seen in Figure 8.27 for the TU50 channel (typical urban, vehicular speed of 50 km/h), the RA100, HT100 and
RA250 GSM channel models. Best performance is achieved via the TU50
channel, which is due to the advantageous 'diversity' effect' introduced by
the impulse response tap at 5 μs in Figure 8.25, since the probability of
both paths having a deep fade simultaneously is fairly low. The worst
performance is experienced via the HT100 channel, where the VE has apparent difficulties in combating excess delays above 15 μs. Interestingly,
the RA100 and RA250 performances are worse than the TU50 integrity,
since the RA models represent virtually single-path conditions with no 'diversity effect'. The BER is in all cases below 1 %, if the E_b/N_0 ratio is in
excess of about 12 dB. This residual BER can then be further reduced by
the GSM concatenated error correction scheme, described in Section 8.5.

The performance of a complete GSM speech channel simulator has been
reported in [22] for the various GSM channel models using vehicular speeds
ranging from 0 km/h (AWGN) through pedestrians walking at 3 km/h to
250 km/h high-speed trains. For slowly walking pedestrians results are
reported both with and without frequency hopping (FH). The concatenated
coded C1 speech BER vs.E_b/N_0 results are reproduced in Figure 8.28,

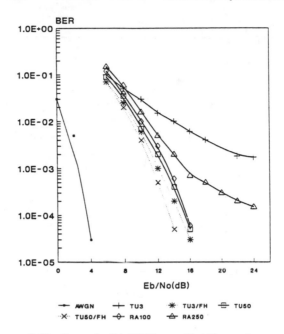

Figure 8.28: Speech C1 BER vs. E_b/N_0 performance

where we observe virtually error free operation for the AWGN channel for E_b/N_0 in excess of 4 dB and for most of the fading channels above 12 dB. When using the TU3 channel the MSs are idleing in deep fades and so the interleaving memory is not sufficiently long to randomise error bursts before channel decoding, which yields a high residual BER. This is seen being effectively combated by FH. The higher residual BER of the RA250 channel is due to the higher Doppler shift and lack of 'diversity effect'. The unprotected C2 bits have a high residual BER in Figure 8.29, which is uneffected by FH. In fact, this residual C2 BER is higher than that of the VE implementations proposed in [19] or [18]. Similar tendencies are recognised as regards to Frame Error Rates (FER) depicted in Figure 8.30. The interference resistance of the system expressed in carrier to interference ratio [C/I (dB)] is characterised in [22], which is again similar to the noise resistance, as seen in Figures 8.31, 8.32 and 8.33. In summary, all reported VE implementations reduce the channel BER to values sufficiently low for the concatenated channel coding/interleaving scheme to remove most of the errors for E_b/N_o and C/I ratios in excess of 12–14 dB, a value providing higher robustness and spectral efficiency than current analogue systems.

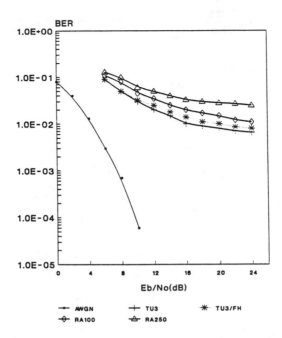

Figure 8.29: Speech C2 BER vs. E_b/N_0 performance

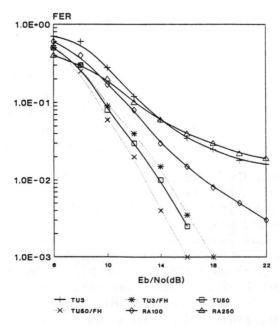

Figure 8.30: Speech FER vs. E_b/N_0 performance

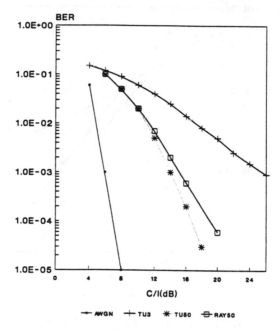

Figure 8.31: Speech C1 BER vs. C/I performance

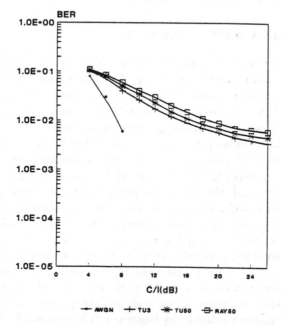

Figure 8.32: Speech C2 BER vs. C/I performance

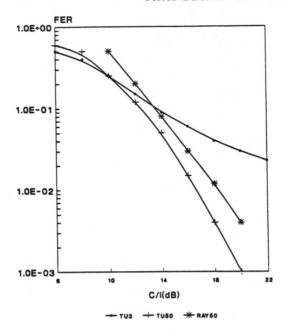

Figure 8.33: Speech FER vs. C/I performance

8.8   Radio Link Control

8.8.1   Link Control Concept

The radio sub-system link control in the GSM system involves procedures necessary for maintaining the link quality and managing traffic distribution, as well as for adaptive RF power control, handover, and to generate radio link failure responses. The call selection and re-selection procedures apply for a MS which is not engaged in communication with a BS, i.e., it is in its idle mode. These procedures allow for the MS to select a cell which provides the highest probability of reliable communications with the serving BS. The adaptive RF power control decreases interference with other co-channel users and, through dense frequency reuse, improves spectral efficiency, whilst maintaining an adequate communications quality. It also facilitates a reduction in power consumption, which is particularly important in hand-held MS's. The handover process maintains a call in progress as the MS moves between cells, or when there is an unacceptable degradation of quality caused by interference, in which case an intra-cell handover to another carrier in the same cell is performed. A radio link failure occurs when a call with an unacceptable voice or data quality cannot be improved either by RF power control or by handover. The reasons for the link failure may be loss of radio coverage or very high interference levels.

The radio sub-system link control procedures rely on measurements of

RXQUAL 0	BER ¡ 0.2%
RXQUAL 1	BER = 0.2% to 0.4%
RXQUAL 2	BER = 0.4% to 0.8%
RXQUAL 3	BER = 0.8% to 1.6%
RXQUAL 4	BER = 1.6% to 3.2%
RXQUAL 5	BER = 3.2% to 6.4%
RXQUAL 6	BER = 6.4% to 12.8%
RXQUAL 7	BER ¿ 12.8%

Table 8.7: Received signal quality vs. channel bit error rate

the received RF signal strength (RXLEV), the received signal quality (RX-QUAL), and the absolute distance between base and mobile stations (DISTANCE). The received RF signal strength measurements are performed on the broadcast control channel (BCCH) carrier which is continuously transmitted by the BS on all timeslots and without variations of the RF level. A MS measures the received signal level from the serving cell and from the BS's in all adjacent cells by tuning and listening to their BCCH carriers. The rms level of the received signal is measured over a dynamic range of -103 to -41 dBm for intervals of one SACCH multiframe (480 ms), with a relative accuracy of ±1 dB within any 20 dB section of this range, and an absolute accuracy of ±4 dB over the range from -103 to -70 dBm under normal conditions. The absolute accuracy over the full dynamic range and under both normal and extreme conditions is ±6 dB. The received signal level is averaged over at least 32 SACCH frames ($\approx$15 s) and mapped to give RXLEV values between 0 and 63, where $RXLEV = 0$ if the received signal level(RSL) is less than $-103\ dBm$, $RXLEV = 1$ if $-103\ dBm \leq RSL < -102\ dBm, \ldots, RXLEV = 63$ if $RSL > -41\ dBm$. The RXLEV parameters are then coded into 6-bit words for transmission to the serving BS via the SACCH.

The received signal quality (RXQUAL) is assessed by estimating the chip error rate, i.e., the BER before channel decoding, using the Viterbi channel equaliser's metrics and/or those of the Viterbi convolutional decoder. Eight values of RXQUAL span the BER range before channel decoding according to Table 8.7.

The absolute distance between base and mobile stations is measured using the "timing advance" parameter. The timing advance is coded as a 6bit number corresponding to a propagation delay from 0 to $63 \cdot 3.69\ \mu s = 232.6\ \mu s$. This allows measurements of an absolute distance from zero to almost 70 km with an accuracy of about 1 km.

The radio link control employs not only the parameters RXLEV, RXQUAL and DISTANCE obtained by the measurements highlighted, but also other parameters transmitted by the BS. A MS needs to identify which surrounding BS it is measuring and the BCCH carrier frequency may not be sufficient for this purpose, since in small cluster sizes the same BCCH frequency may be used in more than one surrounding cell. To avoid ambiguity a 6-bit Base Station Identity Code (BSIC) is transmitted on each BCCH carrier in the SCH. Two other parameters represented by one-bit Boolean flags transmitted in the BCCH data provide additional information about the BS. Namely, PLMN_PERMITTED indicates whether the measured BCCH carrier belongs to a PLMN which the MS is permitted to access. The second flag, CELL_BAR_ACCESS, indicates whether the cell is barred for access by the MS, although it belongs to a permitted PLMN. The parameters BSIC, PLMN_PERMITTED and CELL_BAR_ACCESS, together with the RXLEV, are used in the cell selection and re-selection procedures, as seen in Figure 8.34. A MS in idle mode, i.e., after it has just been switched-on, or after it has lost contact with the network, searches all 124 RF channels and takes readings of RXLEV on each of them. The station then tunes to the carrier with the highest RXLEV and searches for frequency correction bursts (FCB) in order to determine whether or not the carrier is a BCCH carrier. If it is not, then the MS tunes to the next highest carrier, and so on, until it finds a BCCH carrier. The MS then finds a synchronisation burst (SB), synchronises to the BCCH carrier and decodes the parameters BSIC, PLMN_PERMITTED and CELL_BAR_ACCESS from the BCCH data and makes a decision to camp on the cell or to continue the search. The MS may have a BCCH carrier storage option, i.e., store the BCCH carrier frequencies used in the network accessed, in which case the search time would be reduced. The process described is summarised in the flowchart of Figure 8.34.

The RF power control procedures employ RXLEV measurement results. In every SACCH multiframe the BS compares the RXLEV readings reported by the MS, or obtained by the base station, with a set of thresholds. The exact strategy for RF power control is determined by the network operator with the aim of providing an adequate quality of service for speech and data transmissions and keeping interferences low. The criteria for determining the radio link failure are based on the measurements of RXLEV and RXQUAL performed by both the mobile and base stations. The procedures for handling radio link failures result in the re-establishment or the release of the call in progress. The network operator determines the exact criteria employed.

The handover process involves the most complex set of procedures in the radio link control. Handover decisions are based on results of measurements performed both by the base and mobile stations. The base station measures RXLEV, RXQUAL, DISTANCE, and also the interference level in unallocated timeslots, while the MS measures and reports to the BS the values of RXLEV and RXQUAL for the serving cell, and RXLEV for the adjacent cells. When the MS moves away from the BS, the RXLEV and RXQUAL parameters for the serving station become lower, while RXLEV for one of the adjacent cells increases.

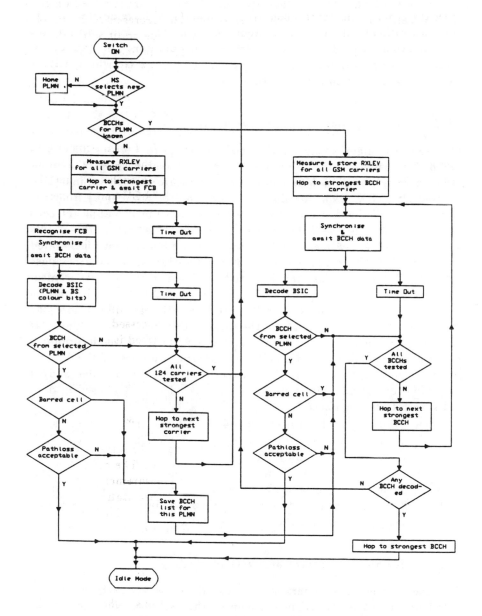

Figure 8.34: Initial cell selection by the MS

8.8.2   A Link Control Algorithm

With the general link control concepts in mind GSM have devised an optional algorithm fulfilling all the system requirements. This algorithm provides an evolutionary basis, from which PLMN operators can start developing their own procedures to meet special local criteria, which will be presented here in a number of steps.

8.8.2.1   BS preprocessing and averaging

The GSM HO algorithm explained with reference to Figures 8.35, 8.36 and 8.37 is based on the evaluation, storage and processing of a number of parameters. The MSs continuously measure the DownLink (DL) received level (RXLEV), downlink received quality (RXQUAL) from the serving cell and the downlink received levels from the nth adjacent cells (RXLEV_NCELL(n)), and report the measured values back to the BS via the SACCH. The new measurement 'samples' are conveniently generated for every new SACCH multiframe of 480 ms duration. If BS power budget (BS-PBGT) control is also implemented, a similar set of values is measured by the BS: UpLink (UL) RXLEV, uplink RXQUAL and RXLEV in unallocated timeslots, representing the interference level. Furthermore, the MS–BS distance is calculated from the timing advance (TA) parameter. It remains for the network operator to resolve how RXQUAL is determined. The options available are to monitor the metric statistics of the Viterbi channel equaliser, that of the Viterbi-type convolutional decoder and/or the code overload rate detected by the external block codes, used in data and speech traffic channels.

In possession of the above mentioned 480 ms based measurement 'samples', the BS has to evaluate their weighted or unweighted averages. Alternatively, median values can be utilised for further decisions, where the extreme outliers are ignored. To have sufficient confidence in the estimates derived from finite-sized measured sample sets, averaging is carried out for at least 32 samples, measured over $32 \cdot 0.48$ s$\approx$15 s durations. The actual timing of the processing is dictated by the OMC.

8.8.2.2   RF power control and HO initiation

Now the averaged parameters are compared in the BS with their associated upper and lower RF power control thresholds, and if any of the UL&DL=(XX) RXLEV_XX & RXQUAL_XX parameters falls outside the required range, the BS attempts to rectify the shortfall by means of RF power control, as seen in Figure 8.35. More explicitly, if any of the four threshold comparisons fails, the BS and MS will attempt to appropriately increase or decrease the transmitted powers to meet the required conditions before initiating HO.

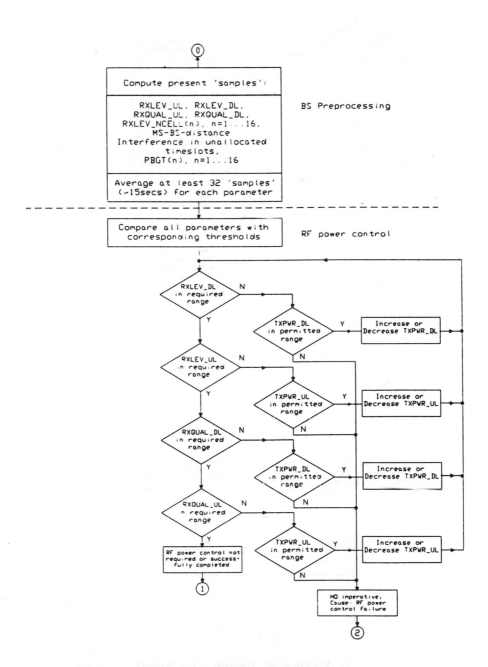

Figure 8.35: Handover preprocessing and RF power control

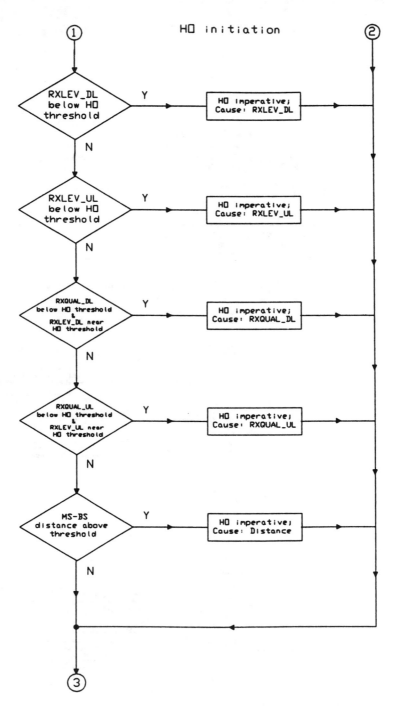

Figure 8.36: Handover initiation

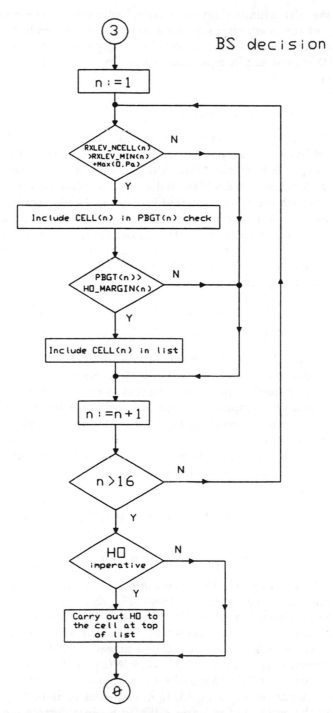

Figure 8.37: BS decision algorithm

In the 'HO initiation' set of threshold comparisons the BS initiates HO if any of the comparisons fails to meet the corresponding condition, as portrayed in Figure 8.36. Observe that the system remembers the cause of the HO request and in most cases when the criteria is not satisfied HO is imperative.

8.8.2.3   Decision algorithm

Upon HO requests due to any of the causes considered in the 'RF power control and HO initiation' phases, the BS sends a message with the 'preferred list of target cells' to the MSC. Alternatively, in case of traffic-motivated HOs the MSC may send a 'HO-candidate enquiry message' to the BS, which responds with the same 'preferred list of target cells' message. This list is compiled using the average received signal levels RXLEV_DL and RXLEV_NCELL(n), as well as a few further system parameters. Also the so-called power budget parameter (PBGT(n)) can be evaluated for each connection taking into account each of the legitimate adjacent cells (n=1...16), using the following equation:

$$
\begin{aligned}
PBGT(n) \;=\; & [min(MS\_TXPWR\_MAX, P) - RXLEV\_DL] \\
- \;& [min(MS\_TXPWR\_MAX(n), P) - RXLEV\_NCELL(n)].
\end{aligned}
$$

Here MS_TXPWR_MAX is the maximum allowed MS transmitted power on a traffic channel in the serving cell to control cellsize, MS_TXPWR_MAX(n) is the same parameter in the nth adjacent channel, while P is the maximum transmitted power capability of the MS. This equation physically evaluates the power budget for each legitimate adjacent cell in contrast to the present serving cell, since the first square bracketed term represents the present pathloss, while the second term the pathloss of the nth candidate serving cell. However, the PBGT(n) parameter is evaluated only for those cells, which satisfy

$$
RXLEV\_NCELL(n) > RXLEV\_MIN(n) + max(0, P_a), \qquad (8.34)
$$

where

$$
P_a = [MS\_TXPWR\_MAX(n) - P], \qquad (8.35)
$$

with P being again the MS's maximum power capability, which is different for vehicle mounted and for handheld MSs, and MS_TXPWR_MAX(n) being the maximum permitted MS power due to coverage area limitation in the nth adjacent cell. In other words, PBGT(n) is evaluated for those candidate target cells, where the received power RXLEV_NCELL(n) exceeds the corresponding minimum RXLEV_MIN(n) by the margin $max(0, P_a)$.

The parameter HO_MARGIN seen in Figure 8.37 in the PBGT(n) comparison is introduced to facilitate a hysteresis in the HO process by requiring the pathloss of adjacent cell n to be considerably more favourable than that of the present serving cell, before HO is requested to it. If

HO_MARGIN=0, there is no hysteresis, while for HO_MARGIN$\neq$0 the HO from cell A to B occurs at a different point from the handover B to A. Clearly, a power budget or pathloss-motivated HO is possible to ensure communications with that BS, which yields the lowest pathloss, even if all the other quality and received power threshold conditions are duly met in the serving cell.

Most of the HO request-causes are self-explanatory, but an interesting case is, when the received signal level RXLEV_UL is high, typically -80...-40 dBm, yet the received signal quality is low. This indicates the probability of high cochannel interference on the uplink, which can be eliminated by intracell HO. If the BS does not support intracell HO, then it sends a HO request to the MSC with the serving cell at the top of its preferred cell list. In some cases the OMC can initiate an intracell HO, for example, due to resource management criteria. It is particularly important to avoid anomalous HO decisions, such as subsequent power increase and decrease commands. Therefore after a power control action the set of samples used in the decision have to be discarded.

8.8.2.4 HO decisions in the MSC

During periods of peak traffic load the number of HO requests is often higher than that of the free traffic channels. In these cases the MSC sorts the HO requests in the following order of priority: RXQUAL, RXLEV, DISTANCE, PBGT. Then requests due to RXQUAL degradation enjoy the highest priority, followed by RXLEV, etc. A further classification principle is the priority order, which can be associated with each adjacent cell, where eight priority levels can be allocated. This allows, for example, umbrella cells to be given low priorities to handle calls only if no other suitable cell can carry the initiated traffic.

8.8.2.5 Handover scenarios

In this subsection a number of examples are used to support our discussions on HO algorithms [24]. Figure 8.38 represents a situation, where the MS is travelling from cell A to cell B, the reference sensitivity level is -102 dBm and the minimum received power level to allow HOs to these cells are RXLEV_MIN(A) and RXLEV_MIN(B), respectively. When the received power levels exceed these thresholds at positions a and b, HO to these cells is possible. Observe that at point b the received signal level from BS_A falls below RXLEV_MIN(A) and therefore HO to BS_B is here imperative, although its received signal level is higher than that from BS_B. When the pathloss from B (PL_B) becomes lower than that from A (PL_A), HO to B is recommended. In this example cell B is probably allocated to an open area with low pathloss exponent and hence, inspite of the lower transmitted power seen, it has a larger coverage than cell A.

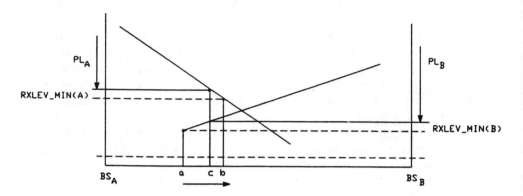

Figure 8.38: Handover without hysteresis

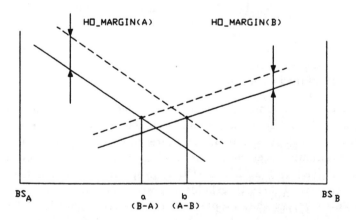

Figure 8.39: Handover with hysteresis

The size of cell B is readily reduced by increasing the minimum expected received level RXLEV_MIN(B). In parallel to this the possible HO region is shrinking and when point a moves past point b as RXLEV_MIN(B) is increased, HO is no longer possible at point c due to lack of received signal power, although the pathloss criterion $PL_B < PL_A$ is met. Conversely, if the size of cell A is reduced by increasing RXLEV_MIN(A), point b keeps moving to the left. When point b moves past point c, HO to cell B becomes imperative due to lack of coverage, before the pathloss criterion $PL_B < PL_A$ becomes true.

In Figure 8.39 the HO hysteresis introduced by the HO_MARGIN is demonstrated. The simple principle is that the received signal level from BS_A must fall significantly, by the HO_MARGIN, below that from BS_B, before HO to cell B is carried out. This ensures that after HO the received level will be by HO_MARGIN(A) dB higher than from the current serving cell A. Observe that the hysteresis area is adjusted by appropriately selecting both HO_MARGINs, where HO from cell A to cell B occurs at point b, while HO from B to A happens at point a.

8.9 Discontinuous Transmission

8.9.1 DTX Concept

The idea of discontinuous transmission (DTX) has long been known in bandwidth and power limited satellite systems, where spectral efficiency is improved using Digital Speech Interpolation (DSI). In mobile radio communications however, GSM is the first system to use voice activity detection (VAD) and DTX to further reduce the MS's power consumption and increase spectral efficiency through reducing interference during silent periods. Assuming an average speech activity of 50 % and a high number of interferers combined with frequency hopping to randomise the interference load, significant spectral efficiency gains can be scored. Due to the reduction in power consumption full DTX operation is mandatory for MSs, but in BSs only receiver DTX functions are compulsory. Earlier adaptive VAD designs were proposed for PCM speech codecs, stationary handsets and indoors background noise [25], [26]. The fundamental problem is how to differentiate between speech and noise, while keeping false noise triggering and speech spurt clipping as low as possible. In vehicle-mounted MSs the severity of the speech/noise recognition problem is aggravated by the excessive non-stationary vehicle background noise. This problem is resolved by deploying a combination of threshold comparisons and spectral domain techniques [27], [28]. Another important associated problem is the introduction of noiseless inactive segments, which is mitigated by introducing comfort noise in these segments at the receiver, which is also addressed in [27], [28].

8.9.2   Voice Activity Detection

The basic function of the VAD is to differentiate between noisy speech
and noise only under very high-noise conditions. Any VAD has to meet a
compromise between the minimisation of false triggering due to high noise
levels and the transmission of low level speech. Fast speech recognition is
crucial to minimise initial talk-spurt clipping, and a short hangover delay
reduces unwanted activity while preventing final talk-spurt clipping. The
specific VAD implementation favoured must be in harmony with the speech
codec selected, but differences between noise and speech properties can be
exploited both in time [25] and frequency domain [27]. In the GSM VAD
a combination of spectral domain and energy differences is utilised in the
decision process.

The VAD's schematic diagram is shown in Figure 8.40. In a first step the
SNR is improved by adaptive noise filtering, the coefficients of which are
determined during noise-only periods. Then the energy of the filtered signal
is compared against an adaptive threshold computed in the 'Threshold
adaptation' block for a speech/noise decision in the 'VAD decision' block.
The adaptive noise filter coefficient- and noise threshold-update must take
place during exclusively noise periods, which is ensured by additionally
checking the signal stationarity and the lack of pitch frequencies by the help
of the 'Periodicity detection' and 'Spectral comparison' blocks. A further
fixed threshold is deployed to ensure that low level noise is not detected as
speech. Finally, a hangover (HGO) mechanism is used to prevent mid-spurt
and end-spurt clipping of speech bursts, as seen in in Figure 8.40.

More specifically, the VAD's fundamental function is to adaptively filter the input signal using the set of filter coefficients a_i, $i = 0 \ldots 8$ during
noise-only periods. The filtered signal's energy P_{VAD} is compared against
an adaptively adjusted threshold Th_{VAD} to derive a speech/noise indicator signal VAD, which after being subjected to hangover imposition, yields
the transmit flag $TXFL$ utilised by the transmitter's DTX handler to en-
able/disable transmissions. The rest of the block diagram is concerned
with the adaptive adjustment of the filter coefficients a_i and that of the
threshold Th_{VAD}.

The 'adaptive block filtering' operation of the 160 input signal samples
$s(n)$ using an 8th order filter a_i yields 168 samples, as follows:

$$s_f(n) = \sum_{i=0}^{8} a_i s(n-i), \; n = 0 \ldots 167, \; 0 \le (n-i) \le 159. \tag{8.36}$$

The energy of the current 20 ms (160 samples) filtered input signal extended
by the filter's 8-sample memory is given by:

$$P_{VAD} = \sum_{n=0}^{167} (\sum_{i=0}^{8} a_i s(n-i))^2, \; 0 \le (n-i) \le 159. \tag{8.37}$$

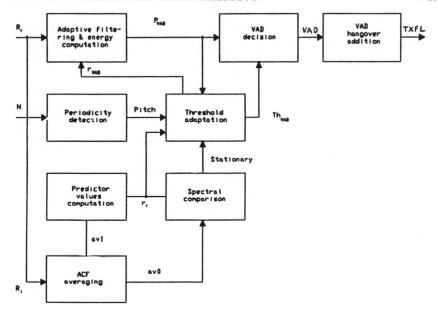

Figure 8.40: Functional block diagram of the VAD

After expanding the operations in the expression of P_{VAD} above and substituting the expressions

$$R_i = \sum_{n=0}^{159} s(n)s(n-i), \quad i = 0\ldots 8, \quad 0 \le (n-i) \le 159 \tag{8.38}$$

and

$$r_i = \sum_{k=0}^{8-i} a_k a_{k+i}, \quad i = 0\ldots 8 \tag{8.39}$$

for the input signal's and the filter coefficient's autocorrelations, respectively, we have:

$$P_{VAD} = r_0 \cdot R_0 + 2 \sum_{i=1}^{8} r_i \cdot R_i. \tag{8.40}$$

The result of the 'VAD decision' is the Boolean flag VAD, which is one if $P_{VAD} > Th_{VAD}$, zero otherwise. The HGO is implemented to prevent the VAD from prematurely curtailing the end of low-energy speech-spurts or removing short mid-speech silent gaps. The principle is that speech is continued to be transmitted for four more 20 ms frames, even if $VAD = 0$ indicates the presence of noise, in case at least three previous 20 ms speech segments were deemed to be present. Should $VAD = 1$ be set during the HGO period, the hangover counter (HOCT) is reset to four.

With the principles of VAD known, we now embark upon the description of the adaptive adjustment of the VAD-threshold, that of the filter

coefficients a_i and their correlations r_i. To get a stationary estimate of
the input signal's statistics, each input signal autocorrelation coefficient R_i
$i = 0 \ldots 8$ is averaged over four frames, i.e., 80 ms to derive the averages
$av0_i(n) = \sum_{j=0}^{3} R_i(n-j)$, $i = 0 \ldots 8, av1_i(n) = av0_i(n-4)$, $i = 0 \ldots 8$,
where n is the 20 ms frame index. The averaged autocorrelation coefficients
$av1_i(n)$ are input to the Schur recursion [12] and the reflection coefficients
$k_i(n)$ are computed exactly, as in the RPE-LTP speech codec. In determining the noise envelope's reflection coefficients, the averages $av1_i(n)$,
$i = 0 \ldots 8$ are used, since $av0_i(n)$ might still contain the end of a speech
burst. In a subsequent step the reflection coefficients k_i are converted to
simple finite impulse response (FIR) LPC filter coefficients a_i, $i = 0 \ldots 8$
and their autocorrelation is computed as follows:

$$r_i = \sum_{k=0}^{8-i} a_k \cdot a_{k+i}, \ i = 1 \ldots 8. \tag{8.41}$$

The LPC filter coefficient autocorrelations r_i, as well as the averaged input
signal autocorrelation coefficients $av0_i$ are then compared using the simple
distance measure d_m defined as:

$$d_m = r_0 \cdot av0_0 + 2 \sum_{i=1}^{8} r_i \cdot av0_i / av0_0 \tag{8.42}$$

to derive a statistical similarity flag called 'Stationary' in the 'Spectral
Comparison' block of Figure 8.40. The spectral distance of the consecutive
20 ms input segments is evaluated by computing $d = (d_m - d_{m-1})$, and
Stationary $= 1$ is set if $d < 0.05$, i.e., the spectrum is deemed stationary,
while *Stationary* $= 0$, if the spectral difference $d \geq 0.05$, i.e., the spectrum
is non-stationary.

The 'Threshold Adaptation' process is based on the input parameters
Stationary, r_{av1} derived so far, as well as on the Boolean flag 'Pitch',
indicating the presence of voiced input and on the energy of the current
adaptive filtered input signal P_{VAD}. This process has two output variables, the VAD decision threshold Th_{VAD} and the updated adaptive filter
coefficient set a_i, $i = 0 \ldots 8$ determining the updated set r_i.

The threshold adaptation updates Th_{VAD} every 20 ms using the flowchart of Figure 8.41 in two basic scenarios. Whenever the signal energy
is very low, i.e., $R_0 < P_{th}$, where the power threshold P_{th} is set by GSM
to 300 000, $Th_{VAD} = P_{LEV} = 80\ 000$ is selected, since any further tests
would be unreliable due to the course quantisation at such low signal level.
If, however, $R_0 \geq P_{th}$ is met, more elaborate tests are performed. If
$(Stat, Pitch) = (1, 0)$ for the current input frame, i.e., the input signal
appears stationary and unvoiced with no pitch periodicity, the adaptation
counter (Adaptcount) is incremented and checked, whether it reached the
value $adp = 8$ to allow threshold update. If not, no further action is taken

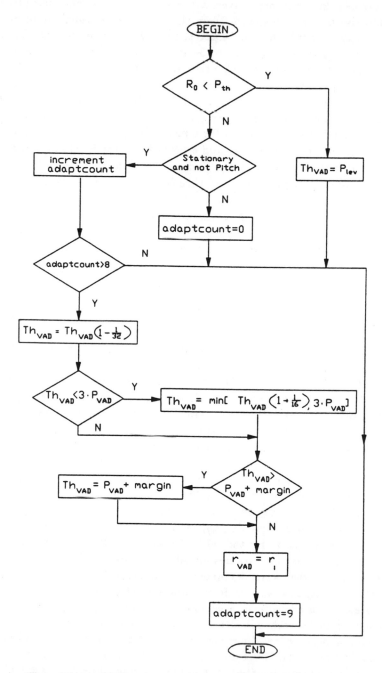

Figure 8.41: VAD adaptive threshold adjustment flowchart

before the next 20 ms frame arrives. Otherwise Th_{VAD} is decreased by the multiplicative factor $(1 - 1/32)$. Then the new Th_{VAD} value is compared with $3 \cdot P_{VAD}$ and in case it is larger than $3 \cdot P_{VAD}$, the threshold really had to be decreased, else it is set to $3 \cdot P_{VAD}$, unless this exceeds a multiplicative increase of $(1+1/16)$. The threshold is not permitted to be higher than $(P_{VAD} + margin)$, where $margin = 8 \cdot 10^6$, and after ensuring this the adaptation counter is set to $adp + 1 = 9$ to allow for continuous threshold adjustments. Finally, by forcing $r_{VAD,i} = r_i$, $i = 0 \ldots 8$, the set of filter coefficient correlations is updated for consecutive power computations.

The last parameter of the threshold adaptation is the Pitch-flag, updated also every 20 ms, which is true if periodic input signal is detected. As seen in the flowchart of Figure 8.41, the threshold Th_{VAD} is only updated, if the input is stationary but not periodic, which is characteristic of noise.

In summary, the GSM VAD strikes a good compromise between lowest possible on-air time, i.e., activity and unobjectionable talk-spurt clipping. The typical channel activities vary from 55 % in quiet locations through 60 % in office noise to 65–70 % in strong airport or railway station noise.

8.9.3   DTX Transmitter Functions

The DTX transmitter's operation is explained by referring to Figure 8.42 and relies on the VAD differentiating between speech and noise. If speech is deemed to be present, i.e., the transmitter state machine is in its speech transmit (SPTX) state, the VAD flag is set to one, while for noise VAD=0. If the VAD stops detecting speech, it does not immediately disable the speech transmission by setting the transmit flag (TXFL) to zero, but first enters the so-called Hangover (HGO) state. The HGO state is designed to prevent negligibly short silence periods from disabling transmissions or to remove final talk spurt clipping, where only a fraction of the frame delivers speech and hence VAD=0 was detected. The HGO delay is of four speech frame durations, i.e., 4·20=80 ms long. Hence the hangover counter (HOCT) is initially set to HOCT=4 and in every subsequent noise frame, where VAD=0, it is decremented by one. When HOCT=0, the HGO delay has elapsed and the 'end of speech' (EOS) flag has to be set to one. Since the last four frames encoded by the speech encoder during the hangover interval were deemed to be noise, their spectral envelope parameters (LARs) as well as RPE subsegment maxima, averaged over four blocks, are used to form a socalled silence identifier (SID) frame to be passed to the speech decoder for comfort noise insertion to 'fill' the subjectively annoying 'deaf' periods introduced by disabled transmissions. Now the first averaged SID frame following the elapse of the hangover is scheduled for transmission, but if it happens to be stolen by the FACCH for example, then the EOS flag has to remind the DTX transmitter to send the subsequent frame instead. However, the frame immediately after the elapse of the HGO is the only one with EOS=1. Clearly, with the

HGO elapsed the system enters the Comfort Noise Update (CNU) state and the SID frames are sent during further silent periods in each SACCH multiframe, i.e., at intervals of 480 ms, whenever it is asked for by the radio subsystem through setting the Noise Update Flag (NUFT) to logical one. In the simplest scenario NUFT=1 is aligned with the timeslots of the SACCH structure. From the CNU state, if VAD=1 is encountered, the DTX transmitter state-machine returns to its speech transmit (SPTX) state, otherwise it enters the Comfort Noise Computation (CNC) state, sets the transmit flag (TXFL) to zero and disables transmissions. Detecting VAD=1 forces the system to SPTX state, while on VAD=0 and NUFT=1 further SID frames have to be transmitted in CNU mode.

8.9.4 DTX Receiver Functions

The DTX receiver's operation is in close cooperation with the entire receiver, since it uses soft and hard decision information from the Viterbi channel equaliser, Viterbi channel decoder and cyclic error detecting block decoder to generate the so-called Bad Frame Indicator (BFI) flag. When the BFI flag signals a corrupted speech or SID frame, the Speech/noise Extrapolation (SE) functions are invoked to improve the perceived link quality. If however, several adjacent frames are damaged, the received signal is gradually muted to zero. The interplay of system elements is completed by the comfort noise generator activated upon reception of SID frames for natural sounding Comfort Noise Insertion (CNI) in inactive speech intervals. The DTX receiver's operation is essentially conducted by the input flags BFI and SID, as evidenced by Figure 8.43.

Firstly, in 'Speech Received(SPRX)' state the SID frame detector decides, whether the received frame is a speech or an SID frame and, after evaluating the received signal quality, forms the pair (BFI, SID). The SID-detector is extremely reliable, since in SID frames all of the 95 C1b FEC coded RPE excitation bits are set to zero at the transmitter and the received frame is only deemed to be an SID sequence rendering SID=1, if at most 15 out of the 95 corresponding bits are non-zero. It will only be used for comfort noise insertion, however, if at most one bit of it is corrupted, i.e. BFI=0, while in case of more than one but less than 16 corrupted bits (BFI,SID)=(1,1) is set, which requires noise extrapolation using gradually muted previous SID frames. The normal operation is described by (BFI=0, SID=0), when an uncorrupted speech frame is received (SPRX). In this situation the received frame is simply decoded by the speech decoder. If a frame with BFI=1 flag has arrived, irrespective of whether speech or noise is deemed to be present, the receiver switches into speech/noise extrapolation mode to improve the subjective link assessment, in case a single speech frame is corrupted or stolen by the FACCH. Several consecutive BFI=1 flags render the receiver to mute its output gradually to zero. When the link-quality improves, BFI=0 is encountered and upon SID=0 the receiver

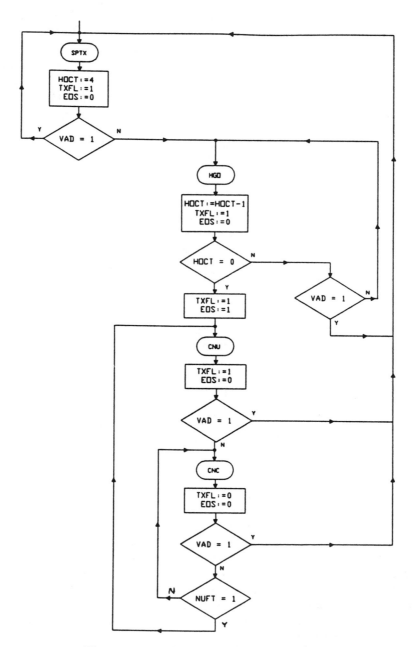

Figure 8.42: DTX transmitter operation

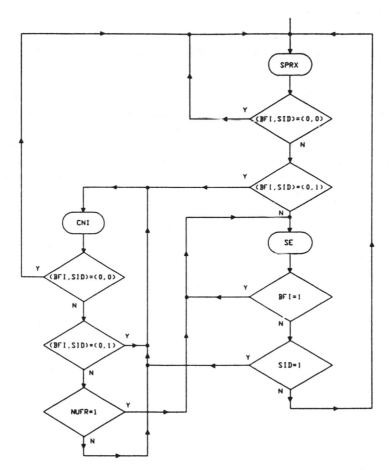

Figure 8.43: DTX receiver operation

returns to its normal speech reception state SPRX.

If, however the pair (BFI=0, SID=1) is detected, the DTX receiver inserts comfort noise to the speech decoder in its CNI state, based on the noise spectral parameters (LARs) received in the last SID frame. These noise spectral LAR parameters are updated by receiving fresh SID frames via the SACCH every 480 ms in each new multiframe. When detecting BFI=0, SID=0, normal speech decoding is invoked, while if this frame happens to be corrupted, i.e., BFI=1 and the receiver's noise update flag NUFR=1 indicates matching alignement with a SACCH time-slot, then again the receiver enters the extrapolation mode (SE). In case several consecutive SID frames are corrupted, i.e. more than one of the 95 zero C1b bits is turned to one, the comfort noise is gradually muted to zero to inform the listener that the link is corrupted. The process described is highlighted also in Figure 8.43.

8.9.5   Comfort Noise Insertion and Speech/Noise Extrapolation

Experiments carried out by GSM have shown that silent gaps inserted by the DTX system are extremely annoying and degrade speech intelligibility. Best subjective and objective results are achieved, if comfort noise of appropriately matched level and spectral envelope is inserted and updated via sending an SID frame at each 480 ms interval through the SACCH, when no speech is transmitted. The 456-bit SID frame is a 'speech-like' frame transmitted every 24th 20 ms speech frame to characterise the current background noise spectral envelope using the Logarithmic Area Ratio (LAR) parameters. The level of the noise is represented by the subsegment maxima computed, but the regularly spaced excitation pulses are set to zero at the transmitter to aid the SID frame recognition at the receiver. The Long Term Predictor (LTP) is disabled by setting its gain to zero, while erratic noise level changes are mitigated by limiting subsequent increases to 50 % of the previous maximum value. Furthermore, the LARs and block maxima are averaged over the last four speech frames, before inclusion in an SID frame. At the receiver the decoded LARs and block maxima are used with locally injected uniformly distributed pseudo-random RPE samples and grid positions to represent the background noise at the transmitter.

Whenever the BFI flag signals a corrupted speech or noise frame, the previous 20 ms frame is input to the speech decoder. This repetition is hardly perceptible, if only one frame is lost in every ten, but becomes inadequate when encountered more frequently. In subsequent corrupted frames therefore their level is gradually muted to zero by decreasing the maximum 64-valued (6-bit logarithmically quantised) subsegment maxima each time by four. Hence it is set to zero in at most 16 subsequent 20 ms speechframes, i.e. in 320 ms.

8.10   Ciphering

The GSM communications security aspects are described in Recommendations 02.09, 02.17, 03.20 and 03.21, while an overview is given in [29]. The GSM security issues center around the Subscriber Identity Module (SIM) received at subscription, which is preferably a removable plug-in module with a Personal Identification Number (PIN). These features facilitate the production of identical handsets with PIN protection against unauthorised use, while allowing GSM access through any GSM handset. The SIM contains, amongst a number of parameters, the International Mobile Subscriber Identity (IMSI), the Individual Subscriber Authentication Key (K_i) and the Authentication Algorithm (A3). On attempting to access the PLMN the MS identifies itself to the network, receives a random number (R), which together with K_i is used to calculate the Signed response (S) by invoking the confidential algorithm (A3): $S=[K_i(A3)R]$. The result S is sent back to the network and compared with the locally computed version to authorise access. In addition to the random number (R) the network sends a key number (K_n) to the MS, which is related to the ciphering key K_c and serves to avoid using different K_c keys at the receiver and transmitter. This key number K_n is then stored by the MS and is included in its first message to the network. Besides S, the MS computes the ciphering key (K_c) using another confidential algorithm (A8) stored in the SIM, and the input parameters K_i and R: $K_c=[K_i(A8)R]$. The ciphering key K_c is also computed in the network and hence no confidential information is sent unprotected via the radio path.

Once authentication is confirmed and both the network and the MS know K_c, the network issues a ciphering mode command and from now on all messages are ciphered at the transmitter and deciphered at the receiver, using the confidential algorithm (A5). Confidentiality is further enhanced by protecting the user's identity, when identification takes place assigning a Temporary Mobile Subscriber Identity (TMSI) valid for a specific location area. This TMSI uniquely describes the IMSI in a specific location area, but outside the area it must be associated with the Location Area Identity (LAI). The network, more precisely the Visitor Location Register (VLR) keeps track of the TMSI-IMSI association and allocates a new TMSI in each new location area update procedure, i.e., in each new VLR.

The following representative example is provided to describe one out of a variety of specific scenarios, where the authentication and ciphering algorithms described are utilised. We assume that the MS associated with a specific TMSI is registered in the VLR. All required MS characteristics are stored in the VLR and identification is based on the LAI and TMSI parameters. As mentioned, authentication is carried out upon each location updating and the set [IMSI, TMSI, K_c, K_n, R, S] is available in the VLR. The process is described with reference to Figure 8.44.

The MS stores its own set of [IMSI, TMSI, LAI, K_i, K_c, K_n] parameters and requests location update via the radio path by sending its $[TMSI_0, LAI_0, K_n]$ parameters to the MSC/BS. The MSC/BS forwards these via the network to the VLR, which issues an authentication request through the network to the MSC/BS and via the radio link to the MS by sending K_n and R. The MS computes K_c and S from K_i and R utilising the algorithms A3 and A8, which are stored in its SIM. The signed response S is transmitted back to the MSC/BS and from there via the network to the VLR, where authentication is performed by comparing the received and locally generated S parameters. The VLR updates the MS's location in its own HLR at the appropriate IMSI entry and assigns a MS Roaming Number (MSRN). The MSRN stored in the MS's HLR is then used by incoming calls to find the MS and route the calls to the appropriate VLR,

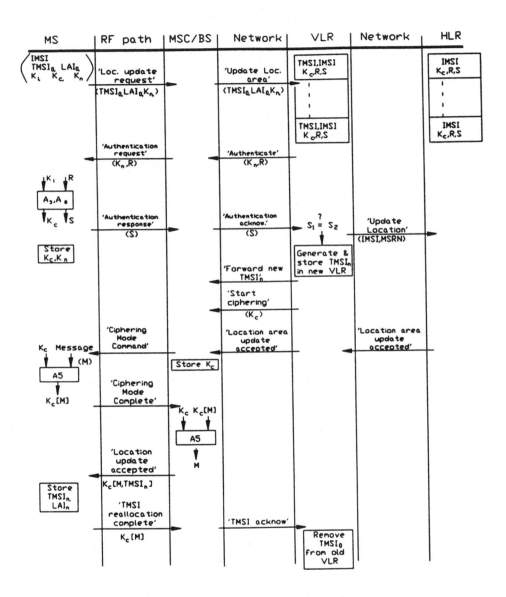

Figure 8.44: Location area update using the confidential algorithms A3,A5 and A8

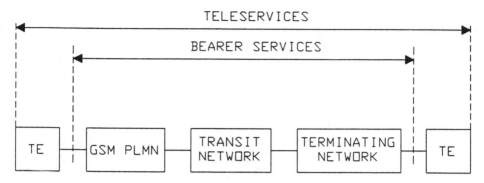

Figure 8.45: Bearer- and Tele-services

where the momentary TMSI and LAI parameters locate and identify the called subscriber.

Simultaneously, the VLR also generates the new $TMSI_n$ and forwards it to the MSC/BS and from now on $TMSI_n$, LAI_n and K_n are used to identify the MS. The HLR acknowledges the location update to the VLR and to the MSC/BS that, in turn, issues a 'ciphering mode command' to the MS. The MS responds with a 'ciphering mode complete' message and therefore ciphers all its messages by the algorthm A5 using K_c, while the MSC/BS deciphers and vice-versa. The MSC/BS informs the MS using a ciphered message that 'location update is accepted' by the system and also sends out the new $TMSI_n$, which is acknowledged by the MS via sending the 'TMSI reallocation complete' message. Finally, this is accepted by the MSC/BS in that it sends a 'channel release' command to the MS and a 'TMSI acknowledge' to the old VLR to discard $TMSI_0$.

8.11 Telecommunication Services

The telecommunication services supported by a GSM PLMN are divided into two broad categories: bearer services and teleservices. The bearer services provide for the transmission of signals between access points (called user-network interfaces in ISDN), while the teleservices provide communications between users according to protocols established by the network operators. The teleservices thus include also the Terminal Equipment (TE) functions, see Figure 8.45.

The bearer services may include more than one transit network. The terminating network may include a GSM PLMN, either the originating one or another one. The terminal equipment (TE) may consist of one or more pieces of equipment—telephone set, Data Terminal Equipment (DTE), teletext terminal, etc. Both bearer services and teleservices are offered together with a set of supplementary services. A supplementary service modifies and/or supplements a basic telecommunication service and consequently it cannot be offered to a customer as a stand alone service.

Bearer Services: The bearer services provide the user with the possibility of gaining access to various forms of communications. For example, information transfer between a user in a GSM PLMN and a user in a terminating network, including the same GSM PLMN, another GSM PLMN and other types of PLMN's. A bearer service involves only low layer attributes (layers 1–3 of the OSI model), such as information transfer, access, interworking, and general attributes. The information transfer capability is concerned with the transfer of different types of information, e.g., digitized speech, through a GSM PLMN and another network, or through a GSM PLMN only. The method of transfer, be it circuit switching or packet transportation is called the transfer mode, while the information transfer rate is the bit rate in circuit mode or the throughput rate in packet mode. The second group of attributes describes the access at the mobile station. The signalling access characterises the protocol on the signalling channel at the access point (V-series protocol, X-series protocol, etc.), while the information access describes the interface according to the protocol used to transfer user information at the access point (V-series interface, X-series interface, etc) and the bit rate in circuit mode, throughput rate in packet mode. The interworking attributes are concerned with the type of the terminating network, such as GSM PLMN, PSTN, ISDN, etc., as well as with the terminal to terminating network interface, e.g., V-series interface, X-series interface, etc. Finally, the general attributes include the supplementary services, quality of service, service interworking, commercial and operational attributes. The following bearer services are supported by a GSM PLMN:

1. Asynchronous 300–9600 bit/s circuit switched data service interworking with the public switched telephone network (PSTN).

2. Circuit switched synchronous data transmission at 300–9600 bit/s, interworking with the PSTN, circuit switched public data networks (CSPDN) and ISDN.

3. Asynchronous 300–9600 bit/s packet assembler/disassembler (PAD) access interworking with the packet switched public data network (PSPDN).

4. Packet switched synchronous 2400–9600 bit/s data service interworking with the PSPDN.

The bearer services can be transparent or non-transparent. In a transparent service the error protection is provided only by Forward Error Correction (FEC). The non-transparent services have the additional protection of Automatic Repeat Request (ARQ) in the radio link protocol, which results in higher data integrity. However, the extra error protection is achieved at the expense of greater transmission delay and reduced throughput.

Teleservices: The teleservices provide the user with the possibility of gaining access to various forms of applications, covering for example:

- Applications involving two terminals, which provide compatible or identical teleservice attributes at an access point in a GSM PLMN and an access point in a terminating network.

- Applications involving a terminal at one access point in a GSM PLMN and a system providing high layer functions (e.g., speech storage system, message handling system, etc) located either within the GSM PLMN or in a terminating network.

A teleservice is characterized by a set of low level attributes and a set of high level attributes. The low level attributes are the same as those used to characterize the bearer services. High level attributes refer to functions and protocols of layers 4 to 7 of the OSI model. They are concerned with the transfer, storage and processing of user messages, provided by a subscriber terminal, a retrieval centre or a network service centre. The high level attributes include a variety of legitimate user information, such as speech, short message, data, videotext, teletext, facsimile, as well as layers 4 to 7 protocol functions, which refer to the layer protocol characteristics of the different teleservices. The teleservices supported by a GSM PLMN are divided into six categories:

1. Transmission of speech information and voice band signalling tones of the PSTN/ISDN.

2. Short message service, which enables a user of a telecommunication network (e.g., PSTN) to send a short alphanumeric message (up to 180 characters) to a mobile subscriber of the GSM network.

3. Message Handling System (MHS) access, providing the transmission of a short message from a message handling system in a fixed network (e.g., paging system) to a mobile station.

4. Videotext access.

5. Teletext transmission.

6. Facsimile transmission.

Supplementary Services: The supplementary services are divided into eight categories:

1. Number identification, entailing five services:

   - calling number identification presentation,

- connected number identification presentation,
- calling number identification restriction,
- connected number identification restriction, and
- malicious calls identification.

The "identification presentation" services provide for the ability to indicate the number of the calling/connected party, while the "identification restriction" services offer to the calling/connected party the ability to restrict presentation of the party's number.

2. Call offering. A group of eight different services: six "call forwarding" services, a "call transfer" service and a "mobile hunting access" service. The call forwarding services permit a called mobile subscriber to have the network send all incoming calls, or just those associated with a specific basic service, addressed to the called mobile subscriber's directory number to another directory number. The call forwarding can be: 1) unconditional, 2) on condition the mobile subscriber is busy, 3) when there is no reply, 4) if the subscriber cannot be reached due to radio congestion, 5) when there is no paging response and 6) whenever the mobile subscriber is not registered. The "call transfer" service enables the served mobile subscriber to transfer an established incoming or outgoing call to a third party. The "mobile hunting access" service can be used only by a mobile PABX and enables all incoming calls to be distributed over a group of accesses, belonging to the mobile PABX.

3. Call completion. This group is divided into three services:

- call waiting,
- call hold, and
- completion of calls to busy subscribers.

The "call waiting" service allows the mobile subscriber to be notified of an incoming call whilst the termination is in a busy state. The subscriber can subsequently either answer, reject or ignore the incoming call. The "call hold" service allows a served mobile subscriber to interrupt communication on an existing call and then subsequently to re-establish communication. The "completion of calls to busy subscribers" service allows a calling mobile subscriber which encounters a busy called subscriber to be notified when the called subscriber becomes unengaged and have the call re-initiated.

4. Multi party. Two services are offered in this group:

- three party, and
- conference calling.

The first one enables a mobile subscriber to establish a three party conversation, while the second service provides the mobile subscriber with the ability to have a multi-connection call, i.e., simultaneous communication between more than two parties.

5. Closed user group service: allows a group of subscribers, connected to the PLMN and/or the ISDN, to intercommunicate only amongst themselves. If required, one or more subscribers may be provided with incoming/outgoing access to subscribers outside this group.

6. Charging, including three services:

   - advice of charge, which allows the subscriber to receive charging information related to the used telecommunication services,

   - freephone service, allowing the served mobile subscriber to be reached with a freephone number and charged for these calls, and

   - reverse charging, which allows a called mobile subscriber to be charged for the usage-based calls.

7. Additional information transfer. A user-to-user signalling service allowing a mobile subscriber to send/receive a limited amount of information to/from another PLMN or ISDN subscriber over the signalling channel.

8. Call restriction. This group of services makes it possible for a mobile subscriber to prevent outgoing or incoming calls. There are seven different types of service:

   - barring of all outgoing calls,

   - barring of all outgoing international calls directed to non-CEPT countries,

   - barring of all outgoing international calls except those directed to the home PLMN country,

   - barring of all outgoing calls when roaming outside the home PLMN country,

   - barring of all incoming calls, and

   - barring of all incoming calls when roaming outside the home PLMN country.

Access to the GSM Network: The different access points in a GSM network are shown in Figure 8.46. At access points 1 and 2 bearer services may be accessed, while at access point 3 teleservices are accessed. All terminal equipment accessing a GSM PLMN interface at one of these access points must meet the specifications of the protocols at that interface.

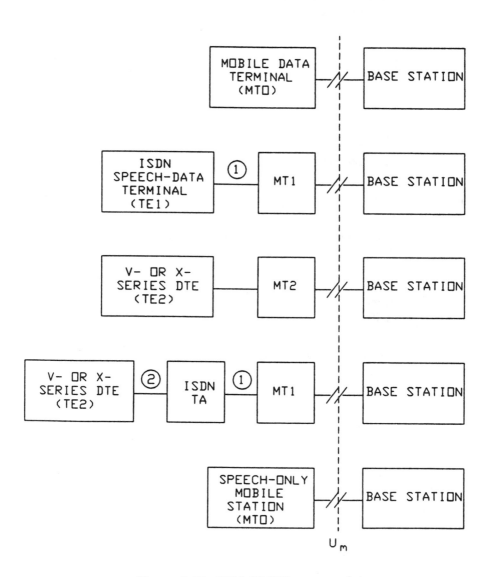

Figure 8.46: GSM PLMN access points

The Mobile Station (MS) is shown to consist of Mobile Termination (MT) and Terminal Equipment (TE). The mobile termination supports functions associated with the management of the radio interface U_m and flow control of user data between interface U_m and access points 1 or 2. These functions include:

— radio transmission termination

— radio channel management

— error protection for information sent across the radio path

— speech encoding/decoding

— flow control and mapping of user data and signalling

— rate adaptation of user data between the radio channel rate and user rates

— multiple terminal support

— mobility management

There are three types of mobile termination: MT0, a fully integrated mobile station, including data terminal and its adaptation functions, MT1 which includes ISDN terminal adaptation functions, and MT2 which includes CCITT V- or X-series terminal adaptation functions.

The terminal equipment may consist of one or more pieces of equipment such as telephone set, Data Terminal Equipment (DTE), teletext terminal, etc. The TE1 type equipment represents an ISDN interface and the TE2 equipment presents a non-ISDN interface, e.g., CCITT V- or X-series interface. A non-ISDN terminal (TE2) may be connected to an MT1 type termination using Terminal Adapter (TA).

Bibliography

[1] *Proceedings of the Nordic Seminar on Digital Land Mobile Radio Communication (DMR), Espoo, Finland,* February 1985.

[2] *Proceedings of the Second Nordic Seminar on Digital Land Mobile Radio Communication (DMRII), Stockholm, Sweden,* October 1986.

[3] *Proceedings of the International Conference on Digital Land Mobile Radio Communication (ICDMC) Venice, Italy,* June/July 1987.

[4] *Proceedings of Digital Cellular Radio Conference, Hagen, FRG,* October 12-14, 1988.

[5] **A. Moloberti.** "Definition of the radio subsystem for the GSM pan-European digital mobile communication system". *Proc. of ICDMC, Venice, Italy,* pp. 37-46, June/July 1987.

[6] **A.W.D. Watson.** "Comparison of the contending multiple access methods for the pan-European mobile radio systems". *IEE Colloquium, Digest No:1986/95,* pp. 2/1-2/6, 7 October 1986.

[7] **E. Natvig.** "Evaluation of six medium bitrate coders for the pan-European digital mobile radio system". *IEEE Journal on Selected Areas in Communications, vol.6, no.2,* pp. 324-334, February 1988.

[8] **D.M. Balston.** "Pan-European cellular radio: or 1991 and all that". *Electronics and Communication Engineering Journal,* pp. 7-13, January/February 1989.

[9] *Group Speciale Mobile (GSM) Recommendation,* April 1988.

[10] **P. Vary, R.J. Sluyter.** "MATS-D speech codec: regular-pulse excitation LPC". *Proc. of the Second Nordic Seminar on Digital Land Mobile Radio Communication (DMRII), Stockholm, Sweden,* pp. 257-261, October, 1986.

[11] **P. Vary, R. Hoffmann.** "Sprachcodec für das europäische Funkfernsprechnetz". *Frequenz 42 (1988) 2/3,* pp. 85-93, 1988.

[12] **J. Schur.** "Über Potenzreihen, die im Innern des Einheitskreises beschränkt sind". *Journal für die reine und angewandte Mathematik, Bd 147,* pp. 205-232, 1917.

[13] W. Webb, L. Hanzo, R. Salami, R. Steele. "Does 16-QAM provide an alternative to a half-rate GSM speech codec?". *Proc. of IEEE-VT Conf., St.-Louis, Missouri, U.S.A.*, May 1991.

[14] K.H.H. Wong. "Transmission of channel coded speech and data over mobile radio channels". *PhD Thesis, Dept. of Electronics and Computer Science, University of Southampton*, 1989.

[15] L. Hanzo, K.H.H. Wong, R. Steele. "Efficient channel coding and interleaving schemes for mobile radio communications". *Proc. of IEE Colloq., Savoy Place, London, U.K.*, 22 February 1988.

[16] L.B. Lopes. "GSM radio link simulation". *IEE Colloquium, University research in Mobile Radio*, pp. 5/1-5/4, 1990.

[17] J.C.S. Cheung, R. Steele. "Modified Viterbi equaliser for mobile radio channels having large multi-path delay". *Electronics Letters, vol.25, no.19*, pp. 1309-1311, 14 Sept., 1989

[18] N.S. Hoult, C.A. Dace, A.P. Cheer. "Implementation of an equaliser for the GSM system". *Proc. of the 5th Int. Conf. on Radio Receivers Associated Systems, Cambridge, U.K.*, 24-26 July, 1990.

[19] R.D'Avella, L. Moreno, M. Sant'Agostino. "An adaptive MLSE receiver for TDMA digital mobile radio". *IEEE Journal on Selected Areas in Communications, vol.7, no.1*, pp. 122-129, January 1989.

[20] J.C.S. Cheung. "Receiver techniques for wideband time division multiple access mobile radio systems". *PhD Mini-Thesis, Univ. of Southampton*, 1990.

[21] J.B. Anderson, T. Aulin, C.E. Sundberg. *Digital phase modulation*, Plenum Press, 1986.

[22] M.R.L. Hodges, S.A. Jensen, P.R. Tattersall. "Laboratory testing of digital cellular radio systems". *BTRL Journal, vol.8, no.1*, pp. 57-66, January 1990.

[23] L. Hanzo, R. Steele, P.M. Fortune. "A subband coding, BCH coding and 16-QAM system for mobile radio speech communications". *IEEE Tr. on VT., Vol. 39*, pp. 327-340, November 1990.

[24] D.J. Targett, H.R. Rast. "Handover-enhanced capabilities of the GSM system". *Proc. of Digital Cellular Radio Conference, Hagen, FRG*, pp. 3C/1-3C/11, October 12-14, 1988.

[25] E. Bacs, L. Hanzo. "A simple real-time adaptive speech detector for SCPC systems". *Proc of ICC'85, Chicago*, pp. 1208-1212, May, 1985.

[26] J.A. Jankowski. "A new digital voice-activated switch". *Comsat Tech. Journal, vol.6, no.1*, pp. 159-170, Spring 1976.

[27] **D.K. Freeman, G. Cosier, C.B. Southcott, I. Boyd.** "The voice
activity detector for the pan-European digital cellular mobile telephone
service". *Proc. of ICASSP'89, Glasgow*, pp. 369-372, 23-26. May, 1989.

[28] **S. Hansen.** "Voice activity detection (VAD) and the operation of discontinuous transmission (DTX) in the GSM system". *Proc. of Digital Cellular
Radio Conference, Hagen, FRG*, pp. 2b/1-2b/14, October 12-14, 1988.

[29] **P.C.J. Arend.** "Security aspects and the implementation in the GSM
system". *IBID.*, pp. 4a/1-4a/7, October 12-14, 1988.

Glossary

A3	Authentication algorithm
A5	Cyphering algorithm
A8	Confidential algorithm to compute the cyphering key
AB	Access burst
ACCH	Associated control channel
ADC	Administration centre
AGCH	Access grant control channel
AUC	Authentication centre
AWGN	Additive gaussian noise
BCCH	Broadcast control channel
BER	Bit error ratio
BFI	Bad frame indicator flag
BN	Bit number
BS	Base station
BS-PBGT	BS power budget: to be evaluated for power budget motivated handovers
BSIC	Base station identifier code
CC	Convolutional codec
CCCH	Commom control channel
CELL_BAR_ACCESS	Boolean flag to indicate, whether the MS is permitted to access the specific traffic cell
CNC	Comfort noise computation
CNI	Comfor noise insertion
CNU	Comfort noise update state in the DTX handler

DB	Dummy burst
DL	Down link
DSI	Digital speech interpolation to improve link efficiency
DTX	Discontinuous transmission for power consumption and interference reduction
EIR	Equipment identity register
EOS	End of speech flag in the DTX handler
FACCH	Fast associated control channel
FCB	Frequency correction burst
FCCH	Frequency correction channel
FEC	Forward error correction
FH	Frequency hopping
FN	TDMA frame number
GMSK	Gaussian minimum shift keying
GP	Guard space
HGO	Hangover in the VAD
HLR	Home location register
HO	Handover
HOCT	Hangover counter in the VAD
HO_MARGIN	Handover margin to facilitate hysteresis
HSN	Hopping sequence number: frequency hopping algorithm's input variable
IMSI	International mobile subscriber identity
ISDN	Integrated services digital network
LAI	Location area identifier
LAR	Logarithmic area ratio
LTP	Long term predictor
MA	Mobile allocation: set of legitimate RF channels, input variable in the frequency hopping algorithm
MAI	Mobile allocation index: output variable of the FH algorithm

MAIO	Mobile allocation index offset: intial RF channel offset, input variable of the FH algorithm
MS	Mobile station
MSC	Mobile switching centre
MSRN	Mobile station roaming number
MS_TXPWR_MAX	Maximum permitted MS transmitted power on a specific traffic channel in a specific traffic cell
MS_TXPWR_MAX(n)	Maximum permitted MS transmitted power on a specific traffic channel in the n-th adjacent traffic cell
NB	Normal burst
NMC	Network management centre
NUFR	Receiver noise update flag
NUFT	Noise update flag to ask for SID frame transmission
OMC	Operation and maintenance centre
PARCOR	Partial correlation
PCH	Paging channel
PCM	Pulse code modulation
PIN	Personal identity number for MSs
PLMN	Public land mobile network
PLMN_PERMITTED	Boolean flag to indicate, whether the MS is permitted to access the specific PLMN
PSTN	Public switched telephone network
QN	Quater bit number
R	Random number in the authentication process
RA	Rural area channel inpulse response
RACH	Random access channel
RF	Radio frequency
RFCH	Radio frequency channel
RFN	Reduced TDMA frame number: equivalent representation of the TDMA frame number, which is used in the synchronisation chanel
RNTABLE	Random number table utilised in the frequency hopping algoirthm

RPE	Regular pulse excited
RPE-LTP	Regular pulse excited codec with long term predictor
RS-232	Serial data transmission standard equivalent to CCITT V24. interface
RXLEV	Received signal level: parameter used in hangovers
RXQUAL	Received signal quality: parameter used in hangovers
S	Signed response in the authentication process
SACCH	Slow associated control channel
SB	Synchronisation burst
SCH	Synchronisation channel
SCPC	Single channel per carrier
SDCCH	Stand-alone dedicated control channel
SE	Speech extrapolation
SID	Silence identifier
SIM	Subscriber identity module in MSs
SPRX	Speech received flag
SPTX	Speech transmit flag in the DTX handler
STP	Short term predictor
TA	Timing advance
TB	Tailing bits
TCH	Traffic channel
TCH/F	Full-rate traffic channel
TCH/F2.4	Full-rate 2.4 kbps data traffic channel
TCH/F4.8	Full-rate 4.8 kbps data traffic channel
TCH/F9.6	Full-rate 9.6 kbps data traffic channel
TCH/FS	Full-rate speech traffic channel
TCH/H	Half-rate traffic channel
TCH/H2.4	Half-rate 2.4 kbps data traffic channel
TCH/H4.8	Half-rate 4.8 kbps data traffic channel
TDMA	Time division multiple access

TMSI	Temporary mobile subscriber identifier
TN	Time slot number
TU	Typical urban channel inpulse response
TXFL	Transmit flag in the DTX handler
UL	Up link
VAD	Voice activity detection
VE	Viterbi equaliser
VLR	Visiting location register

Index